Useful relations

At 298.15 K

RT	2.4790 kJ mol^{-1}	RT/F	25.693 mV
$(RT/F)\ln 10$	59.160 mV	kT/hc	207.225 cm^{-1}
kT	25.693 meV	$V_m^{\ominus}$	2.4790×10^{-2} m^3 mol^{-1} 24.790 dm^3 mol^{-1}

Selected units*

1 N	1 kg m s^{-2}	1 J	1 kg m^2 s^{-2}
1 Pa	1 kg m^{-1} s^{-2}	1 W	1 J s^{-1}
1 V	1 J C^{-1}	1 A	1 C s^{-1}
1 T	1 kg s^{-2} A^{-1}	1 P	10^{-1} kg m^{-1} s^{-1}
1 S	1 Ω^{-1} = 1 A V^{-1}		

* For multiples (milli, mega, etc), see the Resource section

Conversion factors

$\theta/°\mathrm{C} = T/\mathrm{K} - 273.15$*

1 eV	$1.602\,177\times10^{-19}$ J 96.485 kJ mol^{-1} 8065.5 cm^{-1}	1 cal	4.184* J
1 atm	101.325* kPa 760* Torr	1 cm^{-1}	1.9864×10^{-23} J
1 D	$3.335\,64\times10^{-30}$ C m	1 Å	10^{-10} m*

* Exact value

Mathematical relations

$\pi = 3.141\,592\,653\,59\ldots$

$e = 2.718\,281\,828\,46\ldots$

Logarithms and exponentials

$\ln x + \ln y + \ldots = \ln xy\ldots$

$\ln x - \ln y = \ln(x/y)$

$a\ln x = \ln x^a$

$\ln x = (\ln 10)\log x = (2.302\,585\ldots)\log x$

$e^x e^y e^z \ldots = e^{x+y+z+\ldots}$

$e^x/e^y = e^{x-y}$

$(e^x)^a = e^{ax}$

$e^{\pm ix} = \cos x \pm i\sin x$

Series expansions

$$e^x = 1 + x + \frac{x^2}{2!} + \frac{x^3}{3!} + \cdots$$

$$\ln(1+x) = x - \frac{x^2}{2} + \frac{x^3}{3} - \cdots$$

$$\frac{1}{1+x} = 1 - x + x^2 - \qquad \frac{1}{1-x} = 1 + x + x^2 + \cdots$$

$$\sin x = x - \frac{x^3}{3!} + \frac{x^5}{5!} - \cdots \qquad \cos x = 1 - \frac{x^2}{2!} + \frac{x^4}{4!} - \cdots$$

Derivatives; for Integrals, see the *Resource section*

$$d(f+g) = df + dg \qquad d(fg) = f\,dg + g\,df$$

$$d\frac{f}{g} = \frac{1}{g}df - \frac{f}{g^2}dg \qquad \frac{df}{dt} = \frac{df}{dg}\frac{dg}{dt} \quad \text{for} \quad f = f(g(t))$$

$$\left(\frac{\partial y}{\partial x}\right)_z = 1/\left(\frac{\partial x}{\partial y}\right)_z \qquad \left(\frac{\partial y}{\partial x}\right)_z\left(\frac{\partial x}{\partial z}\right)_y\left(\frac{\partial z}{\partial y}\right)_x = -1$$

$$\frac{dx^n}{dx} = nx^{n-1} \qquad \frac{de^{ax}}{dx} = ae^{ax} \qquad \frac{d\ln(ax)}{dx} = \frac{1}{x}$$

$$df = g(x,y)dx + h(x,y)dy \text{ is exact if } \left(\frac{\partial g}{\partial y}\right)_x = \left(\frac{\partial h}{\partial x}\right)_y$$

Greek alphabet*

Α, α	alpha	Ι, ι	iota	Ρ, ρ	rho
Β, β	beta	Κ, κ	kappa	Σ, σ	sigma
Γ, γ	gamma	Λ, λ	lambda	Τ, τ	tau
Δ, δ	delta	Μ, μ	mu	ϒ, υ	upsilon
Ε, ε	epsilon	Ν, ν	nu	Φ, ϕ	phi
Ζ, ζ	zeta	Ξ, ξ	xi	Χ, χ	chi
Η, η	eta	Ο, ο	omicron	Ψ, ψ	psi
Θ, θ	theta	Π, π	pi	Ω, ω	omega

* Oblique versions (α, β, …) are used to denote physical observables.

PERIODIC TABLE OF THE ELEMENTS

Period \ Group	1 I IA	2 II IIA	3 IIIB	4 IVB	5 VB	6 VIB	7 VIIB	8 VIIIB	9 VIIIB	10 VIIIB	11 IB	12 IIB	13 III IIIA	14 IV IVA	15 V VA	16 VI VIA	17 VII VIIA	18 VIII VIIA
1	1 H hydrogen 1.0079 $1s^1$																	2 He helium 4.00 $1s^2$
2	3 Li lithium 6.94 $2s^1$	4 Be beryllium 9.01 $2s^2$											5 B boron 10.81 $2s^22p^1$	6 C carbon 12.01 $2s^22p^2$	7 N nitrogen 14.01 $2s^22p^3$	8 O oxygen 16.00 $2s^22p^4$	9 F fluorine 19.00 $2s^22p^5$	10 Ne neon 20.18 $2s^22p^6$
3	11 Na sodium 22.99 $3s^1$	12 Mg magnesium 24.31 $3s^2$											13 Al aluminium 26.98 $3s^23p^1$	14 Si silicon 28.09 $3s^23p^2$	15 P phosphorus 30.97 $3s^23p^3$	16 S sulfur 32.06 $3s^23p^4$	17 Cl chlorine 35.45 $3s^23p^5$	18 Ar argon 39.95 $3s^23p^6$
4	19 K potassium 39.10 $4s^1$	20 Ca calcium 40.08 $4s^2$	21 Sc scandium 44.96 $3d^14s^2$	22 Ti titanium 47.87 $3d^24s^2$	23 V vanadium 50.94 $3d^34s^2$	24 Cr chromium 52.00 $3d^54s^1$	25 Mn manganese 54.94 $3d^54s^2$	26 Fe iron 55.84 $3d^64s^2$	27 Co cobalt 58.93 $3d^74s^2$	28 Ni nickel 58.69 $3d^84s^2$	29 Cu copper 63.55 $3d^{10}4s^1$	30 Zn zinc 65.41 $3d^{10}4s^2$	31 Ga gallium 69.72 $4s^24p^1$	32 Ge germanium 72.64 $4s^24p^2$	33 As arsenic 74.92 $4s^24p^3$	34 Se selenium 78.96 $4s^24p^4$	35 Br bromine 79.90 $4s^24p^5$	36 Kr krypton 83.80 $4s^24p^6$
5	37 Rb rubidium 85.47 $5s^1$	38 Sr strontium 87.62 $5s^2$	39 Y yttrium 88.91 $4d^15s^2$	40 Zr zirconium 91.22 $4d^25s^2$	41 Nb niobium 92.91 $4d^45s^1$	42 Mo molybdenum 95.94 $4d^55s^1$	43 Tc technetium (98) $4d^55s^2$	44 Ru ruthenium 101.07 $4d^75s^1$	45 Rh rhodium 102.90 $4d^85s^1$	46 Pd palladium 106.42 $4d^{10}$	47 Ag silver 107.87 $4d^{10}5s^1$	48 Cd cadmium 112.41 $4d^{10}5s^2$	49 In indium 114.82 $5s^25p^1$	50 Sn tin 118.71 $5s^25p^2$	51 Sb antimony 121.76 $5s^25p^3$	52 Te tellurium 127.60 $5s^25p^4$	53 I iodine 126.90 $5s^25p^5$	54 Xe xenon 131.29 $5s^25p^6$
6	55 Cs caesium 132.91 $6s^1$	56 Ba barium 137.33 $6s^2$	57 La lanthanum 138.91 $5d^16s^2$	72 Hf hafnium 178.49 $5d^26s^2$	73 Ta tantalum 180.95 $5d^36s^2$	74 W tungsten 183.84 $5d^46s^2$	75 Re rhenium 186.21 $5d^56s^2$	76 Os osmium 190.23 $5d^66s^2$	77 Ir iridium 192.22 $5d^76s^2$	78 Pt platinum 195.08 $5d^96s^1$	79 Au gold 196.97 $5d^{10}6s^1$	80 Hg mercury 200.59 $5d^{10}6s^2$	81 Tl thallium 204.38 $6s^26p^1$	82 Pb lead 207.2 $6s^26p^2$	83 Bi bismuth 208.98 $6s^26p^3$	84 Po polonium (209) $6s^26p^4$	85 At astatine (210) $6s^26p^5$	86 Rn radon (222) $6s^26p^6$
7	87 Fr francium (223) $7s^1$	88 Ra radium (226) $7s^2$	89 Ac actinium (227) $6d^17s^2$	104 Rf rutherfordium (261) $6d^27s^2$	105 Db dubnium (262) $6d^37s^2$	106 Sg seaborgium (263) $6d^47s^2$	107 Bh bohrium (262) $6d^57s^2$	108 Hs hassium (265) $6d^67s^2$	109 Mt meitnerium (266) $6d^77s^2$	110 Ds darmstadtium (271) $6d^87s^2$	111 Rg roentgenium (272) $6d^97s^2$	112 Cn copernicium ? $6d^{10}7s^2$	113 Nh nihonium ? $7s^27p^1$	114 Fl flerovium ? $7s^27p^2$	115 Mc moscovium ? $7s^27p^3$	116 Lv livermorium ? $7s^27p^4$	117 Ts tennessine ? $7s^27p^5$	118 Og oganesson ? $7s^27p^6$

Period															
6	58 Ce cerium 140.12 $4f^15d^16s^2$	59 Pr praseodymium 140.91 $4f^36s^2$	60 Nd neodymium 144.24 $4f^46s^2$	61 Pm promethium (145) $4f^56s^2$	62 Sm samarium 150.36 $4f^66s^2$	63 Eu europium 151.96 $4f^76s^2$	64 Gd gadiolinium 157.25 $4f^75d^16s^2$	65 Tb terbium 158.93 $4f^96s^2$	66 Dy dysprosium 162.50 $4f^{10}6s^2$	67 Ho holmium 164.93 $4f^{11}6s^2$	68 Er erbium 167.26 $4f^{12}6s^2$	69 Tm thulium 168.93 $4f^{13}6s^2$	70 Yb ytterbium 173.04 $4f^{14}6s^2$	71 Lu lutetium 174.97 $5d^16s^2$	Lanthanoids (lanthanides)
7	90 Th thorium 232.04 $6d^27s^2$	91 Pa protactinium 231.04 $5f^26d^17s^2$	92 U uranium 238.03 $5f^36d^17s^2$	93 Np neptunium (237) $5f^46d^17s^2$	94 Pu plutonium (244) $5f^67s^2$	95 Am americium (243) $5f^77s^2$	96 Cm curium (247) $5f^76d^17s^2$	97 Bk berkelium (247) $5f^97s^2$	98 Cf californium (251) $5f^{10}7s^2$	99 Es einsteinium (252) $5f^{11}7s^2$	100 Fm fermium (257) $5f^{12}7s^2$	101 Md mendelevium (258) $5f^{13}7s^2$	102 No nobelium (259) $5f^{14}7s^2$	103 Lr lawrencium (262) $6d^17s^2$	Actinoids (actinides)

Numerical values of molar masses in grams per mole (atomic weights) are quoted to the number of significant figures typical of most naturally occurring samples.

FUNDAMENTAL CONSTANTS

Constant	Symbol	Value		
			Power of 10	Units
Speed of light	c	2.997 924 58*	10^{8}	$\mathrm{m\,s^{-1}}$
Elementary charge	e	1.602 176 634*	10^{-19}	C
Planck's constant	h	6.626 070 15	10^{-34}	J s
	$\hbar = h/2\pi$	1.054 571 817	10^{-34}	J s
Boltzmann's constant	k	1.380 649*	10^{-23}	$\mathrm{J\,K^{-1}}$
Avogadro's constant	N_A	6.022 140 76	10^{23}	$\mathrm{mol^{-1}}$
Gas constant	$R = N_A k$	8.314 462		$\mathrm{J\,K^{-1}\,mol^{-1}}$
Faraday's constant	$F = N_A e$	9.648 533 21	10^{4}	$\mathrm{C\,mol^{-1}}$
Mass				
Electron	m_e	9.109 383 70	10^{-31}	kg
Proton	m_p	1.672 621 924	10^{-27}	kg
Neutron	m_n	1.674 927 498	10^{-27}	kg
Atomic mass constant	m_u	1.660 539 067	10^{-27}	kg
Magnetic constant (vacuum permeability)	μ_0	1.256 637 062	10^{-6}	$\mathrm{J\,s^2\,C^{-2}\,m^{-1}}$
Electric constant (vacuum permittivity)	$\varepsilon_0 = 1/\mu_0 c^2$	8.854 187 813	10^{-12}	$\mathrm{J^{-1}\,C^2\,m^{-1}}$
	$4\pi\varepsilon_0$	1.112 650 056	10^{-10}	$\mathrm{J^{-1}\,C^2\,m^{-1}}$
Bohr magneton	$\mu_B = e\hbar/2m_e$	9.274 010 08	10^{-24}	$\mathrm{J\,T^{-1}}$
Nuclear magneton	$\mu_N = e\hbar/2m_p$	5.050 783 75	10^{-27}	$\mathrm{J\,T^{-1}}$
Proton magnetic moment	μ_p	1.410 606 797	10^{-26}	$\mathrm{J\,T^{-1}}$
g-Value of electron	g_e	2.002 319 304		
Magnetogyric ratio				
Electron	$\gamma_e = g_e e/2m_e$	1.760 859 630	10^{11}	$\mathrm{T^{-1}\,s^{-1}}$
Proton	$\gamma_p = 2\mu_p/\hbar$	2.675 221 674	10^{8}	$\mathrm{T^{-1}\,s^{-1}}$
Bohr radius	$a_0 = 4\pi\varepsilon_0\hbar^2/e^2 m_e$	5.291 772 109	10^{-11}	m
Rydberg constant	$\tilde{R}_\infty = m_e e^4/8h^3 c\varepsilon_0^2$	1.097 373 157	10^{5}	$\mathrm{cm^{-1}}$
	$hc\tilde{R}_\infty/e$	13.605 693 12		eV
Fine-structure constant	$\alpha = \mu_0 e^2 c/2h$	7.297 352 5693	10^{-3}	
	α^{-1}	1.370 359 999 08	10^{2}	
Stefan–Boltzmann constant	$\sigma = 2\pi^5 k^4/15h^3c^2$	5.670 374	10^{-8}	$\mathrm{W\,m^{-2}\,K^{-4}}$
Standard acceleration of free fall	g	9.806 65*		$\mathrm{m\,s^{-2}}$
Gravitational constant	G	6.674 30	10^{-11}	$\mathrm{N\,m^2\,kg^{-2}}$

* Exact value. For current values of the constants, see the National Institute of Standards and Technology (NIST) website.

The two volumes of *Physical Chemistry* available in North America—*Thermodynamics and Kinetics* and *Quantum Chemistry, Spectroscopy, and Statistical Thermodynamics*—contain a selection of Focuses from the international edition of *Physical Chemistry*. This selection has been curated to better reflect the coverage of North American courses, over two semesters. The Focus and Topic numbering from the international edition has been retained, but individual pages have been renumbered to give continuous, unbroken pagination.

Atkins'

PHYSICAL CHEMISTRY

Quantum Chemistry, Spectroscopy, and Statistical Thermodynamics

Twelfth edition

Peter Atkins
Fellow of Lincoln College,
University of Oxford,
Oxford, UK

Julio de Paula
Professor of Chemistry,
Lewis & Clark College,
Portland, Oregon, USA

James Keeler
Associate Professor of Chemistry,
University of Cambridge, and
Walters Fellow in Chemistry at Selwyn College,
Cambridge, UK

OXFORD
UNIVERSITY PRESS

OXFORD
UNIVERSITY PRESS

Great Clarendon Street, Oxford, OX2 6DP,
United Kingdom

Oxford University Press is a department of the University of Oxford.
It furthers the University's objective of excellence in research, scholarship,
and education by publishing worldwide. Oxford is a registered trade mark of
Oxford University Press in the UK and in certain other countries

Eighth edition 2006
Ninth edition 2009
Tenth edition 2014
Eleventh edition 2018

Impression: 1

Published in the United States of America by Oxford University Press
198 Madison Avenue, New York, NY 10016, United States of America

British Library Cataloguing in Publication Data
Data available

Library of Congress Control Number: 2022935397

ISBN 978–0–19–885131–8

Printed in the UK by Bell & Bain Ltd., Glasgow

PREFACE

Our *Physical Chemistry* is continuously evolving in response to users' comments, our own imagination, and technical innovation. The text is mature, but it has been given a new vibrancy: it has become dynamic by the creation of an e-book version with the pedagogical features that you would expect. They include the ability to summon up living graphs, get mathematical assistance in an awkward derivation, find solutions to exercises, get feedback on a multiple-choice quiz, and have easy access to data and more detailed information about a variety of subjects. These innovations are not there simply because it is now possible to implement them: they are there to help students at every stage of their course.

The flexible, popular, and less daunting arrangement of the text into readily selectable and digestible Topics grouped together into conceptually related Focuses has been retained. There have been various modifications of emphasis to match the evolving subject and to clarify arguments either in the light of readers' comments or as a result of discussion among ourselves. We learn as we revise, and pass on that learning to our readers.

Our own teaching experience ceaselessly reminds us that mathematics is the most fearsome part of physical chemistry, and we likewise ceaselessly wrestle with finding ways to overcome that fear. First, there is encouragement to use mathematics, for it is the language of much of physical chemistry. The *How is that done?* sections are designed to show that if you want to make progress with a concept, typically making it precise and quantitative, then you have to deploy mathematics. Mathematics opens doors to progress. Then there is the fine-grained help with the manipulation of equations, with their detailed annotations to indicate the steps being taken. Behind all that are *The chemist's toolkits*, which provide brief reminders of the underlying mathematical techniques. There is more behind them, for the collections of Toolkits available via the e-book take their content further and provide illustrations of how the material is used.

The text covers a very wide area and we have sought to add another dimension: depth. Material that we judge too detailed for the text itself but which provides this depth of treatment, or simply adds material of interest springing form the introductory material in the text, can now be found in enhanced *A deeper look* sections available via the e-book. These sections are there for students and instructors who wish to extend their knowledge and see the details of more advanced calculations.

The main text retains *Examples* (where we guide the reader through the process of answering a question) and *Brief illustrations* (which simply indicate the result of using an equation, giving a sense of how it and its units are used). In this edition a few Exercises are provided at the end of each major section in a Topic along with, in the e-book, a selection of multiple-choice questions. These questions give the student the opportunity to check their understanding, and, in the case of the e-book, receive immediate feedback on their answers. Straightforward Exercises and more demanding Problems appear at the end of each Focus, as in previous editions.

The text is living and evolving. As such, it depends very much on input from users throughout the world. We welcome your advice and comments.

PWA
JdeP
JK

USING THE BOOK

TO THE STUDENT

The twelfth edition of *Atkins' Physical Chemistry* has been developed in collaboration with current students of physical chemistry in order to meet your needs better than ever before. Our student reviewers have helped us to revise our writing style to retain clarity but match the way you read. We have also introduced a new opening section, *Energy: A first look*, which summarizes some key concepts that are used throughout the text and are best kept in mind right from the beginning. They are all revisited in greater detail later. The new edition also brings with it a hugely expanded range of digital resources, including living graphs, where you can explore the consequences of changing parameters, video interviews with practising scientists, video tutorials that help to bring key equations to life in each Focus, and a suite of self-check questions. These features are provided as part of an enhanced e-book, which is accessible by using the access code included in the book.

You will find that the e-book offers a rich, dynamic learning experience. The digital enhancements have been crafted to help your study and assess how well you have understood the material. For instance, it provides assessment materials that give you regular opportunities to test your understanding.

Innovative structure

Short, selectable Topics are grouped into overarching Focus sections. The former make the subject accessible; the latter provides its intellectual integrity. Each Topic opens with the questions that are commonly asked: why is this material important?, what should you look out for as a key idea?, and what do you need to know already?

Resource section

The *Resource section* at the end of the book includes a brief review of two mathematical tools that are used throughout the text: differentiation and integration, including a table of the integrals that are encountered in the text. There is a review of units, and how to use them, an extensive compilation of tables of physical and chemical data, and a set of character tables. Short extracts of most of these tables appear in the Topics themselves: they are there to give you an idea of the typical values of the physical quantities mentioned in the text.

FOCUS 5

SIMPLE MIXTURES

Peter Atkins
Fellow of Lincoln College, University of Oxford, Oxford, UK

Julio de Paula
Professor of Chemistry, Lewis & Clark College, Portland, Oregon, USA

James Keeler
Associate Professor of Chemistry, University of Cambridge, and Walters Fellow in Chemistry at Selwyn College, Cambridge, UK

Mixtures are an essential part of chemistry, either in their own right or as starting materials for chemical reactions. This group of Topics deals with the physical properties of mixtures and shows how to express them in terms of thermodynamic quantities.

5A The thermodynamic description of mixtures

The first Topic in this Focus develops the concept of chemical potential as an example of a partial molar quantity and explores how the chemical potential of a substance is used to describe the physical properties of mixtures. The key idea is that at equilibrium the chemical potential of a species is the same in every phase. By making use of the experimental observations known as Raoult's and Henry's laws, it is possible to express the chemical potential of a substance in terms of its mole fraction in a mixture.

5A.1 Partial molar quantities; 5A.2 The thermodynamics of mixing; 5A.3 The chemical potentials of liquids

5B The properties of solutions

In this Topic, the concept of chemical potential is applied to the discussion of the effect of a solute on certain properties of a solution. These properties include the lowering of the vapour pressure of the solvent, the elevation of its boiling point, the depression of its freezing point, and the origin of osmotic pressure. It is possible to construct a model of a certain class of non-ideal solutions called 'regular solutions', which have properties that diverge from those of ideal solutions.

5B.1 Liquid mixtures; 5B.2 Colligative properties

AVAILABLE IN THE E-BOOK

'Impact on…' sections

'Impact on' sections show how physical chemistry is applied in a variety of modern contexts. They showcase physical chemistry as an evolving subject.

Go to this location in the accompanying e-book to view a list of Impacts.

'A deeper look' sections

These sections take some of the material in the text further and are there if you want to extend your knowledge and see the details of some of the more advanced derivations.

Go to this location in the accompanying e-book to view a list of Deeper Looks.

Group theory tables

A link to comprehensive group theory tables can be found at the end of the accompanying e-book.

The chemist's toolkits

The chemist's toolkits are reminders of the key mathematical, physical, and chemical concepts that you need to understand in order to follow the text.

For a consolidated and enhanced collection of the toolkits found throughout the text, go to this location in the accompanying e-book.

TOPIC 2A Internal energy

➤ Why do you need to know this material?

The First Law of thermodynamics is the foundation of the discussion of the role of energy in chemistry. Wherever the generation or use of energy in physical transformations or chemical reactions is of interest, lying in the background are the concepts introduced by the First Law.

➤ What is the key idea?

The total energy of an isolated system is constant.

➤ What do you need to know already?

This Topic makes use of the discussion of the properties of gases (Topic 1A), particularly the perfect gas law. It builds on the definition of work given in *Energy: A first look*.

A **closed system** has a boundary through which matter cannot be transferred.

Both open and closed systems can exchange energy with their surroundings.

An **isolated system** can exchange neither energy nor matter with its surroundings.

2A.1 Work, heat, and energy

Although thermodynamics deals with the properties of bulk systems, it is enriched by understanding the molecular origins of these properties. What follows are descriptions of work, heat, and energy from both points of view.

Contents

Checklist of concepts

A checklist of key concepts is provided at the end of each Topic, so that you can tick off the ones you have mastered.

Checklist of concepts

- ☐ 1. **Work** is the process of achieving motion against an opposing force.
- ☐ 2. **Energy** is the capacity to do work.
- ☐ 3. **Heat** is the process of transferring energy as a result of

Physical chemistry: people and perspectives

Leading figures in a varity of fields share their unique and varied experiences and careers, and talk about the challenges they faced and their achievements to give you a sense of where the study of physical chemistry can lead.

OXFORD UNIVERSITY PRESS

Physical Chemistry: People and Perspectives

Interview with Sean M. Decatur
President of Kenyon College

PRESENTING THE MATHEMATICS

How is that done?

You need to understand how an equation is derived from reasonable assumptions and the details of the steps involved. This is one role for the *How is that done?* sections. Each one leads from an issue that arises in the text, develops the necessary equations, and arrives at a conclusion. These sections maintain the separation of the equation and its derivation so that you can find them easily for review, but at the same time emphasize that mathematics is an essential feature of physical chemistry.

How is that done? 2B.1 Deriving the relation between enthalpy change and heat transfer at constant pressure

In a typical thermodynamic derivation, as here, a common way to proceed is to introduce successive definitions of the quantities of interest and then apply the appropriate constraints.

Step 1 *Write an expression for $H + \mathrm{d}H$ in terms of the definition of H*

For a general infinitesimal change in the state of the system, U changes to $U + \mathrm{d}U$, p changes to $p + \mathrm{d}p$, and V changes to

The chemist's toolkits

The chemist's toolkits are reminders of the key mathematical, physical, and chemical concepts that you need to understand in order to follow the text. Many of these Toolkits are relevant to more than one Topic, and you can view a compilation of them, with enhancements in the form of more information and brief illustrations, in this section of the accompanying e-book.

The chemist's toolkit 7B.1 Complex numbers

A complex number z has the form $z = x + \mathrm{i}y$, where $\mathrm{i} = \sqrt{-1}$. The complex conjugate of a complex number z is $z^* = x - \mathrm{i}y$. Complex numbers combine together according to the following rules:

Addition and subtraction:

$$(a + \mathrm{i}b) + (c + \mathrm{i}d) = (a + c) + \mathrm{i}(b + d)$$

Annotated equations and equation labels

We have annotated many equations to help you follow how they are developed. An annotation can help you travel across the equals sign: it is a reminder of the substitution used, an approximation made, the terms that have been assumed constant, an integral used, and so on. An annotation can also be a reminder of the significance of an individual term in an expression. We sometimes collect into a small box a collection of numbers or symbols to show how they carry from one line to the next. Many of the equations are labelled to highlight their significance.

$$\ln \mathcal{W} = \ln \frac{N!}{N_0!N_1!N_2!\cdots} \overset{\ln(x/y) = \ln x - \ln y}{=} \ln N! - \ln(N_0!N_1!N_2!\cdots)$$

$$\overset{\ln xy = \ln x + \ln y}{=} \ln N! - \ln N_0! - \ln N_1! - \ln N_2! - \cdots = \ln N! - \sum_i \ln N_i!$$

Checklists of equations

A handy checklist at the end of each topic summarizes the most important equations and the conditions under which they apply. Don't think, however, that you have to memorize every equation in these checklists: they are collected there for ready reference.

Checklist of equations

Property	Equation
Enthalpy	$H = U + pV$
Heat transfer at constant pressure	$\mathrm{d}H = \mathrm{d}q_p$, $\Delta H = q_p$

Video tutorials on key equations

Video tutorials to accompany each Focus dig deeper into some of the key equations used throughout that Focus, emphasizing the significance of an equation, and highlighting connections with material elsewhere in the book.

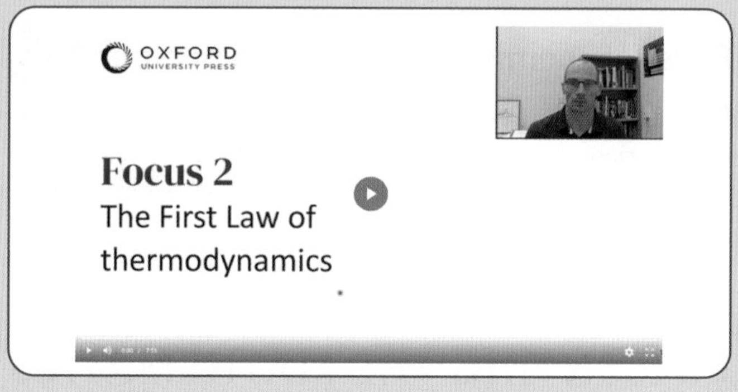

Living graphs

The educational value of many graphs can be heightened by seeing—in a very direct way—how relevant parameters, such as temperature or pressure, affect the plot. You can now interact with key graphs throughout the text in order to explore how they respond as the parameters are changed. These graphs are clearly flagged throughout the book, and you can find links to the dynamic versions in the corresponding location in the e-book.

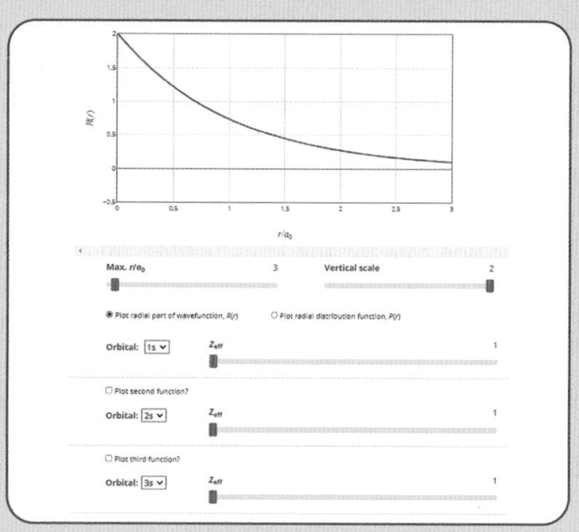

SETTING UP AND SOLVING PROBLEMS

Brief illustrations

A *Brief illustration* shows you how to use an equation or concept that has just been introduced in the text. It shows you how to use data and manipulate units correctly. It also helps you to become familiar with the magnitudes of quantities.

Brief illustration 2B.2

In the reaction 3 $H_2(g) + N_2(g) \rightarrow 2\ NH_3(g)$, 4 mol of gas-phase molecules is replaced by 2 mol of gas-phase molecules, so $\Delta n_g = -2$ mol. Therefore, at 298 K, when $RT = 2.5\ \mathrm{kJ\ mol^{-1}}$, the molar enthalpy and molar internal energy changes taking place in the system are related by

Examples

Worked *Examples* are more detailed illustrations of the application of the material, and typically require you to assemble and deploy several relevant concepts and equations.

Everyone has a different way to approach solving a problem, and it changes with experience. To help in this process, we suggest how you should collect your thoughts and then proceed to a solution. All the worked *Examples* are accompanied by closely related self-tests to enable you to test your grasp of the material after working through our solution as set out in the *Example*.

Example 2B.2 Evaluating an increase in enthalpy with temperature

What is the change in molar enthalpy of N_2 when it is heated from 25 °C to 100 °C? Use the heat capacity information in Table 2B.1.

Collect your thoughts The heat capacity of N_2 changes with temperature significantly in this range, so use eqn 2B.9.

The solution Using $a = 28.58\ \mathrm{J\ K^{-1}\ mol^{-1}}$, $b = 3.77 \times 10^{-3}\ \mathrm{J\ K^{-2}\ mol^{-1}}$,

Self-check questions

This edition introduces self-check questions throughout the text, which can be found at the end of most sections in the e-book. They test your comprehension of the concepts discussed in each section, and provide instant feedback to help you monitor your progress and reinforce your learning. Some of the questions are multiple choice; for them the 'wrong' answers are not simply random numbers but the result of errors that, in our experience, students often make. The feedback from the multiple choice questions not only explains the correct method, but also points out the mistakes that led to the incorrect answer. By working through the multiple-choice questions you will be well prepared to tackle more challenging exercises and problems.

Discussion questions

Discussion questions appear at the end of each Focus, and are organized by Topic. They are designed to encourage you to reflect on the material you have just read, to review the key concepts, and sometimes to think about its implications and limitations.

Exercises and problems

Exercises are provided throughout the main text and, along with Problems, at the end of every Focus. They are all organised by Topic. Exercises are designed as relatively straightforward numerical tests; the Problems are more challenging and typically involve constructing a more detailed answer. For this new edition, detailed solutions are provided in the e-book in the same location as they appear in print.

For the Examples and Problems at the end of each Focus detailed solutions to the odd-numbered questions are provided in the e-book; solutions to the even-numbered questions are available only to lecturers.

FOCUS 1 The properties of gases

To test your understanding of this material, work through the *Exercises*, *Additional exercises*, *Discussion questions*, and *Problems* found throughout this Focus.

Selected solutions can be found at the end of this Focus in the e-book. Solutions to even-numbered questions are available online only to lecturers.

TOPIC 1A The perfect gas

Discussion questions

D1A.1 Explain how the perfect gas equation of state arises by combination of Boyle's law, Charles's law, and Avogadro's principle.

D1A.2 Explain the term 'partial pressure' and explain why Dalton's law is a limiting law.

Additional exercises

E1A.8 Express (i) 22.5 kPa in atmospheres and (ii) 770 Torr in pascals.

E1A.9 Could 25 g of argon gas in a vessel of volume 1.5 dm^3 exert a pressure of 2.0 bar at 30 °C if it behaved as a perfect gas? If not, what pressure would it exert?

E1A.10 A perfect gas undergoes isothermal expansion, which increases its volume by 2.20 dm^3. The final pressure and volume of the gas are 5.04 bar and 4.65 dm^3, respectively. Calculate the original pressure of the gas in (i) bar, (ii) atm.

E1A.11 A perfect gas undergoes isothermal compression, which reduces its volume by 1.80 dm^3. The final pressure and volume of the gas are 1.97 bar and 2.14 dm^3, respectively. Calculate the original pressure of the gas in (i) bar, (ii) torr.

E1A.12 A car tyre (an automobile tire) was inflated to a pressure of 24 lb in^{-2} (1.00 atm = 14.7 lb in^{-2}) on a winter's day when the temperature was −5 °C. What pressure will be found, assuming no leaks have occurred and that the volume is constant, on a subsequent summer's day when the temperature is 35 °C? What complications should be taken into account in practice?

E1A.13 A sample of hydrogen gas was found to have a pressure of 125 kPa when the temperature was 23 °C. What can its pressure be expected to be when the temperature is 11 °C?

E1A.14 A sample of 255 mg of neon occupies 3.00 dm^3 at 122 K. Use the perfect gas law to calculate the pressure of the gas.

E1A.15 A homeowner uses 4.00 × 10^3 m^3 of natural gas in a year to heat a home. Assume that natural gas is all methane, CH_4, and that methane is a perfect gas for the conditions of this problem, which are 1.00 atm and 20 °C. What is the mass of gas used?

E1A.16 At 100 °C and 16.0 kPa, the mass density of phosphorus vapour is 0.6388 kg m^{-3}. What is the molecular formula of phosphorus under these conditions?

E1A.17 Calculate the mass of water vapour present in a room of volume 400 m^3 that contains air at 27 °C on a day when the relative humidity is 60 per cent. *Hint:* Relative humidity is the prevailing partial pressure of water vapour expressed as a percentage of the vapour pressure of water vapour at the same temperature (in this case, 35.6 mbar).

E1A.18 Calculate the mass of water vapour present in a room of volume 250 m^3 that contains air at 23 °C on a day when the relative humidity is 53 per cent (in this case, 28.1 mbar).

E1A.19 Given that the mass density of air at 0.987 bar and 27 °C is 1.146 kg m^{-3}, calculate the mole fraction and partial pressure of nitrogen and oxygen assuming that (i) air consists only of these two gases, (ii) air also contains 1.0 mole per cent Ar.

E1A.20 A gas mixture consists of 320 mg of methane, 175 mg of argon, and 225 mg of neon. The partial pressure of neon at 300 K is 8.87 kPa. Calculate (i) the volume and (ii) the total pressure of the mixture.

E1A.21 The mass density of a gaseous compound was found to be 1.23 kg m^{-3} at 330 K and 20 kPa. What is the molar mass of the compound?

E1A.22 In an experiment to measure the molar mass of a gas, 250 cm^3 of the gas was confined in a glass vessel. The pressure was 152 Torr at 298 K, and after correcting for buoyancy effects, the mass of the gas was 33.5 mg. What is the molar mass of the gas?

E1A.23 The densities of air at −85 °C, 0 °C, and 100 °C are 1.877 g dm^{-3}, 1.294 g dm^{-3}, and 0.946 g dm^{-3}, respectively. From these data, and assuming that air obeys Charles's law, determine a value for the absolute zero of temperature in degrees Celsius.

E1A.24 A certain sample of a gas has a volume of 20.00 dm^3 at 0 °C and 1.000 atm. A plot of the experimental data of its volume against the Celsius temperature, θ, at constant p, gives a straight line of slope 0.0741 dm^3 °C^{-1}. From these data alone (without making use of the perfect gas law), determine the absolute zero of temperature in degrees Celsius.

E1A.25 A vessel of volume 22.4 dm^3 contains 1.5 mol H_2(g) and 2.5 mol N_2(g) at 273.15 K. Calculate (i) the mole fractions of each component, (ii) their partial pressures, and (iii) their total pressure.

Problems

P1A.1 A manometer consists of a U-shaped tube containing a liquid. One side is connected to the apparatus and the other is open to the atmosphere. The pressure p inside the apparatus is given $p = p_{ex} + \rho gh$, where p_{ex} is the external pressure, ρ is the mass density of the liquid in the tube, g = 9.806 m s^{-2} is the acceleration of free fall, and h is the difference in heights of the liquid in the two sides of the tube. (The quantity ρgh is the *hydrostatic pressure* exerted by

Integrated activities

At the end of every Focus you will find questions that span several Topics. They are designed to help you use your knowledge creatively in a variety of ways.

FOCUS 4 Physical transformations of pure substances

Integrated activities

I4.1 Construct the phase diagram for benzene near its triple point at 36 Torr and 5.50 °C from the following data: $\Delta_{fus}H$ = 10.6 kJ mol^{-1}, $\Delta_{vap}H$ = 30.8 kJ mol^{-1}, ρ(s) = 0.891 g cm^{-3}, ρ(l) = 0.879 g cm^{-3}.

I4.2‡ In an investigation of thermophysical properties of methylbenzene R.D. Goodwin (*J. Phys. Chem. Ref. Data* **18**, 1565 (1989)) presented expressions for two coexistence curves. The solid–liquid curve is given by

$$p/\text{bar} = p_3/\text{bar} + 1000(5.60 + 11.727x)x$$

where $x = T/T_3 - 1$ and the triple point pressure and temperature are p_3 = 0.4362 μbar and T_3 = 178.15 K. The liquid–vapour curve is given by

$$\ln(p/\text{bar}) = -10.418/y + 21.157 - 15.996y + 14.015y^2 - 5.0120y^3 + 4.7334(1-y)^{1.70}$$

(c) Plot $T_m/(\Delta_{hb}H_m/\Delta_{hb}S_m)$ for $5 \le N \le 20$. At what value of N does T_m change by less than 1 per cent when N increases by 1?

I4.4‡ A substance as well-known as methane still receives research attention because it is an important component of natural gas, a commonly used fossil fuel. Friend et al. have published a review of thermophysical properties of methane (D.G. Friend, J.F. Ely, and H. Ingham, *J. Phys. Chem. Ref. Data* **18**, 583 (1989)), which included the following vapour pressure data describing the liquid–vapour coexistence curve.

T/K	100	108	110	112	114	120	130	140	150	160	170	190
p/MPa	0.034	0.074	0.088	0.104	0.122	0.192	0.368	0.642	1.041	1.593	2.329	4.521

(a) Plot the liquid–vapour coexistence curve. (b) Estimate the standard boiling point of methane. (c) Compute the standard enthalpy of vaporization of methane (at the standard boiling point), given that the molar volumes of

TAKING YOUR LEARNING FURTHER

'Impact' sections

'*Impact*' sections show you how physical chemistry is applied in a variety of modern contexts. They showcase physical chemistry as an evolving subject. These sections are listed at the beginning of the text, and are referred to at appropriate places elsewhere. You can find a compilation of '*Impact*' sections at the end of the e-book.

A deeper look

These sections take some of the material in the text further. Read them if you want to extend your knowledge and see the details of some of the more advanced derivations. They are listed at the beginning of the text and are referred to where they are relevant. You can find a compilation of Deeper Looks at the end of the e-book.

Group theory tables

If you need character tables, you can find them at the end of the *Resource section*.

TO THE INSTRUCTOR

We have designed the text to give you maximum flexibility in the selection and sequence of Topics, while the grouping of Topics into Focuses helps to maintain the unity of the subject. Additional resources are:

Figures and tables from the book

Lecturers can find the artwork and tables from the book in ready-to-download format. They may be used for lectures without charge (but not for commercial purposes without specific permission).

Key equations

Supplied in Word format so you can download and edit them.

Solutions to exercises, problems, and integrated activities

For the discussion questions, examples, problems, and integrated activities detailed solutions to the even-numbered questions are available to lecturers online, so they can be set as homework or used as discussion points in class.

Lecturer resources are available only to registered adopters of the textbook. To register, simply visit www.oup.com/he/pchem12e_US and follow the appropriate links.

ABOUT THE AUTHORS

Photograph by Natasha Ellis-Knight.

Peter Atkins is a fellow of Lincoln College, Oxford, and emeritus professor of physical chemistry in the University of Oxford. He is the author of over seventy books for students and a general audience. His texts are market leaders around the globe. A frequent lecturer throughout the world, he has held visiting professorships in France, Israel, Japan, China, Russia, and New Zealand. He was the founding chairman of the Committee on Chemistry Education of the International Union of Pure and Applied Chemistry and was a member of IUPAC's Physical and Biophysical Chemistry Division.

Julio de Paula is Professor of Chemistry at Lewis & Clark College. A native of Brazil, he received a B.A. degree in chemistry from Rutgers, The State University of New Jersey, and a Ph.D. in biophysical chemistry from Yale University. His research activities encompass the areas of molecular spectroscopy, photochemistry, and nanoscience. He has taught courses in general chemistry, physical chemistry, biochemistry, inorganic chemistry, instrumental analysis, environmental chemistry, and writing. Among his professional honours are a Christian and Mary Lindback Award for Distinguished Teaching, a Henry Dreyfus Teacher-Scholar Award, and a STAR Award from the Research Corporation for Science Advancement.

Photograph by Nathan Pitt, © University of Cambridge.

James Keeler is Associate Professor of Chemistry, University of Cambridge, and Walters Fellow in Chemistry at Selwyn College. He received his first degree and doctorate from the University of Oxford, specializing in nuclear magnetic resonance spectroscopy. He is presently Head of Department, and before that was Director of Teaching in the department and also Senior Tutor at Selwyn College.

ACKNOWLEDGEMENTS

A book as extensive as this could not have been written without significant input from many individuals. We would like to thank the hundreds of instructors and students who contributed to this and the previous eleven editions:

Scott Anderson, *University of Utah*
Milan Antonijevic, *University of Greenwich*
Elena Besley, *University of Greenwich*
Merete Bilde, *Aarhus University*
Matthew Blunt, *University College London*
Simon Bott, *Swansea University*
Klaus Braagaard Møller, *Technical University of Denmark*
Wesley Browne, *University of Groningen*
Sean Decatur, *Kenyon College*
Anthony Harriman, *Newcastle University*
Rigoberto Hernandez, *Johns Hopkins University*
J. Grant Hill, *University of Sheffield*
Kayla Keller, *Kentucky Wesleyan College*
Kathleen Knierim, *University of Louisiana Lafayette*
Tim Kowalczyk, *Western Washington University*
Kristin Dawn Krantzman, *College of Charleston*
Hai Lin, *University of Colorado Denver*
Mikko Linnolahti, *University of Eastern Finland*
Mike Lyons, *Trinity College Dublin*
Jason McAfee, *University of North Texas*
Joseph McDouall, *University of Manchester*
Hugo Meekes, *Radboud University*
Gareth Morris, *University of Manchester*
David Rowley, *University College London*
Nessima Salhi, *Uppsala University*
Andy S. Sardjan, *University of Groningen*
Trevor Sears, *Stony Brook University*
Gemma Shearman, *Kingston University*
John Slattery, *University of York*
Catherine Southern, *DePaul University*
Michael Staniforth, *University of Warwick*
Stefan Stoll, *University of Washington*
Mahamud Subir, *Ball State University*
Enrico Tapavicza, *CSU Long Beach*
Jeroen van Duifneveldt, *University of Bristol*
Darren Walsh, *University of Nottingham*
Graeme Watson, *Trinity College Dublin*
Darren L. Williams, *Sam Houston State University*
Elisabeth R. Young, *Lehigh University*

Our special thanks also go to the many student reviewers who helped to shape this twelfth edition:

Katherine Ailles, *University of York*
Mohammad Usman Ali, *University of Manchester*
Rosalind Baverstock, *Durham University*
Grace Butler, *Trinity College Dublin*
Kaylyn Cater, *Cardiff University*
Ruth Comerford, *University College Dublin*
Orlagh Fraser, *University of Aberdeen*
Dexin Gao, *University College London*
Suruthi Gnanenthiran, *University of Bath*
Milena Gonakova, *University of the West of England Bristol*
Joseph Ingle, *University of Lincoln*
Jeremy Lee, *University of Durham*
Luize Luse, *Heriot-Watt University*
Zoe Macpherson, *University of Strathclyde*
Sukhbir Mann, *University College London*
Declan Meehan, *Trinity College Dublin*
Eva Pogacar, *Heriot-Watt University*
Pawel Pokorski, *Heriot-Watt University*
Fintan Reid, *University of Strathclyde*
Gabrielle Rennie, *University of Strathclyde*
Annabel Savage, *Manchester Metropolitan University*
Sophie Shearlaw, *University of Strathclyde*
Yutong Shen, *University College London*
Saleh Soomro, *University College London*
Matthew Tully, *Bangor University*
Richard Vesely, *University of Cambridge*
Phoebe Williams, *Nottingham Trent University*

We would also like to thank Michael Clugston for proofreading the entire book, and Peter Bolgar, Haydn Lloyd, Aimee North, Vladimiras Oleinikovas, and Stephanie Smith who all worked alongside James Keeler in the writing of the solutions to the exercises and problems. The multiple-choice questions were developed in large part by Dr Stephanie Smith (Yusuf Hamied Department of Chemistry and Pembroke College, University of Cambridge). These questions and further exercises were integrated into the text by Chloe Balhatchet (Yusuf Hamied Department of Chemistry and Selwyn College, University of Cambridge), who also worked on the living graphs. The solutions to the exercises and problems are taken from the solutions manual for the eleventh edition prepared by Peter Bolgar, Haydn Lloyd, Aimee North, Vladimiras Oleinikovas, Stephanie Smith, and James Keeler, with additional contributions from Chloe Balhatchet.

Last, but by no means least, we acknowledge our two commissioning editors, Jonathan Crowe of Oxford University Press and Jason Noe of OUP USA, and their teams for their assistance, advice, encouragement, and patience. We owe special thanks to Katy Underhill, Maria Bajo Gutiérrez, and Keith Faivre from OUP, who skillfully shepherded this complex project to completion.

BRIEF CONTENTS

FULL CONTENTS

CONVENTIONS

To avoid intermediate rounding errors, but to keep track of values in order to be aware of values and to spot numerical errors, we display intermediate results as *n.nnn*. . . and round the calculation only at the final step.

PHYSICAL CHEMISTRY: PEOPLE AND PERSPECTIVES

To watch these interviews, go to this section of the e-book.

LIST OF TABLES

LIST OF *THE CHEMIST'S TOOLKITS*

LIST OF MATERIAL PROVIDED AS *A DEEPER LOOK*

The list of *A deeper look* material that can be found via the e-book. You will also find references to this material where relevant throughout the book.

LIST OF *IMPACTS*

The list of *Impacts* that can be found via the e-book. You will also find references to this material where relevant throughout the book.

ENERGY A First Look

Much of chemistry is concerned with the transfer and transformation of energy, so right from the outset it is important to become familiar with this concept. The first ideas about energy emerged from **classical mechanics**, the theory of motion formulated by Isaac Newton in the seventeenth century. In the twentieth century classical mechanics gave way to **quantum mechanics**, the theory of motion formulated for the description of small particles, such as electrons, atoms, and molecules. In quantum mechanics the concept of energy not only survived but was greatly enriched, and has come to underlie the whole of physical chemistry.

1 Force

Classical mechanics is formulated in terms of the forces acting on particles, and shows how the paths of particles respond to them by accelerating or changing direction. Much of the discussion focuses on a quantity called the 'momentum' of the particle.

(a) Linear momentum

'Translation' is the motion of a particle through space. The **velocity**, $\boldsymbol{v}$, of a particle is the rate of change of its position. Velocity is a 'vector quantity', meaning that it has both a direction and a magnitude, and is expressed in terms of how fast the particle travels with respect to x-, y-, and z-axes (Fig. 1).

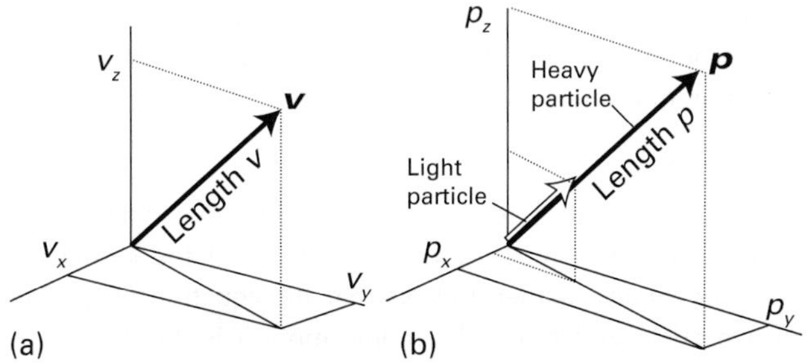

Figure 1 (a) The velocity $\boldsymbol{v}$ is denoted by a vector of magnitude v (the speed) and an orientation that indicates the direction of translational motion. (b) Similarly, the linear momentum $\boldsymbol{p}$ is denoted by a vector of magnitude p and an orientation that corresponds to the direction of motion.

For example, the x-component, v_x, is the particle's rate of change of position along the x-axis:

$$v_x = \frac{dx}{dt} \qquad \text{Component of velocity [definition]} \qquad (1a)$$

Similar expressions may be written for the y- and z-components. The magnitude of the velocity, as represented by the length of the velocity vector, is the **speed**, v. Speed is related to the components of velocity by

$$v = (v_x^2 + v_y^2 + v_z^2)^{1/2} \qquad \text{Speed [definition]} \qquad (1b)$$

The **linear momentum**, $\boldsymbol{p}$, of a particle, like the velocity, is a vector quantity, but takes into account the mass of the particle as well as its speed and direction. Its components are p_x, p_y, and p_z along each axis (Fig. 1b) and its magnitude is p. A heavy particle travelling at a certain speed has a greater linear momentum than a light particle travelling at the same speed. For a particle of mass m, the x-component of the linear momentum is given by

$$p_x = mv_x \qquad \text{Component of linear momentum [definition]} \qquad (2)$$

and similarly for the y- and z-components.

Brief illustration 1

Imagine a particle of mass m attached to a spring. When the particle is displaced from its equilibrium position and then released, it oscillates back and forth about this equilibrium position. This model can be used to describe many features of a chemical bond. In an idealized case, known as the *simple harmonic oscillator*, the displacement from equilibrium $x(t)$ varies with time as

$$x(t) = A\sin 2\pi\nu t$$

In this expression, ν (nu) is the frequency of the oscillation and A is its amplitude, the maximum value of the displacement along the x-axis. The x-component of the velocity of the particle is therefore

$$v_x = \frac{dx}{dt} = \frac{d(A\sin 2\pi\nu t)}{dt} = 2\pi\nu A\cos 2\pi\nu t$$

The x-component of the linear momentum of the particle is

$$p_x = mv_x = 2\pi\nu Am\cos 2\pi\nu t$$

(b) Angular momentum

'Rotation' is the change of orientation in space around a central point (the 'centre of mass'). Its description is very similar to that of translation but with 'angular velocity' taking the place of velocity and 'moment of inertia' taking the place of mass. The **angular velocity**, $\boldsymbol{\omega}$ (omega) is the rate of change of orientation (for example, in radians per second); it is a vector with magnitude ω. The **moment of inertia,** I, is a measure of the mass that is being swung round by the rotational motion. For a particle of mass m moving in a circular path of radius r, the moment of inertia is

$$I = mr^2$$ Moment of inertia [definition] (3a)

For a molecule composed of several atoms, each atom i gives a contribution of this kind, and the moment of inertia around a given axis is

$$I = \sum_i m_i r_i^2 \qquad (3b)$$

where r_i is the perpendicular distance from the mass m_i to the axis. The rotation of a particle is described by its **angular momentum,** $\boldsymbol{J}$, a vector with a length that indicates the rate at which the particle circulates and a direction that indicates the axis of rotation (Fig. 2). The components of angular momentum, J_x, J_y, and J_z, on three perpendicular axes show how much angular momentum is associated with rotation around each axis. The magnitude J of the angular momentum is

$$J = I\omega$$ Magnitude of angular momentum [definition] (4)

Brief illustration 2

A CO_2 molecule is linear, and the length of each CO bond is 116 pm. The mass of each ^{16}O atom is $16.00m_u$, where $m_u = 1.661 \times 10^{-27}$ kg. It follows that the moment of inertia of the molecule around an axis perpendicular to the axis of the molecule and passing through the C atom is

$$\begin{aligned} I &= m_O R^2 + 0 + m_O R^2 = 2m_O R^2 \\ &= 2 \times (16.00 \times 1.661 \times 10^{-27}\ \mathrm{kg}) \times (1.16 \times 10^{-10}\ \mathrm{m})^2 \\ &= 7.15 \times 10^{-46}\ \mathrm{kg\,m^2} \end{aligned}$$

(c) Newton's second law of motion

The central concept of classical mechanics is **Newton's second law of motion**, which states that *the rate of change of momentum is equal to the force acting on the particle.* This law underlies the calculation of the **trajectory** of a particle, a statement about where it is and where it is moving at each moment of time. Like

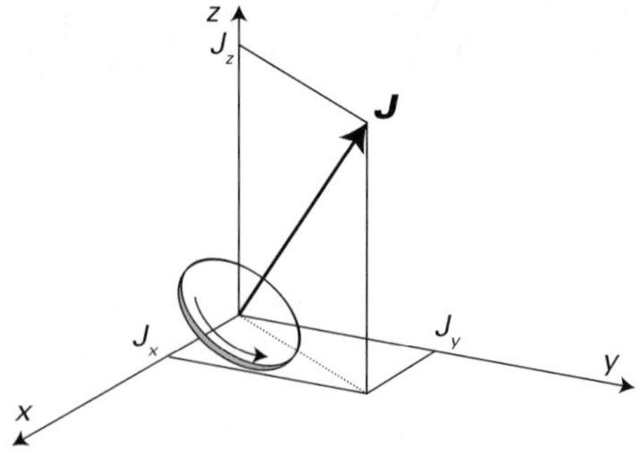

Figure 2 The angular momentum $\boldsymbol{J}$ of a particle is represented by a vector along the axis of rotation and perpendicular to the plane of rotation. The length of the vector denotes the magnitude J of the angular momentum. The direction of motion is clockwise to an observer looking in the direction of the vector.

the velocity and momentum, the **force**, $\boldsymbol{F}$, is a vector quantity with a direction and a magnitude (the 'strength' of the force). Force is reported in newtons, with $1\ \mathrm{N} = 1\ \mathrm{kg\,m\,s^{-2}}$. For motion along the x-axis Newton's second law states that

$$\frac{\mathrm{d}p_x}{\mathrm{d}t} = F_x$$ Newton's second law [in one dimension] (5a)

where F_x is the component of the force acting along the x-axis. Each component of linear momentum obeys the same kind of equation, so the vector $\boldsymbol{p}$ changes with time as

$$\frac{\mathrm{d}\boldsymbol{p}}{\mathrm{d}t} = \boldsymbol{F}$$ Newton's second law [vector form] (5b)

Equation 5 is the **equation of motion** of the particle, the equation that has to be solved to calculate its trajectory.

Brief illustration 3

According to 'Hooke's law', the force acting on a particle undergoing harmonic motion (like that in *Brief illustration* 2) is proportional to the displacement and directed opposite to the direction of motion, so in one dimension

$$F_x = -k_f x$$

where x is the displacement from equilibrium and k_f is the 'force constant', a measure of the stiffness of the spring (or chemical bond). It then follows that the equation of motion of a particle undergoing harmonic motion is $\mathrm{d}p_x/\mathrm{d}t = -k_f x$. Then, because $p_x = mv_x$ and $v_x = \mathrm{d}x/\mathrm{d}t$, it follows that $\mathrm{d}p_x/\mathrm{d}t = m\mathrm{d}v_x/\mathrm{d}t = m\mathrm{d}^2x/\mathrm{d}t^2$. With this substitution, the equation of motion becomes

$$m\frac{\mathrm{d}^2x}{\mathrm{d}t^2} = -kx$$

Equations of this kind, which are called 'differential equations', are solved by special techniques. In most cases in this text, the solutions are simply stated without going into the details of how they are found.

Similar considerations apply to rotation. The change in angular momentum of a particle is expressed in terms of the **torque**, T, a twisting force. The analogue of eqn 5b is then

$$\frac{dJ}{dt} = T \qquad (6)$$

Quantities that describe translation and rotation are analogous, as shown below:

Property	Translation	Rotation
Rate	linear velocity, v	angular velocity, ω
Resistance to change	mass, m	moment of inertia, I
Momentum	linear momentum, p	angular momentum, J
Influence on motion	force, F	torque, T

2 Energy

Energy is a powerful and essential concept in science; nevertheless, its actual nature is obscure and it is difficult to say what it 'is'. However, it can be related to processes that can be measured and can be defined in terms of the measurable process called work.

(a) Work

Work, w, is done in order to achieve motion against an opposing force. The work needed to be done to move a particle through the infinitesimal distance dx against an opposing force F_x is

$$dw_{\text{on the particle}} = -F_x dx \qquad \text{Work [definition]} \qquad (7a)$$

When the force is directed to the left (to negative x), F_x is negative, so for motion to the right (dx positive), the work that must be done to move the particle is positive. With force in newtons and distance in metres, the units of work are joules (J), with $1\,J = 1\,N\,m = 1\,kg\,m^2\,s^{-2}$.

The total work that has to be done to move a particle from x_{initial} to x_{final} is found by integrating eqn 7a, allowing for the possibility that the force may change at each point along the path:

$$w_{\text{on the particle}} = -\int_{x_{\text{initial}}}^{x_{\text{final}}} F_x dx \qquad \text{Work} \qquad (7b)$$

Brief illustration 4

Suppose that when a bond is stretched from its equilibrium value R_e to some arbitrary value R there is a restoring force proportional to the displacement $x = R - R_e$ from the equilibrium length. Then

$$F_x = -k_f(R - R_e) = -k_f x$$

The constant of proportionality, k_f, is the force constant introduced in *Brief illustration* 3. The total work needed to move an atom so that the bond stretches from zero displacement ($x_{\text{initial}} = 0$), when the bond has its equilibrium length, to a displacement $x_{\text{final}} = R_{\text{final}} - R_e$ is

$$w_{\text{on an atom}} = -\int_0^{x_{\text{final}}} (-k_f x)\, dx = k_f \overbrace{\int_0^{x_{\text{final}}} x\, dx}^{\text{Integral A.1}}$$
$$= \tfrac{1}{2}k_f x_{\text{final}}^2 = \tfrac{1}{2}k_f(R_{\text{final}} - R_e)^2$$

(All the integrals required in this book are listed in the *Resource section*.) The work required increases as the square of the displacement: it takes four times as much work to stretch a bond through 20 pm as it does to stretch the same bond through 10 pm.

(b) The definition of energy

Now we get to the core of this discussion. **Energy** *is the capacity to do work*. An object with a lot of energy can do a lot of work; one with little energy can do only little work. Thus, a spring that is compressed can do a lot of work as it expands, so it is said to have a lot of energy. Once the spring is expanded it can do only a little work, perhaps none, so it is said to have only a little energy. The SI unit of energy is the same as that of work, namely the joule, with $1\,J = 1\,kg\,m^2\,s^{-2}$.

A particle may possess two kinds of energy, kinetic energy and potential energy. The **kinetic energy**, E_k, of a particle is the energy it possesses as a result of its motion. For a particle of mass m travelling at a speed v,

$$E_k = \tfrac{1}{2}mv^2 \qquad \text{Kinetic energy [definition]} \qquad (8a)$$

A particle with a lot of kinetic energy can do a lot of work, in the sense that if it collides with another particle it can cause it to move against an opposing force. Because the magnitude of the linear momentum and speed are related by $p = mv$, so $v = p/m$, an alternative version of this relation is

$$E_k = \frac{p^2}{2m} \qquad (8b)$$

It follows from Newton's second law that if a particle is initially stationary and is subjected to a constant force then its linear momentum increases from zero. Because the magnitude of the applied force may be varied at will, the momentum and therefore the kinetic energy of the particle may be increased to any value.

The **potential energy**, E_p or V, of a particle is the energy it possesses as a result of its position. For instance, a stationary weight high above the surface of the Earth can do a lot of work as it falls to a lower level, so is said to have more energy, in this case potential energy, than when it is resting on the surface of the Earth.

This definition can be turned around. Suppose the weight is returned from the surface of the Earth to its original height. The work needed to raise it is equal to the potential energy that it once again possesses. For an infinitesimal change in height, dx, that work is $-F_x dx$. Therefore, the infinitesimal change in potential energy is $dE_p = -F_x dx$. This equation can be rearranged into a relation between the force and the potential energy:

$$F_x = -\frac{dE_p}{dx} \quad \text{or} \quad F_x = -\frac{dV}{dx} \qquad \text{Relation of force to potential energy} \qquad (9)$$

No *universal* expression for the dependence of the potential energy on position can be given because it depends on the type of force the particle experiences. However, there are two very important specific cases where an expression can be given. For a particle of mass m at an altitude h close to the surface of the Earth, the **gravitational potential energy** is

$$E_p(h) = E_p(0) + mgh \qquad \text{Gravitational potential energy [close to surface of the Earth]} \qquad (10)$$

where g is the **acceleration of free fall** (g depends on location, but its 'standard value' is close to 9.81 m s^{-2}). The zero of potential energy is arbitrary. For a particle close to the surface of the Earth, it is common to set $E_p(0) = 0$.

The other very important case (which occurs whenever the structures of atoms and molecules are discussed), is the electrostatic potential energy between two electric charges Q_1 and Q_2 at a separation r in a vacuum. This **Coulomb potential energy** is

$$E_p(r) = \frac{Q_1 Q_2}{4\pi\varepsilon_0 r} \qquad \text{Coulomb potential energy [in a vacuum]} \qquad (11)$$

Charge is expressed in coulombs (C). The constant ε_0 (epsilon zero) is the **electric constant** (or *vacuum permittivity*), a fundamental constant with the value 8.854×10^{-12} C^2 J^{-1} m^{-1}. It is conventional (as in eqn 11) to set the potential energy equal to zero at infinite separation of charges.

The **total energy** of a particle is the sum of its kinetic and potential energies:

$$E = E_k + E_p, \quad \text{or} \quad E = E_k + V \qquad \text{Total energy} \qquad (12)$$

A fundamental feature of nature is that *energy is conserved*; that is, energy can neither be created nor destroyed. Although energy can be transformed from one form to another, its total is constant.

An alternative way of thinking about the potential energy arising from the interaction of charges is in terms of the **potential**, which is a measure of the 'potential' of one charge to affect the potential energy of another charge when the second charge is brought into its vicinity. A charge Q_1 gives rise to a **Coulomb potential** ϕ_1 (phi) such that the potential energy of the interaction with a second charge Q_2 is $Q_2\phi_1(r)$. Comparison of this expression with eqn 11 shows that

$$\phi_1(r) = \frac{Q_1}{4\pi\varepsilon_0 r} \qquad \text{Coulomb potential [in a vacuum]} \qquad (13)$$

The units of potential are joules per coulomb, J C^{-1}, so when the potential is multiplied by a charge in coulombs, the result is the potential energy in joules. The combination joules per coulomb occurs widely and is called a volt (V): 1 V = 1 J C^{-1}.

The language developed here inspires an important alternative energy unit, the **electronvolt** (eV): 1 eV is defined as the potential energy acquired when an electron is moved through a potential difference of 1 V. The relation between electronvolts and joules is

$$1\,\text{eV} = 1.602 \times 10^{-19}\,\text{J}$$

Many processes in chemistry involve energies of a few electronvolts. For example, to remove an electron from a sodium atom requires about 5 eV.

3 Temperature

A key idea of quantum mechanics is that the translational energy of a molecule, atom, or electron that is confined to a region of space, and any rotational or vibrational energy that a molecule possesses, is **quantized**, meaning that it is restricted to certain discrete values. These permitted energies are called **energy levels**. The values of the permitted energies depend on the characteristics of the particle (for instance, its mass) and for translation the extent of the region to which it is confined. The allowed energies are widest apart for particles of small mass confined to small regions of space. Consequently, quantization must be taken into account for electrons bound to nuclei in atoms and molecules. It can be ignored for macroscopic bodies, for which the separation of all kinds of energy levels is so small that for all practical purposes their energy can be varied virtually continuously.

Figure 3 depicts the typical energy level separations associated with rotational, vibrational, and electronic motion. The separation of rotational energy levels (in small molecules, about 10^{-21} J, corresponding to about 0.6 kJ mol^{-1}) is smaller than that of vibrational energy levels (about 10^{-20}–10^{-19} J, or

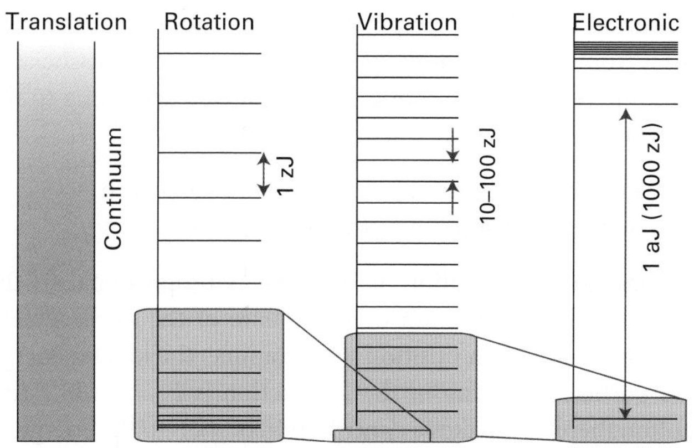

Figure 3 The energy level separations typical of four types of system. (1 zJ = 10^{-21} J; in molar terms, 1 zJ is equivalent to about 0.6 kJ mol^{-1}.)

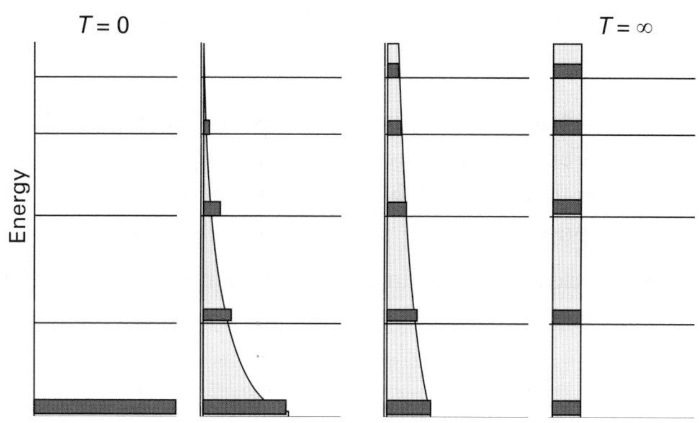

Figure 4 The Boltzmann distribution of populations (represented by the horizontal bars) for a system of five states with different energies as the temperature is raised from zero to infinity. *Interact with the dynamic version of this graph in the e-book.*

6–60 kJ mol^{-1}), which itself is smaller than that of electronic energy levels (about 10^{-18} J, corresponding to about 600 kJ mol^{-1}).

(a) The Boltzmann distribution

The continuous thermal agitation that molecules experience in a sample ensures that they are distributed over the available energy levels. This distribution is best expressed in terms of the occupation of states. The distinction between a state and a level is that a given level may be comprised of several states all of which have the same energy. For instance, a molecule might be rotating clockwise with a certain energy, or rotating counterclockwise with the same energy. One particular molecule may be in a state belonging to a low energy level at one instant, and then be excited into a state belonging to a high energy level a moment later. Although it is not possible to keep track of which state each molecule is in, it is possible to talk about the *average* number of molecules in each state. A remarkable feature of nature is that, for a given array of energy levels, how the molecules are distributed over the states depends on a single parameter, the 'temperature', T.

The **population** of a state is the average number of molecules that occupy it. The populations, whatever the nature of the states (translational, rotational, and so on), are given by a formula derived by Ludwig Boltzmann and known as the **Boltzmann distribution**. According to Boltzmann, the ratio of the populations of states with energies ε_i and ε_j is

$$\frac{N_i}{N_j} = \mathrm{e}^{-(\varepsilon_i-\varepsilon_j)/kT} \qquad \text{Boltzmann distribution} \qquad (14a)$$

where k is **Boltzmann's constant**, a fundamental constant with the value $k = 1.381 \times 10^{-23}$ J K^{-1} and T is the **temperature**, the parameter that specifies the relative populations of states, regardless of their type. Thus, when $T = 0$, the populations of all states other than the lowest state (the 'ground state') of the molecule are zero. As the value of T is increased (the 'temperature is raised'), the populations of higher energy states increase, and the distribution becomes more uniform. This behaviour is illustrated in Fig. 4 for a system with five states of different energy. As predicted by eqn 14a, as the temperature approaches infinity ($T \rightarrow \infty$), the states become equally populated.

In chemical applications it is common to use molar energies, $E_{\mathrm{m},i}$, with $E_{\mathrm{m},i} = N_{\mathrm{A}}\varepsilon_i$, where N_{A} is Avogadro's constant. Then eqn 14a becomes

$$\frac{N_i}{N_j} = \mathrm{e}^{-(E_{\mathrm{m},i}/N_{\mathrm{A}}-E_{\mathrm{m},j}/N_{\mathrm{A}})/kT} = \mathrm{e}^{-(E_{\mathrm{m},i}-E_{\mathrm{m},j})/N_{\mathrm{A}}kT} = \mathrm{e}^{-(E_{\mathrm{m},i}-E_{\mathrm{m},j})/RT} \qquad (14b)$$

where $R = N_{\mathrm{A}}k$. The constant R is known as the 'gas constant'; it appears in expressions of this kind when molar, rather than molecular, energies are specified. Moreover, because it is simply the molar version of the more fundamental Boltzmann constant, it occurs in contexts other than gases.

Brief illustration 5

Methylcyclohexane molecules may exist in one of two conformations, with the methyl group in either an equatorial or axial position. The equatorial form lies 6.0 kJ mol^{-1} lower in energy than the axial form. The relative populations of molecules in the axial and equatorial states at 300 K are therefore

$$\frac{N_{\text{axial}}}{N_{\text{equatorial}}} = \mathrm{e}^{-(E_{\mathrm{m,axial}}-E_{\mathrm{m,equatorial}})/RT}$$
$$= \mathrm{e}^{-(6.0\times10^{3}\,\mathrm{J\,mol^{-1}})/(8.3145\,\mathrm{J\,K^{-1}\,mol^{-1}})\times(300\,\mathrm{K})}$$
$$= 0.090$$

The number of molecules in an axial conformation is therefore just 9 per cent of those in the equatorial conformation.

The important features of the Boltzmann distribution to bear in mind are:

- The distribution of populations is an exponential function of energy and the temperature. As the temperature is increased, states with higher energy become progressively more populated.
- States closely spaced in energy compared to kT are more populated than states that are widely spaced compared to kT.

The energy spacings of translational and rotational states are typically much less than kT at room temperature. As a result, many translational and rotational states are populated. In contrast, electronic states are typically separated by much more than kT. As a result, only the ground electronic state of a molecule is occupied at normal temperatures. Vibrational states are widely separated in small, stiff molecules and only the ground vibrational state is populated. Large and flexible molecules are also found principally in their ground vibrational state, but might have a few higher energy vibrational states populated at normal temperatures.

(b) The equipartition theorem

For gases consisting of non-interacting particles it is often possible to calculate the average energy associated with each type of motion by using the **equipartition theorem**. This theorem arises from a consideration of how the energy levels associated with different kinds of motion are populated according to the Boltzmann distribution. The theorem states that

> At thermal equilibrium, the average value of each quadratic contribution to the energy is $\frac{1}{2}kT$.

A 'quadratic contribution' is one that is proportional to the square of the momentum or the square of the displacement from an equilibrium position. For example, the kinetic energy of a particle travelling in the x-direction is $E_k = p_x^2/2m$. This motion therefore makes a contribution of $\frac{1}{2}kT$ to the energy.

The energy of vibration of atoms in a chemical bond has *two* quadratic contributions. One is the kinetic energy arising from the back and forth motion of the atoms. Another is the potential energy which, for the harmonic oscillator, is $E_p = \frac{1}{2}k_f x^2$ and is a second quadratic contribution. Therefore, the total average energy is $\frac{1}{2}kT + \frac{1}{2}kT = kT$.

The equipartition theorem applies only if many of the states associated with a type of motion are populated. At temperatures of interest to chemists this condition is always met for translational motion, and is usually met for rotational motion. Typically, the separation between vibrational and electronic states is greater than for rotation or translation, and as only a few states are occupied (often only one, the ground state), the equipartition theorem is unreliable for these types of motion.

Checklist of concepts

☐ 1. **Newton's second law of motion** states that the rate of change of momentum is equal to the force acting on the particle.

☐ 2. **Work** is done in order to achieve motion against an opposing force. **Energy** is the capacity to do work.

☐ 3. The **kinetic energy** of a particle is the energy it possesses as a result of its motion.

☐ 4. The **potential energy** of a particle is the energy it possesses as a result of its position.

☐ 5. The total energy of a particle is the sum of its kinetic and potential energies.

☐ 6. The **Coulomb potential energy** between two charges separated by a distance r varies as $1/r$.

☐ 7. The energy levels of confined particles are quantized, as are those of rotating or vibrating molecules.

☐ 8. The **Boltzmann distribution** is a formula for calculating the relative populations of states of various energies.

☐ 9. The **equipartition theorem** states that for a sample at thermal equilibrium the average value of each quadratic contribution to the energy is $\frac{1}{2}kT$.

Checklist of equations

Property	Equation	Comment	Equation number
Component of velocity in *x* direction	$v_x = \mathrm{d}x/\mathrm{d}t$	Definition; likewise for *y* and *z*	1a
Component of linear momentum in *x* direction	$p_x = mv_x$	Definition; likewise for *y* and *z*	2
Moment of inertia	$I = mr^2$	Point particle	3a
	$I = \sum_i m_i r_i^2$	Molecule	3b
Angular momentum	$J = I\omega$		4
Equation of motion	$F_x = \mathrm{d}p_x/\mathrm{d}t$	Motion along *x*-direction	5a
	$\boldsymbol{F} = \mathrm{d}\boldsymbol{p}/\mathrm{d}t$	Newton's second law of motion	5b
	$\boldsymbol{T} = \mathrm{d}\boldsymbol{J}/\mathrm{d}t$	Rotational motion	6
Work opposing a force in the *x* direction	$\mathrm{d}w = -F_x\mathrm{d}x$	Definition	7a
Kinetic energy	$E_\mathrm{k} = \frac{1}{2}mv^2$	Definition; *v* is the speed	8a
Potential energy and force	$F_x = -\mathrm{d}V/\mathrm{d}x$	One dimension	9
Coulomb potential energy	$E_\mathrm{p}(r) = Q_1Q_2/4\pi\varepsilon_0 r$	In a vacuum	11
Coulomb potential	$\phi_1(r) = Q_1/4\pi\varepsilon_0 r$	In a vacuum	13
Boltzmann distribution	$N_i/N_j = \mathrm{e}^{-(\varepsilon_i-\varepsilon_j)/kT}$		14a

Atkins'

PHYSICAL CHEMISTRY

Quantum Chemistry, Spectroscopy, and Statistical Thermodynamics

FOCUS 7

Quantum theory

It was once thought that the motion of atoms and subatomic particles could be expressed using 'classical mechanics', the laws of motion introduced in the seventeenth century by Isaac Newton, for these laws were very successful at explaining the motion of everyday objects and planets. However, a proper description of electrons, atoms, and molecules requires a different kind of mechanics, 'quantum mechanics', which is introduced in this Focus and applied widely throughout the text.

7A The origins of quantum mechanics

Experimental evidence accumulated towards the end of the nineteenth century showed that classical mechanics failed when it was applied to particles as small as electrons. More specifically, careful measurements led to the conclusion that particles may not have an arbitrary energy and that the classical concepts of a particle and wave blend together. This Topic shows how these observations set the stage for the development of the concepts and equations of quantum mechanics in the early twentieth century.

7B Wavefunctions

In quantum mechanics, all the properties of a system are expressed in terms of a wavefunction, which is obtained by solving the equation proposed by Erwin Schrödinger. This Topic focuses on the interpretation of the wavefunction, and specifically what it reveals about the location of a particle.

7C Operators and observables

A central feature of quantum theory is its representation of observables by 'operators', which act on the wavefunction and extract the information it contains. This Topic shows how operators are constructed and used. One consequence of their use is the 'uncertainty principle', one of the most profound departures of quantum mechanics from classical mechanics.

7D Translational motion

Translational motion, motion through space, is one of the fundamental types of motion treated by quantum mechanics. According to quantum theory, a particle constrained to move in a finite region of space is described by only certain wavefunctions and can possess only certain energies. That is, quantization emerges as a natural consequence of solving the Schrödinger equation and the conditions imposed on it. The solutions also expose a number of non-classical features of particles, especially their ability to tunnel into and through regions where classical physics would forbid them to be found.

7E Vibrational motion

This Topic introduces the 'harmonic oscillator', a simple but very important model for the description of vibrations. It shows that the energies of an oscillator are quantized and that an oscillator may be found at displacements that are forbidden by classical physics.

7F Rotational motion

The constraints on the wavefunctions of a body rotating in two and three dimensions result in the quantization of its energy. In addition, because the energy is related to the angular momentum, it follows that angular momentum is also restricted to certain values. The quantization of angular momentum is a very important aspect of the quantum theory of electrons in atoms and of rotating molecules.

7F.1 **Rotation in two dimensions;** 7F.2 **Rotation in three dimensions**

What is an application of this material?

Impact 11, accessed via the e-book, highlights an application of quantum mechanics that is beginning to gain ground as a useful technology. It is based on the expectation that a 'quantum computer' can carry out calculations on many states of a system simultaneously, leading to a new generation of very fast computers. 'Nanoscience' is the study of atomic and molecular assemblies with dimensions ranging from 1 nm to about 100 nm, and 'nanotechnology' is concerned with the incorporation of such assemblies into devices. *Impact* 12, accessed via the e-book, explores quantum mechanical effects that show how the properties of a nanometre-sized assembly depend on its size.

➤ **Go to the e-book for videos that feature the derivation and interpretation of equations, and applications of this material.**

TOPIC 7A The origins of quantum mechanics

➤ **Why do you need to know this material?**

Quantum theory is central to almost every explanation in chemistry. It is used to understand atomic and molecular structure, chemical bonds, and most of the properties of matter.

➤ **What is the key idea?**

Experimental evidence led to the conclusion that energy can be transferred only in discrete amounts, and that the classical concepts of a 'particle' and a 'wave' blend together.

➤ **What do you need to know already?**

You should be familiar with the basic principles of classical mechanics, especially momentum, force, and energy set out in *Energy: A first look*. The discussion of heat capacities of solids makes light use of material in Topic 2A.

The classical mechanics developed by Newton in the seventeenth century is an extraordinarily successful theory for describing the motion of everyday objects and planets. However, late in the nineteenth century scientists started to make observations that could not be explained by classical mechanics. They were forced to revise their entire conception of the nature of matter and replace classical mechanics by a theory that became known as **quantum mechanics**.

7A.1 Energy quantization

Three experiments carried out near the end of the nineteenth century drove scientists to the view that energy can be transferred only in discrete amounts.

7A.1(a) Black-body radiation

The key features of electromagnetic radiation according to classical physics are described in *The chemist's toolkit* 7A.1. It is

The chemist's toolkit 7A.1 Electromagnetic radiation

Electromagnetic radiation consists of oscillating electric and magnetic disturbances that propagate as waves. The two components of an electromagnetic wave are mutually perpendicular and are also perpendicular to the direction of propagation (Sketch 1). Electromagnetic waves travel through a vacuum at a constant speed called the **speed of light**, c, which has the defined value of exactly $2.997\,924\,58 \times 10^8\ \mathrm{m\,s^{-1}}$.

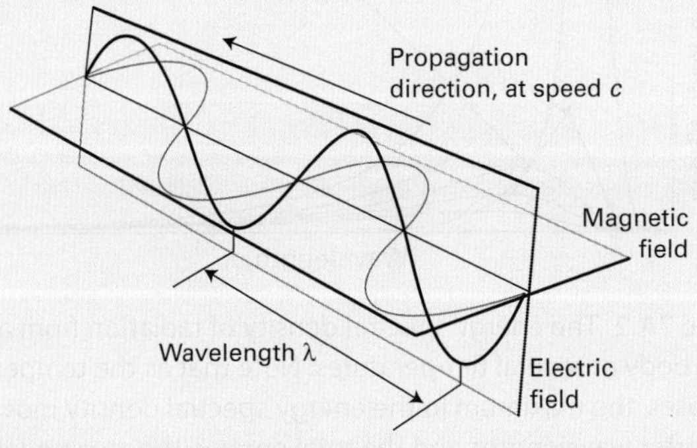

Sketch 1

A wave is characterized by its **wavelength**, λ (lambda), the distance between consecutive peaks of the wave. The classification of electromagnetic radiation according to its wavelength is shown in Sketch 2. Light, which is electromagnetic radiation that is visible to the human eye, has a wavelength in the range 420 nm (violet light) to 700 nm (red light). The properties of a wave may also be expressed in terms of its **frequency**, ν (nu), the number of oscillations in a time interval divided by the duration of the interval. Frequency is reported in hertz, Hz, with $1\ \mathrm{Hz} = 1\ \mathrm{s^{-1}}$ (that is, 1 cycle per second). Light spans the frequency range from 710 THz (violet light) to 430 THz (red light).

The wavelength and frequency of an electromagnetic wave are related by

$$c = \lambda\nu$$

The relation between wavelength and frequency in a vacuum

It is also common to describe a wave in terms of its **wavenumber**, $\tilde{\nu}$ (nu tilde), which is defined as

$$\tilde{\nu} = \frac{1}{\lambda}, \text{ or equivalently } \tilde{\nu} = \frac{\nu}{c}$$

Wavenumber [definition]

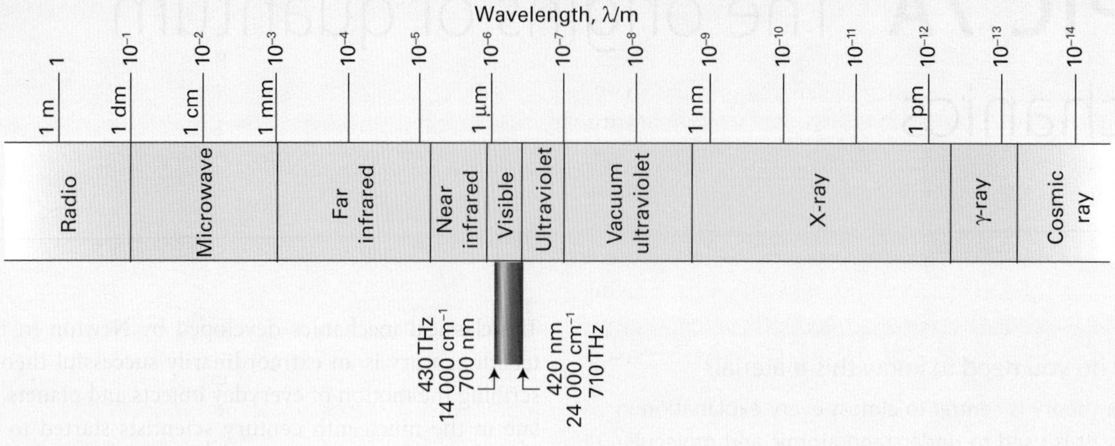

Sketch 2

Thus, wavenumber is the reciprocal of the wavelength and can be interpreted as the number of wavelengths in a given distance. In spectroscopy, for historical reasons, wavenumber is usually reported in units of reciprocal centimetres (cm^{-1}). Visible light therefore corresponds to electromagnetic radiation with a wavenumber of 14000 cm^{-1} (red light) to 24000 cm^{-1} (violet light).

Electromagnetic radiation that consists of a single frequency (and therefore single wavelength) is **monochromatic**, because it corresponds to a single colour. *White light* consists of electromagnetic waves with a continuous, but not uniform, spread of frequencies throughout the visible region of the spectrum.

observed that all objects emit electromagnetic radiation over a range of frequencies with an intensity that depends on the temperature of the object. A familiar example is a heated metal bar that first glows red and then becomes 'white hot' upon further heating. As the temperature is raised, the colour shifts from red towards blue and results in the white glow.

The radiation emitted by hot objects is discussed in terms of a **black body**, a body that emits and absorbs electromagnetic radiation without favouring any wavelengths. A good approximation to a black body is a small hole in an empty container (Fig. 7A.1). Figure 7A.2 shows how the intensity of the radiation from a black body varies with wavelength at several temperatures. At each temperature T there is a wavelength, λ_{max}, at which the intensity of the radiation is a maximum, with T and λ_{max} related by the empirical **Wien's law**:

$$\lambda_{max}T = \text{constant} \qquad \text{Wien's law} \qquad (7A.1)$$

The constant is found to have the value 2.9 mm K. The intensity of the emitted radiation at any temperature declines sharply at short wavelengths (high frequencies). The intensity is effectively a window on to the energy present inside the container,

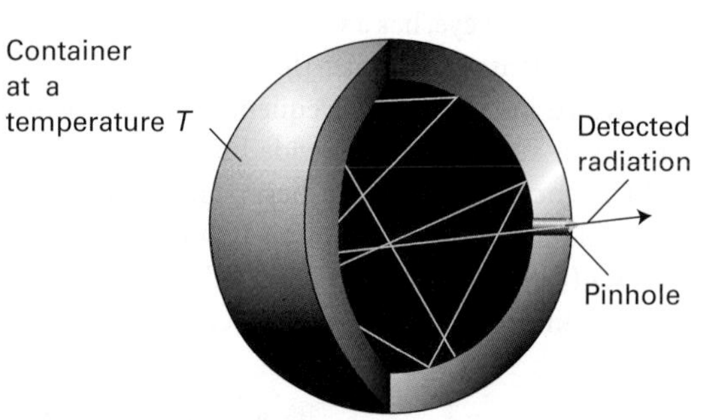

Figure 7A.1 Black-body radiation can be detected by allowing it to leave an otherwise closed container through a pinhole. The radiation is reflected many times within the container and comes to thermal equilibrium with the wall. Radiation leaking out through the pinhole is characteristic of the radiation inside the container.

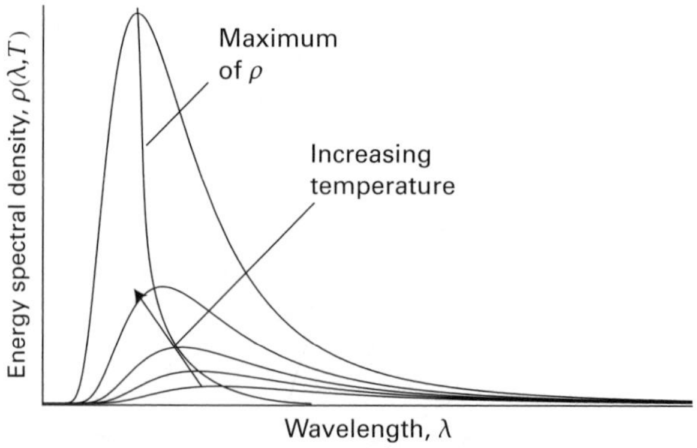

Figure 7A.2 The energy spectral density of radiation from a black body at several temperatures. Note that as the temperature increases, the maximum in the energy spectral density moves to shorter wavelengths and the total energy (the area under the curve) increases.

in the sense that the greater the intensity at a given wavelength, the greater is the energy inside the container due to radiation at that wavelength.

The **energy density**, $\mathcal{E}(T)$, is the total energy inside the container divided by its volume. The **energy spectral density**, $\rho(\lambda,T)$, is defined so that $\rho(\lambda,T)\mathrm{d}\lambda$ is the energy density at temperature T due to the presence of electromagnetic radiation with wavelengths between λ and $\lambda + \mathrm{d}\lambda$. A high energy spectral density at the wavelength λ and temperature T simply means that there is a lot of energy associated with wavelengths lying between λ and $\lambda + \mathrm{d}\lambda$ at that temperature. The energy density is obtained by summing (integrating) the energy spectral density over all wavelengths:

$$\mathcal{E}(T)=\int_0^{\infty}\rho(\lambda,T)\mathrm{d}\lambda \qquad (7A.2)$$

The units of $\mathcal{E}(T)$ are joules per metre cubed ($\mathrm{J\,m^{-3}}$), so the units of $\rho(\lambda,T)$ are $\mathrm{J\,m^{-4}}$. Empirically, the energy density is found to vary as T^4, an observation expressed by the **Stefan–Boltzmann law**:

$$\mathcal{E}(T)=\text{constant}\times T^4 \qquad \text{Stefan–Boltzmann law} \qquad (7A.3)$$

with the constant found to have the value $7.567 \times 10^{-16}\,\mathrm{J\,m^{-3}\,K^{-4}}$.

The container in Fig. 7A.1 emits radiation that can be thought of as oscillations of the electromagnetic field stimulated by the oscillations of electrical charges in the material of the wall. According to classical physics, every oscillator is excited to some extent, and according to the equipartition principle (see *Energy: A first look*) every oscillator, regardless of its frequency, has an average energy of kT. On this basis, the physicist Lord Rayleigh, with minor help from James Jeans, deduced what is now known as the **Rayleigh–Jeans law**:

$$\rho(\lambda,T)=\frac{8\pi kT}{\lambda^4} \qquad \text{Rayleigh–Jeans law} \qquad (7A.4)$$

where k is Boltzmann's constant ($k = 1.381 \times 10^{-23}\,\mathrm{J\,K^{-1}}$).

The Rayleigh–Jeans law is not supported by the experimental measurements. As is shown in Fig. 7A.3, although there is agreement at long wavelengths, it predicts that the energy spectral density (and hence the intensity of the radiation emitted) increases without going through a maximum as the wavelength decreases. That is, the Rayleigh–Jeans law is inconsistent with Wien's law. Equation 7A.4 also implies that the radiation is intense at very short wavelengths and becomes infinitely intense as the wavelength tends to zero. The concentration of radiation at short wavelengths is called the **ultraviolet catastrophe**, and is an unavoidable consequence of classical physics.

In 1900, Max Planck found that the experimentally observed intensity distribution of black-body radiation could be explained by proposing that the energy of each oscillator is limited to discrete values. In particular, Planck assumed that for an electromagnetic oscillator of frequency ν, the permitted energies are integer multiples of $h\nu$:

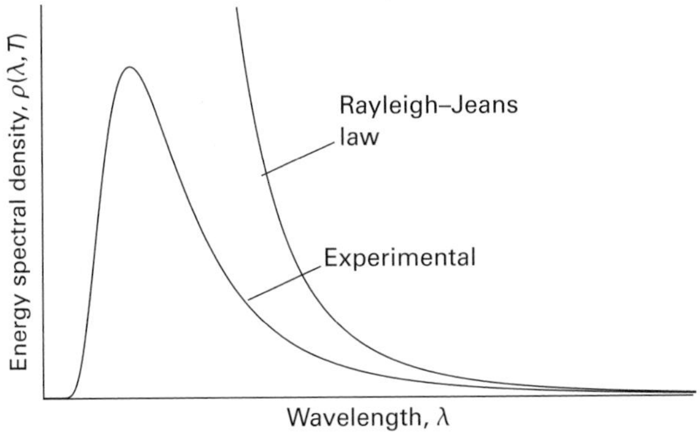

Figure 7A.3 Comparison of the experimental energy spectral density with the prediction of the Rayleigh–Jeans law (eqn 7A.4). The latter predicts an infinite energy spectral density at short wavelengths and infinite overall energy density.

$$E = nh\nu \quad n = 0, 1, 2, \ldots \qquad (7A.5)$$

In this expression h is a fundamental constant now known as **Planck's constant**. The limitation of energies to discrete values is called **energy quantization**. On this basis Planck was able to derive an expression for the energy spectral density which is now called the **Planck distribution**:

$$\rho(\lambda,T)=\frac{8\pi hc}{\lambda^5(\mathrm{e}^{hc/\lambda kT}-1)} \qquad \text{Planck distribution} \qquad (7A.6a)$$

This expression is plotted in Fig. 7A.4 and fits the experimental data very well at all wavelengths. The value of h, which is an undetermined parameter in the theory, can be found by varying its value until the best fit is obtained between the eqn 7A.6a and experimental measurements. The currently accepted value is $h = 6.626 \times 10^{-34}\,\mathrm{J\,s}$.

For short wavelengths, $hc/\lambda kT >> 1$. Because $\mathrm{e}^{hc/\lambda kT} \to \infty$ faster than $\lambda^5 \to 0$, it follows that $\rho \to 0$ as $\lambda \to 0$. Hence, the energy

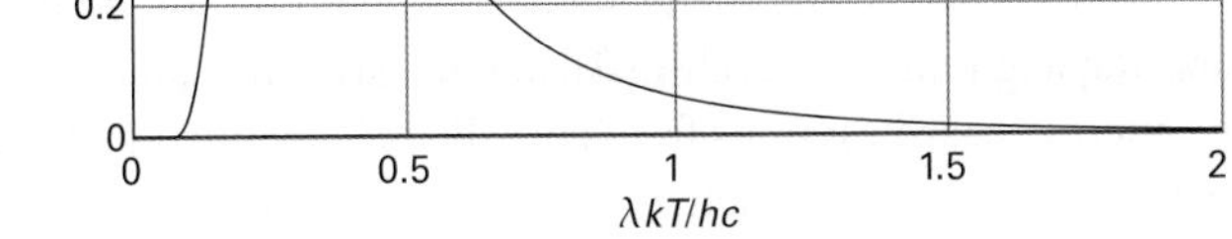

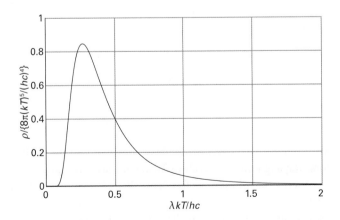

Figure 7A.4 The Planck distribution (eqn 7A.6a) accounts for the experimentally determined energy distribution of black-body radiation. It coincides with the Rayleigh–Jeans distribution at long wavelengths.

spectral density approaches zero at short wavelengths, and so the Planck distribution avoids the ultraviolet catastrophe. For long wavelengths in the sense $hc/\lambda kT << 1$, the denominator in the Planck distribution can be replaced by (see Part 1 of the *Resource section*)

$$e^{hc/\lambda kT}-1=\left(1+\frac{hc}{\lambda kT}+\cdots\right)-1\approx\frac{hc}{\lambda kT}$$

When this approximation is substituted into eqn 7A.6a, the Planck distribution reduces to the Rayleigh–Jeans law, eqn 7A.4. The wavelength at the maximum can be found by differentiation, and is given by $\lambda_{max}T$ = constant, in accord with Wien's law. The value of the constant found in this way is equal to $c_2/5$ ($c_2=hc/k=1.439$ cm K is the **second radiation constant**) and agrees with the experimental value. Wien's law is then written as

$$\lambda_{max}T=c_2/5 \qquad (7A.7)$$

Finally, the total energy density is

$$\mathcal{E}(T)=\int_0^\infty\frac{8\pi hc}{\lambda^5(e^{hc/\lambda kT}-1)}d\lambda=aT^4,\quad \text{with } a=\frac{8\pi^5k^4}{15h^3c^3} \qquad (7A.8)$$

which is finite and agrees with the Stefan–Boltzmann law (eqn 7A.3), also predicting the value of its constant (a) correctly.

Brief illustration 7A.1

Consider eqn 7A.6a with λ_1 = 450 nm (blue light) and λ_2 = 700 nm (red light), and T = 298 K. It follows that

$$\frac{hc}{\lambda_1kT}=\frac{(6.626\times10^{-34}\text{ J s})\times(2.998\times10^8\text{ m s}^{-1})}{(450\times10^{-9}\text{ m})\times(1.381\times10^{-23}\text{ J K}^{-1})\times(298\text{ K})}=107.2\ldots$$

$$\frac{hc}{\lambda_2kT}=\frac{(6.626\times10^{-34}\text{ J s})\times(2.998\times10^8\text{ m s}^{-1})}{(700\times10^{-9}\text{ m})\times(1.381\times10^{-23}\text{ J K}^{-1})\times(298\text{ K})}=68.9\ldots$$

and

$$\frac{\rho(450\text{ nm, }298\text{ K})}{\rho(700\text{ nm, }298\text{ K})}=\left(\frac{700\times10^{-9}\text{ m}}{450\times10^{-9}\text{ m}}\right)^5\times\frac{e^{68.9\ldots}-1}{e^{107.2\ldots}-1}$$
$$=9.11\times(2.30\times10^{-17})=2.10\times10^{-16}$$

At room temperature, the proportion of shorter wavelength radiation is insignificant.

Planck's approach proved successful on account of a key assumption that Planck made about the energy of the oscillators. Rayleigh allowed each oscillator to have the same average energy regardless of its frequency. Planck assumed that the energy has to be a multiple of $h\nu$. As a result, the minimum energy needed to excite oscillators of very high frequency is too large and they remain unexcited. This elimination of the contribution from very high frequency oscillators avoids the ultraviolet catastrophe.

It is sometimes convenient to express the Planck distribution in terms of the frequency. Then $\rho(\nu,T)d\nu$ is the energy density at temperature T due to the presence of electromagnetic radiation with frequencies between ν and $\nu+d\nu$, and

$$\rho(\nu,T)=\frac{8\pi h\nu^3}{c^3(e^{h\nu/kT}-1)} \qquad \text{Planck distribution in terms of frequency} \quad (7A.6b)$$

7A.1(b) Heat capacity

When energy is supplied as heat to a substance its temperature rises; the heat capacity (Topic 2A) is the constant of proportionality between the energy supplied and the temperature rise ($C = dq/dT$ and, at constant volume, $C_{V,m} = (\partial U_m/\partial T)_V$). Experimental measurements made during the nineteenth century had shown that at room temperature the molar heat capacities of many monatomic solids are about $3R$, where R is the gas constant.[1] However, when measurements were made at much lower temperatures it was found that the heat capacity decreased, tending to zero as the temperature approached zero.

Classical physics was unable to explain this temperature dependence. The classical picture of a solid is of atoms oscillating about fixed positions, with the expectation that each oscillating atom will have the same average energy kT. This model predicts that a solid consisting of N atoms, each free to oscillate in three dimensions, will have energy $U = 3NkT$ and hence heat capacity $C_V = (\partial U/\partial T)_V = 3Nk$. The molar heat capacity is therefore predicted to be $3N_Ak$ which, recognizing that $N_Ak=R$, is equal to $3R$ at all temperatures.

In 1905, Einstein suggested applying Planck's hypothesis and supposing that each oscillating atom could have an energy $nh\nu$, where n is an integer and ν is the frequency of the oscillation. Einstein went on to show by using the Boltzmann distribution that each oscillator is unlikely to be excited to high energies and at low temperatures few oscillators can be excited at all. As a consequence, because the oscillators cannot be excited, the heat capacity falls to zero. The quantitative result that Einstein obtained (as shown in Topic 13E) is

$$C_{V,m}(T)=3Rf_E(T),\quad f_E(T)=\left(\frac{\theta_E}{T}\right)^2\left(\frac{e^{\theta_E/2T}}{e^{\theta_E/T}-1}\right)^2 \qquad \text{Einstein formula} \quad (7A.9a)$$

In this expression θ_E is the **Einstein temperature,** $\theta_E=h\nu/k$.

At high temperatures (in the sense $T >> \theta_E$) the exponentials in f_E can be expanded as $e^x=1+x+\cdots$ and higher terms ignored. The result is

$$f_E(T)=\left(\frac{\theta_E}{T}\right)^2\left\{\frac{1+\theta_E/2T+\cdots}{(1+\theta_E/T+\cdots)-1}\right\}^2\approx\left(\frac{\theta_E}{T}\right)^2\left\{\frac{1}{\theta_E/T}\right\}^2\approx1 \qquad (7A.9b)$$

[1] The gas constant occurs in the context of solids because it is actually the more fundamental Boltzmann's constant in disguise: $R=N_Ak$.

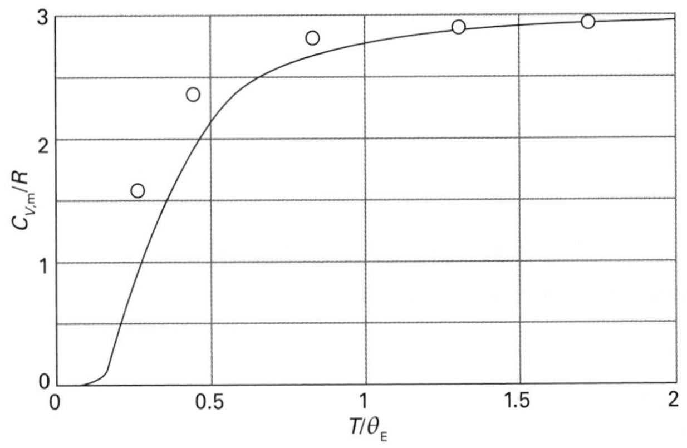

Figure 7A.5 Experimental low-temperature molar heat capacities (open circles) and the temperature dependence predicted on the basis of Einstein's theory (solid line, eqn 7A.9). The theory accounts for the dependence fairly well, but is everywhere too low.

and the classical result ($C_{V,m} = 3R$) is obtained. At low temperatures (in the sense $T << \theta_E$), $e^{\theta_E/T} >> 1$ and

$$f_E(T) \approx \left(\frac{\theta_E}{T}\right)^2 \left(\frac{e^{\theta_E/2T}}{e^{\theta_E/T}}\right)^2 = \left(\frac{\theta_E}{T}\right)^2 e^{-\theta_E/T} \quad (7A.9c)$$

The strongly decaying exponential function goes to zero more rapidly than $1/T^2$ goes to infinity; so $f_E \to 0$ as $T \to 0$, and the heat capacity approaches zero, as found experimentally. The physical reason for this success is that as the temperature is lowered, less energy is available to excite the atomic oscillations. At high temperatures many oscillators are excited into high energy states leading to classical behaviour.

Figure 7A.5 shows the temperature dependence of the heat capacity predicted by the Einstein formula and some experimental data; the value of the Einstein temperature is adjusted to obtain the best fit to the data. The general shape of the curve is satisfactory, but the numerical agreement is in fact quite poor. This discrepancy arises from Einstein's assumption that all the atoms oscillate with the same frequency. A more sophisticated treatment, due to Peter Debye, allows the oscillators to have a range of frequencies from zero up to a maximum. This approach results in much better agreement with the experimental data and there can be little doubt that mechanical motion as well as electromagnetic radiation is quantized.

7A.1(c) Atomic and molecular spectra

The most compelling and direct evidence for the quantization of energy comes from **spectroscopy**, the detection and analysis of the electromagnetic radiation absorbed, emitted, or scattered by a substance. The record of the variation of the intensity of this radiation with frequency (ν), wavelength (λ), or

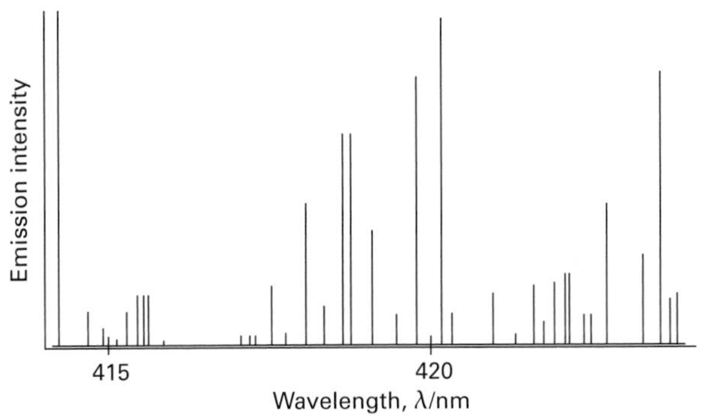

Figure 7A.6 A region of the spectrum of radiation emitted by excited iron atoms consists of radiation at a series of discrete wavelengths (or frequencies).

wavenumber ($\tilde{\nu} = \nu/c$, see *The chemist's toolkit* 7A.1) is called its **spectrum** (from the Latin word for appearance).

An atomic emission spectrum is shown in Fig. 7A.6, and a molecular absorption spectrum is shown in Fig. 7A.7. The obvious feature of both is that radiation is emitted or absorbed at a series of discrete frequencies. This observation can be understood if the energy of the atoms or molecules is also confined to discrete values, because then the energies that a molecule can discard or acquire are also confined to discrete values (Fig. 7A.8). If the energy of an atom or molecule decreases by ΔE, and this energy is carried away as radiation, the frequency of the radiation ν and the change in energy are related by the **Bohr frequency condition**:

$$\Delta E = h\nu \quad \text{Bohr frequency condition} \quad (7A.10)$$

A molecule is said to undergo a **spectroscopic transition**, a change of state, and as a result an emission 'line', a sharply defined peak, appears in the spectrum at frequency ν.

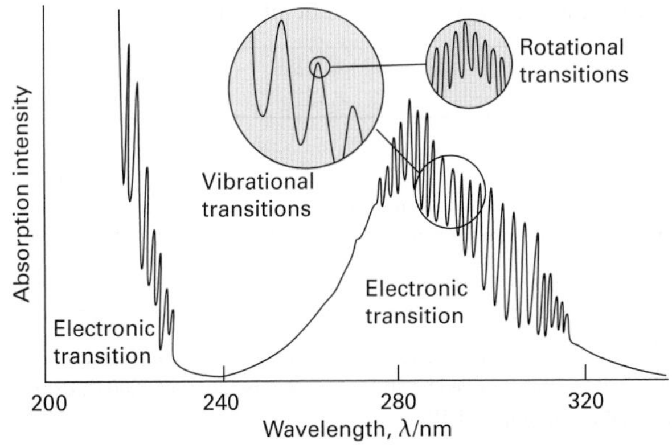

Figure 7A.7 A molecule can change its state by absorbing radiation at definite frequencies. This spectrum is due to the electronic, vibrational, and rotational excitation of sulfur dioxide (SO_2) molecules. The observation of discrete spectral lines suggests that molecules can possess only discrete energies, not an arbitrary energy.

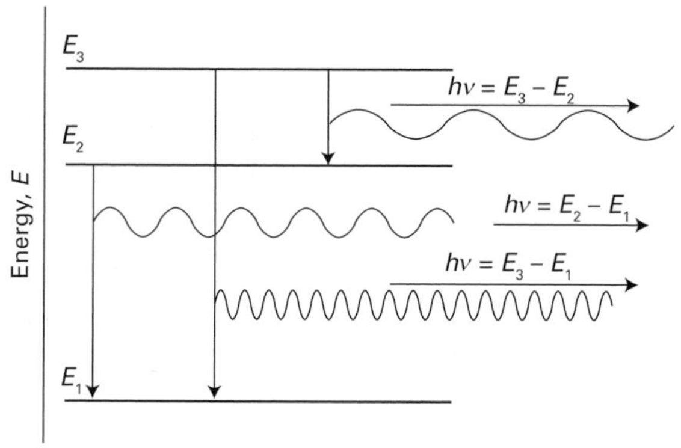

Figure 7A.8 Spectroscopic transitions, such as those shown in Fig. 7A.6, can be accounted for by supposing that an atom (or molecule) emits electromagnetic radiation as it changes from a discrete level of high energy to a discrete level of lower energy. High-frequency radiation is emitted when the energy change is large. Transitions like those shown in Fig. 7A.7 can be explained by supposing that a molecule (or atom) absorbs radiation as it changes from a low-energy level to a higher-energy level.

Brief illustration 7A.2

Atomic sodium produces a yellow glow (as when a sodium salt is introduced into a flame or in some street lamps) resulting from the emission of radiation of 590 nm. The spectroscopic transition responsible for the emission involves electronic energy levels that have a separation given by eqn 7A.10:

$$\Delta E = h\nu = \frac{hc}{\lambda} = \frac{(6.626\times10^{-34}\ \mathrm{J\,s})\times(2.998\times10^{8}\ \mathrm{m\,s^{-1}})}{590\times10^{-9}\ \mathrm{m}}$$
$$= 3.37\times10^{-19}\ \mathrm{J}$$

This energy difference can be expressed in a variety of ways. For instance, multiplication by Avogadro's constant results in an energy separation per mole of atoms, of 203 kJ mol^{-1}, comparable to the energy of a weak chemical bond.

Exercises

E7A.1 Calculate the wavelength and frequency at which the intensity of the radiation is a maximum for a black body at 298 K.

E7A.2 The intensity of the radiation from an object is found to be a maximum at 2000 cm^{-1}. Assuming that the object is a black body, calculate its temperature.

E7A.3 Calculate the molar heat capacity of a monatomic non-metallic solid at 298 K which is characterized by an Einstein temperature of 2000 K. Express your result as a multiple of $3R$.

7A.2 Wave–particle duality

The experiments about to be described show that electromagnetic radiation—which classical physics treats as wave-like—actually also displays the characteristics of particles. Another experiment shows that electrons—which classical physics treats as particles—also display the characteristics of waves. This **wave–particle duality**, the blending together of the characteristics of waves and particles, lies at the heart of quantum mechanics.

7A.2(a) The particle character of electromagnetic radiation

The Planck treatment of black-body radiation introduced the idea that an oscillator of frequency ν can have only the energies 0, $h\nu$, $2h\nu$,.... This quantization leads to the suggestion (and at this stage it is only a suggestion) that the resulting electromagnetic radiation of that frequency can be thought of as consisting of 0, 1, 2,... particles, each particle having an energy $h\nu$. These particles of electromagnetic radiation are now called **photons**. Thus, if an oscillator of frequency ν is excited to its first excited state, then one photon of that frequency is present, if it is excited to its second excited state, then two photons are present, and so on. The observation of discrete emission spectra from atoms and molecules can be pictured as the atom or molecule generating a photon of energy $h\nu$ when it discards an energy of magnitude ΔE, with $\Delta E = h\nu$. Note that one transition generates one photon, not a shower of many photons.

Example 7A.1 Calculating the number of photons

Calculate the number of photons emitted by a 100 W yellow lamp in 1.0 s. Take the wavelength of yellow light as 560 nm, and assume 100 per cent efficiency.

Collect your thoughts Each photon has an energy $h\nu$, so the total number N of photons needed to produce an energy E is $N = E/h\nu$. To use this equation, you need to know the frequency of the radiation (from $\nu = c/\lambda$) and the total energy emitted by the lamp. The latter is given by the product of the power (P, in watts) and the time interval, Δt, for which the lamp is turned on: $E = P\Delta t$ (see *The chemist's toolkit* 2A.1).

The solution The number of photons is

$$N = \frac{E}{h\nu} = \frac{P\Delta t}{h(c/\lambda)} = \frac{\lambda P\Delta t}{hc}$$

Substitution of the data gives (using 1 W = 1 J s^{-1})

$$N = \frac{(5.60\times10^{-7}\ \mathrm{m})\times(100\ \mathrm{J\,s^{-1}})\times(1.0\ \mathrm{s})}{(6.626\times10^{-34}\ \mathrm{J\,s})\times(2.998\times10^{8}\ \mathrm{m\,s^{-1}})} = 2.8\times10^{20}$$

Self-test 7A.1 How many photons does a monochromatic (single frequency) infrared rangefinder of power 1 mW and wavelength 1000 nm emit in 0.1 s?

Answer: 5×10^{14}

So far, the existence of photons is only a suggestion. Experimental evidence for their existence comes from the

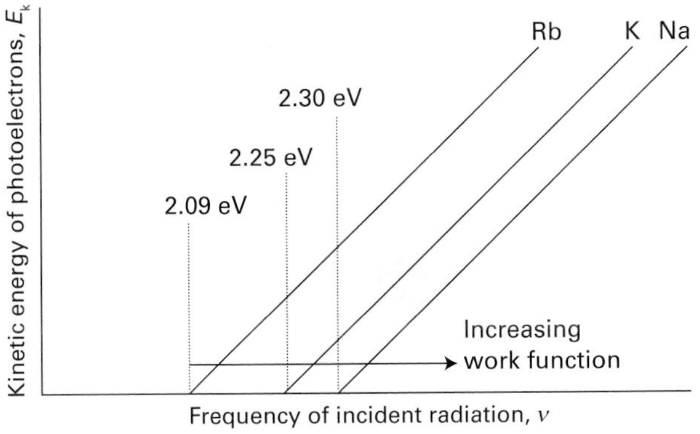

Figure 7A.9 In the photoelectric effect, it is found that no electrons are ejected when the incident radiation has a frequency below a certain value that is characteristic of the metal. Above that value, the kinetic energy of the photoelectrons varies linearly with the frequency of the incident radiation.

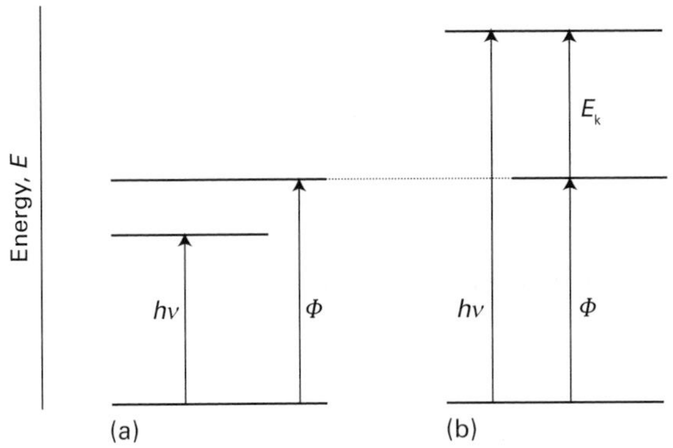

Figure 7A.10 The photoelectric effect can be explained if it is supposed that the incident radiation is composed of photons that have energy proportional to the frequency of the radiation. (a) The energy of the photon is insufficient to drive an electron out of the metal. (b) The energy of the photon is more than enough to eject an electron, and the excess energy is carried away as the kinetic energy of the photoelectron (the ejected electron).

measurement of the energies of electrons produced in the **photoelectric effect**, the ejection of electrons from metals when they are exposed to ultraviolet radiation. The experimental characteristics of the photoelectric effect are as follows:

- No electrons are ejected, regardless of the intensity of the radiation, unless its frequency exceeds a threshold value characteristic of the metal.
- The kinetic energy of the ejected electrons increases linearly with the frequency of the incident radiation but is independent of the intensity of the radiation.
- Even at low radiation intensities, electrons are ejected immediately if the frequency is above the threshold value.

Figure 7A.9 illustrates the first and second characteristics.

These observations strongly suggest that in the photoelectric effect a particle-like projectile collides with the metal and, if the kinetic energy of the projectile is high enough, an electron is ejected. If the projectile is a photon of energy $h\nu$ (ν is the frequency of the radiation), the kinetic energy of the electron is E_k, and the energy needed to remove an electron from the metal, which is called its **work function**, is Φ (uppercase phi), then as illustrated in Fig. 7A.10, the conservation of energy implies that

$$h\nu = E_k + \Phi \quad \text{or} \quad E_k = h\nu - \Phi \qquad \text{Photoelectric effect} \qquad (7A.11)$$

This model explains the three experimental observations:

- Photoejection cannot occur if $h\nu < \Phi$ because the photon brings insufficient energy.
- The kinetic energy of an ejected electron increases linearly with the frequency of the photon.
- When a photon collides with an electron, it gives up all its energy, so electrons should appear as soon as the collisions begin, provided the photons have sufficient energy.

A practical application of eqn 7A.10 is that it provides a technique for the determination of Planck's constant, because the slopes of the lines in Fig. 7A.9 are all equal to h.

Example 7A.2 Calculating the longest wavelength capable of photoejection

A photon of radiation of wavelength 305 nm ejects an electron with a kinetic energy of 1.77 eV from a metal. Calculate the longest wavelength of radiation capable of ejecting an electron from the metal.

Collect your thoughts You can use eqn 7A.11, rearranged into $\Phi = h\nu - E_k$, to compute the work function because you know the frequency of the photon from $\nu = c/\lambda$. The threshold for photoejection is the lowest frequency at which electron ejection occurs without there being any excess energy. This threshold will be reached when the photon energy is equal to the work function. The corresponding wavelength is then found by using $\lambda_{max} = hc/\Phi$.

The solution The work function is

$$\Phi = h\nu - E_k \overset{\nu = c/\lambda}{=} \frac{hc}{\lambda} - E_k$$

$$= \frac{(6.626 \times 10^{-34}\ \text{J s}) \times (2.998 \times 10^{8}\ \text{m s}^{-1})}{305 \times 10^{-9}\ \text{m}}$$

$$- (1.77\ \text{eV}) \times (1.602 \times 10^{-19}\ \text{J eV}^{-1})$$

$$= 3.67... \times 10^{-19}\ \text{J}$$

The longest wavelength that can cause photoejection is therefore

$$\lambda_{\max} = \frac{hc}{\Phi} = \frac{(6.626\times10^{-34}\ \text{J s})\times(2.998\times10^{8}\ \text{m s}^{-1})}{3.67\ldots\times10^{-19}\ \text{J}} = 540\ \text{nm}$$

Self-test 7A.2 When ultraviolet radiation of wavelength 165 nm strikes a certain metal surface, electrons are ejected with a speed of 1.24 Mm s^{-1}. Calculate the speed of electrons ejected by radiation of wavelength 265 nm.

Answer: 735 km s^{-1}

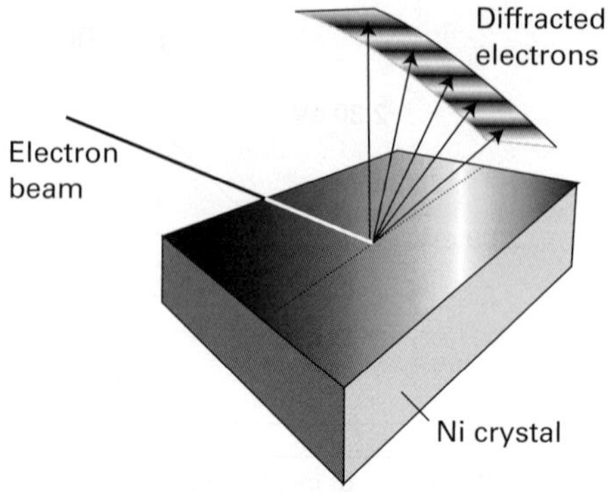

Figure 7A.11 The Davisson–Germer experiment. The scattering of an electron beam from a nickel crystal shows a variation in intensity characteristic of a diffraction experiment in which waves interfere constructively and destructively in different directions.

7A.2(b) The wave character of particles

Although contrary to the long-established wave theory of radiation, the view that radiation consists of particles had been held before, but discarded. No significant scientist, however, had taken the view that matter is wave-like. Nevertheless, experiments carried out in 1927 forced people to consider that possibility. The crucial experiment was performed by Clinton Davisson and Lester Germer, who observed the diffraction of electrons by a crystal (Fig. 7A.11). As remarked in *The chemist's toolkit* 7A.2, diffraction is the interference caused by an object in the path of waves. Davisson and Germer's success was a lucky accident, because a chance rise of temperature caused their polycrystalline sample to anneal, and the ordered planes of atoms then acted as a diffraction grating. The Davisson–Germer experiment, which has since been repeated with other particles (including α particles, molecular hydrogen, and neutrons), shows clearly that particles have wave-like properties. At almost the same time, G.P. Thomson showed that a beam of electrons was diffracted when passed through a thin gold foil.

Some progress towards accounting for wave–particle duality had already been made by Louis de Broglie who, in 1924, suggested that any particle, not only photons, travelling with a linear momentum $p = mv$ (with m the mass and v the speed of the particle) should have in some sense a wavelength given by what is now called the **de Broglie relation**:

$$\lambda = \frac{h}{p} \qquad \text{de Broglie relation} \qquad (7A.12)$$

That is, a particle with a high linear momentum has a short wavelength. Macroscopic bodies have such high momenta even when they are moving slowly (because their mass is so great), that their wavelengths are undetectably small, and the wave-like properties cannot be observed. This undetectability is why classical mechanics can be used to explain the behaviour of macroscopic bodies. It is necessary to invoke quantum mechanics only for microscopic bodies, such as atoms and molecules, in which masses are small.

The chemist's toolkit 7A.2 Diffraction of waves

A characteristic property of waves is that they interfere with one another, which means that they result in a greater amplitude where their displacements add and a smaller amplitude where their displacements subtract (Sketch 1). The former is called 'constructive interference' and the latter 'destructive interference'. The regions of constructive and destructive interference show up as regions of enhanced and diminished intensity. The phenomenon of **diffraction** is the interference caused by an object in the path of waves and occurs when the dimensions of the object are comparable to the wavelength of the radiation. Light waves, with wavelengths of the order of 500 nm, are diffracted by narrow slits.

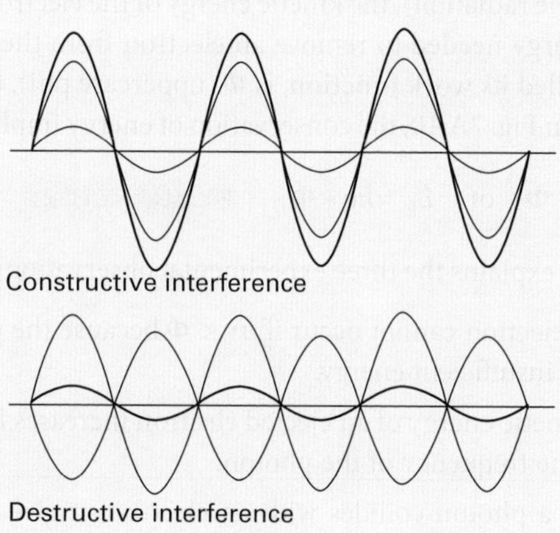

Sketch 1

Example 7A.3 Estimating the de Broglie wavelength

Estimate the wavelength of electrons that have been accelerated from rest through a potential difference of 40 kV. Electrons accelerated in this way are used in the technique of electron diffraction for imaging biological systems and for the determination of the structures of solid surfaces (Topic 19A).

Collect your thoughts To use the de Broglie relation, you need to know the linear momentum, p, of the electrons. To calculate the linear momentum, note that the energy acquired by an electron accelerated through a potential difference $\Delta\phi$ is $e\Delta\phi$, where e is the magnitude of its charge. At the end of the period of acceleration, all the acquired energy is in the form of kinetic energy, $E_k = \frac{1}{2}m_e v^2 = p^2/2m_e$. You can therefore calculate p by setting $p^2/2m_e$ equal to $e\Delta\phi$. For the manipulation of units use $1\,\text{V C} = 1\,\text{J}$ and $1\,\text{J} = 1\,\text{kg m}^2\,\text{s}^{-2}$.

The solution The expression $p^2/2m_e = e\Delta\phi$ implies that $p = (2m_e e\Delta\phi)^{1/2}$ then, from the de Broglie relation $\lambda = h/p$,

$$\lambda = \frac{h}{(2m_e e\Delta\phi)^{1/2}}$$

Substitution of the data and the fundamental constants gives

$$\lambda = \frac{6.626\times10^{-34}\,\text{J s}}{\{2\times(9.109\times10^{-31}\,\text{kg})\times(1.602\times10^{-19}\,\text{C})\times(4.0\times10^{4}\,\text{V})\}^{1/2}} = 6.1\times10^{-12}\,\text{m}$$

or 6.1 pm.

Self-test 7A.3 Calculate the wavelength of (a) a neutron with a translational kinetic energy equal to kT at 300 K, (b) a tennis ball of mass 57 g travelling at $80\,\text{km}\,\text{h}^{-1}$.

Answer: (a) 178 pm, (b) 5.2×10^{-34} m

Exercises

E7A.4 A sodium lamp emits yellow light (550 nm). How many photons does it emit each second if its power is (i) 1.0 W, (ii) 100 W?

E7A.5 The work function of metallic caesium is 2.14 eV. Calculate the kinetic energy and the speed of the electrons ejected by light of wavelength (i) 700 nm, (ii) 300 nm.

E7A.6 To what speed must an electron be accelerated from rest for it to have a de Broglie wavelength of 100 pm? What accelerating potential difference is needed?

Checklist of concepts

- ☐ 1. A **black body** is an object capable of emitting and absorbing all wavelengths of radiation without favouring any wavelength.
- ☐ 2. An electromagnetic field of a given frequency can take up energy only in discrete amounts.
- ☐ 3. Atomic and molecular spectra show that atoms and molecules can take up energy only in discrete amounts.
- ☐ 4. The **photoelectric effect** establishes the view that electromagnetic radiation, regarded in classical physics as wave-like, consists of particles (photons).
- ☐ 5. The diffraction of electrons establishes the view that electrons, regarded in classical physics as particles, are wave-like with a wavelength given by the **de Broglie relation**.
- ☐ 6. **Wave–particle duality** is the recognition that the concepts of particle and wave blend together.

Checklist of equations

Property	Equation	Comment	Equation number
Wien's law	$\lambda_{max}T = c_2/5$	c_2 is the second radiation constant, $c_2 = hc/k$	7A.1, 7A.7
Stefan–Boltzmann law	$\mathcal{E}(T) = aT^4$	$a = 8\pi^5k^4/15h^3c^3$	7A.3, 7A.8
Planck distribution	$\rho(\lambda,T) = 8\pi hc/\{\lambda^5(e^{hc/\lambda kT} - 1)\}$	Black-body radiation	7A.6
	$\rho(\nu,T) = 8\pi h\nu^3/\{c^3(e^{h\nu/kT} - 1)\}$		
Einstein formula for heat capacity of a solid	$C_{V,m}(T) = 3Rf_E(T)$	Einstein temperature: $\theta_E = h\nu/k$	7A.9
	$f_E(T) = (\theta_E/T)^2\{e^{\theta_E/2T}/(e^{\theta_E/T} - 1)\}^2$		
Bohr frequency condition	$\Delta E = h\nu$		7A.10
Photoelectric effect	$E_k = h\nu - \Phi$	Φ is the work function	7A.11
de Broglie relation	$\lambda = h/p$		7A.12

TOPIC 7B Wavefunctions

➤ **Why do you need to know this material?**

Wavefunctions provide the essential foundation for understanding the properties of electrons in atoms and molecules, and are central to explanations in chemistry.

➤ **What is the key idea?**

All the dynamical properties of a system are contained in its wavefunction, which is obtained by solving the Schrödinger equation.

➤ **What do you need to know already?**

You need to be aware of the shortcomings of classical physics that drove the development of quantum theory (Topic 7A).

In classical mechanics an object travels along a definite path or trajectory. In quantum mechanics a particle in a particular state is described by a **wavefunction**, ψ (psi), which is spread out in space, rather than being localized. The wavefunction contains all the dynamical information about the object in that state, such as its position and momentum.

7B.1 The Schrödinger equation

In 1926 Erwin Schrödinger proposed an equation for finding the wavefunctions of any system. The **time-independent Schrödinger equation** for a particle of mass m moving in one dimension with energy E in a system that does not change with time (for instance, its volume remains constant) is

$$-\frac{\hbar^2}{2m}\frac{d^2\psi}{dx^2}+V(x)\psi=E\psi \qquad \text{Time-independent Schrödinger equation} \qquad (7B.1)$$

The constant $\hbar = h/2\pi$ (which is read h-cross or h-bar) is a convenient modification of Planck's constant used widely in quantum mechanics; $V(x)$ is the potential energy of the particle at x. Because the total energy E is the sum of potential and kinetic energies, the first term on the left must be related (in a manner explored later) to the kinetic energy of the particle.

The Schrödinger equation can be regarded as a fundamental postulate of quantum mechanics, but its plausibility can be demonstrated by showing that, for the case of a free particle, it is consistent with the de Broglie relation (Topic 7A). The first step is to note the potential energy $V(x)$ is zero everywhere, so E is equal to the kinetic energy $E = p^2/2m$ and eqn 7B.1 can be written as

$$-\frac{\hbar^2}{2m}\frac{d^2\psi(x)}{dx^2}=\frac{p^2}{2m}\psi(x)$$

A wave of wavelength λ has the form $\psi(x) = \sin(2\pi x/\lambda)$ so the left-hand side of this equation is

$$\begin{aligned}-\frac{\hbar^2}{2m}\frac{d^2\psi(x)}{dx^2} &= -\frac{\hbar^2}{2m}\frac{d^2\sin(2\pi x/\lambda)}{dx^2}\\ &=\frac{\hbar^2}{2m}\left(\frac{2\pi}{\lambda}\right)^2\overbrace{\sin(2\pi x/\lambda)}^{\psi(x)}\\ &=\frac{h^2}{2m\lambda^2}\psi(x)\end{aligned}$$

Comparison of the two preceding equations shows that $p^2/2m = h^2/2m\lambda^2$ and therefore that $p^2 = h^2/\lambda^2$. Therefore $\lambda^2 = h^2/p^2$, which implies the de Broglie relation, $\lambda = h/p$.

7B.2 The Born interpretation

One piece of dynamical information contained in the wavefunction is the location of the particle. Max Born used an analogy with the wave theory of radiation, in which the square of the amplitude of an electromagnetic wave in a region is interpreted as its intensity and therefore (in quantum terms) as a measure of the probability of finding a photon present in the region. The **Born interpretation** of the wavefunction is:

> If the wavefunction of a particle has the value ψ at x, then the probability of finding the particle between x and $x + dx$ is proportional to $|\psi|^2 dx$ (Fig. 7B.1).

The quantity $|\psi|^2 = \psi^\star\psi$, where $\psi^\star$ is the complex conjugate of ψ, allows for the possibility that ψ is complex (see *The chemist's toolkit* 7B.1). If the wavefunction is real, like $\sin(2\pi x/\lambda)$ is, then $|\psi|^2 = \psi^2$.

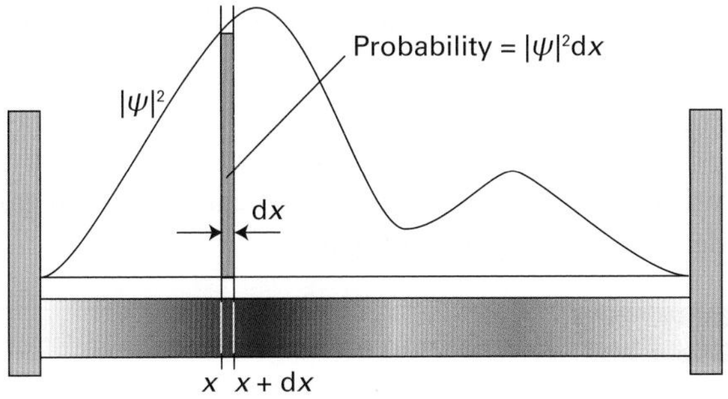

Figure 7B.1 The wavefunction ψ is a probability amplitude in the sense that its square modulus ($\psi^*\psi$ or $|\psi|^2$) is a probability density. The probability of finding a particle in the region between x and $x + dx$ is proportional to $|\psi|^2dx$. Here, the probability density is represented by the density of shading in the superimposed band.

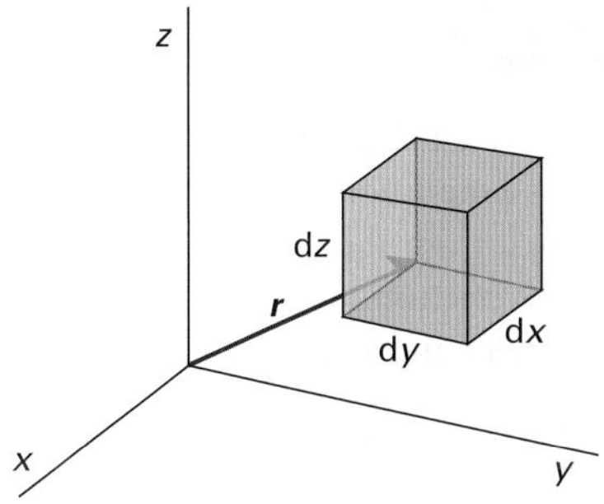

Figure 7B.2 The Born interpretation of the wavefunction in three-dimensional space implies that the probability of finding the particle in the volume element $d\tau = dxdydz$ at some position $\boldsymbol{r}$ is proportional to the product of $d\tau$ and the value of $|\psi(\boldsymbol{r})|^2$ at that position.

The chemist's toolkit 7B.1 Complex numbers

A complex number z has the form $z = x + iy$, where $i = \sqrt{-1}$. The complex conjugate of a complex number z is $z^* = x - iy$. Complex numbers combine together according to the following rules:

Addition and subtraction:

$$(a+ib)+(c+id)=(a+c)+i(b+d)$$

Multiplication:

$$(a+ib)\,(c+id)=(ac-bd)+i(bc+ad)$$

Two special relations are:

$$\text{Modulus: } |z|=(z^*z)^{1/2}=(x^2+y^2)^{1/2}$$

Euler's relation: $e^{i\phi}=\cos\phi+i\sin\phi$, which implies that $e^{i\pi}=-1$, $\cos\phi=\frac{1}{2}(e^{i\phi}+e^{-i\phi})$, and $\sin\phi=-\frac{1}{2}i(e^{i\phi}-e^{-i\phi})$.

Because $|\psi|^2dx$ is a (dimensionless) probability, $|\psi|^2$ is the **probability density**, with the dimensions of 1/length (for a one-dimensional system). The wavefunction ψ itself is called the **probability amplitude**. For a particle free to move in three dimensions (for example, an electron near a nucleus in an atom), the wavefunction depends on the coordinates x, y, and z and is denoted $\psi(\boldsymbol{r})$. In this case the Born interpretation is (Fig. 7B.2):

> If the wavefunction of a particle has the value $\psi(\boldsymbol{r})$ at $\boldsymbol{r}$, then the probability of finding the particle in an infinitesimal volume $d\tau = dxdydz$ at that position is proportional to $|\psi(\boldsymbol{r})|^2d\tau$.

In this case, $|\psi|^2$ has the dimensions of 1/length^3 and the wavefunction itself has dimensions of 1/length$^{3/2}$ (and units such as $m^{-3/2}$).

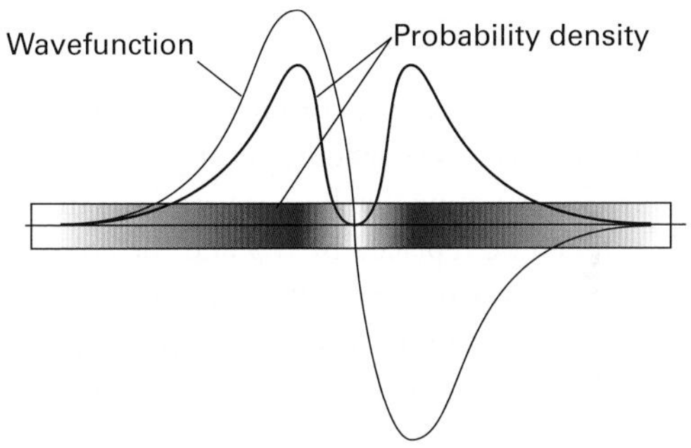

Figure 7B.3 The sign of a wavefunction has no direct physical significance: the positive and negative regions of this wavefunction both correspond to the same probability distribution (as given by the square modulus of ψ and depicted by the density of the shading).

The Born interpretation does away with any worry about the significance of a negative (and, in general, complex) value of ψ because $|\psi|^2$ is always real and nowhere negative. There is no *direct* significance in the negative (or complex) value of a wavefunction: only the square modulus is directly physically significant, and both negative and positive regions of a wavefunction may correspond to a high probability of finding a particle in a region (Fig. 7B.3). However, the presence of positive and negative regions of a wavefunction is of great *indirect* significance, because it gives rise to the possibility of constructive and destructive interference between different wavefunctions.

A wavefunction may be zero at one or more points, and at these locations the probability density is also zero. It is important to distinguish a point at which the wavefunction is zero (for instance, far from the nucleus of a hydrogen atom) from the point at which it passes *through* zero. The latter is called a **node**. A location where the wavefunction approaches zero without actually passing through zero is not a node. Thus, the wavefunction $\sin(2\pi x/\lambda)$ has nodes where the wave passes through zero, but the wavefunction e^{-kx} has no nodes, despite becoming zero as $x \to \infty$.

Brief illustration 7B.1

The wavefunction of an electron in the lowest energy state of a hydrogen atom is proportional to e^{-r/a_0}, where a_0 is a constant and r the distance from the nucleus. The relative probabilities of finding the electron inside a tiny region of volume δV located at the nucleus and a distance a_0 from the nucleus are

$$\frac{P(0)}{P(a_0)} = \frac{\psi(0)^2\delta V}{\psi(a_0)^2\delta V} = \frac{\psi(0)^2}{\psi(a_0)^2} = \frac{e^0}{e^{-2}} = 7.4$$

That is, it is more probable (by a factor of 7) that the electron will be found in a volume element at the nucleus than in a volume element of the same size located at a distance a_0 from the nucleus.

7B.2(a) Normalization

A mathematical feature of the Schrödinger equation is that if ψ is a solution, then so is $N\psi$, where N is any constant. This feature is confirmed by noting that because ψ occurs in every term in eqn 7B.1, it can be replaced by $N\psi$ and the constant factor N cancelled to recover the original equation. This freedom to multiply the wavefunction by a constant factor means that it is always possible to find a **normalization constant**, N, such that rather than the probability density being *proportional* to $|\psi|^2$ it becomes *equal* to $|\psi|^2$.

A normalization constant is found by noting that, for a normalized wavefunction $N\psi$, the probability that a particle is in the region dx is equal to $(N\psi^*)(N\psi)dx$ (N is taken to be real). Furthermore, the sum over all space of these individual probabilities must be 1 (the probability of the particle being somewhere is 1). Expressed mathematically, the latter requirement is

$$N^2\int_{-\infty}^{\infty}\psi^*\psi\,dx = 1 \quad (7B.2)$$

and therefore

$$N = \frac{1}{\left(\int_{-\infty}^{\infty}\psi^*\psi\,dx\right)^{1/2}} \quad (7B.3)$$

Provided this integral has a finite value (that is, the wavefunction is 'square integrable'), the normalization factor can be found and the wavefunction 'normalized' (and specifically 'normalized to 1'). From now on, unless stated otherwise, all wavefunctions are assumed to have been normalized to 1, in which case in one dimension

$$\int_{-\infty}^{\infty}\psi^*\psi\,dx = 1 \quad (7B.4a)$$

and in three dimensions

$$\int_{-\infty}^{\infty}\int_{-\infty}^{\infty}\int_{-\infty}^{\infty}\psi^*\psi\;dx dy dz = 1 \quad (7B.4b)$$

In quantum mechanics it is common to write all such integrals in a short-hand form as

$$\int\psi^*\psi\;d\tau = 1 \quad (7B.4c)$$

where $d\tau$ is the appropriate volume element and the integration is understood as being over all space.

Example 7B.1 Normalizing a wavefunction

Carbon nanotubes are thin hollow cylinders of carbon with diameters between 1 nm and 2 nm, and lengths of several micrometres. According to one simple model, the lowest-energy electrons of the nanotube are described by the wavefunction $\sin(\pi x/L)$, where L is the length of the nanotube. Find the normalized wavefunction.

Collect your thoughts Because the wavefunction is one-dimensional, you need to find the factor N that guarantees that the integral in eqn 7B.4a is equal to 1. The wavefunction is real, so $\psi^* = \psi$. Relevant integrals are found in the *Resource section*.

The solution Write the wavefunction as $\psi = N\sin(\pi x/L)$, where N is the normalization factor. The limits of integration are $x = 0$ to $x = L$ because the space available to the electron spans the length of the tube. It follows that

$$\int\psi^*\psi\,d\tau = N^2\overbrace{\int_0^L \sin^2\frac{\pi x}{L}\,dx}^{\text{Integral T.2}} = \tfrac{1}{2}N^2L$$

For the wavefunction to be normalized, this integral must be equal to 1; that is, $\tfrac{1}{2}N^2L = 1$, and hence

$$N = \left(\frac{2}{L}\right)^{1/2}$$

The normalized wavefunction is therefore

$$\psi = \left(\frac{2}{L}\right)^{1/2}\sin\frac{\pi x}{L}$$

Because L is a length, the dimensions of ψ are 1/length$^{1/2}$, and therefore those of ψ^2 are 1/length, as is appropriate for a probability density in one dimension.

Self-test 7B.1 The wavefunction for the next higher energy level for the electrons in the same tube is $\sin(2\pi x/L)$. Normalize this wavefunction.

Answer: $N = (2/L)^{1/2}$

To calculate the probability of finding the system in a finite region of space the probability density is summed (integrated) over the region of interest. Thus, for a one-dimensional system, the probability P of finding the particle between x_1 and x_2 is given by

$$P = \int_{x_1}^{x_2}|\psi(x)|^2\,dx \quad \text{Probability [one-dimensional region]} \quad (7B.5)$$

Example 7B.2 Evaluating a probability

As seen in *Example* 7B.1, the lowest-energy electrons of a carbon nanotube of length L can be described by the normalized wavefunction $(2/L)^{1/2}\sin(\pi x/L)$. What is the probability of finding the electron between $x = L/4$ and $x = L/2$?

Collect your thoughts Use eqn 7B.5 and the normalized wavefunction to write an expression for the probability of finding the electron in the region of interest. Relevant integrals are given in the *Resource section*.

The solution From eqn 7B.5 the probability is

$$P = \overbrace{\frac{2}{L}\int_{L/4}^{L/2}\sin^2(\pi x/L)\,dx}^{\text{Integral T.2}}$$

It follows that

$$P = \frac{2}{L}\left(\frac{x}{2} - \frac{\sin(2\pi x/L)}{4\pi/L}\right)\Bigg|_{L/4}^{L/2} = \frac{2}{L}\left(\frac{L}{4} - \frac{L}{8} - 0 + \frac{1}{4\pi/L}\right) = 0.409$$

It follows that there is a chance of about 41 per cent that the electron will be found in the region.

Self-test 7B.2 As remarked in *Self-test* 7B.1, the normalized wavefunction of the next higher energy level of the electron in this model of the nanotube is $(2/L)^{1/2}\sin(2\pi x/L)$. What is the probability of finding the electron between $x = L/4$ and $x = L/2$ when it is described by this wavefunction?

Answer: 0.25

7B.2(b) Constraints on the wavefunction

The Born interpretation puts severe restrictions on the acceptability of wavefunctions. The first constraint is that ψ must not be infinite over a finite region; if it were, the Born interpretation would fail. This requirement rules out many possible solutions of the Schrödinger equation, because many mathematically acceptable solutions rise to infinity and are therefore physically unacceptable. The Born interpretation also rules out solutions of the Schrödinger equation that give rise to more than one value of $|\psi|^2$ at a single point. It would be absurd to have more than one value of the probability density for the particle at a point. This restriction is expressed by saying that the wavefunction must be *single-valued*; that is, it must have only one value at each point of space.

The Schrödinger equation itself also implies some mathematical restrictions on the type of functions that can occur. Because it is a second-order differential equation (in the sense that it depends on the second derivative of the wavefunction), $d^2\psi/dx^2$ must be well-defined if the equation is to be applicable everywhere. The second derivative is defined only if the first derivative is continuous. That is (except as specified below), there can be no kinks in the function. In turn, the first derivative is defined only if the function is continuous: no sharp steps are permitted.

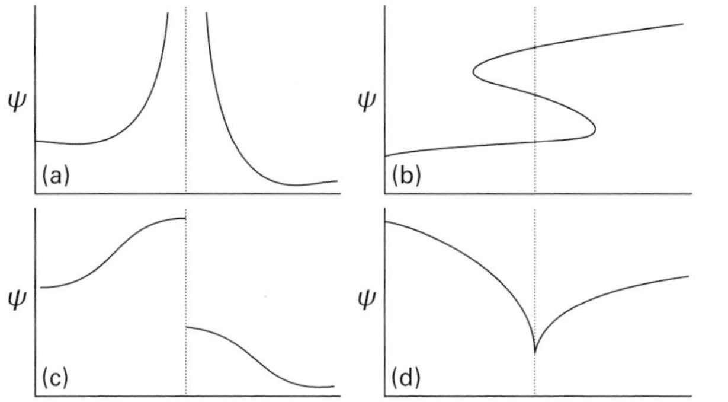

Figure 7B.4 The wavefunction must satisfy stringent conditions for it to be acceptable: (a) unacceptable because it is infinite over a finite region; (b) unacceptable because it is not single-valued; (c) unacceptable because it is not continuous; (d) unacceptable because its slope is discontinuous.

Overall, therefore, the constraints on the wavefunction, which are summarized in Fig. 7B.4, are that it

- must not be infinite over a finite region;
- must be single-valued;
- must be continuous;
- must have a continuous first derivative (slope).

The last of these constraints does not apply if the potential energy has abrupt, infinitely high steps (as in the particle-in-a-box model treated in Topic 7D).

7B.2(c) Quantization

The constraints just noted are so severe that acceptable solutions of the Schrödinger equation do not in general exist for arbitrary values of the energy E. In other words, a particle may possess only certain energies, for otherwise its wavefunction would be physically unacceptable. That is,

> As a consequence of the restrictions on its wavefunction, the energy of a particle is quantized.

These acceptable energies are found by solving the Schrödinger equation for motion of various kinds, and selecting the solutions that conform to the restrictions listed above.

Exercises

E7B.1 A possible wavefunction for an electron in a region of length L (i.e. from $x = 0$ to $x = L$) is $\sin(2\pi x/L)$. (i) Normalize this wavefunction (to 1). (ii) What is the probability of finding the electron in the range dx at $x = L/2$? (iii) At what value or values of x is the probability density a maximum? Locate the positions of any nodes in the wavefunction. You need consider only the range $x = 0$ to $x = L$.

E7B.2 Imagine a particle free to move in the x direction. Which of the following wavefunctions would be acceptable for such a particle? In each case, give your reasons for accepting or rejecting each function. (i) $\psi(x) = x^2$; (ii) $\psi(x) = 1/x$; (iii) $\psi(x) = e^{-x^2}$.

Checklist of concepts

☐ 1. A **wavefunction** is a mathematical function that contains all the dynamical information about a system.

☐ 2. The **Schrödinger equation** is a second-order differential equation used to calculate the wavefunction of a system.

☐ 3. According to the **Born interpretation**, the probability density at a point is proportional to the square of the wavefunction at that point.

☐ 4. A **node** is a point where a wavefunction passes through zero.

☐ 5. A wavefunction is **normalized** if the integral over all space of its square modulus is equal to 1.

☐ 6. A wavefunction must be single-valued, continuous, not infinite over a finite region of space, and (except in special cases) have a continuous slope.

☐ 7. The quantization of energy stems from the constraints that an acceptable wavefunction must satisfy.

Checklist of equations

Property	Equation	Comment	Equation number
The time-independent Schrödinger equation	$-(\hbar^2/2m)(\mathrm{d}^2\psi/\mathrm{d}x^2) + V(x)\psi = E\psi$	One-dimensional system*	7B.1
Normalization	$\int \psi^*\psi\,\mathrm{d}\tau = 1$	Integration over all space	7B.4c
Probability of a particle being between x_1 and x_2	$P = \int_{x_1}^{x_2} \lvert\psi(x)\rvert^2\,\mathrm{d}x$	One-dimensional region	7B.5

* Higher dimensions are treated in Topics 7D, 7F, and 8A.

TOPIC 7C Operators and observables

➤ **Why do you need to know this material?**

To interpret the wavefunction fully it is necessary to be able to extract dynamical information from it. The predictions of quantum mechanics are often very different from those of classical mechanics, and those differences are essential for understanding the structures and properties of atoms and molecules.

➤ **What is the key idea?**

The dynamical information in the wavefunction is extracted by calculating the expectation values of hermitian operators.

➤ **What do you need to know already?**

You need to know that the state of a particle is fully described by a wavefunction (Topic 7B), and that the probability density is proportional to the square modulus of the wavefunction.

A wavefunction contains all the information it is possible to obtain about the dynamical properties of a particle (for example, its location and momentum). The Born interpretation (Topic 7B) provides information about location, but the wavefunction contains other information, which is extracted by using the methods described in this Topic.

7C.1 Operators

The Schrödinger equation can be written in the succinct form

$$\hat{H}\psi = E\psi \qquad \text{Operator form of Schrödinger equation} \qquad (7C.1a)$$

Comparison of this expression with the one-dimensional Schrödinger equation

$$-\frac{\hbar^2}{2m}\frac{d^2\psi}{dx^2} + V(x)\psi = E\psi$$

shows that in one dimension

$$\hat{H} = -\frac{\hbar^2}{2m}\frac{d^2}{dx^2} + V(x) \qquad \text{Hamiltonian operator} \qquad (7C.1b)$$

The quantity $\hat{H}$ (commonly read h-hat) is an **operator**, an expression that carries out a mathematical operation on a function. In this case, the operation is to take the second derivative of ψ, and (after multiplication by $-\hbar^2/2m$) to add the result to the outcome of multiplying ψ by $V(x)$.

The operator $\hat{H}$ plays a special role in quantum mechanics, and is called the **hamiltonian operator** after the nineteenth century mathematician William Hamilton. He developed a form of classical mechanics which, it subsequently turned out, is well suited to the formulation of quantum mechanics. The hamiltonian operator (and commonly simply 'the hamiltonian') is the operator corresponding to the total energy of the system, the sum of the kinetic and potential energies. In eqn 7C.1b the second term on the right is the potential energy, so the first term (the one involving the second derivative) must be the operator for kinetic energy.

In general, an operator acts on a function to produce a new function, as in

$$(\text{operator})(\text{function}) = (\text{new function})$$

In some cases the new function is the same as the original function, perhaps multiplied by a constant. Combinations of operators and functions that have this property are of great importance in quantum mechanics.

Brief illustration 7C.1

When the operator d/dx, which means 'take the derivative of the following function with respect to x', acts on the function sin ax, it generates the new function a cos ax. However, when d/dx operates on e^{-ax} it generates $-ae^{-ax}$, which is the original function multiplied by the constant $-a$.

7C.1(a) Eigenvalue equations

The Schrödinger equation written as in eqn 7C.1a is an **eigenvalue equation**, an equation of the form

$$(\text{operator})(\text{function}) = (\text{constant factor}) \times (\text{same function}) \qquad (7C.2a)$$

In an eigenvalue equation, the action of the operator on the function generates the *same* function, multiplied by a constant. If a general operator is denoted $\hat{\Omega}$ (where Ω is uppercase omega) and the constant factor by ω (lowercase omega), then an eigenvalue equation has the form

$$\hat{\Omega}\psi = \omega\psi \qquad \text{Eigenvalue equation} \qquad (7C.2b)$$

If this relation holds, the function ψ is said to be an **eigenfunction** of the operator $\hat{\Omega}$, and ω is the **eigenvalue** associated with that eigenfunction. With this terminology, eqn 7C.2a can be written

$$(\text{operator})(\text{eigenfunction}) = (\text{eigenvalue})\times(\text{eigenfunction}) \qquad (7C.2c)$$

Equation 7C.1a is therefore an eigenvalue equation in which ψ is an eigenfunction of the hamiltonian and E is the associated eigenvalue. It follows that 'solving the Schrödinger equation' can be expressed as 'finding the eigenfunctions and eigenvalues of the hamiltonian operator for the system'.

Just as the hamiltonian is the operator corresponding to the total energy, there are operators that represent other **observables**, the measurable properties of the system, such as linear momentum or electric dipole moment. For each such operator $\hat{\Omega}$ there is an eigenvalue equation of the form $\hat{\Omega}\psi = \omega\psi$, with the following significance:

> If the wavefunction is an eigenfunction of the operator $\hat{\Omega}$ corresponding to the observable Ω, then the outcome of a measurement of the property Ω will be the eigenvalue corresponding to that eigenfunction.

For instance, if the property of interest, Ω, is the energy, E, then the corresponding operator $\hat{\Omega}$ is the hamiltonian, and the outcome of a measurement of the energy will be the eigenvalue E corresponding to the eigenfunction ψ in the equation $\hat{H}\psi = E\psi$. Quantum mechanics is formulated by constructing the operator corresponding to the observable of interest and then predicting the outcome of a measurement by examining the eigenvalues of the operator.

7C.1(b) The construction of operators

Observables are represented by operators built from the following position and linear momentum operators:

$$\hat{x} = x\times \quad \hat{p}_x = \frac{\hbar}{\mathrm{i}}\frac{\mathrm{d}}{\mathrm{d}x} \qquad \text{Specification of operators} \qquad (7C.3)$$

That is,

- the operator for location along the x-axis is multiplication (of the wavefunction) by x,
- the operator for linear momentum parallel to the x-axis is $\hbar/\mathrm{i}$ times the derivative (of the wavefunction) with respect to x.

The definitions in eqn 7C.3 are used to construct operators for other spatial observables. For example, suppose the potential energy has the form $V(x) = \frac{1}{2}k_f x^2$, where k_f is a constant (this potential energy describes the vibrations of atoms in molecules). Because the operator for x is multiplication by x, by extension the operator for x^2 is multiplication by x and then by x again, or multiplication by x^2. The operator corresponding to $\frac{1}{2}k_f x^2$ is therefore

$$\hat{V}(x) = \tfrac{1}{2}k_f x^2 \times \qquad (7C.4)$$

In practice, the multiplication sign is omitted and multiplication is understood. To construct the operator for kinetic energy, the classical relation between kinetic energy and linear momentum, $E_k = p_x^2/2m$ is used. Then, by using the operator for p_x from eqn 7C.3:

$$\hat{E}_k = \frac{1}{2m}\overbrace{\left(\frac{\hbar}{\mathrm{i}}\frac{\mathrm{d}}{\mathrm{d}x}\right)}^{\hat{p}_x}\overbrace{\left(\frac{\hbar}{\mathrm{i}}\frac{\mathrm{d}}{\mathrm{d}x}\right)}^{\hat{p}_x} = -\frac{\hbar^2}{2m}\frac{\mathrm{d}^2}{\mathrm{d}x^2} \qquad (7C.5)$$

It follows that the operator for the total energy, the hamiltonian operator, is

$$\hat{H} = \hat{E}_k + \hat{V} = -\frac{\hbar^2}{2m}\frac{\mathrm{d}^2}{\mathrm{d}x^2} + \hat{V}(x) \qquad \text{Hamiltonian operator} \qquad (7C.6)$$

where $\hat{V}(x)$ is the operator corresponding to whatever form the potential energy takes, exactly as in eqn 7C.1b.

Example 7C.1 Evaluating an observable

What is the linear momentum of a free particle described by the wavefunctions (a) $\psi(x) = \mathrm{e}^{\mathrm{i}kx}$ and (b) $\psi(x) = \mathrm{e}^{-\mathrm{i}kx}$?

Collect your thoughts You need to operate on ψ with the operator corresponding to linear momentum (eqn 7C.3), and inspect the result. If the outcome is the original wavefunction multiplied by a constant (that is, if the application of the operator results in an eigenvalue equation), then you can identify the constant with the value of the observable.

The solution (a) For $\psi(x) = \mathrm{e}^{\mathrm{i}kx}$,

$$\hat{p}_x\boxed{\psi} = \frac{\hbar}{\mathrm{i}}\frac{\mathrm{d}\boxed{\psi}}{\mathrm{d}x} = \frac{\hbar}{\mathrm{i}}\frac{\mathrm{d}\boxed{\mathrm{e}^{\mathrm{i}kx}}}{\mathrm{d}x} = \frac{\hbar}{\mathrm{i}}\times \mathrm{i}k\boxed{\mathrm{e}^{\mathrm{i}kx}} = \overbrace{+k\hbar}^{\text{Eigenvalue}}\boxed{\psi}$$

This is an eigenvalue equation, with eigenvalue $+k\hbar$. It follows that a measurement of the momentum will give the value $p_x = +k\hbar$.

(b) For $\psi(x) = \mathrm{e}^{-\mathrm{i}kx}$,

$$\hat{p}_x\psi = \frac{\hbar}{\mathrm{i}}\frac{\mathrm{d}\psi}{\mathrm{d}x} = \frac{\hbar}{\mathrm{i}}\frac{\mathrm{d}\mathrm{e}^{-\mathrm{i}kx}}{\mathrm{d}x} = \frac{\hbar}{\mathrm{i}}\times(-\mathrm{i}k)\mathrm{e}^{-\mathrm{i}kx} = \overbrace{-k\hbar}^{\text{Eigenvalue}}\psi$$

Now the eigenvalue is $-k\hbar$, so $p_x = -k\hbar$. In case (a) the momentum is positive, meaning that the particle is travelling in the positive x-direction, whereas in (b) the particle is moving in the opposite direction. This result illustrates a general feature of quantum mechanics: taking the complex conjugate of a wavefunction reverses the direction of travel. An implication is that if the wavefunction is real (such as $\cos(2\pi x/\lambda)$ for a wave with wavelength λ), then taking the complex conjugate leaves the wavefunction unchanged: there is no net direction of travel.

Self-test 7C.1 What is the kinetic energy of a particle described by the wavefunction cos kx?

Answer: $E_k = \hbar^2k^2/2m$

The expression for the kinetic energy operator (eqn 7C.5) reveals an important point about the Schrödinger equation. In mathematics, the second derivative of a function is a measure of its curvature: a large second derivative indicates a sharply curved function (Fig. 7C.1). It follows that a sharply curved wavefunction is associated with a high kinetic energy, and one with a low curvature is associated with a low kinetic energy.

The curvature of a wavefunction in general varies from place to place (Fig. 7C.2). Wherever a wavefunction is sharply curved, its contribution to the total kinetic energy is large. Wherever the wavefunction is not sharply curved, its contribution to the overall kinetic energy is low. The observed kinetic energy of the particle is an average of all the contributions of the kinetic energy from each region. Hence, a particle can be expected to have a high kinetic energy if the average curvature of its wavefunction is high (that is, has sharply curved regions). Locally there can be both positive and negative contributions to the kinetic energy (because the curvature can be either positive, ∪, or negative, ∩), but the average is always positive.

The association of high curvature with high kinetic energy is a valuable guide to the interpretation of wavefunctions and the prediction of their shapes. For example, suppose the wavefunction of a particle with a given total energy and a potential energy that decreases with increasing x is required. Because the difference $E - V = E_k$ increases from left to right, the wavefunction must become more sharply curved by oscillating more rapidly as x increases (Fig. 7C.3). It is therefore likely that the wavefunction will look like the function sketched in the illustration, as more detailed calculation confirms.

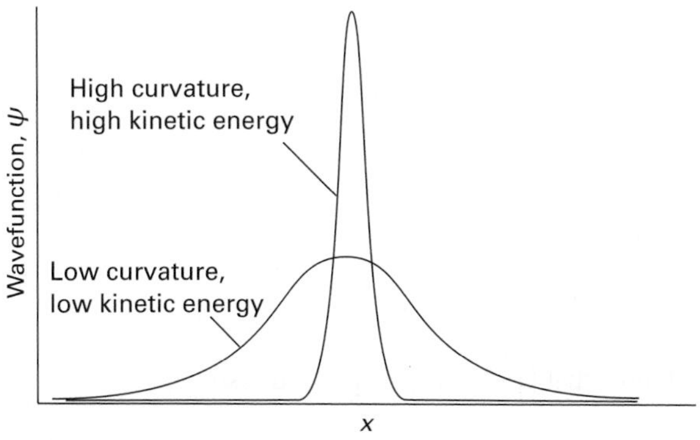

Figure 7C.1 The average kinetic energy of a particle can be inferred from the average curvature of the wavefunction. This figure shows two wavefunctions: the sharply curved function corresponds to a higher kinetic energy than the less sharply curved function.

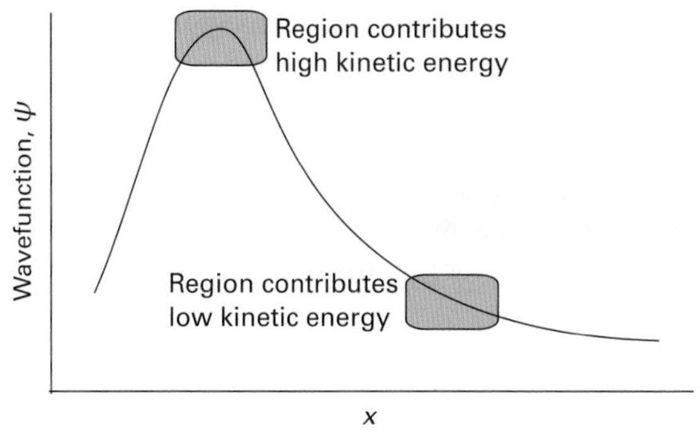

Figure 7C.2 The observed kinetic energy of a particle is an average of contributions from the entire space covered by the wavefunction. Sharply curved regions contribute a high kinetic energy to the average; less sharply curved regions contribute only a small kinetic energy.

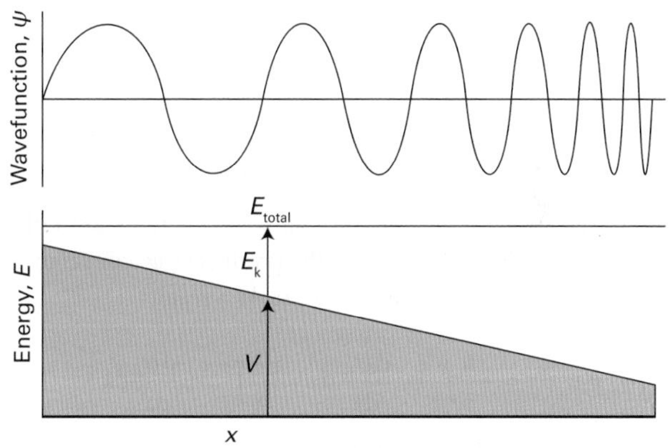

Figure 7C.3 The wavefunction of a particle with a potential energy V that decreases towards the right. As the total energy is constant, the kinetic energy E_k increases to the right, which results in a faster oscillation and hence greater curvature of the wavefunction.

7C.1(c) Hermitian operators

All the quantum mechanical operators that correspond to observables have a very special mathematical property: they are 'hermitian'. A **hermitian operator** is one for which the following relation is true:

$$\int \psi_i^* \hat{\Omega}\psi_j \,\mathrm{d}\tau = \left\{\int \psi_j^* \hat{\Omega}\psi_i \,\mathrm{d}\tau\right\}^* \qquad \text{Hermiticity [definition]} \qquad (7C.7)$$

As stated in Topic 7B, in quantum mechanics $\int \ldots \mathrm{d}\tau$ implies integration over the full range of all relevant spatial variables.

Brief illustration 7C.2

The position operator ($x\,\times$) is hermitian because in this case the order of the factors in the integrand can be changed:

$$\int \psi_i^* x\psi_j \,\mathrm{d}\tau = \int \psi_j x\psi_i^* \mathrm{d}\tau = \left\{\int \psi_j^* x\psi_i \mathrm{d}\tau\right\}^*$$

The final step uses $(\psi^*)^* = \psi$.

The demonstration that the linear momentum operator is hermitian is more involved because the order of functions being differentiated cannot be changed.

How is that done? 7C.1 Proving that the linear momentum operator is hermitian

The task is to show that

$$\int \psi_i^\star \hat{p}_x \psi_j \,\mathrm{d}\tau = \left\{ \int \psi_j^\star \hat{p}_x \psi_i \,\mathrm{d}\tau \right\}^\star$$

with $\hat{p}_x$ given in eqn 7C.3. To do so, use 'integration by parts' (see Part 1 of the *Resource section*) which, when applied to the present case, gives

$$\int \psi_i^\star \hat{p}_x \psi_j \,\mathrm{d}\tau = \frac{\hbar}{\mathrm{i}} \int_{-\infty}^{\infty} \overbrace{\psi_i^\star}^{f} \overbrace{\frac{\mathrm{d}\psi_j}{\mathrm{d}x}}^{\mathrm{d}g/\mathrm{d}x} \mathrm{d}x$$

$$= \frac{\hbar}{\mathrm{i}} \boxed{\overbrace{\overbrace{\psi_i^\star \psi_j}^{fg} \Big|_{-\infty}^{\infty}}^{0}} - \frac{\hbar}{\mathrm{i}} \int_{-\infty}^{\infty} \overbrace{\psi_j}^{g} \overbrace{\frac{\mathrm{d}\psi_i^\star}{\mathrm{d}x}}^{\mathrm{d}f/\mathrm{d}x} \mathrm{d}x$$

The boxed term is zero because all wavefunctions are either zero at $x = \pm\infty$ (see Topic 7B) or the product $\psi_i^\star \psi_j$ converges to the same value at $x = +\infty$ and $x = -\infty$. As a result

$$\int \psi_i^\star \hat{p}_x \psi_j \,\mathrm{d}\tau = -\frac{\hbar}{\mathrm{i}} \int_{-\infty}^{\infty} \psi_j \frac{\mathrm{d}\psi_i^\star}{\mathrm{d}x} \mathrm{d}x = \left\{ \frac{\hbar}{\mathrm{i}} \int_{-\infty}^{\infty} \psi_j^\star \frac{\mathrm{d}\psi_i}{\mathrm{d}x} \mathrm{d}x \right\}^\star$$
$$= \left\{ \int \psi_j^\star \hat{p}_x \psi_i \,\mathrm{d}\tau \right\}^\star$$

as was to be proved. The penultimate line uses $(\psi^\star)^\star = \psi$ and $\mathrm{i}^\star = -\mathrm{i}$.

Hermitian operators are enormously important in quantum mechanics because their eigenvalues are *real*: that is, in eqn 7C.2b $\omega^\star = \omega$. Any measurement must yield a real value because a position, momentum, or an energy cannot be complex or imaginary. Because the outcome of a measurement of an observable is one of the eigenvalues of the corresponding operator, those eigenvalues must be real. It therefore follows that an operator that represents an observable must be hermitian. The proof that their eigenfunctions are real makes use of the definition of hermiticity in eqn 7C.7.

How is that done? 7C.2 Proving that the eigenvalues of hermitian operators are real

Begin by setting ψ_i and ψ_j to be the same, writing them both as ψ. Then eqn 7C.7 becomes

$$\int \psi^\star \hat{\Omega} \psi \,\mathrm{d}\tau = \left\{ \int \psi^\star \hat{\Omega} \psi \,\mathrm{d}\tau \right\}^\star$$

Next suppose that ψ is an eigenfunction of $\hat{\Omega}$ with eigenvalue ω. That is, $\hat{\Omega}\psi = \omega\psi$. Now use this relation in both integrals on the left- and right-hand sides:

$$\int \psi^\star \omega \psi \,\mathrm{d}\tau = \left\{ \int \psi^\star \omega \psi \,\mathrm{d}\tau \right\}^\star$$

The eigenvalue is a constant that can be taken outside the integrals:

$$\omega \boxed{\int \psi^\star \psi \mathrm{d}\tau} = \left\{ \omega \int \psi^\star \psi \mathrm{d}\tau \right\}^\star = \omega^\star \boxed{\int \psi \psi^\star \mathrm{d}\tau}$$

Finally, the (boxed) integrals cancel, leaving $\omega = \omega^\star$. It follows that ω is real.

7C.1(d) Orthogonality

To say that two different functions ψ_i and ψ_j are **orthogonal** means that the integral (over all space) of $\psi_i^\star \psi_j$ is zero:

$$\int \psi_i^\star \psi_j \,\mathrm{d}\tau = 0 \text{ for } i \neq j \qquad \text{Orthogonality [definition]} \quad (7C.8)$$

Functions that are both normalized and mutually orthogonal are called **orthonormal**. Hermitian operators have the important property that

> Eigenfunctions that correspond to different eigenvalues of a hermitian operator are orthogonal.

The proof of this property also follows from the definition of hermiticity (eqn 7C.7).

How is that done? 7C.3 Proving that the eigenfunctions of hermitian operators are orthogonal

Start by supposing that ψ_j is an eigenfunction of $\hat{\Omega}$ with eigenvalue ω_j (that is, $\hat{\Omega}\psi_j = \omega_j \psi_j$) and that ψ_i is an eigenfunction with a different eigenvalue ω_i (that is, $\hat{\Omega}\psi_i = \omega_i \psi_i$, with $\omega_i \neq \omega_j$). Then eqn 7C.7 becomes

$$\int \psi_i^\star \omega_j \psi_j \,\mathrm{d}\tau = \left\{ \int \psi_j^\star \omega_i \psi_i \,\mathrm{d}\tau \right\}^\star$$

The eigenvalues are constants and can be taken outside the integrals; moreover, they are real (being the eigenvalues of hermitian operators), so $\omega_i^\star = \omega_i$. Then

$$\omega_j \int \psi_i^\star \psi_j \,\mathrm{d}\tau = \omega_i \left\{ \int \psi_j^\star \psi_i \,\mathrm{d}\tau \right\}^\star$$

Next, note that $\left\{ \int \psi_j^\star \psi_i \,\mathrm{d}\tau \right\}^\star = \int \psi_j \psi_i^\star \,\mathrm{d}\tau$, so

$$\omega_j \int \psi_i^\star \psi_j \,\mathrm{d}\tau = \omega_i \int \psi_j \psi_i^\star \mathrm{d}\tau, \quad \text{hence} \quad (\omega_j - \omega_i) \int \psi_i^\star \psi_j \,\mathrm{d}\tau = 0$$

The two eigenvalues are different, so $\omega_j - \omega_i \neq 0$; therefore it must be the case that $\int \psi_i^\star \psi_j \,\mathrm{d}\tau = 0$. That is, the two eigenfunctions are orthogonal, as was to be proved.

The hamiltonian operator is hermitian (it corresponds to an observable, the energy, but its hermiticity can be proved specifically). Therefore, if two of its eigenfunctions correspond to different energies, the two functions must be orthogonal. The property of orthogonality is of great importance in quantum mechanics because it eliminates a large number of integrals from calculations. Orthogonality plays a central role in the theory of chemical bonding (Focus 9) and spectroscopy (Focus 11).

Example 7C.2 Verifying orthogonality

Two possible wavefunctions for a particle constrained to be somewhere on the x axis between $x = 0$ and $x = L$ are $\psi_1 = \sin(\pi x/L)$ and $\psi_2 = \sin(2\pi x/L)$. Outside this region the wavefunctions are zero. The wavefunctions correspond to different energies. Verify that the two wavefunctions are mutually orthogonal.

Collect your thoughts To verify the orthogonality of two functions, you need to integrate $\psi_2^*\psi_1 = \sin(2\pi x/L)\sin(\pi x/L)$ over all space, and show that the result is zero. In principle the integral is taken from $x = -\infty$ to $x = +\infty$, but the wavefunctions are zero outside the range $x = 0$ to L so you need integrate only over this range. Relevant integrals are given in the *Resource section*.

The solution To evaluate the integral, use Integral T.5 from the *Resource section* with $a = 2\pi/L$ and $b = \pi/L$:

$$\begin{aligned}\int_0^L \sin(2\pi x/L)\sin(\pi x/L)\mathrm{d}x &= \left\{\frac{\sin(\pi x/L)}{2(\pi/L)} - \frac{\sin(3\pi x/L)}{2(3\pi/L)}\right\}\Bigg|_0^L \\ &= \left\{\frac{\sin \pi}{2(\pi/L)} - \frac{\sin 3\pi}{2(3\pi/L)}\right\} \\ &\quad - \left\{\frac{\sin 0}{2(\pi/L)} - \frac{\sin 0}{2(3\pi/L)}\right\} \\ &= 0\end{aligned}$$

The sine functions have been evaluated by using $\sin n\pi = 0$ for $n = 0, \pm1, \pm2, \ldots$. The two functions are therefore mutually orthogonal.

Self-test 7C.2 The next higher energy level has $\psi_3 = \sin(3\pi x/L)$. Confirm that the functions $\psi_1 = \sin(\pi x/L)$ and $\psi_3 = \sin(3\pi x/L)$ are mutually orthogonal.

Answer: $\int_0^L \sin(3\pi x/L)\sin(\pi x/L)\,\mathrm{d}x = 0$

Exercises

E7C.1 Construct the potential energy operator of a particle with potential energy $V(x) = \frac{1}{2}k_f x^2$, where k_f is a constant.

E7C.2 Identify which of the following functions are eigenfunctions of the operator d/dx: (i) $\cos(kx)$; (ii) e^{ikx}, (iii) kx, (iv) e^{-ax^2}. Give the corresponding eigenvalue where appropriate.

E7C.3 Functions of the form $\sin(n\pi x/L)$, where $n = 1, 2, 3\ldots$, are wavefunctions in a region of length L (between $x = 0$ and $x = L$). Show that the wavefunctions with $n = 2$ and 3 are orthogonal; you will find the necessary integrals in the *Resource section*. (*Hint:* Recall that $\sin(n\pi) = 0$ for integer n.)

7C.2 Superpositions and expectation values

The hamiltonian for a free particle moving in one dimension is

$$\hat{H} = -\frac{\hbar^2}{2m}\frac{\mathrm{d}^2}{\mathrm{d}x^2}$$

The particle is 'free' in the sense that there is no potential to constrain it, hence $V(x) = 0$. It is easily confirmed that $\psi(x) = \cos kx$ is an eigenfunction of this operator, as inferred in Topic 7B (with $k = 2\pi/\lambda$):

$$\hat{H}\psi(x) = -\frac{\hbar^2}{2m}\frac{\mathrm{d}^2}{\mathrm{d}x^2}\cos kx = \frac{k^2\hbar^2}{2m}\cos kx$$

The energy associated with this wavefunction, $k^2\hbar^2/2m$, is therefore well defined, as it is the eigenvalue of an eigenvalue equation. However, the same is not necessarily true of other observables. For instance, $\cos kx$ is not an eigenfunction of the linear momentum operator:

$$\hat{p}_x\psi(x) = \frac{\hbar}{\mathrm{i}}\frac{\mathrm{d}\psi}{\mathrm{d}x} = \frac{\hbar}{\mathrm{i}}\frac{\mathrm{d}\cos kx}{\mathrm{d}x} = -\frac{k\hbar}{\mathrm{i}}\sin kx \qquad (7C.9)$$

This expression is not an eigenvalue equation, because the function on the right ($\sin kx$) is different from that on the left ($\cos kx$).

When the wavefunction of a particle is not an eigenfunction of an operator, the corresponding observable does not have a definite value. However, in the current example the momentum is not completely indefinite because the cosine wavefunction can be written as a **linear combination**, or sum,[1] of e^{ikx} and e^{-ikx} by using the identity $\cos kx = \frac{1}{2}(e^{ikx} + e^{-ikx})$ (see *The chemist's toolkit 7B.1*). As shown in *Example* 7C.1, these two exponential functions are eigenfunctions of $\hat{p}_x$ with eigenvalues $+k\hbar$ and $-k\hbar$, respectively. Therefore each one corresponds to a state of definite but different momentum. The wavefunction $\cos kx$ is said to be a **superposition** of the two individual wavefunctions e^{ikx} and e^{-ikx}, and is written

$$\psi = \underbrace{e^{+ikx}}_{\text{Particle with linear momentum } +k\hbar} + \underbrace{e^{-ikx}}_{\text{Particle with linear momentum } -k\hbar}$$

The interpretation of this superposition is that if many repeated measurements of the momentum are made, then half the

[1] A linear combination is more general than a sum, for it includes weighted sums of the form $ax + by + \cdots$ where $a, b, \ldots$ are constants. A sum is a linear combination with $a = b = \cdots = 1$.

measurements would give the value $p_x = +k\hbar$, and half would give the value $p_x = -k\hbar$. The two values $\pm k\hbar$ occur equally often since e^{ikx} and e^{-ikx} contribute equally to the superposition. All that can be inferred from the wavefunction cos kx about the linear momentum is that the particle it describes is equally likely to be found travelling in the positive and negative x directions, with the same magnitude, $k\hbar$, of the momentum.

A similar interpretation applies to any wavefunction written as a linear combination of eigenfunctions of an operator. In general, a wavefunction can be written as the following linear combination

$$\psi = c_1\psi_1 + c_2\psi_2 + \cdots = \sum_k c_k\psi_k \qquad \text{Linear combination of eigenfunctions} \qquad (7C.10)$$

where the c_k are numerical (possibly complex) coefficients and the ψ_k are different eigenfunctions of the operator $\hat{\Omega}$ corresponding to the observable of interest. The functions ψ_k are said to form a **complete set** in the sense that any arbitrary function can be expressed as a linear combination of them. Then, according to quantum mechanics:

- A single measurement of the observable corresponding to the operator $\hat{\Omega}$ will give one of the eigenvalues corresponding to the ψ_k that contribute to the superposition.
- In a series of measurements, the probability of obtaining a specific eigenvalue is proportional to the square modulus ($|c_k|^2$) of the corresponding coefficient in the linear combination.

The average value of a large number of measurements of an observable Ω is called the **expectation value** of the operator $\hat{\Omega}$, and is written $\langle\Omega\rangle$. For a normalized wavefunction ψ, the expectation value of $\hat{\Omega}$ is calculated by evaluating the integral

$$\langle\Omega\rangle = \int \psi^* \hat{\Omega}\psi \, d\tau \qquad \text{Expectation value [normalized wavefunction, definition]} \qquad (7C.11)$$

This definition can be justified by considering two cases, one where the wavefunction is an eigenfunction of the operator $\hat{\Omega}$ and another where the wavefunction is a superposition of that operator's eigenfunctions.

How is that done? 7C.4 Justifying the expression for the expectation value of an operator

If the wavefunction ψ is an eigenfunction of $\hat{\Omega}$ with eigenvalue ω (so $\hat{\Omega}\psi = \omega\psi$),

$$\langle\Omega\rangle = \int \psi^* \overbrace{\hat{\Omega}\psi}^{\omega\psi} d\tau = \int \psi^* \omega\psi \, d\tau \overset{\omega\text{ a constant}}{=} \omega \int \psi^*\psi \, d\tau \overset{\psi\text{ normalized}}{=} \omega$$

The interpretation of this expression is that, because the wavefunction is an eigenfunction of $\hat{\Omega}$, each observation of the property Ω results in the same value ω; the average value of all the observations is therefore ω.

Now suppose the (normalized) wavefunction is the linear combination of two eigenfunctions of the operator $\hat{\Omega}$, each of which is individually normalized to 1. Then

$$\begin{aligned}\langle\Omega\rangle &= \int (c_1\psi_1 + c_2\psi_2)^* \hat{\Omega}(c_1\psi_1 + c_2\psi_2)d\tau \\ &= \int (c_1\psi_1 + c_2\psi_2)^* (c_1\overbrace{\hat{\Omega}\psi_1}^{\omega_1\psi_1} + c_2\overbrace{\hat{\Omega}\psi_2}^{\omega_2\psi_2})d\tau \\ &= \int (c_1\psi_1 + c_2\psi_2)^* (c_1\omega_1\psi_1 + c_2\omega_2\psi_2)d\tau \\ &= c_1^*c_1\omega_1 \overbrace{\int \psi_1^*\psi_1 d\tau}^{1} + c_2^*c_2\omega_2 \overbrace{\int \psi_2^*\psi_2 d\tau}^{1} \\ &\quad + c_1^*c_2\omega_2 \overbrace{\int \psi_1^*\psi_2 d\tau}^{0} + c_2^*c_1\omega_1 \overbrace{\int \psi_2^*\psi_1 d\tau}^{0}\end{aligned}$$

The first two integrals on the right are both equal to 1 because the wavefunctions ψ_1 and ψ_2 are individually normalized. Because ψ_1 and ψ_2 correspond to different eigenvalues of a hermitian operator, they are orthogonal, so the third and fourth integrals on the right are zero. Therefore

$$\langle\Omega\rangle = |c_1|^2\omega_1 + |c_2|^2\omega_2$$

The interpretation of this expression is that in a series of measurements each individual measurement yields either ω_1 or ω_2, but that the probability of ω_1 occurring is $|c_1|^2$, and likewise the probability of ω_2 occurring is $|c_2|^2$. The average is the sum of the two eigenvalues, but with each weighted according to the probability that it will occur in a measurement:

$$\begin{aligned}\text{average} &= (\text{probability of } \omega_1 \text{ occurring}) \times \omega_1 \\ &\quad + (\text{probability of } \omega_2 \text{ occurring}) \times \omega_2\end{aligned}$$

The expectation value therefore predicts the result of taking a series of measurements, each of which gives an eigenvalue, and then taking the weighted average of these values. This conclusion justifies the form of eqn 7C.11.

Example 7C.3 Calculating an expectation value

Calculate the average value of the position of an electron in the lowest energy state of a one-dimensional box of length L, with the (normalized) wavefunction $\psi = (2/L)^{1/2}\sin(\pi x/L)$ inside the box and zero outside it.

Collect your thoughts The average value of the position is the expectation value of the operator corresponding to position, which is multiplication by x. To evaluate $\langle x\rangle$, you need to evaluate the integral in eqn 7C.11 with $\hat{\Omega} = \hat{x} = x\times$.

The solution The expectation value of position is

$$\langle x\rangle = \int_0^L \psi^*\hat{x}\psi \, dx \quad \text{with } \psi = \left(\frac{2}{L}\right)^{1/2} \sin\frac{\pi x}{L} \quad \text{and } \hat{x} = x\times$$

The integral is restricted to the region $x = 0$ to $x = L$ because outside this region the wavefunction is zero. Use Integral T.11 from the *Resources section* to obtain

$$\langle x\rangle = \frac{2}{L}\overbrace{\int_0^L x\sin^2\frac{\pi x}{L}\,\mathrm{d}x}^{\text{Integral T.11}} = \frac{2}{L}\times\boxed{\frac{L^2}{4}} = \tfrac{1}{2}L$$

This result means that if a very large number of measurements of the position of the electron are made, then the mean value will be at the centre of the box.

Self-test 7C.3 Evaluate the mean square position, $\langle x^2\rangle$, of the electron; you will need Integral T.12 from the *Resource section*.

Answer: $L^2\left(\frac{1}{3}-\frac{1}{2\pi^2}\right) = 0.217L^2$

The mean kinetic energy of a particle in one dimension is the expectation value of the operator given in eqn 7C.5. Therefore,

$$\langle E_\mathrm{k}\rangle = \int_{-\infty}^{\infty}\psi^\star \hat{E}_\mathrm{k}\psi\,\mathrm{d}x = -\frac{\hbar^2}{2m}\int_{-\infty}^{\infty}\psi^\star\frac{\mathrm{d}^2\psi}{\mathrm{d}x^2}\mathrm{d}x \qquad (7\mathrm{C}.12)$$

This conclusion confirms the previous assertion that the kinetic energy is related to the average of the curvature of the wavefunction: a large contribution to the observed value comes from regions where the wavefunction is sharply curved (so $\mathrm{d}^2\psi/\mathrm{d}x^2$ is large) and the wavefunction itself is large (so that $\psi^\star$ is large there too).

Exercises

E7C.4 An electron in a region of length L is described by the normalized wavefunction $\psi(x) = (2/L)^{1/2}\sin(2\pi x/L)$ in the range $x = 0$ to $x = L$; outside this range the wavefunction is zero. Evaluate $\langle x^3\rangle$. The necessary integrals will be found in the *Resource section*.

E7C.5 An electron in a one-dimensional region of length L is described by the normalized wavefunction $\psi(x) = (2/L)^{1/2}\sin(2\pi x/L)$ in the range $x = 0$ to $x = L$; outside this range the wavefunction is zero. The expectation value of the momentum of the electron is found from eqn 7C.11, which in this case is

$$\langle p_x\rangle = \frac{2}{L}\int_0^L \sin(2\pi x/L)\hat{p}_x\sin(2\pi x/L)\mathrm{d}x$$

Evaluate the differential and then the integral, and hence find $\langle p_x\rangle$. The necessary integrals will be found in the *Resource section*.

7C.3 The uncertainty principle

The wavefunction $\psi = \mathrm{e}^{ikx}$ is an eigenfunction of $\hat{p}_x$ with eigenvalue $+k\hbar$: in this case the wavefunction describes a particle with a definite state of linear momentum. Where, though, is the particle? The probability density is proportional to $\psi^\star\psi$, so if the particle is described by the wavefunction e^{ikx} the probability density is proportional to $(\mathrm{e}^{ikx})^\star\mathrm{e}^{ikx} = \mathrm{e}^{-ikx}\mathrm{e}^{ikx} = \mathrm{e}^{-ikx+ikx} = \mathrm{e}^0 = 1$. In other words, the probability density is the same for all values of x: the location of the particle is completely unpredictable. In summary, if the momentum of the particle is known precisely, it is not possible to predict its location.

This conclusion is an example of the consequences of the **Heisenberg uncertainty principle**, one of the most celebrated results of quantum mechanics:

> It is impossible to specify simultaneously, with arbitrary precision, both the linear momentum and the position of a particle.

Note that the uncertainty principle also implies that if the position is known precisely, then the momentum cannot be predicted. The argument runs as follows.

Suppose the particle is known to be at a definite location, then its wavefunction must be large there and zero everywhere else (Fig. 7C.4). Such a wavefunction can be created by superimposing a large number of harmonic (sine and cosine) functions, or, equivalently, a number of e^{ikx} functions (because $\mathrm{e}^{ikx} = \cos kx + \mathrm{i}\sin kx$). In other words, a sharply localized wavefunction, called a **wavepacket**, can be created by forming a linear combination of wavefunctions that correspond to many different linear momenta.

The superposition of a few harmonic functions gives a wavefunction that spreads over a range of locations (Fig. 7C.5). However, as the number of wavefunctions in the superposition increases, the wavepacket becomes sharper on account of the more complete interference between the positive and negative regions of the individual waves. When an infinite number of components are used, the wavepacket is a sharp, infinitely narrow spike, which corresponds to perfect localization of the particle. Now the particle is perfectly localized but all information about its momentum has been lost. A measurement of the momentum will give a result corresponding to any one of the infinite number of waves in the superposition, and which one it will give is unpredictable. Hence, if the location of the particle is known precisely (implying that its wavefunction is a superposition of an infinite number of momentum eigenfunctions), then its momentum is completely unpredictable.

The quantitative version of the uncertainty principle is

$$\Delta p_q \Delta q \geq \tfrac{1}{2}\hbar \qquad \text{Heisenberg uncertainty principle} \qquad (7\mathrm{C}.13\mathrm{a})$$

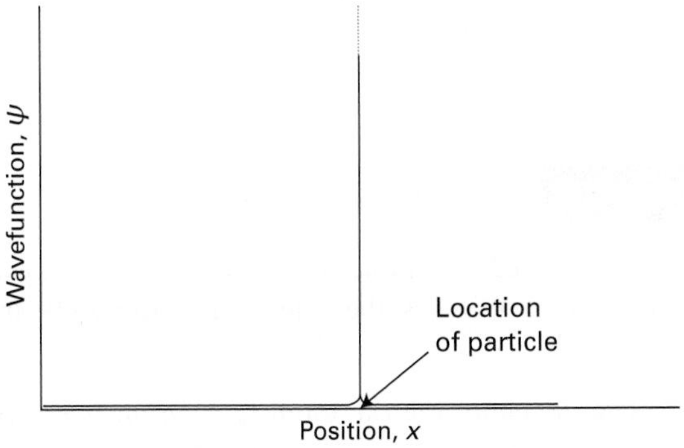

Figure 7C.4 The wavefunction of a particle at a well-defined location is a sharply spiked function that has zero amplitude everywhere except at the position of the particle.

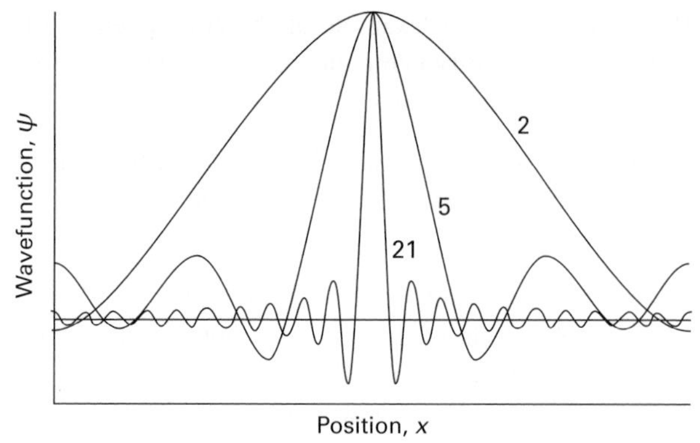

Figure 7C.5 The wavefunction of a particle with an ill-defined location can be regarded as a superposition of several wavefunctions of definite wavelength that interfere constructively in one place but destructively elsewhere. As more waves are used in the superposition (as given by the numbers attached to the curves), the location becomes more precise at the expense of uncertainty in the momentum of the particle. An infinite number of waves are needed in the superposition to construct the wavefunction of the perfectly localized particle.

In this expression Δp_q is the 'uncertainty' in the linear momentum parallel to the axis q, and Δq is the uncertainty in position along that axis. These 'uncertainties' are given by the root-mean-square deviations of the observables from their mean values:

$$\Delta p_q = (\langle p_q^2\rangle - \langle p_q\rangle^2)^{1/2} \quad \Delta q = (\langle q^2\rangle - \langle q\rangle^2)^{1/2} \qquad (7\text{C.13b})$$

If there is complete certainty about the position of the particle ($\Delta q = 0$), then the only way that eqn 7C.13a can be satisfied is for $\Delta p_q = \infty$, which implies complete uncertainty about the momentum. Conversely, if the momentum parallel to an axis is known exactly ($\Delta p_q = 0$), then the position along that axis must be completely uncertain ($\Delta q = \infty$).

The p and q that appear in eqn 7C.13a refer to the same direction in space. Therefore, whereas simultaneous specification of the position on the x-axis and momentum parallel to the x-axis are restricted by the uncertainty relation, simultaneous location of position on x and motion parallel to y or z are not restricted.

Example 7C.4 Using the uncertainty principle

Suppose the speed of a projectile of mass 1.0 g is known to within 1 μm s^{-1}. What is the minimum uncertainty in its position?

Collect your thoughts You can estimate Δp from $m\Delta v$, where Δv is the uncertainty in the speed; then use eqn 7C.13a to estimate the *minimum* uncertainty in position, Δq, by using it in the form $\Delta p\Delta q = \frac{1}{2}\hbar$ rearranged into $\Delta q = \frac{1}{2}\hbar/\Delta p$. You will need to use $1\,\text{J} = 1\,\text{kg}\,\text{m}^2\,\text{s}^{-2}$.

The solution The minimum uncertainty in position is

$$\Delta q = \frac{\hbar}{2m\Delta v}$$

$$= \frac{1.055\times10^{-34}\,\text{J s}}{2\times(1.0\times10^{-3}\,\text{kg})\times(1\times10^{-6}\,\text{m s}^{-1})} = 5\times10^{-26}\,\text{m}$$

This uncertainty is completely negligible for all practical purposes. However, if the mass is that of an electron, then the same uncertainty in speed implies an uncertainty in position far larger than the diameter of an atom (the analogous calculation gives $\Delta q = 60$ m).

Self-test 7C.4 Estimate the minimum uncertainty in the speed of an electron in a one-dimensional region of length $2a_0$, the approximate diameter of a hydrogen atom, where a_0 is the Bohr radius, 52.9 pm.

Answer: 500 km s^{-1}

The Heisenberg uncertainty principle is more general than even eqn 7C.13a suggests. It applies to any pair of observables, called **complementary observables**, for which the corresponding operators $\hat{\Omega}_1$ and $\hat{\Omega}_2$ have the property

$$\hat{\Omega}_1\hat{\Omega}_2\psi \neq \hat{\Omega}_2\hat{\Omega}_1\psi \qquad \text{Complementarity of observables} \qquad (7\text{C.14})$$

The term on the left implies that $\hat{\Omega}_2$ acts first, then $\hat{\Omega}_1$ acts on the result, and the term on the right implies that the operations are performed in the opposite order. When the effect of two operators applied in succession depends on their order (as this equation implies), they do not **commute**. The different outcomes of the effect of applying $\hat{\Omega}_1$ and $\hat{\Omega}_2$ in a different order are expressed by introducing the **commutator** of the two operators, which is defined as

$$[\hat{\Omega}_1, \hat{\Omega}_2] = \hat{\Omega}_1\hat{\Omega}_2 - \hat{\Omega}_2\hat{\Omega}_1 \qquad \text{Commutator [definition]} \qquad (7\text{C.15})$$

By using the definitions of the operators for position and momentum, an explicit value of this commutator can be found.

How is that done? 7C.5 Evaluating the commutator of position and momentum

You need to consider the effect of $\hat{x}\hat{p}_x$ (that is, the effect of $\hat{p}_x$ followed by the effect on the outcome of multiplication by x) on an arbitrary wavefunction ψ, which need not be an eigenfunction of either operator.

$$\hat{x}\hat{p}_x\psi = x\times\frac{\hbar}{\text{i}}\frac{\text{d}\psi}{\text{d}x}$$

Then you need to consider the effect of $\hat{p}_x\hat{x}$ on the same function (that is, the effect of multiplication by x followed by the effect of $\hat{p}_x$ on the outcome):

d(fg)/dx = (df/dx)g + f(dg/dx)

$$\hat{p}_x\hat{x}\psi = \frac{\hbar}{\text{i}}\frac{\text{d}(x\psi)}{\text{d}x} = \frac{\hbar}{\text{i}}\left(\psi + x\frac{\text{d}\psi}{\text{d}x}\right)$$

The second expression is different from the first, so $\hat{p}_x\hat{x}\psi \neq \hat{x}\hat{p}_x\psi$ and therefore the two operators do not commute. You can infer the value of the commutator from the difference of the two expressions:

$$[\hat{x},\hat{p}_x]\psi = \hat{x}\hat{p}_x\psi - \hat{p}_x\hat{x}\psi = -\frac{\hbar}{\mathrm{i}}\psi = \mathrm{i}\hbar\psi, \quad \text{so } [\hat{x},\hat{p}_x]\psi = \mathrm{i}\hbar\psi$$

This relation is true for any wavefunction ψ, so the commutator is

$$[\hat{x},\hat{p}_x] = \mathrm{i}\hbar \qquad (7\text{C}.16)$$

Commutator of position and momentum operators

The commutator in eqn 7C.16 is of such central significance in quantum mechanics that it is taken as a fundamental distinction between classical mechanics and quantum mechanics. In fact, this commutator may be taken as a postulate of quantum mechanics. Classical mechanics supposed, falsely as is now known, that the position and momentum of a particle could be specified simultaneously with arbitrary precision. However, quantum mechanics shows that position and momentum are complementary, and that a choice must be made: position can be specified, but at the expense of momentum, or momentum can be specified, but at the expense of position.

Exercises

E7C.6 Calculate the minimum uncertainty in the speed of a ball of mass 500 g that is known to be within 1.0 μm of a certain point on a bat. What is the minimum uncertainty in the position of a bullet of mass 5.0 g that is known to have a speed somewhere between 350.000 01 m s^{-1} and 350.000 00 m s^{-1}?

E7C.7 The speed of a certain proton is 0.45 Mm s^{-1}. If the uncertainty in its momentum is to be reduced to 0.0100 per cent, what uncertainty in its location must be tolerated?

7C.4 The postulates of quantum mechanics

The principles of quantum theory can be summarized as a series of postulates, which will form the basis for chemical applications of quantum mechanics throughout the text.

The wavefunction: All dynamical information is contained in the wavefunction ψ for the system, which is a mathematical function found by solving the appropriate Schrödinger equation for the system.

The Born interpretation: If the wavefunction of a particle has the value ψ at some position r, then the probability of finding the particle in an infinitesimal volume $\mathrm{d}\tau = \mathrm{d}x\mathrm{d}y\mathrm{d}z$ at that position is proportional to $|\psi(r)|^2\mathrm{d}\tau$.

Acceptable wavefunctions: An acceptable wavefunction must be single-valued, continuous, not infinite over a finite region of space, and (except in special cases) have a continuous slope.

Observables: Observables, Ω, are represented by hermitian operators, $\hat{\Omega}$, built from the position and momentum operators specified in eqn 7C.3.

Observations and expectation values: A single measurement of the observable represented by the operator $\hat{\Omega}$ gives one of the eigenvalues of $\hat{\Omega}$. If the wavefunction is not an eigenfunction of $\hat{\Omega}$, the average of many measurements is given by the expectation value, $\langle\Omega\rangle$, defined in eqn 7C.11.

The Heisenberg uncertainty principle: It is impossible to specify simultaneously, with arbitrary precision, both the linear momentum and the position of a particle and, more generally, any pair of observables represented by operators that do not commute.

Checklist of concepts

☐ 1. The Schrödinger equation is an **eigenvalue equation.**

☐ 2. An **operator** carries out a mathematical operation on a function, in general transforming it into a new function.

☐ 3. The **hamiltonian operator** is the operator corresponding to the total energy of the system, the sum of the kinetic and potential energies.

☐ 4. The wavefunction corresponding to a specific energy is an **eigenfunction** of the hamiltonian operator.

☐ 5. Two different functions are **orthogonal** if the integral (over all space) of their product is zero.

☐ 6. **Hermitian operators** have real eigenvalues and orthogonal eigenfunctions.

☐ 7. **Observables** are represented by hermitian operators.

☐ 8. **Orthonormal** functions are sets of functions that are both normalized and mutually orthogonal.

☐ 9. When the system is not described by a single eigenfunction of an operator, it may be expressed as a **superposition** of such eigenfunctions.

☐ 10. The mean value of a series of observations is given by the **expectation value** of the corresponding operator.

☐ 11. The **uncertainty principle** restricts the precision with which complementary observables may be specified and measured simultaneously.

☐ 12. **Complementary observables** are observables for which the corresponding operators do not commute.

Checklist of equations

Property	Equation	Comment	Equation number
Eigenvalue equation	$\hat{\Omega}\psi = \omega\psi$	ψ eigenfunction; ω eigenvalue	7C.2b
Hermiticity	$\int \psi_i^* \hat{\Omega}\psi_j \,d\tau = \left\{\int \psi_j^* \hat{\Omega}\psi_i \,d\tau\right\}^*$	Hermitian operators have real eigenvalues and orthogonal eigenfunctions	7C.7
Orthogonality	$\int \psi_i^* \psi_j \,d\tau = 0$ for $i \neq j$	Integration over all space	7C.8
Expectation value	$\langle \Omega \rangle = \int \psi^* \hat{\Omega}\psi \,d\tau$	Definition; assumes ψ normalized	7C.11
Heisenberg uncertainty principle	$\Delta p_q \Delta q \geq \frac{1}{2}\hbar$	For position and momentum	7C.13a
Commutator of two operators	$[\hat{\Omega}_1, \hat{\Omega}_2] = \hat{\Omega}_1\hat{\Omega}_2 - \hat{\Omega}_2\hat{\Omega}_1$	The observables are complementary if $[\hat{\Omega}_1, \hat{\Omega}_2] \neq 0$	7C.15
	Special case: $[\hat{x}, \hat{p}_x] = i\hbar$		7C.16

TOPIC 7D Translational motion

> ➤ Why do you need to know this material?
>
> The application of quantum theory to translational motion reveals the origin of quantization and non-classical features, such as tunnelling and zero-point energy. This material is important for the discussion of atoms and molecules that are free to move within a restricted volume, such as a gas in a container.
>
> ➤ What is the key idea?
>
> The translational energy of a particle confined to a finite region of space is quantized, and under certain conditions particles can pass into and through classically forbidden regions.
>
> ➤ What do you need to know already?
>
> You should know that the wavefunction is the solution of the Schrödinger equation (Topic 7B). In one instance you need to be familiar with the techniques of deriving dynamical properties from the wavefunction by using the operators corresponding to the observables (Topic 7C).

Translation, motion through space, is one of the basic types of motion. Quantum mechanics, however, shows that translation can have several non-classical features, such as its confinement to discrete energies and passage into and through classically forbidden regions.

7D.1 Free motion in one dimension

A free particle is unconstrained by any potential, which may be taken to be zero everywhere. In one dimension $V(x) = 0$ everywhere, so the Schrödinger equation becomes (Topic 7B)

$$-\frac{\hbar^2}{2m}\frac{d^2\psi(x)}{dx^2} = E\psi(x) \qquad \text{Free motion [one dimension]} \qquad (7D.1)$$

The most straightforward way to solve this simple second-order differential equation is to take the known general form of solutions of equations of this kind, and then show that it does indeed satisfy eqn 7D.1.

How is that done? 7D.1 Finding the solutions to the Schrödinger equation for a free particle in one dimension

The general solution of a second-order differential equation of the kind in eqn 7D.1 is

$$\psi_k(x) = Ae^{ikx} + Be^{-ikx}$$

where k, A, and B are constants. You can verify that $\psi_k(x)$ is a solution of eqn 7D.1 by substituting it into the left-hand side of the equation, evaluating the derivatives, and then confirming that you have generated the right-hand side. Because $de^{\pm ax}/dx = \pm ae^{\pm ax}$, the left-hand side becomes

$$-\frac{\hbar^2}{2m}\frac{d^2}{dx^2}\overbrace{(Ae^{ikx}+Be^{-ikx})}^{\psi_k(x)} = -\frac{\hbar^2}{2m}\{A(ik)^2e^{ikx}+B(-ik)^2e^{-ikx}\}$$

$$= \overbrace{\boxed{\frac{k^2\hbar^2}{2m}}}^{E_k}\overbrace{(Ae^{ikx}+Be^{-ikx})}^{\psi_k(x)}$$

The left-hand side is therefore equal to a constant × $\psi_k(x)$, which is the same as the term on the right-hand side of eqn 7D.1 provided the constant, the boxed term, is identified with E. The value of the energy depends on the value of k, so henceforth it will be written E_k. The wavefunctions and energies of a free particle are therefore

$$\psi_k(x) = Ae^{ikx} + Be^{-ikx} \qquad E_k = \frac{k^2\hbar^2}{2m} \qquad \text{Wavefunctions and energies [one dimension]} \qquad (7D.2)$$

The wavefunctions in eqn 7D.2 are continuous, have continuous slope everywhere, are single-valued, and do not go to infinity: they are therefore acceptable wavefunctions for all values of k. Because k can take any value, the energy can take any non-negative value, including zero. As a result, *the translational energy of a free particle is not quantized.*

As seen in Topic 7C in general a wavefunction can be written as a superposition (a linear combination) of the eigenfunctions of an operator. The wavefunctions of eqn 7D.2 can be recognized as superpositions of the two functions $e^{\pm ikx}$ which are eigenfunctions of the linear momentum operator with eigenvalues $\pm k\hbar$ (Topic 7C). These eigenfunctions correspond to states with definite linear momentum:

$$\psi_k(x) = \underbrace{Ae^{+ikx}}_{\substack{\text{Particle with linear}\\ \text{momentum}+k\hbar}} + \underbrace{Be^{-ikx}}_{\substack{\text{Particle with linear}\\ \text{momentum}-k\hbar}}$$

According to the interpretation given in Topic 7C, if a system is described by the wavefunction $\psi_k(x)$, then repeated measurements of the momentum will give $+k\hbar$ (that is, the particle travelling in the positive x-direction) with a probability proportional to A^2, and $-k\hbar$ (that is, the particle travelling in the negative x-direction) with a probability proportional to B^2. Only if A or B is zero does the particle have a definite momentum of $-k\hbar$ or $+k\hbar$, respectively.

Brief illustration 7D.1

Suppose an electron emerges from an accelerator moving towards positive x with kinetic energy 1.0 eV (1 eV = 1.602×10^{-19} J). The wavefunction for such a particle is given by eqn 7D.2 with $B = 0$ because the momentum is definitely in the positive x-direction. The value of k is found by rearranging the expression for the energy in eqn 7D.2 into

$$k = \left(\frac{2m_eE_k}{\hbar^2}\right)^{1/2} = \left(\frac{2\times(9.109\times10^{-31}\text{ kg})\times(1.6\times10^{-19}\text{ J})}{(1.055\times10^{-34}\text{ J s})^2}\right)^{1/2}$$
$$= 5.1\times10^{9}\text{ m}^{-1}$$

or 5.1 nm^{-1} (with 1 nm = 10^{-9} m). Therefore, the wavefunction is $\psi(x) = Ae^{5.1ix/\text{nm}}$.

So far, the motion of the particle has been confined to the x-axis. In general, the linear momentum is a vector (see *Energy: A first look*) directed along the line of travel of the particle. Then $\boldsymbol{p} = \boldsymbol{k}\hbar$, the magnitude of the vector is $p = k\hbar$, and its component on each axis is $p_q = k_q\hbar$, with $q = x, y$, or z. The wavefunction for each component is proportional to e^{ik_qq} and overall equal to $e^{i(k_xx+k_yy+k_zz)}$.[1]

Exercises

E7D.1 Evaluate the linear momentum and kinetic energy of a free electron described by the wavefunction e^{ikx} with $k = 3$ nm^{-1}.

E7D.2 Write the wavefunction for a particle of mass 2.0 g travelling to the left with kinetic energy 20 J.

7D.2 Confined motion in one dimension

Consider a **particle in a box** in which a particle of mass m is confined to a region of one-dimensional space between two impenetrable walls. The potential energy is zero inside the box

[1] In terms of scalar products, this overall wavefunction would be written $e^{i\boldsymbol{k}\cdot\boldsymbol{r}}$.

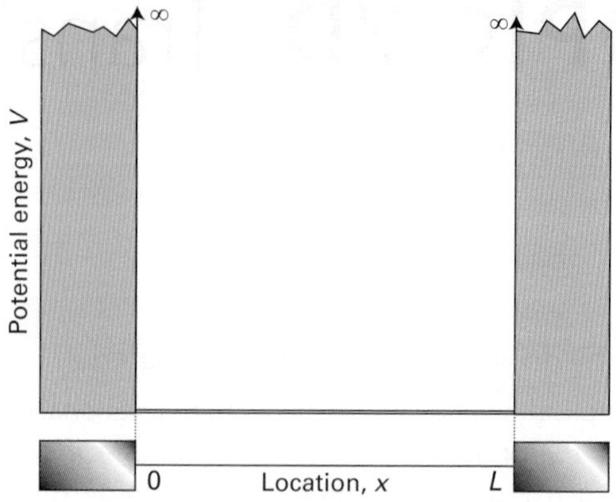

Figure 7D.1 The potential energy for a particle in a one-dimensional box. The potential is zero between $x = 0$ and $x = L$, and then rises to infinity outside this region, resulting in impenetrable walls which confine the particle.

but rises abruptly to infinity at the walls located at $x = 0$ and $x = L$ (Fig. 7D.1). When the particle is between the walls, the Schrödinger equation is the same as for a free particle (eqn 7D.1), so the general solutions given in eqn 7D.2 are also the same. However, it will prove convenient to rewrite the wavefunction in terms of sines and cosines by using $e^{\pm ikx} = \cos kx \pm i \sin kx$ (*The chemist's toolkit* 7B.1)

$$\begin{aligned}\psi_k(x) &= Ae^{ikx} + Be^{-ikx}\\ &= A(\cos kx + i\sin kx) + B(\cos kx - i\sin kx)\\ &= (A+B)\cos kx + i(A-B)\sin kx\end{aligned}$$

The constants i(A – B) and $A + B$ can be denoted C and D, respectively, in which case

$$\psi_k(x) = C\sin kx + D\cos kx \qquad (7D.3)$$

Outside the box the wavefunctions must be zero as the particle will not be found in a region where its potential energy would be infinite:

$$\text{For } x < 0 \text{ and } x > L, \psi_k(x) = 0 \qquad (7D.4)$$

7D.2(a) The acceptable solutions

One of the requirements placed on a wavefunction is that it must be continuous. It follows that since the wavefunction is zero when $x < 0$ (where the potential energy is infinite) the wavefunction must be zero *at* $x = 0$ which is the point where the potential energy rises to infinity. Likewise, the wavefunction is zero where $x > L$ and so must be zero at $x = L$ where the potential energy also rises to infinity. These two restrictions are the **boundary conditions**, or constraints on the function:

$$\psi_k(0) = 0 \text{ and } \psi_k(L) = 0 \qquad \text{Boundary conditions} \qquad (7D.5)$$

Now it is necessary to show that the requirement that the wavefunction must satisfy these boundary conditions implies that only certain wavefunctions are acceptable, and that as a consequence only certain energies are allowed.

How is that done? 7D.2 Showing that the boundary conditions lead to quantized levels

You need to start from the general solution and then explore the consequences of imposing the boundary conditions.

Step 1 *Apply the boundary conditions*

At $x = 0$, $\psi_k(0) = C \sin 0 + D \cos 0 = D$ (because $\sin 0 = 0$ and $\cos 0 = 1$). One boundary condition is $\psi_k(0) = 0$, so it follows that $D = 0$.

At $x = L$, $\psi_k(L) = C \sin kL$. The boundary condition $\psi_k(L) = 0$ therefore requires that $\sin kL = 0$, which in turn requires that $kL = n\pi$ with $n = 1, 2, \ldots$. Although $n = 0$ also satisfies the boundary condition it is ruled out because the wavefunction would be $C \sin 0 = 0$ for all values of x, and the particle would be found nowhere. Negative integral values of n also satisfy the boundary condition, but simply result in a change of sign of the wavefunction (because $\sin(-\theta) = -\sin\theta$). It therefore follows that the wavefunctions that satisfy the two boundary conditions are $\psi_k(x) = C \sin(n\pi x/L)$ with $n = 1, 2, \ldots$ and $k = n\pi/L$.

Step 2 *Normalize the wavefunctions*

To normalize the wavefunction, write it as $N \sin(n\pi x/L)$ and require that the integral of the square of the wavefunction over all space is equal to 1. The wavefunction is zero outside the range $0 \le x \le L$, so the integration needs to be carried out only inside this range:

$$\int_0^L \psi^2 \mathrm{d}x = N^2 \overbrace{\left(\int_0^L \sin^2 \frac{n\pi x}{L}\mathrm{d}x\right)}^{\text{Integral T.2}} = N^2 \times \frac{L}{2} = 1, \text{ so } N = \left(\frac{2}{L}\right)^{1/2}$$

Step 3 *Identify the allowed energies*

According to eqn 7D.2, $E_k = k^2\hbar^2/2m$, but because k is limited to the values $k = n\pi/L$ with $n = 1, 2, \ldots$ the energies are restricted to the values

$$E_k = \frac{k^2\hbar^2}{2m} = \frac{(n\pi/L)^2(h/2\pi)^2}{2m} = \frac{n^2h^2}{8mL^2}$$

At this stage it is sensible to replace the label k by the label n, and to label the wavefunctions and energies as $\psi_n(x)$ and E_n. The normalized wavefunctions and energies are therefore

$$\psi_n(x) = \left(\frac{2}{L}\right)^{1/2} \sin\left(\frac{n\pi x}{L}\right) \qquad E_n = \frac{n^2h^2}{8mL^2}, \; n = 1, 2, \ldots \qquad (7D.6)$$

Particle in a one-dimensional box

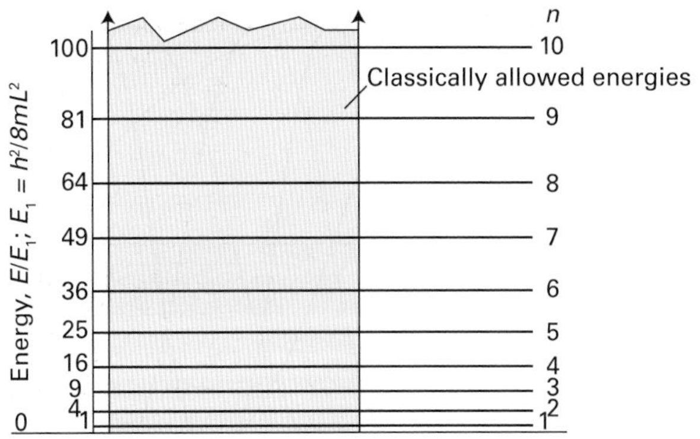

Figure 7D.2 The energy levels for a particle in a box. Note that the energy levels increase as n^2, and that their separation increases as the quantum number increases. Classically, the particle is allowed to have any value of the energy shown as a tinted area.

The fact that n is restricted to positive integer values implies that the energy of the particle in a one-dimensional box is quantized. This quantization arises from the boundary conditions that ψ must satisfy. This conclusion is general:

> The need to satisfy boundary conditions implies that only certain wavefunctions are acceptable, and hence restricts the eigenvalues to discrete values.

The integer n that has been used to label the wavefunctions and energies is an example of a 'quantum number'. In general, a **quantum number** is an integer (in some cases, Topic 8B, a half-integer) that labels the state of the system. For a particle in a one-dimensional box there are an infinite number of acceptable solutions, and the quantum number n specifies the one of interest (Fig. 7D.2).[2] As well as acting as a label, a quantum number can often be used to calculate the value of a property, such as the energy corresponding to the state, as in eqn 7D.6.

7D.2(b) The properties of the wavefunctions

Figure 7D.3 shows some of the wavefunctions of a particle in a one-dimensional box. The points to note are as follows.

- The wavefunctions are all sine functions with the same maximum amplitude but different wavelengths; the wavelength gets shorter as n increases.
- Shortening the wavelength results in a sharper average curvature of the wavefunction and therefore an increase in the kinetic energy of the particle (recall that, as $V = 0$ inside the box, the energy is entirely kinetic).
- The number of nodes (the points where the wavefunction passes through zero) also increases as n increases; the wavefunction ψ_n has $n - 1$ nodes.

[2] You might object that the wavefunctions have a discontinuous slope at the edges of the box, and so do not qualify as acceptable according to the criteria in Topic 7B. This is a rare instance where the requirement does not apply, because the potential energy suddenly jumps to an infinite value.

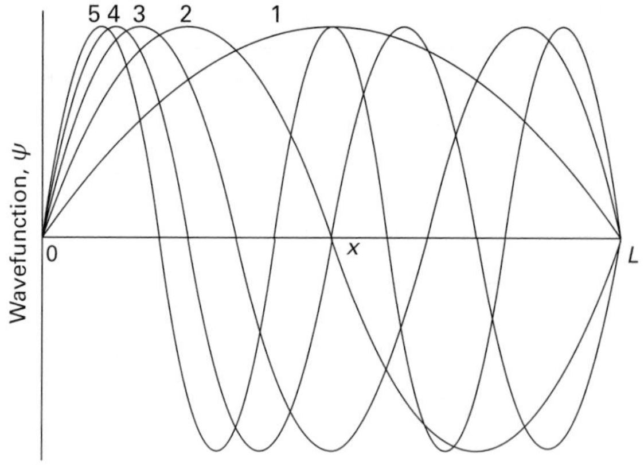

Figure 7D.3 The first five normalized wavefunctions of a particle in a box. As the energy increases the wavelength decreases, and successive functions possess one more half wave. The wavefunctions are zero outside the box.

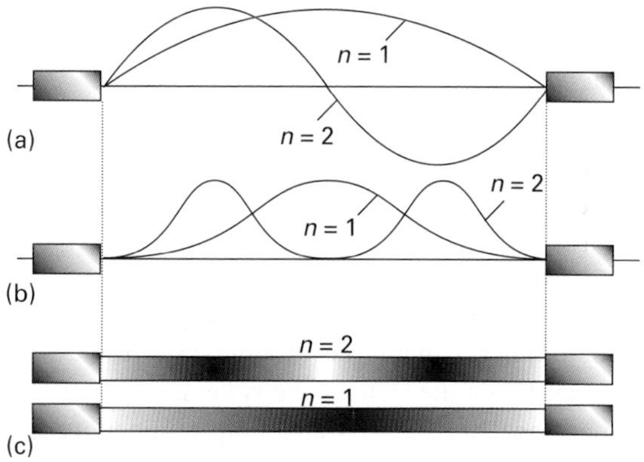

Figure 7D.4 (a) The first two wavefunctions for a particle in a box, (b) the corresponding probability densities, and (c) a representation of the probability density in terms of the darkness of shading. *Interact with the dynamic version of this graph in the e-book.*

The probability density for a particle in a one-dimensional box is

$$\psi_n^2(x) = \frac{2}{L}\sin^2\left(\frac{n\pi x}{L}\right) \qquad (7D.7)$$

and varies with position. The non-uniformity in the probability density is pronounced when n is small (Fig. 7D.4). The maxima in the probability density give the locations at which the particle has the greatest probability of being found.

Brief illustration 7D.2

As explained in Topic 7B, the total probability of finding the particle in a specified region is the integral of $\psi(x)^2\mathrm{d}x$ over that region. Therefore, the probability of finding the particle with $n = 1$ in a region between $x = 0$ and $x = L/2$ is

$$P = \int_0^{L/2}\psi_1^2\mathrm{d}x = \frac{2}{L}\overbrace{\int_0^{L/2}\sin^2\left(\frac{\pi x}{L}\right)\mathrm{d}x}^{\text{Integral T.2}} = \frac{2}{L}\left[\frac{x}{2} - \frac{1}{4\pi/L}\sin\left(\frac{2\pi x}{L}\right)\right]_0^{L/2}$$

$$= \frac{2}{L}\left(\frac{L}{4} - \frac{1}{4\pi/L}\overbrace{\sin\pi}^{0}\right) = \tfrac{1}{2}$$

The result should not be a surprise, because the probability density is symmetrical around $x = L/2$. The probability of finding the particle between $x = 0$ and $x = L/2$ must therefore be half of the probability of finding the particle between $x = 0$ and $x = L$, which is 1.

As n increases the separation between the maxima in the probability density $\psi_n^2(x)$ decreases (Fig. 7D.5). For a macroscopic object this separation is very much smaller than the dimensions of the object. Therefore, when averaged over the dimensions of the object, the probability density becomes constant. This behaviour of the probability density at high quantum numbers reflects the classical result that a particle bouncing between the walls of a container spends equal times at all points. This conclusion is an example of the **correspondence principle**, which states that as high quantum numbers are reached, the classical result emerges from quantum mechanics.

7D.2(c) The properties of the energy

The linear momentum of a particle in a box is not well defined because the wavefunction $\sin kx$ is not an eigenfunction of the linear momentum operator. However, because $\sin kx = (\mathrm{e}^{ikx} - \mathrm{e}^{-ikx})/2\mathrm{i}$,

$$\psi_n(x) = \left(\frac{2}{L}\right)^{1/2}\sin\left(\frac{n\pi x}{L}\right) = \frac{1}{2\mathrm{i}}\left(\frac{2}{L}\right)^{1/2}\left(\mathrm{e}^{\mathrm{i}n\pi x/L} - \mathrm{e}^{-\mathrm{i}n\pi x/L}\right) \qquad (7D.8)$$

It follows that, if repeated measurements are made of the linear momentum, half will give the value $+n\pi\hbar/L$ and half will give the value $-n\pi\hbar/L$. This conclusion is the quantum mechanical

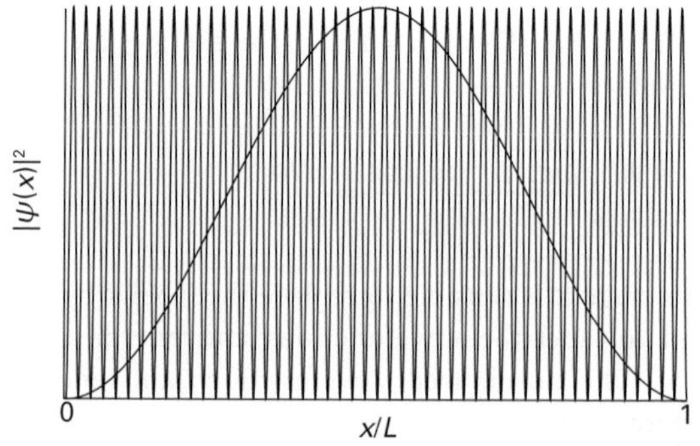

Figure 7D.5 The probability density $\psi^2(x)$ for a large quantum number (here $n = 50$ compared with $n = 1$). For high n the probability density is nearly uniform when averaged over several oscillations.

version of the classical picture in which the particle bounces back and forth in the box, spending equal times travelling to the left and to the right.

Because n cannot be zero, the lowest energy that the particle may possess is not zero (as allowed by classical mechanics, corresponding to a stationary particle) but

$$E_1 = \frac{h^2}{8mL^2} \qquad \text{Zero-point energy} \qquad (7D.9)$$

This lowest, irremovable energy is called the **zero-point energy**. The physical origin of the zero-point energy can be explained in two ways:

- The Heisenberg uncertainty principle states that $\Delta p_x \Delta x \geq \frac{1}{2}\hbar$. For a particle confined to a box, Δx has a finite value, therefore Δp_x cannot be zero, as that would violate the uncertainty principle. Therefore the kinetic energy cannot be zero.
- If the wavefunction is to be zero at the walls, but smooth, continuous, and not zero everywhere, then it must be curved, and curvature in a wavefunction implies the possession of kinetic energy.

Brief illustration 7D.3

The lowest energy of an electron in a region of length 100 nm is given by eqn 7D.6 with $n = 1$:

$$E_1 = \frac{1^2 \times (6.626\times10^{-34}\ \mathrm{J\,s})^2}{8\times(9.109\times10^{-31}\ \mathrm{kg})\times(100\times10^{-9}\ \mathrm{m})^2} = 6.02\times10^{-24}\ \mathrm{J}$$

where $1\ \mathrm{J} = 1\ \mathrm{kg\,m^2\,s^{-2}}$ has been used. The energy E_1 can be expressed as 6.02 yJ ($1\ \mathrm{yJ} = 10^{-24}\ \mathrm{J}$).

The separation between adjacent energy levels with quantum numbers n and $n + 1$ is

$$E_{n+1} - E_n = \frac{(n+1)^2h^2}{8mL^2} - \frac{n^2h^2}{8mL^2} = (2n+1)\frac{h^2}{8mL^2} \qquad (7D.10)$$

This separation decreases as the length of the container increases, and is very small when the container has macroscopic dimensions. The separation of adjacent levels becomes zero when the walls are infinitely far apart. Atoms and molecules free to move in normal laboratory-sized vessels may therefore be treated as though their translational energy is not quantized.

Example 7D.1 Estimating an absorption wavelength

β-Carotene (**1**) is a linear polyene in which 10 single and 11 double bonds alternate along a chain of 22 carbon atoms. The delocalized π system can be approximated as a one-dimensional box, with electrons having energies given by eqn 7D.6. If each CC bond length is taken to be 140 pm, the length of the molecular box in β-carotene is $L = 2.94$ nm. Estimate the wavelength of the light absorbed by this molecule when it undergoes a transition from its ground state to the next higher excited state.

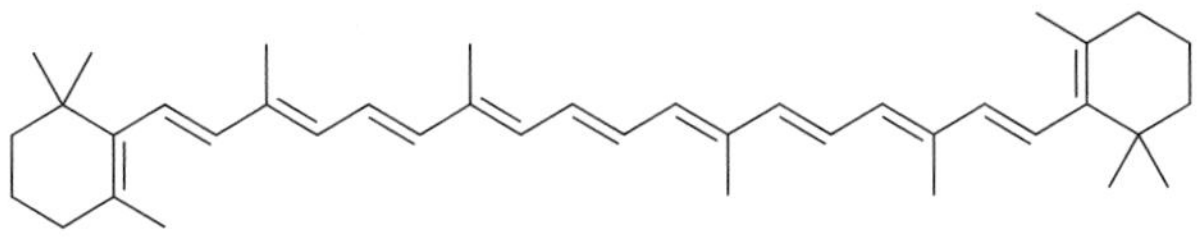

1 β-Carotene

Collect your thoughts Recall that each C atom in the π system contributes one p electron to the π-orbitals and two electrons occupy each of the energy levels. Use eqn 7D.10 to calculate the energy separation between the highest occupied and the lowest unoccupied levels, and convert that energy to a wavelength by using the Bohr frequency condition (eqn 7A.9, $\Delta E = h\nu$).

The solution There are 22 C atoms in the π system, so in the ground state each of the levels up to $n = 11$ is occupied by two electrons. The first excited state is one in which an electron is promoted from $n = 11$ to $n = 12$. The energy difference between the ground and first excited state is therefore

$$\begin{aligned}\Delta E &= E_{12} - E_{11}\\ &= (2\times11+1)\frac{(6.626\times10^{-34}\ \mathrm{J\,s})^2}{8\times(9.109\times10^{-31}\ \mathrm{kg})\times(2.94\times10^{-9}\ \mathrm{m})^2}\\ &= 1.60\ldots\times10^{-19}\ \mathrm{J}\end{aligned}$$

or 0.160 aJ. It follows from the Bohr frequency condition ($\Delta E = h\nu$) that the frequency of radiation required to cause this transition is

$$\nu = \frac{\Delta E}{h} = \frac{1.60\ldots\times10^{-19}\ \mathrm{J}}{6.626\times10^{-34}\ \mathrm{J\,s}} = 2.42\times10^{14}\ \mathrm{s^{-1}}$$

or 242 THz ($1\ \mathrm{THz} = 10^{12}\ \mathrm{Hz}$), corresponding to a wavelength $\lambda = c/\nu = 1240$ nm.

Comment. The experimental value is 603 THz (λ = 497 nm), corresponding to radiation in the visible range of the electromagnetic spectrum. The model is too crude to expect quantitative agreement, but the calculation at least predicts a wavelength in the right general range.

Self-test 7D.1 Estimate a typical nuclear excitation energy in electronvolts ($1\ \mathrm{eV} = 1.602\times10^{-19}$ J) by calculating the first excitation energy of a proton confined to a one-dimensional box with a length equal to the diameter of a nucleus (approximately 1×10^{-15} m, or 1 fm).

Answer: 0.6 GeV, where $1\ \mathrm{GeV} = 10^9\ \mathrm{eV}$

Exercises

E7D.3 Calculate the energy separations in joules, kilojoules per mole, electronvolts, and reciprocal centimetres between the levels (i) $n = 2$ and $n = 1$, (ii) $n = 6$ and $n = 5$ of an electron in a box of length 1.0 nm.

E7D.4 Calculate the probability that a particle will be found between $0.49L$ and $0.51L$ in a box of length L for (i) ψ_1, (ii) ψ_2. You may assume that the wavefunction is constant in this range, so the probability is $\psi^2\delta x$.

E7D.5 For a particle in a box of length L sketch the wavefunction corresponding to the state with the lowest energy and on the same graph sketch the corresponding probability density. Without evaluating any integrals, explain why the expectation value of x is equal to $L/2$.

7D.3 Confined motion in two and more dimensions

Now consider a rectangular two-dimensional region, between 0 and L_1 along x, and between 0 and L_2 along y. Inside this region the potential energy is zero, but at the edges it rises to infinity (Fig. 7D.6). As in the one-dimensional case, the wave-function can be expected to be zero at the edges of this region (at $x = 0$ and L_1, and at $y = 0$ and L_2), and to be zero outside the region. Inside the region the particle has contributions to its kinetic energy from its motion along both the x and y directions, and so the Schrödinger equation has two kinetic energy terms, one for each axis. For a particle of mass m the equation is

$$-\frac{\hbar^2}{2m}\left(\frac{\partial^2\psi}{\partial x^2}+\frac{\partial^2\psi}{\partial y^2}\right)=E\psi \qquad (7D.11)$$

Equation 7D.11 is a *partial* differential equation, and the resulting wavefunctions are functions of both x and y, denoted $\psi(x,y)$.

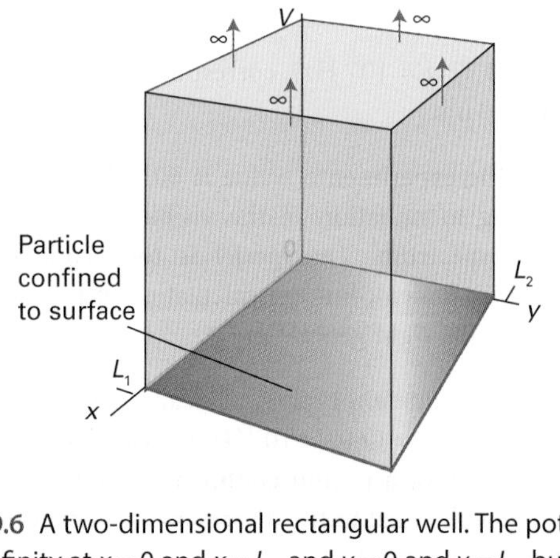

Figure 7D.6 A two-dimensional rectangular well. The potential goes to infinity at $x = 0$ and $x = L_1$, and $y = 0$ and $y = L_2$, but in between these values the potential is zero. The particle is confined to this rectangle by impenetrable walls.

7D.3(a) Energy levels and wavefunctions

The procedure for finding the allowed wavefunctions and energies involves starting with the two-dimensional Schrödinger equation, and then applying the 'separation of variables' technique to turn it into two separate one-dimensional equations.

How is that done? 7D.3 Constructing the wavefunctions for a particle in a two-dimensional box

The 'separation of variables' technique, which is explained and used here, is used in several cases in quantum mechanics.

Step 1 *Apply the separation of variables technique*

First, recognize the presence of two operators, each of which acts on functions of only x or y:

$$\hat{H}_x=-\frac{\hbar^2}{2m}\frac{\partial^2}{\partial x^2} \qquad \hat{H}_y=-\frac{\hbar^2}{2m}\frac{\partial^2}{\partial y^2}$$

Equation 7D.11, which is

$$\overbrace{-\frac{\hbar^2}{2m}\frac{\partial^2}{\partial x^2}}^{\hat{H}_x}\psi\overbrace{-\frac{\hbar^2}{2m}\frac{\partial^2}{\partial y^2}}^{\hat{H}_y}\psi=E\psi$$

then becomes

$$\hat{H}_x\psi+\hat{H}_y\psi=E\psi$$

Now suppose that the wavefunction ψ can be expressed as the product of two functions, $\psi(x,y) = X(x)Y(y)$, one depending only on x and the other depending only on y. This assumption is the central step of the procedure, and does not work for all partial differential equations: that it works here must be demonstrated. With this substitution the preceding equation becomes

$$\hat{H}_xX(x)Y(y)+\hat{H}_yX(x)Y(y)=EX(x)Y(y)$$

Then, because $\hat{H}_x$ operates on (takes the second derivatives with respect to x of) $X(x)$, and likewise for $\hat{H}_y$ and $Y(y)$, this equation is the same as

$$Y(y)\hat{H}_xX(x)+X(x)\hat{H}_yY(y)=EX(x)Y(y)$$

Division of both sides by $X(x)Y(y)$ then gives

$$\overbrace{\frac{1}{X(x)}\hat{H}_xX(x)}^{\text{Depends only on }x}+\overbrace{\frac{1}{Y(y)}\hat{H}_yY(y)}^{\text{Depends only on }y}=\overbrace{E}^{\text{A constant}}$$

If x is varied, only the first term can change; but the other two terms do not change, so the first term must be a constant for the equality to remain true. The same is true of the second term when y is varied. Therefore, denoting these constants as E_X and E_Y,

$$\frac{1}{X(x)}\hat{H}_x X(x) = E_X, \text{ so } \hat{H}_x X(x) = E_X X(x)$$

$$\frac{1}{Y(y)}\hat{H}_y Y(y) = E_Y, \text{ so } \hat{H}_y Y(y) = E_Y Y(y)$$

with $E_X + E_Y = E$. The procedure has successfully separated the partial differential equation into two ordinary differential equations, one in x and the other in y.

Step 2 *Recognize the two ordinary differential equations*

Each of the two equations is identical to the Schrödinger equation for a particle in a one-dimensional box, one for the coordinate x and the other for the coordinate y. The boundary conditions are also essentially the same (that the wavefunction must be zero at the walls). Consequently, the two solutions are

$$X_{n_1}(x) = \left(\frac{2}{L_1}\right)^{1/2} \sin\left(\frac{n_1\pi x}{L_1}\right) \qquad E_{X,n_1} = \frac{n_1^2 h^2}{8mL_1^2}$$

$$Y_{n_2}(y) = \left(\frac{2}{L_2}\right)^{1/2} \sin\left(\frac{n_2\pi y}{L_2}\right) \qquad E_{Y,n_2} = \frac{n_2^2 h^2}{8mL_2^2}$$

with each of n_1 and n_2 taking the values 1, 2,… independently.

Step 3 *Assemble the complete wavefunction*

Inside the box, which is when $0 \le x \le L_1$ and $0 \le y \le L_2$, the wavefunction is the product $X_{n_1}(x)Y_{n_2}(y)$, and is given by eqn 7D.12a below. Inside the material of the walls, the wavefunction is zero. The energies are the sum $E_{X,n_1} + E_{Y,n_2}$. The two quantum numbers take the values $n_1 = 1, 2,\ldots$ and $n_2 = 1, 2, \ldots$ independently. Overall, therefore,

$$\psi_{n_1,n_2}(x,y) = \frac{2}{(L_1L_2)^{1/2}} \sin\left(\frac{n_1\pi x}{L_1}\right) \sin\left(\frac{n_2\pi y}{L_2}\right)$$

Wavefunctions [two dimensions] (7D.12a)

$$E_{n_1,n_2} = \left(\frac{n_1^2}{L_1^2} + \frac{n_2^2}{L_2^2}\right)\frac{h^2}{8m}$$

Energy levels [two dimensions] (7D.12b)

Some of the wavefunctions are plotted as contours in Fig. 7D.7. They are the two-dimensional versions of the wavefunctions shown in Fig. 7D.3. Whereas in one dimension the wavefunctions resemble states of a vibrating string with ends fixed, in two dimensions the wavefunctions correspond to vibrations of a rectangular plate with fixed edges.

Brief illustration 7D.4

Consider an electron confined to a square cavity of side L (that is $L_1 = L_2 = L$), and in the state with quantum numbers $n_1 = 1$, $n_2 = 2$. Because the probability density is

$$\psi_{1,2}^2(x,y) = \frac{4}{L^2}\sin^2\left(\frac{\pi x}{L}\right)\sin^2\left(\frac{2\pi y}{L}\right)$$

the most probable locations correspond to $\sin^2(\pi x/L) = 1$ and $\sin^2(2\pi y/L) = 1$, or $(x,y) = (L/2,L/4)$ and $(L/2,3L/4)$. The least probable locations (the nodes, where the wavefunction passes through zero) correspond to zeroes in the probability density within the box, which occur along the line $y = L/2$.

There is one other two-dimensional case that can be solved exactly (but is not separable into x- and y-solutions): a particle confined to an equilateral triangular area. The solutions are set out in *A deeper look* 7D.1, available to read in the e-book accompanying this text.

A three-dimensional box can be treated in the same way as a planar rectangular region: the wavefunctions are products of three terms and the energy is a sum of three terms. As before, each term is analogous to that for the one-dimensional case. Overall, therefore,

$$\psi_{n_1,n_2,n_3}(x,y,z) = \left(\frac{8}{L_1L_2L_3}\right)^{1/2} \sin\left(\frac{n_1\pi x}{L_1}\right)\sin\left(\frac{n_2\pi y}{L_2}\right)\sin\left(\frac{n_3\pi z}{L_3}\right)$$

Wavefunctions [three dimensions] (7D.13a)

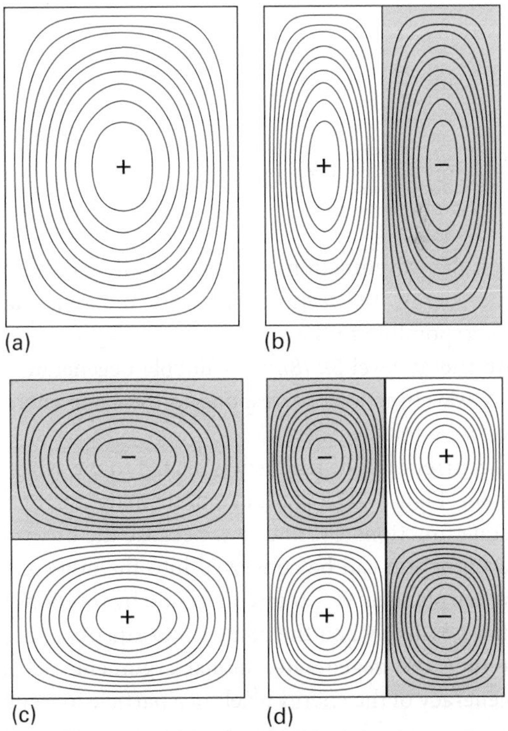

Figure 7D.7 The wavefunctions for a particle confined to a rectangular surface depicted as contours of equal amplitude. (a) $n_1 = 1, n_2 = 1$, the state of lowest energy; (b) $n_1 = 2, n_2 = 1$; (c) $n_1 = 1, n_2 = 2$; (d) $n_1 = 2, n_2 = 2$. *Interact with the dynamic version of this graph in the e-book.*

for $0 \le x \le L_1, 0 \le y \le L_2, 0 \le z \le L_3$

$$E_{n_1,n_2,n_3} = \left(\frac{n_1^2}{L_1^2}+\frac{n_2^2}{L_2^2}+\frac{n_3^2}{L_3^2}\right)\frac{h^2}{8m} \quad \text{Energy levels [three dimensions]} \quad (7D.13b)$$

The quantum numbers n_1, n_2, and n_3 are all positive integers 1, 2,... that can be chosen independently. The system has a zero-point energy, the value of $E_{1,1,1}$.

7D.3(b) Degeneracy

A special feature of the solutions arises when a two-dimensional box is not merely rectangular but square, with $L_1 = L_2 = L$. Then the wavefunctions and their energies are

$$\psi_{n_1,n_2}(x,y) = \frac{2}{L}\sin\left(\frac{n_1\pi x}{L}\right)\sin\left(\frac{n_2\pi y}{L}\right) \quad \text{for } 0 \le x \le L, 0 \le y \le L$$
$$\psi_{n_1,n_2}(x,y) = 0 \quad \text{outside box} \qquad \text{Wavefunctions [square]} \quad (7D.14a)$$

$$E_{n_1,n_2} = (n_1^2+n_2^2)\frac{h^2}{8mL^2} \quad \text{Energy levels [square]} \quad (7D.14b)$$

Consider the cases $n_1 = 1, n_2 = 2$ and $n_1 = 2, n_2 = 1$:

$$\psi_{1,2} = \frac{2}{L}\sin\left(\frac{\pi x}{L}\right)\sin\left(\frac{2\pi y}{L}\right) \qquad E_{1,2} = (1^2+2^2)\frac{h^2}{8mL^2} = \frac{5h^2}{8mL^2}$$

$$\psi_{2,1} = \frac{2}{L}\sin\left(\frac{2\pi x}{L}\right)\sin\left(\frac{\pi y}{L}\right) \qquad E_{2,1} = (2^2+1^2)\frac{h^2}{8mL^2} = \frac{5h^2}{8mL^2}$$

Although the wavefunctions are different, they correspond to the same energy. The technical term for different wavefunctions corresponding to the same energy is **degeneracy**, and in this case energy level $5h^2/8mL^2$ is 'doubly degenerate'. In general, if N wavefunctions correspond to the same energy, then that level is 'N-fold degenerate'.

The occurrence of degeneracy is related to the symmetry of the system. Figure 7D.8 shows contour diagrams of the two degenerate functions $\psi_{1,2}$ and $\psi_{2,1}$. Because the box is square, one wavefunction can be converted into the other simply by rotating the plane by 90°. Interconversion by rotation through 90° is not possible when the plane is not square, and $\psi_{1,2}$ and $\psi_{2,1}$ are then not degenerate. Similar arguments account for the degeneracy of the energy levels of a particle in a cubic box. Other examples of degeneracy occur in quantum mechanical systems (for instance, in the hydrogen atom, Topic 8A), and all of them can be traced to the symmetry properties of the system.

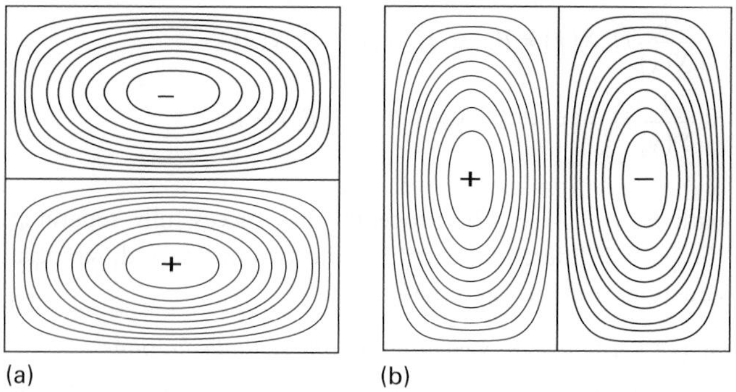

Figure 7D.8 Two of the wavefunctions for a particle confined to a geometrically square well: (a) $n_1 = 1, n_2 = 2$; (b) $n_1 = 2, n_2 = 1$. The two functions correspond to the same energy and are said to be degenerate. Note that one wavefunction can be converted into the other by rotation of the plane by 90°: degeneracy is always a consequence of symmetry.

Brief illustration 7D.5

The energy of a particle in a two-dimensional square box of side L in the energy level with $n_1 = 1, n_2 = 7$ is

$$E_{1,7} = (1^2+7^2)\frac{h^2}{8mL^2} = \frac{50h^2}{8mL^2}$$

The level with $n_1 = 7$ and $n_2 = 1$ has the same energy. Thus, at first sight the energy level $50h^2/8mL^2$ is doubly degenerate. However, in certain systems there may be levels that are not apparently related by symmetry but have the same energy and are said to be 'accidentally' degenerate. Such is the case here, for the level with $n_1 = 5$ and $n_2 = 5$ also has energy $50h^2/8mL^2$. The level is therefore actually three-fold degenerate. Accidental degeneracy is also encountered in the hydrogen atom (Topic 8A) and can always be traced to a 'hidden' symmetry, one that is not immediately obvious.

Exercises

E7D.6 An electron is confined to a square well of length L. What would be the length of the box such that the zero-point energy of the electron is equal to its rest mass energy, m_ec^2? Express your answer in terms of the parameter $\lambda_C = h/m_ec$, the 'Compton wavelength' of the electron.

E7D.7 For a particle in a square box of side L, at what position (or positions) is the probability density a maximum if the wavefunction has $n_1 = 2, n_2 = 2$? Also, describe the position of any node or nodes in the wavefunction.

E7D.8 Consider a particle in a cubic box. What is the degeneracy of the level that has an energy three times that of the lowest level?

7D.4 Tunnelling

A new quantum-mechanical feature appears when the potential energy does not rise abruptly to infinity at the walls (Fig. 7D.9). Consider the case in which there are two regions where the potential energy is zero separated by a barrier where it rises to a finite value, V_0. Suppose the energy of the particle is less than V_0. A particle arriving from the left of the barrier has an oscillating wavefunction but inside the barrier the wavefunction decays rather than oscillates (this feature is justified below). Provided the barrier is not too wide the wavefunction emerges to the right, but with reduced amplitude; it then continues to oscillate once it is back in a region where it has zero potential energy. As a result of this behaviour the particle has a non-zero probability of passing through the barrier. That is forbidden classically because a particle cannot have a potential energy that exceeds its total energy. The ability of a particle to penetrate into, and possibly pass through, a classically forbidden region is called **tunnelling**.

The Schrödinger equation can be used to calculate the probability of tunnelling of a particle of mass m incident from the left on a rectangular potential energy barrier of width W. On the left of the barrier ($x < 0$) the wavefunctions are those of a particle with $V = 0$, so from eqn 7D.2,

$$\psi = Ae^{ikx} + Be^{-ikx} \qquad k\hbar = (2mE)^{1/2} \qquad \text{Wavefunction to left of barrier} \qquad (7D.15)$$

The Schrödinger equation for the region representing the barrier ($0 \le x \le W$), where the potential energy has the constant value V_0, is

$$-\frac{\hbar^2}{2m}\frac{d^2\psi(x)}{dx^2} + V_0\psi(x) = E\psi(x) \qquad (7D.16)$$

Provided $E < V_0$ the general solutions of eqn 7D.16 are

$$\psi = Ce^{\kappa x} + De^{-\kappa x} \qquad \kappa\hbar = \{2m(V_0 - E)\}^{1/2} \qquad \text{Wavefunction inside barrier} \qquad (7D.17)$$

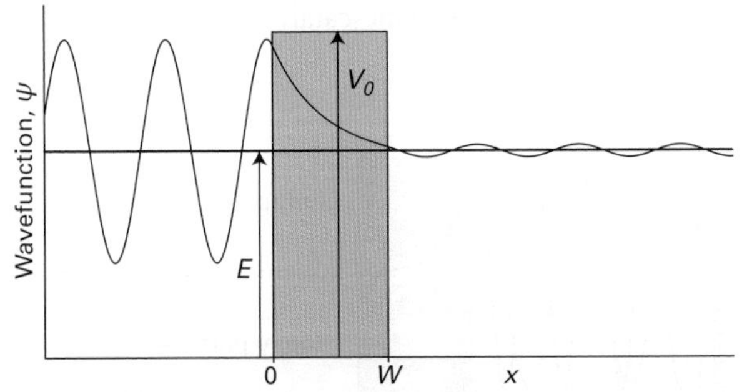

Figure 7D.9 The wavefunction for a particle encountering a potential barrier. Provided that the barrier is neither too wide nor too tall, the wavefunction will be non-zero as it exits to the right. *Interact with the dynamic version of this graph in the e-book.*

as can be verified by substituting this solution into the left-hand side of eqn 7D.16. The important feature to note is that the two exponentials in eqn 7D.17 are now real functions, as distinct from the complex, oscillating functions for the region where $V = 0$. To the right of the barrier ($x > W$), where $V = 0$ again, the wavefunctions are

$$\psi = A'e^{ikx} \qquad k\hbar = (2mE)^{1/2} \qquad \text{Wavefunction to right of barrier} \qquad (7D.18)$$

Note that to the right of the barrier, the particle can be moving only to the right and therefore only the term e^{ikx} contributes as it corresponds to a particle with positive linear momentum (moving to the right).

The complete wavefunction for a particle incident from the left consists of (Fig. 7D.10):

- an incident wave (Ae^{ikx} corresponds to positive linear momentum);
- a wave reflected from the barrier (Be^{-ikx} corresponds to negative linear momentum, motion to the left);
- the exponentially changing amplitude inside the barrier (eqn 7D.17);
- an oscillating wave (eqn 7D.18) representing the propagation of the particle to the right after tunnelling through the barrier successfully.

The probability that a particle is travelling towards positive x (to the right) on the left of the barrier ($x < 0$) is proportional to $|A|^2$, and the probability that it is travelling to the right after passing through the barrier ($x > W$) is proportional to $|A'|^2$. The ratio of these two probabilities, $|A'|^2/|A|^2$, which expresses the probability of the particle tunnelling through the barrier, is called the **transmission probability**, T.

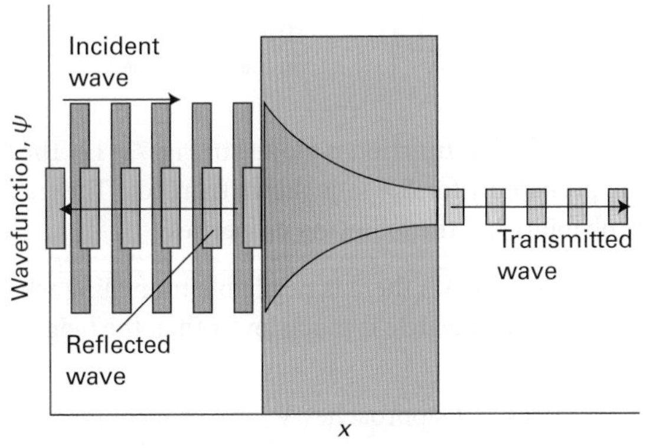

Figure 7D.10 When a particle is incident on a barrier from the left, the wavefunction consists of a wave representing linear momentum to the right, a reflected component representing momentum to the left, a varying but not oscillating component inside the barrier, and a (weak) wave representing motion to the right on the far side of the barrier.

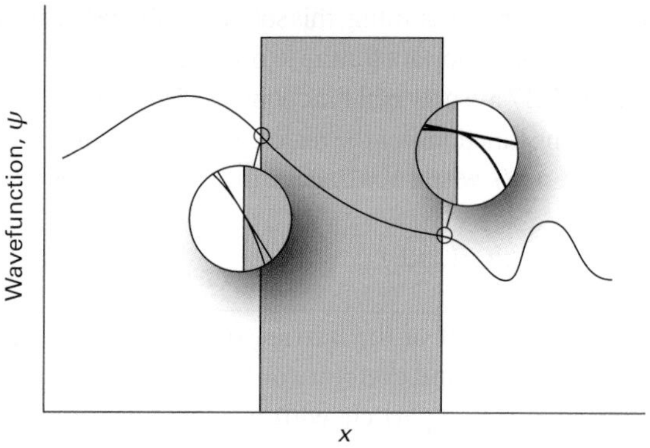

Figure 7D.11 The wavefunction and its slope must be continuous at the edges of the barrier. The conditions for continuity enable the wavefunctions at the junctions of the three zones to be connected and hence relations between the coefficients that appear in the solutions of the Schrödinger equation to be obtained.

The values of the coefficients A, B, C, and D are found by applying the usual criteria of acceptability to the wavefunction. Because an acceptable wavefunction must be continuous at the edges of the barrier (at $x=0$ and $x=W$)

$$\begin{aligned}&\text{at } x=0: A+B=C+D\\&\text{at } x=W: Ce^{\kappa W}+De^{-\kappa W}=A'e^{ikW}\end{aligned} \qquad (7D.19a)$$

Their slopes (their first derivatives) must also be continuous at these positions (Fig. 7D.11):

$$\begin{aligned}&\text{at } x=0: ikA-ikB=\kappa C-\kappa D\\&\text{at } x=W: \kappa Ce^{\kappa W}-\kappa De^{-\kappa W}=ikA'e^{ikW}\end{aligned} \qquad (7D.19b)$$

After lengthy algebraic manipulations of these four equations, the transmission probability turns out to be

$$T=\left\{1+\frac{(e^{\kappa W}-e^{-\kappa W})^2}{16\varepsilon(1-\varepsilon)}\right\}^{-1} \qquad \text{Transmission probability [rectangular barrier]} \qquad (7D.20a)$$

where $\varepsilon = E/V_0$. This function is plotted in Fig. 7D.12. The transmission probability for $E > V_0$ is shown there too. The transmission probability has the following properties:

- $T \approx 0$ for $E << V_0$: there is negligible tunnelling when the energy of the particle is much lower than the height of the barrier;
- T increases as E approaches V_0: the probability of tunnelling increases as the energy of the particle rises to match the height of the barrier;
- T approaches 1 for $E > V_0$, but the fact that it does not immediately reach 1 means that there is a probability of the particle being reflected by the barrier even though according to classical mechanics it can pass over it;
- $T \approx 1$ for $E >> V_0$, as expected classically: the barrier is invisible to the particle when its energy is much higher than the barrier.

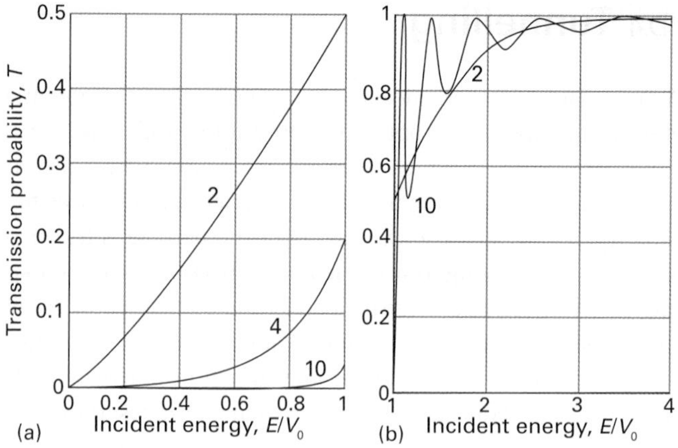

Figure 7D.12 The transmission probabilities T for passage through a rectangular potential barrier. The horizontal axis is the energy of the incident particle expressed as a multiple of the barrier height. The curves are labelled with the value of $W(2mV_0)^{1/2}/\hbar$. (a) $E < V_0$; (b) $E > V_0$.

For high, wide barriers (in the sense that $\kappa W >> 1$), eqn 7D.20a simplifies to

$$T \approx 16\varepsilon(1-\varepsilon)e^{-2\kappa W} \qquad \text{Rectangular potential barrier; } \kappa W >> 1 \qquad (7D.20b)$$

The transmission probability decreases exponentially with the thickness of the barrier and with $m^{1/2}$ (because $\kappa \propto m^{1/2}$). It follows that particles of low mass are more able to tunnel through barriers than heavy ones (Fig. 7D.13). Tunnelling is very important for electrons and muons ($m_\mu \approx 207m_e$), and moderately important for protons ($m_p \approx 1840m_e$); for heavier particles it is less important.

A number of effects in chemistry depend on the ability of the proton to tunnel more readily than the deuteron. The very rapid equilibration of proton transfer reactions is also a manifestation of the ability of protons to tunnel through barriers and transfer quickly from an acid to a base. Tunnelling of protons between acidic and basic groups is also an important feature of the mechanism of some enzyme-catalysed reactions.

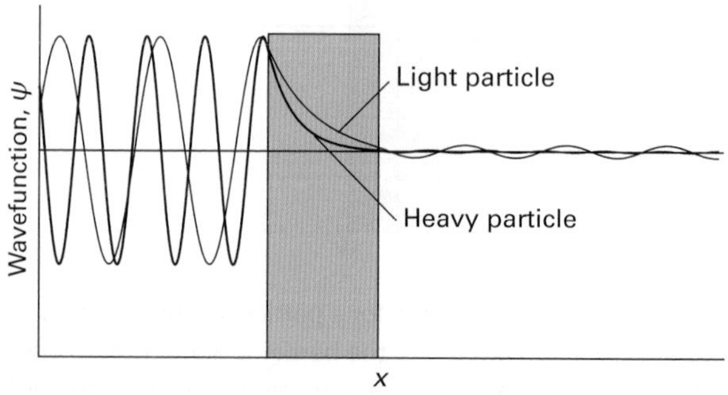

Figure 7D.13 The wavefunction of a heavy particle decays more rapidly inside a barrier than that of a light particle. Consequently, a light particle has a greater probability of tunnelling through the barrier.

Brief illustration 7D.6

Suppose that a proton of an acidic hydrogen atom is confined to an acid that can be represented by a barrier of height 2.000 eV and length 100 pm. The probability that a proton with energy 1.995 eV (corresponding to 0.3195 aJ) can escape from the acid is computed using eqn 7D.20a, with $\varepsilon = E/V_0 = 1.995\,\text{eV}/2.000\,\text{eV} = 0.9975$ and $V_0 - E = 0.005\,\text{eV}$ (corresponding to 8.0×10^{-22} J). The quantity κ is given by eqn 7D.17:

$$\kappa = \frac{\{2\times(1.67\times10^{-27}\ \text{kg})\times(8.0\times10^{-22}\ \text{J})\}^{1/2}}{1.055\times10^{-34}\ \text{J s}}$$
$$=1.54\cdots\times10^{10}\ \text{m}^{-1}$$

It follows that

$$\kappa W = (1.54\ldots\times10^{10}\ \text{m}^{-1})\times(100\times10^{-12}\ \text{m}) = 1.54\ldots$$

Equation 7D.20a then yields

$$T = \left\{1+\frac{(\text{e}^{1.54\ldots}-\text{e}^{-1.54\ldots})^2}{16\times0.9975\times(1-0.9975)}\right\}^{-1}$$
$$=1.97\times10^{-3}$$

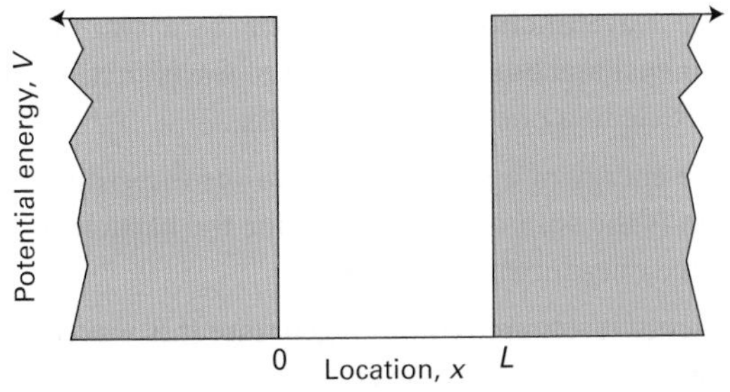

Figure 7D.14 A potential well with a finite depth.

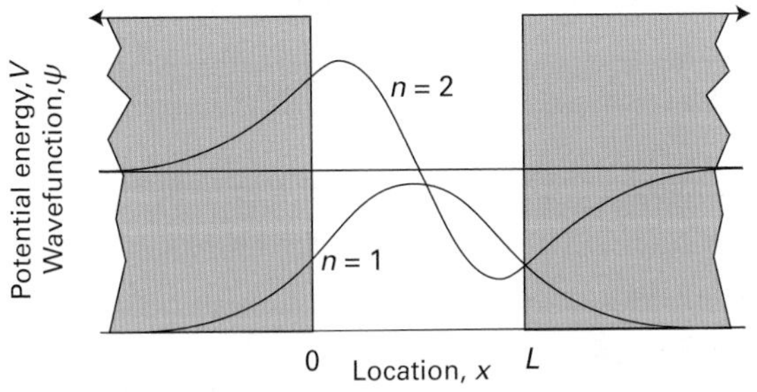

Figure 7D.15 Wavefunctions of the lowest two bound levels for a particle in the potential well shown in Fig. 7D.14.

A problem related to tunnelling is that of a particle in a square-well potential of finite depth (Fig. 7D.14). Inside the well the potential energy is zero and the wavefunctions oscillate as they do for a particle in an infinitely deep box. At the edges, the potential energy rises to a finite value V_0. If $E < V_0$ the wavefunction decays as it penetrates into the walls, just as it does when it enters a barrier. The wavefunctions are found by ensuring, as in the discussion of the potential barrier, that they and their slopes are continuous at the edges of the potential. The two lowest energy solutions are shown in Fig. 7D.15.

For a potential well of finite depth, there are a finite number of wavefunctions with energy less than V_0: they are referred to as **bound states**, in the sense that the particle is mainly confined to the well. Detailed consideration of the Schrödinger equation for the problem shows that the number of bound states is equal to N, with

$$N-1<\frac{(8mV_0)^{1/2}L}{h}<N \qquad (7\text{D}.21)$$

where V_0 is the depth of the well and L is its width. This relation shows that the deeper and wider the well, the greater the number of bound states. As the depth becomes infinite, so the number of bound states also becomes infinite, as for the particle in a box treated earlier in this Topic.

Exercise

E7D.9 Suppose that the junction between two semiconductors can be represented by a barrier of height 2.0 eV and length 100 pm. Calculate the transmission probability of an electron with energy 1.5 eV.

Checklist of concepts

☐ 1. The translational energy of a free particle is not quantized.

☐ 2. The need to satisfy **boundary conditions** implies that only certain wavefunctions are acceptable and restricts observables, specifically the energy, to discrete values.

☐ 3. A **quantum number** is an integer (in certain cases, a half-integer) that labels the state of the system.

☐ 4. A particle in a box possesses a **zero-point energy**, an irremovable minimum energy.

☐ 5. The **correspondence principle** states that the quantum mechanical result with high quantum numbers should agree with the predictions of classical mechanics.

☐ 6. The wavefunction for a particle in a two- or three-dimensional box is the product of wavefunctions for the particle in a one-dimensional box.

☐ 7. The energy of a particle in a two- or three-dimensional box is the sum of the energies for the particle in two or three one-dimensional boxes.

☐ 8. Energy levels are ***N*-fold degenerate** if *N* wavefunctions correspond to the same energy.

☐ 9. The occurrence of degeneracy is a consequence of the symmetry of the system.

☐ 10. **Tunnelling** is penetration into or through a classically forbidden region.

☐ 11. The probability of tunnelling decreases with an increase in the height and width of the potential barrier.

☐ 12. Light particles are more able to tunnel through barriers than heavy ones.

Checklist of equations

Property	Equation	Comment	Equation number
Free particle:		All values of k allowed	7D.2
Wavefunctions	$\psi_k = A\mathrm{e}^{\mathrm{i}kx} + B\mathrm{e}^{-\mathrm{i}kx}$		
Energies	$E_k = k^2\hbar^2/2m$		
Particle in a box			
One dimension:			
Wavefunctions	$\psi_n(x) = (2/L)^{1/2}\sin(n\pi x/L),\ 0 \le x \le L$ $\psi_n(x) = 0,\ x < 0$ and $x > L$	$n = 1, 2, \ldots$	7D.6
Energies	$E_n = n^2h^2/8mL^2$		
Two dimensions*:			
Wavefunctions	$\psi_{n_1,n_2}(x,y) = X_{n_1}(x)Y_{n_2}(y)$ $X_{n_1}(x) = (2/L_1)^{1/2}\sin(n_1\pi x/L_1),\ 0 \le x \le L_1$ $Y_{n_2}(y) = (2/L_2)^{1/2}\sin(n_2\pi y/L_2),\ 0 \le y \le L_2$	$n_1, n_2 = 1, 2, \ldots$	7D.12a
Energies	$E_{n_1,n_2} = (n_1^2/L_1^2 + n_2^2/L_2^2)h^2/8m$		7D.12b
Transmission probability	$T = \{1 + (\mathrm{e}^{\kappa W} - \mathrm{e}^{-\kappa W})^2/16\varepsilon(1-\varepsilon)\}^{-1}$	Rectangular potential barrier	7D.20a
	$T \approx 16\varepsilon(1-\varepsilon)\mathrm{e}^{-2\kappa W}$	High, wide rectangular barrier	7D.20b

* Solutions for three dimensions can be written analogously.

TOPIC 7E Vibrational motion

➤ Why do you need to know this material?

Molecular vibration plays a role in the interpretation of thermodynamic properties, such as heat capacities (Topics 2A and 13E), and of the rates of chemical reactions (Topic 18C). The measurement and interpretation of the vibrational frequencies of molecules is the basis of infrared spectroscopy (Topics 11C and 11D).

➤ What is the key idea?

The energy of vibrational motion is quantized, with the separation of energy levels large for strong restoring forces and light masses.

➤ What do you need to know already?

You should know how to formulate the Schrödinger equation for a given potential energy. You should also be familiar with the concepts of tunnelling (Topic 7D) and the expectation value of an observable (Topic 7C).

Atoms in molecules and solids vibrate around their equilibrium positions as bonds stretch, compress, and bend. As seen in *Energy: A first look*, the simplest model for this kind of motion is the 'harmonic oscillator', which is considered in detail in this Topic.

7E.1 The harmonic oscillator

In classical mechanics a **harmonic oscillator** is a particle of mass m that experiences a 'Hooke's law' restoring force proportional to its displacement, x, from the equilibrium position. For a one-dimensional system,

$$F_x = -k_f x \qquad \text{Hooke's law} \qquad (7E.1)$$

where k_f is the **force constant**, which characterizes the strength of the restoring force and is expressed in newtons per metre ($N\,m^{-1}$). As seen in *Energy: A first look*, the displacement varies with time as

$$x(t) = A\sin 2\pi\nu t \qquad \nu = \frac{1}{2\pi}\left(\frac{k_f}{m}\right)^{1/2} \qquad (7E.2)$$

That is, the position of the particle oscillates *harmonically* (as a sine function) with frequency ν (units: Hz). The *angular frequency* of the oscillator is $\omega = 2\pi\nu$ (units: radians per second). It follows that the angular frequency of a classical harmonic oscillator is $\omega = (k_f/m)^{1/2}$.

The potential energy V is related to force by $F = -\mathrm{d}V/\mathrm{d}x$ (see *Energy: A first look*), so the potential energy corresponding to a Hooke's law restoring force is

$$V(x) = \tfrac{1}{2}k_f x^2 \qquad \text{Parabolic potential energy} \qquad (7E.3)$$

This form of potential energy is called a 'harmonic potential energy' or a 'parabolic potential energy' (Fig. 7E.1). As the particle moves away from the equilibrium position its potential energy increases and so its kinetic energy, and hence its speed, decreases. At some point all the energy is potential and the particle comes to rest at a turning point. The particle then accelerates back towards and through the equilibrium position. The greatest probability of finding the particle is where it is moving most slowly, which is close to the turning points.

The turning point, x_{tp}, of a classical oscillator occurs when its potential energy $\tfrac{1}{2}k_f x^2$ is equal to its total energy, so

$$x_{tp} = \pm\left(\frac{2E}{k_f}\right)^{1/2} \qquad \text{Turning point [classical harmonic oscillator]} \qquad (7E.4)$$

The turning point increases with the total energy: in classical terms, the amplitude of the swing of a pendulum or the displacement of a mass on a spring increases.

The Schrödinger equation for mass m experiencing a harmonic potential energy is

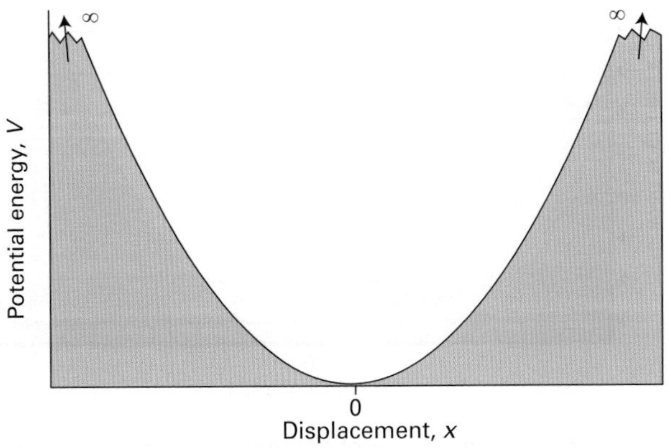

Figure 7E.1 The potential energy for a harmonic oscillator is the parabolic function $V(x) = \tfrac{1}{2}k_f x^2$, where x is the displacement from equilibrium. The larger the force constant k_f the steeper the curve and narrower the curve becomes.

$$-\frac{\hbar^2}{2m}\frac{d^2\psi(x)}{dx^2}+\tfrac{1}{2}k_f x^2\psi(x)=E\psi(x)$$ Schrödinger equation [harmonic oscillator] (7E.5)

The potential energy becomes infinite at $x = \pm\infty$, and so the wavefunction is zero at these limits. However, as the potential energy rises smoothly rather than abruptly to infinity, as it does for a particle in a box, the wavefunction decreases smoothly towards zero rather than becoming zero abruptly. The boundary conditions $\psi(\pm\infty) = 0$ imply that only some solutions of the Schrödinger equation are acceptable, and therefore that the energy of the oscillator is quantized.

7E.1(a) The energy levels

Equation 7E.5 is a standard form of differential equation and its solutions are well known to mathematicians.[1] The energies permitted by the boundary conditions are

$$E_v=\left(v+\tfrac{1}{2}\right)\hbar\omega \qquad \omega=(k_f/m)^{1/2} \qquad v=0,1,2,\ldots$$ Energy levels (7E.6)

where v is the **vibrational quantum number**. Note that the energies depend on ω, which has the same dependence on the mass and the force constant as the angular frequency of a classical oscillator (eqn 7E.2) and is high when the force constant is large and the mass small. The separation of adjacent levels is

$$E_{v+1}-E_v=\hbar\omega \quad (7E.7)$$

for all v. The energy levels therefore form a uniform ladder with spacing $\hbar\omega$ (Fig. 7E.2). The energy separation $\hbar\omega$ is negligibly small for macroscopic objects (with large mass) but significant for objects with mass similar to that of an atom.

The energy of the lowest level, with $v = 0$, is not zero:

$$E_0=\tfrac{1}{2}\hbar\omega$$ Zero-point energy (7E.8)

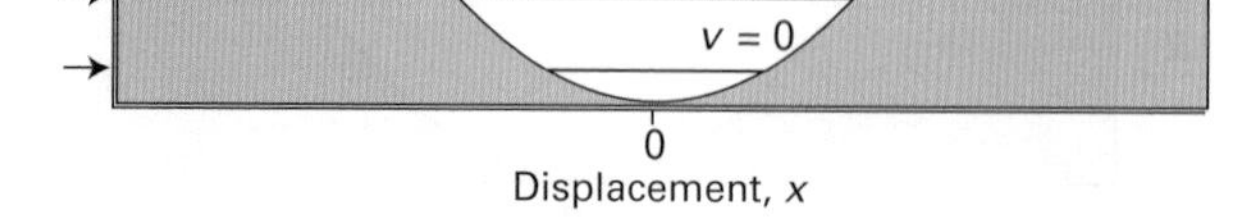

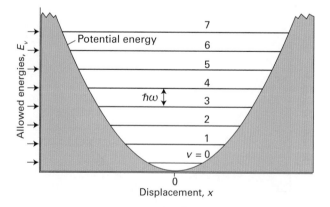

Figure 7E.2 The energy levels of a harmonic oscillator are evenly spaced with separation $\hbar\omega$, where $\omega = (k_f/m)^{1/2}$. Even in its lowest energy state, an oscillator has an energy greater than zero.

[1] For details, see our *Molecular quantum mechanics*, Oxford University Press, Oxford (2011).

The physical reason for the existence of this zero-point energy is the same as for the particle in a box (Topic 7D). The particle is confined, so its position is not completely uncertain. It follows that its momentum, and hence its kinetic energy, cannot be zero. A classical interpretation of the zero-point energy is that the quantum oscillator is never completely at rest and therefore has kinetic energy; moreover, because its motion samples the potential energy away from the equilibrium position, it also has non-zero potential energy.

The model of a particle oscillating in a parabolic potential is used to describe the vibrational motion of a diatomic molecule A–B (and, with elaboration, Topic 11D, polyatomic molecules). In this case both atoms move as the bond between them is stretched and compressed and the mass m is replaced by the **effective mass**, μ, given by

$$\mu=\frac{m_A m_B}{m_A+m_B}$$ Effective mass [diatomic molecule] (7E.9)

When A is much heavier than B, m_B can be neglected in the denominator and the effective mass is $\mu \approx m_B$, the mass of the lighter atom. In this case, only the light atom moves and the heavy atom acts as a stationary anchor.

Brief illustration 7E.1

The effective mass of $^1H^{35}Cl$ is

$$\mu=\frac{m_H m_{Cl}}{m_H+m_{Cl}}=\frac{1.0078m_u\times 34.9688m_u}{1.0078m_u+34.9688m_u}=0.9796m_u$$

which is close to the mass of the hydrogen atom. The force constant of the bond is $k_f = 516.3\,N\,m^{-1}$. It follows from eqn 7E.6 and $1\,N = 1\,kg\,m\,s^{-2}$, with μ in place of m, that

$$\omega=\left(\frac{k_f}{\mu}\right)^{1/2}=\left(\frac{516.3\,N\,m^{-1}}{0.9796\times(1.66054\times10^{-27}\,kg)}\right)^{1/2}$$
$$=5.634\times10^{14}\,s^{-1}$$

or (after division by 2π) 89.67 THz. The separation of adjacent levels (eqn 7E.7) is

$$E_{v+1}-E_v=\hbar\omega=(1.054\,57\times10^{-34}\,J\,s)\times(5.634\times10^{14}\,s^{-1})$$
$$=5.941\times10^{-20}\,J$$

or 59.41 zJ, about 0.37 eV. This energy separation corresponds to 36 kJ mol^{-1}, which is chemically significant. The zero-point energy (eqn 7E.8) of this molecular oscillator is 29.71 zJ, which corresponds to 0.19 eV, or 18 kJ mol^{-1}.

7E.1(b) The wavefunctions

The acceptable solutions of eqn 7E.5, all have the form

$$\psi(x)=N\times(\text{polynomial in } x)$$
$$\times(\text{bell-shaped Gaussian function})$$

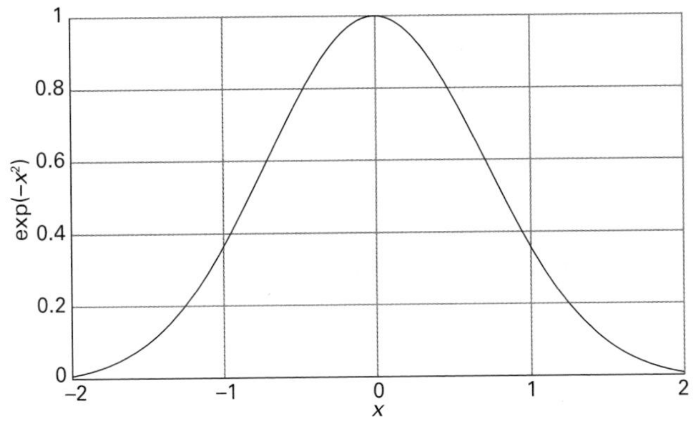

Figure 7E.3 The graph of the Gaussian function, $f(x)=e^{-x^2}$.

where N is a normalization constant. A Gaussian function is a bell-shaped function of the form e^{-x^2} (Fig. 7E.3). The precise form of the wavefunctions is

$$\psi_v(x)=N_v H_v(y)e^{-y^2/2},\ y=\frac{x}{\alpha},\ \alpha=\left(\frac{\hbar^2}{mk_f}\right)^{1/4}$$

Wavefunctions (7E.10)

The factor $H_v(y)$ is a **Hermite polynomial**; the form of these polynomials are listed in Table 7E.1. Note that the first few Hermite polynomials are rather simple: for instance, $H_0(y)=1$ and $H_1(y)=2y$. Hermite polynomials, which are members of a class of functions called 'orthogonal polynomials', have a wide range of important properties which allow a number of quantum mechanical calculations to be done with relative ease. For example, a useful 'recursion' relation is

$$H_{v+1}-2yH_v+2vH_{v-1}=0 \tag{7E.11}$$

and an important integral is

$$\int_{-\infty}^{\infty}H_{v'}H_v e^{-y^2}\,dy=\begin{cases}0 & \text{if } v'\neq v\\ \pi^{1/2}2^v v! & \text{if } v'=v\end{cases} \tag{7E.12}$$

The wavefunction for the ground state, which has $v=0$, is

$$\psi_0(x)=N_0e^{-y^2/2}=N_0e^{-x^2/2\alpha^2}$$

Ground-state wavefunction (7E.13a)

Table 7E.1 The Hermite polynomials

v	$H_v(y)$
0	1
1	$2y$
2	$4y^2-2$
3	$8y^3-12y$
4	$16y^4-48y^2+12$
5	$32y^5-160y^3+120y$
6	$64y^6-480y^4+720y^2-120$

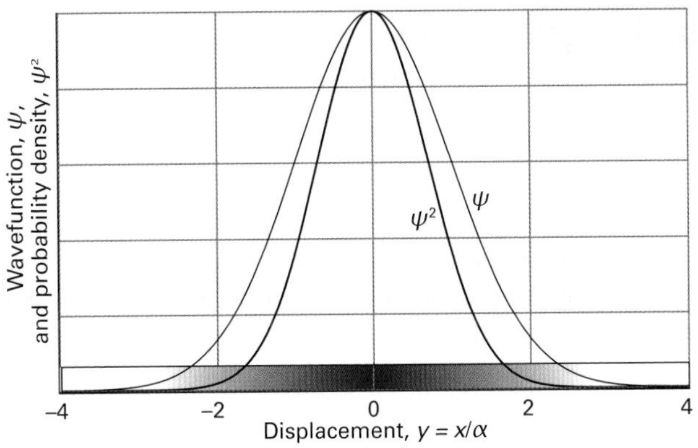

Figure 7E.4 The normalized wavefunction and probability density (shown also by shading) for the lowest energy state of a harmonic oscillator.

and the corresponding probability density is

$$\psi_0^2(x)=N_0^2e^{-y^2}=N_0^2e^{-x^2/\alpha^2}$$

Ground-state probability density [harmonic oscillator] (7E.13b)

The wavefunction and the probability density are shown in Fig. 7E.4. The probability density has its maximum value at $x=0$, the equilibrium position, but is spread about this position. The curvature is consistent with the kinetic energy being non-zero; the spread is consistent with the potential energy also being non-zero, so jointly resulting in a zero-point energy. The wavefunction for the first excited state, $v=1$, is

$$\psi_1(x)=N_1 2ye^{-y^2/2}=N_1\left(\frac{2}{\alpha}\right)xe^{-x^2/2\alpha^2}$$

First excited-state wavefunction [harmonic oscillator] (7E.14)

This function has a node at zero displacement ($x=0$), and the probability density has maxima at $x=\pm\alpha$ (Fig. 7E.5).

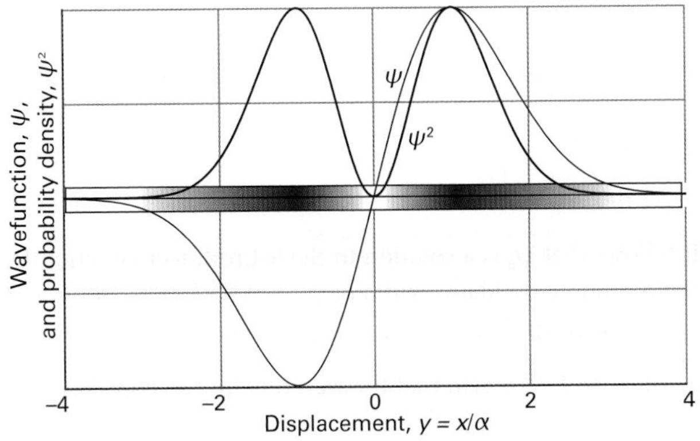

Figure 7E.5 The normalized wavefunction and probability density (shown also by shading) for the first excited state of a harmonic oscillator.

Example 7E.1 Confirming that a wavefunction is a solution of the Schrödinger equation

Confirm that the ground-state wavefunction (eqn 7E.11a) is a solution of the Schrödinger equation (eqn 7E.5).

Collect your thoughts You need to substitute the wavefunction given in eqn 7E.13a into eqn 7E.5 and see that the left-hand side generates the right-hand side of the equation. Use the definition of α in eqn 7E.10.

The solution First, evaluate the second derivative of the ground-state wavefunction by differentiating it twice in succession, starting with

$$\frac{\mathrm{d}}{\mathrm{d}x}N_0\mathrm{e}^{-x^2/2\alpha^2} = -N_0\left(\frac{x}{\alpha^2}\right)\mathrm{e}^{-x^2/2\alpha^2}$$

and then

$$\frac{\mathrm{d}^2}{\mathrm{d}x^2}N_0\mathrm{e}^{-x^2/2\alpha^2} = -\frac{\mathrm{d}}{\mathrm{d}x}\overbrace{N_0\left(\frac{x}{\alpha^2}\right)}^{f}\overbrace{\mathrm{e}^{-x^2/2\alpha^2}}^{g}$$

d(fg)/dx = fdg/dx + gdf/dx

$$= -\frac{N_0}{\alpha^2}\mathrm{e}^{-x^2/2\alpha^2} + N_0\left(\frac{x}{\alpha^2}\right)^2\mathrm{e}^{-x^2/2\alpha^2}$$

$$= -(1/\alpha^2)\psi_0 + (x^2/\alpha^4)\psi_0$$

Now substitute this expression and $\alpha^2 = (\hbar^2/mk_\mathrm{f})^{1/2}$ into the left-hand side of eqn 7E.5, which first becomes

$$-\frac{\hbar^2}{2m}\left(-\frac{1}{\alpha^2}\right)\psi_0 - \frac{\hbar^2}{2m}\left(\frac{x^2}{\alpha^4}\right)\psi_0 + \tfrac{1}{2}k_\mathrm{f}x^2\psi_0 = E\psi_0$$

and then

$$\overbrace{\frac{\hbar^2}{2m}\left(\frac{mk_\mathrm{f}}{\hbar^2}\right)^{1/2}}^{(\hbar/2)(k_\mathrm{f}/m)^{1/2}}\psi_0 \boxed{-\overbrace{\frac{\hbar^2}{2m}\left(\frac{mk_\mathrm{f}}{\hbar^2}\right)}^{k_\mathrm{f}/2}x^2\psi_0 + \tfrac{1}{2}k_\mathrm{f}x^2\psi_0} = E\psi_0$$

Therefore (keeping track of the boxed terms)

$$\frac{\hbar}{2}\left(\frac{k_\mathrm{f}}{m}\right)^{1/2}\psi_0 \boxed{-\tfrac{1}{2}k_\mathrm{f}x^2\psi_0 + \tfrac{1}{2}k_\mathrm{f}x^2\psi_0} = E\psi_0$$

The boxed terms cancel, leaving

$$\frac{\hbar}{2}\overbrace{\left(\frac{k_\mathrm{f}}{m}\right)^{1/2}}^{\omega}\psi_0 = E\psi_0$$

It follows that ψ_0 is a solution to the Schrödinger equation for the harmonic oscillator with energy $E = \tfrac{1}{2}\hbar(k_\mathrm{f}/m)^{1/2} = \tfrac{1}{2}\hbar\omega$, in accord with eqn 7E.8 for the zero-point energy.

Self-test 7E.1 Confirm that the wavefunction in eqn 7E.13 is a solution of eqn 7E.5 and evaluate its energy.

Answer: Yes, with $E_1 = \tfrac{3}{2}\hbar\omega$

The shapes of several of the wavefunctions are shown in Fig. 7E.6 and the corresponding probability densities are shown in Fig. 7E.7. These probability densities show that, as the quantum number increases, the positions of highest probability migrate towards the classical turning points. This behaviour is another example of the correspondence principle (Topic 7D) in which at high quantum numbers the classical behaviour emerges from the quantum behaviour.

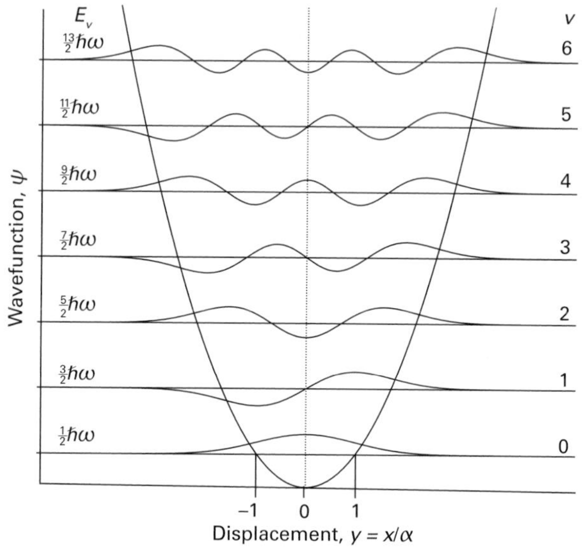

Figure 7E.6 The normalized wavefunctions for the first seven states of a harmonic oscillator. Note that the number of nodes is equal to v. The wavefunctions with even v are symmetric about $y = 0$, and those with odd v are anti-symmetric. The wavefunctions are shown superimposed on the potential energy function, and the horizontal axis for each wavefunction is set at the corresponding energy. The classical turning points occur where these horizontal lines cross the potential energy curve. *Interact with the dynamic version of this graph in the e-book.*

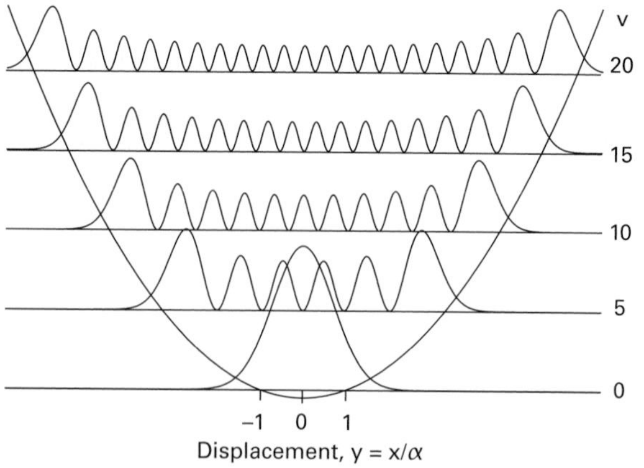

Figure 7E.7 The probability densities for the states of a harmonic oscillator with $v = 0, 5, 10, 15$, and 20. Note how the regions of highest probability density move towards the turning points of the classical motion as v increases.

The wavefunctions have the following features:

- The Gaussian function decays quickly to zero as the displacement in either direction increases, so all the wavefunctions approach zero at large displacements: the particle is unlikely to be found at large displacements.
- The wavefunction oscillates between the classical turning points but decays without oscillating outside them.
- The exponent y^2 is proportional to $x^2 \times (mk_f)^{1/2}$, so the wavefunctions decay more rapidly for large masses and strong restoring forces (stiff springs).
- As v increases, the Hermite polynomials become larger at large displacements (as x^v), so the wavefunctions grow large before the Gaussian function damps them down to zero: as a result, the wavefunctions spread over a wider range as v increases (Fig. 7E.6).

Example 7E.2 Normalizing a harmonic oscillator wavefunction

Find the normalization constant for the harmonic oscillator wavefunctions.

Collect your thoughts A wavefunction is normalized (to 1) by evaluating the integral of $|\psi|^2$ over all space and then finding the normalization factor from eqn 7B.3 ($N = 1/(\int\psi^*\psi\, d\tau)^{1/2}$). The normalized wavefunction is then equal to $N\psi$. In this one-dimensional problem, the volume element is dx and the integration is from $-\infty$ to $+\infty$. The wavefunctions are expressed in terms of the dimensionless variable $y = x/\alpha$, so begin by expressing the integral in terms of y by using $dx = \alpha dy$. The integral required is given by eqn 7E.12.

The solution The unnormalized wavefunction is

$$\psi_v(x) = H_v(y)e^{-y^2/2}$$

It follows from eqn 7E.12 that

$$\int_{-\infty}^{\infty}\psi_v^*\psi_v\, dx = \alpha\int_{-\infty}^{\infty}\psi_v^*\psi_v\, dy = \alpha\int_{-\infty}^{\infty}H_v^2(y)e^{-y^2}dy = \alpha\pi^{1/2}2^v v!$$

where $v! = v(v-1)(v-2)\ldots 1$ and $0! \equiv 1$. Therefore,

$$N_v = \left(\frac{1}{\alpha\pi^{1/2}2^v v!}\right)^{1/2} \quad \text{Normalization constant} \quad (7E.15)$$

Note that N_v is different for each value of v.

Self-test 7E.2 Confirm, by explicit evaluation of the integral, that ψ_0 and ψ_1 are orthogonal.

Answer: Show that $\int_{-\infty}^{\infty}\psi_0^*\psi_1 dx = 0$ by using the information in Table 7E.1 and eqn 7E.13.

Exercises

E7E.1 Calculate the zero-point energy of a harmonic oscillator consisting of a particle of mass 2.33×10^{-26} kg and force constant 155 N m^{-1}.

E7E.2 For a certain harmonic oscillator of effective mass 1.33×10^{-25} kg, the difference in adjacent energy levels is 4.82 zJ. Calculate the force constant of the oscillator.

E7E.3 Calculate the wavelength of the photon needed to excite a transition between neighbouring energy levels of a harmonic oscillator of effective mass equal to that of a proton ($1.0078m_u$) and force constant 855 N m^{-1}.

7E.2 Properties of the harmonic oscillator

The average value of a property is calculated by evaluating the expectation value of the corresponding operator (eqn 7C.11, $\langle\Omega\rangle = \int\psi^*\hat{\Omega}\psi\, d\tau$ for a normalized wavefunction). For a harmonic oscillator,

$$\langle\Omega\rangle_v = \int_{-\infty}^{\infty}\psi_v^*\hat{\Omega}\psi_v\, dx \quad (7E.16)$$

When the explicit wavefunctions are substituted, the integrals might look fearsome, but the Hermite polynomials have many features that simplify the calculation.

7E.2(a) Mean values

Equation 7E.16 can be used to calculate the mean displacement, $\langle x\rangle$, and the mean square displacement, $\langle x^2\rangle$, for a harmonic oscillator in a state with quantum number v.

How is that done? 7E.1 Calculating the mean values of x and x^2 for a harmonic oscillator

The evaluation of the integrals needed to compute $\langle x\rangle$ and $\langle x^2\rangle$ is simplified by recognizing the symmetry of the problem and using the special properties of the Hermite polynomials.

Step 1 *Use a symmetry argument to find the mean displacement*

The mean displacement $\langle x\rangle$ is expected to be zero because the probability density of the oscillator is symmetrical about zero; that is, there is equal probability of positive and negative displacements.

Step 2 *Confirm the result by examining the necessary integral*

More formally, the mean value of x, which is the expectation value of x, is

$$\langle x\rangle_v = \int_{-\infty}^{\infty}\psi_v^* x\psi_v\, dx = N_v^2\int_{-\infty}^{\infty}(H_v e^{-y^2/2})x(H_v e^{-y^2/2})dx$$

$x = \alpha y \quad dx = \alpha dy$

$$= \alpha^2 N_v^2\int_{-\infty}^{\infty}\underbrace{y(H_v e^{-y^2/2})^2}_{\text{An odd function}}dy$$

The integrand is an odd function because when $y \to -y$ it changes sign (the squared term does not change sign, but the term y does). The integral of an odd function over a symmetrical range is necessarily zero, so

$$\langle x\rangle_v = 0 \text{ for all } v \qquad \text{Mean displacement} \quad (7E.17a)$$

Step 3 *Find the mean square displacement*

The mean square displacement, the expectation value of x^2, is

$$\langle x^2\rangle_v = N_v^2\int_{-\infty}^{\infty}(H_v e^{-y^2/2})x^2(H_v e^{-y^2/2})dx$$

$x = \alpha y \quad dx = \alpha dy$

$$= \alpha^3 N_v^2\int_{-\infty}^{\infty}(H_v e^{-y^2/2})y^2(H_v e^{-y^2/2})dy$$

You can develop the factor y^2H_v by using the recursion relation (eqn 7E.11) rearranged into $yH_v = vH_{v-1} + \frac{1}{2}H_{v+1}$. After multiplying this expression by y it becomes

$$y^2H_v = vyH_{v-1} + \tfrac{1}{2}yH_{v+1}$$

Now use the recursion relation (with v replaced by $v-1$ or $v+1$) again for both yH_{v-1} and yH_{v+1}:

$$yH_{v-1} = (v-1)H_{v-2} + \tfrac{1}{2}H_v$$
$$yH_{v+1} = (v+1)H_v + \tfrac{1}{2}H_{v+2}$$

It follows that

$$y^2H_v = vyH_{v-1} + \tfrac{1}{2}\boxed{yH_{v+1}} = v\{(v-1)H_{v-2} + \tfrac{1}{2}H_v\} + \tfrac{1}{2}\{\boxed{(v+1)H_v + \tfrac{1}{2}H_{v+2}}\}$$
$$= v(v-1)H_{v-2} + \left(v+\tfrac{1}{2}\right)H_v + \tfrac{1}{4}H_{v+2}$$

Substitution of this result into the integral gives

$$\langle x^2\rangle_v = \alpha^3N_v^2\int_{-\infty}^{\infty}(H_v e^{-y^2/2})\overbrace{\left\{v(v-1)H_{v-2} + \left(v+\tfrac{1}{2}\right)H_v + \tfrac{1}{4}H_{v+2}\right\}}^{y^2H_v}e^{-y^2/2}dy$$
$$= \alpha^3N_v^2 v(v-1)\overbrace{\int_{-\infty}^{\infty}H_vH_{v-2}e^{-y^2}dy}^{0}$$
$$+\alpha^3N_v^2\left(v+\tfrac{1}{2}\right)\overbrace{\int_{-\infty}^{\infty}H_vH_ve^{-y^2}dy}^{\pi^{1/2}2^v v!} + \tfrac{1}{4}\alpha^3N_v^2\overbrace{\int_{-\infty}^{\infty}H_vH_{v+2}e^{-y^2}dy}^{0}$$
$$= \alpha^3N_v^2\left(v+\tfrac{1}{2}\right)\pi^{1/2}2^v v!$$

Each of the three integrals is evaluated by making use of eqn 7E.12. Therefore, after noting the expression for N_v in eqn 7E.15,

$$\langle x^2\rangle_v = \alpha^3\left(v+\tfrac{1}{2}\right)\pi^{1/2}2^v v! \times \overbrace{\frac{1}{\alpha\pi^{1/2}2^v v!}}^{N_v^2} = \left(v+\tfrac{1}{2}\right)\alpha^2$$

Finally, with, $\alpha^2 = (\hbar^2/mk_f)^{1/2}$

$$\langle x^2\rangle_v = \left(v+\tfrac{1}{2}\right)\frac{\hbar}{(mk_f)^{1/2}} \qquad \text{Mean square displacement} \quad (7E.17b)$$

The result for $\langle x\rangle_v$ shows that the oscillator is equally likely to be found on either side of $x = 0$ (like a classical oscillator). The result for $\langle x^2\rangle_v$ shows that the mean square displacement increases with v. This increase is apparent from the probability densities in Fig. 7E.7, and corresponds to the amplitude of a classical harmonic oscillator becoming greater as its energy increases.

The mean potential energy of an oscillator, which is the expectation value of $V = \frac{1}{2}k_f x^2$, can now be calculated:

$$\langle V\rangle_v = \tfrac{1}{2}\langle k_f x^2\rangle_v = \tfrac{1}{2}k_f\langle x^2\rangle_v = \tfrac{1}{2}\left(v+\tfrac{1}{2}\right)\hbar\left(\frac{k_f}{m}\right)^{1/2}$$

or

$$\langle V\rangle_v = \tfrac{1}{2}\left(v+\tfrac{1}{2}\right)\hbar\omega \qquad \text{Mean potential energy} \quad (7E.18a)$$

The total energy in the state with quantum number v is $\left(v+\frac{1}{2}\right)\hbar\omega$, so it follows that

$$\langle V\rangle_v = \tfrac{1}{2}E_v \qquad \text{Mean kinetic energy} \quad (7E.18b)$$

The total energy is the sum of the potential and kinetic energies, $E_v = \langle V\rangle_v + \langle E_k\rangle_v$, so the mean kinetic energy of the oscillator is

$$\langle E_k\rangle_v = E_v - \langle V\rangle_v = E_v - \tfrac{1}{2}E_v = \tfrac{1}{2}E_v \qquad \text{Mean kinetic energy} \quad (7E.18c)$$

The result that the mean potential and kinetic energies of a harmonic oscillator are equal (and therefore that both are equal to half the total energy) is a special case of the **virial theorem**:

> If the potential energy of a particle has the form $V = ax^b$, then its mean potential and kinetic energies are related by

$$2\langle E_k\rangle = b\langle V\rangle \qquad \text{Virial theorem} \quad (7E.19)$$

For a harmonic oscillator $b = 2$, so $\langle E_k\rangle_v = \langle V\rangle_v$. The virial theorem is a short cut to the establishment of a number of useful results, and it is used elsewhere (for example, in Topic 8A).

7E.2(b) Tunnelling

A quantum oscillator may be found at displacements with $V > E$, which are forbidden by classical physics because they correspond to negative kinetic energy. That is, a harmonic oscillator can tunnel into classically forbidden displacements. As shown in *Example* 7E.3, for the lowest energy state of the harmonic oscillator, there is about an 8 per cent chance of finding the oscillator at classically forbidden displacements in either direction. As illustrated in the following *Example*, these tunnelling probabilities are independent of the force constant and mass of the oscillator (as explained in the *Example*).

Example 7E.3 Calculating the tunnelling probability for the harmonic oscillator

Calculate the probability that the ground-state harmonic oscillator will be found in a classically forbidden region.

Collect your thoughts Find the expression for the classical turning point, x_{tp}, where the kinetic energy goes to zero, by equating the potential energy to the total energy of the harmonic oscillator. You can then calculate the probability of finding the oscillator at a displacement beyond x_{tp} by integrating $\psi^2 dx$ between x_{tp} and infinity

$$P = \int_{x_{tp}}^{\infty} \psi_v^2 dx$$

By symmetry, the probability of the particle being found in the classically forbidden region from $-x_{tp}$ to $-\infty$ is the same.

The solution The substitution of E_v into eqn 7E.4 for the classical turning point x_{tp} gives

$$x_{tp} = \pm\left(\frac{2E_v}{k_f}\right)^{1/2}$$

The variable of integration in the integral P is best expressed in terms of $y = x/\alpha$ with $\alpha = (\hbar^2/mk_f)^{1/4}$. With these substitutions, and also using $E_v = \left(v+\tfrac{1}{2}\right)\hbar\omega$, the turning points are given by

$$y_{tp} = \frac{x_{tp}}{\alpha} = \left\{\frac{2\left(v+\tfrac{1}{2}\right)\hbar\omega}{\alpha^2 k_f}\right\}^{1/2} \overset{\omega=(k_f/m)^{1/2}}{=} (2v+1)^{1/2}$$

For the state of lowest energy ($v = 0$), $y_{tp} = 1$ and the probability of being beyond that point is

$$P = \int_{x_{tp}}^{\infty} \psi_0^2 dx \overset{dx=\alpha dy}{=} \alpha\int_1^{\infty} \psi_0^2 dy = \alpha N_0^2 \int_1^{\infty} e^{-y^2} dy$$

with

$$N_0 = \left(\frac{1}{\alpha\pi^{1/2}2^0 0!}\right)^{1/2} \overset{2^0=1;\ 0!=1}{=} \left(\frac{1}{\alpha\pi^{1/2}}\right)^{1/2}$$

Therefore

$$P = \frac{1}{\pi^{1/2}}\int_1^{\infty} e^{-y^2} dy$$

The integral must be evaluated numerically (by using mathematical software), and is equal to 0.139.... It follows that $P = 0.079$. That is, in 7.9 per cent of a large number of observations of an oscillator in the state with quantum number $v = 0$, the particle will be found beyond the (positive) classical turning point. It will be found with the same probability at negative forbidden displacements. The total probability of finding the oscillator in a classically forbidden region when it is in its ground state is about 16 per cent.

Comment. Note that the result is independent of both k_f and m. As seen in Topic 7D, tunnelling is most important for particles of low mass, so how can this conclusion be justified physically? The answer lies in the fact that the turning point comes closer to the equilibrium point as either the mass or the force constant increases. The amplitude of the wavefunction at x_{tp} therefore becomes greater, and although the tunnelling falls off more rapidly, it starts from a higher value at x_{tp} itself. The effects balance, and the net outcome is independence of the penetration probability on mass and force constant.

Self-test 7E.3 Calculate the probability that a harmonic oscillator in the state with quantum number $v = 1$ will be found at a classically forbidden extension. You will need to use mathematical software to evaluate the integral.

Answer: $P = 0.056$

The probability of finding the oscillator in classically forbidden regions decreases quickly with increasing v, and vanishes entirely as v approaches infinity, as is expected from the correspondence principle. Macroscopic oscillators (such as pendulums) are in states with very high quantum numbers, so the tunnelling probability is wholly negligible and classical mechanics is reliable. Molecules, however, are normally in their vibrational ground states, and for them the probability is very significant and classical mechanics is misleading.

Exercises

E7E.4 Sketch the form of the wavefunctions for the harmonic oscillator with quantum numbers $v = 0$ and 1. Use a symmetry argument to explain why these two wavefunctions are orthogonal (do not evaluate any integrals).

E7E.5 How many nodes are there in the wavefunction of a harmonic oscillator with (i) $v = 3$; (ii) $v = 4$?

E7E.6 At what displacements is the probability density a maximum for a state of a harmonic oscillator with $v = 1$? (Express your answers in terms of the coordinate y.)

Checklist of concepts

☐ **1.** The energy levels of a quantum mechanical harmonic oscillator are evenly spaced.

☐ **2.** The wavefunctions of a quantum mechanical harmonic oscillator are products of a **Hermite polynomial** and a Gaussian (bell-shaped) function.

☐ **3.** A quantum mechanical harmonic oscillator has **zero-point energy**, an irremovable minimum energy.

☐ **4.** The probability of finding a quantum mechanical harmonic oscillator at classically forbidden displacements is significant for the ground vibrational state ($v = 0$) but decreases quickly with increasing v.

Checklist of equations

Property	Equation	Comment	Equation number
Energy levels	$E_v = \left(v+\tfrac{1}{2}\right)\hbar\omega \quad \omega = (k_f/m)^{1/2}$	$v = 0, 1, 2,\ldots$	7E.6
Zero-point energy	$E_0 = \tfrac{1}{2}\hbar\omega$		7E.8
Wavefunctions	$\psi_v(x) = N_v H_v(y)e^{-y^2/2}$ $y = x/\alpha,\ \alpha = (\hbar^2/mk_f)^{1/4}$	$v = 0, 1, 2,\ldots$	7E.10
Normalization constant	$N_v = (1/\alpha\pi^{1/2}2^v v!)^{1/2}$		7E.15
Mean displacement	$\langle x\rangle_v = 0$		7E.17a
Mean square displacement	$\langle x^2\rangle_v = \left(v+\tfrac{1}{2}\right)\hbar/(mk_f)^{1/2}$		7E.17b
Virial theorem	$2\langle E_k\rangle = b\langle V\rangle$	$V = ax^b$	7E.19

TOPIC 7F Rotational motion

> ➤ Why do you need to know this material?
>
> Angular momentum is central to the description of the electronic structure of atoms and molecules and the interpretation of molecular spectra.
>
> ➤ What is the main idea?
>
> The energy, angular momentum, and orientation of the angular momentum of a rotating body are quantized.
>
> ➤ What do you need to know already?
>
> You should be aware of the postulates of quantum mechanics and the role of boundary conditions (Topics 7C and 7D). Background information on the description of rotation is found in *Energy: A first look*.

Rotational motion is encountered in many aspects of chemistry, including the electronic structures of atoms, because electrons orbit (in a quantum mechanical sense) around nuclei and spin on their axis. Molecules also rotate; transitions between their rotational states affect the appearance of spectra and their detection gives valuable information about the structures of molecules.

7F.1 Rotation in two dimensions

Consider a particle of mass m constrained to move in a circular path (a 'ring') of radius r in the xy-plane (Fig. 7F.1). The particle might not be an actual body but could represent the centre of a molecule or other object constrained to rotate in a plane, such as the group of three hydrogen atoms in a methyl group freely rotating about its axis.

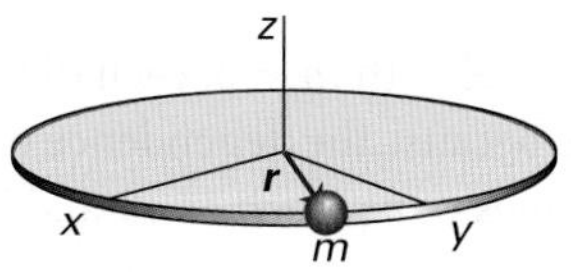

Figure 7F.1 A particle on a ring is free to move in the xy-plane around a circular path of radius r.

The Schrödinger equation for motion around the ring is

$$-\frac{\hbar^2}{2m}\left(\frac{\partial^2}{\partial x^2}+\frac{\partial^2}{\partial y^2}\right)\psi(x,y)=E\psi(x,y) \qquad (7F.1)$$

The equation is best expressed in cylindrical coordinates r and ϕ with $z=0$ (*The chemist's toolkit* 7F.1) because they reflect the symmetry of the system. In cylindrical coordinates

$$\frac{\partial^2}{\partial x^2}+\frac{\partial^2}{\partial y^2}=\boxed{\frac{\partial^2}{\partial r^2}+\frac{1}{r}\frac{\partial}{\partial r}}+\frac{1}{r^2}\frac{\partial^2}{\partial \phi^2} \qquad (7F.2)$$

However, because the radius of the path is fixed, the (boxed) derivatives with respect to r can be discarded. Only the last term in eqn 7F.2 then survives and the Schrödinger equation becomes

$$-\frac{\hbar^2}{2mr^2}\frac{\mathrm{d}^2\psi(\phi)}{\mathrm{d}\phi^2}=E\psi(\phi) \qquad (7F.3a)$$

The partial derivative has been replaced by a complete derivative because ϕ is now the only variable. The term mr^2 is the

> **The chemist's toolkit 7F.1** Cylindrical coordinates
>
> For systems with cylindrical symmetry it is best to work in **cylindrical coordinates** r, ϕ, and z (Sketch 1), with
>
> $$x=r\cos\phi \qquad y=r\sin\phi$$
>
> and where
>
> $$0\le r\le\infty \qquad 0\le\phi\le 2\pi \qquad -\infty\le z\le+\infty$$
>
> The volume element is
>
> $$\mathrm{d}\tau=r\mathrm{d}r\mathrm{d}\phi\mathrm{d}z$$
>
> For motion in a plane, $z=0$ and the volume element is
>
> $$\mathrm{d}\tau=r\mathrm{d}r\mathrm{d}\phi$$
>
> 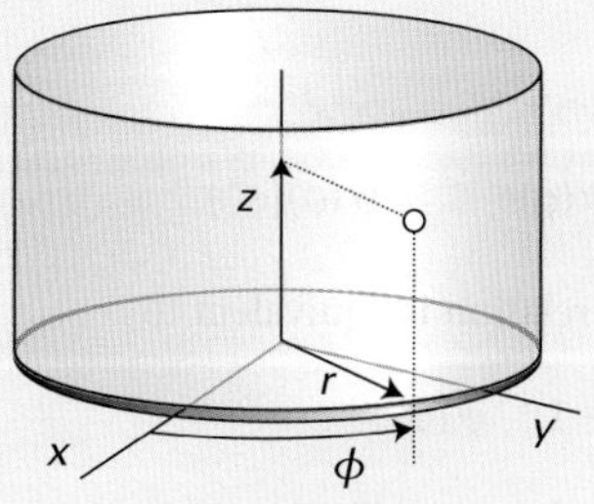
>
>
> **Sketch 1**

moment of inertia, $I = mr^2$, and so the Schrödinger equation becomes

$$-\frac{\hbar^2}{2I}\frac{d^2\psi(\phi)}{d\phi^2} = E\psi(\phi) \qquad \text{Schrödinger equation [particle on a ring]} \qquad (7F.3b)$$

7F.1(a) The solutions of the Schrödinger equation

The most straightforward way of finding the solutions of eqn 7F.3b is to take the known general solution to a second-order differential equation of this kind and show that it does indeed satisfy the equation. Then find the allowed solutions and energies by imposing the relevant boundary conditions.

How is that done? 7F.1 Finding the solutions of the Schrödinger equation for a particle on a ring

A solution of eqn 7F.3b is

$$\psi(\phi) = e^{im_l\phi}$$

where, as yet, m_l is an arbitrary dimensionless number (the notation is explained later). This solution is not the most general, which would be $\psi(\phi) = Ae^{im_l\phi} + Be^{-im_l\phi}$, but is sufficiently general for the present purpose.

Step 1 *Verify that the function satisfies the equation*

To verify that $\psi(\phi)$ is a solution note that

$$\frac{d^2}{d\phi^2}e^{im_l\phi} = \frac{d}{d\phi}(im_l)e^{im_l\phi} = (im_l)^2 e^{im_l\phi} = -m_l^2 \overbrace{e^{im_l\phi}}^{\psi} = -m_l^2\psi$$

Then

$$-\frac{\hbar^2}{2I}\frac{d^2\psi}{d\phi^2} = -\frac{\hbar^2}{2I}\boxed{(-m_l^2\psi)} = \frac{m_l^2\hbar^2}{2I}\psi$$

which has the form constant × ψ, so the proposed wavefunction is indeed a solution and the corresponding energy is $m_l^2\hbar^2/2I$.

Step 2 *Impose the appropriate boundary conditions*

The requirement that a wavefunction must be single-valued implies the existence of a **cyclic boundary condition**, the requirement that the wavefunction must be the same after a complete revolution: $\psi(\phi + 2\pi) = \psi(\phi)$ (Fig. 7F.2). In this case

$$\begin{aligned}\psi(\phi + 2\pi) &= e^{im_l(\phi+2\pi)} = e^{im_l\phi}e^{2\pi i m_l} \\ &= \psi(\phi)e^{2\pi i m_l} = \psi(\phi)(e^{i\pi})^{2m_l}\end{aligned}$$

As $e^{i\pi} = -1$, this relation is equivalent to

$$\psi(\phi + 2\pi) = (-1)^{2m_l}\psi(\phi)$$

The cyclic boundary condition $\psi(\phi + 2\pi) = \psi(\phi)$ requires $(-1)^{2m_l} = 1$; this requirement is satisfied for any positive or negative integer value of m_l, including 0.

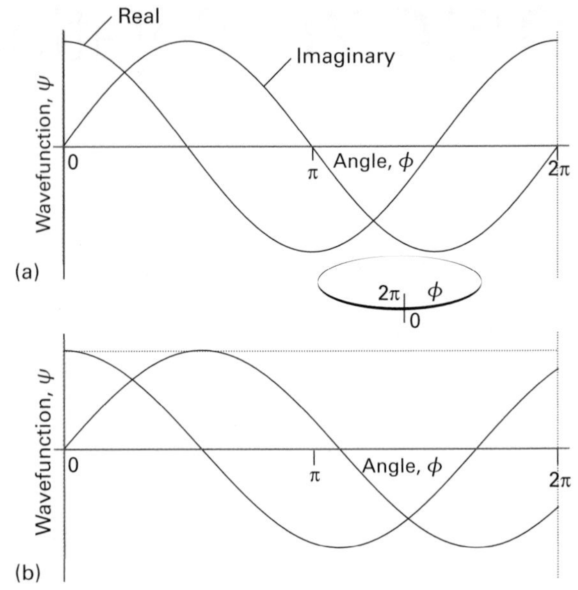

Figure 7F.2 Two possible solutions of the Schrödinger equation for a particle on a ring. The circumference has been opened out into a straight line; the points at $\phi = 0$ and 2π are identical. The solution in (a), $e^{i\phi} = \cos\phi + i\sin\phi$, is acceptable because after a complete revolution the wavefunction has the same value. The solution in (b), $e^{0.9i\phi} = \cos(0.9\phi) + i\sin(0.9\phi)$ is unacceptable because its value, both for the real and imaginary parts, is not the same at $\phi = 0$ and 2π.

Step 3 *Normalize the wavefunction*

A wavefunction is normalized by finding the normalization constant N given by eqn 7B.3, $N = 1/\left(\int\psi^*\psi d\tau\right)^{1/2}$. In this case, the wavefunction depends only on the angle ϕ, the 'volume element' is $d\phi$, and the range of integration is from $\phi = 0$ to 2π, so the normalization constant is

$$N = \frac{1}{\left(\int_0^{2\pi}\psi^*\psi d\phi\right)^{1/2}} = \frac{1}{\left(\int_0^{2\pi}\underbrace{e^{-im_l\phi}e^{im_l\phi}}_{1} d\phi\right)^{1/2}} = \frac{1}{(2\pi)^{1/2}}$$

The normalized wavefunctions and corresponding energies are labelled with the integer m_l, which is playing the role of a quantum number:

$$\psi_{m_l}(\phi) = \frac{e^{im_l\phi}}{(2\pi)^{1/2}}, \qquad E_{m_l} = \frac{m_l^2\hbar^2}{2I}, \quad m_l = 0, \pm1, \pm2, \ldots \qquad \text{Wavefunctions and energy levels of a particle on a ring} \qquad (7F.4)$$

Apart from the level with $m_l = 0$, each of the energy levels is doubly degenerate because the dependence of the energy on m_l^2 means that two values of m_l (such as +1 and −1) correspond to the same energy.[1]

[1] When quoting the value of m_l, it is good practice always to give the sign, even if m_l is positive. Thus, write $m_l = +1$, not $m_l = 1$.

7F.1(b) Quantization of angular momentum

For a particle travelling on a circular path of radius r about the z-axis, and therefore confined to the xy-plane, the angular momentum is represented by a vector that is perpendicular to the plane. Its only component is

$$l_z = \pm pr \tag{7F.5}$$

where p is the magnitude of the linear momentum in the xy-plane at any instant. When $l_z > 0$, the particle travels in a clockwise direction as viewed from below (**1**); when $l_z < 0$, the motion is anticlockwise (**2**). However, not all values of l_z are allowed and it can be shown that, like rotational energy, angular momentum is quantized.

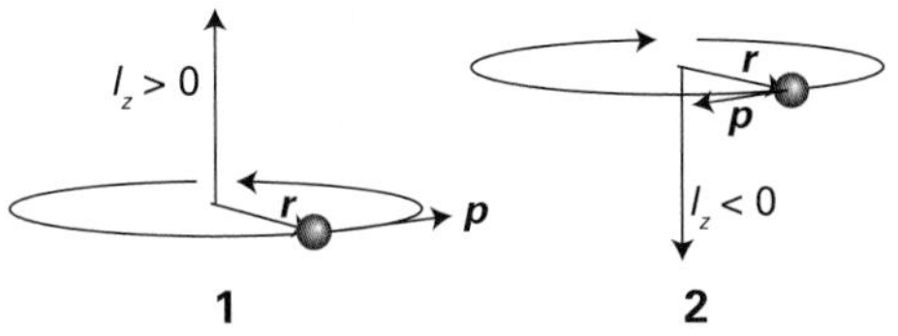

How is that done? 7F.2 Showing that angular momentum is quantized

As explained in Topic 7C, the outcome of a measurement of a property is one of the eigenvalues of the corresponding operator. The first step is therefore to identify the operator corresponding to angular momentum, and then to identify its eigenvalues.

Step 1 *Construct the operator for angular momentum*

Because the particle is confined to the xy-plane, its angular momentum is directed along the z-axis, so only this component need be considered. As you may recall from introductory physics, the z-component of the angular momentum is $l_z = xp_y - yp_x$. The corresponding operator is formed by replacing x, y, p_x, and p_y by their corresponding operators (Topic 7C; $\hat{q} = q\times$ and $\hat{p}_q = (\hbar/\mathrm{i})\partial/\partial q$, with $q = x$ and y), which gives

$$\hat{l}_z = \frac{\hbar}{\mathrm{i}}\left(x\frac{\partial}{\partial y} - y\frac{\partial}{\partial x}\right) \quad \text{Operator for } l_z \tag{7F.6a}$$

When expressed in cylindrical coordinates this operator becomes

$$\hat{l}_z = \frac{\hbar}{\mathrm{i}}\frac{\mathrm{d}}{\mathrm{d}\phi} \tag{7F.6b}$$

Step 2 *Verify that the wavefunctions are eigenfunctions of this operator*

To decide whether the wavefunctions in eqn 7F.4 are eigenfunctions of $\hat{l}_z$, allow it to act on the wavefunction (ignoring the normalization factor):

$$\hat{l}_z\psi_{m_l} = \frac{\hbar}{\mathrm{i}}\frac{\mathrm{d}}{\mathrm{d}\phi}\mathrm{e}^{\mathrm{i}m_l\phi} = \frac{\hbar}{\mathrm{i}}\mathrm{i}m_l\overbrace{\mathrm{e}^{\mathrm{i}m_l\phi}}^{\psi_{m_l}} = m_l\hbar\psi_{m_l}$$

The wavefunction is an eigenfunction of the angular momentum, with the eigenvalue $m_l\hbar$. In summary,

$$\hat{l}_z\psi_{m_l}(\phi) = m_l\hbar\psi_{m_l}(\phi) \quad m_l = 0, \pm1, \pm2, \ldots \quad \text{Eigenfunctions of } \hat{l}_z \tag{7F.7}$$

Because m_l is confined to discrete values, the z-component of angular momentum is quantized. In accord with the significance of the orientation of l_z in **1** and **2**, when m_l is positive, the z-component of angular momentum is positive (clockwise rotation when seen from below); when m_l is negative, the z-component of angular momentum is negative (anticlockwise when seen from below).

The important features of the results so far are:

- The energies are quantized because m_l is confined to integer values.
- The occurrence of m_l as its square means that the energy of rotation is independent of the sense of rotation (the sign of m_l), as expected physically.
- Apart from the state with $m_l = 0$, all the energy levels are doubly degenerate; rotation can be clockwise or anticlockwise with the same energy.
- When $m_l = 0$, $E_0 = 0$. That is, there is no zero-point energy: the particle can be stationary.
- As m_l increases, the wavefunctions oscillate with shorter wavelengths and so have greater curvature, corresponding to increasing kinetic energy (Fig. 7F.3).
- As pointed out in Topic 7D, a wavefunction that is complex represents a direction of motion, and taking its complex conjugate reverses the direction. For two states with the same value of $|m_l|$, the wavefunctions with $m_l > 0$ and $m_l < 0$ are each other's complex conjugate. They correspond to motion in opposite directions.

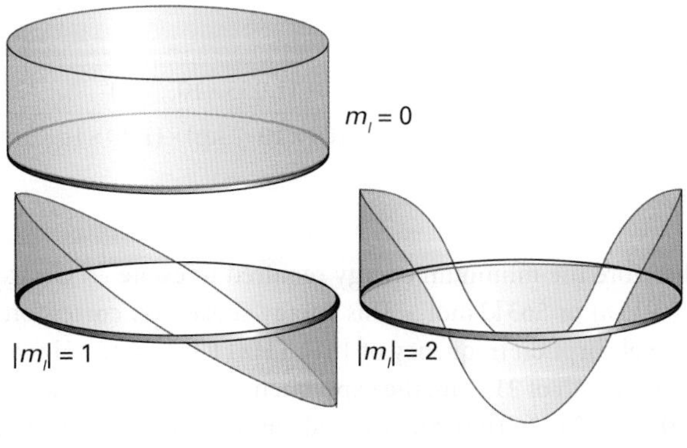

Figure 7F.3 The real parts of the wavefunctions of a particle on a ring. As the energy increases, so do the number of nodes and the curvature.

The probability density predicted by the wavefunctions of eqn 7F.4 is uniform around the ring:

$$\psi_{m_l}^{\star}(\phi)\psi_{m_l}(\phi) = \left(\frac{e^{im_l\phi}}{(2\pi)^{1/2}}\right)^{\star}\left(\frac{e^{im_l\phi}}{(2\pi)^{1/2}}\right) = \left(\frac{e^{-im_l\phi}}{(2\pi)^{1/2}}\right)\left(\frac{e^{im_l\phi}}{(2\pi)^{1/2}}\right) = \frac{1}{2\pi}$$

Angular momentum and angular position are a pair of complementary observables (in the sense defined in Topic 7C), and the inability to specify them simultaneously with arbitrary precision is another example of the uncertainty principle. In this case, the z-component of angular momentum is known exactly (as $m_l\hbar$) but the location of the particle on the ring is completely unknown, as implied by the uniform probability density.

Example 7F.1 Using the particle-on-a-ring model

The particle-on-a-ring is a crude but illustrative model of cyclic, conjugated molecular systems. Treat the π electrons in benzene as particles freely moving over a circular ring of six carbon atoms and calculate the minimum energy required for the excitation of a π electron. The carbon–carbon bond length in benzene is 140 pm.

Collect your thoughts Each carbon atom contributes one π electron, so there are six electrons to accommodate. Each state is occupied by two electrons, so only the $m_l = 0, +1$, and -1 states are occupied (with the last two being degenerate). The minimum energy required for excitation corresponds to a transition of an electron from the $m_l = +1$ (or -1) state to the $m_l = +2$ (or -2) state. Use eqn 7F.4, and the mass of the electron, to calculate the energies of the states. A hexagon can be inscribed inside a circle with a radius equal to the side of the hexagon, so take $r = 140$ pm.

The solution From eqn 7F.4, the energy separation between the states with $m_l = +1$ and $m_l = +2$ is

$$\Delta E = E_{+2} - E_{+1} = (4-1)\times\frac{(1.055\times10^{-34}\ \mathrm{J\,s})^2}{2\times(9.109\times10^{-31}\ \mathrm{kg})\times(1.40\times10^{-10}\ \mathrm{m})^2}$$
$$= 9.35\times10^{-19}\ \mathrm{J}$$

Therefore the minimum energy required to excite an electron is 0.935 aJ or 563 kJ mol^{-1}. This energy separation corresponds to an absorption frequency of 1410 THz (1 THz = 10^{12} Hz) and a wavelength of 213 nm; the experimental value for a transition of this kind is 260 nm. Such a crude model cannot be expected to give quantitative agreement, but the value is at least of the right order of magnitude.

Self-test 7F.1 Use the particle-on-a-ring model to calculate the energy separation between the $m_l = +3$ and $m_l = +4$ states of π electrons in coronene, $C_{24}H_{12}$ (**3**). Assume that the radius of the ring is three times the carbon–carbon bond length in benzene and that the electrons are confined to the periphery of the molecule.

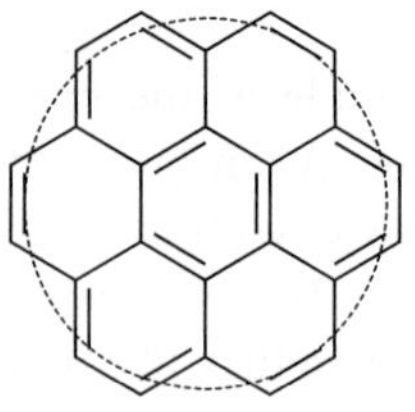

3 Coronene
(model ring dotted)

Answer: $\Delta E = 0.0147$ zJ or 8.83 J mol^{-1}

Exercises

E7F.1 Calculate the minimum excitation energy (i.e. the difference in energy between the first excited state and the ground state) of a proton constrained to rotate in a circle of radius 100 pm around a fixed point.

E7F.2 The wavefunction for the motion of a particle on a ring is of the form $\psi = Ne^{im_l\phi}$. Evaluate the normalization constant, N.

7F.2 Rotation in three dimensions

Now consider a particle of mass m that is free to move anywhere on the surface of a sphere of radius r. As for two-dimensional rotation the 'particle' may be an electron in motion around a nucleus or a point in a rotating molecule.

7F.2(a) The wavefunctions and energy levels

The potential energy of the particle is the same at all angles and for convenience may be taken to be zero (in an atom, the potential due to the nucleus is constant at a given radius). The Schrödinger equation is therefore

$$-\frac{\hbar^2}{2m}\left(\frac{\partial^2}{\partial x^2}+\frac{\partial^2}{\partial y^2}+\frac{\partial^2}{\partial z^2}\right)\psi = E\psi \qquad \text{(7F.8a)}$$

The sum of the three second derivatives is the 'laplacian' and denoted ∇^2 (and read 'del-squared'). Then the equation becomes

$$-\frac{\hbar^2}{2m}\nabla^2\psi = E\psi \qquad \text{(7F.8b)}$$

To take advantage of the symmetry of the problem it is appropriate to change to spherical polar coordinates (*The chemist's toolkit* 7F.2) when the laplacian becomes

$$\nabla^2 = \frac{1}{r}\frac{\partial^2}{\partial r^2}r + \frac{1}{r^2}\Lambda^2$$

The derivatives with respect to θ and ϕ have been collected in Λ^2, which is called the 'legendrian' and is given by

$$\Lambda^2 = \frac{1}{\sin^2\theta}\frac{\partial^2}{\partial\phi^2} + \frac{1}{\sin\theta}\frac{\partial}{\partial\theta}\sin\theta\frac{\partial}{\partial\theta}$$

In the present case, r is fixed, so the derivatives with respect to r in the laplacian can be ignored and only the term Λ^2/r^2 survives. The Schrödinger equation then becomes

$$-\frac{\hbar^2}{2m}\frac{1}{r^2}\Lambda^2\psi(\theta,\phi) = E\psi(\theta,\phi)$$

The term mr^2 in the denominator can be recognized as the moment of inertia, I, of the particle, so the Schrödinger equation is

$$-\frac{\hbar^2}{2I}\Lambda^2\psi(\theta,\phi) = E\psi(\theta,\phi) \quad \text{Schrödinger equation [particle on a sphere]} \quad (7F.9)$$

There are two cyclic boundary conditions to fulfil. The first is the same as for the two-dimensional case, where the wavefunction must join up on completing a circuit around the equator, as specified by the angle ϕ. The second is a similar requirement that the wavefunction must join up on encircling over the poles, as specified by the angle θ. Although θ sweeps from 0 only to π, as ϕ sweeps from 0 to 2π the entire spherical surface is covered, the wavefunctions must match everywhere, and the boundary condition on θ is effectively cyclic. These two conditions are illustrated in Fig. 7F.4. Once again, it can be shown that the need to satisfy them leads to the conclusion that both the energy and the angular momentum are quantized.

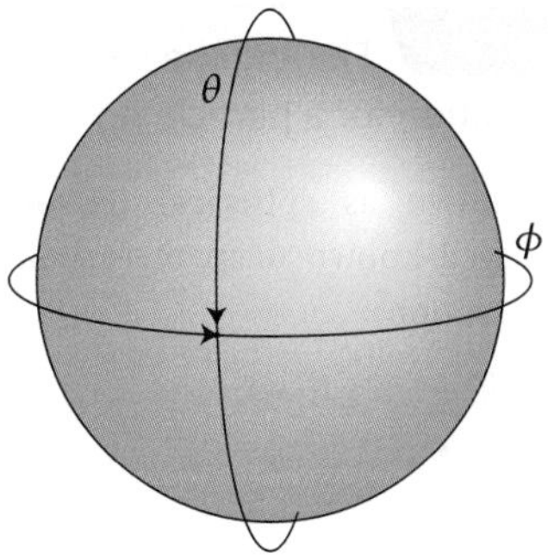

Figure 7F.4 The wavefunction of a particle on the surface of a sphere must reproduce itself on passing round the equator and over the poles. This double requirement leads to wavefunctions described by two quantum numbers.

The chemist's toolkit 7F.2 Spherical polar coordinates

The mathematics of systems with spherical symmetry (such as atoms) is often greatly simplified by using **spherical polar coordinates** (Sketch 1): r, the distance from the origin (the radius), θ, the colatitude, and ϕ, the azimuth. The ranges of these coordinates are (with angles in radians, Sketch 2): $0 \le r \le +\infty$, $0 \le \theta \le \pi$, $0 \le \phi \le 2\pi$.

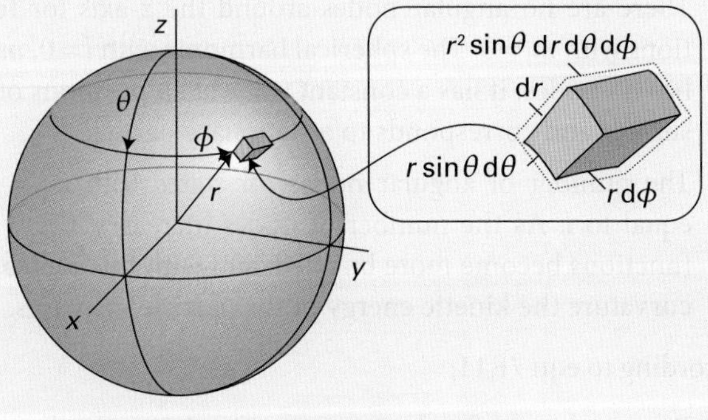

Sketch 1

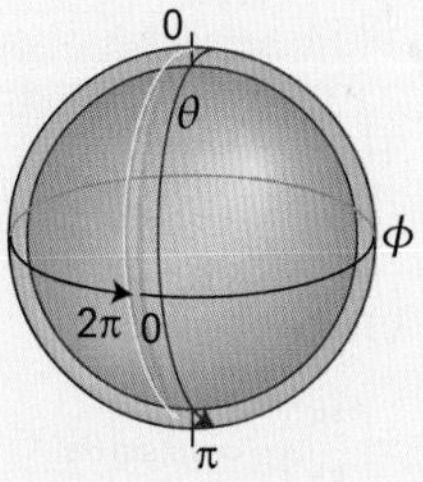

Sketch 2

Cartesian and polar coordinates are related by

$$x = r\sin\theta\cos\phi \quad y = r\sin\theta\sin\phi \quad z = r\cos\theta$$

The volume element in Cartesian coordinates is $d\tau = dxdydz$, and in spherical polar coordinates it becomes

$$d\tau = r^2\sin\theta\, drd\theta d\phi$$

An integral of a function $f(r,\theta,\phi)$ over all space in polar coordinates therefore has the form

$$\int f d\tau = \int_{r=0}^{\infty}\int_{\theta=0}^{\pi}\int_{\phi=0}^{2\pi} f(r,\theta,\phi)r^2\sin\theta\, drd\theta d\phi$$

How is that done? 7F.3 Finding the solutions of the Schrödinger equation for a particle on a sphere

The functions known as **spherical harmonics**, $Y_{l,m_l}(\theta,\phi)$ (Table 7F.1), are well known to mathematicians and are the solutions of the equation[2]

$$\Lambda^2 Y_{l,m_l}(\theta,\phi) = -l(l+1)Y_{l,m_l}(\theta,\phi), \quad l = 0, 1, 2, \ldots, \quad m_l = 0, \pm 1, \ldots, \pm l \qquad (7F.10)$$

These functions satisfy the two boundary conditions and are normalized in the sense that

$$\int_{\phi=0}^{2\pi}\int_{\theta=0}^{\pi} Y_{l,m_l}(\theta,\phi)^* Y_{l,m_l}(\theta,\phi)\sin\theta\,\mathrm{d}\theta\,\mathrm{d}\phi = 1$$

Step 1 *Show that the spherical harmonics solve the Schrödinger equation*

The spherical harmonics are substituted for the wavefunctions in eqn 7F.9 to give

$$-\frac{\hbar}{2I}\overbrace{\Lambda^2 Y_{l,m_l}(\theta,\phi)}^{-l(l+1)Y_{l,m_l}} = \overbrace{l(l+1)\frac{\hbar^2}{2I}}^{E}Y_{l,m_l}(\theta,\phi)$$

The spherical harmonics are therefore solutions of the Schrödinger equation with energies $E = l(l+1)\hbar^2/2I$. Note that the energies depend only on l but not on m_l.

Step 2 *Show that the wavefunctions are also eigenfunctions of the z-component of angular momentum*

The operator for the z-component of angular momentum is $\hat{l}_z = (\hbar/\mathrm{i})\partial/\partial\phi$. From Table 7F.1 note that each spherical harmonic is of the form $Y_{l,m_l}(\theta,\phi) = \mathrm{e}^{\mathrm{i}m_l\phi}f(\theta)$. It then follows that

$$\hat{l}_z Y_{l,m_l}(\theta,\phi) = \hat{l}_z \overbrace{\mathrm{e}^{\mathrm{i}m_l\phi}f(\theta)}^{Y_{l,m_l}(\theta,\phi)} = \frac{\hbar}{\mathrm{i}}\frac{\partial}{\partial\phi}\mathrm{e}^{\mathrm{i}m_l\phi}f(\theta) = m_l\hbar\times\mathrm{e}^{\mathrm{i}m_l\phi}f(\theta)$$
$$= m_l\hbar\times Y_{l,m_l}(\theta,\phi)$$

Therefore, the $Y_{l,m_l}(\theta,\phi)$ are eigenfunctions of $\hat{l}_z$ with eigenvalues $m_l\hbar$.

In summary, the $Y_{l,m_l}(\theta,\phi)$ are solutions to the Schrödinger equation for a particle on a sphere, with the corresponding energies given by

$$E_{l,m_l} = l(l+1)\frac{\hbar^2}{2I} \quad l = 0, 1, 2, \ldots \quad m_l = 0, \pm 1, \ldots \pm l \qquad (7F.11)$$

Energy levels [particle on a sphere]

Table 7F.1 The spherical harmonics

l	m_l	$Y_{l,m_l}(\theta,\phi)$
0	0	$\left(\frac{1}{4\pi}\right)^{1/2}$
1	0	$\left(\frac{3}{4\pi}\right)^{1/2}\cos\theta$
	±1	$\mp\left(\frac{3}{8\pi}\right)^{1/2}\sin\theta\,\mathrm{e}^{\pm\mathrm{i}\phi}$
2	0	$\left(\frac{5}{16\pi}\right)^{1/2}(3\cos^2\theta - 1)$
	±1	$\mp\left(\frac{15}{8\pi}\right)^{1/2}\cos\theta\sin\theta\,\mathrm{e}^{\pm\mathrm{i}\phi}$
	±2	$\left(\frac{15}{32\pi}\right)^{1/2}\sin^2\theta\,\mathrm{e}^{\pm2\mathrm{i}\phi}$
3	0	$\left(\frac{7}{16\pi}\right)^{1/2}(5\cos^3\theta - 3\cos\theta)$
	±1	$\mp\left(\frac{21}{64\pi}\right)^{1/2}(5\cos^2\theta - 1)\sin\theta\,\mathrm{e}^{\pm\mathrm{i}\phi}$
	±2	$\left(\frac{105}{32\pi}\right)^{1/2}\sin^2\theta\cos\theta\,\mathrm{e}^{\pm2\mathrm{i}\phi}$
	±3	$\mp\left(\frac{35}{64\pi}\right)^{1/2}\sin^3\theta\,\mathrm{e}^{\pm3\mathrm{i}\phi}$

[2] See the first section of *A deeper look* 7F.1, available to read in the e-book accompanying this text, for details of how the separation of variables procedure is used to find the form of the spherical harmonics.

The integers l and m_l are now identified as quantum numbers: l is the **orbital angular momentum quantum number** (the name is justified in the following section) and m_l is the **magnetic quantum number** (for reasons related to early spectroscopic observations in magnetic fields). The energy is specified by l alone, but for each value of l there are $2l+1$ values of m_l, so each energy level is $(2l + 1)$-fold degenerate. Each wavefunction is also an eigenfunction of $\hat{l}_z$ and therefore corresponds to a definite value, $m_l\hbar$, of the z-component of the angular momentum.

Figure 7F.5 shows a representation of the spherical harmonics for $l = 0$ to 4 and $m_l = 0$ in each case. The labels denoting the different signs of the wavefunction emphasizes the location of the angular nodes (the positions at which the wavefunction passes through zero). Note that:

- There are no angular nodes around the z-axis for functions with $m_l = 0$. The spherical harmonic with $l = 0$, $m_l = 0$ has no nodes: it has a constant value at all positions of the surface and corresponds to a stationary particle.
- The number of angular nodes for states with $m_l = 0$ is equal to l. As the number of nodes increases, the wavefunctions become more buckled, and with this increasing curvature the kinetic energy of the particle increases.

According to eqn 7F.11,

- Because l is confined to non-negative integral values, the energy is quantized.
- The energies are independent of the value of m_l, because the energy is independent of the direction and orientation of the rotational motion.

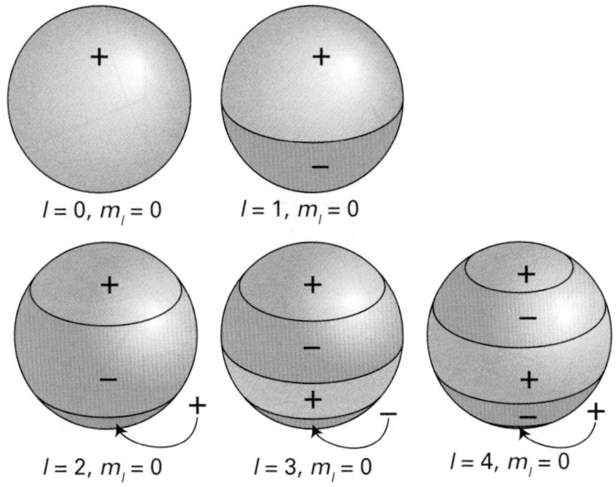

Figure 7F.5 A representation of the wavefunctions of a particle on the surface of a sphere that emphasizes the location of angular nodes. Note that the number of nodes increases as the value of l increases. All these wavefunctions correspond to $m_l = 0$; the wavefunctions do not depend on the angle ϕ.

- There are $2l + 1$ different wavefunctions (one for each value of m_l) that correspond to the same energy, so it follows that a level with quantum number l is $(2l + 1)$-fold degenerate.
- There is no zero-point energy: $E_{0,0} = 0$.

Example 7F.2 Using the rotational energy levels

The particle on a sphere is a good starting point for developing a model for the rotation of a diatomic molecule. Treat the rotation of $^1H^{127}I$ as a hydrogen atom rotating around a stationary I atom (this is a good first approximation as the I atom is so heavy it hardly moves). The bond length is 160 pm. Evaluate the energies and degeneracies of the lowest four rotational energy levels of $^1H^{127}I$. What is the frequency of the transition between the lowest two rotational levels?

Collect your thoughts The moment of inertia is $I = m_{^1H}R^2$ with $R = 160$ pm; the rotational energies are given in eqn 7F.11.[3] A transition between two rotational levels can be brought about by the emission or absorption of a photon with a frequency given by the Bohr frequency condition (Topic 7A, $h\nu = \Delta E$).

The solution The moment of inertia is

$$I = \overbrace{(1.675\times10^{-27}\ \text{kg})}^{m_{^1H}} \times \overbrace{(1.60\times10^{-10}\ \text{m})^2}^{R^2} = 4.29\times10^{-47}\ \text{kg m}^2$$

It follows that

$$\frac{\hbar^2}{2I} = \frac{(1.055\times10^{-34}\ \text{J s})^2}{2\times(4.29\times10^{-47}\ \text{kg m}^2)} = 1.30\times10^{-22}\ \text{J}$$

or 0.130 zJ. Draw up the following table:

l	E/zJ	Degeneracy
0	0	1
1	0.260	3
2	0.780	5
3	1.56	7

The energy separation between the two lowest rotational energy levels ($l = 0$ and 1) is 2.60×10^{-22} J, which corresponds to a photon of frequency

$$\nu = \frac{\Delta E}{h} = \frac{2.60\times10^{-22}\ \text{J}}{6.626\times10^{-34}\ \text{J s}} = 3.92\times10^{11}\ \overbrace{\text{s}^{-1}}^{\text{Hz}} = 392\ \text{GHz}$$

Comment. Radiation of this frequency belongs to the microwave region of the electromagnetic spectrum, so microwave spectroscopy is used to study molecular rotations (Topic 11B.2). Because the transition frequencies depend on the moment of inertia and frequencies can be measured with great precision, microwave spectroscopy is a very precise technique for the determination of bond lengths.

Self-test 7F.2 What is the frequency of the transition between the lowest two rotational levels in $^2H^{127}I$? (Assume that the bond length is the same as for $^1H^{127}I$ and that the iodine atom is stationary.)

Answer: 196 GHz

7F.2(b) Angular momentum

Orbital angular momentum is the angular momentum of a particle around a fixed point in space. It is represented by a vector $\boldsymbol{l}$, a quantity with both magnitude and direction. The components of $\boldsymbol{l}$ when it lies in a general orientation in three dimensions are

$$l_x = yp_z - zp_y \qquad l_y = zp_x - xp_z \qquad l_z = xp_y - yp_x \qquad \text{Components of } \boldsymbol{l} \quad (7F.12)$$

where p_x is the component of the linear momentum in the x-direction at any instant, and likewise p_y and p_z in the other directions. The square of the magnitude of the angular momentum vector is given by

$$l^2 = l_x^2 + l_y^2 + l_z^2 \qquad \text{Square of the magnitude of } \boldsymbol{l} \quad (7F.13)$$

Classically, the kinetic energy of a particle circulating on a ring is $E_k = J^2/2I$ (see *Energy: A first look*), where J, in the notation used there, is the magnitude of the angular momentum.

[3] As seen in Topic 11B.1, the rotational energy levels of a molecule are usually denoted by the angular momentum quantum number J rather than l. To keep this example in line with the current discussion, we use l to describe the angular momentum of the H atom.

By comparing this relation with eqn 7F.11, it follows that the square of the magnitude of the orbital angular momentum is $l(l+1)\hbar^2$, so the magnitude of the orbital angular momentum is

$$\text{Magnitude} = \{l(l+1)\}^{1/2}\hbar, \qquad l = 0, 1, 2, \ldots$$

Magnitude of orbital angular momentum (7F.14)

The spherical harmonics are also eigenfunctions of $\hat{l}_z$ and the corresponding eigenvalues give the z-component of the orbital angular momentum:

$$z\text{-Component} = m_l\hbar, \qquad m_l = 0, \pm1, \ldots \pm l$$

z-Component of orbital angular momentum (7F.15)

So, both the magnitude and the z-component of orbital angular momentum are quantized.

Brief illustration 7F.1

The lowest four rotational energy levels of any object rotating in three dimensions correspond to $l = 0, 1, 2, 3$. The following table can be constructed by using eqns 7F.14 and 7F.15.

l	Magnitude of orbital angular momentum/ $\hbar$	Degeneracy	z-Component of orbital angular momentum/ $\hbar$
0	0	1	0
1	$2^{1/2}$	3	$0, \pm1$
2	$6^{1/2}$	5	$0, \pm1, \pm2$
3	$12^{1/2}$	7	$0, \pm1, \pm2, \pm3$

7F.2(c) The vector model

The result that m_l is confined to the values $0, \pm1, \ldots \pm l$ for a given value of l means that the component of angular momentum about the z-axis—the contribution to the total angular momentum of rotation around that axis—may take only $2l + 1$ values. If the orbital angular momentum is represented by a vector of length $\{l(l+1)\}^{1/2}$, it follows that this vector must be oriented so that its projection on the z-axis is m_l and that it can have only $2l + 1$ orientations rather than the continuous range of orientations of a rotating classical body (Fig. 7F.6). The remarkable implication is that

The orientation of a rotating body is quantized.

The quantum mechanical result that a rotating body may not take up an arbitrary orientation with respect to some specified axis (such as an axis defined by the direction of an externally applied electric or magnetic field) is called **space quantization**.

The preceding discussion has referred to the z-component of orbital angular momentum and there has been no reference to the x- and y-components. The reason for this omission is found by examining the operators for the three components, each one being given by a term like that in eqn 7F.6a:

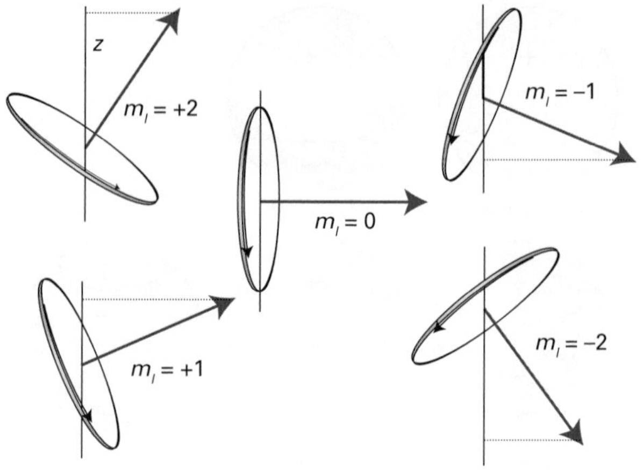

Figure 7F.6 The permitted orientations of angular momentum when $l = 2$. This representation is too specific because the azimuthal orientation of the vector (its angle around z) is indeterminate.

$$\hat{l}_x = \frac{\hbar}{\mathrm{i}}\left(y\frac{\partial}{\partial z} - z\frac{\partial}{\partial y}\right)$$
$$\hat{l}_y = \frac{\hbar}{\mathrm{i}}\left(z\frac{\partial}{\partial x} - x\frac{\partial}{\partial z}\right)$$
$$\hat{l}_z = \frac{\hbar}{\mathrm{i}}\left(x\frac{\partial}{\partial y} - y\frac{\partial}{\partial x}\right)$$

Orbital angular momentum operators (7F.16)

Each of these expressions can be derived in the same way as eqn 7F.6a by converting the classical expressions for the components of the angular momentum in eqn 7F.12 into their quantum mechanical operator equivalents. The commutation relations among the three operators (Problem P7F.9), are

$$[\hat{l}_x, \hat{l}_y] = \mathrm{i}\hbar\hat{l}_z \qquad [\hat{l}_y, \hat{l}_z] = \mathrm{i}\hbar\hat{l}_x \qquad [\hat{l}_z, \hat{l}_x] = \mathrm{i}\hbar\hat{l}_y$$

Angular momentum commutation relations (7F.17)

The three operators do not commute, so they represent complementary observables (Topic 7C). It is possible to have precise knowledge of only one of the components of the angular momentum, so if l_z is specified exactly (as in the preceding discussion), neither l_x nor l_y can be specified.

The operator for the square of the magnitude of the angular momentum in eqn 7F.6b commutes with all three components (Problem P7F.11):

$$[\hat{l}^2, \hat{l}_q] = 0 \qquad q = x, y, \text{ and } z$$

Commutators of angular momentum operators (7F.18)

It follows that both the square magnitude and one component, commonly the z-component, of the angular momentum can be specified simultaneously and precisely. The illustration in Fig. 7F.6, which is summarized in Fig. 7F.7(a), therefore gives a false impression of the state of the system, because it suggests

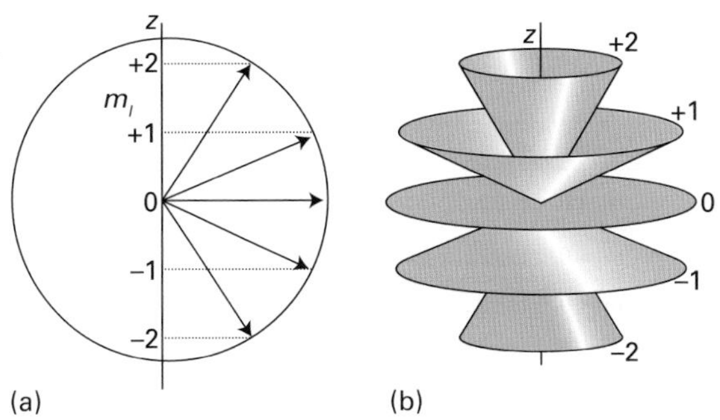

Figure 7F.7 (a) A summary of Fig. 7F.6. However, because the azimuthal angle of the vector around the z-axis is indeterminate, a better representation is as in (b), where each vector lies at an unspecified azimuthal angle on its cone.

definite values for the x- and y-components too. A better picture must reflect the impossibility of specifying l_x and l_y if l_z is known.

The **vector model** of angular momentum uses pictures like that in Fig. 7F.7(b). The cones are drawn with side $\{l(l+1)\}^{1/2}$ units, and represent the magnitude of the angular momentum. Each cone has a definite projection (of m_l units) on to the z-axis, representing the precisely known value of l_z. The projections of the vector on to the x- and y-axes, which give the values of l_x and l_y, are indefinite: the vector representing angular momentum can be thought of as lying with its tip on any point on the mouth of the cone. At this stage it should not be thought of as sweeping round the cone; that aspect of the model is added when the picture is allowed to convey more information (Topics 8B and 8C).

Brief illustration 7F.2

If the wavefunction of a rotating molecule is given by the spherical harmonic $Y_{3,+2}$ then the angular momentum can be represented by a cone

- with a side of length $12^{1/2}$ (representing the magnitude of $12^{1/2}\hbar$); and
- with a projection of +2 on the z-axis (representing the z-component of $+2\hbar$).

Exercises

E7F.3 The moment of inertia of a CH_4 molecule is 5.27×10^{-47} kg m^2. What is the minimum energy needed to start it rotating?

E7F.4 What is the magnitude of the angular momentum of a CH_4 molecule when it is rotating with its minimum energy?

E7F.5 Draw scale vector diagrams to represent the states (i) $l=1$, $m_l=+1$, (ii) $l=2$, $m_l=0$.

Checklist of concepts

☐ 1. The energy and angular momentum for a particle rotating in two- or three-dimensions are quantized; quantization results from the requirement that the wavefunction satisfies **cyclic boundary conditions**.

☐ 2. All energy levels of a particle rotating in two dimensions are doubly degenerate except for the lowest level ($m_l=0$).

☐ 3. A rotating particle does not have a zero-point energy.

☐ 4. It is impossible to specify simultaneously the angular momentum and location of a particle with arbitrary precision.

☐ 5. For a particle rotating in three dimensions, the cyclic boundary conditions imply that the magnitude and z-component of the angular momentum are quantized.

☐ 6. **Space quantization** refers to the quantum mechanical result that a rotating body may not take up an arbitrary orientation with respect to some specified axis.

☐ 7. The three components of orbital angular momentum are mutually complementary observables.

☐ 8. Only the magnitude of the orbital angular momentum and one of its components can be specified simultaneously with arbitrary precision.

☐ 9. In the **vector model** of angular momentum, the orbital angular momentum is represented by a cone with a side of length $\{l(l+1)\}^{1/2}$ and a projection of m_l on the z-axis. The vector can be thought of as lying with its tip on an indeterminate point on the mouth of the cone.

Checklist of equations

Property	Equation	Comment	Equation number
Wavefunction of particle on a ring	$\psi_{m_l}(\phi) = e^{im_l\phi}/(2\pi)^{1/2}$	$m_l = 0, \pm1, \pm2,\ldots$	7F.4
Energy of particle on a ring	$E_{m_l} = m_l^2\hbar^2/2I$	$m_l = 0, \pm1, \pm2,\ldots, I = mr^2$	7F.4
z-Component of angular momentum of particle on a ring	$m_l\hbar$	$m_l = 0, \pm1, \pm2,\ldots$	7F.7
Wavefunction of particle on a sphere	$\psi(\theta,\phi) = Y_{l,m_l}(\theta,\phi)$	Y is a spherical harmonic (Table 7F.1)	
Energy of particle on a sphere	$E_{l\,m_l} = l(l+1)\hbar^2/2I$	$l = 0, 1, 2,\ldots$	7F.11
Magnitude of angular momentum	$\{l(l+1)\}^{1/2}\hbar$	$l = 0, 1, 2,\ldots$	7F.14
z-Component of angular momentum	$m_l\hbar$	$m_l = 0, \pm1, \pm2,\ldots \pm l$	7F.15
Angular momentum commutation relations	$[\hat{l}_x,\hat{l}_y] = i\hbar\hat{l}_z,\ [\hat{l}_y,\hat{l}_z] = i\hbar\hat{l}_x,\ [\hat{l}_z,\hat{l}_x] = i\hbar\hat{l}_y$		7F.17
	$[\hat{l}^2,\hat{l}_q] = 0,\ q = x,y,z$		7F.18

FOCUS 7 Quantum theory

To test your understanding of this material, work through the *Exercises, Additional exercises, Discussion questions,* and *Problems* found throughout this Focus.

Selected solutions can be found at the end of this Focus in the e-book. Solutions to even-numbered questions are available online only to lecturers.

TOPIC 7A The origins of quantum mechanics

Discussion questions

D7A.1 Summarize the evidence that led to the introduction of quantum mechanics.

D7A.2 Explain how Planck's introduction of quantization accounted for the properties of black-body radiation.

D7A.3 Explain how Einstein's introduction of quantization accounted for the properties of heat capacities at low temperatures.

D7A.4 Explain the meaning and summarize the consequences of wave–particle duality.

Additional exercises

E7A.7 Calculate the wavelength and frequency at which the intensity of the radiation is a maximum for a black body at 2.7 K.

E7A.8 The intensity of the radiation from an object is found to be a maximum at 282 GHz (1 GHz = 10^9 Hz). Assuming that the object is a black body, calculate its temperature.

E7A.9 Calculate the molar heat capacity of a monatomic non-metallic solid at 500 K which is characterized by an Einstein temperature of 300 K. Express your result as a multiple of $3R$.

E7A.10 Calculate the energy of the quantum involved in the excitation of (i) an electronic oscillation of period 1.0 fs, (ii) a molecular vibration of period 10 fs, (iii) a pendulum of period 1.0 s. Express the results in joules and kilojoules per mole.

E7A.11 Calculate the energy of the quantum involved in the excitation of (i) an electronic oscillation of period 2.50 fs, (ii) a molecular vibration of period 2.21 fs, (iii) a balance wheel of period 1.0 ms. Express the results in joules and kilojoules per mole.

E7A.12 Calculate the energy of a photon and the energy per mole of photons for radiation of wavelength (i) 600 nm (red), (ii) 550 nm (yellow), (iii) 400 nm (blue).

E7A.13 Calculate the energy of a photon and the energy per mole of photons for radiation of wavelength (i) 200 nm (ultraviolet), (ii) 150 pm (X-ray), (iii) 1.00 cm (microwave).

E7A.14 Calculate the speed to which a stationary H atom would be accelerated if it absorbed a photon of wavelength (i) 600 nm (red), (ii) 550 nm (yellow), (iii) 400 nm (blue).

E7A.15 Calculate the speed to which a stationary ^{4}He atom (mass $4.0026m_u$) would be accelerated if it absorbed a photon of wavelength (i) 200 nm (ultraviolet), (ii) 150 pm (X-ray), (iii) 1.00 cm (microwave).

E7A.16 A laser used to read CDs emits red light of wavelength 700 nm. How many photons does it emit each second if its power is (i) 0.10 W, (ii) 1.0 W?

E7A.17 The work function of metallic rubidium is 2.09 eV. Calculate the kinetic energy and the speed of the electrons ejected by light of wavelength (i) 650 nm, (ii) 195 nm.

E7A.18 A glow-worm of mass 5.0 g emits red light (650 nm) with a power of 0.10 W entirely in the backward direction. To what speed will it have accelerated after 10 y if released into free space and assumed to live?

E7A.19 A photon-powered spacecraft of mass 10.0 kg emits radiation of wavelength 225 nm with a power of 1.50 kW entirely in the backward direction. To what speed will it have accelerated after 10.0 y if released into free space?

E7A.20 To what speed must a proton be accelerated from rest for it to have a de Broglie wavelength of 100 pm? What accelerating potential difference is needed?

E7A.21 To what speed must an electron be accelerated for it to have a de Broglie wavelength of 3.0 cm?

E7A.22 To what speed must a proton be accelerated for it to have a de Broglie wavelength of 3.0 cm?

E7A.23 The 'fine-structure constant', α, plays a special role in the structure of matter; its approximate value is 1/137. What is the de Broglie wavelength of an electron travelling at αc, where c is the speed of light?

E7A.24 Calculate the linear momentum of photons of wavelength 350 nm. At what speed does a hydrogen molecule need to travel for it to have the same linear momentum?

E7A.25 Calculate the de Broglie wavelength of (i) a mass of 1.0 g travelling at 1.0 cm s^{-1}; (ii) the same, travelling at 100 km s^{-1}; (iii) a He atom travelling at 1000 m s^{-1} (a typical speed at room temperature).

E7A.26 Calculate the de Broglie wavelength of an electron accelerated from rest through a potential difference of (i) 100 V; (ii) 1.0 kV; (iii) 100 kV.

Problems

P7A.1 Calculate the energy density in the range 650 nm to 655 nm inside a cavity at (a) 25 °C, (b) 3000 °C. For this relatively small range of wavelength it is acceptable to approximate the integral of the energy spectral density $\rho(\lambda, T)$ between λ_1 and λ_2 by $\rho(\lambda,T) \times (\lambda_2 - \lambda_1)$.

P7A.2 Calculate the energy density in the range 1000 cm^{-1} to 1010 cm^{-1} inside a cavity at (a) 25 °C, (b) 4 K.

P7A.3 Demonstrate that the Planck distribution reduces to the Rayleigh–Jeans law at long wavelengths.

P7A.4 The wavelength λ_{max} at which the Planck distribution is a maximum can be found by solving $d\rho(\lambda, T)/dT = 0$. Differentiate $\rho(\lambda,T)$ with respect to T and show that the condition for the maximum can be expressed as $xe^x - 5(e^x - 1) = 0$, where $x = hc/\lambda kT$. There are no analytical solutions to this equation, but a numerical approach gives $x = 4.965$ as a solution. Use this result to confirm *Wien's law*, that $\lambda_{max}T$ is a constant, deduce an expression for the constant, and compare it to the value quoted in the text.

P7A.5 For a black body, the temperature and the wavelength of the emission maximum, λ_{max}, are related by Wien's law, $\lambda_{max}T = hc/4.965k$; see Problem 7A.4. Values of λ_{max} from a small pinhole in an electrically heated container were determined at a series of temperatures, and the results are given below. Deduce the value of Planck's constant.

θ/°C	1000	1500	2000	2500	3000	3500
λ_{max}/nm	2181	1600	1240	1035	878	763

P7A.6‡ Solar energy strikes the top of the Earth's atmosphere at 1361 W m^{-2}. About 30 per cent of this energy is reflected directly back into space. The Earth–atmosphere system absorbs the remaining energy and re-radiates it into space as black-body radiation at $5.672 \times 10^{-8}(T/\text{K})^4$ W m^{-2}, where T is the temperature. Assuming that the arrangement has come to equilibrium, what is the average black-body temperature of the Earth? Calculate the wavelength at which the black-body radiation from the Earth is at a maximum.

P7A.7 The total energy density of black-body radiation is found by integrating the energy spectral density over all wavelengths, eqn 7A.2. Evaluate this integral for the Planck distribution. This is most easily done by making the substitution $x = hc/\lambda kT$; you will need the integral $\int_0^\infty \{x^3/(e^x - 1)\}dx = \pi^4/15$. Hence deduce the *Stefan–Boltzmann law* that the total energy density of black-body radiation is proportional to T^4, and find the constant of proportionality.

P7A.8‡ Prior to Planck's derivation of the distribution law for black-body radiation, Wien found empirically a closely related distribution function which is very nearly but not exactly in agreement with the experimental results, namely $\rho(\lambda,T) = (a/\lambda^5)e^{-b/\lambda kT}$. This formula shows small deviations from Planck's at long wavelengths. (a) Find a form of the Planck distribution which is appropriate for short wavelengths (*Hint:* consider the behaviour of the term $e^{hc/\lambda kT} - 1$ in this limit). (b) Compare your expression from part (a) with Wien's empirical formula and hence determine the constants a and b. (c) Integrate Wien's empirical expression for $\rho(\lambda,T)$ over all wavelengths and show that the result is consistent with the Stefan–Boltzmann law (*Hint:* to compute the integral use the substitution $x = hc/\lambda kT$ and then refer to the *Resource section*). (d) Show that Wien's empirical expression is consistent with Wien's law.

P7A.9‡ The temperature of the Sun's surface is approximately 5800 K. On the assumption that the human eye evolved to be most sensitive at the wavelength of light corresponding to the maximum in the Sun's radiant energy distribution, identify the colour of light to which the eye is the most sensitive.

P7A.10 The Einstein frequency is often expressed in terms of an equivalent temperature θ_E, where $\theta_E = h\nu/k$. Confirm that θ_E has the dimensions of temperature, and express the criterion for the validity of the high-temperature form of the Einstein equation in terms of θ_E. Evaluate θ_E for (a) diamond, for which $\nu = 46.5$ THz, and (b) for copper, for which $\nu = 7.15$ THz. Use these values to calculate the molar heat capacity of each substance at 25 °C, expressing your answers as multiples of $3R$.

TOPIC 7B Wavefunctions

Discussion questions

D7B.1 Describe how a wavefunction summarizes the dynamical properties of a system and how those properties may be predicted.

D7B.2 Explain the relation between probability amplitude, probability density, and probability.

D7B.3 Identify the constraints that the Born interpretation puts on acceptable wavefunctions.

Additional exercises

E7B.3 A possible wavefunction for an electron in a region of length L is $\sin(3\pi x/L)$. Normalize this wavefunction (to 1).

E7B.4 Normalize (to 1) the wavefunction e^{-ax^2} in the range $-\infty \le x \le \infty$, with $a > 0$. Refer to the *Resource section* for the necessary integral.

E7B.5 Normalize (to 1) the wavefunction e^{-ax} in the range $0 \le x \le \infty$, with $a > 0$.

E7B.6 Which of the following functions can be normalized (in all cases the range for x is from $x = -\infty$ to ∞, and a is a positive constant): (i) e^{-ax^2}; (ii) e^{-ax}. Which of these functions are acceptable as wavefunctions?

E7B.7 Which of the following functions can be normalized (in all cases the range for x is from $x = -\infty$ to ∞, and a is a positive constant): (i) $\sin(ax)$; (ii) $\cos(ax)\,e^{-x^2}$? Which of these functions are acceptable as wavefunctions?

E7B.8 A possible wavefunction for an electron confined to a region of length L is $\sin(3\pi x/L)$. What is the probability of finding the electron in the range dx at $x = L/6$?

E7B.9 A possible wavefunction for an electron confined to a region of length L is $\sin(2\pi x/L)$. What is the probability of finding the electron between $x = L/4$ and $x = L/2$?

E7B.10 A possible wavefunction for an electron confined to a region of length L is $\sin(3\pi x/L)$. What is the probability of finding the electron between $x = 0$ and $x = L/3$?

E7B.11 What are the dimensions of a wavefunction that describes a particle free to move in both the x and y directions?

‡ These problems were supplied by Charles Trapp and Carmen Giunta.

E7B.12 The wavefunction for a particle free to move between $x = 0$ and $x = L$ is $(2/L)^{1/2}\sin(\pi x/L)$; confirm that this wavefunction has the expected dimensions.

E7B.13 Imagine a particle confined to move on the circumference of a circle ('a particle on a ring'), such that its position can be described by an angle ϕ in the range 0–2π. Which of the following wavefunctions would be acceptable for such a particle? In each case, give your reasons for accepting or rejecting each function. (i) $\cos\phi$; (ii) $\sin\phi$; (iii) $\cos(0.9\phi)$.

E7B.14 A possible wavefunction for an electron confined to a region of length L is $\sin(3\pi x/L)$. At what value or values of x is the probability density a maximum? Locate the position or positions of any nodes in the wavefunction. You need consider only the range $x = 0$ to $x = L$.

Problems

P7B.1 Imagine a particle confined to move on the circumference of a circle ('a particle on a ring'), such that its position can be described by an angle ϕ in the range 0 to 2π. Find the normalizing factor for the wavefunctions: (a) $e^{i\phi}$ and (b) $e^{im_l\phi}$, where m_l is an integer.

P7B.2 Imagine a particle confined to move on the circumference of a circle ('a particle on a ring'), such that its position can be described by an angle ϕ in the range 0 to 2π. Find the normalizing factor for the wavefunctions: (a) $\cos\phi$; (b) $\sin m_l\phi$, where m_l is an integer.

P7B.3 A particle is confined to a two-dimensional region with $0 \le x \le L_x$ and $0 \le y \le L_y$. Normalize (to 1) the functions (a) $\sin(\pi x/L_x)\sin(\pi y/L_y)$ and (b) $\sin(\pi x/L)\sin(\pi y/L)$ for the case $L_x = L_y = L$.

P7B.4 Normalize (to 1) the wavefunction $e^{-ax^2}e^{-by^2}$ for a system in two dimensions with $a > 0$ and $b > 0$, and with x and y both allowed to range from $-\infty$ to ∞. Refer to the *Resource section* for relevant integrals.

P7B.5 Suppose that in a certain system a particle free to move along one dimension (with $0 \le x \le \infty$) is described by the unnormalized wavefunction $\psi(x)e^{-ax}$ with $a = 2\ \text{m}^{-1}$. What is the probability of finding the particle at a distance $x \ge 1$ m? (*Hint:* You will need to normalize the wavefunction before using it to calculate the probability.)

P7B.6 Suppose that in a certain system a particle free to move along x (without constraint) is described by the unnormalized wavefunction $\psi(x) = e^{-ax^2}$ with $a = 0.2\ \text{m}^{-2}$. Use mathematical software to calculate the probability of finding the particle at $x \ge 1$ m.

P7B.7 A normalized wavefunction for a particle confined between 0 and L in the x direction is $\psi = (2/L)^{1/2}\sin(\pi x/L)$. Suppose that $L = 10.0$ nm. Calculate the probability that the particle is (a) between $x = 4.95$ nm and 5.05 nm, (b) between $x = 1.95$ nm and 2.05 nm, (c) between $x = 9.90$ nm and 10.00 nm, (d) between $x = 5.00$ nm and 10.00 nm.

P7B.8 A normalized wavefunction for a particle confined between 0 and L in the x direction, and between 0 and L in the y direction (that is, to a square of side L) is $\psi = (2/L)\sin(\pi x/L)\sin(\pi y/L)$. The probability of finding the particle between x_1 and x_2 along x, and between y_1 and y_2 along y is

$$P = \int_{y=y_1}^{y=y_2}\int_{x=x_1}^{x=x_2} \psi^2 \mathrm{d}x\,\mathrm{d}y$$

Calculate the probability that the particle is: (a) between $x = 0$ and $x = L/2$, $y = 0$ and $y = L/2$ (i.e. in the bottom left-hand quarter of the square); (b) between $x = L/4$ and $x = 3L/4$, $y = L/4$ and $y = 3L/4$ (i.e. a square of side $L/2$ centred on $x = y = L/2$).

P7B.9 The normalized ground-state wavefunction of a hydrogen atom is $\psi(r) = (1/\pi a_0^3)^{1/2}e^{-r/a_0}$ where $a_0 = 53$ pm (the Bohr radius) and r is the distance from the nucleus. (a) Calculate the probability that the electron will be found somewhere within a small sphere of radius 1.0 pm centred on the nucleus. (b) Now suppose that the same sphere is located at $r = a_0$. What is the probability that the electron is inside it? You may approximate the probability of being in a small volume δV at position r by $\psi(r)^2\delta V$.

P7B.10 Atoms in a chemical bond vibrate around the equilibrium bond length. An atom undergoing vibrational motion is described by the wavefunction $\psi(x) = Ne^{-x^2/2a^2}$, where a is a constant and $-\infty \le x \le \infty$. (a) Find the normalizing factor N. (b) Use mathematical software to calculate the probability of finding the particle in the range $-a \le x \le a$ (the result will be expressed in terms of the 'error function', erf(x)).

P7B.11 Atoms in a chemical bond vibrate around the equilibrium bond length. Suppose that an atom undergoing vibrational motion is described by the wavefunction $\psi(x) = Nxe^{-x^2/2a^2}$. Where is the most probable location of the atom?

TOPIC 7C Operators and observables

Discussion questions

D7C.1 How may the curvature of a wavefunction be interpreted?

D7C.2 Describe the relation between operators and observables in quantum mechanics.

D7C.3 Use the properties of wavepackets to account for the uncertainty relation between position and linear momentum.

Additional exercises

E7C.8 Construct the potential energy operator of a particle with potential energy $V(x) = D_e(1 - e^{-ax})^2$, where D_e and a are constants.

E7C.9 Identify which of the following functions are eigenfunctions of the operator $\mathrm{d}^2/\mathrm{d}x^2$: (i) $\cos(kx)$; (ii) e^{ikx}, (iii) kx, (iv) e^{-ax^2}. Give the corresponding eigenvalue where appropriate.

E7C.10 Functions of the form $\sin(n\pi x/L)$, where $n = 1, 2, 3\ldots$, are wavefunctions in a region of length L (between $x = 0$ and $x = L$). Show that the wavefunctions with $n = 2$ and 4 are orthogonal.

E7C.11 Functions of the form $\cos(n\pi x/L)$, where $n = 1, 3, 5\ldots$, can be used to model the wavefunctions of particles confined to the region between $x = -L/2$ and $x = +L/2$. The integration is limited to the range $-L/2$ to $+L/2$ because the wavefunction is zero outside this range. Show that the wavefunctions are orthogonal for $n = 1$ and 3. You will find the necessary integral in the *Resource section*.

E7C.12 For the same system as in Exercise E7C.11 show that the wavefunctions with $n = 3$ and 5 are orthogonal.

E7C.13 Imagine a particle confined to move on the circumference of a circle ('a particle on a ring'), such that its position can be described by an angle ϕ in the range 0–2π. The wavefunctions for this system are of the form $\psi_m(\phi) = e^{im_l\phi}$ with m_l an integer. Show that the wavefunctions with $m_l = +1$ and $+2$ are orthogonal. (*Hint:* Note that $(e^{ix})^* = e^{-ix}$, and that $e^{ix} = \cos x + i \sin x$.)

E7C.14 For the same system as in Exercise E7C.13 show that the wavefunctions with $m_l = +1$ and -2 are orthogonal.

E7C.15 Find $\langle x \rangle$ when the wavefunction is $\psi(x) = (2/L)^{1/2}\sin(\pi x/L)$ for a particle confined between $x = 0$ and $x = L$.

E7C.16 Find $\langle p_x \rangle$ for the case where the normalized wavefunction is $\psi(x) = (2/L)^{1/2}\sin(\pi x/L)$ for a particle confined between $x = 0$ and $x = L$.

E7C.17 Imagine a particle confined to move on the circumference of a circle ('a particle on a ring'), such that its position can be described by an angle ϕ in the range 0–2π. The *normalized* wavefunctions for this system are of the form $\psi_{m_l}(\phi) = (1/2\pi)^{1/2} e^{im_l\phi}$ with m_l an integer. The expectation value of a quantity represented by the operator $\hat{\Omega}$ is given by

$$\Omega_{m_l} = \int_0^{2\pi} \psi^*_{m_l}(\phi)\hat{\Omega}\psi_{m_l}(\phi)\mathrm{d}\phi$$

where $\psi_{m_l}(\phi)$ are the normalized wavefunctions. Compute the expectation value of the position, specified by the angle ϕ, for the case $m_l = +1$, and then for the general case of integer m_l.

E7C.18 For the system described in Exercise E7C.17, evaluate the expectation value of the angular momentum represented by the operator $(\hbar/i)\mathrm{d}/\mathrm{d}\phi$ for the case $m_l = +1$, and then for the general case of integer m_l.

E7C.19 An electron is confined to a linear region with a length of the same order as the diameter of an atom (about 100 pm). Calculate the minimum uncertainties in its position and speed.

E7C.20 The speed of a certain electron is 995 km s^{-1}. If the uncertainty in its momentum is to be reduced to 0.0010 per cent, what uncertainty in its location must be tolerated?

Problems

P7C.1 Identify which of the following functions are eigenfunctions of the inversion operator $\hat{i}$, which has the effect of making the replacement $x \to -x$: (a) $x^3 - kx$, (b) $\cos kx$, (c) $x^2 + 3x - 1$. Identify the eigenvalue of $\hat{i}$ when relevant.

P7C.2 An electron in a one-dimensional region of length L is described by the wavefunction $\psi_n(x) = \sin(n\pi x/L)$, where $n = 1, 2,\ldots$, in the range $x = 0$ to $x = L$; outside this range the wavefunction is zero. The orthogonality of these wavefunctions is confirmed by considering the integral

$$I = \int_0^L \sin(n\pi x/L)\sin(m\pi x/L)\mathrm{d}x$$

(a) Use the identity $\sin A \sin B = \frac{1}{2}\{\cos(A - B) - \cos(A + B)\}$ to rewrite the integrand as a sum of two terms. (b) Consider the case $n = 2$, $m = 1$, and make separate sketch graphs of the two terms identified in part (a) in the range $x = 0$ to $x = L$. (c) Make use of the properties of the cosine function to argue that the area enclosed between the curves and the x axis is zero in both cases, and hence that the integral is zero. (d) Generalize the argument for the case of arbitrary n and m ($n \neq m$).

P7C.3 Confirm that the kinetic energy operator, $-(\hbar^2/2m)\mathrm{d}^2/\mathrm{d}x^2$, is hermitian. (*Hint:* Use the same approach as in the text, but because a second derivative is involved you will need to integrate by parts twice; you may assume that the derivatives of the wavefunctions go to zero as $x \to \pm\infty$.)

P7C.4 The operator corresponding to the angular momentum of a particle is $(\hbar/i)\mathrm{d}/\mathrm{d}\phi$, where ϕ is an angle. For such a system the criterion for an operator $\hat{\Omega}$ to be hermitian is

$$\int_0^{2\pi} \psi_i^*(\phi)\hat{\Omega}\psi_j(\phi)\mathrm{d}\phi = \left[\int_0^{2\pi} \psi_j^*(\phi)\hat{\Omega}\psi_i(\phi)\mathrm{d}\phi\right]^*$$

Show that $(\hbar/i)\mathrm{d}/\mathrm{d}\phi$ is a hermitian operator. (*Hint:* Use the same approach as in the text; recall that the wavefunction must be single-valued, so $\psi_i(\phi) = \psi_i(\phi + 2\pi)$.)

P7C.5 (a) Show that the sum of two hermitian operators $\hat{A}$ and $\hat{B}$ is also a hermitian operator. (*Hint:* Start by separating the appropriate integral into two terms, and then apply the definition of hermiticity.) (b) Show that the product of a hermitian operator with itself is also a hermitian operator. Start by considering the integral

$$I = \int \psi_i^* \hat{\Omega}\hat{\Omega}\psi_j \,\mathrm{d}\tau$$

Recall that $\hat{\Omega}\psi_j$ is simply another function, so the integral can be thought of as

$$I = \int \psi_i^* \hat{\Omega} \overbrace{(\hat{\Omega}\psi_j)}^{\text{a function}} \,\mathrm{d}\tau$$

Now apply the definition of hermiticity and complete the proof.

P7C.6 Calculate the expectation value of the linear momentum p_x of a particle described by the following normalized wavefunctions (in each case N is the appropriate normalizing factor, which you do not need to find): (a) Ne^{ikx}, (b) $N\cos kx$, (c) Ne^{-ax^2}, where in each one x ranges from $-\infty$ to $+\infty$.

P7C.7 A particle freely moving in one dimension x with $0 \le x \le \infty$ is in a state described by the normalized wavefunction $\psi(x) = a^{1/2}e^{-ax/2}$, where a is a constant. Evaluate the expectation value of the position operator.

P7C.8 The normalized wavefunction of an electron in a linear accelerator is $\psi = (\cos\chi)e^{ikx} + (\sin\chi)e^{-ikx}$, where χ (chi) is a parameter. (a) What is the probability that the electron will be found with a linear momentum (a) $+k\hbar$, (b) $-k\hbar$? (c) What form would the wavefunction have if it were 90 per cent certain that the electron had linear momentum $+k\hbar$? (d) Evaluate the kinetic energy of the electron.

P7C.9 (a) Show that the expectation value of a hermitian operator is real. (*Hint:* Start from the definition of the expectation value and then apply the definition of hermiticity to it.) (b) Show that the expectation value of an operator that can be written as the square of a hermitian operator is positive. (*Hint:* Start from the definition of the expectation value for the operator $\hat{\Omega}\hat{\Omega}$; recognize that $\hat{\Omega}\psi$ is a function, and then apply the definition of hermiticity.)

P7C.10 Suppose the wavefunction of an electron in a one-dimensional region is a linear combination of $\cos nx$ functions. (a) Use mathematical software or a spreadsheet to construct superpositions of cosine functions as

$$\psi(x) = \frac{1}{N}\sum_{k=1}^{N}\cos(k\pi x)$$

where the constant $1/N$ (not a normalization constant) is introduced to keep the superpositions with the same overall magnitude. Set $x = 0$ at the centre of the screen and build the superposition there; consider the range $x = -1$ to $+1$. (b) Explore how the probability density $\psi^2(x)$ changes with the value of N. (c) Evaluate the root-mean-square location of the packet, $\langle x^2 \rangle^{1/2}$. (d) Determine the probability that a given momentum will be observed.

P7C.11 A particle is in a state described by the normalized wavefunction $\psi(x) = (2a/\pi)^{1/4}e^{-ax^2}$, where a is a constant and $-\infty \le x \le \infty$. (a) Calculate the expectation values $\langle x \rangle$, $\langle x^2 \rangle$, $\langle p_x \rangle$, and $\langle p_x^2 \rangle$; the necessary integrals will be found in the *Resource section*. (b) Use these results to calculate $\Delta p_x = \{\langle p_x^2 \rangle - \langle p_x \rangle^2\}^{1/2}$ and $\Delta x = \{\langle x^2 \rangle - \langle x \rangle^2\}^{1/2}$. (c) Hence verify that the value of the product $\Delta p_x \Delta x$ is consistent with the predictions from the uncertainty principle.

P7C.12 A particle is in a state described by the normalized wavefunction $\psi(x) = a^{1/2}e^{-ax/2}$, where a is a constant and $0 \le x \le \infty$. Evaluate the expectation value of the commutator of the position and momentum operators.

P7C.13 Evaluate the commutators of the operators (a) d/dx and $1/x$, (b) d/dx and x^2. (*Hint:* Follow the procedure in the text by considering, for case (a), $(d/dx)(1/x)\psi$ and $(1/x)(d/dx)\psi$; recall that ψ is a function of x, so it will be necessary to use the product rule to evaluate some of the derivatives.)

P7C.14 Evaluate the commutators of the operators $\hat{a}$ and $\hat{a}^+$ where $\hat{a} = (\hat{x} + i\hat{p}_x)/2^{1/2}$ and $\hat{a}^+ = (\hat{x} - i\hat{p}_x)/2^{1/2}$.

P7C.15 Evaluate the commutators (a) $[\hat{H},\hat{p}_x]$ and (b) $[\hat{H},\hat{x}]$ where $\hat{H} = \hat{p}_x^2/2m + \hat{V}(x)$. Choose (i) $V(x) = V_0$, a constant, (ii) $V(x) = \frac{1}{2}k_f x^2$. (*Hint:* See the hint for *Problem* P7C.13.)

TOPIC 7D Translational motion

Discussion questions

D7D.1 Explain the physical origin of quantization for a particle confined to the interior of a one-dimensional box.

D7D.2 Describe the features of the solution of the particle in a one-dimensional box that appear in the solutions of the particle in two- and three-dimensional boxes. What feature occurs in the two- and three-dimensional box that does not occur in the one-dimensional box?

D7D.3 Explain the physical origin of quantum mechanical tunnelling. Why is tunnelling more likely to contribute to the mechanisms of electron transfer and proton transfer processes than to mechanisms of group transfer reactions, such as AB + C → A + BC (where A, B, and C are large molecular groups)?

Additional exercises

E7D.10 Evaluate the linear momentum and kinetic energy of a free proton described by the wavefunction e^{-ikx} with $k = 5\ \text{nm}^{-1}$.

E7D.11 Write the wavefunction for a particle of mass 1.0 g travelling to the right at $10\ \text{m s}^{-1}$.

E7D.12 Calculate the energy separations in joules, kilojoules per mole, electronvolts, and reciprocal centimetres between the levels (i) $n = 3$ and $n = 2$, (ii) $n = 7$ and $n = 6$ of an electron in a box of length 1.50 nm.

E7D.13 For a particle in a one-dimensional box, show that the wavefunctions ψ_1 and ψ_2 are orthogonal. The necessary integrals will be found in the *Resource section*.

E7D.14 For a particle in a one-dimensional box, show that the wavefunctions ψ_1 and ψ_3 are orthogonal.

E7D.15 Calculate the probability that a particle will be found between $0.65L$ and $0.67L$ in a box of length L for the case where the wavefunction is (i) ψ_1, (ii) ψ_2. You may make the same approximation as in *Exercise* E7D.4.

E7D.16 Without evaluating any integrals, state the value of the expectation value of x for a particle in a box of length L for the case where the wavefunction has $n = 2$. Explain how you arrived at your answer.

E7D.17 For a particle in a box of length L sketch the wavefunction corresponding to the state with $n = 1$ and on the same graph sketch the corresponding probability density. Without evaluating any integrals, explain why for this wavefunction the expectation value of x^2 is not equal to $(L/2)^2$.

E7D.18 For a particle in a box of length L sketch the wavefunction corresponding to the state with $n = 1$ and on the same graph sketch the corresponding probability density. For this wavefunction, explain whether you would expect the expectation value of x^2 to be greater than or less than the square of the expectation value of x.

E7D.19 An electron is confined to a cubic box of side of L. What would be the length of the box such that the zero-point energy of the electron is equal to its rest mass energy, $m_e c^2$? Express your answer in terms of the parameter $\lambda_C = h/m_e c$, the 'Compton wavelength' of the electron.

E7D.20 For a particle in a box of length L and in the state with $n = 3$, at what positions is the probability density a maximum? At what positions is the probability density zero?

E7D.21 For a particle in a box of length L and in the state with $n = 5$, at what positions is the probability density a maximum? At what positions is the probability density a minimum?

E7D.22 For a particle in a box of length L, write the expression for the energy levels, E_n, and then write a similar expression E'_n for the energy levels when the length of the box has increased to $1.1L$ (that is, an increase by 10 per cent). Calculate $(E'_n - E_n)/E_n$, the fractional change in the energy that results from extending the box.

E7D.23 For a particle in a cubical box of side L, write the expression for the energy levels, E_{n_1,n_2,n_3}, and then write a similar expression E'_{n_1,n_2,n_3} for the energy levels when the side of the box has decreased to $0.9L$ (that is, a decrease by 10 per cent). Calculate $(E'_{n_1,n_2,n_3} - E_{n_1,n_2,n_3})/E_{n_1,n_2,n_3}$, the fractional change in the energy that results from expanding the box.

E7D.24 Find an expression for the value of n of a particle of mass m in a one-dimensional box of length L such that the separation between neighbouring levels is equal to the mean energy of thermal motion $\left(\frac{1}{2}kT\right)$. Calculate the value of n for the case of a helium atom in a box of length 1 cm at 298 K.

E7D.25 Find an expression for the value of n of a particle of mass m in a one-dimensional box of length L such that the energy of the level is equal to the mean energy of thermal motion $\left(\frac{1}{2}kT\right)$. Calculate the value of n for the case of an argon atom in a box of length 0.1 cm at 298 K.

E7D.26 For a particle in a square box of side L, at what position (or positions) is the probability density a maximum if the wavefunction has $n_1 = 1$, $n_2 = 3$? Also, describe the position of any node or nodes in the wavefunction.

E7D.27 For a particle in a rectangular box with sides of length $L_1 = L$ and $L_2 = 2L$, find a state that is degenerate with the state $n_1 = n_2 = 2$. (*Hint:* You will need to experiment with some possible values of n_1 and n_2.) Is this degeneracy associated with symmetry?

E7D.28 For a particle in a rectangular box with sides of length $L_1 = L$ and $L_2 = 2L$, find a state that is degenerate with the state $n_1 = 2$, $n_2 = 8$. Would you expect there to be any degenerate states for a rectangular box with $L_1 = L$ and $L_2 = \sqrt{2}L$? Explain your reasoning.

E7D.29 Consider a particle in a cubic box. What is the degeneracy of the level that has an energy $\frac{14}{3}$ times that of the lowest level?

E7D.30 Suppose that a proton of an acidic hydrogen atom is confined to an acid that can be represented by a barrier of height 2.0 eV and length 100 pm. Calculate the probability that a proton with energy 1.5 eV can escape from the acid.

Problems

P7D.1 Calculate the separation between the two lowest levels for an O_2 molecule in a one-dimensional container of length 5.0 cm. At what value of n does the energy of the molecule reach $\frac{1}{2}kT$ at 300 K, and what is the separation of this level from the one immediately below?

P7D.2 A nitrogen molecule is confined in a cubic box of volume 1.00 m^3. (i) Assuming that the molecule has an energy equal to $\frac{3}{2}kT$ at $T = 300$ K, what is the value of $n = (n_x^2 + n_y^2 + n_z^2)^{1/2}$ for this molecule? (ii) What is the energy separation between the levels n and $n + 1$? (iii) What is the de Broglie wavelength of the molecule?

P7D.3 Calculate the expectation values of x and x^2 for a particle in the state with $n = 1$ in a one-dimensional square-well potential.

P7D.4 Calculate the expectation values of p_x and p_x^2 for a particle in the state with $n = 2$ in a one-dimensional square-well potential.

P7D.5 When β-carotene (**1**) is oxidized *in vivo*, it breaks in half and forms two molecules of retinal (vitamin A), which is a precursor to the pigment in the retina responsible for vision. The conjugated system of retinal consists of 11 C atoms and one O atom. In the ground state of retinal, each level up to $n = 6$ is occupied by two electrons. Assuming an average internuclear distance of 140 pm, calculate (a) the separation in energy between the ground state and the first excited state in which one electron occupies the state with $n = 7$, and (b) the frequency of the radiation required to produce a transition between these two states. (c) Using your results, choose among the words in parentheses to generate a rule for the prediction of frequency shifts in the absorption spectra of linear polyenes:

> The absorption spectrum of a linear polyene shifts to (higher/lower) frequency as the number of conjugated atoms (increases/decreases).

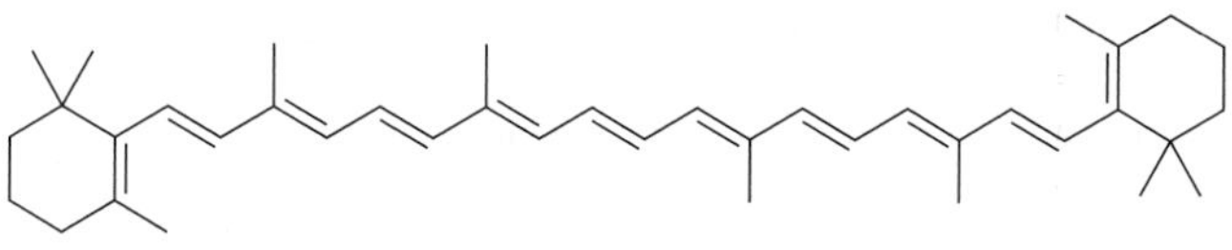

1 β-Carotene

P7D.6 Consider a particle of mass m confined to a one-dimensional box of length L and in a state with normalized wavefunction ψ_n. (a) Without evaluating any integrals, explain why $\langle x \rangle = L/2$. (b) Without evaluating any integrals, explain why $\langle p_x \rangle = 0$. (c) Derive an expression for $\langle x^2 \rangle$ (the necessary integrals will be found in the *Resource section*). (d) For a particle in a box the energy is given by $E_n = n^2h^2/8mL^2$ and, because the potential energy is zero, all of this energy is kinetic. Use this observation and, without evaluating any integrals, explain why $\langle p_x^2 \rangle = n^2h^2/4L^2$.

P7D.7 Consider a particle of mass m confined to a one-dimensional box of length L and in a state with normalized wavefunction ψ_n. According to Topic 7C, the uncertainty in the position is $\Delta x = \{\langle x^2 \rangle - \langle x \rangle^2\}^{1/2}$ and for the linear momentum it is $\Delta p_x = \{\langle p_x^2 \rangle - \langle p_x \rangle^2\}^{1/2}$. (a) For this system find expressions for $\langle x \rangle$, $\langle x^2 \rangle$, $\langle p_x \rangle$, and $\langle p_x^2 \rangle$; hence find expressions for Δx and Δp_x. (b) Go on to find an expression for the product $\Delta x \Delta p_x$. (c) Show that for $n = 1$ and $n = 2$ the result from (b) is in accord with the Heisenberg uncertainty principle, and infer that this is also true for $n \geq 1$.

P7D.8‡ A particle is confined to move in a one-dimensional box of length L. If the particle is behaving classically, then it simply bounces back and forth in the box, moving with a constant speed. (a) Explain why the probability density, $P(x)$, for the classical particle is $1/L$. (*Hint:* What is the total probability of finding the particle in the box?) (b) Explain why the average value of x^n is $\langle x^n \rangle = \int_0^L P(x)x^n dx$. (c) By evaluating such an integral, find $\langle x \rangle$ and $\langle x^2 \rangle$. (d) For a quantum particle $\langle x \rangle = L/2$ and $\langle x^2 \rangle = L^2\left(\frac{1}{3} - 1/2n^2\pi^2\right)$. Compare these expressions with those you have obtained in part (c), recalling that the correspondence principle states that, for very large values of the quantum numbers, the predictions of quantum mechanics approach those of classical mechanics.

P7D.9 (a) Set up the Schrödinger equation for a particle of mass m in a three-dimensional rectangular box with sides L_1, L_2, and L_3. Show that the Schrödinger equation is separable. (b) Show that the wavefunction and the energy are defined by three quantum numbers. (c) Specialize the result from part (b) to an electron moving in a cubic box of side $L = 5$ nm and draw an energy diagram resembling Fig. 7D.2 and showing the first 15 energy levels. Note that each energy level might be degenerate. (d) Compare the energy level diagram from part (c) with the energy level diagram for an electron in a one-dimensional box of length $L = 5$ nm. Are the energy levels more or less sparsely distributed in the cubic box than in the one-dimensional box?

P7D.10 In the text the one-dimensional particle-in-a-box problem involves confining the particle to the range from $x = 0$ to $x = L$. This problem explores a similar situation in which the potential energy is zero between $x = -L/2$ and $x = +L/2$, and infinite elsewhere. (a) Identify the boundary conditions that apply in this case. (b) Show that $\cos(kx)$ is a solution of the Schrödinger equation for the region with zero potential energy, find the values of k for which the boundary conditions are satisfied, and hence derive an expression for the corresponding energies. Sketch the three wavefunctions with the lowest energies. (c) Repeat the process, but this time with the trial wavefunction $\sin(k'x)$. (d) Compare the complete set of energies you have obtained in parts (b) and (c) with the energies for the case where the particle is confined between 0 and L: are they the same? (e) Normalize the wavefunctions (the necessary integrals are in the *Resource section*). (f) Without evaluating any integrals, explain why $\langle x \rangle = 0$ for both sets of wavefunctions.

P7D.11 Many biological electron transfer reactions, such as those associated with biological energy conversion, may be visualized as arising from electron tunnelling between protein-bound co-factors, such as cytochromes, quinones, flavins, and chlorophylls. This tunnelling occurs over distances that are often greater than 1.0 nm, with sections of protein separating electron donor from acceptor. For a specific combination of donor and acceptor, the rate of electron tunnelling is proportional to the transmission probability, with $\kappa \approx 7$ nm^{-1} (eqn 7D.17). By what factor does the rate of electron tunnelling between two co-factors increase as the distance between them changes from 2.0 nm to 1.0 nm? You may assume that the barrier is such that eqn 7D.20b is appropriate.

P7D.12 Derive eqn 7D.20a, the expression for the transmission probability and show that when $\kappa W >> 1$ it reduces to eqn 7D.20b. The derivation proceeds by requiring that the wavefunction and its first derivative are continuous at the edges of the barrier, as expressed by eqns 7D.19a and 7D.19b.

P7D.13‡ A particle of mass m moves in one dimension in a region divided into three zones: zone 1 has $V = 0$ for $-\infty < x \leq 0$; zone 2 has $V = V_2$ for $0 \leq x \leq W$; zone 3 has $V = V_3$ for $W \leq x < \infty$. In addition, $V_3 < V_2$. In zone 1 the wavefunction is $A_1e^{ik_1x} + B_1e^{-ik_1x}$; the term e^{ik_1x} represents the wave incident on the barrier V_2, and the term e^{-ik_1x} represents the reflected wave. In zone 2 the wavefunction is $A_2e^{k_2x} + B_2e^{-k_2x}$. In zone 3 the wavefunction has only a forward component, $A_3e^{ik_3x}$, which represents a particle that has traversed the barrier. Consider a case in which the energy of the particle E is greater than V_3 but less than V_2, so that zone 2 represents a barrier. The transmission probability, T, is the ratio of the square modulus of the zone 3 amplitude to the square modulus of the incident amplitude, that is, $T = |A_3|^2/|A_1|^2$. (a) Derive an expression for T by imposing the requirement that the wavefunction and its slope must be continuous at the zone boundaries. You can simplify the calculation by assuming from the outset that $A_1 = 1$. (b) Show that this equation for T reduces to eqn 7D.20b in the high, wide barrier limit when $V_1 = V_3 = 0$. (c) Draw a graph of the probability of proton tunnelling when $V_3 = 0$, $W = 50$ pm, and $E = 10$ kJ mol^{-1} in the barrier range $E < V_2 < 2E$.

P7D.14 A potential barrier of height V extends from $x = 0$ to positive x. Inside this barrier the normalized wavefunction is $\psi = Ne^{-\kappa x}$. Calculate (a) the probability that the particle is inside the barrier and (b) the average penetration depth of the particle into the barrier.

P7D.15 Use mathematical software or a spreadsheet for the following procedures:

(a) Plot the probability density for a particle in a box with $n = 1, 2, \ldots 5$, and $n = 50$. How do your plots illustrate the correspondence principle?

(b) Plot the transmission probability T against E/V for passage by (i) a hydrogen molecule, (ii) a proton, and (iii) an electron through a barrier of height V.

(c) Use mathematical software to generate three-dimensional plots of the wavefunctions for a particle confined to a rectangular surface with (i) $n_1 = 1, n_2 = 1$, the state of lowest energy, (ii) $n_1 = 1, n_2 = 2$, (iii) $n_1 = 2, n_2 = 1$, and (iv) $n_1 = 2, n_2 = 2$. Deduce a rule for the number of nodal lines in a wavefunction as a function of the values of n_1 and n_2.

TOPIC 7E Vibrational motion

Discussion questions

D7E.1 Describe the variation with the mass and force constant of the separation of the vibrational energy levels of a harmonic oscillator.

D7E.2 In what ways does the quantum mechanical description of a harmonic oscillator merge with its classical description at high quantum numbers?

D7E.3 To what quantum mechanical principle can you attribute the existence of a zero-point vibrational energy?

Additional exercises

E7E.7 Calculate the zero-point energy of a harmonic oscillator consisting of a particle of mass 5.16×10^{-26} kg and force constant 285 N m^{-1}.

E7E.8 For a certain harmonic oscillator of effective mass 2.88×10^{-25} kg, the difference in adjacent energy levels is 3.17 zJ. Calculate the force constant of the oscillator.

E7E.9 Calculate the wavelength of the photon needed to excite a transition between neighbouring energy levels of a harmonic oscillator of effective mass equal to that of an oxygen atom ($15.9949m_u$) and force constant 544 N m^{-1}.

E7E.10 Sketch the form of the wavefunctions for the harmonic oscillator with quantum numbers $v = 1$ and 2. Use a symmetry argument to explain why these two wavefunctions are orthogonal (do not evaluate any integrals).

E7E.11 Assuming that the vibrations of a $^{35}Cl_2$ molecule are equivalent to those of a harmonic oscillator with a force constant $k_f = 329$ N m^{-1}, what is the zero-point energy of vibration of this molecule? Use $m(^{35}\mathrm{Cl}) = 34.9688m_u$.

E7E.12 Assuming that the vibrations of a $^{14}N_2$ molecule are equivalent to those of a harmonic oscillator with a force constant $k_f = 2293.8$ N m^{-1}, what is the zero-point energy of vibration of this molecule? Use $m(^{14}\mathrm{N}) = 14.0031m_u$.

E7E.13 The classical turning points of a harmonic oscillator occur at the displacements at which all of the energy is potential energy; that is, when $E_v = \frac{1}{2}k_f x_{tp}^2$. For a particle of mass m_u undergoing harmonic motion with force constant $k_f = 1000$ N m^{-1}, calculate the energy of the state with $v = 0$ and hence find the separation between the classical turning points. Repeat the calculation for an oscillator with $k_f = 100$ N m^{-1}.

E7E.14 The classical turning points of a harmonic oscillator occur at the displacements at which all of the energy is potential energy; that is, when $E_v = \frac{1}{2}k_f x_{tp}^2$. For a particle of mass m_u undergoing harmonic motion with force constant $k_f = 1000$ N m^{-1}, calculate the energy of the state with $v = 1$ and hence find the separation between the classical turning points. Express your answer as a percentage of a typical bond length of 110 pm.

E7E.15 How many nodes are there in the wavefunction of a harmonic oscillator with (i) $v = 5$; (ii) $v = 35$?

E7E.16 Locate the nodes of a harmonic oscillator wavefunction with $v = 2$. (Express your answers in terms of the coordinate y.)

E7E.17 Locate the nodes of the harmonic oscillator wavefunction with $v = 3$.

E7E.18 At what displacements is the probability density a maximum for a state of a harmonic oscillator with $v = 3$?

Problems

P7E.1 If the vibration of a diatomic A–B is modelled using a harmonic oscillator, the vibrational frequency is given by $\omega = (k_f/\mu)^{1/2}$, where μ is the effective mass, $\mu = m_A m_B/(m_A + m_B)$. If atom A is substituted by an isotope (for example ^{2}H substituted for ^{1}H), then to a good approximation the force constant remains the same. Why? (*Hint:* Is there any change in the number of charged species?) (a) Show that when an isotopic substitution is made for atom A, such that its mass changes from m_A to $m_{A'}$, the vibrational frequency of A′–B, $\omega_{A'B}$, can be expressed in terms of the vibrational frequency of A–B, ω_{AB} as $\omega_{A'B} = \omega_{AB}(\mu_{AB}/\mu_{A'B})^{1/2}$, where μ_{AB} and $\mu_{A'B}$ are the effective masses of A–B and A′–B, respectively. (b) The vibrational frequency of ^{1}H^{35}Cl is 5.63×10^{14} s^{-1}. Calculate the vibrational frequency of (i) ^{2}H^{35}Cl and (ii) ^{1}H^{37}Cl. Use integer relative atomic masses.

P7E.2 Consider the case where in the diatomic molecule A–B the mass of B is much greater than that of A. (a) Show that for an isotopic substitution of A, the ratio of vibrational frequencies is $\omega_{A'B} \approx \omega_{AB}(m_A/m_{A'})^{1/2}$. (b) Use this expression to calculate the vibrational frequency of ^{2}H^{35}Cl (the vibrational frequency of ^{1}H^{35}Cl is 5.63×10^{14} s^{-1}). (c) Compare your answer with the value obtained in the *Problem* P7E.1. (d) In organic molecules it is commonly observed that the C–H stretching frequency is reduced by a factor of around 0.7 when ^{1}H is substituted by ^{2}H: rationalize this observation.

P7E.3 The vibrational frequency of $^1H_2$ is 131.9 THz. What is the vibrational frequency of $^2H_2$ and of $^3H_2$? Use integer relative atomic masses for this estimate.

P7E.4 The force constant for the bond in CO is 1857 N m^{-1}. Calculate the vibrational frequencies (in Hz) of ^{12}C^{16}O, ^{13}C^{16}O, ^{12}C^{18}O, and ^{13}C^{18}O. Use integer relative atomic masses for this estimate.

P7E.5 In infrared spectroscopy it is common to observe a transition from the $v = 0$ to $v = 1$ vibrational level. If this transition is modelled as a harmonic oscillator, the energy of the photon involved is $\hbar\omega$, where ω is the vibrational frequency. (a) Show that the wavenumber of the radiation corresponding to photons of this energy, $\tilde{\nu}$, is given by $\tilde{\nu} = \omega/2\pi c$, where c is the speed of light.

(b) The vibrational frequency of $^{1}H^{35}Cl$ is $\omega = 5.63 \times 10^{14}\ s^{-1}$; calculate $\tilde{\nu}$. (c) Derive an expression for the force constant k_f in terms of $\tilde{\nu}$. (d) For $^{12}C^{16}O$ the $\upsilon = 0 \rightarrow 1$ transition is observed at 2170 cm^{-1}. Calculate the force constant and estimate the wavenumber at which the corresponding absorption occurs for $^{14}C^{16}O$. Use integer relative atomic masses for this estimate.

P7E.6 The following data give the wavenumbers (wavenumbers in cm^{-1}) of the $\upsilon = 0 \rightarrow 1$ transition of a number of diatomic molecules. Calculate the force constants of the bonds and arrange them in order of increasing stiffness. Use integer relative atomic masses.

$^{1}H^{35}Cl$	$^{1}H^{81}Br$	$^{1}H^{127}I$	$^{12}C^{16}O$	$^{14}N^{16}O$
2990	2650	2310	2170	1904

P7E.7 Carbon monoxide binds strongly to the Fe^{2+} ion of the haem (heme) group of the protein myoglobin. Estimate the vibrational frequency of CO bound to myoglobin by making the following assumptions: the atom that binds to the haem group is immobilized, the protein is infinitely more massive than either the C or O atom, the C atom binds to the Fe^{2+} ion, and binding of CO to the protein does not alter the force constant of the CO bond. The wavenumber of the $\upsilon = 0 \rightarrow 1$ transition of $^{12}C^{16}O$ is 2170 cm^{-1}.

P7E.8 Carbon monoxide binds strongly to the Fe^{2+} ion of the haem (heme) group of the protein myoglobin. It is possible to estimate the vibrational frequency of CO bound to myoglobin by assuming that the atom that binds to the haem group is immobilized, and that the protein is infinitely more massive than either the C or O atom. Suppose that you have at your disposal a supply of myoglobin, a suitable buffer in which to suspend the protein, $^{12}C^{16}O$, $^{13}C^{16}O$, $^{12}C^{18}O$, $^{13}C^{18}O$, and an infrared spectrometer. Assuming that isotopic substitution does not affect the force constant of the CO bond, describe a set of experiments that: (a) proves which atom, C or O, binds to the haem group of myoglobin, and (b) allows for the determination of the force constant of the CO bond for myoglobin-bound carbon monoxide.

P7E.9 A function of the form e^{-gx^2} is a solution of the Schrödinger equation for the harmonic oscillator (eqn 7E.5), provided that g is chosen correctly. In this problem you will find the correct form of g. (a) Start by substituting $\psi = e^{-gx^2}$ into the left-hand side of eqn 7E.5 and evaluating the second derivative. (b) You will find that in general the resulting expression is not of the form constant × ψ, implying that ψ is not a solution to the equation. However, by choosing the value of g such that the terms in x^2 cancel one another, a solution is obtained. Find the required form of g and hence the corresponding energy. (c) Confirm that the function so obtained is indeed the ground state of the harmonic oscillator, as quoted in eqn 7E.13a, and that it has the energy expected from eqn 7E.6.

P7E.10 Write the normalized form of the ground state wavefunction of the harmonic oscillator in terms of the variable y and the parameter α. (a) Write the integral you would need to evaluate to find the mean displacement $\langle y \rangle$, and then use a symmetry argument to explain why this integral is equal to 0. (b) Calculate $\langle y^2 \rangle$ (the necessary integral will be found in the *Resource section*). (c) Repeat the process for the first excited state.

P7E.11 The expectation value of the kinetic energy of a harmonic oscillator is most easily found by using the virial theorem, but in this Problem you will find it directly by evaluating the expectation value of the kinetic energy operator with the aid of the properties of the Hermite polynomials given in eqn 7E.12. (a) Write the kinetic energy operator $\hat{T}$ in terms of x and show that it can be rewritten in terms of the variable y (introduced in eqn 7E.10) and the frequency ω as

$$\hat{T} = -\tfrac{1}{2}\hbar\omega\frac{d^2}{dy^2}$$

The expectation value of this operator for an harmonic oscillator wavefunction with quantum number υ is

$$\langle T \rangle_\upsilon = -\tfrac{1}{2}\hbar\omega\alpha N_\upsilon^2\int_{-\infty}^{\infty} H_\upsilon e^{-y^2/2}\frac{d^2}{dy^2}H_\upsilon e^{-y^2/2}dy$$

where N_υ is the normalization constant (eqn 7E.10) and α is defined in eqn 7E.7 (the term α arises from $dx = \alpha dy$). (b) Evaluate the second derivative and then use the property $H_\upsilon'' - 2yH_\upsilon' + 2\upsilon H_\upsilon = 0$; where the prime indicates a derivative, to rewrite the derivatives in terms of the H_υ (you should be able to eliminate all the derivatives). (c) Now proceed as in the text, in which terms of the form yH_υ are rewritten by using the property $H_{\upsilon+1} - 2yH_\upsilon + 2\upsilon H_{\upsilon-1} = 0$; you will need to apply this twice. (d) Finally, evaluate the integral using the properties of the integrals of the Hermite polynomials given in Table 7E.1 and so obtain the result quoted in the text.

P7E.12 Calculate the values of $\langle x^3 \rangle_\upsilon$ and $\langle x^4 \rangle_\upsilon$ for a harmonic oscillator by using the properties of the Hermite polynomials given in Table 7E.1; follow the approach used in the text.

P7E.13 Use the same approach as in *Example* 7E.3 to calculate the probability that a harmonic oscillator in the first excited state will be found in the classically forbidden region. You will need to use mathematical software to evaluate the appropriate integral. Compare the result you obtain with that for the ground state and comment on the difference.

P7E.14 Use the same approach as in *Example* 7E.3 to calculate the probability that a harmonic oscillator in the states $\upsilon = 0, 1, \ldots 7$ will be found in the classically forbidden region. You will need to use mathematical software to evaluate the final integrals. Plot the probability as a function of υ and interpret the result in terms of the correspondence principle.

P7E.15 The intensities of spectroscopic transitions between the vibrational states of a molecule are proportional to the square of the integral $\int \psi_{\upsilon'} x \psi_\upsilon dx$ over all space. Use the relations between Hermite polynomials given in eqn 7E.12 to show that the only permitted transitions are those for which $\upsilon' = \upsilon \pm 1$ and evaluate the integral in these cases.

P7E.16 The potential energy of the rotation of one CH_3 group relative to its neighbour in ethane can be expressed as $V(\phi) = V_0 \cos 3\phi$. Show that for small displacements the motion of the group is harmonic and derive an expression for the energy of excitation from $\upsilon = 0$ to $\upsilon = 1$. (*Hint:* Use a series expansion for cos 3ϕ.) What do you expect to happen to the energy levels and wavefunctions as the excitation increases to high quantum numbers?

P7E.17 (a) Without evaluating any integrals, explain why you expect $\langle x \rangle_\upsilon = 0$ for all states of a harmonic oscillator. (b) Use a physical argument to explain why $\langle p_x \rangle_\upsilon = 0$. (c) Equation 7E.18c gives $\langle E_k \rangle_\upsilon = \tfrac{1}{2}E_\upsilon$. Recall that the kinetic energy is given by $p^2/2m$ and hence find an expression for $\langle p_x^2 \rangle_\upsilon$. (d) Note from Topic 7C that the uncertainty in the position, Δx, is given by $\Delta x = (\langle x^2 \rangle - \langle x \rangle^2)^{1/2}$ and likewise for the momentum $\Delta p_x = (\langle p_x^2 \rangle - \langle p_x \rangle^2)^{1/2}$. Find expressions for Δx and Δp_x (the expression for $\langle x^2 \rangle_\upsilon$ is given in the text). (e) Hence find an expression for the product $\Delta x \Delta p_x$ and show that the Heisenberg uncertainty principle is satisfied. (f) For which state is the product $\Delta x \Delta p_x$ a minimum?

P7E.18 Use mathematical software or a spreadsheet to gain some insight into the origins of the nodes in the harmonic oscillator wavefunctions by plotting the Hermite polynomials $H_\upsilon(y)$ for $\upsilon = 0$ through 5.

TOPIC 7F Rotational motion

Discussion questions

D7F.1 Discuss the physical origin of quantization of energy for a particle confined to motion on a ring.

D7F.2 Describe the features of the solution of the particle on a ring that appear in the solution of the particle on a sphere. What concept applies to the latter but not to the former?

D7F.3 Describe the vector model of angular momentum in quantum mechanics. What features does it capture?

Additional exercises

E7F.6 The rotation of a molecule can be represented by the motion of a particle moving over the surface of a sphere. Calculate the magnitude of its angular momentum when $l = 1$ and the possible components of the angular momentum along the z-axis. Express your results as multiples of $\hbar$.

E7F.7 The rotation of a molecule can be represented by the motion of a particle moving over the surface of a sphere with angular momentum quantum number $l = 2$. Calculate the magnitude of its angular momentum and the possible components of the angular momentum along the z-axis. Express your results as multiples of $\hbar$.

E7F.8 For a particle on a ring, how many nodes are there in the real part, and in the imaginary part, of the wavefunction for (i) $m_l = 0$ and (ii) $m_l = +3$? In both cases, find the values of ϕ at which any nodes occur.

E7F.9 For a particle on a ring, how many nodes are there in the real part, and in the imaginary part of the wavefunction for (i) $m_l = +1$ and (ii) $m_l = +2$? In both cases, find the values of ϕ at which any nodes occur.

E7F.10 The wavefunction for the motion of a particle on a ring can also be written $\psi = N\cos(m_l\phi)$, where m_l is integer. Evaluate the normalization constant, N.

E7F.11 By considering the integral $\int_0^{2\pi} \psi_{m_l}^* \psi_{m_l'} \mathrm{d}\phi$, where $m_l \neq m_l'$, confirm that wavefunctions for a particle in a ring with different values of the quantum number m_l are mutually orthogonal.

E7F.12 By considering the integral $\int_0^{2\pi} \cos m_l\phi \cos m_l'\phi \,\mathrm{d}\phi$, where $m_l \neq m_l'$, confirm that the wavefunctions $\cos m_l\phi$ and $\cos m_l'\phi$ for a particle on a ring are orthogonal. (*Hint:* To evaluate the integral, first apply the identity $\cos A \cos B = \frac{1}{2}\{\cos(A+B) + \cos(A-B)\}$.)

E7F.13 Consider a proton constrained to rotate in a circle of radius 100 pm around a fixed point. For this arrangement, calculate the value of $|m_l|$ corresponding to a rotational energy equal to the classical average energy at 25 °C (which is equal to $\frac{1}{2}kT$).

E7F.14 The moment of inertia of an SF_6 molecule is $3.07 \times 10^{-45}\,\mathrm{kg\,m^2}$. What is the minimum energy needed to start it rotating?

E7F.15 The moment of inertia of a CH_4 molecule is $5.27 \times 10^{-47}\,\mathrm{kg\,m^2}$. Calculate the energy needed to excite a CH_4 molecule from a state with $l = 1$ to a state with $l = 2$.

E7F.16 The moment of inertia of an SF_6 molecule is $3.07 \times 10^{-45}\,\mathrm{kg\,m^2}$. Calculate the energy needed to excite an SF_6 molecule from a state with $l = 2$ to a state with $l = 3$.

E7F.17 What is the magnitude of the angular momentum of an SF_6 molecule when it is rotating with its minimum energy?

E7F.18 Draw the vector diagram for all the permitted states of a particle with $l = 6$.

E7F.19 How many angular nodes are there for the spherical harmonic $Y_{3,0}$ and at which values of θ do they occur?

E7F.20 Based on the pattern of nodes in Fig. 7F.5, how many angular nodes do you expect there to be for the spherical harmonic $Y_{4,0}$? Does it have a node at $\theta = 0$?

E7F.21 Consider the real part of the spherical harmonic $Y_{1,+1}$. At which values of ϕ do angular nodes occur? These angular nodes can also be described as planes: identify the positions of the corresponding planes (for example, the angular node with $\phi = 0$ is the xz-plane). Do the same for the imaginary part.

E7F.22 Consider the real part of the spherical harmonic $Y_{2,+2}$. At which values of ϕ do angular nodes occur? Identify the positions of the corresponding planes. Repeat the process for the imaginary part.

E7F.23 What is the degeneracy of a molecule rotating with $J = 3$?

E7F.24 What is the degeneracy of a molecule rotating with $J = 4$?

E7F.25 Draw diagrams to scale, and similar to Fig. 7F.7a, representing the states (i) $l = 1$, $m_l = -1, 0, +1$, (ii) $l = 2$ and all possible values of m_l.

E7F.26 Draw diagrams to scale, and similar to Fig. 7F.7a, representing the states (i) $l = 0$, (ii) $l = 3$ and all possible values of m_l.

E7F.27 Derive an expression for the angle between the vector representing angular momentum l with z-component $m_l = +l$ (that is, its maximum value) and the z-axis. What is this angle for $l = 1$ and for $l = 5$?

E7F.28 Derive an expression for the angle between the vector representing angular momentum l with z-component $m_l = +l$ and the z-axis. What value does this angle take in the limit that l becomes very large? Interpret your result in the light of the correspondence principle.

Problems

P7F.1 The particle on a ring is a useful model for the motion of electrons around the porphyrin ring (**2**), the conjugated macrocycle that forms the structural basis of the haem (heme) group and the chlorophylls. The group may be modelled as a circular ring of radius 440 pm, with 22 electrons in the conjugated system moving along its perimeter. In the ground state of the molecule each state is occupied by two electrons. (a) Calculate the energy and angular momentum of an electron in the highest occupied level. (b) Calculate the frequency of radiation that can induce a transition between the highest occupied and lowest unoccupied levels.

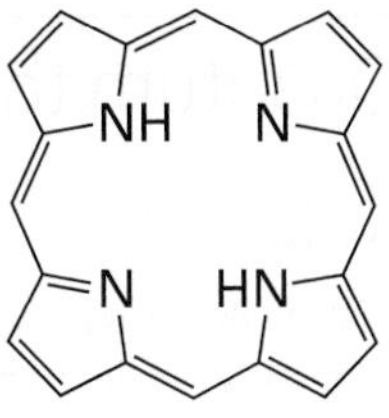

2 Porphyrin ring

P7F.2 Consider the following wavefunctions (i) $e^{i\phi}$, (ii) $e^{-2i\phi}$, (iii) $\cos\phi$, and (iv) $(\cos\chi)e^{i\phi} + (\sin\chi)e^{-i\phi}$ each of which describes a particle on a ring. (a) Decide whether or not each wavefunction is an eigenfunction of the operator $\hat{l}_z$ for the z-component of the angular momentum ($\hat{l}_z = (\hbar/i)(d/d\phi)$); where the function is an eigenfunction, give the eigenvalue. (b) For the functions that are not eigenfunctions, calculate the expectation value of l_z (you will first need to normalize the wavefunction). (c) Repeat the process but this time for the kinetic energy, for which the operator is $-(\hbar^2/2I)(d^2/d\phi^2)$. (d) Which of these wavefunctions describe states of definite angular momentum, and which describe states of definite kinetic energy?

P7F.3 Is the Schrödinger equation for a particle on an elliptical ring of semi-major axes a and b separable? (*Hint:* Although r varies with angle ϕ, the two are related by $r^2 = a^2\sin^2\phi + b^2\cos^2\phi$.)

P7F.4 Calculate the energies of the first four rotational levels of $^1H^{127}I$ free to rotate in three dimensions; use for its moment of inertia $I = \mu R^2$, with $\mu = m_H m_I/(m_H + m_I)$ and $R = 160$ pm. Use integer relative atomic masses for this estimate.

P7F.5 Consider the three spherical harmonics (a) $Y_{0,0}$, (b) $Y_{2,-1}$, and (c) $Y_{3,+3}$. (a) For each spherical harmonic, substitute the explicit form of the function taken from Table 7F.1 into the left-hand side of eqn 7F.8 (the Schrödinger equation for a particle on a sphere) and confirm that the function is a solution of the equation; give the corresponding eigenvalue (the energy) and show that it agrees with eqn 7F.10. (b) Likewise, show that each spherical harmonic is an eigenfunction of $\hat{l}_z = (\hbar/i)(d/d\phi)$ and give the eigenvalue in each case.

P7F.6 Confirm that $Y_{1,+1}$, taken from Table 7F.1, is normalized. You will need to integrate $Y^*_{1,+1}Y_{1,+1}$ over all space using the relevant volume element:

$$\int_{\theta=0}^{\pi}\int_{\phi=0}^{2\pi} Y^*_{1,+1}Y_{1,+1}\,\overbrace{\sin\theta\,d\theta\,d\phi}^{\text{volume element}}$$

P7F.7 Confirm that $Y_{1,0}$ and $Y_{1,+1}$, taken from Table 7F.1, are orthogonal. You will need to integrate $Y^*_{1,+1}Y_{1,+1}$ over all space using the relevant volume element:

$$\int_{\theta=0}^{\pi}\int_{\phi=0}^{2\pi} Y^*_{1,0}Y_{1,+1}\,\overbrace{\sin\theta\,d\theta\,d\phi}^{\text{volume element}}$$

(*Hint:* A useful result for evaluating the integral is $(d/d\theta)\sin^3\theta = 3\sin^2\theta\cos\theta$.)

P7F.8 (a) Show that $\psi = c_1Y_{l,m_l} + c_2Y_{l,m_l'}$ is an eigenfunction of Λ^2 with eigenvalue $-l(l+1)$; c_1 and c_2 are arbitrary coefficients. (*Hint:* Apply Λ^2 to ψ and use the properties given in eqn 7F.9.) (b) The spherical harmonics $Y_{1,+1}$ and $Y_{1,-1}$ are complex functions (see Table 7F.1), but as they are degenerate eigenfunctions of Λ^2, any linear combination of them is also an eigenfunction, as was shown in part (a). Show that the combinations $\psi_a = -Y_{1,+1} + Y_{1,-1}$ and $\psi_b = i(Y_{1,+1} + Y_{1,-1})$ are real. (c) Show that ψ_a and ψ_b are orthogonal (you will need to integrate using the relevant volume element, $\sin\theta d\theta d\phi$). (d) Normalize ψ_a and ψ_b. (e) Identify the angular nodes in these two functions and the planes to which they correspond. (f) Is ψ_a an eigenfunction of $\hat{l}_z$? Discuss the significance of your answer.

P7F.9 In this problem you will establish the commutation relations, given in eqn 7E.14, between the operators for the x-, y-, and z-components of angular momentum, which are defined in eqn 7F.13. In order to manipulate the operators correctly it is helpful to imagine that they are acting on some arbitrary function f: it does not matter what f is, and at the end of the proof it is simply removed. Consider $[\hat{l}_x,\hat{l}_y] = \hat{l}_x\hat{l}_y - \hat{l}_y\hat{l}_x$. Consider the effect of the first term on some arbitrary function f and evaluate

$$\hat{l}_x\hat{l}_y f = -\hbar^2\left(\overbrace{y\frac{\partial}{\partial z}}^{A} - \overbrace{z\frac{\partial}{\partial y}}^{B}\right)\left(\overbrace{z\frac{\partial f}{\partial x}}^{C} - \overbrace{x\frac{\partial f}{\partial z}}^{D}\right)$$

The next step is to multiply out the parentheses, and in doing so care needs to be taken over the order of operations. (b) Repeat the procedure for the other term in the commutator, $\hat{l}_y\hat{l}_x f$. (c) Combine the results from parts (a) and (b) so as to evaluate $\hat{l}_x\hat{l}_y f - \hat{l}_y\hat{l}_x f$; you should find that many of the terms cancel. Confirm that the final expression you have is indeed $i\hbar\hat{l}_z f$, where $\hat{l}_z$ is given in eqn 7F.13. (d) The definitions in eqn 7F.13 are related to one another by cyclic permutation of the x, y, and z. That is, by making the permutation $x \to y$, $y \to z$, and $z \to x$, you can move from one definition to the next: confirm that this is so. (e) The same cyclic permutation can be applied to the commutators of these operators. Start with $[\hat{l}_x,\hat{l}_y] = i\hbar\hat{l}_z$ and show that cyclic permutation generates the other two commutators in eqn 7F.14.

P7F.10 Show that $\hat{l}_z$ and $\hat{l}^2$ both commute with the hamiltonian for a hydrogen atom. What is the significance of this result? Begin by noting that $\hat{l}^2 = \hat{l}_x^2 + \hat{l}_y^2 + \hat{l}_z^2$. Then show that $[\hat{l}_z,\hat{l}_q^2] = [\hat{l}_z,\hat{l}_q]\hat{l}_q + \hat{l}_q[\hat{l}_z,\hat{l}_q]$ and then use the angular momentum commutation relations in eqn 7F.14.

P7F.11 Starting from the definition of the operator $\hat{l}_z$ given in eqn 7F.13, show that in spherical polar coordinates it can be expressed as $\hat{l}_z = -i\hbar\partial/\partial\phi$. (*Hint:* You will need to express the Cartesian coordinates in terms of the spherical polar coordinates; refer to *The chemist's toolkit* 7F.2).

P7F.12 A particle confined within a spherical cavity is a starting point for the discussion of the electronic properties of spherical metal nanoparticles. Here, you are invited to show in a series of steps that the $l = 0$ energy levels of an electron in a spherical cavity of radius R are quantized and given by $E_n = n^2h^2/8m_eR^2$. (a) The hamiltonian for a particle free to move inside a spherical cavity of radius a is

$$\hat{H} = -\frac{\hbar^2}{2m}\nabla^2 \quad \text{with } \nabla^2 = \frac{1}{r}\frac{\partial^2}{\partial r^2}r + \frac{1}{r^2}\Lambda^2$$

Show that the Schrödinger equation is separable into radial and angular components. That is, begin by writing $\psi(r,\theta,\phi) = R(r)Y(\theta,\phi)$, where $R(r)$ depends only on the distance of the particle from the centre of the sphere, and $Y(\theta,\phi)$ is a spherical harmonic. Then show that the Schrödinger equation can be separated into two equations, one for $R(r)$, the radial equation, and the other for $Y(\theta,\phi)$, the angular equation. (b) Consider the case $l = 0$. Show by differentiation that the solution of the radial equation has the form

$$R(r) = (2\pi a)^{-1/2}\frac{\sin(n\pi r/a)}{r}$$

(c) Now go on to show (by acknowledging the appropriate boundary conditions) that the allowed energies are given by $E_n = n^2h^2/8ma^2$. With substitution of m_e for m and of R for a, this is the equation given above for the energy.

FOCUS 7 Quantum theory

Integrated activities

I7.1‡ A star too small and cold to shine has been found by S. Kulkarni et al. (*Science*, 1478 (1995)). The spectrum of the object shows the presence of methane which, according to the authors, would not exist at temperatures much above 1000 K. The mass of the star, as determined from its gravitational effect on a companion star, is roughly 20 times the mass of Jupiter. The star is considered to be a brown dwarf, the coolest ever found.

(a) Derive an expression for $\Delta_rG^\ominus$ for $CH_4(g) \to C(\text{graphite}) + 2\,H_2(g)$ at temperature T. Proceed by using data from the tables in the *Resource section* to find $\Delta_rH^\ominus$ and $\Delta_rS^\ominus$ at 298 K and then convert these values to an arbitrary temperature T by using heat capacity data, also from the tables (assume that the heat capacities do not vary with temperature). (b) Find the temperature above which $\Delta_rG^\ominus$ becomes positive. (The solution to the relevant equation cannot be found analytically, so use mathematical software to find a numerical solution or plot a graph). Does your result confirm the assertion that methane could not exist at temperatures much above 1000 K? (c) Assume the star behaves as a black body at 1000 K, and calculate the wavelength at which the radiation from it is maximum. (d) Estimate the fraction of the energy density of the star that it emitted in the visible region of the spectrum (between 420 nm and 700 nm). (You may assume that over this wavelength range $\Delta\lambda$ it is acceptable to approximate the integral of the Planck distribution by $\rho(\lambda,T)\Delta\lambda$.)

I7.2 Describe the features that stem from nanometre-scale dimensions that are not found in macroscopic objects.

I7.3 Explain why the particle in a box and the harmonic oscillator are useful models for quantum mechanical systems: what chemically significant systems can they be used to represent?

I7.4 Suppose that 1.0 mol of perfect gas molecules all occupy the lowest energy level of a cubic box. (a) How much work must be done to change the volume of the box by ΔV? (b) Would the work be different if the molecules all occupied a state $n \neq 1$? (c) What is the relevance of this discussion to the expression for the expansion work discussed in Topic 2A? (d) Can you identify a distinction between adiabatic and isothermal expansion?

I7.5 Evaluate $\Delta x = (\langle x^2\rangle - \langle x\rangle^2)^{1/2}$ and $\Delta p_x = (\langle p_x^2\rangle - \langle p_x\rangle^2)^{1/2}$ for the ground state of (a) a particle in a box of length L and (b) a harmonic oscillator. Discuss these quantities with reference to the uncertainty principle.

I7.6 Repeat Problem I7.5 for (a) a particle in a box and (b) a harmonic oscillator in a general quantum state (n and v, respectively).

FOCUS 8

Atomic structure and spectra

This Focus discusses the use of quantum mechanics to describe and investigate the 'electronic structure' of atoms, the arrangement of electrons around their nuclei. The concepts are of central importance for understanding the properties of atoms and molecules, and hence have extensive chemical applications.

8A Hydrogenic atoms

This Topic uses the principles of quantum mechanics introduced in Focus 7 to describe the electronic structure of a 'hydrogenic atom', a one-electron atom or ion of general atomic number Z. Hydrogenic atoms are important because their Schrödinger equations can be solved exactly. Moreover, they provide a set of concepts that are used to describe the structures of many-electron atoms and molecules. Solving the Schrödinger equation for an electron in an atom involves the separation of the wavefunction into angular and radial parts and the resulting wavefunctions are the hugely important 'atomic orbitals' of hydrogenic atoms.

8B Many-electron atoms

A 'many-electron atom' is an atom or ion with more than one electron. Examples include all neutral atoms other than H; so even He, with only two electrons, is a many-electron atom. This Topic uses hydrogenic atomic orbitals to describe the structures of many-electron atoms. Then, in conjunction with the concept of 'spin' and the 'Pauli exclusion principle', it describes the origin of the periodicity of atomic properties and the structure of the periodic table.

8C Atomic spectra

The spectra of many-electron atoms are more complicated than that of hydrogen. Similar principles apply, but Coulombic and magnetic interactions between the electrons give rise to a variety of energy differences, which are summarized by constructing 'term symbols'. These symbols act as labels that display the total orbital and spin angular momentum of a many-electron atom and are used to express the selection rules that govern their spectroscopic transitions.

What is an application of this material?

Impact 13, accessed via the e-book, focuses on the use of atomic spectroscopy to examine stars. By analysing their spectra it is possible to determine the composition of their outer layers and the surrounding gases and to determine features of their physical state.

➤ **Go to the e-book for videos that feature the derivation and interpretation of equations, and applications of this material.**

TOPIC 8A Hydrogenic atoms

➤ Why do you need to know this material?

An understanding of the structure of hydrogenic atoms is central to the description of all other atoms, the periodic table, and bonding. All accounts of the structures of molecules are based on the language and concepts introduced here.

➤ What is the key idea?

Atomic orbitals are one-electron wavefunctions for atoms; they are labelled by three quantum numbers that specify the energy and angular momentum of the electron.

➤ What do you need to know already?

You need to be aware of the concept of a wavefunction (Topic 7B) and its interpretation. You also need to know how to set up a Schrödinger equation and how boundary conditions result in only certain solutions being acceptable (Topics 7D and 7F). The solutions make use of the properties of orbital angular momentum (Topic 7F).

When an electric discharge is passed through gaseous hydrogen, the H_2 molecules are dissociated and the energetically excited H atoms that are produced emit electromagnetic radiation at a number of discrete frequencies (and therefore discrete wavenumbers), producing a spectrum of a series of 'lines' (Fig. 8A.1).

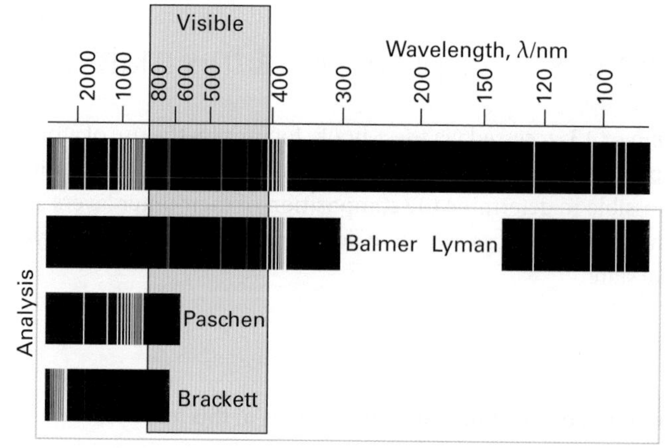

Figure 8A.1 The spectrum of atomic hydrogen. Both the observed spectrum and its resolution into overlapping series are shown. Note that the Balmer series lies in the visible region.

The Swedish spectroscopist Johannes Rydberg noted (in 1890) that the wavenumbers of all the lines are given by the expression

$$\tilde{\nu} = \tilde{R}_{\mathrm{H}}\left(\frac{1}{n_1^2} - \frac{1}{n_2^2}\right) \quad \text{Spectral lines of a hydrogen atom} \quad (8A.1)$$

with $n_1 = 1$ (the *Lyman series*), 2 (the *Balmer series*), and 3 (the *Paschen series*), and that in each case $n_2 = n_1 + 1, n_1 + 2, \ldots$. The constant $\tilde{R}_{\mathrm{H}}$ is now called the **Rydberg constant** for the hydrogen atom and is found empirically to have the value 109 677 cm^{-1}.

8A.1 The structure of hydrogenic atoms

Consider a **hydrogenic atom**, an atom or ion of arbitrary atomic number Z but having a single electron. Hydrogen itself is an example (with $Z = 1$). The Coulomb potential energy of an electron in a hydrogenic atom of atomic number Z and therefore nuclear charge Ze is (see *Energy: A first look*)

$$V(r) = -\frac{Ze^2}{4\pi\varepsilon_0 r} \quad (8A.2)$$

where r is the distance of the electron from the nucleus and ε_0 is the electric constant (formerly the 'vacuum permittivity'). The hamiltonian for the entire atom, which consists of an electron and a nucleus of mass m_{N}, is therefore

$$\hat{H} = \hat{E}_{\mathrm{k,electron}} + \hat{E}_{\mathrm{k,nucleus}} + \hat{V}(r) = -\frac{\hbar^2}{2m_{\mathrm{e}}}\nabla_{\mathrm{e}}^2 - \frac{\hbar^2}{2m_{\mathrm{N}}}\nabla_{\mathrm{N}}^2 - \frac{Ze^2}{4\pi\varepsilon_0 r} \quad \text{Hamiltonian for a hydrogenic atom} \quad (8A.3)$$

The subscripts e and N on ∇^2 indicate differentiation with respect to the electron or nuclear coordinates.

8A.1(a) The separation of variables

Physical intuition suggests that the full Schrödinger equation ought to separate into two equations, one for the motion of the atom as a whole through space and the other for the motion of the electron relative to the nucleus. The Schrödinger equation for the internal motion of the electron relative to the nucleus is[1]

$$-\frac{\hbar^2}{2\mu}\nabla^2\psi - \frac{Ze^2}{4\pi\varepsilon_0 r}\psi = E\psi \qquad \frac{1}{\mu} = \frac{1}{m_e} + \frac{1}{m_N}$$ Schrödinger equation for a hydrogenic atom (8A.4)

where differentiation is now with respect to the coordinates of the electron relative to the nucleus. The quantity μ is called the **reduced mass**. The reduced mass is very similar to the electron mass because m_N, the mass of the nucleus, is much larger than the mass of an electron, so $1/\mu \approx 1/m_e$ and therefore $\mu \approx m_e$. In all except the most precise work, the reduced mass can be replaced by m_e.

Because the potential energy is centrosymmetric (independent of angle), the equation for the wavefunction is expected to be separable into radial and angular components, as in

$$\psi(r,\theta,\phi) = R(r)Y(\theta,\phi) \qquad (8A.5)$$

with $R(r)$ the **radial wavefunction** and $Y(\theta,\phi)$ the **angular wavefunction**. The equation does separate, and the two contributions to the wavefunction are solutions of two equations:

$$\Lambda^2 Y = -l(l+1)Y \qquad (8A.6a)$$

$$-\frac{\hbar^2}{2\mu}\left(\frac{d^2R}{dr^2} + \frac{2}{r}\frac{dR}{dr}\right) + V_{eff}R = ER \qquad (8A.6b)$$

where

$$V_{eff}(r) = -\frac{Ze^2}{4\pi\varepsilon_0 r} + \frac{l(l+1)\hbar^2}{2\mu r^2} \qquad (8A.6c)$$

Equation 8A.6a is the same as the Schrödinger equation for a particle free to move at constant radius around a central point (Topic 7F). The allowed solutions are the spherical harmonics (Table 7F.1), and are specified by the quantum numbers l and m_l. As explained in Topic 7F, the orbital angular quantum number l specifies the magnitude of the orbital angular momentum and the magnetic quantum number m_l specifies the z-component of that momentum (see below). Equation 8A.6b is called the **radial wave equation**. The radial wave equation describes the motion of a particle of mass μ in a one-dimensional region $0 \le r < \infty$ where the potential energy is $V_{eff}(r)$.

8A.1 (b) The radial solutions

Some features of the shapes of the radial wavefunctions can be anticipated by examining the form of $V_{eff}(r)$. The first term in eqn 8A.6c is the Coulomb potential energy of the electron in the field of the nucleus. The second term stems from what

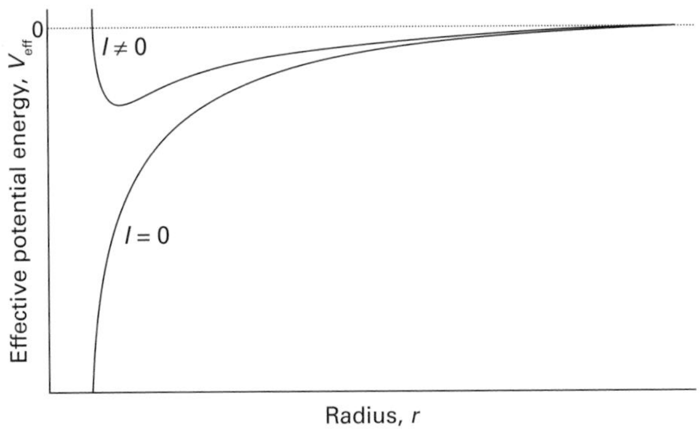

Figure 8A.2 The effective potential energy of an electron in the hydrogen atom. When the electron has zero orbital angular momentum, the effective potential energy is the Coulombic potential energy. When the electron has non-zero orbital angular momentum, the centrifugal effect gives rise to a positive contribution which is very large close to the nucleus. The $l = 0$ and $l \neq 0$ wavefunctions are therefore very different near the nucleus.

in classical physics would be called the centrifugal force arising from the angular momentum of the electron around the nucleus. The consequences are:

- When $l = 0$, the electron has no angular momentum, and the effective potential energy is purely Coulombic; the force exerted on the electron is attractive at all radii (Fig. 8A.2).
- When $l \neq 0$, the centrifugal term gives a positive contribution to the effective potential energy, corresponding to a repulsive force at all radii.
- When the electron is close to the nucleus ($r \approx 0$), the centrifugal contribution to the potential energy, which is proportional to $1/r^2$, dominates the Coulombic contribution, which is proportional to $1/r$; the net result is an effective repulsion of the electron from the nucleus.

It follows that the two effective potential energies, the one for $l = 0$ and the one for $l \neq 0$, are qualitatively very different close to the nucleus. However, they are similar at large distances because the centrifugal contribution tends to zero more rapidly (as $1/r^2$) than the Coulombic contribution (as $1/r$). Therefore, the solutions with $l = 0$ and $l \neq 0$ are expected to be quite different near the nucleus but similar far away from it.

Two features of the radial wavefunction are important:

- Close to the nucleus the radial wavefunction is proportional to $(r/a_0)^l$ with a_0 the Bohr radius, 52.9 pm (see below), and $r/a_0 < 1$. The higher the orbital angular momentum, the closer $(r/a_0)^l$ remains to zero and it is less likely that the electron will be found there (Fig. 8A.3).
- Far from the nucleus all radial wavefunctions approach zero exponentially.

[1] See the first section of *A deeper look* 8A.1, available to read in the e-book accompanying this text, for full details of this separation procedure and then the second section for the calculations that lead to eqn 8A.6.

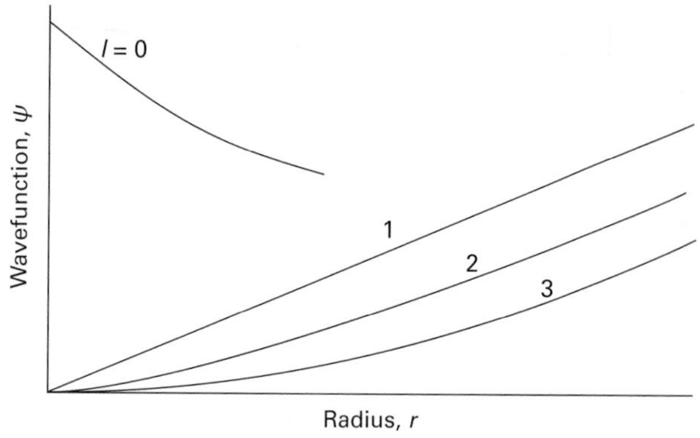

Figure 8A.3 Close to the nucleus, orbitals with $l = 1$ are proportional to r, orbitals with $l = 2$ are proportional to r^2, and orbitals with $l = 3$ are proportional to r^3. Electrons are progressively excluded from the neighbourhood of the nucleus as l increases. An orbital with $l = 0$ has a finite, non-zero value at the nucleus.

The detailed solution of the radial equation for the full range of radii shows how the form r^l close to the nucleus blends into the exponentially decaying form at great distances. It turns out that the two regions are bridged by a polynomial in r and that

$$R(r) = \overbrace{r^l}^{\text{Dominant close to the nucleus}} \times \overbrace{(\text{polynomial in } r)}^{\text{Bridges the two ends of the function}} \times \overbrace{(\text{decaying exponential in } r)}^{\text{Dominant far from the nucleus}} \quad (8A.7)$$

The radial wavefunction therefore has the form

$$R(r) = r^l L(r)\mathrm{e}^{-r}$$

with various constants and where $L(r)$ is the bridging polynomial. That polynomial has the following features:

- Close to the nucleus ($r \approx 0$) the polynomial is a constant and $\mathrm{e}^{-r} \approx 1$, so $R(r) \propto r^l$.
- Far from the nucleus the dominant term in the polynomial is proportional to r^{n-l-1}, where n is an integer; so, regardless of the value of l, all the wavefunctions of a given value of n are proportional to $r^{n-1}\mathrm{e}^{-r}$ and decay exponentially to zero in the same way.[2]

The detailed solution also shows that, for the wavefunction to be acceptable, the value of n that appears in the polynomial can take only positive integral values, and specifically $n = 1, 2,\ldots$ and the two other quantum numbers l and m_l are confined to $l = 0, 1, 2,\ldots, n-1$ and $m_l = 0, \pm1, \pm2,\ldots, \pm l$. The allowed energies are:

$$E_{n,l,m_l} = -\frac{\mu e^4}{32\pi^2\varepsilon_0^2\hbar^2} \times \frac{Z^2}{n^2} \qquad \text{Bound-state energies} \quad (8A.8)$$

[2] Note that exponential functions e^{-x} always dominate simple powers, x^n.

Note that the energies are independent of l and m_l, so each level is n^2-fold degenerate.

So far, only the general form of the radial wavefunctions has been given. It is now time to show how they depend on various fundamental constants and the atomic number of the atom. They are most simply written in terms of the dimensionless quantity ρ (rho), where

$$\rho = \frac{2Zr}{na} \quad a = \frac{m_\mathrm{e}}{\mu}a_0 \quad a_0 = \frac{4\pi\varepsilon_0\hbar^2}{m_\mathrm{e}e^2} \quad (8A.9)$$

The **Bohr radius**, a_0, has the value 52.9 pm; it is so called because the same quantity appeared in Bohr's early model of the hydrogen atom as the radius of the electron orbit of lowest energy. In practice, because $m_\mathrm{e} << m_\mathrm{N}$ (so $m_\mathrm{e}/\mu \approx 1$) there is so little difference between a and a_0 that it is safe to use a_0 in the definition of ρ for all atoms (even for ^{1}H, $a = 1.0005a_0$). In terms of these quantities and with the various quantum numbers displayed, the radial wavefunctions for an electron with quantum numbers n and l are the (real) functions

$$R_{n,l}(r) = N_{n,l}\rho^l L_{n,l}(\rho)\mathrm{e}^{-\rho/2} \qquad \text{Radial wavefunctions} \quad (8A.10)$$

where $L_{n,l}(\rho)$ is an *associated Laguerre polynomial*. These polynomials have quite simple forms, such as 1, ρ, and $2 - \rho$ (they can be picked out in Table 8A.1). The factor $N_{n,l}$ ensures that the radial wavefunction is normalized to 1 in the sense that

$$\int_0^\infty R_{n,l}(r)^2 r^2 \mathrm{d}r = 1 \quad (8A.11)$$

where the r^2 comes from the volume element in spherical coordinates; see *The chemist's toolkit* 7F.2. Specifically, the components of eqn 8A.10 can be interpreted as follows:

- The exponential factor ensures that the wavefunction approaches zero far from the nucleus.

Table 8A.1 Hydrogenic radial wavefunctions

n	l	$R_{n,l}(r)$
1	0	$2\left(\frac{Z}{a}\right)^{3/2}\mathrm{e}^{-\rho/2}$
2	0	$\frac{1}{8^{1/2}}\left(\frac{Z}{a}\right)^{3/2}(2-\rho)\mathrm{e}^{-\rho/2}$
2	1	$\frac{1}{24^{1/2}}\left(\frac{Z}{a}\right)^{3/2}\rho\mathrm{e}^{-\rho/2}$
3	0	$\frac{1}{243^{1/2}}\left(\frac{Z}{a}\right)^{3/2}(6-6\rho+\rho^2)\mathrm{e}^{-\rho/2}$
3	1	$\frac{1}{486^{1/2}}\left(\frac{Z}{a}\right)^{3/2}(4-\rho)\rho\mathrm{e}^{-\rho/2}$
3	2	$\frac{1}{2430^{1/2}}\left(\frac{Z}{a}\right)^{3/2}\rho^2\mathrm{e}^{-\rho/2}$

$\rho = (2Z/na)r$ with $a = 4\pi\varepsilon_0\hbar^2/\mu e^2$. For an infinitely heavy nucleus (or one that may be assumed to be), $\mu = m_\mathrm{e}$ and $a = a_0$, the Bohr radius.

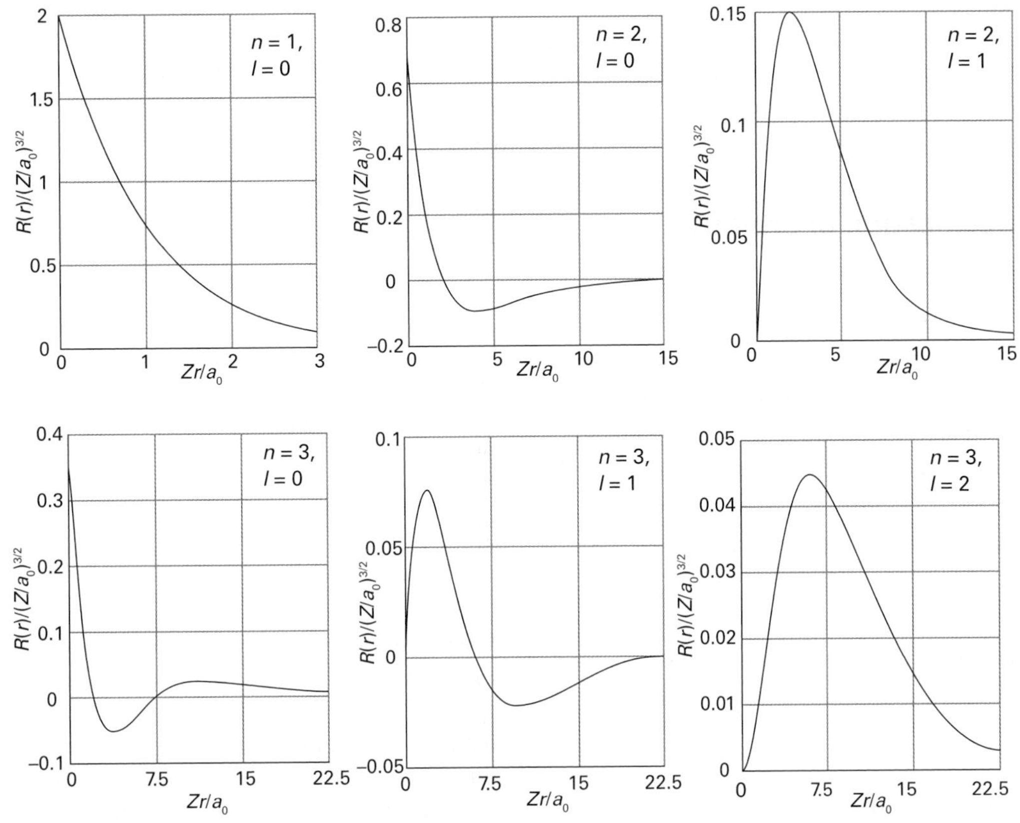

Figure 8A.4 The radial wavefunctions of the first few states of hydrogenic atoms of atomic number Z. Note that the orbitals with $l = 0$ have a non-zero and finite value at the nucleus. The horizontal scales are different in each case: as the principal quantum number increases, so too does the size of the orbital. *Interact with the dynamic version of this graph in the e-book.*

- The factor ρ^l ensures that (provided $l > 0$) the wavefunction vanishes at the nucleus.
- The associated Laguerre polynomial is a function that in general oscillates from positive to negative values and accounts for the presence of **radial nodes.** The zero at $r = 0$ is not a radial node because the radial wavefunction does not pass through zero at that point (because r cannot be negative).

Expressions for some radial wavefunctions are given in Table 8A.1 and illustrated in Fig. 8A.4. Finally, with the form of the radial wavefunction established, the total wavefunction, eqn 8A.5, in full dress becomes

$$\psi_{n,l,m_l}(r,\theta,\phi) = R_{n,l}(r)Y_{l,m_l}(\theta,\phi) \qquad (8A.12)$$

Brief illustration 8A.1

To calculate the probability density at the nucleus for an electron with $n = 1$, $l = 0$, and $m_l = 0$, evaluate ψ at $r = 0$:

$$\psi_{1,0,0}(0,\theta,\phi) = R_{1,0}(0)Y_{0,0}(\theta,\phi) = 2\left(\frac{Z}{a_0}\right)^{3/2}\left(\frac{1}{4\pi}\right)^{1/2}$$

The probability density is therefore

$$\psi^2_{1,0,0}(0,\theta,\phi) = \frac{Z^3}{\pi a_0^3}$$

which evaluates to 2.15×10^{-6} pm^{-3} when $Z = 1$.

Exercises

E8A.1 The wavefunction for the ground state of a hydrogen atom is Ne^{-r/a_0}. Evaluate the normalization constant N.

E8A.2 Evaluate the probability density at the nucleus of an electron with $n = 2$, $l = 0$, $m_l = 0$.

E8A.3 By differentiation of the 2s radial wavefunction, show that it has two extrema in its amplitude, and locate them.

8A.2 Atomic orbitals and their energies

An **atomic orbital** is a one-electron wavefunction for an electron in an atom, and for hydrogenic atoms has the form specified in eqn 8A.12. Each hydrogenic atomic orbital is defined by three quantum numbers, designated n, l, and m_l. An electron described by one of the wavefunctions in eqn 8A.12 is said to 'occupy' that orbital. For example, an electron described by the wavefunction $\psi_{1,0,0}$ is said to 'occupy' the orbital with $n = 1$, $l = 0$, and $m_l = 0$.

8A.2(a) The specification of orbitals

Each of the three quantum numbers specifies a different attribute of the orbital:

- The **principal quantum number**, n, specifies the energy of the orbital (through eqn 8A.8); it takes the values $n = 1, 2, 3, \ldots$.
- The **orbital angular momentum quantum number**, l, specifies the magnitude of the angular momentum of the electron as $\{l(l+1)\}^{1/2}\hbar$, with $l = 0, 1, 2, \ldots, n-1$.
- The **magnetic quantum number**, m_l, specifies the z-component of the angular momentum as $m_l\hbar$, with $m_l = 0, \pm1, \pm2, \ldots, \pm l$.

Note how the value of the principal quantum number controls the maximum value of l, and how the value of l controls the range of values of m_l.

8A.2(b) The energy levels

The energy levels predicted by eqn 8A.8 are depicted in Fig. 8A.5. Some important features are:

- The energies, and also the separation of neighbouring levels, are proportional to Z^2, so the levels are four times as wide apart (and the ground state four times lower in energy) in He^+ ($Z = 2$) than in H ($Z = 1$).
- All the energies given by eqn 8A.8 are negative. They refer to the **bound states** of the atom, in which the energy of the atom is lower than that of the infinitely separated, stationary electron and nucleus (which corresponds to the zero of energy).
- There are also solutions of the Schrödinger equation with positive energies. These solutions correspond to **unbound states** of the electron, the states to which an electron is raised when it is ejected from the atom by a high-energy collision or photon. The energies of the unbound electron are not quantized and form the continuum states of the atom.

Equation 8A.8, which can be written as

$$E_{n,l,m_l} = -\frac{hcZ^2\tilde{R}_N}{n^2}, \quad \tilde{R}_N = \frac{\mu e^4}{8\varepsilon_0^2 ch^3} \qquad \text{Bound-state energies} \quad (8A.13)$$

is consistent with the spectroscopic result summarized by eqn 8A.1, with the Rydberg constant for the atom identified as

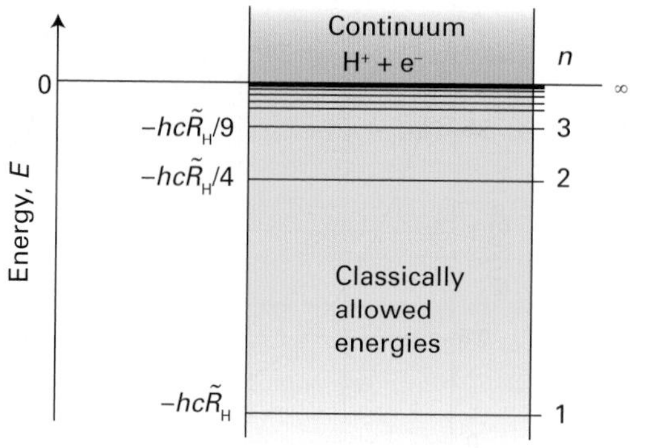

Figure 8A.5 The energy levels of a hydrogen atom. The values are relative to an infinitely separated, stationary electron and a proton.

$$\tilde{R}_N = \frac{\mu}{m_e} \times \tilde{R}_\infty \quad \tilde{R}_\infty = \frac{m_e e^4}{8\varepsilon_0^2 h^3 c} \qquad \text{Rydberg constant} \quad (8A.14)$$

where μ is the reduced mass of the atom and $\tilde{R}_\infty$ is the **Rydberg constant**; the constant $\tilde{R}_N$ is the value that constant takes for a specified atom N (not nitrogen!), such as hydrogen, when N is replaced by H and μ takes the appropriate value. Insertion of the values of the fundamental constants into the expression for $\tilde{R}_H$ gives almost exact agreement with the experimental value for hydrogen. The only discrepancies arise from the neglect of relativistic corrections (in simple terms, the increase of mass with speed), which the non-relativistic Schrödinger equation ignores.

Brief illustration 8A.2

The value of $\tilde{R}_\infty$ is 109 737 cm^{-1}. The reduced mass of a hydrogen atom with $m_p = 1.672\,62 \times 10^{-27}$ kg and $m_e = 9.109\,38 \times 10^{-31}$ kg is

$$\mu = \frac{m_e m_p}{m_e + m_p} = \frac{(9.109\,38\times10^{-31}\,\text{kg})\times(1.672\,62\times10^{-27}\,\text{kg})}{(9.109\,38\times10^{-31}\,\text{kg})+(1.672\,62\times10^{-27}\,\text{kg})}$$
$$= 9.104\,42\times10^{-31}\,\text{kg}$$

It then follows that

$$\tilde{R}_H = \frac{9.104\,42\times10^{-31}\,\text{kg}}{9.109\,38\times10^{-31}\,\text{kg}}\times109\,737\,\text{cm}^{-1} = 109\,677\,\text{cm}^{-1}$$

and that the ground state of the electron ($n = 1$, $l = 0$, $m_l = 0$) lies at

$$E_{1,0,0} = -hc\tilde{R}_H = -(6.626\,08\times10^{-34}\,\text{J s})$$
$$\times(2.997\,945\times10^{10}\,\text{cm s}^{-1})\times(109\,677\,\text{cm}^{-1})$$
$$= -2.178\,70\times10^{-18}\,\text{J}$$

or −2.178 70 aJ. This energy corresponds to −13.598 eV.

8A.2(c) Ionization energies

The **ionization energy**, I, of an element is the minimum energy required to remove an electron from the **ground state**, the state of lowest energy, of one of its atoms in the gas phase. Because the ground state of hydrogen is the state with $n = 1$, with energy $E_1 = -hc\tilde{R}_H$ and the atom is ionized when the electron has been excited to the level corresponding to $n = \infty$ (see Fig. 8A.5), the energy that must be supplied is

$$I = hc\tilde{R}_H \tag{8A.15}$$

The value of I is 2.179 aJ (1 aJ = 10^{-18} J), which corresponds to 13.60 eV.[3]

Example 8A.1 Measuring an ionization energy spectroscopically

The emission spectrum of atomic hydrogen shows lines at 82 259, 97 492, 102 824, 105 292, 106 632, and 107 440 cm^{-1}, which correspond to transitions to the same lower state from successive upper states with $n = 2, 3, \ldots$. Determine the ionization energy of the lower state.

Collect your thoughts The spectroscopic determination of ionization energies depends on the identification of the 'series limit', the wavenumber at which the series terminates and becomes a continuum. If the upper state lies at an energy $-hc\tilde{R}_H/n^2$, then the wavenumber of the photon emitted when the atom makes a transition to the lower state, with energy E_{lower}, is

$$\tilde{\nu} = -\frac{\tilde{R}_H}{n^2} - \frac{E_{lower}}{hc} \overset{I=-E_{lower}}{=} -\frac{\tilde{R}_H}{n^2} + \frac{I}{hc}$$

A plot of the wavenumbers against $1/n^2$ should give a straight line of slope $-\tilde{R}_H$ and intercept I/hc. Use software to calculate a least-squares fit of the data in order to obtain a result that reflects the precision of the data.

The solution The wavenumbers are plotted against $1/n^2$ in Fig. 8A.6. From the (least-squares) intercept, it follows that $I/hc = 109\,679\ cm^{-1}$, so the ionization energy is

$$\begin{aligned} I &= hc \times (109\,677\ cm^{-1}) \\ &= (6.626\,08 \times 10^{-34}\ J\,s) \times (2.997\,945 \times 10^{10}\ cm\,s^{-1}) \\ &\quad \times (109\,677\ cm^{-1}) \\ &= 2.1787 \times 10^{-18}\ J \end{aligned}$$

or 2.1787 aJ, corresponding to 1312.1 kJ mol^{-1} (the negative of the value of E calculated in *Brief illustration* 8A.2).

[3] Ionization energies are sometimes referred to as *ionization potentials*. That is incorrect, but not uncommon. If the term is used at all, it should denote the electrical potential difference through which an electron must be moved for the change in its potential energy to be equal to the ionization energy, and reported in volts: the ionization energy of hydrogen is 13.60 eV; its ionization potential is 13.60 V.

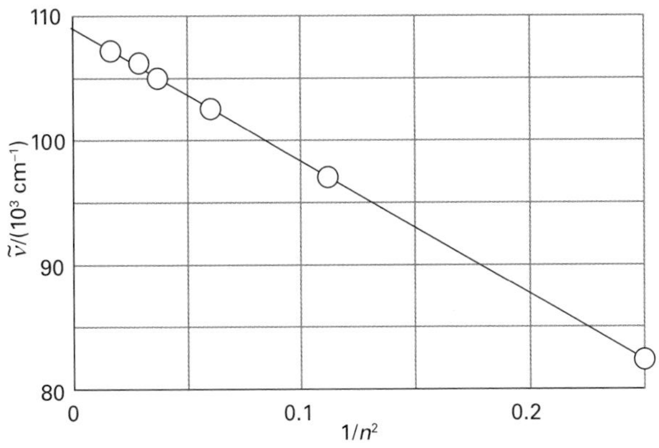

Figure 8A.6 The plot of the data in *Example* 8A.1 used to determine the ionization energy of an atom (in this case, of H).

Self-test 8A.1 The emission spectrum of atomic deuterium shows lines at 15 238, 20 571, 23 039, and 24 380 cm^{-1}, which correspond to transitions from successive upper states with $n = 3, 4, \ldots$ to the same lower state. Determine (a) the ionization energy of the lower state, (b) the ionization energy of the ground state, (c) the mass of the deuteron (by expressing the Rydberg constant in terms of the reduced mass of the electron and the deuteron, and solving for the mass of the deuteron).

Answer: (a) 328.1 kJ mol^{-1}, (b) 1312.4 kJ mol^{-1}, (c) 2.8×10^{-27} kg, a result very sensitive to $\tilde{R}_D$

8A.2(d) Shells and subshells

All the orbitals of a given value of n are said to form a single **shell** of the atom. In a hydrogenic atom (and only in a hydrogenic atom), all orbitals of given n, and therefore belonging to the same shell, have the same energy. It is common to refer to successive shells by letters:

$n =$ 1 2 3 4…
K L M N… *Specification of shells*

Thus, all the orbitals of the shell with $n = 2$ form the L shell of the atom, and so on.

The orbitals with the same value of n but different values of l are said to form a **subshell** of a given shell. These subshells are also generally referred to by letters:

$l =$ 0 1 2 3 4 5 6…
s p d f g h i… *Specification of subshells*

All orbitals of the same subshell have the same energy in all kinds of atoms, not only hydrogenic atoms. After $l = 3$ the letters run alphabetically (j is not used because in some languages i and j are not distinguished). Figure 8A.7 is a version of Fig. 8A.5 which shows the subshells explicitly. Because l can range from 0 to $n - 1$, giving n values in all, it follows that there

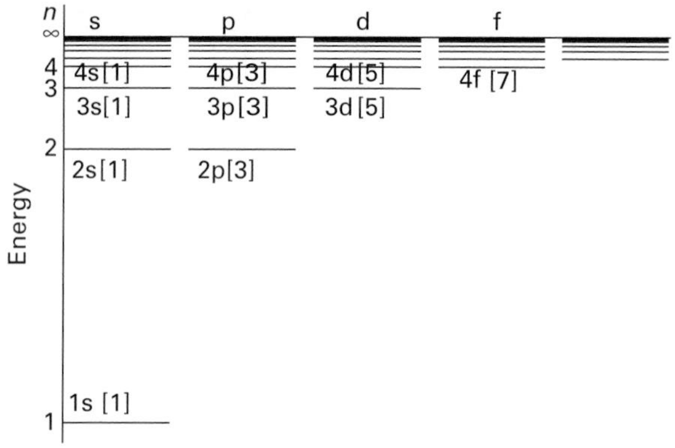

Figure 8A.7 The energy levels of a hydrogenic atom showing the subshells and (in square brackets) the numbers of orbitals in each subshell. All orbitals of a given shell have the same energy.

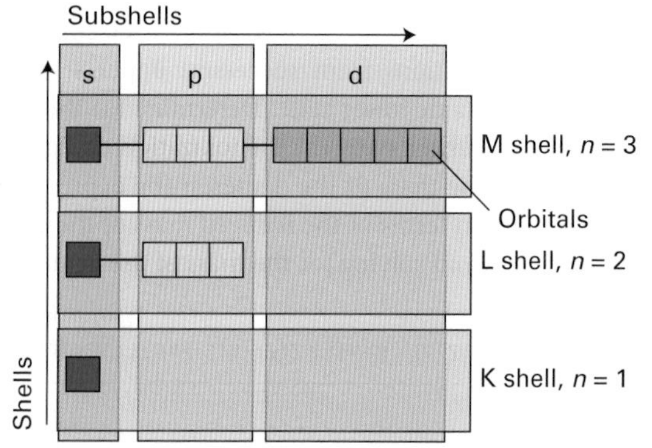

Figure 8A.8 The organization of orbitals (small squares) into subshells (characterized by l) and shells (characterized by n).

are n subshells of a shell with principal quantum number n. The organization of orbitals in the shells is summarized in Fig. 8A.8. The number of orbitals in a shell of principal quantum number n is n^2, so in a hydrogenic atom each energy level is n^2-fold degenerate.

Brief illustration 8A.3

When $n = 1$ there is only one subshell, that with $l = 0$, and that subshell contains only one orbital, with $m_l = 0$ (the only value of m_l permitted). When $n = 2$, there are four orbitals, one in the s subshell with $l = 0$ and $m_l = 0$, and three in the $l = 1$ subshell with $m_l = +1, 0, -1$. When $n = 3$ there are nine orbitals (one with $l = 0$, three with $l = 1$, and five with $l = 2$).

8A.2(e) s Orbitals

The orbital occupied in the ground state is the one with $n = 1$ (and therefore with $l = 0$ and $m_l = 0$, the only possible values of these quantum numbers when $n = 1$). From Table 8A.1 and with $Y_{0,0} = (1/4\pi)^{1/2}$ (Table 7F.1) it follows that (for $Z = 1$):

$$\psi = \frac{1}{(\pi a_0^3)^{1/2}} e^{-r/a_0} \tag{8A.16}$$

This wavefunction has the following features:

- It is independent of angle and has the same value at all points of constant radius; that is, the 1s orbital (the s orbital with $n = 1$, and in general ns) is 'spherically symmetrical'.
- It decays exponentially from a maximum value of $1/(\pi a_0^3)^{1/2}$ at the nucleus (at $r = 0$), so that the probability density of the electron is greatest at the nucleus itself.

The general form of the ground-state wavefunction can be understood by considering the contributions of the potential and kinetic energies to the total energy of the electron in the atom. The more compact the orbital, the shorter the average distance of the electron from the nucleus, and therefore the lower (more negative) its average potential energy. However, the more compact the wavefunction, the greater its curvature and hence the greater its average kinetic energy. This dependence of energy on the compactness of the wavefunction is illustrated in Fig 8A.9. The most compact wavefunction has the lowest potential energy but the highest kinetic energy and a high energy overall. The least compact wavefunction has a low kinetic energy but a high potential energy and a high energy overall. The wavefunction that lies between these two has a moderately high kinetic energy, a moderately low potential energy, and a total energy that is lowest of the three. It corresponds to the observed ground-state wavefunction.

One way of depicting the probability density of the electron is to represent $|\psi|^2$ by the density of shading (Fig. 8A.10). A simpler procedure is to show only the **boundary surface**, the

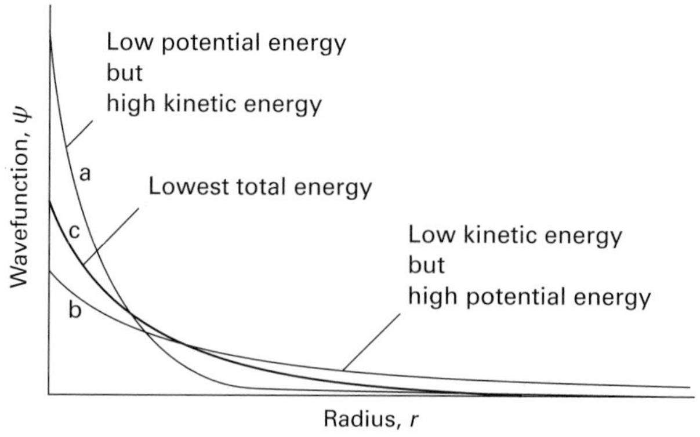

Figure 8A.9 The balance of kinetic and potential energies that accounts for the structure of the ground state of hydrogenic atoms. (a) The sharply curved but localized orbital has high mean kinetic energy, but low mean potential energy; (b) the mean kinetic energy is low, but the potential energy is not very favourable; (c) the compromise of moderate kinetic energy and moderately favourable potential energy.

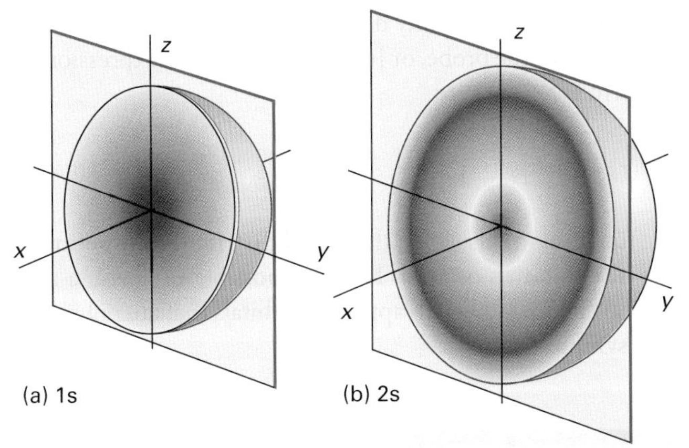

Figure 8A.10 Representations of cross-sections through the (a) 1s and (b) 2s hydrogenic atomic orbitals in terms of their electron probability densities (as represented by the density of shading).

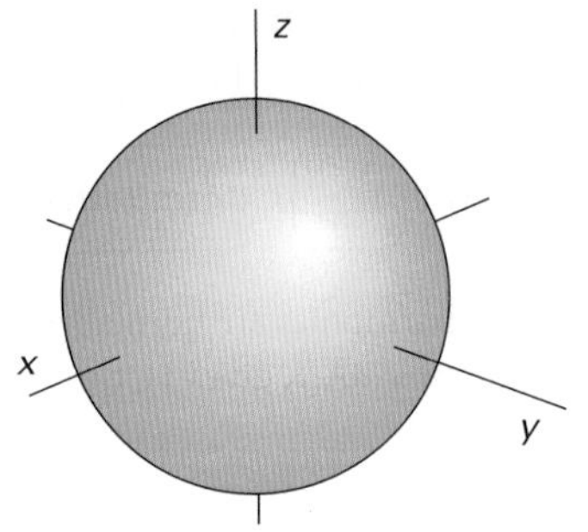

Figure 8A.11 The boundary surface of a 1s orbital, within which there is a 90 per cent probability of finding the electron. All s orbitals have spherical boundary surfaces.

surface that mirrors the shape of the orbital and captures a high proportion (typically about 90 per cent) of the electron probability. Formally, computer software typically constructs an **isosurface**, a surface of constant probability density that essentially captures a high proportion of the probability and mirrors the characteristic shape of the orbital. Most of the orbital depictions in this text are notional: they will be referred to as 'boundary surfaces' but apart from those for s orbitals they are only general indications of their shapes. For the 1s orbital, the boundary surface is a sphere centred on the nucleus (Fig. 8A.11).

Example 8A.2 Calculating the mean radius of an orbital

Calculate the mean radius of a hydrogenic 1s orbital.

Collect your thoughts The mean radius is the expectation value

$$\langle r\rangle = \int \psi^{*} r\psi \,\mathrm{d}\tau = \int r|\psi|^2 \mathrm{d}\tau$$

where the wavefunction is normalized. You need to evaluate the integral by using the wavefunctions given in Table 8A.1 and $\mathrm{d}\tau = r^2\mathrm{d}r \sin\theta \,\mathrm{d}\theta \,\mathrm{d}\phi$ (*The chemist's toolkit* 7F.2). The angular parts of the wavefunction (Table 7F.1) are normalized in the sense that

$$\int_{\theta=0}^{\pi}\int_{\phi=0}^{2\pi} \left|Y_{l,m_l}\right|^2 \sin\theta \,\mathrm{d}\theta\mathrm{d}\phi = 1$$

The relevant integral over r is given in the *Resource section*.

The solution With the wavefunction written in the form $\psi = RY$, the integration (with the integral over the angular variables, which is equal to 1, in the boxes) is

$$\langle r\rangle = \int_0^{\infty} \boxed{\int_0^{\pi}\int_0^{2\pi}} rR_{n,l}^2 \boxed{\left|Y_{l,m_l}\right|^2} r^2\,\mathrm{d}r \boxed{\sin\theta \,\mathrm{d}\theta\mathrm{d}\phi} = \int_0^{\infty} r^3 R_{n,l}^2 \mathrm{d}r$$

For a 1s orbital

$$R_{1,0} = 2\left(\frac{Z}{a_0}\right)^{3/2} \mathrm{e}^{-Zr/a_0}$$

Hence

$$\langle r\rangle = \frac{4Z^3}{a_0^3}\overbrace{\int_0^{\infty} r^3 \mathrm{e}^{-2Zr/a_0}\mathrm{d}r}^{\text{Integral E.3}} = \frac{4Z^3}{a_0^3}\times\frac{3!}{(2Z/a_0)^4} = \frac{3a_0}{2Z}$$

Self-test 8A.2 Evaluate the mean radius of a 3s orbital. *Hint:* Use mathematical software.

Answer: $27a_0/2Z$

All s orbitals are spherically symmetric, but differ in the number of radial nodes. For example, the 1s, 2s, and 3s orbitals have 0, 1, and 2 radial nodes, respectively. In general, an ns orbital has $n - 1$ radial nodes. As n increases, the radius of the spherical boundary surface that essentially captures a given fraction of the probability also increases.

Brief illustration 8A.4

The radial nodes of a 2s orbital lie at the locations where the associated Laguerre polynomial factor (Table 8A.1) is equal to zero. In this case the factor is simply $2 - \rho$ so there is a node at $\rho = 2$. For a 2s orbital, $\rho = Zr/a_0$, so the radial node occurs at $r = 2a_0/Z$ (see Fig. 8A.4).

8A.2(f) Radial distribution functions

The wavefunction yields, through the value of $|\psi|^2$, the probability of finding an electron in any region. As explained in Topic 7B, $|\psi|^2$ is a probability *density* (dimensions: 1/volume) and can be interpreted as a (dimensionless) probability when multiplied by the (infinitesimal) volume of interest. Imagine a probe with a fixed volume $d\tau$, and sensitive to electrons, that can move around near the nucleus of a hydrogenic atom. Because the probability density in the ground state of the atom is proportional to e^{-2Zr/a_0}, the reading from the detector decreases exponentially as the probe is moved out along any radius but is constant if the probe is moved on a circle of constant radius (Fig. 8A.12).

Now consider the total probability of finding the electron anywhere between the two walls of a spherical shell of thickness dr at a radius r. The sensitive volume of the probe is now the volume of the shell (Fig. 8A.13), which is $4\pi r^2 dr$ (the product of its surface area, $4\pi r^2$, and its thickness, dr). Note that the volume probed increases with distance from the nucleus and is zero at the nucleus itself, when $r = 0$. The probability that the electron will be found between the inner and outer surfaces of

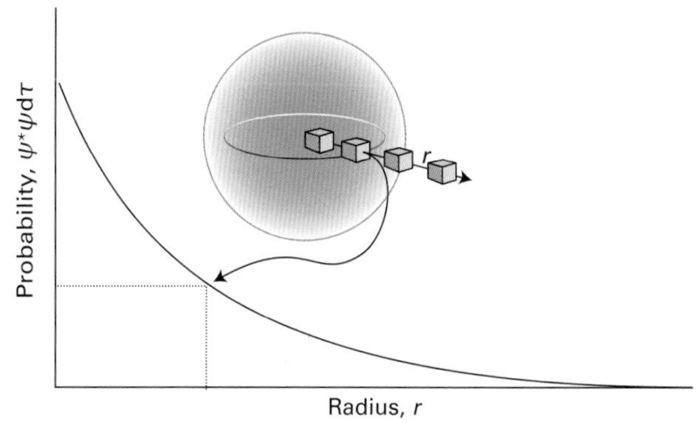

Figure 8A.12 For a 1s orbital a constant-volume electron-sensitive detector (the small cube) gives its greatest reading at the nucleus, and a smaller reading elsewhere. The same reading is obtained anywhere on a circle of given radius at any orientation: the s orbital is spherically symmetrical.

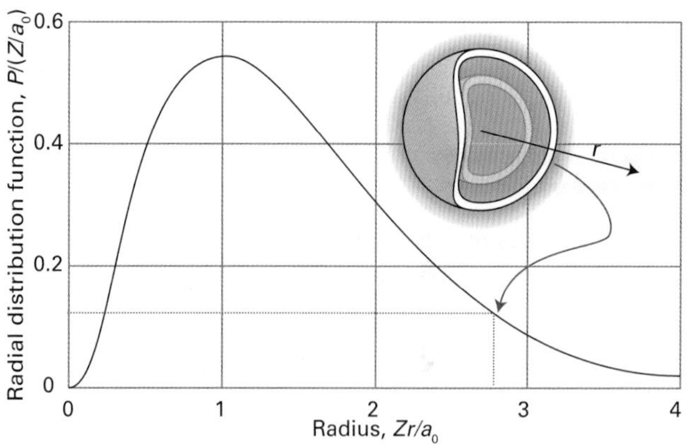

Figure 8A.13 The radial distribution function $P(r)$ is the probability density that the electron will be found anywhere in a shell of radius r; the probability itself is $P(r)dr$, where dr is the thickness of the shell. For a 1s electron in hydrogen, $P(r)$ is a maximum when r is equal to the Bohr radius a_0. The value of $P(r)dr$ is equivalent to the reading that a detector shaped like a spherical shell of thickness dr would give as its radius is varied.

this shell is the probability density at the radius r multiplied by the volume of the probe, or $|\psi(r)|^2 \times 4\pi r^2 dr$. This expression has the form $P(r)dr$, where

$$P(r) = 4\pi r^2 |\psi(r)|^2 \qquad \text{Radial distribution function [s orbitals only]} \qquad (8A.17a)$$

The function $P(r)$ is called the **radial distribution function** (in this case, for an s orbital). It is also possible to devise a more general expression which applies to orbitals that are not spherically symmetrical.

How is that done? 8A.1 Deriving the general form of the radial distribution function

The probability of finding an electron in a volume element $d\tau$ when its wavefunction is $\psi = RY$ is $|RY|^2 d\tau$ with $d\tau = r^2 dr\, \sin\theta d\theta d\phi$. The total probability of finding the electron at any angle in a shell of radius r and thickness dr is the integral of this probability over the entire surface, and is written $P(r)dr$; so

$$P(r)dr = \boxed{\int_0^{\pi}\int_0^{2\pi}} R(r)^2 \boxed{|Y_{l,m_l}|^2} r^2 dr \boxed{\sin\theta d\theta d\phi}$$

Because the spherical harmonics are normalized to 1 (the boxed integration, as in *Example* 8A.2, gives 1), the final result is

$$P(r) = r^2 R(r)^2 \qquad \text{Radial distribution function [general form]} \qquad (8A.17b)$$

The radial distribution function is a probability density in the sense that, when it is multiplied by dr, it gives the probability of finding the electron anywhere between the two walls of a spherical shell of thickness dr at the radius r. For a 1s orbital,

$$P(r) = \frac{4Z^3}{a_0^3} r^2 e^{-2Zr/a_0} \qquad (8A.18)$$

This expression can be interpreted as follows:

- Because $r^2 = 0$ at the nucleus, $P(0) = 0$. The volume of the shell is zero when $r = 0$ so the probability of finding the electron in the shell is zero.
- As $r \to \infty$, $P(r) \to 0$ on account of the exponential term. The wavefunction has fallen to zero at great distances from the nucleus and there is little probability of finding the electron even in a large shell.
- The increase in r^2 and the decrease in the exponential factor means that P passes through a maximum at an intermediate radius (see Fig. 8A.13); it marks the most probable radius at which the electron will be found regardless of direction.

Example 8A.3 Calculating the most probable radius

Calculate the most probable radius, r_{mp}, at which an electron will be found when it occupies a 1s orbital of a hydrogenic atom of atomic number Z, and tabulate the values for the one-electron species from H to Ne^{9+}.

Collect your thoughts You need to find the radius at which the radial distribution function of the hydrogenic 1s orbital has a maximum value by solving $dP/dr = 0$. If there are several maxima, you should choose the one corresponding to the greatest amplitude.

The solution The radial distribution function is given in eqn 8A.18. It follows that

$$\frac{dP}{dr} = \frac{4Z^3}{a_0^3}\left(2r - \frac{2Zr^2}{a_0}\right)e^{-2Zr/a_0} = \frac{8rZ^3}{a_0^3}\left(1 - \frac{Zr}{a_0}\right)e^{-2Zr/a_0}$$

This function is zero other than at $r = 0$ where the term in parentheses is zero, which is at

$$r_{mp} = \frac{a_0}{Z}$$

Then, with $a_0 = 52.9$ pm, the most probable radii are

	H	He^+	Li^{2+}	Be^{3+}	B^{4+}	C^{5+}	N^{6+}	O^{7+}	F^{8+}	Ne^{9+}
r_{mp}/pm	52.9	26.5	17.6	13.2	10.6	8.82	7.56	6.61	5.88	5.29

Notice how the 1s orbital is drawn towards the nucleus as the nuclear charge increases. At uranium the most probable radius is only 0.58 pm, almost 100 times closer than for hydrogen. (On a scale where $r_{mp} = 10$ cm for H, $r_{mp} = 1$ mm for U.) However, extending this result to very heavy atoms neglects important relativistic effects that complicate the calculation.

Self-test 8A.3 Find the most probable distance of a 2s electron from the nucleus in a hydrogenic atom. *Hint:* Use mathematical software.

Answer: $(3+5^{1/2})a_0/Z = 5.24a_0/Z$; this value reflects the expansion of the atom as its energy increases.

8A.2(g) p Orbitals

All three 2p orbitals have $l = 1$, and therefore the same magnitude of angular momentum; they are distinguished by different values of m_l, the quantum number that specifies the component of angular momentum around a chosen axis (conventionally taken to be the z-axis). The orbital with $m_l = 0$, for instance, has zero angular momentum around the z-axis. Its angular variation is given by the spherical harmonic $Y_{1,0}$, which is proportional to $\cos\theta$ (see Table 7F.1). Therefore, the probability density, which is proportional to $\cos^2\theta$, has its maximum value on either side of the nucleus along the z-axis (at $\theta = 0$ and $180°$, where $\cos^2\theta = 1$). Specifically, the wavefunction of a 2p orbital with $m_l = 0$ is

$$\psi_{2,1,0} = R_{2,1}(r)Y_{1,0}(\theta,\phi) = \frac{1}{4(2\pi)^{1/2}}\left(\frac{Z}{a_0}\right)^{5/2}\boxed{r\cos\theta}e^{-Zr/2a_0} = \boxed{r\cos\theta}f(r) \qquad (8A.19a)$$

where $f(r)$ is a function only of r. Because in spherical polar coordinates $z = r\cos\theta$ (*The chemist's toolkit* 7F.2), this wavefunction may also be written

$$\psi_{2,1,0} = zf(r) \qquad (8A.19b)$$

All p orbitals with $m_l = 0$ and any value of n have wavefunctions of this form, but $f(r)$ depends on the value of n. This way of writing the orbital is the origin of the name 'p_z orbital': the general form of its boundary surface is shown in Fig. 8A.14. The wavefunction is zero everywhere in the xy-plane, where $z = 0$, so the xy-plane is a **nodal plane** of the orbital: the wavefunction changes sign on going from one side of the plane to the other.

The wavefunctions of 2p orbitals with $m_l = \pm 1$ are

$$\psi_{2,1,\pm1} = R_{2,1}(r)Y_{1,\pm1}(\theta,\phi) = \mp\frac{1}{8\pi^{1/2}}\left(\frac{Z}{a_0}\right)^{5/2} r\sin\theta\, e^{\pm i\phi}\, e^{-Zr/2a_0} = \mp\frac{1}{2^{1/2}}r\sin\theta\, e^{\pm i\phi}f(r) \qquad (8A.20)$$

In Topic 7D it is explained that a particle described by a complex wavefunction has net motion. In the present case, the functions correspond to non-zero angular momentum about the z-axis: $e^{+i\phi}$ corresponds to clockwise rotation when viewed from below, and $e^{-i\phi}$ corresponds to counterclockwise rotation (from the same viewpoint). They have zero amplitude where $\theta = 0$ and $180°$ (along the z-axis) and maximum amplitude at $90°$,

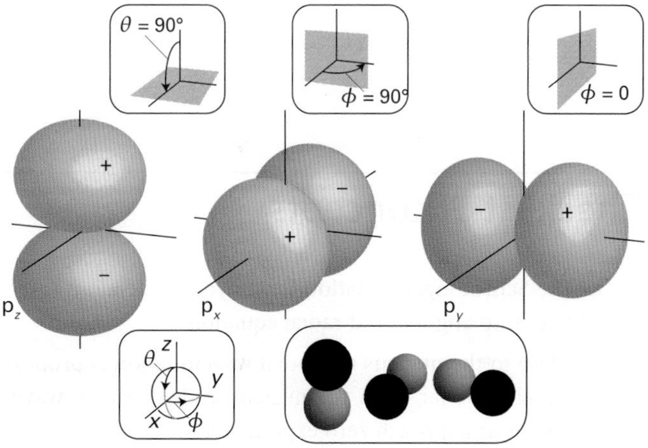

Figure 8A.14 The general shapes of the boundary surfaces of 2p orbitals. A nodal plane passes through the nucleus and separates the two lobes of each orbital. The dark and light lobes denote regions of opposite sign of the wavefunction. The angles that specify the nodal planes are shown. All p orbitals have boundary surfaces like those shown here.

which is in the xy-plane. To draw the functions it is usual to represent them by forming the linear combinations

$$\psi_{2p_x} = \frac{1}{2^{1/2}}(-\psi_{2,1,+1} + \psi_{2,1,-1}) \overset{e^{i\phi}+e^{-i\phi}=2\cos\phi}{=} r\sin\theta\cos\phi f(r) = xf(r)$$

$$\psi_{2p_y} = \frac{i}{2^{1/2}}(\psi_{2,1,+1} + \psi_{2,1,-1}) \overset{e^{i\phi}-e^{-i\phi}=2i\sin\phi}{=} r\sin\theta\sin\phi f(r) = yf(r) \quad (8A.21)$$

These linear combinations correspond to zero orbital angular momentum around the z-axis, as they are superpositions of states with equal and opposite values of m_l. The p_x orbital has the same shape as a p_z orbital, but it is directed along the x-axis (see Fig. 8A.14); the p_y orbital is similarly directed along the y-axis. The wavefunction of any p orbital of a given shell can be written as a product of x, y, or z and the same function f (which depends on the value of n).

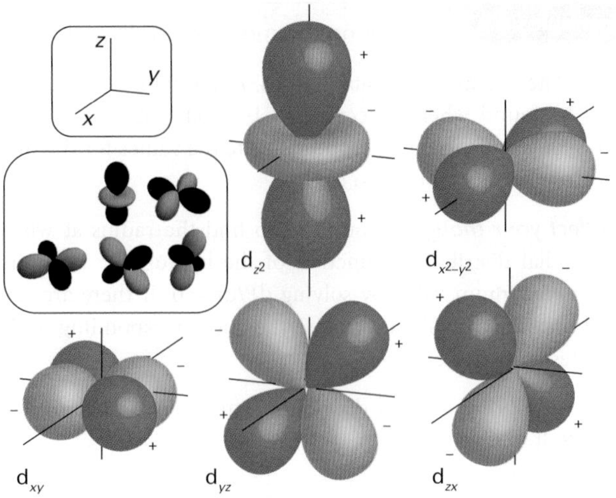

Figure 8A.15 The general form of the boundary surfaces of 3d orbitals. The purple and yellow areas (which are mirrored in the insert) denote regions of opposite sign of the wavefunction. All d orbitals have boundary surfaces like those shown here.

8A.2(h) d Orbitals

When $n = 3$, l can be 0, 1, or 2. As a result, this shell consists of one 3s orbital, three 3p orbitals, and five 3d orbitals. Each value of the quantum number $m_l = 0, \pm1, \pm2$ corresponds to a different value of the component of angular momentum about the z-axis. As for the p orbitals, d orbitals with opposite values of m_l (and hence opposite senses of motion around the z-axis) may be combined in pairs to give real wavefunctions, and the boundary surfaces of the resulting functions are shown in Fig. 8A.15. The real linear combinations have the following forms, with the function $f(r)$ depending on the value of n:

$$\psi_{d_{xy}} = xyf(r) \quad \psi_{d_{yz}} = yzf(r) \quad \psi_{d_{zx}} = zxf(r)$$
$$\psi_{d_{x^2-y^2}} = \tfrac{1}{2}(x^2-y^2)f(r) \quad \psi_{d_{z^2}} = \frac{1}{12^{1/2}}(3z^2-r^2)f(r) \quad (8A.22)$$

These linear combinations give rise to the notation d_{xy}, d_{yz}, etc. for the d-orbitals. With the exception of the d_{z^2} orbital, each combination has two angular nodes that divide the orbital into four lobes. For the d_{z^2} orbital, the two angular nodes combine to give a conical surface that separates the main lobes from a smaller toroidal component encircling the nucleus.

Exercises

E8A.4 What subshells and orbitals are available in the M shell?

E8A.5 What is the orbital angular momentum (as multiples of $\hbar$) of an electron in the orbitals (i) 1s, (ii) 3s, (iii) 3d? Give the numbers of angular and radial nodes in each case.

E8A.6 Write down the expression for the radial distribution function of a 2p electron in a hydrogenic atom of atomic number Z and identify the radius at which the electron is most likely to be found.

Checklist of concepts

☐ 1. The Schrödinger equation for a hydrogenic atom separates into angular and radial equations.

☐ 2. Close to the nucleus the radial wavefunction is proportional to r^l; far from the nucleus all hydrogenic wavefunctions approach zero exponentially.

☐ 3. An **atomic orbital** is a one-electron wavefunction for an electron in an atom.

☐ 4. An atomic orbital is specified by the values of the **quantum numbers** n, l, and m_l.

☐ 5. The energies of the bound states of hydrogenic atoms are proportional to $-Z^2/n^2$.

☐ 6. The **ionization energy** of an element is the minimum energy required to remove an electron from the ground state of one of its atoms.

☐ 7. Orbitals of a given value of n form a **shell** of an atom, and within that shell orbitals of the same value of l form **subshells**.

☐ 8. Orbitals of the same shell all have the same energy in hydrogenic atoms; orbitals of the same subshell of a shell are degenerate in all types of atoms.

☐ 9. **s Orbitals** are spherically symmetrical and have nonzero probability density at the nucleus.

☐ 10. A **radial distribution function** is the probability density for the distribution of the electron as a function of distance from the nucleus.

☐ 11. There are three **p orbitals** in a given subshell; each one has one angular node.

☐ 12. There are five **d orbitals** in a given subshell; each one has two angular nodes.

Checklist of equations

Property	Equation	Comment	Equation number
Wavenumbers of the spectral lines of a hydrogen atom	$\tilde{\nu} = \tilde{R}_H(1/n_1^2 - 1/n_2^2)$	$\tilde{R}_H$ is the Rydberg constant for hydrogen (expressed as a wavenumber)	8A.1
Bohr radius	$a_0 = 4\pi\varepsilon_0\hbar^2/m_e e^2$	$a_0 = 52.9$ pm	8A.9
Wavefunctions of hydrogenic atoms	$\psi_{n,l,m_l}(r,\theta,\phi) = R_{n,l}(r)Y_{l,m_l}(\theta,\phi)$	Y_{l,m_l} are spherical harmonics	8A.12
Energies of hydrogenic atoms	$E_{n,l,m_l} = -hcZ^2\tilde{R}_N/n^2$, $\tilde{R}_N = \mu e^4/8\varepsilon_0^2 ch^3$	$\tilde{R}_N \approx \tilde{R}_\infty$, the Rydberg constant; $\mu = m_e m_N/(m_e + m_N)$	8A.13
Radial distribution function	$P(r) = r^2 R(r)^2$	$P(r) = 4\pi r^2\psi^2$ for s orbitals	8A.17b

TOPIC 8B Many-electron atoms

➤ Why do you need to know this material?

Many-electron atoms are the building blocks of all compounds, and to understand their properties, including their ability to participate in chemical bonding, it is essential to understand their electronic structure. Moreover, a knowledge of that structure explains the structure of the periodic table and all that it summarizes.

➤ What is the key idea?

Electrons occupy the orbitals that result in the lowest energy of the atom, subject to the requirements of the Pauli exclusion principle.

➤ What do you need to know already?

This Topic builds on the account of the structure of hydrogenic atoms (Topic 8A), especially their shell structure.

A **many-electron atom** (or *polyelectron atom*) is an atom with more than one electron. The Schrödinger equation for a many-electron atom is complicated because all the electrons interact with one another. One very important consequence of these interactions is that orbitals of the same value of n but different values of l are no longer degenerate. Moreover, even for a helium atom, with just two electrons, it is not possible to find analytical expressions for the orbitals and energies, so it is necessary to use various approximations.

8B.1 The orbital approximation

The wavefunction of a many-electron atom is a very complicated function of the coordinates of all the electrons, written as $\Psi(\boldsymbol{r}_1,\boldsymbol{r}_2,\ldots)$, where $\boldsymbol{r}_i$ is the vector from the nucleus to electron i (uppercase psi, Ψ, is commonly used to denote a many-electron wavefunction). The **orbital approximation** states that a reasonable first approximation to this exact wavefunction is obtained by thinking of each electron as occupying its 'own' orbital, and writing

$$\Psi(\boldsymbol{r}_1,\boldsymbol{r}_2,\ldots)=\psi(\boldsymbol{r}_1)\psi(\boldsymbol{r}_2)\ldots \qquad \text{Orbital approximation} \qquad (8B.1)$$

The individual orbitals can be assumed to resemble the hydrogenic orbitals based on nuclei with charges modified by the presence of all the other electrons in the atom. This assumption can be justified if, to a first approximation, electron–electron interactions are ignored.

How is that done? 8B.1 Justifying the orbital approximation

Consider a system in which the hamiltonian for the energy is the sum of two contributions, one for electron 1 and the other for electron 2: $\hat{H}=\hat{H}_1+\hat{H}_2$. In an actual two-electron atom (such as a helium atom, with $Z=2$), there is an additional term (proportional to $1/r_{12}$, where r_{12} is the distance between the two electrons) corresponding to their interaction:

$$\hat{H}=\overbrace{-\frac{\hbar^2}{2m_e}\nabla_1^2-\frac{2e^2}{4\pi\varepsilon_0 r_1}}^{\hat{H}_1}\overbrace{-\frac{\hbar^2}{2m_e}\nabla_2^2-\frac{2e^2}{4\pi\varepsilon_0 r_2}}^{\hat{H}_2}+\frac{e^2}{4\pi\varepsilon_0 r_{12}}$$

In the orbital approximation the final term is ignored. Then the task is to show that if $\psi(\boldsymbol{r}_1)$ is an eigenfunction of $\hat{H}_1$ with energy E_1, and $\psi(\boldsymbol{r}_2)$ is an eigenfunction of $\hat{H}_2$ with energy E_2, then the product $\Psi(\boldsymbol{r}_1,\boldsymbol{r}_2)=\psi(\boldsymbol{r}_1)\psi(\boldsymbol{r}_2)$ is an eigenfunction of the combined hamiltonian $\hat{H}$. To do so write

$$\begin{aligned}\hat{H}\Psi(\boldsymbol{r}_1,\boldsymbol{r}_2)&=(\hat{H}_1+\hat{H}_2)\psi(\boldsymbol{r}_1)\psi(\boldsymbol{r}_2)\\&=\overbrace{\hat{H}_1\psi(\boldsymbol{r}_1)\psi(\boldsymbol{r}_2)}^{\psi(\boldsymbol{r}_2)\hat{H}_1\psi(\boldsymbol{r}_1)}+\overbrace{\hat{H}_2\psi(\boldsymbol{r}_1)\psi(\boldsymbol{r}_2)}^{\psi(\boldsymbol{r}_1)\hat{H}_2\psi(\boldsymbol{r}_2)}\\&=\psi(\boldsymbol{r}_2)\overbrace{\hat{H}_1\psi(\boldsymbol{r}_1)}^{E_1\psi(\boldsymbol{r}_1)}+\psi(\boldsymbol{r}_1)\overbrace{\hat{H}_2\psi(\boldsymbol{r}_2)}^{E_2\psi(\boldsymbol{r}_2)}\\&=\psi(\boldsymbol{r}_2)E_1\psi(\boldsymbol{r}_1)+\psi(\boldsymbol{r}_1)E_2\psi(\boldsymbol{r}_2)=(E_1+E_2)\psi(\boldsymbol{r}_1)\psi(\boldsymbol{r}_2)\\&=E\Psi(\boldsymbol{r}_1,\boldsymbol{r}_2)\end{aligned}$$

where $E=E_1+E_2$, which is the desired result. Note how each hamiltonian operates on only its 'own' wavefunction. If the electrons interact (as they do in fact), then the term in $1/r_{12}$ must be included, and the proof fails. Therefore, this description is only approximate, but it is a useful model for discussing the chemical properties of atoms and is the starting point for more sophisticated descriptions of atomic structure.

The orbital approximation can be used to express the electronic structure of an atom by reporting its **configuration**, a statement of its occupied orbitals (usually, but not necessarily, in its ground state). Thus, as the ground state of a hydrogenic atom consists of the single electron in a 1s orbital, its configuration is reported as $1s^1$ (read 'one-ess-one').

A He atom has two electrons. The first electron occupies a 1s hydrogenic orbital, but because $Z = 2$ that orbital is more compact than in H itself. The second electron joins the first in the 1s orbital, so the electron configuration of the ground state of He is $1s^2$ (read 'one-ess-two').

Brief illustration 8B.1

According to the orbital approximation, each electron in He occupies a hydrogenic 1s orbital of the kind given in Topic 8A. Anticipating (see below) that the electrons experience an effective nuclear charge $Z_{\text{eff}}e$ rather than the actual charge on the nucleus with $Z = 2$ (specifically, as seen later, a charge $1.69e$ rather than $2e$), then the two-electron wavefunction of the atom is

$$\Psi(r_1,r_2) = \overbrace{\frac{Z_{\text{eff}}^{3/2}}{(\pi a_0^3)^{1/2}}\mathrm{e}^{-Z_{\text{eff}}r_1/a_0}}^{\psi_{1s}(r_1)} \times \overbrace{\frac{Z_{\text{eff}}^{3/2}}{(\pi a_0^3)^{1/2}}\mathrm{e}^{-Z_{\text{eff}}r_2/a_0}}^{\psi_{1s}(r_2)}$$

$$= \frac{Z_{\text{eff}}^3}{\pi a_0^3}\mathrm{e}^{-Z_{\text{eff}}(r_1+r_2)/a_0}$$

There is nothing particularly mysterious about a two-electron wavefunction: in this case it is a simple exponential function of the distances of the two electrons from the nucleus.

Exercise

E8B.1 Construct the wavefunction for an excited state of the He atom with configuration $1s^1 2s^1$. Use $Z_{\text{eff}} = 2$ for the 1s electron and $Z_{\text{eff}} = 1$ for the 2s electron.

8B.2 The Pauli exclusion principle

It is tempting to suppose that the electronic configurations of the atoms of successive elements with atomic numbers $Z = 3, 4, \ldots$, and therefore with Z electrons, are simply $1s^Z$. That, however, is not the case. The reason lies in two aspects of nature: that electrons possess 'spin' and that they must obey the very fundamental 'Pauli principle'.

8B.2(a) Spin

The quantum mechanical property of electron **spin**, the possession of an intrinsic angular momentum, was identified by an experiment performed by Otto Stern and Walther Gerlach in 1921, who shot a beam of silver atoms through an inhomogeneous magnetic field (Fig. 8B.1). The idea behind the experiment was that each atom possesses a certain electronic angular momentum and (because moving charges generate a magnetic field) as a result behaves like a small bar magnet aligned with the direction of the angular momentum vector. As the atoms pass through the inhomogeneous magnetic field they are deflected, with the deflection depending on the relative orientation of the applied magnetic field and the atomic magnet.

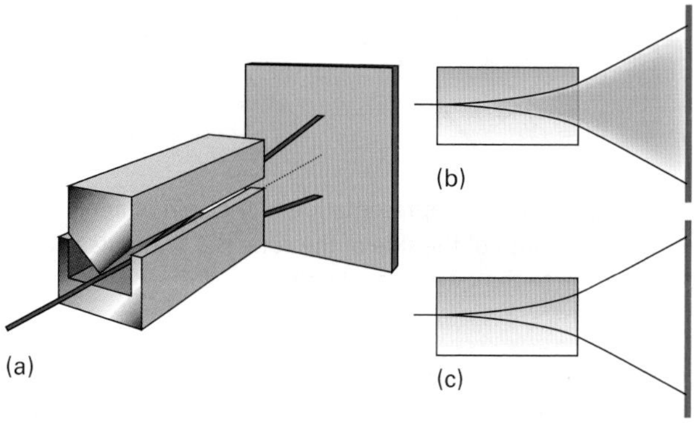

Figure 8B.1 (a) The experimental arrangement for the Stern–Gerlach experiment: the magnet provides an inhomogeneous field. (b) The classically expected result. (c) The observed outcome using silver atoms.

The classical expectation is that the electronic angular momentum, and hence the resulting magnet, can be oriented in any direction. Each atom would be deflected into a direction that depends on the orientation and the beam should spread out into a broad band as it emerges from the magnetic field. In contrast, the expectation from quantum mechanics is that the angular momentum, and hence the atomic magnet, has only discrete orientations (Topic 7F). Each of these orientations results in the atoms being deflected in a specific direction, so the beam should split into a number of sharp bands, each corresponding to a different orientation of the angular momentum of the electrons in the atom.

In their first experiment, Stern and Gerlach appeared to confirm the classical prediction. However, the experiment is difficult because collisions between the atoms in the beam blur the bands. When they repeated the experiment with a beam of very low intensity (so that collisions were less frequent), they observed discrete bands, and so confirmed the quantum prediction. However, Stern and Gerlach observed *two* bands of Ag atoms in their experiment. This observation seems to conflict with one of the predictions of quantum mechanics, because an angular momentum l gives rise to $2l + 1$ orientations, which is equal to 2 only if $l = \frac{1}{2}$, contrary to the requirement that l is an integer. The conflict was resolved by the suggestion that the angular momentum they were observing was not due to orbital angular momentum (the motion of an electron around the atomic nucleus) but arose instead from the rotation of the electron about its own axis, its 'spin'.

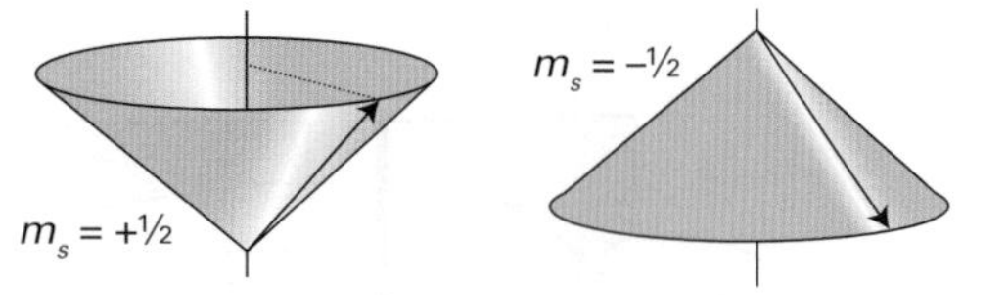

Figure 8B.2 The vector representation of the spin of an electron. The length of the side of the cone is $3^{1/2}/2$ units and the projections on to the z-axis are $\pm\frac{1}{2}$ units.

The spin of an electron does not have to satisfy the same boundary conditions as those for a particle circulating through space around a central point, so the quantum number for spin angular momentum is subject to different restrictions. The **spin quantum number** s is used in place of the orbital angular momentum quantum number l (Topic 7F; like l, s is a non-negative number) and m_s, the **spin magnetic quantum number**, is used in place of m_l for the projection on the z-axis. The magnitude of the spin angular momentum is $\{s(s+1)\}^{1/2}\hbar$ and the component $m_s\hbar$ is restricted to the $2s+1$ values $m_s = s, s-1, \ldots, -s$. To account for Stern and Gerlach's observation, $s=\frac{1}{2}$ and $m_s = \pm\frac{1}{2}$.[1]

The detailed analysis of the spin of a particle is sophisticated and shows that the property should not be taken to be an actual spinning motion. It is better to regard 'spin' as an intrinsic property like mass and charge: every electron has exactly the same value and the magnitude of the spin angular momentum of an electron cannot be changed. However, the picture of an actual spinning motion can be very useful when used with care. In the vector model of angular momentum (Topic 7F), the spin may lie in two different orientations (Fig. 8B.2). One orientation corresponds to $m_s = +\frac{1}{2}$ (this state is often denoted α or ↑); the other orientation corresponds to $m_s = -\frac{1}{2}$ (this state is denoted β or ↓).

Other elementary particles have characteristic spin. For example, protons and neutrons are spin-$\frac{1}{2}$ particles (i.e. $s=\frac{1}{2}$). Because the masses of a proton and a neutron are so much greater than the mass of an electron, yet they all have the same spin angular momentum, the classical picture would be of these two particles spinning much more slowly than an electron. Some mesons, another variety of fundamental particle, are spin-1 particles (i.e. $s = 1$), as are some atomic nuclei, but for our purposes the most important spin-1 particle is the photon. The importance of photon spin in spectroscopy is explained in Topic 11A; nuclear spin is the basis of nuclear magnetic resonance (Topic 12A).

Brief illustration 8B.2

The magnitude of the spin angular momentum, like any angular momentum, is $\{s(s+1)\}^{1/2}\hbar$. For any spin-$\frac{1}{2}$ particle, not only electrons, this angular momentum is $\left(\frac{3}{4}\right)^{1/2}\hbar = 0.866\hbar$, or 9.13×10^{-35} J s. The component on the z-axis is $m_s\hbar$, which for a spin-$\frac{1}{2}$ particle is $\pm\frac{1}{2}\hbar$, or $\pm5.27\times10^{-35}$ J s.

Particles with half-integral spin are called **fermions** and those with integral spin (including 0) are called **bosons**. Thus, electrons and protons are fermions; photons are bosons. It is a very deep feature of nature that all the elementary particles that constitute matter are fermions whereas the elementary particles that transmit the forces that bind fermions together are all bosons. Photons, for example, transmit the electromagnetic force that binds together electrically charged particles. Matter, therefore, is an assembly of fermions held together by forces conveyed by bosons.

8B.2(b) The Pauli principle

With the concept of spin established, it is possible to resume discussion of the electronic structures of atoms. Lithium, with $Z = 3$, has three electrons. The first two occupy a 1s orbital drawn even more closely than in He around the more highly charged nucleus. The third electron, however, does not join the first two in the 1s orbital because that configuration is forbidden by the **Pauli exclusion principle**:

> No more than two electrons may occupy any given orbital, and if two do occupy one orbital, then their spins must be paired.

Electrons with paired spins, denoted ↑↓, have zero net spin angular momentum because the spin of one electron is cancelled by the spin of the other. Specifically, one electron has $m_s = +\frac{1}{2}$ the other has $m_s = -\frac{1}{2}$. In the vector model they are orientated on their respective cones so that the resultant spin is zero (Fig. 8B.3). The exclusion principle is the key to the structure of complex atoms, to chemical periodicity, and to molecular structure. It was proposed by Wolfgang Pauli in 1925 when he

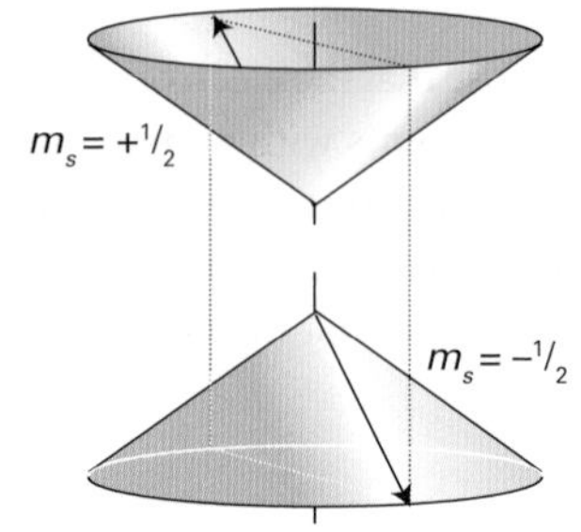

Figure 8B.3 Electrons with paired spins have zero resultant spin angular momentum. They can be represented by two vectors that lie at an indeterminate position on the cones shown here, but wherever one lies on its cone, the other points in the opposite direction; their resultant is zero.

[1] You will sometimes see the quantum number s used in place of m_s, and written $s = \pm\frac{1}{2}$. That is wrong: like l, s is never negative and denotes the magnitude of the spin angular momentum. For the z-component, use m_s.

was trying to account for the absence of some lines in the spectrum of helium. Later he was able to derive a very general form of the principle from theoretical considerations.

The Pauli *exclusion* principle is a special case of a general statement called the **Pauli principle**:

> When the labels of any two identical fermions are exchanged, the total wavefunction changes sign; when the labels of any two identical bosons are exchanged, the sign of the total wavefunction remains the same.

By 'total wavefunction' is meant the entire wavefunction, including the spin of the particles. It is possible to show that the Pauli principle implies the Pauli exclusion principle.

How is that done? 8B.2 Justifying the Pauli exclusion principle

Consider the wavefunction for two electrons, $\Psi(1,2)$. The Pauli principle implies that it is a fact of nature (which has its roots in the theory of relativity) that the wavefunction must change sign if the labels 1 and 2 are interchanged wherever they occur in the function:

$$\Psi(2,1) = -\Psi(1,2) \tag{8B.2}$$

Suppose the two electrons occupy the same orbital ψ, then in the orbital approximation the overall *spatial* wavefunction is $\psi(r_1)\psi(r_2)$, which for simplicity will be denoted $\psi(1)\psi(2)$. To apply the Pauli principle, it is necessary to consider the *total* wavefunction, the wavefunction including spin. The rest of the argument uses the following steps.

Step 1 *Write the total wavefunctions*

For each electron the spin wavefunction can be α, corresponding to $m_s = +\frac{1}{2}$, or β, corresponding to $m_s = -\frac{1}{2}$. For two electrons there are four possible spin wavefunctions: both α, denoted $\alpha(1)\alpha(2)$, both β, denoted $\beta(1)\beta(2)$, and one α and the other β, denoted either $\alpha(1)\beta(2)$ or $\alpha(2)\beta(1)$. Because it is not possible to know which electron is α and which is β, in the last case it is appropriate to express the spin contribution to the total wavefunction as the (normalized) linear combinations[2]

$$\begin{aligned}\sigma_+(1,2) &= \frac{1}{2^{1/2}}\{\alpha(1)\beta(2)+\beta(1)\alpha(2)\}\\ \sigma_-(1,2) &= \frac{1}{2^{1/2}}\{\alpha(1)\beta(2)-\beta(1)\alpha(2)\}\end{aligned} \tag{8B.3}$$

These combinations allow one spin to be α and the other β with equal probability; the former corresponds to parallel spins (the individual spins do not cancel) and the latter to paired spins (the individual spins cancel). The total wavefunction of the system is therefore the product of the orbital part and one of the four spin contributions:

$$\begin{aligned}&\psi(1)\psi(2)\alpha(1)\alpha(2) \quad \psi(1)\psi(2)\beta(1)\beta(2)\\ &\psi(1)\psi(2)\sigma_+(1,2) \quad \psi(1)\psi(2)\sigma_-(1,2)\end{aligned} \tag{8B.4}$$

Step 2 *Apply the Pauli principle*

The Pauli principle states that for a wavefunction to be acceptable (for electrons), it must change sign when the electrons are exchanged. In each case, exchanging the labels 1 and 2 converts $\psi(1)\psi(2)$ into $\psi(2)\psi(1)$, which is the same, because the order of multiplying the functions does not change the value of the product. The same is true of $\alpha(1)\alpha(2)$ and $\beta(1)\beta(2)$. Therefore, $\psi(1)\psi(2)\alpha(1)\alpha(2)$ and $\psi(1)\psi(2)\beta(1)\beta(2)$ are not allowed, because they do not change sign. When the labels are exchanged the combination $\sigma_+(1,2)$ becomes

$$\sigma_+(2,1) = \frac{1}{2^{1/2}}\{\alpha(2)\beta(1)+\beta(2)\alpha(1)\} = \sigma_+(1,2)$$

because the central term is simply the original function written in a different order. The product $\psi(1)\psi(2)\sigma_+(1,2)$ is therefore also disallowed. Finally, consider $\sigma_-(1,2)$:

$$\begin{aligned}\sigma_-(2,1) &= \frac{1}{2^{1/2}}\{\alpha(2)\beta(1)-\beta(2)\alpha(1)\}\\ &= -\frac{1}{2^{1/2}}\{\alpha(1)\beta(2)-\beta(1)\alpha(2)\} = -\sigma_-(1,2)\end{aligned}$$

The combination $\psi(1)\psi(2)\sigma_-(1,2)$ therefore does change sign (it is 'antisymmetric') and is acceptable.

Step 3 *Summarize the results*

In summary, only one of the four possible states is allowed by the Pauli principle: the one that survives has paired α and β spins. This is the content of the Pauli exclusion principle.

The Pauli exclusion principle (but not the more general Pauli principle) is irrelevant when the orbitals occupied by the electrons are different, and both electrons may then have, but need not have, the same spin state. In each case the overall wavefunction must still be antisymmetric and must satisfy the Pauli principle itself.

Now returning to lithium, Li ($Z = 3$), the third electron cannot enter the 1s orbital because that orbital is already full: the K shell (the shell with $n = 1$, Topic 8A) is complete and the two electrons form a **closed shell**, a shell in which all the orbitals are fully occupied. Because a similar closed shell is characteristic of the He atom, it is commonly denoted [He]. The third electron cannot enter the K shell and must occupy the next available orbital, which is one with $n = 2$ and hence belonging to the L shell (which consists of the four orbitals with $n = 2$). It is now necessary to decide whether the next available orbital is the 2s orbital or a 2p orbital, and therefore whether the lowest energy configuration of the atom is $[\text{He}]2s^1$ or $[\text{He}]2p^1$.

[2] A stronger justification for taking these linear combinations is that they correspond to eigenfunctions of the total spin operators S^2 and S_z, with $M_S = 0$ and, respectively, $S = 1$ and 0.

Exercise

E8B.2 How many electrons can occupy subshells with $l = 3$?

8B.3 The building-up principle

Unlike in hydrogenic atoms, the 2s and 2p orbitals (and, in general, the subshells of a given shell) do not have the same energy in many-electron atoms.

8B.3(a) Penetration and shielding

An electron in a many-electron atom experiences a Coulombic repulsion from all the other electrons present. If the electron is at a distance r from the nucleus, it experiences an average repulsion that can be represented by a point negative charge located at the nucleus and equal in magnitude to the total charge of all the other electrons within a sphere of radius r (Fig. 8B.4). This property is a conclusion of classical electrostatics, where the effect of a spherical distribution of charge can be represented by a point charge of the same magnitude located at its centre. The effect of this point negative charge is to reduce the full charge of the nucleus from Ze to $Z_{\text{eff}}e$, the **effective nuclear charge**. In everyday parlance, Z_{eff} itself is commonly referred to as the 'effective nuclear charge'. The electron is said to experience a **shielded** nuclear charge, and the difference between Z and Z_{eff} is called the **shielding constant**, σ:

$$Z_{\text{eff}} = Z - \sigma \qquad \text{Nuclear shielding} \qquad (8B.5)$$

The electrons do not actually 'block' the full Coulombic attraction of the nucleus: the shielding constant is simply a way of expressing the net outcome of the nuclear attraction and the electronic repulsions in terms of a single equivalent charge at the centre of the atom.

The shielding constant is different for s and p electrons because they have different radial distribution functions and therefore respond to the other electrons in the atom to different extents (Fig. 8B.5). An s electron has a greater **penetration** through inner shells than a p electron, in the sense that an s electron is more likely to be found close to the nucleus than a p electron of the same shell. Because only electrons inside the sphere defined by the location of the electron of interest contribute to shielding, an s electron experiences less shielding than a p electron. Consequently, as a result of the combined effects of penetration and shielding, an s electron is more tightly bound than a p electron of the same shell. Similarly, a d electron penetrates less than a p electron of the same shell (recall that a d orbital is proportional to $(r/a_0)^2$ close to the nucleus, whereas a p orbital is proportional to r/a_0, so the amplitude of a d orbital is smaller there than that of a p orbital), and therefore experiences more shielding.

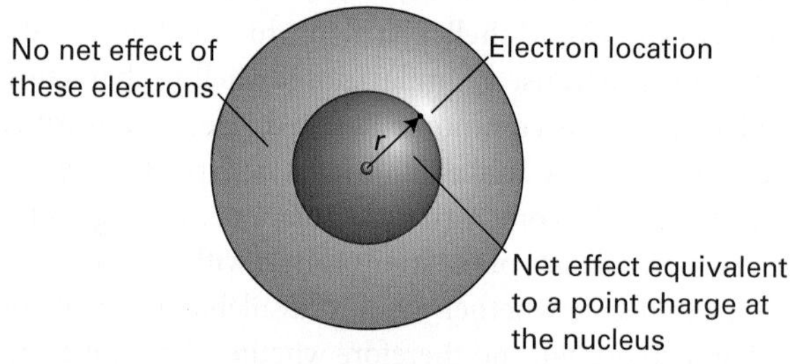

Figure 8B.4 An electron at a distance r from the nucleus experiences a Coulombic repulsion from all the electrons within a sphere of radius r. This repulsion is equivalent to that from a point negative charge located on the nucleus. The negative charge reduces the effective nuclear charge of the nucleus from Ze to $Z_{\text{eff}}e$.

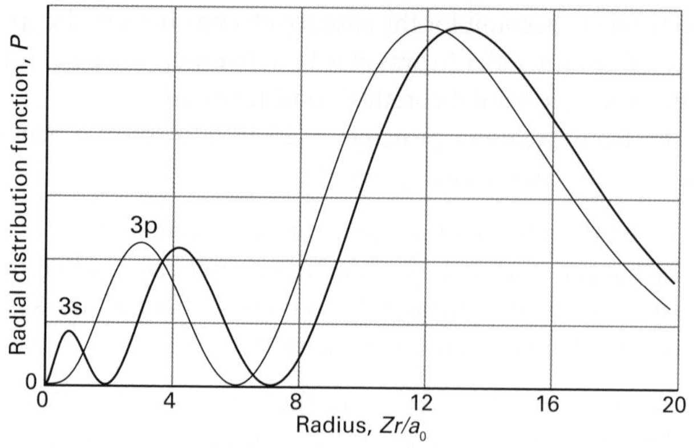

Figure 8B.5 An electron in an s orbital (here a 3s orbital) is more likely to be found close to the nucleus than an electron in a p orbital of the same shell (note the closeness of the innermost peak of the 3s orbital to the nucleus at $r = 0$). Hence an s electron experiences less shielding and is more tightly bound than a p electron of the same shell.

Shielding constants for different types of electrons in atoms have been calculated from wavefunctions obtained by numerical solution of the Schrödinger equation, leading to the effective nuclear charges shown in Table 8B.1. In general, valence-shell s electrons do experience higher effective nuclear charges than p electrons, although there are some discrepancies.

Table 8B.1 Effective nuclear charge*

Element	Z	Orbital	Z_{eff}
He	2	1s	1.6875
C	6	1s	5.6727
		2s	3.2166
		2p	3.1358

* More values are given in the *Resource section*.

Brief illustration 8B.3

The effective nuclear charge for 1s, 2s, and 2p electrons in a carbon atom are 5.6727, 3.2166, and 3.1358, respectively. The radial distribution functions for these orbitals (Topic 8A) are generated by forming $P(r) = r^2R(r)^2$, where $R(r)$ is the radial wavefunction, which is given in Table 8A.1. The three radial distribution functions, taking into account the effective nuclear

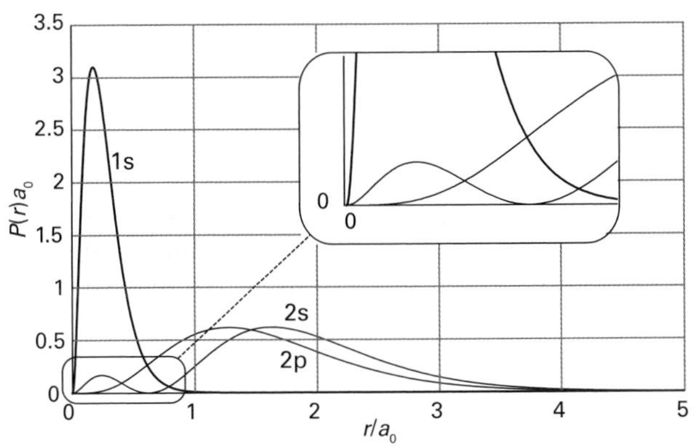

Figure 8B.6 The radial distribution functions for electrons in a carbon atom, as calculated in *Brief illustration* 8B.3.

charges, are plotted in Fig. 8B.6. As can be seen (especially in the magnified view close to the nucleus), the s orbital has greater penetration than the p orbital. The average radii of the 2s and 2p orbitals are 99 pm and 84 pm, respectively, which shows that the average distance of a 2s electron from the nucleus is greater than that of a 2p orbital. To account for the lower energy of the 2s orbital, the extent of penetration is more important than the average distance from the nucleus.

The consequence of penetration and shielding is that the energies of subshells of a shell in a many-electron atom (those with the same values of n but different values of l) in general lie in the order s < p < d < f. The individual orbitals of a given subshell (those with the same value of l but different values of m_l) remain degenerate because they all have the same radial characteristics and so experience the same effective nuclear charge.

To complete the Li story, consider that, because the shell with $n = 2$ consists of two subshells, with the 2s subshell lower in energy than the 2p subshell, the third electron occupies the 2s orbital (the only orbital in that subshell). This occupation results in the ground-state configuration $1s^2 2s^1$, with the central nucleus surrounded by a complete helium-like shell of two 1s electrons, and around that a more diffuse 2s electron. The electrons in the outermost shell of an atom in its ground state are called the **valence electrons** because they are largely responsible for the chemical bonds that the atom forms (and 'valence', as explained in Focus 9, refers to the ability of an atom to form bonds). Thus, the valence electron in Li is a 2s electron and its other two electrons belong to its core.

8B.3(b) Hund's rules

The extension of the argument used to account for the structures of H, He, and Li is called the **building-up principle**, or the *Aufbau principle*, from the German word for 'building up', and should be familiar from introductory courses. In brief, imagine the bare nucleus of atomic number Z, and then feed into the orbitals Z electrons in succession. The order of occupation, following the shells and their subshells arranged in order of increasing energy, is

1s 2s 2p 3s 3p 4s 3d 4p 5s 4d 5p 6s

Each orbital may accommodate up to two electrons.

Brief illustration 8B.4

Consider the carbon atom, for which $Z = 6$ and there are six electrons to accommodate. Two electrons enter and fill the 1s orbital, two enter and fill the 2s orbital, leaving two electrons to occupy the orbitals of the 2p subshell. Hence the ground-state configuration of C is $1s^2 2s^2 2p^2$, or more succinctly $[He]2s^2 2p^2$, with [He] the helium-like $1s^2$ core.

It is possible to be more precise about the configuration of a carbon atom than in *Brief illustration* 8B.4. The last two electrons are expected to occupy different 2p orbitals because they are then farther apart on average and repel each other less than if they were in the same orbital. Thus, one electron can be thought of as occupying the $2p_x$ orbital and the other the $2p_y$ orbital (the x, y, z designation is arbitrary, and it would be equally valid to use the complex forms of these orbitals), and the lowest energy configuration of the atom is $[He]2s^2 2p_x^1 2p_y^1$. The same rule applies whenever degenerate orbitals of a subshell are available for occupation. Thus, another rule of the building-up principle is:

> Electrons occupy different orbitals of a given subshell before doubly occupying any one of them.

For instance, nitrogen ($Z = 7$) has the ground-state configuration $[He]2s^2 2p_x^1 2p_y^1 2p_z^1$, and only at oxygen ($Z = 8$) is a 2p orbital doubly occupied, giving $[He]2s^2 2p_x^2 2p_y^1 2p_z^1$.

When electrons occupy orbitals singly it is necessary to invoke **Hund's maximum multiplicity rule**:

> An atom in its ground state adopts a configuration with the greatest number of unpaired electrons.

The explanation of Hund's rule is subtle, but it reflects the quantum mechanical property of **spin correlation**. In essence, the effect of spin correlation is to allow the atom to shrink slightly when the spins are parallel, so the electron–nucleus interaction is improved. As a consequence, in the ground state of the carbon atom, the two 2p electrons have parallel spins, all three 2p electrons in the N atoms have parallel spins, and the two 2p electrons in different orbitals in the O atom have parallel spins (the two in the $2p_x$ orbital are necessarily paired). The effect can be explained by considering the Pauli principle and showing that electrons with parallel spins behave as if they have a tendency to stay apart, and hence repel each other less.

How is that done? 8B.3 Exploring the origins of spin correlation

Suppose electron 1 is in orbital a and described by a wavefunction $\psi_a(r_1)$, and electron 2 is in orbital b with wavefunction $\psi_b(r_2)$. Then, in the orbital approximation, the joint spatial wavefunction of the electrons is the product $\Psi = \psi_a(r_1)\psi_b(r_2)$. However, this wavefunction is not acceptable, because it suggests that it is possible to know which electron is in which orbital. According to quantum mechanics, the correct description is either of the two following wavefunctions:

$$\Psi_{\pm} = \frac{1}{2^{1/2}}\{\psi_a(r_1)\psi_b(r_2) \pm \psi_b(r_1)\psi_a(r_2)\}$$

According to the Pauli principle, because Ψ_+ is symmetrical under interchange of the electrons, it must be combined with an antisymmetric spin wavefunction (the one denoted σ_-). That combination corresponds to a spin-paired state. Conversely, Ψ_- is antisymmetric, so it must be combined with one of the three symmetric spin wavefunctions. These three symmetric states correspond to electrons with parallel spins (see Topic 8C for an explanation of this point).

Now consider the behaviour of the two wavefunctions $\Psi_\pm$ when one electron approaches another, and $r_1 = r_2$. As a result, Ψ_- vanishes, which means that there is zero probability of finding the two electrons at the same point in space when they have parallel spins. In contrast, the wavefunction Ψ_+ does not vanish when the two electrons are at the same point in space. Because the two electrons have different relative spatial distributions depending on whether their spins are parallel or not, it follows that their Coulombic interaction is different, and hence that the two states described by these wavefunctions have different energies, with the spin-parallel state lower in energy than the spin-paired state.

Neon, with $Z = 10$, has the configuration [He]$2s^22p^6$, which completes the L shell. This closed-shell configuration is denoted [Ne], and acts as a core for subsequent elements. The next electron must enter the 3s orbital and begin a new shell, so an Na atom, with $Z = 11$, has the configuration [Ne]$3s^1$. Like lithium with the configuration [He]$2s^1$, sodium has a single s electron outside a complete core. This analysis hints at the origin of chemical periodicity. The L shell is completed by eight electrons, so the element with $Z = 11$ (Na) should have similar properties to the element with $Z = 3$ (Li). Likewise, Mg ($Z = 12$) should be similar to Be ($Z = 4$), and so on, up to the noble gases He ($Z = 2$), Ne ($Z = 10$), and Ar ($Z = 18$).

At potassium ($Z = 19$) the next orbital in line for occupation is 4s: this orbital is brought below 3d by the effects of penetration and shielding, and the ground state configuration is [Ar]$4s^1$. Calcium ($Z = 20$) is likewise [Ar]$4s^2$. At this stage the five 3d orbitals are in line for occupation, but there are complications arising from the energy changes arising from the interaction of the electrons in the valence shell, and penetration arguments alone are no longer reliable.

Calculations of the type discussed in Topic 8B show that for the atoms from scandium to zinc the energies of the 3d orbitals are always lower than the energy of the 4s orbital, in spite of the greater penetration of a 4s electron. However, spectroscopic results show that Sc has the configuration [Ar]$3d^14s^2$, not [Ar]$3d^3$ nor [Ar]$3d^24s^1$. To understand this observation, consider the nature of electron–electron repulsions in 3d and 4s orbitals. Because the average distance of a 3d electron from the nucleus is less than that of a 4s electron, two 3d electrons are so close together that they repel each other more strongly than two 4s electrons do, and $3d^2$ and $3d^3$ configurations are disfavoured. As a result, Sc has the configuration [Ar]$3d^14s^2$ rather than the two alternatives, for then the strong electron–electron repulsions in the 3d orbitals are minimized. The total energy of the atom is lower despite the cost of allowing electrons to populate the high energy 4s orbital (Fig. 8B.7). The effect just described is generally true for scandium to zinc, so their electron configurations are of the form [Ar]$3d^n4s^2$, where $n = 1$ for scandium and $n = 10$ for zinc. Two notable exceptions, which are observed experimentally, are Cr, with electron configuration [Ar]$3d^54s^1$, and Cu, with electron configuration [Ar]$3d^{10}4s^1$. At gallium, these complications disappear and the building-up principle is used in the same way as in preceding periods. Now the 4s and 4p subshells constitute the valence shell, and the period terminates with krypton. Because 18 electrons have intervened since argon, this row is the first 'long period' of the periodic table.

At this stage it becomes apparent that sequential occupation of the orbitals in successive shells results in periodic similarities in the electronic configurations. This periodicity of structure accounts for the formulation of the **periodic table**.

The vertical columns of the periodic table are called **groups** and (in the modern convention) numbered from 1 to 18. Successive rows of the periodic table are called **periods**, the

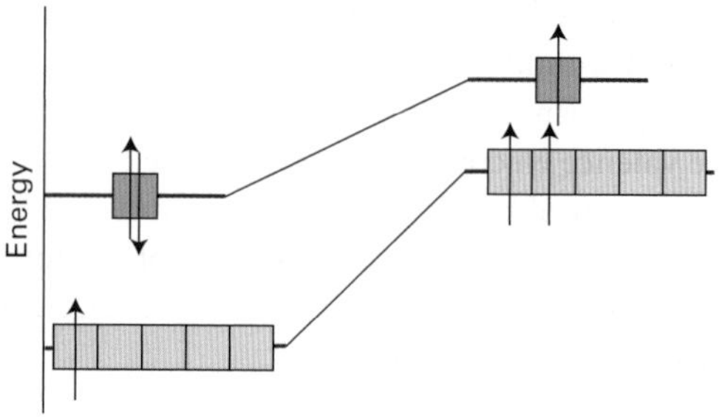

Figure 8B.7 Strong electron–electron repulsions in the 3d orbitals are minimized in the ground state of Sc if the atom has the configuration [Ar]$3d^14s^2$ (shown on the left) instead of [Ar]$3d^24s^1$ (shown on the right). The total energy of the atom is lower when it has the [Ar]$3d^14s^2$ configuration despite the cost of populating the high energy 4s orbital.

number of the period being equal to the principal quantum number of the valence shell.

The periodic table is divided into s, p, d, and f **blocks**, according to the subshell that is last to be occupied in the formulation of the electronic configuration of the atom. The members of the d block (specifically the members of Groups 3–11 in the d block) are also known as the **transition metals**; those of the f block (which is not divided into numbered groups) are sometimes called the **inner transition metals**. The upper row of the f block (Period 6) consists of the **lanthanoids** (still commonly the 'lanthanides') and the lower row (Period 7) consists of the **actinoids** (still commonly the 'actinides').

The configurations of cations of elements in the s, p, and d blocks of the periodic table are derived by removing electrons from the ground-state configuration of the neutral atom in a specific order. First, remove valence p electrons, then valence s electrons, and then as many d electrons as are necessary to achieve the specified charge. The configurations of anions of the p-block elements are derived by continuing the building-up procedure and adding electrons to the neutral atom until the configuration of the next noble gas has been reached.

Brief illustration 8B.5

Because the configuration of vanadium is [Ar]$3d^34s^2$, the V^{2+} cation has the configuration [Ar]$3d^3$. It is reasonable to remove the more energetic 4s electrons in order to form the cation, but it is not obvious why the [Ar]$3d^3$ configuration is preferred in V^{2+} over the [Ar]$3d^14s^2$ configuration, which is found in the isoelectronic Sc atom. Calculations show that the energy difference between [Ar]$3d^3$ and [Ar]$3d^14s^2$ depends on Z_{eff}. As Z_{eff} increases, transfer of a 4s electron to a 3d orbital becomes more favourable because the electron–electron repulsions are compensated by attractive interactions between the nucleus and the electrons in the spatially compact 3d orbital. Indeed, calculations reveal that, for a sufficiently large Z_{eff}, [Ar]$3d^3$ is lower in energy than [Ar]$3d^14s^2$. This conclusion explains why V^{2+} has a [Ar]$3d^3$ configuration and also accounts for the observed [Ar]$4s^03d^n$ configurations of the M^{2+} cations of Sc through Zn.

8B.3(c) Atomic and ionic radii

The **atomic radius** of an element is half the distance between the centres of neighbouring atoms in a solid (such as Cu) or, for non-metals, in a homonuclear molecule (such as H_2 or S_8). As seen in Table 8B.2 and Fig. 8B.8, atomic radii tend to decrease from left to right across a period of the periodic table, and increase down each group. These general trends are justified as follows:

- The decrease across a period can be traced to the increase in nuclear charge, which draws the electrons in closer to the nucleus. The increase in nuclear charge is partly cancelled by the increase in the number of electrons, but because electrons are spread over a region of space, one electron does not fully shield one nuclear charge, so the increase in nuclear charge dominates.
- The increase in atomic radius down a group (despite the increase in nuclear charge) is explained by the fact that the valence shells of successive periods correspond to higher principal quantum numbers. That is, successive periods correspond to the start and then completion of successive (and more distant) shells of the atom that surround each other like the successive layers of an onion. The need to occupy a more distant shell leads to a larger atom despite the increased nuclear charge.

A modification of the increase down a group is encountered in Period 6, for the radii of the atoms in the d block and in the following atoms of the p block are not as large as would be expected by simple extrapolation down the group. The reason can be traced to the fact that in Period 6 the f orbitals are in the process of being occupied. An f electron is a very inefficient

Table 8B.2 Atomic radii of main-group elements, *r*/pm

Li 157	Be 112	B 88	C 77	N 74	O 66	F 64
Na 191	Mg 160	Al 143	Si 118	P 110	S 104	Cl 99
K 235	Ca 197	Ga 153	Ge 122	As 121	Se 117	Br 114
Rb 250	Sr 215	In 167	Sn 158	Sb 141	Te 137	I 133
Cs 272	Ba 224	Tl 171	Pb 175	Bi 182	Po 167	

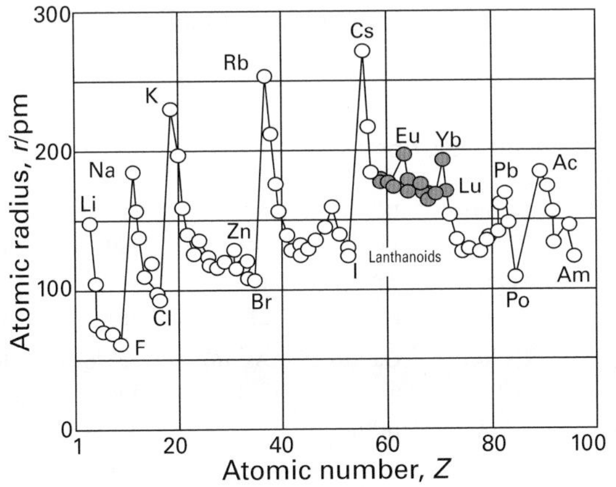

Figure 8B.8 The variation of atomic radius through the periodic table. Note the contraction of radius following the lanthanoids in Period 6 (following Lu, lutetium).

shielder of nuclear charge (for reasons connected with its radial extension), and as the atomic number increases from La to Lu, there is a considerable contraction in radius. By the time the d block resumes (at hafnium, Hf), the poorly shielded but considerably increased nuclear charge has drawn in the surrounding electrons, and the atoms are compact. They are so compact, that the metals in this region of the periodic table (iridium to lead) are very dense. The reduction in radius below that expected by extrapolation from preceding periods is called the **lanthanide contraction**.

The **ionic radius** of an element is its share of the distance between neighbouring ions in an ionic solid. That is, the distance between the centres of a neighbouring cation and anion is the sum of the two ionic radii. The size of the 'share' leads to some ambiguity in the definition. One common definition sets the ionic radius of O^{2-} equal to 140 pm, but there are other scales, and care must be taken not to mix them. Ionic radii also vary with the number of counterions (ions of opposite charge) around a given ion; unless otherwise stated, the values in this text have been corrected to correspond to an environment of six counterions.

When an atom loses one or more valence electrons to form a cation, the remaining atomic core is smaller than the parent atom. Therefore, a cation is invariably smaller than its parent atom. For example, the atomic radius of Na, with the configuration $[Ne]3s^1$, is 191 pm, but the ionic radius of Na^+, with the configuration [Ne], is only 102 pm (Table 8B.3). Like atomic radii, cation radii increase down each group because electrons are occupying shells with higher principal quantum numbers.

An anion is larger than its parent atom because the electrons added to the valence shell repel one another. Without a compensating increase in the nuclear charge, which would draw the electrons closer to the nucleus and each other, the ion expands. The variation in anion radii shows the same trend as that for atoms and cations, with the smallest anions at the upper right of the periodic table, close to fluorine (Table 8B.3).

Table 8B.3 Ionic radii, r/pm*

$Li^+(4)$ 59	$Be^{2+}(4)$ 27	$B^{3+}(4)$ 12	N^{3-} 171	$O^{2-}(6)$ 140	$F^-(6)$ 133
$Na^+(6)$ 102	$Mg^{2+}(6)$ 72	$Al^{3+}(6)$ 53	P^{3-} 212	$S^{2-}(6)$ 184	$Cl^-(6)$ 181
$K^+(6)$ 138	$Ca^{2+}(6)$ 100	$Ga^{3+}(6)$ 62	$As^{3-}(6)$ 222	$Se^{2-}(6)$ 198	$Br^-(6)$ 196
$Rb^+(6)$ 149	$Sr^{2+}(6)$ 116	$In^{3+}(6)$ 79		$Te^{2-}(6)$ 221	$I^-(6)$ 220
$Cs^+(6)$ 167	$Ba^{2+}(6)$ 136	$Tl^{3+}(6)$ 88			

* Numbers in parentheses are the *coordination numbers* of the ions, the numbers of species (for example, counterions, solvent molecules) around the ions. Values for ions without a coordination number stated are estimates. More values are given in the *Resource section*.

Brief illustration 8B.6

The Ca^{2+}, K^+, and Cl^- ions have the configuration [Ar]. However, their radii differ because they have different nuclear charges. The Ca^{2+} ion has the largest nuclear charge, so it has the strongest attraction for the electrons and the smallest radius. The Cl^- ion has the lowest nuclear charge of the three ions and, as a result, the largest radius.

8B.3(d) Ionization energies and electron affinities

The minimum energy necessary to remove an electron from a many-electron atom in the gas phase is the **first ionization energy**, I_1, of the element. The **second ionization energy**, I_2, is the minimum energy needed to remove a second electron (from the singly charged cation). The variation of the first ionization energy through the periodic table is shown in Fig. 8B.9 and some numerical values are given in Table 8B.4.

The **electron affinity**, E_{ea}, is the energy released when an electron attaches to a gas-phase atom (Table 8B.5). In a common, logical (given its name), but not universal convention (which is adopted here), the electron affinity is positive if energy is released when the electron attaches to the atom. That is, $E_{ea} > 0$ implies that electron attachment is exothermic.

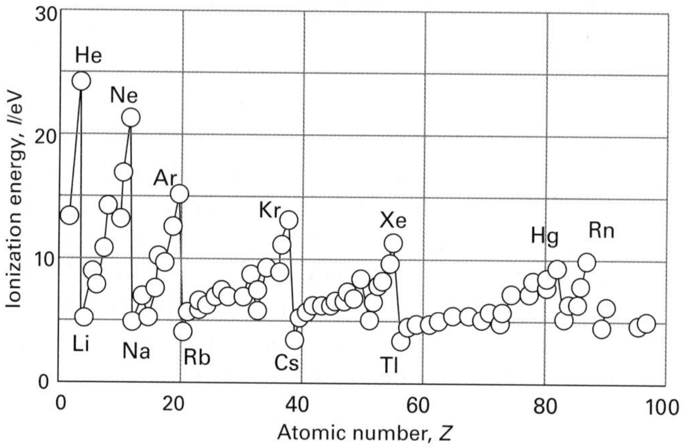

Figure 8B.9 The first ionization energies of the elements plotted against atomic number.

Table 8B.4 First and second ionization energies*

Element	I_1/(kJ mol^{-1})	I_2/(kJ mol^{-1})
H	1312	
He	2372	5251
Mg	738	1451
Na	496	4562

* More values are given in the *Resource section*.

Table 8B.5 Electron affinities, $E_a/(\text{kJ mol}^{-1})$*

Element	$E_a/(\text{kJ mol}^{-1})$	Element	$E_a/(\text{kJ mol}^{-1})$
Cl	349		
F	322		
H	73		
O	141	O^-	−844

* More values are given in the *Resource section*.

As will be familiar from introductory chemistry, ionization energies and electron affinities show periodicities. The former is more regular and concentrated on here:

- Lithium has a low first ionization energy because its outermost electron is well shielded from the nucleus by the core ($Z_{eff} = 1.3$, compared with $Z = 3$).
- The ionization energy of Be ($Z = 4$) is greater but that of B is lower because in the latter the outermost electron occupies a 2p orbital and is less strongly bound than if it had been a 2s electron.
- The ionization energy increases from B to N on account of the increasing effective nuclear charge.
- However, the ionization energy of O is less than would be expected by simple extrapolation. The explanation is that at oxygen a 2p orbital must become doubly occupied, and the electron–electron repulsions are increased above what would be expected by simple extrapolation along the row. In addition, the loss of a 2p electron results in a configuration with a half-filled subshell (like that of N), which is an arrangement of low energy, so the energy of $O^+ + e^-$ is lower than might be expected, and the ionization energy is correspondingly low too. (The kink is less pronounced in the next row, between phosphorus and sulfur, because their orbitals are more diffuse.)
- The values for O, F, and Ne fall roughly on the same line, the increase of their ionization energies reflecting the increasing attraction of the more highly charged nuclei for the outermost electrons.
- The outermost electron in sodium ($Z = 11$) is 3s. It is far from the nucleus, and the latter's charge is shielded by the compact, complete neon-like core, with the result that $Z_{eff} \approx 2.5$. As a result, the ionization energy of Na is substantially lower than that of Ne ($Z = 10$, $Z_{eff} \approx 5.8$).
- The periodic cycle starts again along this row, and the variation of the ionization energy can be traced to similar reasons.

Electron affinities are greatest close to fluorine, for the incoming electron enters a vacancy in a compact valence shell and can interact strongly with the nucleus. The attachment of an electron to an anion (as in the formation of O^{2-} from O^-) is invariably endothermic, so E_{ea} is negative. The incoming electron is repelled by the charge already present. Electron affinities are also small, and may be negative, when an electron enters an orbital that is far from the nucleus (as in the heavier alkali metal atoms) or is forced by the Pauli principle to occupy a new shell (as in the noble gas atoms).

Exercises

E8B.3 Write the ground-state electron configurations of the d-metals from scandium to zinc.

E8B.4 Write the electronic configuration of the Ni^{2+} ion.

E8B.5 Consider the atoms of the Period 2 elements of the periodic table. Predict which element has the lowest first ionization energy.

8B.4 Self-consistent field orbitals

The preceding treatment of the electronic configuration of many-electron species is only approximate because of the complications introduced by electron–electron interactions. However, computational techniques are available that give reliable approximate solutions for the wavefunctions and energies. The techniques were originally introduced by D.R. Hartree (before computers were available) and then modified by V. Fock to take into account the Pauli principle correctly. In broad outline, the **Hartree–Fock self-consistent field** (HF-SCF) procedure is as follows.

Start with an idea of the structure of the atom as suggested by the building-up principle. In the Ne atom, for instance, the principle suggests the configuration $1s^22s^22p^6$ with the orbitals approximated by hydrogenic atomic orbitals with the appropriate effective nuclear charges. Now consider one of the 2p electrons. A Schrödinger equation can be written for this electron by ascribing to it a potential energy due to the nuclear attraction and the average repulsion from the other electrons. Although the equation is for the 2p orbital, that repulsion, and therefore the equation, depends on the wavefunctions of all the other occupied orbitals in the atom. To solve the equation, guess an approximate form of the wavefunctions of all the other orbitals and then solve the Schrödinger equation for the 2p orbital. The procedure is then repeated for the 1s and 2s orbitals. This sequence of calculations gives the form of the 2p, 2s, and 1s orbitals, and in general they will differ from the set used to start the calculation. These improved orbitals can be used in another cycle of calculation, and a second improved set of orbitals and a better energy are obtained. The recycling continues until the orbitals and energies obtained are insignificantly different from those used at the start of the current cycle. The solutions are then self-consistent and accepted as solutions of the problem.

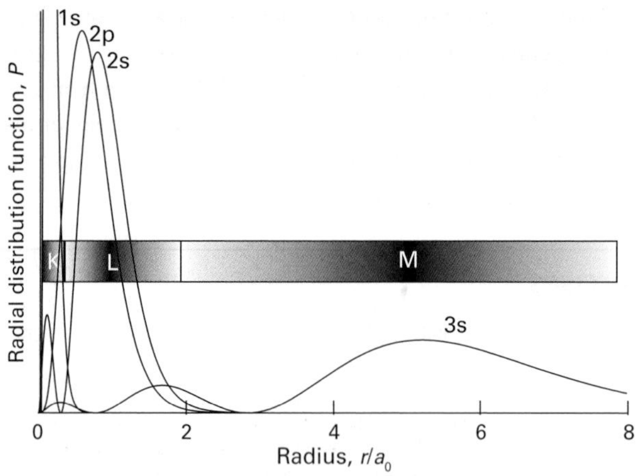

Figure 8B.10 The radial distribution functions for the orbitals of Na based on HF-SCF calculations. Note the shell-like structure, with the 3s orbital outside the inner concentric K and L shells.

Figure 8B.10 shows plots of some of the HF-SCF radial distribution functions for sodium. They show the grouping of electron density into shells, as was anticipated by the early chemists, and the differences of penetration as discussed above. These calculations therefore support the qualitative discussions that are used to explain chemical periodicity. They also considerably extend that discussion by providing detailed wavefunctions and precise energies.

Checklist of concepts

- ☐ 1. In the **orbital approximation**, each electron is regarded as being described by its own wavefunction; the overall wavefunction of a many-electron atom is the product of the orbital wavefunctions.
- ☐ 2. The **configuration** of an atom is the statement of its occupied orbitals.
- ☐ 3. The **Pauli exclusion principle**, a special case of the Pauli principle, limits to two the number of electrons that can occupy a given orbital.
- ☐ 4. In many-electron atoms, s orbitals lie at a lower energy than p orbitals of the same shell due to the combined effects of **penetration** and **shielding**.
- ☐ 5. The **building-up principle** is a procedure for predicting the ground state electron configuration of an atom.
- ☐ 6. Electrons occupy different orbitals of a given subshell before doubly occupying any one of them.
- ☐ 7. An atom in its ground state adopts a configuration with the greatest number of unpaired electrons.
- ☐ 8. The **atomic radius** of an element is half the distance between the centres of neighbouring atoms in a solid or in a homonuclear molecule.
- ☐ 9. The **ionic radius** of an element is its share of the distance between neighbouring ions in an ionic solid.
- ☐ 10. The **first ionization energy** is the minimum energy necessary to remove an electron from a many-electron atom in the gas phase.
- ☐ 11. The **second ionization energy** is the minimum energy needed to remove an electron from a singly charged cation.
- ☐ 12. The **electron affinity** is the energy released when an electron attaches to a gas-phase atom.
- ☐ 13. The atomic radius, ionization energy, and electron affinity vary periodically through the periodic table.
- ☐ 14. In the **Hartree–Fock self-consistent field** (HF-SCF) procedure the Schrödinger equation is solved numerically and iteratively until the solutions no longer change (to within certain criteria).

Checklist of equations

Property	Equation	Comment	Equation number
Orbital approximation	$\Psi(r_1,r_2,\ldots) = \psi(r_1)\psi(r_2)\ldots$		8B.1
Effective nuclear charge	$Z_{\text{eff}} = Z - \sigma$	The charge is this number times e	8B.5

TOPIC 8C Atomic spectra

➤ **Why do you need to know this material?**

A knowledge of the energies of electrons in atoms is essential for understanding many chemical properties and chemical bonding.

➤ **What is the key idea?**

The frequency and wavenumber of radiation emitted or absorbed when atoms undergo electronic transitions provide detailed information about their electronic energy states.

➤ **What do you need to know already?**

This Topic draws on knowledge of the energy levels of hydrogenic atoms (Topic 8A) and the configurations of many-electron atoms (Topic 8B). In places, it uses the properties of angular momentum (Topic 7F).

The general idea behind atomic spectroscopy is straightforward: lines in the spectrum (in either emission or absorption) occur when the electron distribution in an atom undergoes a **transition**, a change of state, in which its energy changes by ΔE. This transition leads to the emission or is accompanied by the absorption of a photon of frequency $\nu = |\Delta E|/h$ and wavenumber $\tilde{\nu} = |\Delta E|/hc$. In spectroscopy, transitions are said to take place between two **terms**. Broadly speaking, a term is simply another name for the energy level of an atom, but as this Topic progresses its full significance will become clear.

8C.1 The spectra of hydrogenic atoms

Not all transitions between the possible terms are observed. Spectroscopic transitions are **allowed**, if they can occur, or **forbidden**, if they cannot occur. A **selection rule** is a statement about which transitions are allowed.

The origin of selection rules can be identified by considering transitions in hydrogenic atoms. A photon has an intrinsic spin angular momentum corresponding to $s = 1$ (Topic 8B). Because total angular momentum is conserved in a transition, the angular momentum of the electron must change to compensate for the angular momentum carried away by the photon. Thus, an electron in a d orbital ($l = 2$) cannot make a transition into an s orbital ($l = 0$) because the photon cannot carry away enough angular momentum. Similarly, an s electron cannot make a transition to another s orbital, because there would then be no change in the angular momentum of the electron to make up for the angular momentum carried away by the photon. A more formal treatment of selection rules requires mathematical manipulation of the wavefunctions for the initial and final states of the atom.

How is that done? 8C.1 Identifying selection rules

The underlying classical idea behind a spectroscopic transition is that, for an atom or molecule to be able to interact with the electromagnetic field and absorb or create a photon of frequency ν, it must possess, at least transiently, a dipole oscillating at that frequency. The consequences of this idea are explored in the following steps.

Step 1 *Write an expression for the transition dipole moment*

The transient dipole is expressed quantum mechanically as the **transition dipole moment**, μ_{fi}, between the initial and final states i and f, where[1]

$$\mu_{fi} = \int \psi_f^* \hat{\mu} \psi_i \,d\tau \qquad (8C.1)$$

and $\hat{\mu}$ is the electric dipole moment operator. For a one-electron atom $\hat{\mu}$ is multiplication by $-e\boldsymbol{r}$. Because $\boldsymbol{r}$ is a vector with components x, y, and z, $\hat{\mu}$ is also a vector, with components $\mu_x = -ex$, $\mu_y = -ey$, and $\mu_z = -ez$. If the transition dipole moment is zero, then the transition is forbidden; the transition is allowed if the transition dipole moment is non-zero.

Step 2 *Formulate the integrand in terms of spherical harmonics*

To evaluate a transition dipole moment, consider each component in turn. For example, for the z-component,

$$\mu_{z,fi} = -e\int \psi_f^* z \psi_i \,d\tau$$

In spherical polar coordinates (see *The chemist's toolkit* 7F.2) $z = r\cos\theta$. It is convenient to express $\cos\theta$ in terms of spherical harmonics. From Table 7F.1 it is seen that $Y_{1,0} = (3/4\pi)^{1/2}\cos\theta$, so it follows that $\cos\theta = (4\pi/3)^{1/2}Y_{1,0}$. Therefore z can be written as $z = r\cos\theta = (4\pi/3)^{1/2} rY_{1,0}$. The wavefunctions for the initial

[1] See our *Physical chemistry: Quanta, matter, and change* (2014) for a detailed development of the form of eqn 8C.1.

and final states are atomic orbitals of the form $R_{n,l}(r)Y_{l,m_l}(\theta,\phi)$ (Topic 8A). With these substitutions the integral becomes

$$\int\psi_{\mathrm{f}}^{\star}z\psi_{\mathrm{i}}\mathrm{d}\tau =$$

$$\int_0^{\infty}\int_0^{\pi}\int_0^{2\pi}\overbrace{R_{n_{\mathrm{f}},l_{\mathrm{f}}}\boxed{Y^{\star}_{l_{\mathrm{f}},m_{l,\mathrm{f}}}}}^{\psi_{\mathrm{f}}^{\star}}\overbrace{\left(\frac{4\pi}{3}\right)^{1/2}r\boxed{Y_{1,0}}}^{z}\overbrace{R_{n_{\mathrm{i}},l_{\mathrm{i}}}\boxed{Y_{l_{\mathrm{i}},m_{l,\mathrm{i}}}}}^{\psi_{\mathrm{i}}}\overbrace{r^2\mathrm{d}r\boxed{\sin\theta\mathrm{d}\theta\mathrm{d}\phi}}^{\mathrm{d}\tau}$$

This multiple integral is the product of three factors, an integral over r and two integrals (boxed) over the angles, so the factors on the right can be grouped as follows:

$$\int\psi_{\mathrm{f}}^{\star}z\psi_{\mathrm{i}}\mathrm{d}\tau =$$

$$\left(\frac{4\pi}{3}\right)^{1/2}\int_0^{\infty}R_{n_{\mathrm{f}},l_{\mathrm{f}}}r^3R_{n_{\mathrm{i}},l_{\mathrm{i}}}\mathrm{d}r\boxed{\int_0^{\pi}\int_0^{2\pi}Y^{\star}_{l_{\mathrm{f}},m_{l,\mathrm{f}}}Y_{1,0}Y_{l_{\mathrm{i}},m_{l,\mathrm{i}}}\sin\theta\mathrm{d}\theta\mathrm{d}\phi}$$

Step 3 *Evaluate the angular integral*

It follows from the properties of the spherical harmonics that the integral

$$I=\int_0^{\pi}\int_0^{2\pi}Y^{\star}_{l_{\mathrm{f}},m_{l,\mathrm{f}}}Y_{l,m}Y_{l_{\mathrm{i}},m_{l,\mathrm{i}}}\sin\theta\,\mathrm{d}\theta\mathrm{d}\phi$$

is zero unless $l_{\mathrm{f}} = l_{\mathrm{i}} \pm l$ and $m_{l,\mathrm{f}} = m_{l,\mathrm{i}} + m$. Because in the present case $l = 1$ and $m = 0$, the angular integral, and hence the z-component of the transition dipole moment, is zero unless $\Delta l = l_{\mathrm{f}} - l_{\mathrm{i}} = \pm 1$ and $\Delta m_l = m_{l,\mathrm{f}} - m_{l,\mathrm{i}} = 0$, which is a part of the set of selection rules. The same procedure, but considering the x- and y-components, results in the complete set of rules:

$$\Delta l=\pm 1 \quad \Delta m_l=0,\pm 1 \qquad \text{Selection rules for hydrogenic atoms} \qquad (8C.2)$$

The principal quantum number n can change by any amount consistent with the value of Δl for the transition, because it does not relate directly to the angular momentum.

Brief illustration 8C.1

To identify the orbitals to which a 4d electron may make radiative transitions, first identify the value of l and then apply the selection rule for this quantum number. Because $l = 2$, the final orbital must have $l = 1$ or 3. Thus, an electron may make a transition from a 4d orbital to any np orbital (subject to $\Delta m_l = 0, \pm 1$) and to any nf orbital (subject to the same rule). However, it cannot undergo a transition to any other orbital, such as an ns or an nd orbital.

The selection rules and the atomic energy levels jointly account for the structure of a **Grotrian diagram** (Fig. 8C.1), which summarizes the energies of the terms and the transitions between them. In some versions, the thicknesses of the transition lines in the diagram denote their relative intensities in the spectrum.

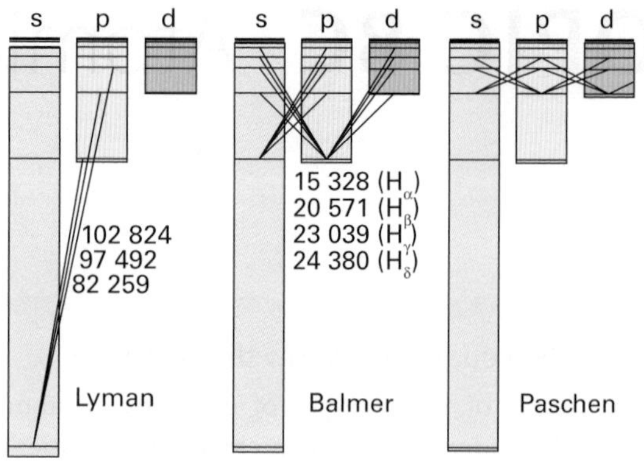

Figure 8C.1 A Grotrian diagram that summarizes the appearance and analysis of the spectrum of atomic hydrogen. The wavenumbers of some transitions (in cm^{-1}) are indicated.

Exercises

E8C.1 Identify the transition responsible for the shortest and longest wavelength lines in the Lyman series.

E8C.2 Calculate the wavelength, frequency, and wavenumber of the $n = 2 \rightarrow n = 1$ transition in He^+.

E8C.3 Which of the following transitions are allowed in the electronic emission spectrum of a hydrogenic atom: (i) 2s → 1s, (ii) 2p → 1s, (iii) 3d → 2p?

8C.2 The spectra of many-electron atoms

The spectra of atoms rapidly become very complicated as the number of electrons increases, in part because their energy levels, their terms, are not given solely by the energies of the orbitals but depend on the interactions between the electrons.

8C.2(a) Singlet and triplet terms

Consider the energy levels of a He atom, with its two electrons. The ground-state configuration is $1s^2$, and an excited configuration is one in which an electron has been promoted into a different orbital to give, for instance, the configuration $1s^12s^1$. The two electrons need not be paired because they occupy different orbitals. According to Hund's maximum multiplicity rule (Topic 8B), the state of the atom with the spins parallel lies lower in energy than the state in which they are paired. Both states are permissible, correspond to different terms, and can contribute to the spectrum of the atom.

Parallel and antiparallel (paired) spins differ in their total spin angular momentum. In the paired case, the two spin momenta cancel, and there is zero net spin (as depicted in

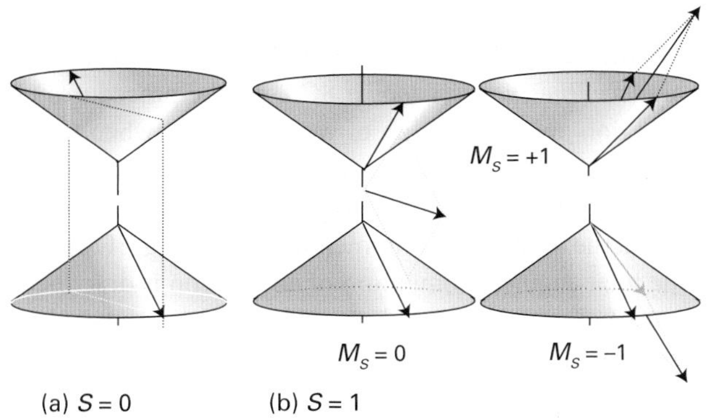

Figure 8C.2 (a) Electrons with paired spins have zero resultant spin angular momentum ($S = 0$). They can be represented by two vectors that lie at an indeterminate position on the cones shown here, but wherever one lies on its cone, the other points in the opposite direction; their resultant is zero. (b) When two electrons have parallel spins, they have a non-zero total spin angular momentum ($S = 1$). There are three ways of achieving this resultant, which are shown by these vector representations. Note that, whereas two paired spins are precisely antiparallel, two 'parallel' spins are not strictly parallel. The notation S, M_S is explained later.

Fig. 8C.2a). Its state is the one denoted σ_- in the discussion of the Pauli principle (Topic 8B):

$$\sigma_-(1,2) = \frac{1}{2^{1/2}}\{\alpha(1)\beta(2) - \beta(1)\alpha(2)\} \tag{8C.3a}$$

The angular momenta of two parallel spins add to give a non-zero total spin. As illustrated in Fig. 8C.2(b), there are three ways of achieving non-zero total spin. The wavefunctions of the three spin states are the symmetric combinations introduced in Topic 8B:

$$\alpha(1)\alpha(2)$$

$$\sigma_+(1,2) = \frac{1}{2^{1/2}}\{\alpha(1)\beta(2) + \beta(1)\alpha(2)\} \tag{8C.3b}$$

$$\beta(1)\beta(2)$$

The state of the He atom in which the two electrons are paired and their spins are described by eqn 8C.3a gives rise to a **singlet term**. The alternative arrangement, in which the spins are parallel and are described by any of the three expressions in eqn 8C.3b, gives rise to a **triplet term**. The fact that the parallel arrangement of spins in the triplet term of the $1s^12s^1$ configuration of the He atom lies lower in energy than the antiparallel arrangement, the singlet term, can now be expressed by saying that the triplet term of the $1s^12s^1$ configuration of He lies lower in energy than the singlet term. This is a general conclusion and applies to other atoms (and molecules):

> For states arising from the same configuration, the triplet term generally lies lower than the singlet term.

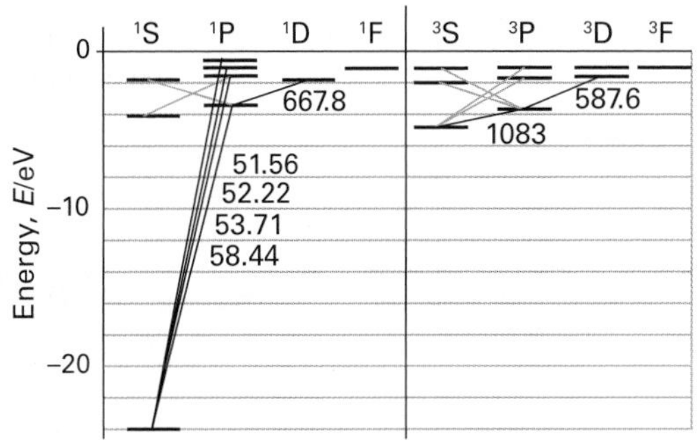

Figure 8C.3 Some of the transitions responsible for the spectrum of atomic helium. The labels give the wavelengths (in nanometres) of the transitions. The labels along the top of the diagram are term symbols introduced in Section 8C.2(c).

The origin of the energy difference lies in the effect of spin correlation on the Coulombic interactions between electrons, as in the case of Hund's maximum multiplicity rule for ground-state configurations (Topic 8B): electrons with parallel spins tend to avoid each other. Because the Coulombic interaction between electrons in an atom is strong, the difference in energies between singlet and triplet terms of the same configuration can be large. The singlet and triplet terms of the configuration $1s^12s^1$ of He, for instance, differ by 6421 cm^{-1} (corresponding to 0.80 eV).

The spectrum of atomic helium is more complicated than that of atomic hydrogen, but there are two simplifying features. One is that the only excited configurations to consider are of the form $1s^1nl^1$; that is, only one electron is excited. Excitation of two electrons requires an energy greater than the ionization energy of the atom, so the He^+ ion is formed instead of the doubly excited atom. Second, and as seen later in this Topic, no radiative transitions take place between singlet and triplet terms because the relative orientation of the two electron spins cannot change during a transition. Thus, there is a spectrum arising from transitions between singlet terms (including the ground state) and between triplet terms, but not between the two. Spectroscopically, helium behaves like two distinct species. The Grotrian diagram for helium in Fig. 8C.3 shows the two sets of transitions.

8C.2(b) Spin–orbit coupling

An electron has a magnetic moment that arises from its spin. Similarly, an electron with orbital angular momentum (that is, an electron in an orbital with $l > 0$) is in effect a circulating current, and possesses a magnetic moment that arises from its orbital momentum. The interaction of the spin magnetic moment with the magnetic moment arising from the orbital angular momentum is called **spin–orbit coupling**. The strength of the coupling, and its effect on the energy levels of the atom, depend on the relative orientations of the spin and orbital magnetic

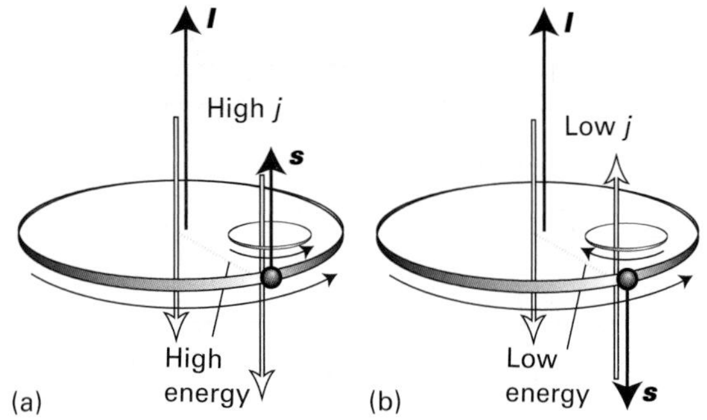

Figure 8C.4 Spin–orbit coupling is a magnetic interaction between spin and orbital magnetic moments; the black arrows show the direction of the angular momentum and the open arrows show the direction of the associated magnetic moments. When the angular momenta are parallel, as in (a), the magnetic moments are aligned unfavourably; when they are opposed, as in (b), the interaction is favourable. This magnetic coupling is the cause of the splitting of a term into levels.

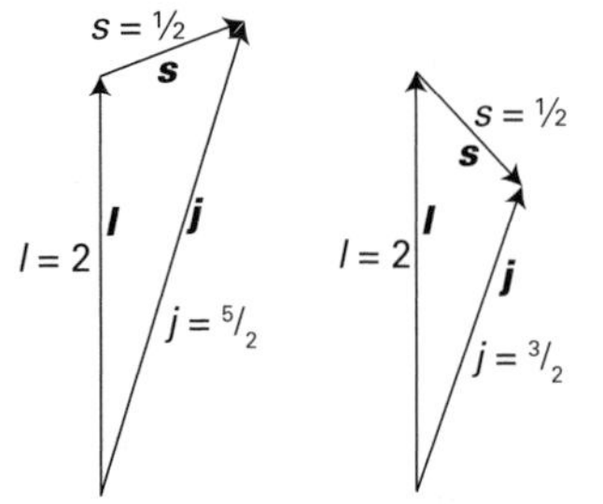

Figure 8C.5 The coupling of the spin and orbital angular momenta of a d electron ($l = 2$) gives two possible values of j depending on the relative orientations of the spin and orbital angular momenta of the electron.

moments, and therefore on the relative orientations of the two angular momenta (Fig. 8C.4).

One way of expressing the dependence of the spin–orbit interaction on the relative orientation of the spin and orbital momenta is to say that it depends on the total angular momentum of the electron, the vector sum of its spin and orbital momenta (see *The chemist's toolkit* 8C.1 for the procedures used for addition and subtraction of vectors). Thus, when the spin and orbital angular momenta are nearly parallel, the total angular momentum is high; when the two angular momenta are opposed, the total angular momentum is low (Fig. 8C.5).

The total angular momentum of an electron is described by the quantum numbers j and m_j, with $j = l + \frac{1}{2}$ or $j = l - \frac{1}{2}$. The different values of j that can arise for a given value of l label the **levels** of a term. For $l = 0$, the only permitted value is $j = \frac{1}{2}$ (the total angular momentum is the same as the spin angular momentum because there is no other source of angular momentum in the atom). When $l = 1$, j may be either $\frac{3}{2}$ (the spin and orbital angular momenta are in the same sense) or $\frac{1}{2}$ (the spin and angular momenta are in opposite senses).

Brief illustration 8C.2

To identify the levels that may arise from the configurations (a) d^1 and (b) s^1, identify the value of l and then the possible values of j. (a) For a d electron, $l = 2$ and there are two levels in the configuration, one with $j = 2 + \frac{1}{2} = \frac{5}{2}$ and the other with $j = 2 - \frac{1}{2} = \frac{3}{2}$. (b) For an s electron $l = 0$, so only one level is possible, and $j = \frac{1}{2}$.

With a little work, it is possible to incorporate the effect of spin–orbit coupling on the energies of the levels.

How is that done? 8C.2 Deriving an expression for the energy of spin–orbit interaction

Classically, the energy of a magnetic moment $\boldsymbol{\mu}$ in a magnetic field $\mathcal{B}$ is equal to their scalar product $-\boldsymbol{\mu}\cdot\mathcal{B}$. Follow these steps to arrive at an expression for the spin–orbit interaction energy.

The chemist's toolkit 8C.1 Combining vectors

In three dimensions, the vectors $\boldsymbol{u}$ (with components u_x, u_y, and u_z) and $\boldsymbol{v}$ (with components v_x, v_y, and v_z) have the general form:

$$\boldsymbol{u} = u_x\boldsymbol{i} + u_y\boldsymbol{j} + u_z\boldsymbol{k} \quad \boldsymbol{v} = v_x\boldsymbol{i} + v_y\boldsymbol{j} + v_z\boldsymbol{k}$$

where $\boldsymbol{i}$, $\boldsymbol{j}$, and $\boldsymbol{k}$ are **unit vectors**, vectors of magnitude 1, pointing along the positive directions on the x, y, and z axes. The operations of addition, subtraction, and multiplication are as follows:

1. Addition:

$$\boldsymbol{v} + \boldsymbol{u} = (v_x + u_x)\boldsymbol{i} + (v_y + u_y)\boldsymbol{j} + (v_z + u_z)\boldsymbol{k}$$

2. Subtraction:

$$\boldsymbol{v} - \boldsymbol{u} = (v_x - u_x)\boldsymbol{i} + (v_y - u_y)\boldsymbol{j} + (v_z - u_z)\boldsymbol{k}$$

3. Multiplication (the 'scalar product'):

$$\boldsymbol{v}\cdot\boldsymbol{u} = v_x u_x + v_y u_y + v_z u_z$$

The scalar product of a vector with itself gives the square magnitude of the vector:

$$\boldsymbol{v}\cdot\boldsymbol{v} = v_x^2 + v_y^2 + v_z^2 = v^2$$

Step 1 *Write an expression for the energy of interaction*

If the magnetic field arises from the orbital angular momentum of the electron, it is proportional to $\boldsymbol{l}$; if the magnetic moment μ is that of the electron spin, then it is proportional to $\boldsymbol{s}$. It follows that the energy of interaction is proportional to the scalar product $\boldsymbol{s}\cdot\boldsymbol{l}$:

$$\text{Energy of interaction} = -\boldsymbol{\mu}\cdot\boldsymbol{\mathcal{B}} \propto \boldsymbol{s}\cdot\boldsymbol{l}$$

Step 2 *Express the scalar product in terms of the magnitudes of the vectors*

Note that the total angular momentum is the vector sum of the spin and orbital momenta: $\boldsymbol{j} = \boldsymbol{l} + \boldsymbol{s}$. The magnitude of the vector $\boldsymbol{j}$ is calculated by evaluating

$$\overbrace{\boldsymbol{j}\cdot\boldsymbol{j}}^{j^2} = (\boldsymbol{l}+\boldsymbol{s})\cdot(\boldsymbol{l}+\boldsymbol{s}) = \overbrace{\boldsymbol{l}\cdot\boldsymbol{l}}^{l^2} + \overbrace{\boldsymbol{s}\cdot\boldsymbol{s}}^{s^2} + 2\boldsymbol{s}\cdot\boldsymbol{l}$$

so

$$j^2 = l^2 + s^2 + 2\boldsymbol{s}\cdot\boldsymbol{l}$$

That is,

$$\boldsymbol{s}\cdot\boldsymbol{l} = \tfrac{1}{2}(j^2 - l^2 - s^2)$$

This equation is a classical result.

Step 3 *Replace the classical magnitudes by their quantum mechanical versions*

To derive the quantum mechanical version of this expression, replace all the quantities on the right with their quantum-mechanical values, which are of the form $j(j+1)\hbar^2$, etc (Topic 7F):

$$\boldsymbol{s}\cdot\boldsymbol{l} = \tfrac{1}{2}\{j(j+1) - l(l+1) - s(s+1)\}\hbar^2$$

Then, by inserting this expression into the formula for the energy of interaction ($E \propto \boldsymbol{s}\cdot\boldsymbol{l}$) and writing the constant of proportionality as $hc\tilde{A}/\hbar^2$, obtain an expression for the energy in terms of the quantum numbers and the **spin–orbit coupling constant**, $\tilde{A}$ (a wavenumber):

$$E_{l,s,j} = \tfrac{1}{2}hc\tilde{A}\{j(j+1) - l(l+1) - s(s+1)\} \qquad (8\text{C}.4)$$

Spin–orbit interaction energy

Brief illustration 8C.3

The unpaired electron in the ground state of an alkali metal atom has $l = 0$, so $j = \frac{1}{2}$. Because the orbital angular momentum is zero in this state, the spin–orbit coupling energy is zero (as is confirmed by setting $j = s$ and $l = 0$ in eqn 8C.4). When the electron is excited to an orbital with $l = 1$, it has orbital angular momentum and can give rise to a magnetic field that interacts with its spin. In this configuration the electron can have $j = \frac{3}{2}$ or $j = \frac{1}{2}$, and the energies of these levels are

$$E_{1,1/2,3/2} = \tfrac{1}{2}hc\tilde{A}\{\tfrac{3}{2}\times\tfrac{5}{2} - 1\times 2 - \tfrac{1}{2}\times\tfrac{3}{2}\} = \tfrac{1}{2}hc\tilde{A}$$

$$E_{1,1/2,1/2} = \tfrac{1}{2}hc\tilde{A}\{\tfrac{1}{2}\times\tfrac{3}{2} - 1\times 2 - \tfrac{1}{2}\times\tfrac{3}{2}\} = -hc\tilde{A}$$

The corresponding energies are shown in Fig. 8C.6. Note that the barycentre (the 'centre of gravity') of the levels is unchanged, because there are four states of energy $\frac{1}{2}hc\tilde{A}$ and two of energy $-hc\tilde{A}$.

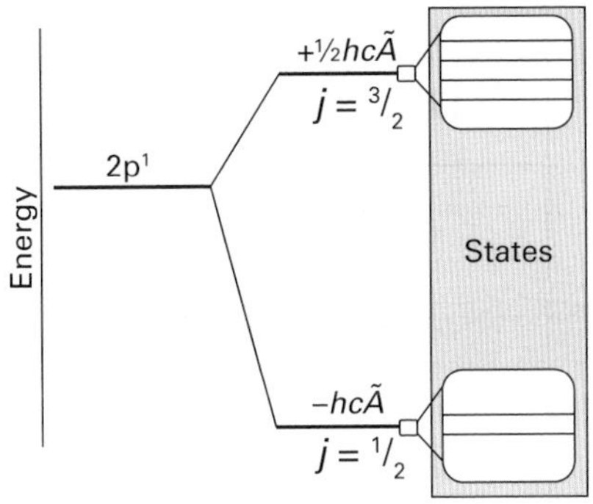

Figure 8C.6 The levels of a $2p^1$ configuration arising from spin–orbit coupling. Note that the low-j level lies below the high-j level in energy. The number of states in a level with quantum number j is $2j + 1$.

The strength of the spin–orbit coupling depends on the nuclear charge. To understand why this is so, imagine riding on the orbiting electron and seeing a charged nucleus apparently orbiting around you (like the Sun rising and setting). As a result, you find yourself at the centre of a ring of current. The greater the nuclear charge, the greater is this current, and therefore the stronger is the magnetic field you detect. Because the spin magnetic moment of the electron interacts with this orbital magnetic field, it follows that the greater the nuclear charge, the stronger is the spin–orbit interaction. It turns out that the coupling increases sharply with atomic number (as Z^4) because not only is the current greater but the electron is drawn closer to the nucleus. Whereas the coupling is only weak in H (giving rise to shifts of energy levels of no more than about $0.4\,\text{cm}^{-1}$), in heavy atoms like Pb it is very strong (giving shifts of the order of thousands of reciprocal centimetres).

Two spectral lines are observed when the p electron of an electronically excited alkali metal atom undergoes a transition into a lower s orbital. One line is due to a transition starting in a $j = \frac{3}{2}$ level of the upper term; the other line is due to a transition starting in the $j = \frac{1}{2}$ level of the same term. The two lines are jointly an example of the **fine structure** of a spectrum, the structure due to spin-orbit coupling. Fine structure can be seen in the emission spectrum from sodium vapour excited by an electric discharge or in a flame. The yellow line at 589 nm (close to $17\,000\,\text{cm}^{-1}$) is actually a doublet composed of one line at 589.76 nm ($16\,956.2\,\text{cm}^{-1}$) and another at 589.16 nm

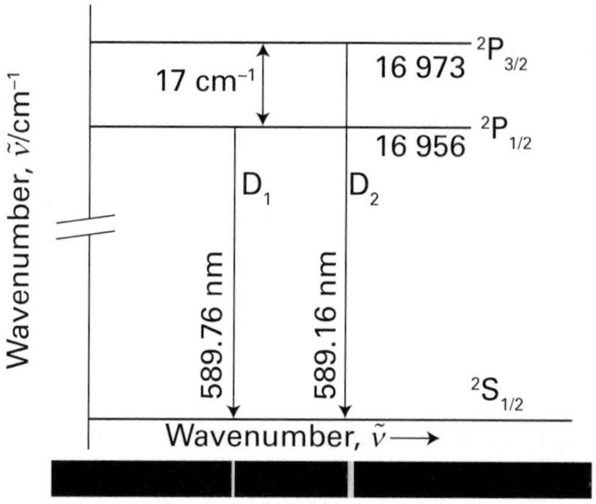

Figure 8C.7 The energy-level diagram for the formation of the sodium D lines; the j values are the subscripts of term symbols such as $^2P_{1/2}$. The splitting of the spectral lines (by 17 cm^{-1}) reflects the splitting of the levels of the ^2P term.

(16 973.4 cm^{-1}); the components of this doublet are the 'D lines' of the spectrum (Fig. 8C.7). Therefore, in Na, the spin–orbit coupling affects the energies by about 17 cm^{-1}.

Example 8C.1 Analysing a spectrum for the spin–orbit coupling constant

The origin of the D lines in the spectrum of atomic sodium is shown in Fig. 8C.7. Calculate the spin–orbit coupling constant for the upper configuration of the Na atom.

Collect your thoughts It follows from Fig. 8C.7 that the splitting of the lines is equal to the energy separation of the $j = \frac{3}{2}$ and $\frac{1}{2}$ levels of the excited configuration. You need to express this separation in terms of $\tilde{A}$ by using eqn 8C.4.

The solution The two levels are split by

$$\Delta\tilde{\nu} = \left(E_{1,\frac{1}{2},\frac{3}{2}} - E_{1,\frac{1}{2},\frac{1}{2}}\right)/hc = \tfrac{1}{2}\tilde{A}\left\{\tfrac{3}{2}\left(\tfrac{3}{2}+1\right) - \tfrac{1}{2}\left(\tfrac{1}{2}+1\right)\right\} = \tfrac{3}{2}\tilde{A}$$

The experimental value of $\Delta\tilde{\nu}$ is 17.2 cm^{-1}. Therefore

$$\tilde{A} = \tfrac{2}{3} \times (17.2\,\text{cm}^{-1}) = 11.5\,\text{cm}^{-1}$$

The same calculation repeated for the atoms of other alkali metals gives Li: 0.23 cm^{-1}, K: 38.5 cm^{-1}, Rb: 158 cm^{-1}, Cs: 370 cm^{-1}. Note the increase of $\tilde{A}$ with atomic number (but more slowly than Z^4 for these many-electron atoms).

Self-test 8C.1 The configuration … $4p^6 5d^1$ of rubidium has two levels at 25 700.56 cm^{-1} and 25 703.52 cm^{-1} above the ground state. What is the spin–orbit coupling constant in this excited state?

Answer: 1.18 cm^{-1}

8C.2(c) Term symbols

The discussion so far has used expressions such as 'the $j = \frac{3}{2}$ level of a doublet term with $l = 1$'. A **term symbol**, which is a symbol looking like $^2P_{3/2}$ or $^3D_2$, gives the total spin, total orbital angular momentum, and total overall angular momentum, in a succinct way.

A term symbol gives three pieces of information:

- The letter (P or D in the examples) indicates the total orbital angular momentum quantum number, L.
- The left superscript in the term symbol (the 2 in $^2P_{3/2}$) gives the multiplicity of the term.
- The right subscript on the term symbol (the 3/2 in $^2P_{3/2}$) is the value of the total angular momentum quantum number, J, and labels the level of the term.

The meaning of these statements can be discussed in the light of the contributions to the energies summarized in Fig. 8C.8.

When several electrons are present, it is necessary to judge how their individual orbital angular momenta add together to augment or oppose each other. The **total orbital angular momentum quantum number**, L, gives the magnitude of the angular momentum through $\{L(L+1)\}^{1/2}\hbar$. It has $2L+1$ orientations distinguished by the quantum number M_L, which can take the values $0, \pm1, \ldots, \pm L$. Similar remarks apply to the **total spin quantum number**, S, and the quantum number M_S, and the **total angular momentum quantum number**, J, and the quantum number M_J.

The value of L (a non-negative integer) is obtained by coupling the individual orbital angular momenta by using the **Clebsch–Gordan series**:

$$L = l_1 + l_2, l_1 + l_2 - 1, \ldots, |l_1 - l_2| \qquad \text{Clebsch-Gordan series} \quad (8C.5)$$

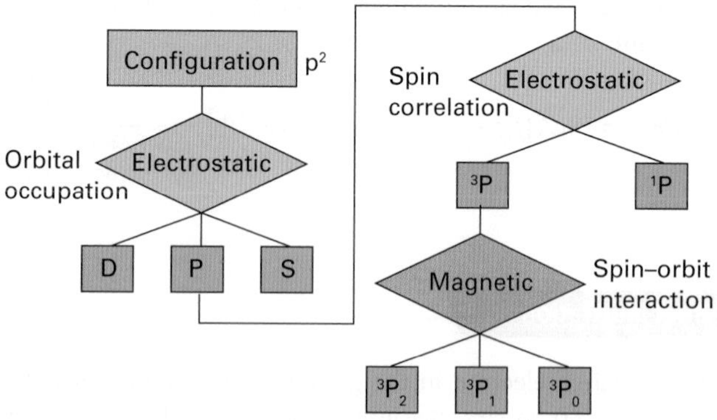

Figure 8C.8 A summary of the types of interaction that are responsible for the various kinds of splitting of energy levels in atoms. For light (low Z) atoms, magnetic interactions are small, but in heavy (high Z) atoms they may dominate the electrostatic (charge–charge) interactions.

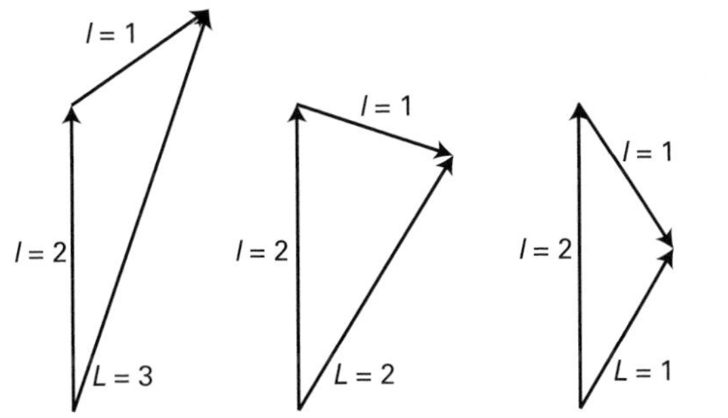

Figure 8C.9 The total orbital angular momenta of a p electron and a d electron correspond to $L = 3$, 2, and 1 and reflect the different relative orientations of the two momenta.

The modulus signs are attached to $l_1 - l_2$ to ensure that L is non-negative. The maximum value, $L = l_1 + l_2$, is obtained when the two orbital angular momenta are in nearly the same direction; the lowest value, $|l_1 - l_2|$, is obtained when they are nearly in opposite directions. Note the 'nearly': the magnitudes of the component vectors and the resultant do not in general allow the vectors to be exactly in the same or opposite directions. The intermediate values represent possible intermediate relative orientations of the two momenta (Fig. 8C.9). For two p electrons (for which $l_1 = l_2 = 1$), $L = 2, 1, 0$. The code for converting the value of L into a letter is the same as for the s, p, d, f,... designation of orbitals, but uses uppercase Roman letters:[2]

L:	0	1	2	3	4	5	6...
	S	P	D	F	G	H	I...

Thus, a p^2 configuration has $L = 2, 1, 0$ and gives rise to D, P, and S terms. The terms differ in energy on account of the different spatial distribution of the electrons and the consequent differences in repulsion between them.[3]

A closed shell has zero orbital angular momentum because all the individual orbital angular momenta sum to zero. Therefore, when working out term symbols, only the electrons of the unfilled shell need to be considered. In the case of a single electron outside a closed shell, the value of L is the same as the value of l; so the configuration $[Ne]3s^1$ has only an S term.

Example 8C.2 Deriving the total orbital angular momentum of a configuration

Find the terms that can arise from the configurations (a) d^2, (b) p^3.

Collect your thoughts Use the Clebsch–Gordan series and begin by finding the minimum value of L (so that you know where the series terminates). When there are more than two electrons to couple together, you need to use two series in succession: first to couple two electrons, and then to couple the third to each combined state, and so on.

The solution (a) The minimum value is $|l_1 - l_2| = |2 - 2| = 0$. Therefore,

$$L = 2+2, 2+2-1, \ldots, 0 = 4, 3, 2, 1, 0$$

corresponding to G, F, D, P, and S terms, respectively. (b) Coupling two p electrons gives a minimum value of $|1 - 1| = 0$. Therefore,

$$L' = 1+1, 1+1-1, \ldots, 0 = 2, 1, 0$$

Now couple $l_3 = 1$ with $L' = 2$, to give $L = 3, 2, 1$; with $L' = 1$, to give $L = 2, 1, 0$; and with $L' = 0$, to give $L = 1$. The overall result is

$$L = 3, 2, 2, 1, 1, 1, 0$$

giving one F, two D, three P, and one S term.

Self-test 8C.2 Repeat the question for the configurations (a) f^1d^1 and (b) d^3.

Answer: (a) H, G, F, D, P; (b) I, 2H, 3G, 4F, 5D, 3P, S

When there are several electrons to be taken into account, their total spin angular momentum quantum number, S (a non-negative integer or half-integer), must be assessed. Once again the Clebsch–Gordan series is used, but now in the form

$$S = s_1 + s_2, s_1 + s_2 - 1, \ldots, |s_1 - s_2| \qquad (8C.6)$$

to decide on the value of S, noting that each electron has $s = \frac{1}{2}$. For two electrons the possible values of S are 1 and 0 (Fig. 8C.10). If there are three electrons, the total spin angular momentum is obtained by coupling the third spin to each of the values of S for the first two spins, which results in $S = \frac{3}{2}$ and $\frac{1}{2}$.

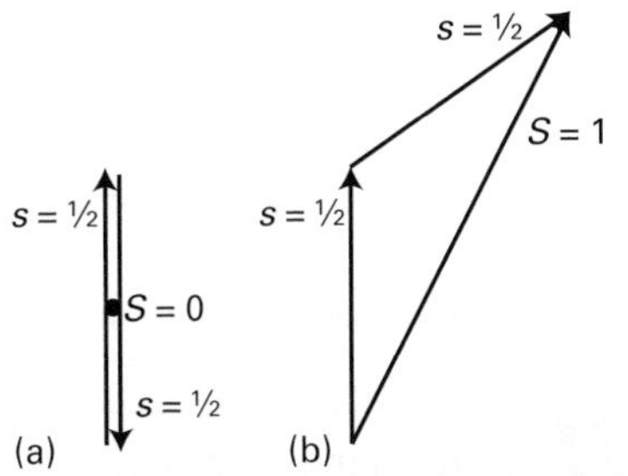

Figure 8C.10 For two electrons (each of which has $s = \frac{1}{2}$), only two total spin states are permitted ($S = 0, 1$). (a) The state with $S = 0$ can have only one value of M_S ($M_S = 0$) and gives rise to a singlet term; (b) the state with $S = 1$ can have any of three values of M_S (+1, 0, −1) and gives rise to a triplet term. The vector representations of the $S = 0$ and 1 states are shown in Fig. 8C.2.

[2] The convention of using lowercase letters to label orbitals and uppercase letters to label overall states applies throughout spectroscopy, not just to atoms.

[3] Throughout this discussion of atomic spectroscopy, distinguish italic S, the total spin quantum number, from Roman S, the term label.

The **multiplicity** of a term is the value of $2S+1$. When $S=0$ (as for a closed shell, like $1s^2$) the electrons are all paired and there is no net spin: this arrangement gives a singlet term, ^1S. A lone electron has $S=s=\frac{1}{2}$, so a configuration such as $[Ne]3s^1$ can give rise to a doublet term, ^2S. Likewise, the configuration $[Ne]3p^1$ is a doublet, ^2P. When there are two unpaired (parallel spin) electrons $S=1$, so $2S+1=3$, giving a triplet term, such as ^3D. The relative energies of singlets and triplets are discussed earlier in the Topic, where it is seen that their energies differ on account of spin correlation.

As already explained, the quantum number j gives the relative orientation of the spin and orbital angular momenta of a single electron. The **total angular momentum quantum number**, J (a non-negative integer or half-integer), does the same for several electrons. If there is a single electron outside a closed shell, $J=j$, with j either $l+\frac{1}{2}$ or $|l-\frac{1}{2}|$. The $[Ne]3s^1$ configuration has $j=\frac{1}{2}$ (because $l=0$ and $s=\frac{1}{2}$), so the ^2S term has a single level, denoted $^2S_{1/2}$. The $[Ne]3p^1$ configuration has $l=1$; therefore $j=\frac{3}{2}$ and $\frac{1}{2}$; the ^2P term therefore has two levels, $^2P_{3/2}$ and $^2P_{1/2}$. These levels lie at different energies on account of the spin–orbit interaction.

If there are several electrons outside a closed shell it is necessary to consider the coupling of all the spins and all the orbital angular momenta. This complicated problem can be simplified when the spin–orbit coupling is weak (for atoms of low atomic number), by using the **Russell–Saunders coupling** scheme. This scheme is based on the view that, if spin–orbit coupling is weak, then it is effective only when all the orbital momenta are operating cooperatively. That is, all the orbital angular momenta of the electrons couple to give a total L, and all the spins are similarly coupled to give a total S. Only at this stage do the two kinds of momenta couple through the spin–orbit interaction to give a total J. The permitted values of J are given by the Clebsch–Gordan series

$$J=L+S, L+S-1, \ldots, |L-S| \qquad (8C.7)$$

For example, in the case of the ^3D term of the configuration $[Ne]2p^13p^1$, the permitted values of J are 3, 2, 1 (because ^3D has $L=2$ and $S=1$), so the term has three levels, $^3D_3$, $^3D_2$, and $^3D_1$.

When $L \geq S$, the multiplicity is equal to the number of levels. For example, a ^2P term $\left(L=1>S=\frac{1}{2}\right)$ has the two levels $^2P_{3/2}$ and $^2P_{1/2}$, and ^3D ($L=2>S=1$) has the three levels $^3D_3$, $^3D_2$, and $^3D_1$. However, this is not the case when $L<S$: the term ^2S $\left(L=0<S=\frac{1}{2}\right)$, for example, has only the one level $^2S_{1/2}$.

Example 8C.3 Deriving term symbols

Write the term symbols arising from the ground-state configurations of (a) Na and (b) F, and (c) the excited configuration $1s^22s^22p^13p^1$ of C.

Collect your thoughts Begin by writing the configurations, but ignore inner closed shells. Then couple the orbital momenta to find L and the spins to find S. Next, couple L and S to find J. Finally, express the term as $^{2S+1}\{L\}_J$, where $\{L\}$ is the appropriate letter. For F, for which the valence configuration is $2p^5$, treat the single gap in the closed-shell $2p^6$ configuration as a single spin-$\frac{1}{2}$ particle.

The solution (a) For Na, the configuration is $[Ne]3s^1$, and consider only the single 3s electron. Because $L=l=0$ and $S=s=\frac{1}{2}$, the only possible value is $J=\frac{1}{2}$. Hence the term symbol is $^2S_{1/2}$. (b) For F, the configuration is $[He]2s^22p^5$, which can be treated as $[Ne]2p^{-1}$ (where the notation $2p^{-1}$ signifies the absence of a 2p electron). Hence $L=l=1$, and $S=s=\frac{1}{2}$. Two values of J are possible: $J=\frac{3}{2}, \frac{1}{2}$. Hence, the term symbols for the two levels are $^2P_{3/2}$ and $^2P_{1/2}$. (c) This is a two-electron problem, and $l_1=l_2=1$, $s_1=s_2=\frac{1}{2}$. It follows that $L=2, 1, 0$ and $S=1, 0$. The terms are therefore ^3D and ^1D, ^3P and ^1P, and ^3S and ^1S. For ^3D, $L=2$ and $S=1$; hence $J=3, 2, 1$ and the levels are $^3D_3$, $^3D_2$, and $^3D_1$. For ^1D, $L=2$ and $S=0$, so the single level is $^1D_2$. The three levels of ^3P are $^3P_2$, $^3P_1$, and $^3P_0$, and there is just one level for the singlet, $^1P_1$. For the ^3S term there is only one level, $^3S_1$ (because $J=1$ only), and the singlet term is $^1S_0$.

Comment. Fewer terms arise from a configuration like … $2p^2$ or … $3p^2$ than from a configuration like … $2p^13p^1$ because the Pauli exclusion principle forbids parallel arrangements of spins when two electrons occupy the same orbital. The analysis of the terms arising in such cases requires more detail than given here.

Self-test 8C.3 Identify the terms arising from the configurations (a) $2s^12p^1$, (b) $2p^13d^1$.

Answer: (a) $^3P_2$, $^3P_1$, $^3P_0$, $^1P_1$; (b) $^3F_4$, $^3F_3$, $^3F_2$, $^1F_3$, $^3D_3$, $^3D_2$, $^3D_1$, $^1D_2$, $^3P_2$, $^3P_1$, $^3P_0$, $^1P_1$

Russell–Saunders coupling fails when the spin–orbit coupling is large (in heavy atoms, those with high Z). In that case, the individual spin and orbital momenta of the electrons are coupled into individual j values; then these momenta are combined into a grand total, J, given by a Clebsch–Gordan series. This scheme is called ***jj*-coupling**. For example, in a p^2 configuration, the individual values of j are $\frac{3}{2}$ and $\frac{1}{2}$ for each electron. If the spin and the orbital angular momentum of each electron are coupled together strongly, it is best to consider each electron as a particle with angular momentum $j=\frac{3}{2}$ or $\frac{1}{2}$. These individual total momenta then couple as follows:

j_1	j_2	J
$\frac{3}{2}$	$\frac{3}{2}$	3, 2, 1, 0
$\frac{3}{2}$	$\frac{1}{2}$	2, 1
$\frac{1}{2}$	$\frac{3}{2}$	2, 1
$\frac{1}{2}$	$\frac{1}{2}$	1, 0

For heavy atoms, in which *jj*-coupling is appropriate, it is best to discuss their energies by using these quantum numbers.

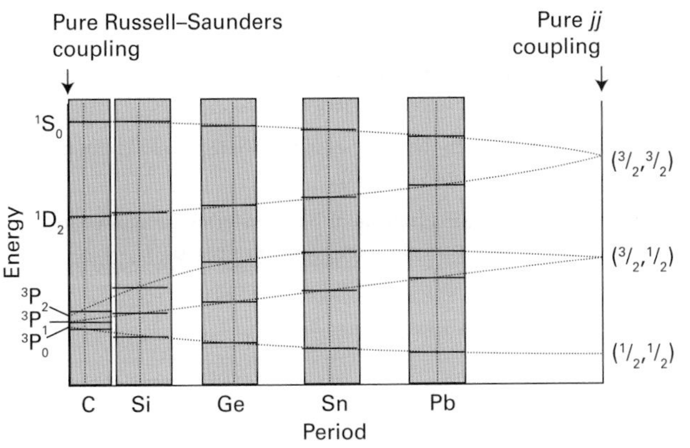

Figure 8C.11 The correlation diagram for some of the states of a two-electron system. All atoms lie between the two extremes, but the heavier the atom, the closer it lies to the pure *jj*-coupling case.

Although *jj*-coupling should be used for assessing the energies of heavy atoms, the term symbols derived from Russell–Saunders coupling can still be used as labels. To see why this procedure is valid, it is useful to examine how the energies of the atomic states change as the spin–orbit coupling increases in strength. Such a **correlation diagram** is shown in Fig. 8C.11. It shows that there is a correspondence between the low spin–orbit coupling (Russell–Saunders coupling) and high spin–orbit coupling (*jj*-coupling) schemes, so the labels derived by using the Russell–Saunders scheme can be used to label the states of the *jj*-coupling scheme.

8C.2(d) Hund's rules and term symbols

Hund's rules for predicting the ground state configuration of a many-electron atom are introduced in Topic 8B. With the notation introduced in this Topic, they can be extended and expressed by referring to term symbols. As already remarked, the terms arising from a given configuration differ in energy because they represent different relative orientations of the angular momenta of the electrons and therefore different spatial distributions. The terms arising from the ground-state configuration of an atom (and less reliably from other configurations) can be put into the order of increasing energy by using the following version of the rules, which summarize the preceding discussion:

1. For a given configuration, the term of greatest multiplicity lies lowest in energy.

As discussed in Topic 8B, this rule is a consequence of spin correlation, the quantum-mechanical tendency of electrons with parallel spins to stay apart from one another.

2. For a given multiplicity, the term with the highest value of L lies lowest in energy.

This rule can be explained classically by noting that two electrons have a high orbital angular momentum if they circulate in the same direction, in which case they can stay apart. If they circulate in opposite directions, they meet. Thus, a D term is expected to lie lower in energy than an S term of the same multiplicity.

3. For atoms with less than half-filled shells, the level with the lowest value of J lies lowest in energy; for more than half-filled shells, the highest value of J lies lowest.

This rule arises from considerations of spin–orbit coupling. Thus, for a state of low J, the orbital and spin angular momenta lie in opposite directions, and so too do the corresponding magnetic moments. In classical terms the magnetic moments are then antiparallel, with the N pole of one close to the S pole of the other, which is a low-energy arrangement.

8C.2(e) Selection rules

Any state of the atom, and any spectral transition, can be specified by using term symbols. For example, the transitions giving rise to the yellow sodium doublet (which are shown in Fig. 8C.7) are

$$3p^1\,{}^2P_{3/2} \rightarrow 3s^1\,{}^2S_{1/2} \quad 3p^1\,{}^2P_{1/2} \rightarrow 3s^1\,{}^2S_{1/2}$$

By convention, the upper term precedes the lower. The corresponding absorptions are therefore denoted ${}^2P_{3/2} \leftarrow {}^2S_{1/2}$ and ${}^2P_{1/2} \leftarrow {}^2S_{1/2}$. (The configurations have been omitted.)

As seen in Section 8C.1, selection rules arise from the conservation of angular momentum during a transition and from the fact that a photon has a spin of 1. They can therefore be expressed in terms of the term symbols, because the latter carry information about angular momentum. A detailed analysis leads to the following rules:

$$\begin{aligned}&\Delta S = 0,\\ &\Delta L = 0, \pm 1,\ \Delta l = \pm 1,\\ &\Delta J = 0, \pm 1 \text{ but } J = 0 \leftarrow\!|\!\rightarrow J = 0\end{aligned} \qquad \text{Selection rules for light atoms} \qquad (8C.8)$$

where the symbol $\leftarrow\!|\!\rightarrow$ denotes a forbidden transition. The rule about ΔS (no change of overall spin) stems from the fact that electromagnetic radiation interacts with all the spins in an atom equally and therefore cannot twist individual spins into new relative orientations. The rules about ΔL and Δl express the fact that the orbital angular momentum of an individual electron must change (so $\Delta l = \pm 1$), but whether or not this results in an overall change of orbital momentum depends on the coupling.

The selection rules given above apply when Russell–Saunders coupling is valid (in light atoms, those of low Z). If labelling the terms of heavy atoms with symbols like ^{3}D, then the selection rules progressively fail as the atomic number increases because the quantum numbers S and L become ill defined as *jj*-coupling becomes more appropriate. As explained above, Russell–Saunders term symbols are only a convenient way of labelling the terms of heavy atoms: they do not bear any direct relation to the actual angular momenta of the electrons in a heavy atom. For this reason, transitions between singlet and triplet states (for which $\Delta S = \pm 1$), while forbidden in light atoms, are allowed in heavy atoms.

Exercises

E8C.4 What atomic terms are possible for the electron configuration ns^1nd^1? Which term is likely to lie lowest in energy?

E8C.5 Calculate the shifts in the energies of the two terms of a d^1 configuration that can arise from spin–orbit coupling.

E8C.6 Which of the following transitions between terms are allowed in the electronic emission spectrum of a many-electron atom: (i) ${}^3D_2 \rightarrow {}^3P_1$, (ii) ${}^3P_2 \rightarrow {}^1S_0$, (iii) ${}^3F_4 \rightarrow {}^3D_3$?

Checklist of concepts

☐ 1. Two electrons with paired spins in a configuration give rise to a **singlet term**; if their spins are parallel, they give rise to a **triplet term**.

☐ 2. The orbital and spin angular momenta interact magnetically.

☐ 3. **Spin–orbit coupling** results in the levels of a term having different energies.

☐ 4. Fine structure in a spectrum is due to transitions to different levels of a term.

☐ 5. A **term symbol** specifies the angular momentum states of an atom.

☐ 6. Angular momenta are combined into a resultant by using the **Clebsch–Gordan series**.

☐ 7. The **multiplicity** of a term is the value of $2S + 1$.

☐ 8. The total angular momentum in light atoms is obtained on the basis of **Russell–Saunders coupling**; in heavy atoms, ***jj*-coupling** is used.

☐ 9. The term with the maximum multiplicity lies lowest in energy.

☐ 10. For a given multiplicity, the term with the highest value of L lies lowest in energy.

☐ 11. For atoms with less than half-filled shells, the level with the lowest value of J lies lowest in energy; for more than half-filled shells, the highest value of J lies lowest.

☐ 12. Selection rules for light atoms include the fact that changes of total spin do not occur.

Checklist of equations

Property	Equation	Comment	Equation number
Spin–orbit interaction energy	$E_{l,s,j} = \frac{1}{2}hc\tilde{A}\{j(j+1) - l(l+1) - s(s+1)\}$		8C.4
Clebsch–Gordan series	$J = j_1 + j_2, j_1 + j_2 - 1, \ldots, \lvert j_1 - j_2 \rvert$	J, j denote any kind of angular momentum	8C.5
Selection rules	$\Delta S = 0$, $\Delta L = 0, \pm 1$, $\Delta l = \pm 1$, $\Delta J = 0, \pm 1$, but $J = 0 \leftarrow\mid\rightarrow J = 0$	Light atoms	8C.8

FOCUS 8 Atomic structure and spectra

To test your understanding of this material, work through the *Exercises, Additional exercises, Discussion questions*, and *Problems* found throughout this Focus.

Selected solutions can be found at the end of this Focus in the e-book. Solutions to even-numbered questions are available online only to lecturers.

TOPIC 8A Hydrogenic atoms

Discussion questions

D8A.1 Describe the separation of variables procedure as it is applied to simplify the description of a hydrogenic atom free to move through space.

D8A.2 List and describe the significance of the quantum numbers needed to specify the internal state of a hydrogenic atom.

D8A.3 Explain the significance of (a) a boundary surface and (b) the radial distribution function for hydrogenic orbitals.

Additional exercises

E8A.7 State the orbital degeneracy of the levels in a hydrogen atom that have energy (i) $-hc\tilde{R}_H$; (ii) $-\frac{1}{9}hc\tilde{R}_H$; (iii) $-\frac{1}{25}hc\tilde{R}_H$.

E8A.8 State the orbital degeneracy of the levels in a hydrogenic atom (Z in parentheses) that have energy (i) $-4hc\tilde{R}_N$, (2); (ii) $-\frac{1}{4}hc\tilde{R}_N$ (4), and (iii) $-hc\tilde{R}_N$ (5).

E8A.9 The wavefunction for the 2s orbital of a hydrogen atom is $N(2-r/a_0)e^{-r/2a_0}$. Evaluate the normalization constant N.

E8A.10 Evaluate the probability density at the nucleus of an electron with $n=3$, $l=0$, $m_l=0$.

E8A.11 By differentiation of the 3s radial wavefunction, show that it has three extrema in its amplitude, and locate them.

E8A.12 At what radius does the probability density of an electron in the H atom fall to 50 per cent of its maximum value?

E8A.13 At what radius in the H atom does the radial distribution function of the ground state have (i) 50 per cent, (ii) 75 per cent of its maximum value?

E8A.14 Locate the radial nodes in the 3s orbital of a hydrogenic atom.

E8A.15 Locate the radial nodes in the 4p orbital of a hydrogenic atom. You need to know that, in the notation of eqn 8A.10, $L_{4,1}(\rho) \propto 20 - 10\rho + \rho^2$, with $\rho = \frac{1}{2}Zr/a_0$.

E8A.16 The wavefunction of one of the d orbitals is proportional to $\cos\theta \sin\theta \cos\phi$. At what angles does it have nodal planes?

E8A.17 The wavefunction of one of the d orbitals is proportional to $\sin^2\theta \sin 2\phi$. At what angles does it have nodal planes?

E8A.18 Write down the expression for the radial distribution function of a 2s electron in a hydrogenic atom of atomic number Z and identify the radius at which it is a maximum. *Hint:* Use mathematical software.

E8A.19 Write down the expression for the radial distribution function of a 3s electron in a hydrogenic atom of atomic number Z and identify the radius at which the electron is most likely to be found. *Hint:* Use mathematical software.

E8A.20 Write down the expression for the radial distribution function of a 3p electron in a hydrogenic atom of atomic number Z and identify the radius at which the electron is most likely to be found. *Hint:* Use mathematical software.

E8A.21 What subshells and orbitals are available in the N shell?

E8A.22 What is the orbital angular momentum (as multiples of $\hbar$) of an electron in the orbitals (i) 4d, (ii) 2p, (iii) 3p? Give the numbers of angular and radial nodes in each case.

E8A.23 Locate the radial nodes of each of the 2p orbitals of a hydrogenic atom of atomic number Z.

E8A.24 Locate the radial nodes of each of the 3d orbitals of a hydrogenic atom of atomic number Z.

Problems

P8A.1 At what point (not radius) is the probability density a maximum for the 2p electron?

P8A.2 Show by explicit integration that (a) hydrogenic 1s and 2s orbitals, (b) $2p_x$ and $2p_y$ orbitals are mutually orthogonal.

P8A.3 The value of $\tilde{R}_\infty$ is 109 737 cm^{-1}. What is the energy of the ground state of a deuterium atom? Take $m_D = 2.013\,55m_u$.

P8A.4 Predict the ionization energy of Li^{2+} given that the ionization energy of He^+ is 54.36 eV.

P8A.5 Explicit expressions for hydrogenic orbitals are given in Tables 7F.1 (for the angular component) and 8A.1 (for the radial component). (a) Verify both that the $3p_x$ orbital is normalized (to 1) and that $3p_x$ and $3d_{xy}$ are mutually orthogonal. *Hint:* It is sufficient to show that the functions $e^{i\phi}$ and $e^{2i\phi}$ are mutually orthogonal. (b) Identify the positions of both the radial nodes and nodal planes of the 3s, $3p_x$, and $3d_{xy}$ orbitals. (c) Calculate the mean radius of the 3s orbital. *Hint:* Use mathematical software. (d) Draw a graph of the radial distribution function for the three orbitals (of part (b)) and discuss the significance of the graphs for interpreting the properties of many-electron atoms.

P8A.6 Determine whether the p_x and p_y orbitals are eigenfunctions of l_z. If not, does a linear combination exist that is an eigenfunction of l_z?

P8A.7 The 'size' of an atom is sometimes considered to be measured by the radius of a sphere within which there is a 90 per cent probability of finding the electron in the outermost occupied orbital. Calculate the 'size' of a hydrogen atom in its ground state according to this definition. Go on to explore how the 'size' varies as the definition is changed to other percentages, and plot your conclusion.

P8A.8 Some atomic properties depend on the average value of $1/r$ rather than the average value of r itself. Evaluate the expectation value of $1/r$ for (a) a hydrogenic 1s orbital, (b) a hydrogenic 2s orbital, (c) a hydrogenic 2p orbital. (d) Does $\langle 1/r\rangle = 1/\langle r\rangle$?

P8A.9 One of the most famous of the obsolete theories of the hydrogen atom was proposed by Niels Bohr. It has been replaced by quantum mechanics, but by a remarkable coincidence (not the only one where the Coulomb potential is concerned), the energies it predicts agree exactly with those obtained from the Schrödinger equation. In the Bohr atom, an electron travels in a circle around the nucleus. The Coulombic force of attraction ($Ze^2/4\pi\varepsilon_0 r^2$) is balanced by the centrifugal effect of the orbital motion. Bohr proposed that the angular momentum is limited to integral values of $\hbar$. When the two forces are balanced, the atom remains in a stationary state until it makes a spectral transition. Calculate the energies of a hydrogenic atom using the Bohr model.

P8A.10 The Bohr model of the atom is specified in Problem 8A.9. (a) What features of it are untenable according to quantum mechanics? (b) How does the ground state of the Bohr atom differ from the actual ground state? (c) Is there an experimental distinction between the Bohr and quantum mechanical models of the ground state?

P8A.11 Atomic units of length and energy may be based on the properties of a particular atom. The usual choice is that of a hydrogen atom, with the unit of length being the Bohr radius, a_0, and the unit of energy being the 'hartree', E_h, which is equal to twice the (negative of the) energy of the 1s orbital (specifically, and more precisely, $E_h = 2hc\tilde{R}_\infty$). Positronium consists of an electron and a positron (same mass, opposite charge) orbiting round their common centre of mass. If the positronium atom (e^+,e^-) were used instead, with analogous definitions of units of length and energy, what would be the relation between these two sets of atomic units?

TOPIC 8B Many-electron atoms

Discussion questions

D8B.1 Describe the orbital approximation for the wavefunction of a many-electron atom. What are the limitations of the approximation?

D8B.2 Outline the electron configurations of many-electron atoms in terms of their location in the periodic table.

D8B.3 Describe and account for the variation of first ionization energies along Period 2 of the periodic table. Would you expect the same variation in Period 3?

D8B.4 Describe the self-consistent field procedure for calculating the form of the orbitals and the energies of many-electron atoms.

Additional exercises

E8B.6 Construct the wavefunction for an excited state of the He atom with configuration $1s^13s^1$. Use $Z_{eff} = 2$ for the 1s electron and $Z_{eff} = 1$ for the 3s electron.

E8B.7 How many electrons can occupy subshells with $l = 5$?

E8B.8 Write the ground-state electron configurations of the d-metals from yttrium to cadmium.

E8B.9 Write the electronic configuration of the O^{2-} ion.

E8B.10 Consider the atoms of the Period 2 elements of the periodic table. Predict which element has the lowest second ionization energy.

Problems

P8B.1 In 1976 it was mistakenly believed that the first of the 'superheavy' elements had been discovered in a sample of mica. Its atomic number was believed to be 126. What is the most probable distance of the innermost electrons from the nucleus of an atom of this element? (In such elements, relativistic effects are very important, but ignore them here.)

P8B.2 Why is the electronic configuration of the yttrium atom $[Kr]4d^15s^2$ and that of the silver atom $[Kr]4d^{10}5s^1$?

P8B.3 The d-metals iron, copper, and manganese form cations with different oxidation states. For this reason, they are found in many oxidoreductases and in several proteins of oxidative phosphorylation and photosynthesis. Explain why many d-metals form cations with different oxidation states.

P8B.4 One important function of atomic and ionic radius is in regulating the uptake of oxygen by haemoglobin, for the change in ionic radius that accompanies the conversion of Fe(II) to Fe(III) when O_2 attaches triggers a conformational change in the protein. Which do you expect to be larger: Fe^{2+} or Fe^{3+}? Why?

P8B.5 Thallium, a neurotoxin, is the heaviest member of Group 13 of the periodic table and is found most usually in the +1 oxidation state. Aluminium, which causes anaemia and dementia, is also a member of the group but its chemical properties are dominated by the +3 oxidation state. Examine this issue by plotting the first, second, and third ionization energies for the Group 13 elements against atomic number. Explain the trends you observe. *Hints*: The third ionization energy, I_3, is the minimum energy needed to remove an electron from the doubly charged cation: $E^{2+}(g) \rightarrow E^{3+}(g) + e^-(g)$, $I_3 = E(E^{3+}) - E(E^{2+})$.

TOPIC 8C Atomic spectra

Discussion questions

D8C.1 Discuss the origin of the series of lines in the emission spectrum of hydrogen. What region of the electromagnetic spectrum is associated with each of the series shown in Fig. 8C.1?

D8C.2 Specify and account for the selection rules for transitions in (a) hydrogenic atoms, and (b) many-electron atoms.

D8C.3 Explain the origin of spin–orbit coupling and how it affects the appearance of a spectrum.

D8C.4 Why does the spin–orbit coupling constant depend so strongly on the atomic number?

Additional exercises

E8C.7 The Pfund series has $n_1 = 5$. Identify the transition responsible for the shortest and longest wavelength lines in the Pfund series.

E8C.8 Calculate the wavelength, frequency, and wavenumber of the $n = 5 \rightarrow n = 4$ transition in Li^{2+}.

E8C.9 Which of the following transitions are allowed in the electronic emission spectrum of a hydrogenic atom: (i) 5d → 2s, (ii) 5p → 3s, (iii) 6p → 4f?

E8C.10 Identify the levels of the configuration p^1.

E8C.11 Identify the levels of the configuration f^1.

E8C.12 What are the permitted values of *j* for (i) a d electron, (ii) an f electron?

E8C.13 What are the permitted values of *j* for (i) a p electron, (ii) an h electron?

E8C.14 An electron in two different states of an atom is known to have $j = \frac{3}{2}$ and $\frac{1}{2}$. What is its orbital angular momentum quantum number in each case?

E8C.15 What are the allowed total angular momentum quantum numbers of a composite system in which $j_1 = 5$ and $j_2 = 3$?

E8C.16 What information does the term symbol $^1D_2$ provide about the angular momentum of an atom?

E8C.17 What information does the term symbol $^3F_4$ provide about the angular momentum of an atom?

E8C.18 Suppose that an atom has (i) 2, (ii) 3 electrons in different orbitals. What are the possible values of the total spin quantum number *S*? What is the multiplicity in each case?

E8C.19 Suppose that an atom has (i) 4, (ii) 5, electrons in different orbitals. What are the possible values of the total spin quantum number *S*? What is the multiplicity in each case?

E8C.20 What are the possible values of the total spin quantum numbers *S* and M_S for the Ni^{2+} ion?

E8C.21 What are the possible values of the total spin quantum numbers *S* and M_S for the V^{2+} ion?

E8C.22 What atomic terms are possible for the electron configuration np^1nd^1? Which term is likely to lie lowest in energy?

E8C.23 What values of *J* may occur in the terms (i) ^1S, (ii) ^2P, (iii) ^3P? How many states (distinguished by the quantum number M_J) belong to each level?

E8C.24 What values of *J* may occur in the terms (i) ^3D, (ii) ^4D, (iii) ^2G? How many states (distinguished by the quantum number M_J) belong to each level?

E8C.25 Give the possible term symbols for (i) Li [He]$2s^1$, (ii) Na [Ne]$3p^1$.

E8C.26 Give the possible term symbols for (i) Zn [Ar]$3d^{10}4s^2$, (ii) Br [Ar]$3d^{10}4s^24p^5$.

E8C.27 Calculate the shifts in the energies of the two terms in an f^1 configuration that can arise from spin–orbit coupling.

E8C.28 Which of the following transitions between terms are allowed in the electronic emission spectrum of a many-electron atom: (i) $^2P_{3/2} \rightarrow {}^2S_{1/2}$, (ii) $^3P_0 \rightarrow {}^3S_1$, (iii) $^3D_3 \rightarrow {}^1P_1$?

Problems

P8C.1 The *Humphreys series* is a group of lines in the spectrum of atomic hydrogen. It begins at 12 368 nm and has been traced to 3281.4 nm. What are the transitions involved? What are the wavelengths of the intermediate transitions?

P8C.2 A series of lines involving a common level in the spectrum of atomic hydrogen lies at 656.46 nm, 486.27 nm, 434.17 nm, and 410.29 nm. What is the wavelength of the next line in the series? What is the ionization energy of the atom when it is in the lower state of the transitions?

P8C.3 The distribution of isotopes of an element may yield clues about the nuclear reactions that occur in the interior of a star. Show that it is possible to use spectroscopy to confirm the presence of both $^4He^+$ and $^3He^+$ in a star by calculating the wavenumbers of the $n = 3 \rightarrow n = 2$ and of the $n = 2 \rightarrow n = 1$ transitions for each ionic isotope.

P8C.4 The Li^{2+} ion is hydrogenic and has a Lyman series at 740 747 cm^{-1}, 877 924 cm^{-1}, 925 933 cm^{-1}, and beyond. Show that the energy levels are of the form $-hc\tilde{R}_{Li}/n^2$ and find the value of $\tilde{R}_{Li}$ for this ion. Go on to predict the wavenumbers of the two longest-wavelength transitions of the Balmer series of the ion and find its ionization energy.

P8C.5 A series of lines in the spectrum of neutral Li atoms rise from transitions between $1s^22p^1\,{}^2P$ and $1s^2nd^1\,{}^2D$ and occur at 610.36 nm, 460.29 nm, and 413.23 nm. The d orbitals are hydrogenic. It is known that the transition from the ^2P to the ^2S term (which arises from the ground-state configuration $1s^22s^1$) occurs at 670.78 nm. Calculate the ionization energy of the ground-state atom.

P8C.6‡ W.P. Wijesundera et al. (*Phys. Rev.* A **51**, 278 (1995)) attempted to determine the electron configuration of the ground state of lawrencium, element 103. The two contending configurations are [Rn]$5f^{14}7s^27p^1$ and [Rn]$5f^{14}6d7s^2$. Write down the term symbols for each of these configurations, and identify the lowest level within each configuration. Which level would be lowest according to a simple estimate of spin–orbit coupling?

P8C.7 An emission line from K atoms is found to have two closely spaced components, one at 766.70 nm and the other at 770.11 nm. Account for this observation, and deduce what information you can.

‡ These problems were supplied by Charles Trapp and Carmen Giunta.

P8C.8 Calculate the mass of the deuteron given that the first line in the Lyman series of ^1H lies at 82 259.098 cm^{-1} whereas that of ^2H lies at 82 281.476 cm^{-1}. Calculate the ratio of the ionization energies of ^1H and ^2H.

P8C.9 Positronium consists of an electron and a positron (same mass, opposite charge) orbiting round their common centre of mass. The broad features of the spectrum are therefore expected to be hydrogen-like, the differences arising largely from the mass differences. Predict the wavenumbers of the first three lines of the Balmer series of positronium. What is the binding energy of the ground state of positronium?

P8C.10 The *Zeeman effect* is the modification of an atomic spectrum by the application of a strong magnetic field. It arises from the interaction between applied magnetic fields and the magnetic moments due to orbital and spin angular momenta (recall the evidence provided for electron spin by the Stern–Gerlach experiment, Topic 8B). To gain some appreciation for the so-called *normal Zeeman effect*, which is observed in transitions involving singlet states, consider a p electron, with $l = 1$ and $m_l = 0, \pm 1$. In the absence of a magnetic field, these three states are degenerate. When a field of magnitude $\mathcal{B}$ is present, the degeneracy is removed and it is observed that the state with $m_l = +1$ moves up in energy by $\mu_B\mathcal{B}$, the state with $m_l = 0$ is unchanged, and the state with $m_l = -1$ moves down in energy by $\mu_B\mathcal{B}$, where $\mu_B = e\hbar/2m_e = 9.274 \times 10^{-24}$ J T^{-1} is the 'Bohr magneton'. Therefore, a transition between a $^1S_0$ term and a $^1P_1$ term consists of three spectral lines in the presence of a magnetic field where, in the absence of the magnetic field, there is only one. (a) Calculate the splitting in reciprocal centimetres between the three spectral lines of a transition between a $^1S_0$ term and a $^1P_1$ term in the presence of a magnetic field of 2 T (where 1 T = 1 kg s^{-2} A^{-1}). (b) Compare the value you calculated in (a) with typical optical transition wavenumbers, such as those for the Balmer series of the H atom. Is the line splitting caused by the normal Zeeman effect relatively small or relatively large?

P8C.11 Some of the selection rules for hydrogenic atoms were derived in the text. Complete the derivation by considering the x- and y-components of the electric dipole moment operator.

P8C.12 Hydrogen is the most abundant element in all stars. However, neither absorption nor emission lines due to neutral hydrogen are found in the spectra of stars with effective temperatures higher than 25 000 K. Account for this observation.

FOCUS 8 Atomic structure and spectra

Integrated activities

I8.1 An electron in the ground-state He^+ ion undergoes a transition to a state specified by the quantum numbers $n = 4$, $l = 1$, $m_l = +1$. (a) Describe the transition using term symbols. (b) Calculate the wavelength, frequency, and wavenumber of the transition. (c) By how much does the mean radius of the electron change due to the transition? You need to know that the mean radius of a hydrogenic orbital is

$$r_{n,l,m_l} = \frac{n^2 a_0}{Z}\left\{1 + \frac{1}{2}\left[1 - \frac{l(l+1)}{n^2}\right]\right\}$$

I8.2‡ Highly excited atoms have electrons with large principal quantum numbers. Such *Rydberg atoms* have unique properties and are of interest to astrophysicists. (a) For hydrogen atoms with large n, derive a relation for the separation of energy levels. (b) Calculate this separation for $n = 100$; also calculate the average radius (see the preceding activity), and the ionization energy. (c) Could a thermal collision with another hydrogen atom ionize this Rydberg atom? (d) What minimum velocity of the second atom is required? (e) Sketch the likely form of the radial wavefunction for a 100s orbital.

I8.3‡ Stern–Gerlach splittings of atomic beams are small and require either large magnetic field gradients or long magnets for their observation. For a beam of atoms with zero orbital angular momentum, such as H or Ag, the deflection is given by $x = \pm(\mu_B L^2/4E_k)\mathrm{d}\mathcal{B}/\mathrm{d}z$, where $\mu_B = e\hbar/2m_e = 9.274 \times 10^{-24}$ J T^{-1} is the 'Bohr magneton', L is the length of the magnet, E_k is the average kinetic energy of the atoms in the beam, and $\mathrm{d}\mathcal{B}/\mathrm{d}z$ is the magnetic field gradient across the beam. Calculate the magnetic field gradient required to produce a splitting of 1.00 mm in a beam of Ag atoms from an oven at 1000 K with a magnet of length 50 cm.

FOCUS 9

Molecular structure

The concepts developed in Focus 8, particularly those of orbitals, can be extended to a description of the electronic structures of molecules. There are two principal quantum mechanical theories of molecular electronic structure: 'valence-bond theory' is centred on the concept of the shared electron pair; 'molecular orbital theory' treats electrons as being distributed over all the nuclei in a molecule.

Prologue The Born–Oppenheimer approximation

The starting point for the theories discussed here and the interpretation of spectroscopic results (Focus 11) is the 'Born-Oppenheimer approximation', which separates the relative motions of nuclei and electrons in a molecule.

9A Valence-bond theory

The key concept of this Topic is the wavefunction for a shared electron pair, which is then used to account for the structures of a wide variety of molecules. The theory introduces the concepts of σ and π bonds, promotion, and hybridization, which are used widely in chemistry.

9B Molecular orbital theory: the hydrogen molecule-ion

In molecular orbital theory the concept of an atomic orbital is extended to that of a 'molecular orbital', which is a wavefunction that spreads over all the atoms in a molecule. This Topic focuses on the hydrogen molecule-ion, setting the scene for the application of the theory to more complicated molecules.

9C Molecular orbital theory: homonuclear diatomic molecules

The principles established for the hydrogen molecule-ion are extended to other homonuclear diatomic molecules and ions. The principal differences are that all the valence-shell atomic orbitals must be included and that they give rise to a more varied collection of molecular orbitals. The building-up principle for atoms is extended to the occupation of molecular orbitals and used to predict the electronic configurations of molecules and ions.

9D Molecular orbital theory: heteronuclear diatomic molecules

The molecular orbital theory of heteronuclear diatomic molecules introduces the possibility that the atomic orbitals on the two atoms contribute unequally to the molecular orbital. As a result, the molecule is polar. The polarity can be expressed in terms of the concept of electronegativity. This Topic shows how quantum mechanics is used to calculate the energy and form of a molecular orbital arising from the overlap of different atomic orbitals.

9E Molecular orbital theory: polyatomic molecules

Molecular orbital theory applies equally well to larger molecules; however, the calculations involved can no longer be done by hand but require the use of sophisticated computer programs. A flavour of how molecular orbitals are calculated comes from using the 'Hückel method' to find the orbitals of planar conjugated polyenes. Severe approximations are necessary in order to make the calculations tractable by simple methods but the method sets the scene for more sophisticated procedures. The latter have given rise to the huge and vibrant field of computational theoretical chemistry in which elaborate computations are used to predict molecular properties. This Topic describes briefly how those calculations are formulated and displayed.

9E.1 **The Hückel approximation;** 9E.2 **Applications;** 9E.3 **Computational chemistry**

9F Computational chemistry

Molecular orbital theory is the foundation of current computational methods that can predict molecular structure and many physical and chemical properties of substances as simple as diatomic molecules and as complex as biopolymers. This Topic introduces the main ideas behind these methods, again using the hydrogen molecule as an example. Initially the focus is on the formulation of the 'Hartree–Fock equations', which guide the calculation of molecular orbitals. The 'density functional' approach, which offers greater precision and computational efficiency, is also described.

9F.1 **The central challenge;** 9F.2 **The Hartree–Fock formalism;** 9F.3 **The Roothaan equations;** 9F.4 **Evaluation and approximation of the integrals;** 9F.5 **Density functional theory**

What is an application of this material?

The concepts introduced in this chapter pervade the whole of chemistry and are encountered throughout the text. Two biochemical aspects are discussed here. In *Impact* 14, accessed via the e-book, simple concepts are used to account for the reactivity of small molecules that occur in organisms. *Impact* 15, accessed via the e-book, provides a glimpse of the contribution of computational chemistry to the explanation of the thermodynamic and spectroscopic properties of several biologically significant molecules.

➤ **Go to the e-book for videos that feature the derivation and interpretation of equations, and applications of this material.**

PROLOGUE The Born–Oppenheimer approximation

All theories of molecular structure make the same simplification at the outset. Whereas the Schrödinger equation for a hydrogen atom can be solved exactly, an exact solution is not possible for any molecule because even the simplest molecule consists of three particles (two nuclei and one electron). Therefore, it is common to adopt the **Born–Oppenheimer approximation** in which it is supposed that the nuclei, being so much heavier than an electron, move relatively slowly and may be treated as stationary while the electrons move in their field. That is, the nuclei are assumed to be fixed at arbitrary locations, and the Schrödinger equation is then solved for the wavefunction of the electrons alone.

To use the Born–Oppenheimer approximation for a diatomic molecule, the nuclear separation is set at a chosen value, the Schrödinger equation for the electrons is then solved and the energy calculated. Then a different separation is selected, the calculation repeated, and so on for other values of the separation. In this way the variation of the energy of the molecule with bond length is explored, and a **molecular potential energy curve** is obtained (see the illustration). It is called a potential energy curve because the kinetic energy of the stationary nuclei is zero.

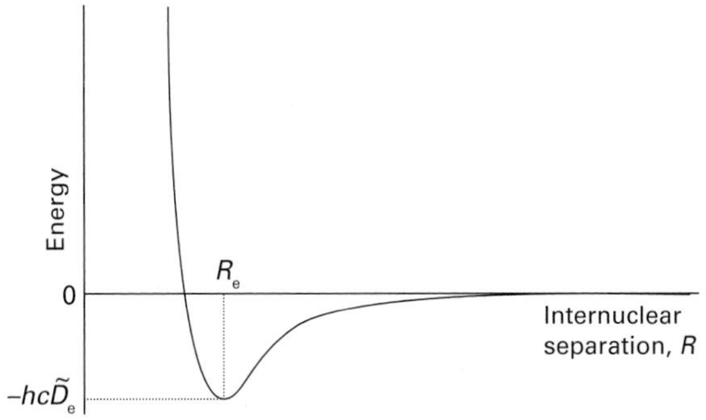

A molecular potential energy curve. The equilibrium bond length corresponds to the energy minimum.

Once the curve has been calculated or determined experimentally (by using the spectroscopic techniques described in Focus 11), it is possible to identify two important parameters. The first is the **equilibrium bond length**, R_e, the internuclear separation at the minimum of the curve. The second is the depth, $hc\tilde{D}_e$, of the minimum below the energy of the infinitely widely separated atoms (when R is very much larger than R_e).

When more than one molecular parameter is changed in a polyatomic molecule, such as its various bond lengths and angles, a potential energy *surface* is obtained. The overall equilibrium shape of the molecule corresponds to the global minimum of the surface.

TOPIC 9A Valence-bond theory

➤ **Why do you need to know this material?**

The language introduced by valence-bond theory is used throughout chemistry, especially in the description of the properties and reactions of organic compounds.

➤ **What is the key idea?**

A bond forms when an electron in an atomic orbital on one atom pairs its spin with that of an electron in an atomic orbital on another atom.

➤ **What do you need to know already?**

You need to know about atomic orbitals (Topic 8A) and the concepts of normalization and orthogonality (Topics 7B and 7C). This Topic also makes use of the Pauli principle (Topic 8B).

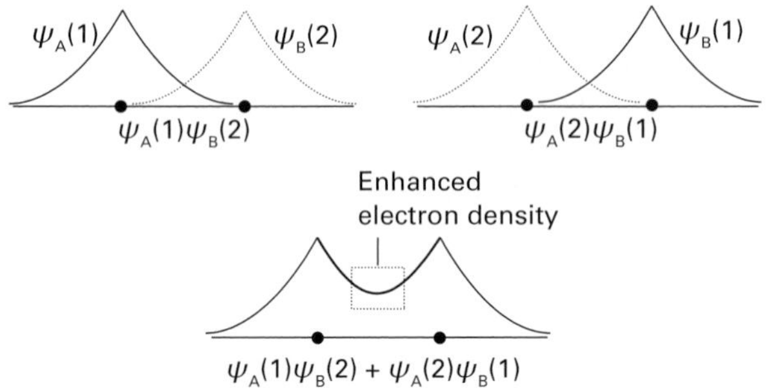

Figure 9A.1 It is very difficult to represent valence-bond wavefunctions because they refer to two electrons simultaneously. However, this illustration is an attempt. The atomic orbital for electron 1 is represented by the full line, and that of electron 2 is represented by the dotted line. The left illustration represents $\psi_A(1)\psi_B(2)$ and the right illustration represents the contribution $\psi_A(2)\psi_B(1)$. When the two contributions are superimposed, there is interference between them, resulting in an enhanced (two-electron) density in the internuclear region.

Valence-bond theory (VB theory) begins by considering the chemical bond in molecular hydrogen, H_2. The basic concepts are first extended to other diatomic molecules and ions, and then to polyatomic molecules.

9A.1 Diatomic molecules

Consider two widely separated H atoms, each with one electron. Electron 1 is in the H1s atomic orbital on atom A, $\psi_{\mathrm{H1s_A}}$, and electron 2 is in the H1s atomic orbital on atom B, $\psi_{\mathrm{H1s_B}}$. The spatial wavefunction describing this arrangement is

$$\Psi(1,2)=\psi_{\mathrm{H1s_A}}(\boldsymbol{r}_1)\psi_{\mathrm{H1s_B}}(\boldsymbol{r}_2) \qquad (9\mathrm{A}.1)$$

For simplicity, this wavefunction will be written $\Psi(1,2) = \psi_A(1)\psi_B(2)$. When the atoms are close together, it is not possible to know whether it is electron 1 or electron 2 that is on A. An equally valid description is therefore $\Psi(1,2) = \psi_A(2)\psi_B(1)$, in which electron 2 is on A and electron 1 is on B. When two outcomes are equally probable in quantum mechanics, the true state of the system is described as a superposition of the wavefunctions for each possibility (Topic 7C). Therefore a better description of the molecule than either wavefunction alone is one of the (unnormalized) linear combinations $\Psi(1,2) = \psi_A(1)\psi_B(2) \pm \psi_A(2)\psi_B(1)$. The combination with lower energy turns out to be the one with a + sign, so the valence-bond wavefunction of the electrons in an H_2 molecule is

$$\Psi(1,2)=\psi_A(1)\psi_B(2)+\psi_A(2)\psi_B(1) \qquad \text{A valence-bond wavefunction} \qquad (9\mathrm{A}.2)$$

The reason why this linear combination has a lower energy than either the separate atoms or the linear combination with a negative sign can be traced to the constructive interference between the wave patterns represented by the terms $\psi_A(1)\psi_B(2)$ and $\psi_A(2)\psi_B(1)$, and the resulting enhancement of the probability density of the electrons in the internuclear region (Fig. 9A.1).

Brief illustration 9A.1

The wavefunction for an H1s orbital ($Z = 1$) is given in Topic 8A as $(\pi a_0^3)^{-1/2}\mathrm{e}^{-r/a_0}$. If the distance of electron 1 from nucleus A is written r_{A1}, and likewise for the other electron and nucleus, the wavefunction in eqn 9A.2 can be written

$$\Psi(1,2) = \overbrace{\frac{1}{(\pi a_0^3)^{1/2}} e^{-r_{A1}/a_0}}^{\psi_A(1)} \times \overbrace{\frac{1}{(\pi a_0^3)^{1/2}} e^{-r_{B2}/a_0}}^{\psi_B(2)} + \overbrace{\frac{1}{(\pi a_0^3)^{1/2}} e^{-r_{A2}/a_0}}^{\psi_A(2)} \times \overbrace{\frac{1}{(\pi a_0^3)^{1/2}} e^{-r_{B1}/a_0}}^{\psi_B(1)}$$

$$= \frac{1}{\pi a_0^3}\{e^{-(r_{A1}+r_{B2})/a_0} + e^{-(r_{A2}+r_{B1})/a_0}\}$$

The electron distribution described by the wavefunction in eqn 9A.2 is called a σ **bond** (Fig. 9A.2). A σ bond has cylindrical symmetry around the internuclear axis, and is so called because, when viewed along the internuclear axis, it resembles a pair of electrons in an s orbital (and σ is the Greek equivalent of s).

A chemist's picture of a covalent bond is one in which the spins of two electrons pair as the atomic orbitals overlap. It can be shown that the origin of the role of spin is that the wavefunction in eqn 9A.2 can be formed only by two spin-paired electrons.

How is that done? 9A.1 Establishing the origin of electron pairs in VB theory

The Pauli principle requires the overall wavefunction of two electrons, the wavefunction including spin, to change sign when the labels of the electrons are interchanged (Topic 8B). The overall VB wavefunction for two electrons is

$$\Psi(1,2) = \{\psi_A(1)\psi_B(2) + \psi_A(2)\psi_B(1)\}\sigma(1,2)$$

where σ represents the spin component of the wavefunction. When the labels 1 and 2 are interchanged, this wavefunction becomes

$$\Psi(2,1) = \{\psi_A(2)\psi_B(1) + \psi_A(1)\psi_B(2)\}\sigma(2,1)$$
$$= \{\psi_A(1)\psi_B(2) + \psi_A(2)\psi_B(1)\}\sigma(2,1)$$

The Pauli principle requires that $\Psi(2,1) = -\Psi(1,2)$, which is satisfied only if $\sigma(2,1) = -\sigma(1,2)$. The combination of two spins that has this property is

$$\sigma_-(1,2) = \frac{1}{2^{1/2}}\{\alpha(1)\beta(2) - \beta(1)\alpha(2)\}$$

which corresponds to paired electron spins (Topic 8B). Therefore, the state of lower energy (and hence the formation of a chemical bond) is achieved if the electron spins are paired. Spin pairing is not an end in itself: it is necessary in order to achieve the lowest energy wavefunction.

The VB description of H_2 can extended to other homonuclear diatomic molecules, such as N_2. The starting point for the discussion of N_2 is the valence electron configuration of each atom, which is $2s^2 2p_x^1 2p_y^1 2p_z^1$. It is conventional to take the z-axis to be the internuclear axis in a linear molecule, so each atom is imagined as having a $2p_z$ orbital pointing towards a $2p_z$ orbital on the other atom, with the $2p_x$ and $2p_y$ orbitals perpendicular to the axis. A σ bond is then formed by spin pairing between the two electrons in the two $2p_z$ orbitals (Fig. 9A.3). Its spatial wavefunction is given by eqn 9A.2, but now ψ_A and ψ_B stand for the two $2p_z$ orbitals.

The remaining N2p orbitals ($2p_x$ and $2p_y$) cannot merge to give σ bonds as they do not have cylindrical symmetry around the internuclear axis. Instead, they merge to form two 'π bonds'. A π **bond** arises from the spin pairing of electrons in two p orbitals that approach side-by-side (Fig. 9A.3). It is so called because, viewed along the internuclear axis, a π bond resembles a pair of electrons in a p orbital (and π is the Greek equivalent of p).[1]

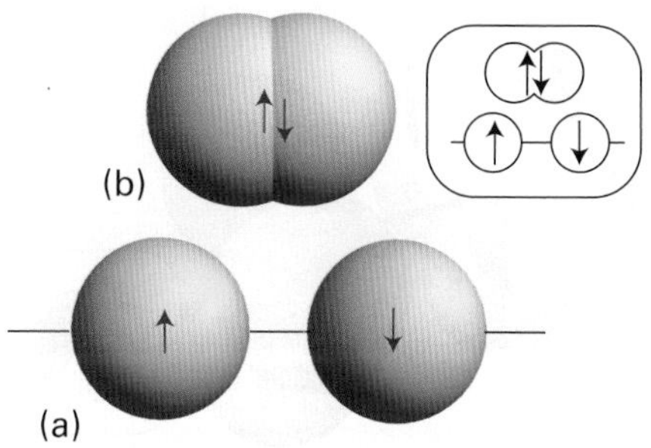

Figure 9A.2 The orbital overlap and spin pairing between electrons in two s orbitals that results in the formation of a σ bond.

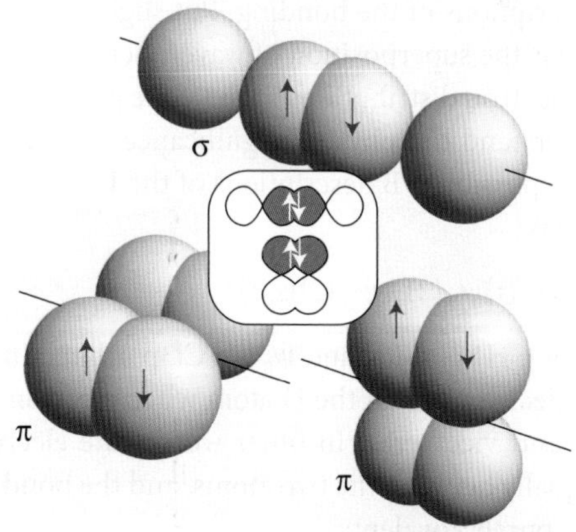

Figure 9A.3 (a) Orbital overlap and spin pairing between electrons in two p orbitals pointing along the internuclear axis (the z-axis) results in the formation of a σ bond. (b) A π bond results from orbital overlap and spin pairing between electrons in p orbitals with their axes perpendicular to the internuclear axis. Each bond has two lobes of electron density separated by a nodal plane.

[1] π bonds can also be formed from d orbitals in the appropriate orientation.

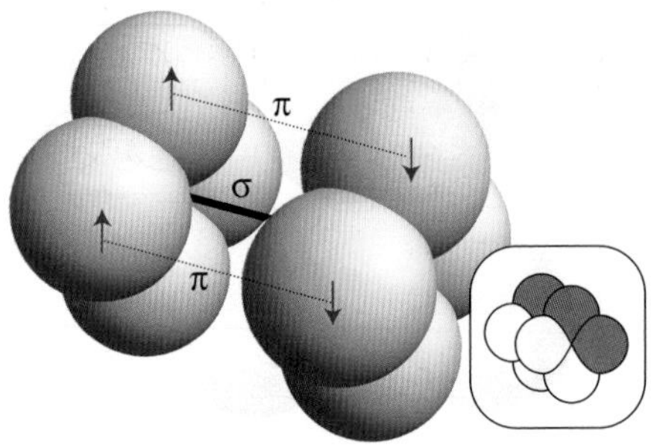

Figure 9A.4 The dinitrogen molecule has one σ bond and two π bonds.

There are two π bonds in N_2, one formed by spin pairing in two neighbouring $2p_x$ orbitals and the other by spin pairing in two neighbouring $2p_y$ orbitals. The overall bonding pattern in N_2 is therefore a σ bond plus two π bonds (Fig. 9A.4), which is consistent with the Lewis structure :N≡N: for dinitrogen.

Exercise

E9A.1 Write the valence-bond wavefunction for the single bond in HF.

9A.2 Resonance

For a given molecule there may be more than one plausible way of describing the bonding using electron pair bonds, and in VB theory this possibility is taken into account by introducing the concept of **resonance**. The idea is that the VB wavefunction can be a superposition of the wavefunctions that correspond to different descriptions of the bonding. Put slightly more formally, resonance is the superposition of wavefunctions that represent different electron distributions in the same nuclear framework.

To understand the physical significance of resonance, consider some possible VB descriptions of the HCl molecule. One possibility is

$$\Psi_{\text{H-Cl}} = \psi_A(1)\psi_B(2) + \psi_A(2)\psi_B(1)$$

where ψ_A is a H1s orbital and ψ_B is a Cl3p orbital. In this wavefunction electron 1 is on the H atom when electron 2 is on the Cl atom, and vice versa. In other words, the electron pair is shared equally between the two atoms, and the bond can be described as purely covalent.

Two other possibilities involve both electrons being on the Cl atom, which is the ionic form H^+Cl^-, or both electrons being on the H atom, which is the ionic form H^-Cl^+

$$\Psi_{\text{H}^+\text{Cl}^-} = \psi_B(1)\psi_B(2) \qquad \Psi_{\text{H}^-\text{Cl}^+} = \psi_A(1)\psi_A(2)$$

Chemical intuition indicates that the latter possibility is much less likely than the former, so it will be ignored from now on. A better description of the wavefunction for the molecule is as a superposition of the covalent and ionic descriptions, written as $\Psi_{\text{HCl}} = \Psi_{\text{H-Cl}} + \lambda\Psi_{\text{H}^+\text{Cl}^-}$ with λ (lambda) a numerical coefficient.

In general, the VB wavefunction, taking into account resonance, is written

$$\Psi = \Psi_{\text{covalent}} + \lambda\Psi_{\text{ionic}} \tag{9A.3}$$

where Ψ_{covalent} is the two-electron wavefunction for the purely covalent form of the bond and Ψ_{ionic} is the two-electron wavefunction for the ionic form of the bond. In this case, where one structure is pure covalent and the other pure ionic, it is called **ionic–covalent resonance**. The interpretation of the (unnormalized) wavefunction, which is called a **resonance hybrid**, is that if the molecule is inspected, then the probability that it would be found with an ionic structure is proportional to λ^2. If $\lambda^2 << 1$, the covalent description is dominant. If $\lambda^2 >> 1$, the ionic description is dominant. Resonance is not a flickering between the contributing states: it is a blending of their characteristics. It is only a mathematical device for achieving a closer approximation to the true wavefunction of the molecule than that represented by any single contributing electronic structure alone.

A systematic way of calculating the value of λ is provided by the **variation principle**:

> If an arbitrary wavefunction is used to calculate the energy, then the value calculated is never less than the true energy.

(This principle is derived and used in Topic 9C.) The arbitrary wavefunction is called the **trial wavefunction**. The principle implies that if the energy, the expectation value of the hamiltonian, is calculated for various trial wavefunctions with different values of the parameter λ, then the best value of λ is the one that results in the lowest energy. The ionic contribution to the resonance is then proportional to λ^2.

Brief illustration 9A.2

Consider a bond described by eqn 9A.3. If the lowest energy is reached when $\lambda = 0.1$, then the best description of the bond in the molecule is a resonance structure described by the wavefunction $\Psi = \Psi_{\text{covalent}} + 0.1\Psi_{\text{ionic}}$. This wavefunction implies that the probabilities of finding the molecule in its covalent and ionic forms are in the ratio 100:1 (because $0.1^2 = 0.01$).

Exercise

E9A.2 Write the valence-bond wavefunction for the resonance hybrid HF ↔ H^+F^- ↔ H^-F^+ (allow for different contributions of each structure).

9A.3 Polyatomic molecules

Each σ bond in a polyatomic molecule is formed by the spin pairing of electrons in atomic orbitals with cylindrical symmetry around the relevant internuclear axis. Likewise, π bonds are formed by pairing electrons that occupy atomic orbitals of the appropriate symmetry.

Brief illustration 9A.3

The VB description of H_2O is as follows. The valence-electron configuration of an O atom is $2s^2 2p_x^2 2p_y^1 2p_z^1$. The two unpaired electrons in the O2p orbitals can each pair with an electron in an H1s orbital, and each combination results in the formation of a σ bond (each bond has cylindrical symmetry about the respective O–H internuclear axis). Because the $2p_y$ and $2p_z$ orbitals lie at 90° to each other, the two σ bonds also lie at 90° to each other (Fig. 9A.5). Therefore, H_2O is predicted to be an angular molecule, which it is. However, the theory predicts a bond angle of 90°, whereas the actual bond angle is 104.5°.

Resonance plays an important role in the VB description of polyatomic molecules. One of the most famous examples of resonance is in the VB description of benzene, where the wavefunction of the molecule is written as a superposition of the many-electron wavefunctions of the two covalent Kekulé structures:

$\Psi = \Psi($ 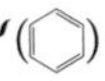 $) + \Psi($  $)$ (9A.4)

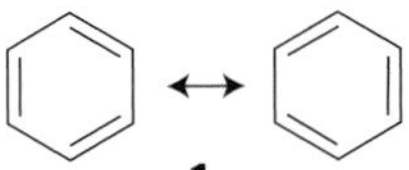

1

The two contributing structures have identical energies, so they contribute equally to the superposition. In this case the effect of resonance (which is represented by a double-headed arrow (**1**)) is to distribute double-bond character around the ring and to make the lengths and strengths of all the carbon–carbon bonds identical. The wavefunction is improved by allowing resonance because it allows the electrons to adjust into a distribution of lower energy. This lowering is called the **resonance stabilization** of the molecule and, in the context of VB theory, is largely responsible for the unusual stability of aromatic rings.

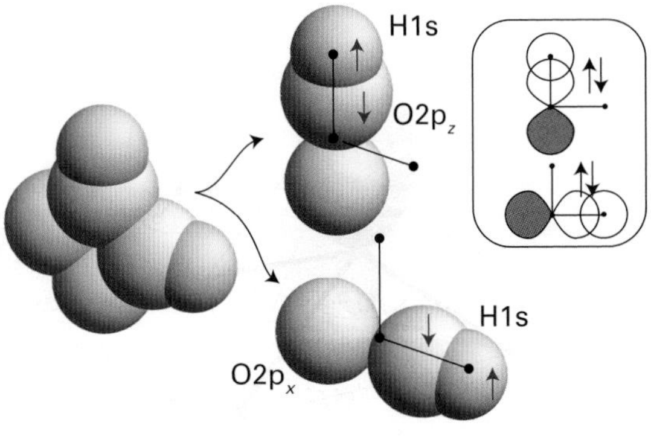

Figure 9A.5 In a primitive view of the structure of an H_2O molecule, each bond is formed by the overlap and spin pairing of an H1s electron and an O2p electron.

Resonance always lowers the energy, and the lowering is greatest when the contributing structures have similar energies. The wavefunction of benzene is improved still further, and the calculated energy of the molecule is lowered still further, if ionic–covalent resonance is also considered, by allowing a small admixture of ionic structures, such as (**2**).

2

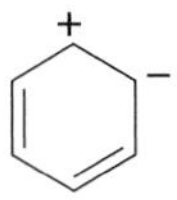

9A.3(a) Promotion

A deficiency of this initial formulation of VB theory is its inability to account for the fact that carbon commonly forms four bonds (it is 'tetravalent'). The ground-state configuration of carbon is $2s^2 2p_x^1 2p_y^1$, which suggests that a carbon atom should be capable of forming only two bonds, not four.

This deficiency is overcome by allowing for **promotion**, the excitation of an electron to an orbital of higher energy. In carbon, for example, the promotion of a 2s electron to a 2p orbital can be thought of as leading to the configuration $2s^1 2p_x^1 2p_y^1 2p_z^1$, with four unpaired electrons in separate orbitals. These electrons may pair with four electrons in orbitals provided by four other atoms (such as four H1s orbitals if the molecule is CH_4), and hence form four σ bonds. Although energy is required to promote the electron, it is more than recovered by the promoted atom's ability to form four bonds in place of the two bonds of the unpromoted atom.

Promotion, and the formation of four bonds, is a characteristic feature of carbon because the promotion energy is quite small: the promoted electron leaves a doubly occupied 2s orbital and enters a vacant 2p orbital, hence significantly relieving the electron–electron repulsion it experiences in the ground state. However, it is important to remember that promotion is not a 'real' process in which an atom somehow becomes excited and then forms bonds: rather, it is a notional contribution to the overall energy change that occurs when bonds form.

Brief illustration 9A.4

Sulfur can form six bonds (an 'expanded octet'), as in the molecule SF_6. Because the ground-state electron configuration of sulfur is $[Ne]3s^2 3p^4$, in one version of VB theory this bonding pattern requires the promotion of a 3s electron and a 3p electron to two different 3d orbitals, which are nearby in energy, to produce the notional configuration $[Ne]3s^1 3p^3 3d^2$. Now all six of the valence electrons are in different orbitals and capable of bond formation with six electrons provided by six F atoms. In an alternative approach, the molecule is regarded as a resonance hybrid of structures of the form $(SF_4)^{2+}(F^-)_2$, without the need to invoke d-orbitals. Calculations suggest the latter model is closer to the truth.

9A.3(b) Hybridization

The description of the bonding in CH_4 (and other alkanes) is still incomplete because it implies the presence of three σ bonds of one type (formed from H1s and C2p orbitals) and a fourth σ bond of a distinctly different character (formed from H1s and C2s). This problem is overcome by realizing that the electron density distribution in the promoted atom can be described in an alternative way in which each electron occupies a **hybrid orbital** formed from a combination of the C2s and C2p orbitals of the same atom. If these atomic orbitals are combined in a suitable way the result is four hybrid orbitals in a tetrahedral arrangement.

How is that done? 9A.2 Constructing tetrahedral hybrid orbitals

Each tetrahedral bond can be regarded as directed to one corner of a unit cube (3). Suppose that each hybrid can be written in the form $h = as + b_x p_x + b_y p_y + b_z p_z$, where the orbital designations represent the corresponding atomic wavefunctions. The hybrid h_1 that points to the corner with coordinates (1,1,1) must have equal contributions from all three p orbitals, so the three b coefficients can be set equal to each other and $h_1 = as + b(p_x + p_y + p_z)$. The other three hybrids have the same composition (they are equivalent, apart from their direction in space), but are orthogonal to h_1. This orthogonality is achieved by choosing different signs for the p orbitals but the same overall composition. For instance, choosing $h_2 = as + b(-p_x - p_y + p_z)$, the orthogonality condition is

$$\begin{aligned}\int h_1 h_2 \mathrm{d}\tau &= \int \{as + b(p_x + p_y + p_z)\}\{as + b(-p_x - p_y + p_z)\}\mathrm{d}\tau \\ &= a^2 \overbrace{\int s^2 \mathrm{d}\tau}^{1} - b^2 \overbrace{\int p_x^2 \mathrm{d}\tau}^{1} - b^2 \overbrace{\int p_y^2 \mathrm{d}\tau}^{1} + b^2 \overbrace{\int p_z^2 \mathrm{d}\tau}^{1} \\ &\quad - ab \overbrace{\int sp_x \mathrm{d}\tau}^{0} - \cdots - b^2 \overbrace{\int p_x p_y \mathrm{d}\tau}^{0} + \cdots \\ &= a^2 - b^2 - b^2 + b^2 = a^2 - b^2 = 0\end{aligned}$$

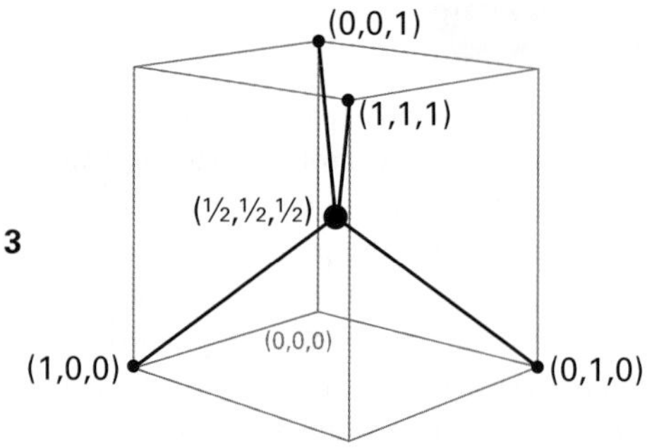

3

The values of the integrals come from the fact that the atomic orbitals are normalized and mutually orthogonal (Topic 7C). It follows that a solution is $a = b$ and that the two hybrid orbitals are $h_1 = s + p_x + p_y + p_z$ and $h_2 = s - p_x - p_y + p_z$. The alternative solution, $a = -b$, simply corresponds to choosing different absolute phases for the p orbitals. A similar argument but with $h_3 = as + b(-p_x + p_y - p_z)$ and $h_4 = as + b(p_x - p_y - p_z)$ leads to two other hybrids. In summary,

$$\begin{aligned} h_1 &= s + p_x + p_y + p_z & h_2 &= s - p_x - p_y + p_z \\ h_3 &= s - p_x + p_y - p_z & h_4 &= s + p_x - p_y - p_z \end{aligned} \qquad (9A.5)$$

sp³ hybrid orbitals

As a result of the interference between the component orbitals, each hybrid orbital consists of a large lobe pointing in the direction of one corner of a regular tetrahedron (Fig. 9A.6). The angle between the axes of the hybrid orbitals is the tetrahedral angle, $\arccos(-\frac{1}{3}) = 109.47°$. Because each hybrid is built from one s orbital and three p orbitals, it is called an **sp^3 hybrid orbital**.

It is now straightforward to see how the VB description of the CH_4 molecule leads to a tetrahedral molecule containing four equivalent C–H bonds. Each hybrid orbital of the promoted C atom contains a single unpaired electron; an H1s electron can pair with each one, giving rise to a σ bond pointing to a corner of a tetrahedron. For example, the (unnormalized) two-electron wavefunction for the bond formed by the hybrid orbital h_1 and the H1s orbital is

$$\Psi(1,2) = h_1(1)\psi_{\mathrm{H1s}}(2) + h_1(2)\psi_{\mathrm{H1s}}(1) \qquad (9A.6)$$

As for H_2, to achieve this wavefunction, the two electrons it describes must be paired. Because each sp^3 hybrid orbital has the same composition, all four σ bonds are identical apart from their orientation in space (Fig. 9A.7).

A hybrid orbital has enhanced amplitude in the internuclear region, which arises from the constructive interference between the s orbital and the positive lobes of the p orbitals. As a result, the bond strength is greater than for a bond formed from an s or p orbital alone. This increased bond strength is another factor that helps to repay the promotion energy.

The hybridization of N atomic orbitals always results in the formation of N hybrid orbitals, which may either form bonds or may contain **lone pairs** of electrons, pairs of electrons that

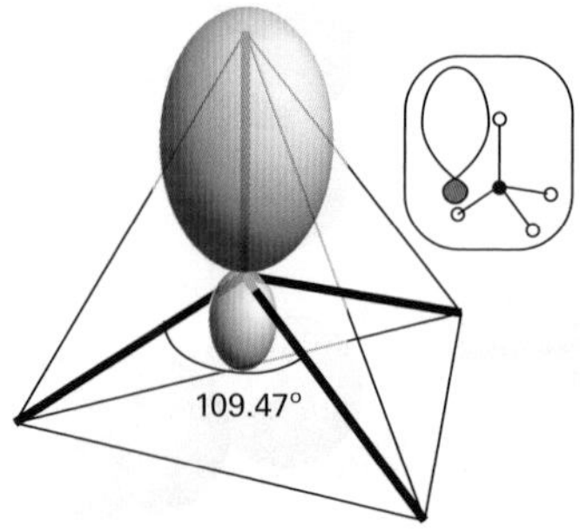

Figure 9A.6 An sp^3 hybrid orbital formed from the superposition of s and p orbitals on the same atom. There are four such hybrids: each one points towards the corner of a regular tetrahedron.

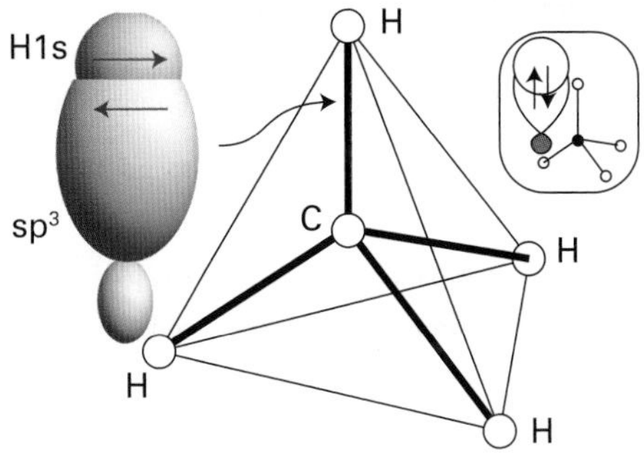

Figure 9A.7 Each sp^3 hybrid orbital forms a σ bond by overlap with an H1s orbital located at the corner of the tetrahedron. This model is consistent with the equivalence of the four bonds in CH_4.

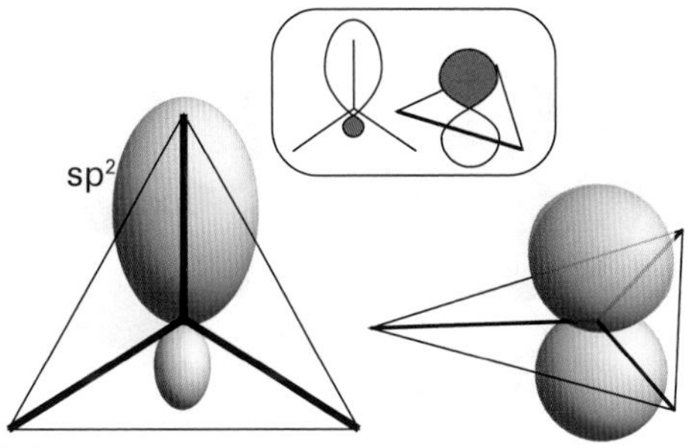

Figure 9A.8 (a) An s orbital and two p orbitals can be hybridized to form three equivalent orbitals that point towards the corners of an equilateral triangle. (b) The remaining, unhybridized p orbital is perpendicular to the plane.

do not participate directly in bond formation (but may influence the shape of the molecule).

Brief illustration 9A.5

To accommodate the observed bond angle of 104.5° in H_2O in VB theory it is necessary to suppose that the oxygen 2s and three 2p orbitals hybridize. As a first approximation, suppose they hybridize to form four equivalent sp^3 orbitals. Two of the hybrids are occupied by electron pairs, and so become lone pairs. The remaining two pairs of electrons form two O–H bonds at 109.5°. The actual hybridization will be slightly different to account for the observed bond angle not being exactly the tetrahedral angle.

Hybridization is also used to describe the structure of an ethene molecule, $H_2C{=}CH_2$, and the torsional rigidity of double bonds. An ethene molecule is planar, with HCH and HCC bond angles close to 120°. To reproduce the σ bonding structure, each C atom is regarded as being promoted to a $2s^12p^3$ configuration. However, instead of using all four orbitals to form hybrids, **sp^2 hybrid orbitals** are formed:

$$\begin{aligned} h_1 &= s + 2^{1/2}p_y \\ h_2 &= s + \left(\tfrac{3}{2}\right)^{1/2}p_x - \left(\tfrac{1}{2}\right)^{1/2}p_y \\ h_3 &= s - \left(\tfrac{3}{2}\right)^{1/2}p_x - \left(\tfrac{1}{2}\right)^{1/2}p_y \end{aligned} \qquad sp^2\ \text{hybrid orbitals} \qquad (9A.7)$$

These hybrids lie in a plane and point towards the corners of an equilateral triangle at 120° to each other (Fig. 9A.8). The third 2p orbital ($2p_z$) is not included in the hybridization; it lies along an axis perpendicular to the plane formed by the hybrids. The different signs of the coefficients, as well as ensuring that the hybrids are mutually orthogonal, also ensure that constructive interference takes place in different regions of space, so giving the patterns in the illustration.

The sp^2-hybridized C atoms each form three σ bonds by spin pairing with either a hybrid orbital on the other C atom or with H1s orbitals. The σ framework therefore consists of C–H and C–C σ bonds at 120° to each other. When the two CH_2 groups lie in the same plane, each electron in the two unhybridized p orbitals can pair and form a π bond (Fig. 9A.9). The formation of this π bond locks the framework into the planar arrangement, because any rotation of one CH_2 group relative to the other leads to a weakening of the π bond (and consequently an increase in energy of the molecule).

A similar description applies to ethyne, HC≡CH, a linear molecule. Now the C atoms are **sp hybridized**, and the σ bonds are formed using hybrid atomic orbitals of the form

$$h_1 = s + p_z \qquad h_2 = s - p_z \qquad \text{sp hybrid orbitals} \qquad (9A.8)$$

These two hybrids lie along the internuclear axis (the *z*-axis). The electrons in them pair either with an electron in the corresponding hybrid orbital on the other C atom or with an electron in one of the H1s orbitals. Electrons in the two remaining p orbitals on each atom, which are perpendicular to the molecular axis, pair to form two perpendicular π bonds (Fig. 9A.10).

Other hybridization schemes, particularly those involving d orbitals, are often invoked in VB descriptions of molecular structure to be consistent with other molecular geometries (Table 9A.1).

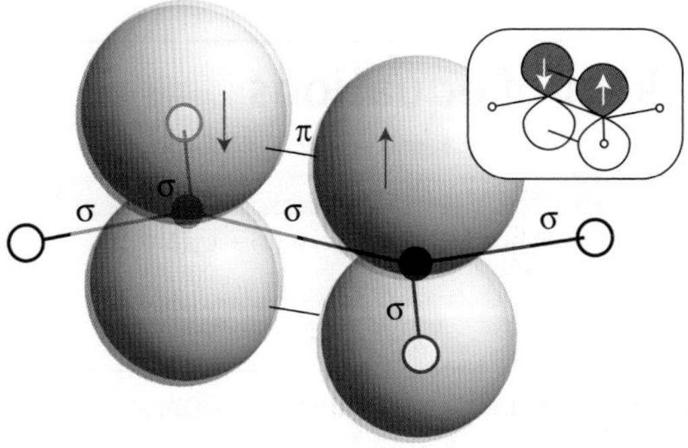

Figure 9A.9 A representation of the structure of a double bond in ethene; only the π bond is shown explicitly.

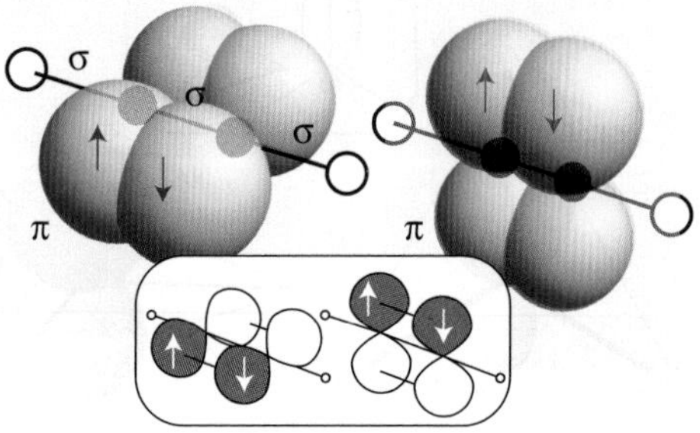

Figure 9A.10 A representation of the structure of a triple bond in ethyne; only the π bonds are shown explicitly.

Table 9A.1 Some hybridization schemes

Coordination number	Arrangement	Composition
2	Linear	sp, pd, sd
	Angular	sd
3	Trigonal planar	sp^2, p^2d
	Unsymmetrical planar	spd
	Trigonal pyramidal	pd^2
4	Tetrahedral	sp^3, sd^3
	Irregular tetrahedral	spd^2, p^3d, pd^3
	Square planar	p^2d^2, sp^2d
5	Trigonal bipyramidal	sp^3d, spd^3
	Tetragonal pyramidal	sp^2d^2, sd^4, pd^4, p^3d^2
	Pentagonal planar	p^2d^3
6	Octahedral	sp^3d^2
	Trigonal prismatic	spd^4, pd^5
	Trigonal antiprismatic	p^3d^3

Brief illustration 9A.6

Consider an octahedral molecule, such as SF_6. The promotion of sulfur's electrons as in *Brief illustration* 9A.4, followed by the model that invokes sp^3d^2 hybridization results in six equivalent hybrid orbitals pointing towards the corners of a regular octahedron.

Exercises

E9A.3 Describe the structure of a P_2 molecule in valence-bond terms. Why is P_4 a more stable form of molecular phosphorus than P_2?

E9A.4 Account for the ability of nitrogen to form four bonds, as in NH_4^+.

E9A.5 Describe the bonding in 1,3-butadiene using hybrid orbitals.

Checklist of concepts

☐ 1. A bond forms when an electron in an atomic orbital on one atom pairs its spin with that of an electron in an atomic orbital on another atom.

☐ 2. A **σ bond** has cylindrical symmetry around the internuclear axis.

☐ 3. **Resonance** is the superposition of structures with different electron distributions but the same nuclear arrangement.

☐ 4. A **π bond** has symmetry like that of a p orbital perpendicular to the internuclear axis.

☐ 5. **Promotion** is the notional excitation of an electron to an empty orbital to enable the formation of additional bonds.

☐ 6. **Hybridization** is the blending together of atomic orbitals on the same atom to achieve the appropriate directional properties and enhanced overlap.

Checklist of equations

Property	Equation	Comment	Equation number
Valence-bond wavefunction	$\Psi = \psi_A(1)\psi_B(2) + \psi_A(2)\psi_B(1)$	Spins must be paired*	9A.2
Resonance	$\Psi = \Psi_{\text{covalent}} + \lambda\Psi_{\text{ionic}}$	Ionic–covalent resonance	9A.3
Hybridization	$h = as + bp + \cdots$	All atomic orbitals on the same atom; specific forms in the text	9A.5, 9A.7, and 9A.8

* The spin contribution is $\sigma_-(1,2) = (1/2^{1/2})\{\alpha(1)\beta(2) - \beta(1)\alpha(2)\}$

TOPIC 9B Molecular orbital theory: the hydrogen molecule-ion

> ➤ **Why do you need to know this material?**
>
> Molecular orbital theory is the basis of almost all descriptions of chemical bonding and computational methods, for both individual molecules and solids.
>
> ➤ **What is the key idea?**
>
> Molecular orbitals are wavefunctions that spread over all the atoms in a molecule and are commonly represented as linear combinations of atomic orbitals.
>
> ➤ **What do you need to know already?**
>
> You need to be familiar with the shapes of atomic orbitals (Topic 8A) and how an energy is calculated from a wavefunction (Topic 7C). The entire discussion is within the framework of the Born–Oppenheimer approximation (see the *Prologue* for this Focus).

In **molecular orbital theory** (MO theory), electrons do not belong to particular bonds but spread throughout the entire molecule. This theory has been more fully developed than valence-bond theory (Topic 9A) and provides the language that is widely used in modern discussions of bonding. This Topic introduces the essential features of the theory by applying it to the very simplest molecular species, the hydrogen molecule-ion, H_2^+, which has only one electron. These ideas are used in the other Topics in this Focus to describe the structures of progressively more complex systems.

9B.1 Linear combinations of atomic orbitals

The hamiltonian for the single electron in H_2^+ is

$$\hat{H} = -\frac{\hbar^2}{2m_e}\nabla_1^2 + V \qquad V = -\frac{e^2}{4\pi\varepsilon_0}\left(\frac{1}{r_{A1}} + \frac{1}{r_{B1}} - \frac{1}{R}\right) \qquad (9B.1)$$

where r_{A1} and r_{B1} are the distances of the electron from the two nuclei A and B (**1**) and R is the distance between the two nuclei. In the expression for V, the first two terms in parentheses are the attractive contribution from the interaction between the electron and the nuclei; the remaining term is the repulsive interaction between the nuclei. The collection of fundamental constants $e^2/4\pi\varepsilon_0$ occurs widely throughout this Focus and is denoted j_0.

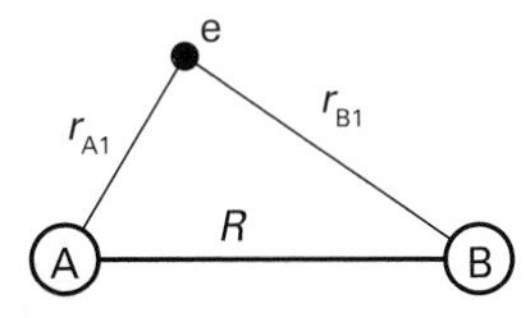

1

The one-electron wavefunctions obtained by solving the Schrödinger equation $\hat{H}\psi = E\psi$ are called **molecular orbitals**. A molecular orbital ψ gives, through the value of $|\psi|^2$, the distribution of the electron in the molecule. A molecular orbital is like an atomic orbital, but spreads throughout the molecule.

9B.1(a) The construction of linear combinations

The Schrödinger equation for H_2^+ can be solved analytically. However, the resulting wavefunctions are very complicated and the approach cannot be extended to larger molecules. Here a simpler approach is adopted which, although it is more approximate, can be extended readily to other molecules.

If the two nuclei in H_2^+ are far apart, then the electron is found either in the H1s atomic orbital of atom A, ψ_A, or in the H1s atomic orbital of atom B, ψ_B. The two atomic orbitals are identical so there is equal probability of the electron being found in each. Therefore the wavefunction for the electron in the molecule (the molecular orbital) can be written $\psi = \psi_A + \psi_B$ because in this wavefunction the two atomic orbitals appear with equal weight. An equally acceptable alternative would be to write the molecular orbital as $\psi = \psi_A - \psi_B$. The probability density depends on the square of the wavefunction, so the minus sign does not affect the conclusion that the electron has equal probability of being found in either orbital. As the separation of the nuclei is reduced, it is assumed that the molecular orbitals retain these forms.

The two molecular orbitals $\psi = \psi_A + \psi_B$ and $\psi = \psi_A - \psi_B$ are just the simplest example of the outcome of a procedure known as the **linear combination of atomic orbitals** (LCAO) in which a molecular orbital is constructed from a superposition of atomic orbitals. An approximate molecular orbital formed in this way is called an **LCAO-MO**. For H_2^+ the two LCAO-MOs formed from the two H1s atomic orbitals are:

$$\psi_\pm = N_\pm(\psi_A \pm \psi_B) \qquad \text{Linear combination of atomic orbitals} \qquad (9B.2)$$

where $N_\pm$ is a normalization factor. Because they are constructed from s atomic orbitals both molecular orbitals have cylindrical symmetry around the internuclear axis and are therefore described as **σ orbitals**. More precisely, this classification indicates that the orbital has zero orbital angular momentum around the internuclear axis. Figure 9B.1 shows two different representations of the molecular orbital ψ_+. Figures 9B.2 and 9B.3 are three-dimensional representations of ψ_+ and ψ_- in which surfaces of constant amplitude are plotted.

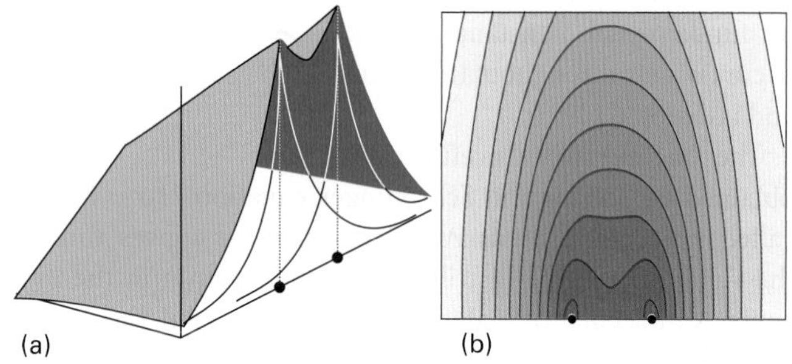

Figure 9B.1 (a) The amplitude of the molecular orbital ψ_+ of the hydrogen molecule-ion in a plane containing the two nuclei and (b) a contour representation of the amplitude.

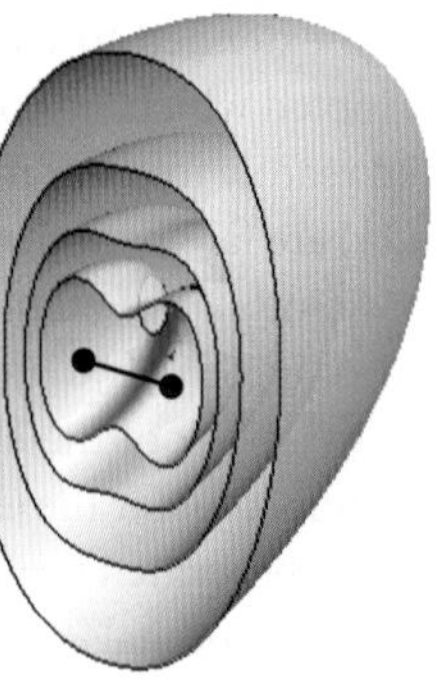

Figure 9B.2 Surfaces of constant amplitude of the wavefunction ψ_+ of the hydrogen molecule-ion.

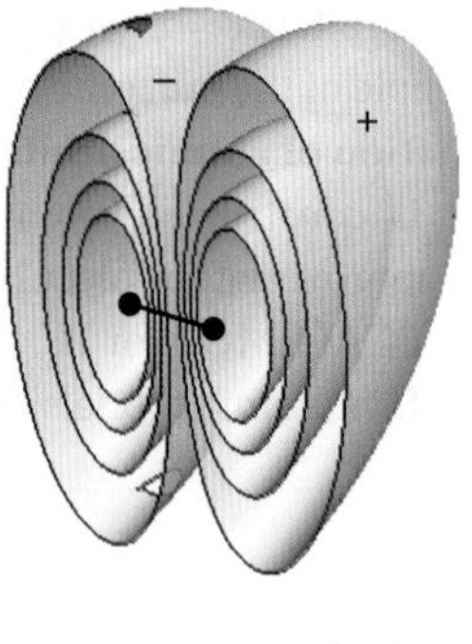

Figure 9B.3 Surfaces of constant amplitude of the wavefunction ψ_- of the hydrogen molecule-ion.

Example 9B.1 Normalizing a molecular orbital

Normalize the molecular orbital ψ_+ in eqn 9B.2.

Collect your thoughts You need to find the factor N_+ such that $\int\psi_+^*\psi_+\mathrm{d}\tau = 1$, where the integration is over the whole of space. To proceed, you should substitute the LCAO-MO into this integral and make use of the fact that the atomic orbitals are individually normalized and (in this case) real.

The solution Substitution of the wavefunction gives

$$\begin{aligned}\int\psi_+^*\psi_+\mathrm{d}\tau &= N_+^2\int(\psi_A+\psi_B)^2\mathrm{d}\tau\\ &= N_+^2\left(\overbrace{\int\psi_A^2\mathrm{d}\tau}^{1}+\overbrace{\int\psi_B^2\mathrm{d}\tau}^{1}+2\overbrace{\int\psi_A\psi_B\mathrm{d}\tau}^{S}\right)\\ &= 2(1+S)N_+^2\end{aligned}$$

where $S = \int\psi_A\psi_B\mathrm{d}\tau$ and has a value that depends on the nuclear separation (this 'overlap integral' will play a significant role later). For the integral to be equal to 1,

$$N_+ = \frac{1}{\{2(1+S)\}^{1/2}}$$

For H_2^+ at its equilibrium bond length $S \approx 0.59$, so $N_+ = 0.56$.

Self-test 9B.1 Normalize the orbital ψ_- in eqn 9B.2 and evaluate N_- for $S = 0.59$.

Answer: $N_- = 1/\{2(1-S)\}^{1/2}$, so $N_- = 1.10$

9B.1(b) Bonding orbitals

According to the Born interpretation, the probability density of the electron at each point in H_2^+ is proportional to the square modulus of its wavefunction at that point. The probability density corresponding to the wavefunction ψ_+ in eqn 9B.2 is

$$\psi_+^2 \propto \psi_A^2 + \psi_B^2 + 2\psi_A\psi_B \qquad \text{Bonding probability density} \qquad (9B.3)$$

This probability density is plotted in Fig. 9B.4. An important feature becomes apparent in the internuclear region, where

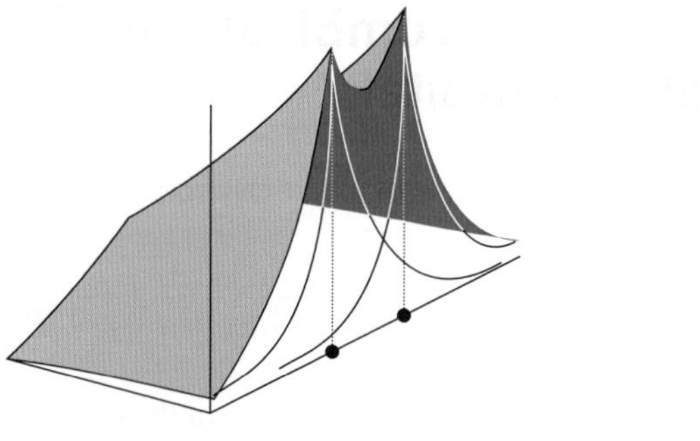

Figure 9B.4 The electron density resulting from the wavefunction ψ_+ of the hydrogen molecule-ion shown in Fig. 9B.2. Note the accumulation of electron density in the internuclear region.

both atomic orbitals have similar amplitudes. According to eqn 9B.3, the total probability density is proportional to the sum of:

- ψ_A^2, the probability density if the electron were confined to atom A;
- ψ_B^2, the probability density if the electron were confined to atom B;
- $2\psi_A\psi_B$, an extra contribution to the density from both atomic orbitals.

The last contribution, the **overlap density**, is crucial, because it represents an enhancement of the probability of finding the electron in the internuclear region. The enhancement can be traced to the constructive interference of the two atomic orbitals: each has a positive amplitude in the internuclear region, so the total amplitude is greater there than if the electron were confined to a single atom. This observation is summarized as

> Bonds form as a result of the build-up of electron density where atomic orbitals overlap and interfere constructively.

The conventional explanation of this observation is based on the notion that accumulation of electron density between the nuclei puts the electron in a position where it interacts strongly with both nuclei. Hence, the energy of the molecule is lower than that of the separate atoms, where each electron can interact strongly with only one nucleus.

This conventional explanation, however, has been called into question, because shifting an electron away from a nucleus into the internuclear region *raises* its potential energy. The modern (and still controversial) explanation does not emerge from the simple LCAO treatment given here. It seems that, at the same time as the electron shifts into the internuclear region, the atomic orbitals shrink. This orbital shrinkage improves the electron–nucleus attraction more than it is decreased by the migration to the internuclear region, so there is a net lowering of potential energy. The kinetic energy of the electron is also modified because the curvature of the wavefunction is changed, but the change in kinetic energy is dominated by the change in potential energy.

Throughout the following discussion the strength of chemical bonds is ascribed to the accumulation of electron density in the internuclear region. In molecules more complicated than H_2^+ the true source of energy lowering may be this accumulation of electron density or some indirect but related effect.

The σ orbital just described is an example of a **bonding orbital**, an orbital which, if occupied, helps to bind two atoms together. An electron that occupies a σ orbital is called a **σ electron**. With this notation, the configuration of the ground state of H_2^+ is written σ^1.

The energy E_σ of the σ orbital is:[1]

$$E_\sigma = E_{\rm H1s} + \frac{j_0}{R} - \frac{j+k}{1+S} \qquad \text{Energy of bonding orbital} \qquad (9B.4)$$

[1] For a derivation of eqn 9B.4, see *A deeper look* 9B.1, available to read in the e-book accompanying this text.

where $E_{\rm H1s}$ is the energy of a H1s orbital, j_0/R is the potential energy of repulsion between the two nuclei (recall that $j_0 = e^2/4\pi\varepsilon_0$), and

$$S = \int \psi_A\psi_B\,d\tau = \left\{1 + \frac{R}{a_0} + \frac{1}{3}\left(\frac{R}{a_0}\right)^2\right\} e^{-R/a_0} \qquad (9B.5a)$$

$$j = j_0\int \frac{\psi_A^2}{r_B}\,d\tau = \frac{j_0}{R}\left\{1 - \left(1 + \frac{R}{a_0}\right)e^{-2R/a_0}\right\} \qquad (9B.5b)$$

$$k = j_0\int \frac{\psi_A\psi_B}{r_B}\,d\tau = \frac{j_0}{a_0}\left(1 + \frac{R}{a_0}\right)e^{-R/a_0} \qquad (9B.5c)$$

Note that

$$\frac{j_0}{a_0} = \frac{e^2}{4\pi\varepsilon_0 a_0} = \frac{e^2}{4\pi\varepsilon_0} \times \frac{\pi m_e e^2}{\varepsilon_0 h^2} = \frac{m_e e^4}{4\varepsilon_0^2 h^2} = 2hc\tilde{R}_\infty \qquad (9B.5d)$$

The numerical value of $2hc\tilde{R}_\infty$ (when expressed in electronvolts) is 27.21 eV. Figure 9B.5 shows how the integrals S, j, and k vary with the internuclear separation R. These integrals are interpreted as follows:

- All three integrals are positive and decline towards zero at large internuclear separations (S and k on account of the exponential term, j on account of the factor $1/R$). The integral S is discussed in more detail in Topic 9C.
- The integral j is a measure of the interaction between one nucleus and the electron density centred on the other nucleus.
- The integral k is a measure of the interaction between one nucleus and the excess electron density in the internuclear region arising from overlap.

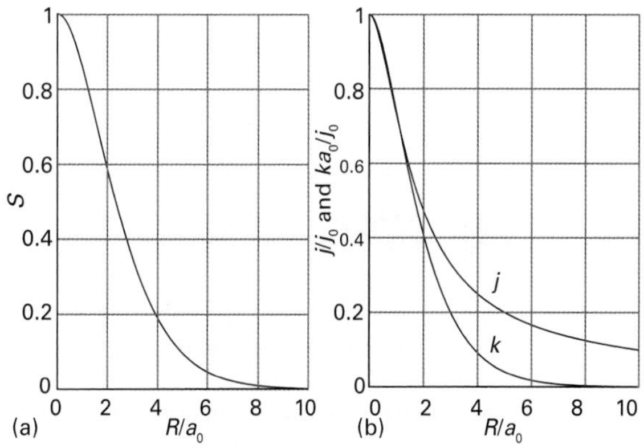

Figure 9B.5 The dependence of the integrals (a) S, (b) j and k on the internuclear distance, each calculated for H_2^+.

Brief illustration 9B.1

It turns out (see below) that the minimum value of E_σ occurs at $R = 2.49a_0$. At this separation

$$S=\left\{1+2.49+\frac{2.49^2}{3}\right\}e^{-2.49}=0.461$$

$$j=\frac{j_0/a_0}{2.49}(1-3.49e^{-4.98})=0.392j_0/a_0$$

$$k=\frac{j_0}{a_0}(1+2.49)e^{-2.49}=0.289j_0/a_0$$

Therefore, with $j_0/a_0 = 27.21$ eV, $j = 10.7$ eV, and $k = 7.87$ eV. The energy separation between the bonding MO and the H1s atomic orbital (being cautious with rounding) is

$$E_\sigma - E_{\mathrm{H1s}} = \frac{j_0}{R}-\frac{j+k}{1+S} = \frac{27.21\text{ eV}}{2.49}-\frac{(10.7\text{ eV})+(7.87\text{ eV})}{1+0.461} = -1.78\text{ eV}$$

Figure 9B.6 shows a plot of $E_\sigma - E_{\mathrm{H1s}}$ against R, that is a plot of the energy of the orbital relative to that of the separated atoms. The energy of the σ orbital decreases as the internuclear separation is decreased from large values because electron density accumulates in the internuclear region as the constructive interference between the atomic orbitals increases (Fig. 9B.7). However, at small separations there is too little space between the nuclei for significant accumulation of electron density there. In addition, the potential energy of nucleus–nucleus repulsion (which is proportional to $1/R$) becomes large. As a result, the energy of the molecular orbital rises at short distances, resulting in a minimum in the potential energy curve. The depth of this minimum is denoted $hc\tilde{D}_e$. From the expressions given in eqns 9B.4 and 9B.5 it is possible to compute that $R_e = 2.49a_0 = 132$ pm and $hc\tilde{D}_e = 1.76$ eV (171 kJ mol^{-1}). The experimental values are 106 pm and 2.6 eV, so this simple LCAO-MO description of the molecule, while inaccurate, is not absurdly wrong.

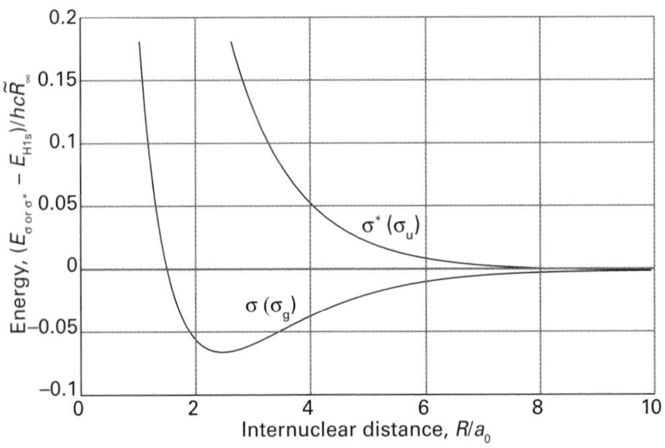

Figure 9B.6 The calculated molecular potential energy curves for a hydrogen molecule-ion showing the variation of the energies of the bonding and antibonding orbitals as the internuclear distance is changed. The energy E_σ is that of the σ orbital and E_{σ^*} is that of σ*.

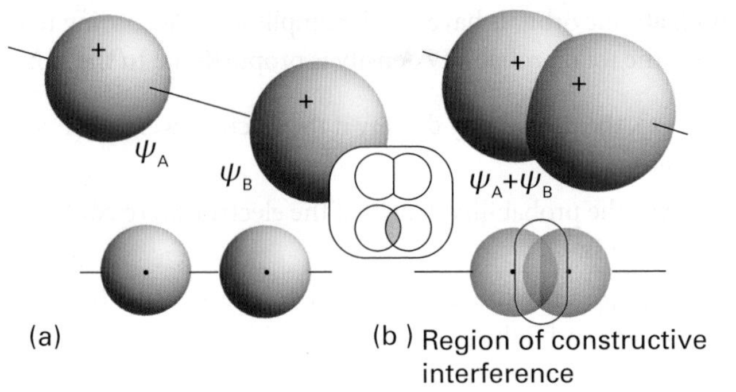

Figure 9B.7 A representation of the constructive interference that occurs when two H1s orbitals overlap and form a bonding σ orbital. In (a) the atomic orbitals are far apart and so do not interact; in (b) the orbitals are close enough to interact significantly.

9B.1(c) Antibonding orbitals

The linear combination ψ_- in eqn 9B.2 has higher energy than ψ_+. The orbital has cylindrical symmetry about the internuclear access so is classified as a σ orbital. For now it is labelled σ*, with the star indicating that it is an antibonding orbital. This orbital has a nodal plane perpendicular to the internuclear axis and passing through the mid-point of the bond where ψ_A and ψ_B cancel exactly (Figs. 9B.8 and 9B.9).

The probability density is

$$\psi_-^2 \propto \psi_A^2+\psi_B^2-2\psi_A\psi_B \quad \text{Antibonding probability density} \quad (9B.6)$$

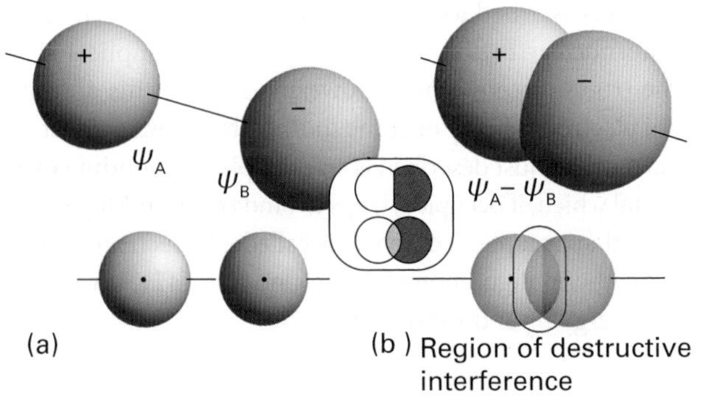

Figure 9B.8 A representation of the destructive interference that occurs when two H1s orbitals overlap and form an antibonding σ* orbital.

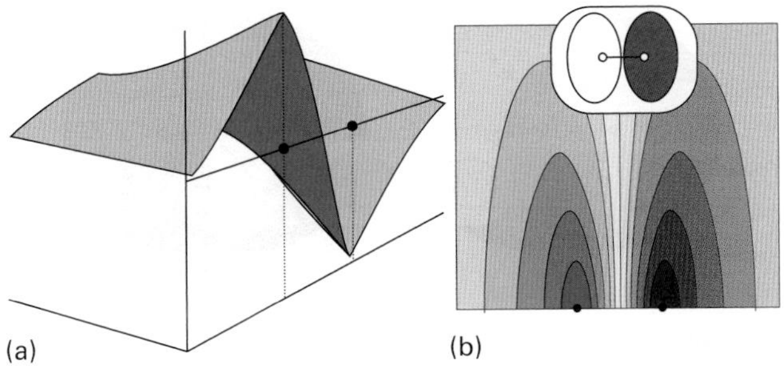

Figure 9B.9 (a) The amplitude of the antibonding molecular orbital ψ_- in a hydrogen molecule-ion in a plane containing the two nuclei and (b) a contour representation of the amplitude. Note the internuclear nodal plane.

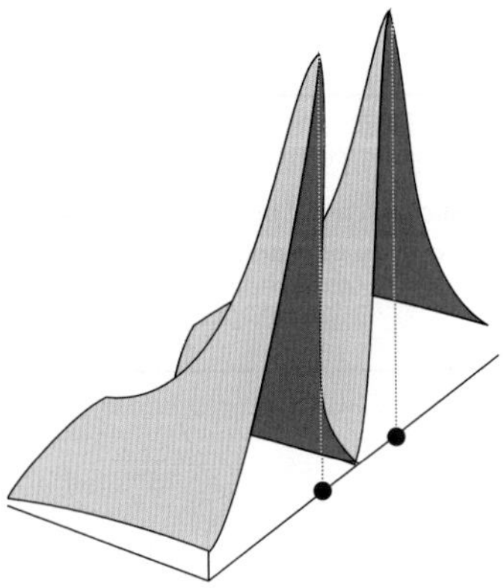

Figure 9B.10 The electron density calculated by forming the square of the wavefunction used to construct Fig. 9B.9. Note the reduction of electron density in the internuclear region.

There is a reduction in the probability density between the nuclei due to the term $-2\psi_A\psi_B$ (Fig. 9B.10); in physical terms, there is destructive interference where the two atomic orbitals overlap. The σ* orbital is an example of an **antibonding orbital**. If such an orbital is occupied it contributes to a reduction in the cohesion between two atoms and helps to raise the energy of the molecule relative to the separated atoms.

The energy E_{σ^*} of the σ* antibonding orbital is[2]

$$E_{\sigma^*} = E_{\mathrm{H1s}} + \frac{j_0}{R} - \frac{j-k}{1-S} \tag{9B.7}$$

where the integrals S, j, and k are the same as in eqn 9B.5. The variation of E_{σ^*} with R is shown in Fig. 9B.6. It is evident from this plot that if the electron is in the antibonding orbital the energy of the molecule is raised above that of the separated atoms. In other words, the antibonding electron has a destabilizing effect. This effect is partly due to the fact that an antibonding electron is excluded from the internuclear region and hence is distributed largely outside the bonding region. In effect, whereas a bonding electron pulls two nuclei together, an antibonding electron pulls the nuclei apart (Fig. 9B.11).

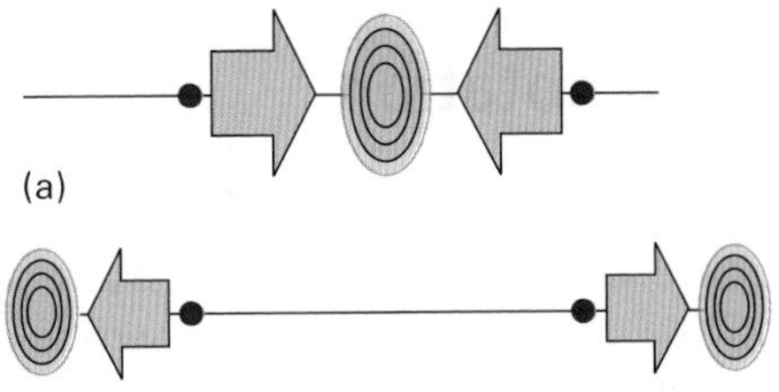

Figure 9B.11 A partial explanation of the origin of bonding and antibonding effects. (a) In a bonding orbital, the nuclei are attracted to the accumulation of electron density in the internuclear region. (b) In an antibonding orbital, the nuclei are attracted to an accumulation of electron density outside the internuclear region.

Figure 9B.6 also shows another feature drawn on later: $|E_{\sigma^*} - E_{\mathrm{H1s}}| > |E_\sigma - E_{\mathrm{H1s}}|$, which indicates that *the antibonding orbital is more antibonding than the bonding orbital is bonding.* This important conclusion stems in part from the presence of the nucleus–nucleus repulsion (j_0/R): this contribution raises the energy of both molecular orbitals.

Brief illustration 9B.2

At the minimum of the bonding orbital energy $R = 2.49a_0$, and, from *Brief illustration* 9B.1, $S = 0.461$, $j = 10.7\,\mathrm{eV}$, and $k = 7.87\,\mathrm{eV}$. It follows that at that separation, the energy of the antibonding orbital relative to that of a hydrogen atom 1s orbital is

$$\begin{aligned} E_{\sigma^*} - E_{\mathrm{H1s}} &= \frac{j_0}{R} - \frac{j-k}{1-S} \\ &= \frac{27.21\,\mathrm{eV}}{2.49} - \frac{(10.7\,\mathrm{eV})-(7.87\,\mathrm{eV})}{1-0.461} \\ &= +5.75\,\mathrm{eV} \end{aligned}$$

At this separation the antibonding orbital lies 5.75 eV above the energy of the separated atoms, whereas the bonding orbital lies 1.76 eV below this energy. As commented on above, the antibonding orbital is more antibonding that the bonding orbital is bonding.

Exercises

E9B.1 Normalize to 1 the molecular orbital $\psi = \psi_A + \lambda\psi_B$ in terms of the parameter λ and the overlap integral S. Assume that ψ_A and ψ_B are normalized to 1.

E9B.2 Derive an expression for the energy separation of σ and σ* orbitals in H_2^+ in terms of j, k, and S. Simplify your expression for the case $|S| \ll 1$. How do you interpret this result?

[2] This result is obtained by applying the strategy in *A deeper look* 9B.1, available to read in the e-book accompanying this text.

9B.2 Orbital notation

For homonuclear diatomic molecules (molecules consisting of two atoms of the same element, such as N_2), it proves helpful to label a molecular orbital according to its **inversion symmetry**. This label indicates the behaviour of the wavefunction when it is inverted through the centre of the molecule.[3] Thus, any point on the bonding σ orbital that is projected through the centre of the molecule and out an equal distance on the other side leads to an identical value (and sign) of the wavefunction (Fig. 9B.12). This so-called **gerade symmetry** (from the German word for 'even') is denoted by a subscript g, as in σ_g. The same procedure applied to the antibonding σ* orbital results in the same amplitude but opposite sign of the wavefunction. This **ungerade symmetry** ('odd symmetry') is denoted by a subscript u, as in σ_u.

The inversion symmetry classification is not applicable to heteronuclear diatomic molecules (diatomic molecules formed by atoms from two different elements, such as CO) because these molecules do not have a centre of inversion.

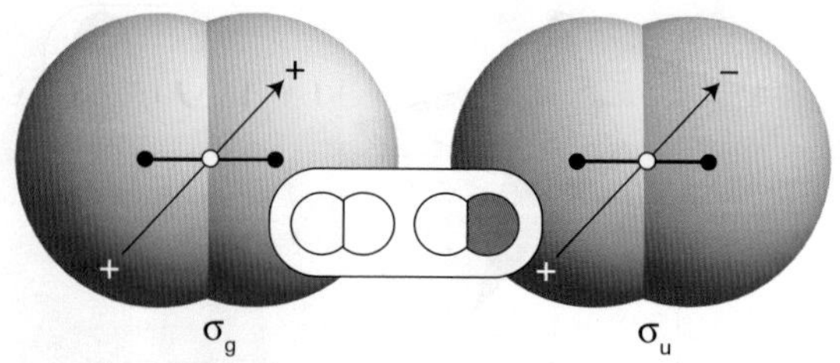

Figure 9B.12 The inversion symmetry of an orbital is even (g) if its wavefunction is unchanged under inversion through the centre of inversion (the open circle) of the molecule, but odd (u) if the wavefunction changes sign. Heteronuclear diatomic molecules do not have a centre of inversion, so for them the g, u classification is irrelevant.

Exercise

E9B.3 Identify the g or u character of bonding and antibonding π orbitals formed by side-by-side overlap of p atomic orbitals.

Checklist of concepts

- ☐ 1. A **molecular orbital** is constructed from a linear combination of atomic orbitals.
- ☐ 2. A **bonding orbital** arises from the constructive overlap of neighbouring atomic orbitals.
- ☐ 3. An **antibonding orbital** arises from the destructive overlap of neighbouring atomic orbitals.
- ☐ 4. σ **Orbitals** have cylindrical symmetry and zero orbital angular momentum around the internuclear axis.
- ☐ 5. A molecular orbital in a homonuclear diatomic molecule is labelled 'gerade' (g) or 'ungerade' (u) according to its behaviour under **inversion symmetry**.

Checklist of equations

Property	Equation	Comment	Equation number
Linear combination of atomic orbitals	$\psi_\pm = N_\pm(\psi_A \pm \psi_B)$	Homonuclear diatomic molecule	9B.2
Energies of σ orbitals formed from two 1s atomic orbitals	$E_\sigma = E_{H1s} + j_0/R - (j+k)/(1+S)$		9B.4
	$E_{\sigma^*} = E_{H1s} + j_0/R - (j-k)/(1-S)$		9B.7
Molecular integrals	$S = \int \psi_A \psi_B \mathrm{d}\tau$		9B.5a
	$j = j_0 \int (\psi_A^2/r_B)\mathrm{d}\tau$		9B.5b
	$k = j_0 \int (\psi_A \psi_B/r_B)\mathrm{d}\tau$		9B.5c

[3] More formally this is an inversion through a centre of inversion, as is discussed in Topic 10A.

TOPIC 9C Molecular orbital theory: homonuclear diatomic molecules

➤ Why do you need to know this material?

To be useful molecular orbital theory needs to be extended to cover molecules with more than one electron.

➤ What is the key idea?

Molecular orbitals are formed from all available valence-shell orbitals and the building-up principle is used to identify the configuration of lowest energy.

➤ What do you need to know already?

You need to be familiar with the discussion of the bonding and antibonding molecular orbitals (Topic 9B) and the building-up principle for atoms (Topic 8B).

The ground electronic configurations of many-electron atoms are found using hydrogenic atomic orbitals and the building-up principle (Topic 8B). A similar approach is used for many-electron homonuclear diatomic molecules, but in this case using the molecular orbitals resembling those for the one-electron hydrogen molecule-ion (Topic 9B).

9C.1 Electron configurations

The starting point of the molecular orbital theory (MO theory) of bonding in diatomic molecules (and ions) is the construction of molecular orbitals as linear combinations of the available atomic orbitals. Once the molecular orbitals have been formed, a building-up principle, like that for atoms (Topic 8B), is used to establish their ground-state electron configurations:

- The electrons are accommodated in the molecular orbitals so as to achieve the lowest overall energy subject to the constraint of the Pauli exclusion principle that no more than two electrons may occupy a single orbital (and then their spins must be paired).
- If several degenerate molecular orbitals are available, electrons are added singly to each individual orbital before any one orbital is doubly occupied (because that minimizes electron–electron repulsions).
- According to Hund's maximum multiplicity rule, if two electrons do occupy different degenerate orbitals, then a lower energy is obtained if their spins are parallel.

9C.1(a) MO energy level diagrams

Consider H_2, the simplest many-electron diatomic molecule. Each H atom contributes a 1s orbital (as in H_2^+), which combine to form bonding σ_g and antibonding σ_u orbitals, as explained in Topic 9B. At the equilibrium nuclear separation these orbitals have the energies shown in Fig. 9C.1, which is called a **molecular orbital energy level diagram**. Note that from two atomic orbitals two molecular orbitals are built. In general, from *N* atomic orbitals *N* molecular orbitals can be built.

There are two electrons to accommodate, and both can enter the σ_g orbital by pairing their spins, as required by the Pauli principle. The ground-state configuration is therefore σ_g^2 and the bond consists of an electron pair in a bonding σ orbital. This approach shows that an electron pair, which was the focus of Lewis's account of chemical bonding, represents the maximum number of electrons that can enter a bonding molecular orbital.

A straightforward extension of this argument explains why helium does not form diatomic molecules. Each He atom contributes a 1s orbital, so σ_g and σ_u molecular orbitals can be constructed. Although these orbitals differ in detail from those in H_2, their general shapes are the same and the same qualitative energy level diagram can be used in the discussion. There are four electrons to accommodate. Two can enter the σ_g orbital, but then it is full, and the next two must enter the σ_u orbital (Fig. 9C.2). The ground electronic configuration of He_2 is therefore $\sigma_g^2\sigma_u^2$. Recall from the discussion in Topic 9B that the σ_u orbital is higher in energy above the separate atoms

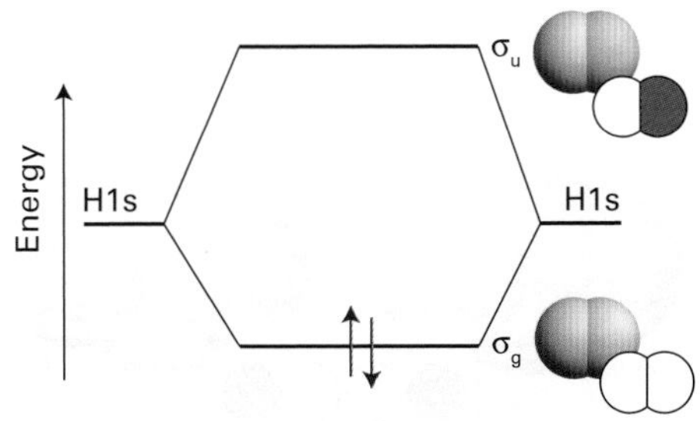

Figure 9C.1 A molecular orbital energy level diagram for orbitals constructed from the overlap of H1s orbitals. The energies of the atomic orbitals are indicated by the lines at the outer edges of the diagram, and the energies of the molecular orbitals are shown in the middle. The ground electronic configuration of H_2 is obtained by accommodating the two electrons in the lowest available orbital (the bonding orbital, σ_g).

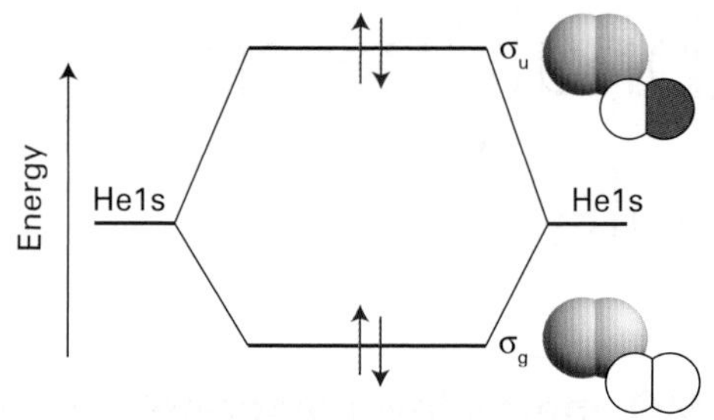

Figure 9C.2 The ground-state electronic configuration of the hypothetical four-electron molecule He_2 (at an arbitrary internuclear separation) has two electrons in the bonding orbital σ_g and two in the antibonding orbital σ_u. The molecule has a higher energy than the separated atoms, and so is unstable with respect to dissociation.

than the σ_g is lower. Therefore, the He_2 molecule has a higher energy than the separated atoms, so it is unstable relative to them and dihelium does not form.

9C.1(b) σ Orbitals and π orbitals

The concepts introduced so far also apply to homonuclear diatomics in general. In the elementary treatment used here, only the orbitals of the valence shell are used to form molecular orbitals so, for molecules formed with atoms from Period 2 elements, only the 2s and 2p atomic orbitals are considered.

A general principle of MO theory is that

> All orbitals of the appropriate symmetry contribute to a molecular orbital.

Thus, σ orbitals are built by forming linear combinations of all atomic orbitals that have cylindrical symmetry about the internuclear axis. These orbitals include the 2s orbitals on each atom and the $2p_z$ orbitals on the two atoms (Fig. 9C.3; the *z*-axis on each atom lies along the internuclear axis and points towards the neighbouring atom). The general form of the σ orbitals that may be formed is therefore

$$\psi = c_{A2s}\psi_{A2s} + c_{B2s}\psi_{B2s} + c_{A2p_z}\psi_{A2p_z} + c_{B2p_z}\psi_{B2p_z} \tag{9C.1}$$

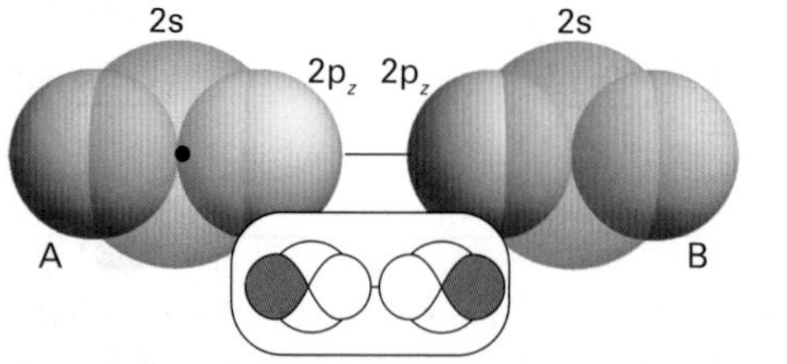

Figure 9C.3 According to molecular orbital theory, σ orbitals are built from all orbitals that have the appropriate symmetry. In homonuclear diatomic molecules of Period 2, that means that two 2s and two $2p_z$ orbitals should be used. From these four orbitals, four molecular orbitals can be built.

From these four atomic orbitals four molecular orbitals of σ symmetry can be formed by an appropriate choice of the coefficients *c*.

Because the 2s and 2p orbitals on each atom have such different energies, they may be treated separately (this approximation is removed later). That is, the four σ molecular orbitals fall approximately into two sets, one consisting of two molecular orbitals formed from the 2s orbitals

$$\psi = c_{A2s}\psi_{A2s} + c_{B2s}\psi_{B2s} \tag{9C.2a}$$

and another consisting of two orbitals formed from the $2p_z$ orbitals

$$\psi = c_{A2p_z}\psi_{A2p_z} + c_{B2p_z}\psi_{B2p_z} \tag{9C.2b}$$

In a homonuclear diatomic molecule the energies of the 2s orbitals on atoms A and B are the same. Their coefficients are therefore the same, $c_{A2s} = c_{B2s}$, or simply differ in sign, $c_{A2s} = -c_{B2s}$. The same is true of the $2p_z$ orbitals on each atom. Therefore, the two sets of orbitals have the form $\psi_{A2s} \pm \psi_{B2s}$ and $\psi_{A2p_z} \pm \psi_{B2p_z}$, and in each case there is a bonding combination and an antibonding combination.

The molecular orbitals arising from 2s and $2p_z$ orbitals are classified as σ_g or σ_u in the same way as in Topic 9B. Orbitals with the same label are distinguished by numbering them, starting with the lowest in energy, as $1\sigma_g$, $2\sigma_g$, and so on. Therefore, the σ bonding orbital formed from the 2s orbitals is denoted $1\sigma_g$ and the σ antibonding orbital formed from the same atomic orbitals is denoted $1\sigma_u$.

The two $2p_z$ orbitals directed along the internuclear axis also overlap strongly. They may interfere either constructively to give a bonding σ_g orbital or destructively to give an antibonding σ_u orbital (Fig. 9C.4). If it is supposed that the 2p atomic orbitals lie significantly higher in energy than the 2s orbitals, then the bonding and antibonding orbitals arising from the overlap of the $2p_z$ lie higher in energy than the $1\sigma_g$ and $1\sigma_u$ orbitals, and so are labelled $2\sigma_g$ and $2\sigma_u$, respectively.

Now consider the $2p_x$ and $2p_y$ orbitals of each atom. These orbitals are perpendicular to the internuclear axis and overlap broadside-on when the atoms are close together. This overlap

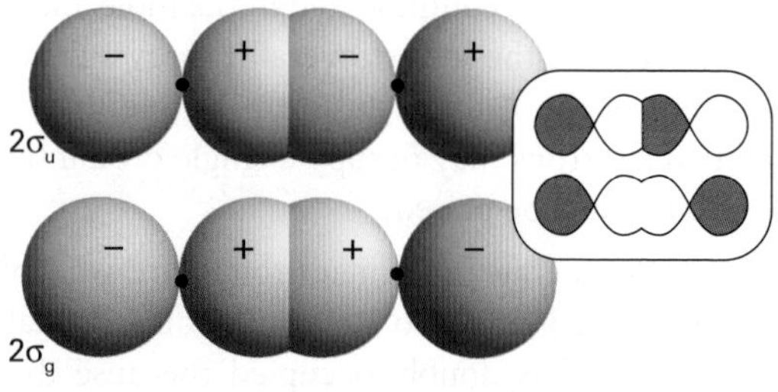

Figure 9C.4 A representation of the form of the bonding and antibonding σ orbitals built from the overlap of p orbitals. These illustrations are schematic.

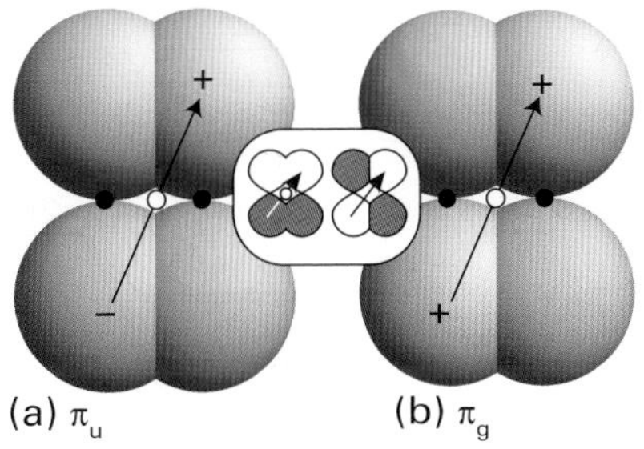

Figure 9C.5 The parity of (a) π bonding and (b) π antibonding molecular orbitals. The open circle is the centre of inversion.

may be constructive, resulting in a bonding π orbital, or destructive, resulting in an antibonding π orbital (Fig. 9C.5). The notation π is the analogue of p in atoms: when viewed along the axis of the molecule, a π orbital looks like a p orbital and has one unit of orbital angular momentum around the internuclear axis.

The two neighbouring $2p_x$ orbitals overlap to give a bonding and antibonding π_x orbital, and the two $2p_y$ orbitals overlap to give two π_y orbitals. The π_x and π_y bonding orbitals are degenerate; so too are their antibonding partners. As seen in Fig. 9C.5, a bonding π orbital has odd parity (u) and the antibonding π orbital has even parity (g). The lower two doubly degenerate orbitals are therefore labelled $1\pi_u$ and their higher energy antibonding partners are labelled $1\pi_g$.

9C.1(c) The overlap integral

As in the discussion of the hydrogen molecule-ion, the lowering of energy that results from constructive interference between neighbouring atomic orbitals (and the raising of energy that results from destructive interference) correlates with the extent of overlap of the orbitals. As explained in Topic 9B, the extent to which two atomic orbitals overlap is measured by the **overlap integral**, S:

$$S = \int \psi_A^* \psi_B \, d\tau \qquad \text{Overlap integral [definition]} \quad (9C.3)$$

If the amplitude of atomic orbital ψ_A on A is small in all the regions of space where the amplitude of atomic orbital ψ_B on B is large, the product $\psi_A\psi_B$ will be small everywhere. Therefore, the overlap integral—the sum of these products—is small (Fig. 9C.6a). In contrast, if the amplitudes of the two orbitals are both large in the same region of space, the product $\psi_A\psi_B$ will be large and the overlap integral will be significant (Fig. 9C.6b). If the two atomic orbitals are identical and are normalized (for instance, 1s orbitals on the same nucleus), then $S = 1$ when the internuclear separation is zero. In some cases, simple formulas can be given for overlap integrals (Table 9C.1); they are plotted in Fig. 9C.7.

Now consider the arrangement in which an s orbital spreads into the same region of space as a p_x orbital of a different atom

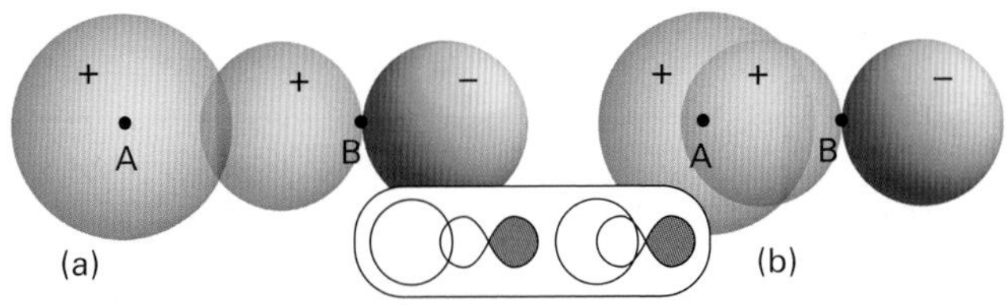

Figure 9C.6 (a) The two orbitals are far apart so in regions of space where the wavefunction of one is large, the other is small. The overlap integral S is therefore small. (b) When the orbitals are closer together, there are regions of space in which both wavefunctions are large and so the overlap integral is significant. Note that the overlap integral will decrease as the two atoms approach more closely than shown here, because the region of negative amplitude of the p orbital starts to overlap the positive amplitude of the s orbital. When the centres of the atoms coincide, $S = 0$.

Table 9C.1 Overlap integrals between hydrogenic orbitals*

Orbitals	Overlap integral, S
1s,1s	$\left(1+\eta+\frac{1}{3}\eta^2\right)e^{-\eta}$
2s,2s	$\left(1+\frac{1}{2}\eta+\frac{1}{12}\eta^2+\frac{1}{240}\eta^4\right)e^{-\eta/2}$
$2p_x$,$2p_x$ (π)	$\left(1+\frac{1}{2}\eta+\frac{1}{10}\eta^2+\frac{1}{120}\eta^3\right)e^{-\eta/2}$
$2p_z$,$2p_z$ (σ)	$-\left(1+\frac{1}{2}\eta+\frac{1}{20}\eta^2-\frac{1}{60}\eta^3-\frac{1}{240}\eta^4\right)e^{-\eta/2}$

* $\eta = ZR/a_0$

(Fig. 9C.8). The integral over the region where the product of the wavefunctions is positive exactly cancels the integral over the region where the product is negative, so overall $S = 0$ exactly. Therefore, there is no net overlap between the s and p_x orbitals in this arrangement.

The extent of overlap as measured by the overlap integral is suggestive of the contribution that different kinds of orbital overlap makes to bond formation, but the value of the integral must be treated with caution. Thus, the overlap integral for broadside overlap of $2p_x$ or $2p_y$ orbitals is typically greater than that for the overlap of $2p_z$ orbitals, suggesting weaker overlap in σ bonding than in π bonding.

However, the constructive overlap in the region between the nuclei and on the axis is greater in σ interactions, and its effect on bonding is more important than the overall extent of overlap. As a result, the separation of $1\pi_u$ and $1\pi_g$ orbitals is likely to be smaller than the separation of $2\sigma_g$ and $2\sigma_u$ orbitals in the same molecule. The relative energies of these orbitals is therefore likely to be as shown in Fig. 9C.9, and electrons occupying π_u orbitals are likely to be less effective at bonding than those occupying the σ_g orbitals derived from the same p orbitals.

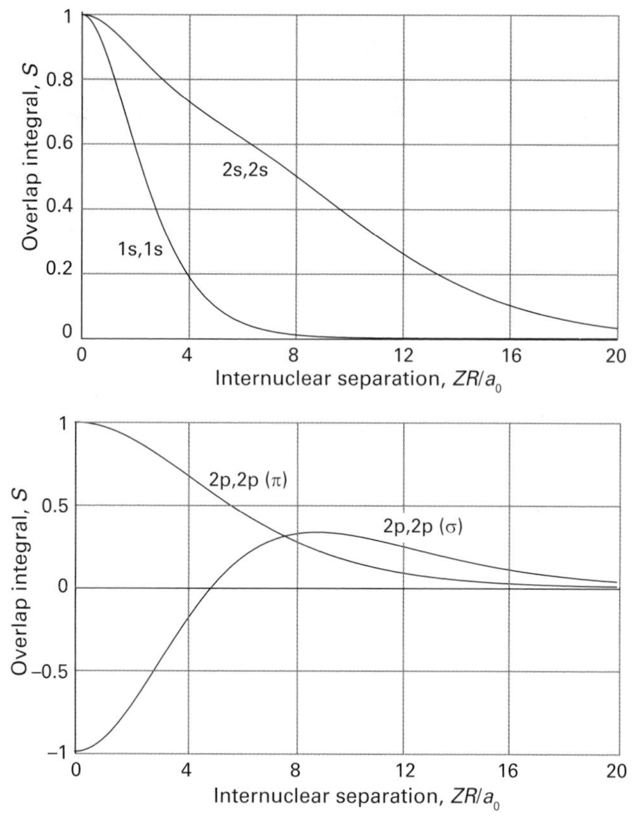

Figure 9C.7 The variation of the overlap integral, S, between two hydrogenic orbitals with the internuclear separation. A negative value of S corresponds to separations at which the contribution to the overlap of the positive region of one 2p orbital with the negative lobe of the other 2p orbital outweighs that from the regions where both have the same sign.

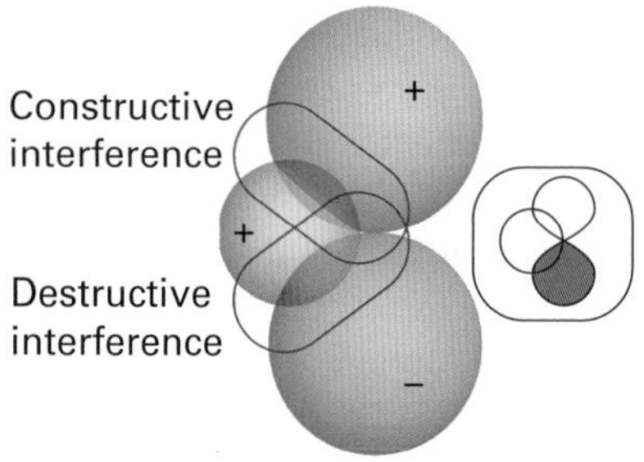

Figure 9C.8 A p orbital in the orientation shown here has zero net overlap ($S=0$) with the s orbital at all internuclear separations.

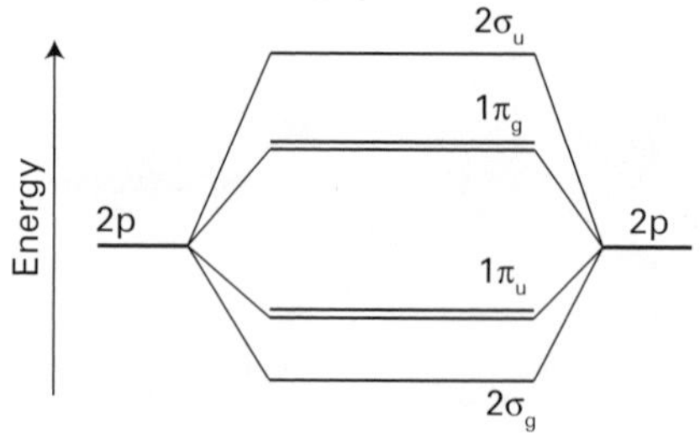

Figure 9C.9 As explained in the text, the separation of $1\pi_u$ and $1\pi_g$ orbitals is likely to be smaller than the separation of $2\sigma_g$ and $2\sigma_u$ orbitals in the same molecule, leading to the relative energies shown here.

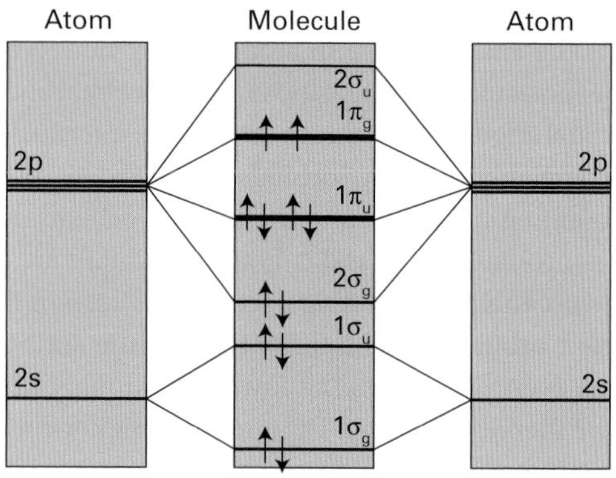

Figure 9C.10 The molecular orbital energy level diagram for homonuclear diatomic molecules. The lines in the middle are an indication of the energies of the molecular orbitals that can be formed by overlap of atomic orbitals. Energy increases upwards. As remarked in the text, this diagram is appropriate for O_2 (the configuration shown) and F_2.

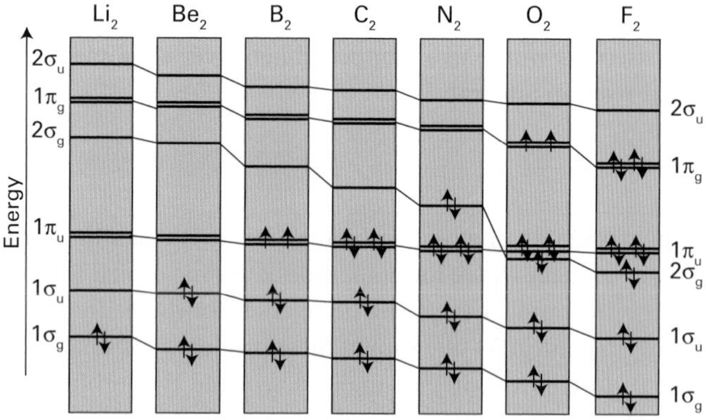

Figure 9C.11 The variation of the orbital energies of Period 2 homonuclear diatomics.

9C.1(d) Period 2 diatomic molecules

To construct the molecular orbital energy level diagram for Period 2 homonuclear diatomic molecules, eight molecular orbitals are formed from the eight valence shell orbitals (four from each atom). The ordering suggested by the discussion above is shown in Fig. 9C.10. However, remember that this scheme assumes that the 2s and $2p_z$ orbitals contribute to different sets of σ molecular orbitals. In fact all four atomic orbitals have the same symmetry around the internuclear axis and contribute jointly to the four σ orbitals. Hence, there is no guarantee that this order of energies will be found, and detailed calculation shows that the order varies along Period 2 (Fig. 9C.11). The order shown in Fig. 9C.12 is appropriate as far as N_2, and Fig. 9C.10 is appropriate for O_2 and F_2. The relative order is

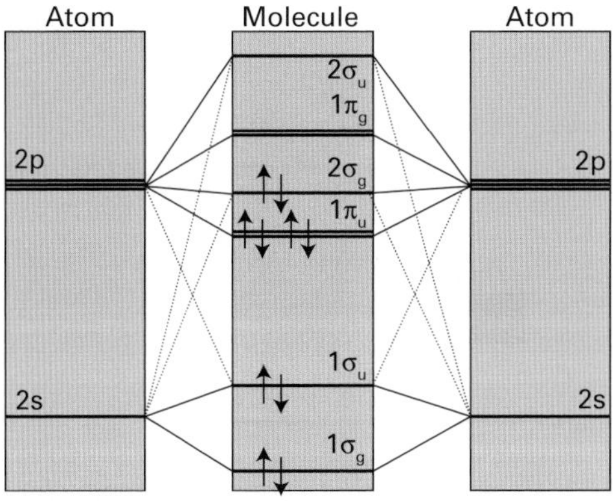

Figure 9C.12 An alternative molecular orbital energy level diagram for homonuclear diatomic molecules. Energy increases upwards. As remarked in the text, this diagram is appropriate for Period 2 homonuclear diatomics up to and including N_2 (the configuration shown).

controlled by the energy separation of the 2s and 2p orbitals in the atoms, which increases across the period. The change in the order of the $1\pi_u$ and $2\sigma_g$ orbitals occurs at about N_2.

With the molecular orbital energy level diagram established, the probable ground-state configurations of the molecules are deduced by adding the appropriate number of electrons to the orbitals and following the building-up rules. Anionic species (such as the peroxide ion, O_2^{2-}) need more electrons than the parent neutral molecules; cationic species (such as O_2^+) need fewer.

Consider N_2, which has 10 valence electrons. Two electrons fill the $1\sigma_g$ orbital (with paired spins), and two more fill the $1\sigma_u$ orbital. There are two $1\pi_u$ orbitals, so four electrons can be accommodated in them. The last two electrons enter the $2\sigma_g$ orbital. Therefore, the ground-state configuration of N_2 is $1\sigma_g^2 1\sigma_u^2 1\pi_u^4 2\sigma_g^2$, shown in Fig. 9C.12. It is sometimes helpful to include an asterisk to denote an antibonding orbital, in which case this configuration would be denoted $1\sigma_g^2 1\sigma_u^{*2} 1\pi_u^4 2\sigma_g^2$.

A measure of the net bonding in a diatomic molecule is its **bond order**, b:

$$b = \tfrac{1}{2}(N - N^*) \qquad \text{Bond order [definition]} \qquad (9C.4)$$

where N is the number of electrons in bonding orbitals and N^* is the number of electrons in antibonding orbitals.

Brief illustration 9C.1

Each electron pair in a bonding orbital increases the bond order by 1 and each pair in an antibonding orbital decreases b by 1. For H_2, $b = 1$, corresponding to a single bond, H–H, between the two atoms. In He_2, $b = 0$, and there is no bond. In N_2, $b = \tfrac{1}{2}(8-2) = 3$. This bond order accords with the Lewis structure of the molecule (:N≡N:).

The ground-state electron configuration of O_2, with 12 valence electrons, is found using the ordering of orbitals in Fig. 9C.10, and is $1\sigma_g^2 1\sigma_u^2 2\sigma_g^2 1\pi_u^4 1\pi_g^2$ (or $1\sigma_g^2 1\sigma_u^{*2} 2\sigma_g^2 1\pi_u^4 1\pi_g^{*2}$). The bond order is $b = \tfrac{1}{2}(8-4) = 2$. According to the building-up principle the two $1\pi_g$ electrons occupy two different orbitals: one will enter $1\pi_{g,x}$ and the other will enter $1\pi_{g,y}$. Because the electrons are in different orbitals they can have parallel spins, which corresponds to a lower energy than if they were paired. Therefore, an O_2 molecule is predicted to have a net spin angular momentum with $S = 1$ and, in the language introduced in Topic 8C, to be in a triplet state. As electron spin is the source of a magnetic moment, oxygen is also predicted to be paramagnetic, a substance that tends to be drawn into a magnetic field (Topic 15C). This prediction, which VB theory does not make, is confirmed by experiment.

An F_2 molecule has two more electrons than an O_2 molecule. Its configuration is therefore $1\sigma_g^2 1\sigma_u^{*2} 2\sigma_g^2 1\pi_u^4 1\pi_g^{*4}$ and $b = 1$, so F_2 is a singly bonded molecule, in agreement with its Lewis structure. The hypothetical molecule dineon, Ne_2, has two more electrons than F_2: its configuration is $1\sigma_g^2 1\sigma_u^{*2} 2\sigma_g^2 1\pi_u^4 1\pi_g^{*4} 2\sigma_u^{*2}$ and $b = 0$. The bond order of zero is consistent with the fact that neon occurs as a monatomic gas.

The bond order is a useful parameter for discussing the characteristics of bonds, because it correlates with bond length and bond strength. For bonds between atoms of a given pair of elements:

- The greater the bond order, the shorter the bond.
- The greater the bond order, the greater the bond strength.

Table 9C.2 lists some typical bond lengths in diatomic and polyatomic molecules. The strength of a bond is measured by its bond dissociation energy, and Table 9C.3 lists some experimental values.

Brief illustration 9C.2

From Fig. 9C.12, the electron configurations and bond orders of N_2 and N_2^+ are

N_2	$1\sigma_g^2 1\sigma_u^{*2} 1\pi_u^4 2\sigma_g^2$	$b = 3$
N_2^+	$1\sigma_g^2 1\sigma_u^{*2} 1\pi_u^4 2\sigma_g^1$	$b = 2\tfrac{1}{2}$

The cation has the smaller bond order, so you should expect it to have the smaller dissociation energy. The experimental dissociation energies are 942 kJ mol^{-1} for N_2 and 842 kJ mol^{-1} for N_2^+.

Table 9C.2 Bond lengths*

Bond	Order	R_e/pm
HH	1	74.14
NN	3	109.76
HCl	1	127.45
CH	1	*114*
CC	1	*154*
	2	*134*
	3	*120*

* More values will be found in the *Resource section*. Numbers in italics are mean values for polyatomic molecules.

Table 9C.3 Bond dissociation energies, $N_A hc\tilde{D}_0$*

Bond	Order	$N_A hc\tilde{D}_0$/(kJ mol^{-1})
HH	1	432.1
NN	3	941.7
HCl	1	427.7
CH	1	*435*
CC	1	*368*
	2	*720*
	3	*962*

* More values will be found in the *Resource section*. Numbers in italics are mean values for polyatomic molecules.

Exercises

E9C.1 Give the ground-state electron configurations, identify the highest occupied molecular orbital (the HOMO), and state the bond orders of (i) Li_2, (ii) Be_2, and (iii) C_2.

E9C.2 From the ground-state electron configurations of B_2 and C_2, predict which molecule should have the greater dissociation energy.

E9C.3 Which has the higher dissociation energy, F_2 or F_2^+?

9C.2 Photoelectron spectroscopy

So far, molecular orbitals have been regarded as purely theoretical constructs, but is there experimental evidence for their existence? A particularly useful technique for probing the energies of molecular orbitals is **photoelectron spectroscopy** (PES). In this technique the molecules are irradiated with high-energy photons of known energy. The photons are sufficiently energetic to cause electrons to be ejected from some orbitals. By measuring the energy of the **photoelectron**, the ejected electron, it is possible to infer the energy of the orbital from which it has been ejected.

Energy is conserved when a photon ionizes a sample. Therefore, the sum of the ionization energy, I, of the molecule and the kinetic energy of the photoelectron (moving at speed v) must be equal to the energy of the incident photon, $h\nu$ (Fig. 9C.13):

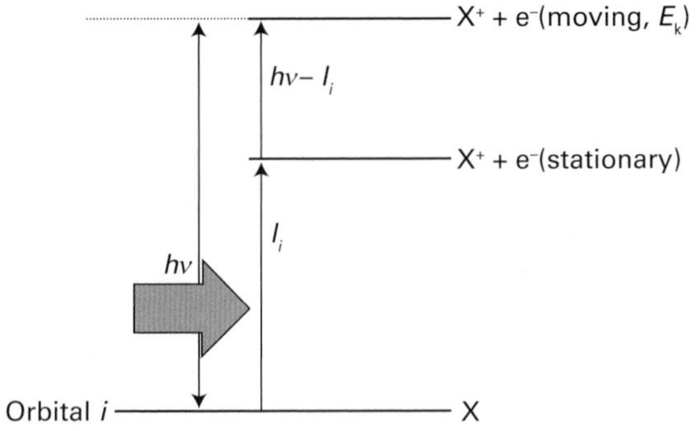

Figure 9C.13 An incoming photon carries an energy $h\nu$; an energy I_i is needed to remove an electron from an orbital i, and the difference appears as the kinetic energy of the electron.

$$h\nu = \tfrac{1}{2}m_e v^2 + I \qquad (9C.5)$$

This equation can be refined in two ways. First, photoelectrons may originate from one of a number of different orbitals, each one having a different ionization energy. Hence, a series of photoelectrons with different kinetic energies will be obtained, each one satisfying $h\nu = \tfrac{1}{2}m_e v^2 + I_i$, where I_i is the ionization energy for ejection of an electron from an orbital i. Therefore, by measuring the kinetic energies of the photoelectrons, and knowing the frequency ν, these ionization energies can be determined.

Photoelectron spectra are interpreted in terms of an approximation called **Koopmans' theorem**, which states that the ionization energy I_i is equal to the orbital energy of the ejected electron (formally: $I_i = -\varepsilon_i$). That is, the ionization energy can be identified with the energy of the orbital from which it is ejected. The theorem is only an approximation because it ignores the fact that the remaining electrons adjust their distributions when ionization occurs.

The ionization energies of molecules are several electronvolts, even for valence electrons, so it is essential to work in at least the ultraviolet region of the spectrum and with wavelengths of less than about 200 nm. Much work has been done with radiation generated by a discharge through helium: the He(I) line ($1s^1 2p^1 \rightarrow 1s^2$) lies at 58.43 nm, corresponding to a photon energy of 21.22 eV. Its use gives rise to the technique of **ultraviolet photoelectron spectroscopy** (UPS). When core electrons are being studied, photons of even higher energy are needed to expel them: X-rays are used, and the technique is denoted XPS.

The kinetic energies of the photoelectrons are measured using an electrostatic deflector that produces different deflections in the paths of the photoelectrons as they pass between charged plates (Fig. 9C.14). As the field strength between the plates is increased, electrons of different speeds, and therefore kinetic energies, reach the detector. The electron flux can be recorded and plotted against kinetic energy to obtain the photoelectron spectrum (Fig. 9C.15).

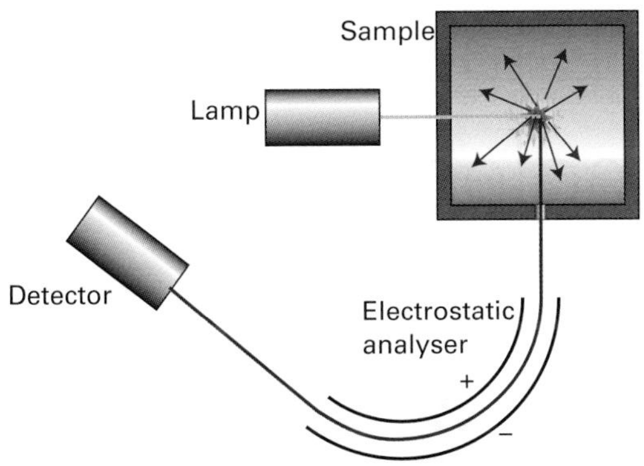

Figure 9C.14 A photoelectron spectrometer consists of a source of ionizing radiation (such as a helium discharge lamp for UPS and an X-ray source for XPS), an electrostatic analyser, and an electron detector. The deflection of the electron path caused by the analyser depends on the speed of the electrons.

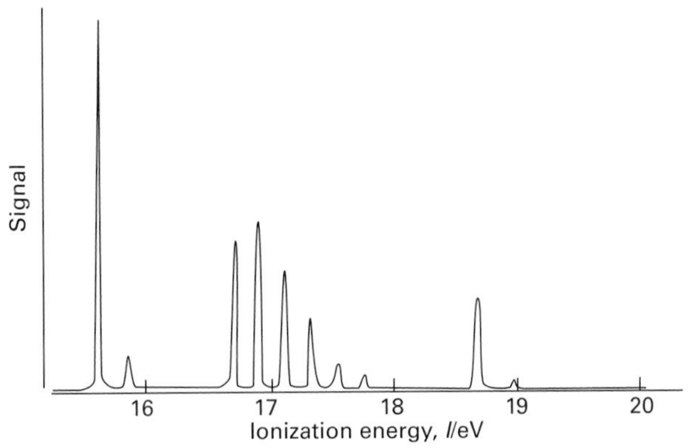

Figure 9C.15 The photoelectron spectrum of N_2 recorded using He(I) radiation.

Brief illustration 9C.3

The photoelectrons of highest kinetic energy ejected from N_2 in a spectrometer using He(I) radiation have kinetic energies of 5.63 eV. Photons of helium(I) radiation have energy 21.22 eV, so it follows that 21.22 eV = 5.63 eV + I_i, and therefore I_i = 15.59 eV. This ionization energy is the energy needed to remove an electron from the occupied molecular orbital with the highest energy of the N_2 molecule, the $2\sigma_g$ bonding orbital. Photoelectrons are also detected at 4.53 eV, corresponding to an ionization energy of 16.7 eV. The likely origin of these electrons is the $1\pi_u$ orbital.

Photoejection commonly results in cations that are excited vibrationally. Because different energies are needed to excite different vibrational states of the ion, the photoelectrons appear with different kinetic energies. The result is **vibrational fine structure**, a progression of lines with a spacing in energy that corresponds to the vibrational frequency of the molecular ion. This fine structure occurs between 16.7 eV and 18 eV in the photoelectron spectrum of N_2 shown in Fig. 9C.15.

Exercises

E9C.4 What is the speed of a photoelectron ejected from an orbital with ionization energy 12.0 eV by a photon of radiation of wavelength 100 nm?

E9C.5 Predict the form of the ultraviolet photoelectron spectrum of H_2. Speculate on how the corresponding spectrum of H_2^+ might differ from that of H_2.

Checklist of concepts

☐ 1. Molecular orbitals are constructed as linear combinations of all valence orbitals of the appropriate symmetry.

☐ 2. As a first approximation, σ orbitals are constructed separately from valence s and p orbitals.

☐ 3. π **Orbitals** are constructed from the sideways overlap of p orbitals.

☐ 4. An **overlap integral** is a measure of the extent of orbital overlap.

☐ 5. According to the building-up principle, electrons occupy the available molecular orbitals so as to achieve the lowest total energy subject to the Pauli exclusion principle.

☐ 6. If electrons occupy different orbitals, the lowest energy is obtained if their spins are parallel.

☐ 7. The greater the **bond order** of a molecule or ion between the same two atoms, the shorter and stronger is the bond.

☐ 8. **Photoelectron spectroscopy** is a technique for determining the energies of electrons in molecular orbitals.

Checklist of equations

Property	Equation	Comment	Equation number
Overlap integral	$S=\int\psi_A^*\psi_B\mathrm{d}\tau$	Integration over all space	9C.3
Bond order	$b=\frac{1}{2}(N-N^*)$	N and N^* are the numbers of electrons in bonding and antibonding orbitals, respectively	9C.4
Photoelectron spectroscopy	$h\nu=\frac{1}{2}m_e v^2+I$	Interpret I as I_i, the ionization energy from orbital i	9C.5

TOPIC 9D Molecular orbital theory: heteronuclear diatomic molecules

> ➤ **Why do you need to know this material?**
>
> Many diatomic molecules are built from different elements and to understand their structure and properties it is necessary to take into account the difference in the atomic orbitals used to form their molecular orbitals.
>
> ➤ **What is the key idea?**
>
> The molecular orbitals in heteronuclear diatomic molecules have unequal contributions from the atomic orbitals from which they are constructed, resulting in polarization of the bond.
>
> ➤ **What do you need to know already?**
>
> You need to be familiar with the molecular orbitals of homonuclear diatomic molecules (Topic 9C) and the concepts of normalization and orthogonality (Topic 7C). This Topic makes light use of determinants (*The chemist's toolkit* 9D.1).

The electrons in a covalent bond in a heteronuclear diatomic species are not distributed equally over the atoms: it is energetically favourable for the electron pair to be found closer to one atom than to the other. This imbalance results in a **polar bond**, a bond in which the bonding electron density is shared unequally between the bonded atoms. The bond in HF, for instance, is polar, with the bonding electron density greater near the F atom than the H atom. The accumulation of bonding electron density near the F atom results in that atom having a net negative charge, which is called a **partial negative charge** and denoted δ–. There is a matching **partial positive charge**, δ+, on the H atom (Fig. 9D.1).

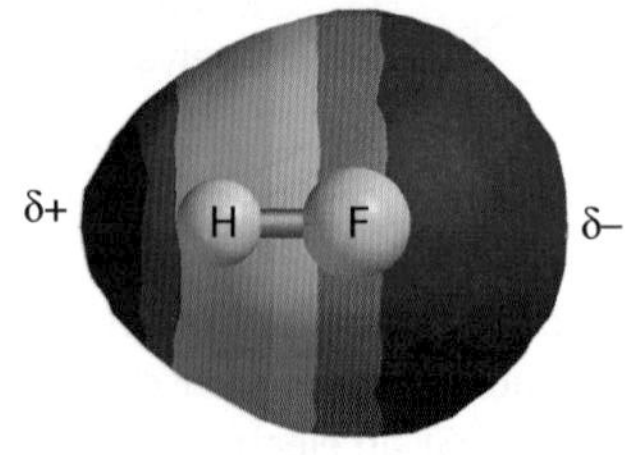

Figure 9D.1 The electron density of the molecule HF, computed using one of the methods described in Topic 9F. Different colours show the variation in the electrostatic potential and hence the net charge, with blue representing the region with largest partial positive charge, and red the region with largest partial negative charge.

9D.1 Polar bonds and electronegativity

The description of polar bonds is a straightforward extension of the molecular orbital theory of homonuclear diatomic molecules (Topic 9C). The principal difference stems from the fact that the atomic orbitals on the two atoms have different energies and spatial extents.

A polar bond consists of two electrons in a bonding molecular orbital of the form

$$\psi = c_A\psi_A + c_B\psi_B \qquad \text{Wavefunction of a polar bond} \qquad (9D.1)$$

with unequal coefficients. The interpretation of this wavefunction is that the probability that the electron will be found on atom A is proportional to $|c_A|^2$, and the probability that it will be found on atom B is proportional to $|c_B|^2$. A nonpolar bond has $|c_A|^2 = |c_B|^2$, and a pure ionic bond has one coefficient equal to zero (so the species A^+B^- would have $c_A = 0$ and $c_B = 1$). The atomic orbital with the lower energy makes the larger contribution to the bonding molecular orbital. The opposite is true of the antibonding orbital, for which the dominant contribution comes from the atomic orbital with higher energy.

The distribution of partial charges in bonds is commonly discussed in terms of the **electronegativity**, χ (chi), of the elements involved. The electronegativity is a parameter introduced by Linus Pauling as a measure of the power of an atom in a bond to attract electrons to itself. Pauling used valence-bond arguments to suggest that an appropriate numerical scale of electronegativities could be defined in terms of bond dissociation energies and proposed that the difference in electronegativities could be expressed as

$$|\chi_A - \chi_B| = \{hc\tilde{D}_0(\text{AB})/\text{eV} - \tfrac{1}{2}[hc\tilde{D}_0(\text{AA})/\text{eV} + hc\tilde{D}_0(\text{BB})/\text{eV}]\}^{1/2}$$

$$\text{Pauling electronegativity [definition]} \qquad (9D.2)$$

where $hc\tilde{D}_0(\text{XY})$ is the dissociation energy of an X–Y bond. This expression gives differences of electronegativities; to

Table 9D.1 Pauling electronegativities*

Element	χ_P
H	2.2
C	2.6
N	3.0
O	3.4
F	4.0
Cl	3.2
Cs	0.79

* More values are given in the *Resource section*.

establish an absolute scale Pauling chose individual values that gave the best match to the values obtained from eqn 9D.2. Electronegativities based on this definition are called **Pauling electronegativities** (Table 9D.1). The most electronegative elements are found in Groups 15, 16, and 17, especially in Periods 2 and 3. It is found that the greater the difference in electronegativities, the greater the polar character of the bond. The difference for HF, for instance, is 1.8; a C–H bond, which is commonly regarded as almost nonpolar, has an electronegativity difference of 0.4.

Brief illustration 9D.1

The bond dissociation energies of H_2, Cl_2, and HCl are 4.52 eV, 2.51 eV, and 4.47 eV, respectively. From eqn 9D.2,

$$|\chi_{\text{Pauling}}(\text{H}) - \chi_{\text{Pauling}}(\text{Cl})| = \{4.47 - \tfrac{1}{2}(4.52 + 2.51)\}^{1/2} = 0.98 \approx 1.0$$

The spectroscopist Robert Mulliken proposed an alternative definition of electronegativity. Mulliken argued that an element is likely to be highly electronegative if it has a high ionization energy (so it will not release electrons readily) and a high electron affinity (so it is energetically favourable to acquire electrons). The **Mulliken electronegativity scale** is therefore based on the definition

$$\chi = \tfrac{1}{2}(I + E_{\text{ea}})/\text{eV} \quad \text{Mulliken electronegativity [definition]} \quad (9\text{D}.3)$$

where I is the ionization energy of the element and E_{ea} is its electron affinity. The greater the value of the Mulliken electronegativity the greater is the contribution of that atom to the electron distribution in the bond. There is one word of caution: the values of I and E_{ea} in eqn 9D.3 are strictly those for a special 'valence state' of the atom, not a true spectroscopic state, but that complication is ignored here. The Mulliken and Pauling scales are approximately in line with each other. A reasonably reliable conversion between the two is

$$\chi_{\text{Pauling}} = 1.35\chi_{\text{Mulliken}}^{1/2} - 1.37 \quad (9\text{D}.4)$$

Exercise

E9D.1 The bond dissociation energy of O–H is 467 kJ mol^{-1}, that of H–H is 432 kJ mol^{-1}, and that of O–O is 146 kJ mol^{-1}. Calculate $|\chi_{\text{Pauling}}(\text{O}) - \chi_{\text{Pauling}}(\text{H})|$; note that the bond energies must be converted to eV to evaluate this difference in electronegativities.

9D.2 The variation principle

The systematic way of finding the coefficients in the linear combinations used to build molecular orbitals is provided by the **variation principle**:

> If an arbitrary wavefunction is used to calculate the energy, the value calculated is never less than the true ground-state energy.

It can be justified by setting up an arbitrary 'trial function' and showing that the corresponding energy is not less than the true ground-state energy (it might be the same).

How is that done? 9D.1 Justifying the variation principle

Any arbitrary function can be expressed as a linear combination of the eigenfunctions ψ_n of the exact hamiltonian for a molecule. In the present case, consider a normalized trial wavefunction written as a linear combination $\psi_{\text{trial}} = \sum_n c_n\psi_n$ and suppose that the ψ_n are themselves normalized and mutually orthogonal.

Step 1 *Write an expression for the difference between the calculated and true energy*

The energy associated with the normalized trial function is the expectation value of the hamiltonian

$$E = \int \psi_{\text{trial}}^* \hat{H} \psi_{\text{trial}} \, \mathrm{d}\tau$$

The lowest energy of the system (the energy of the ground state) is E_0, the eigenvalue of $\hat{H}$ corresponding to ψ_0. Consider the difference $E - E_0$, which can be expressed as follows

$$E - E_0 = \int \psi_{\text{trial}}^* \hat{H} \psi_{\text{trial}} \, \mathrm{d}\tau - E_0 \overbrace{\int \psi_{\text{trial}}^* \psi_{\text{trial}} \, \mathrm{d}\tau}^{1}$$

because the trial wavefunction is normalized. Bringing the constant E_0 into the second integral leads to

$$\begin{aligned} E - E_0 &= \int \psi_{\text{trial}}^* \hat{H} \psi_{\text{trial}} \, \mathrm{d}\tau - \int \psi_{\text{trial}}^* E_0 \psi_{\text{trial}} \, \mathrm{d}\tau \\ &= \int \boxed{\psi_{\text{trial}}^*} (\hat{H} - E_0) \boxed{\psi_{\text{trial}}} \, \mathrm{d}\tau \end{aligned}$$

Replacing the trial wavefunction with the linear combination of eigenfunctions of the hamiltonian gives

$$\begin{aligned} E - E_0 &= \int \left(\sum_n \boxed{c_n^* \psi_n^*}\right) (\hat{H} - E_0) \left(\sum_{n'} \boxed{c_{n'} \psi_{n'}}\right) \mathrm{d}\tau \\ &= \sum_{n,n'} c_n^* c_{n'} \int \psi_n^* (\hat{H} - E_0) \psi_{n'} \, \mathrm{d}\tau \end{aligned}$$

where the coefficients are gathered outside the integral and the two sums become a double sum over all possible values of n and n'.

Step 2 *Simplify the expression*

Because $\int \psi_n^* \hat{H}\psi_{n'}\mathrm{d}\tau = E_{n'}\int \psi_n^*\psi_{n'}\mathrm{d}\tau$ and $\int \psi_n^* E_0\psi_{n'}\mathrm{d}\tau = E_0\int \psi_n^*\psi_{n'}\mathrm{d}\tau$, write

$$\int \psi_n^*(\hat{H}-E_0)\psi_{n'}\mathrm{d}\tau = (E_{n'}-E_0)\int \psi_n^*\psi_{n'}\mathrm{d}\tau$$

It follows that

$$E - E_0 = \sum_{n,n'} c_n^* c_{n'}(E_{n'}-E_0)\overbrace{\int \psi_n^*\psi_{n'}\mathrm{d}\tau}^{1 \text{ if } n'=n,\ 0 \text{ otherwise}}$$

Step 3 *Analyse the final expression*

The eigenfunctions ψ_n are orthogonal, so only terms with $n'=n$ contribute to this sum. Because each eigenfunction is normalized, each surviving integral is 1. Consequently

$$E - E_0 = \sum_n \overbrace{c_n^* c_n}^{\geq 0}\overbrace{(E_n - E_0)}^{\geq 0} \geq 0$$

The quantity $c_n^* c_n$ is necessarily real and greater than or equal to zero, and because E_0 is the lowest energy, $E_n - E_0$ is also greater than or equal to zero. It follows that the product of the two terms on the right is greater than or equal to zero. Therefore, $E \geq E_0$, as asserted.

The variation principle is the basis of all modern molecular structure calculations. The principle implies that, if the coefficients in the trial wavefunction are varied until the lowest energy is achieved (by evaluating the expectation value of the hamiltonian for the wavefunction in each case), then those coefficients will be the best for that particular form of trial function. A lower energy might be obtained with a more complicated wavefunction, such as by taking a linear combination of several atomic orbitals on each atom. However, for a molecular orbital constructed from a given **basis set**, a given set of atomic orbitals, the variation principle gives the optimum molecular orbital of that kind.

9D.2(a) The procedure

The practical application of the variation principle can be illustrated by applying it to the trial wavefunction in eqn 9D.1, where the coefficients define the trial function.

How is that done? 9D.2 Applying the variation principle to a heteronuclear diatomic molecule

The trial wavefunction in eqn 9D.1, $\psi = c_A\psi_A + c_B\psi_B$, is real but not normalized because at this stage the coefficients can take arbitrary values. Because it is real, write $\psi^* = \psi$, and to normalize it multiply it by $N = 1/(\int \psi^*\psi\,\mathrm{d}\tau)^{1/2} = 1/(\int \psi^2\mathrm{d}\tau)^{1/2}$. From now on $N\psi$ is used as the trial function. Then follow these steps.

Step 1 *Write an expression for the energy*

The energy is the expectation value of the hamiltonian. Using the normalized real trial function $N\psi$ this expectation value is

$$E = N^2\int \psi\hat{H}\psi\,\mathrm{d}\tau = \frac{\int \psi\hat{H}\psi\,\mathrm{d}\tau}{\int \psi^2\mathrm{d}\tau} \qquad (9D.5)$$

The denominator is

$$\begin{aligned}\int \psi^2\mathrm{d}\tau &= \int (c_A\psi_A + c_B\psi_B)^2\mathrm{d}\tau \\ &= c_A^2\overbrace{\int \psi_A^2\mathrm{d}\tau}^{1} + c_B^2\overbrace{\int \psi_B^2\mathrm{d}\tau}^{1} + 2c_Ac_B\overbrace{\int \psi_A\psi_B\mathrm{d}\tau}^{S} \\ &= c_A^2 + c_B^2 + 2c_Ac_BS\end{aligned}$$

because the individual atomic orbitals are normalized to 1 and the third integral is the overlap integral S (eqn 9C.3, $S = \int \psi_A\psi_B\mathrm{d}\tau$). The numerator is

$$\begin{aligned}\int \psi\hat{H}\psi\,\mathrm{d}\tau &= \int (c_A\psi_A + c_B\psi_B)\hat{H}(c_A\psi_A + c_B\psi_B)\,\mathrm{d}\tau \\ &= c_A^2\overbrace{\int \psi_A\hat{H}\psi_A\mathrm{d}\tau}^{\alpha_A} + c_B^2\overbrace{\int \psi_B\hat{H}\psi_B\mathrm{d}\tau}^{\alpha_B} + c_Ac_B\overbrace{\int \psi_A\hat{H}\psi_B\mathrm{d}\tau}^{\beta} \\ &\quad + c_Ac_B\overbrace{\int \psi_B\hat{H}\psi_A\mathrm{d}\tau}^{\beta}\end{aligned}$$

The significance of the quantities α_A, α_B, and β (which are all energies) is discussed shortly.[1] The hamiltonian is hermitian (Topic 7C), therefore the third and fourth integrals are equal and

$$\int \psi\hat{H}\psi\,\mathrm{d}\tau = c_A^2\alpha_A + c_B^2\alpha_B + 2c_Ac_B\beta$$

At this point the complete expression for E is

$$E = \frac{c_A^2\alpha_A + c_B^2\alpha_B + 2c_Ac_B\beta}{c_A^2 + c_B^2 + 2c_Ac_BS}$$

Step 2 *Minimize the energy*

Now search for values of the coefficients in the trial function that minimize the value of E. The procedure is a standard problem in calculus, and is solved by finding the coefficients for which

$$\frac{\partial E}{\partial c_A} = 0 \qquad \frac{\partial E}{\partial c_B} = 0$$

The derivative is most easily calculated by first multiplying both sides of the expression for E by the denominator of the quotient to give

$$E(c_A^2 + c_B^2 + 2c_Ac_BS) = c_A^2\alpha_A + c_B^2\alpha_B + 2c_Ac_B\beta$$

[1] Note that here α and β are nothing to do with electron spin: here they symbolize two contributions to the energy.

The derivative with respect to c_A is then found, being careful to note that E is a function of c_A so the term on the left must be differentiated as a product. Therefore

$$\overbrace{\overbrace{\frac{\partial E}{\partial c_A}(c_A^2+c_B^2+2c_Ac_BS)}^{(\partial u/\partial x)v}+\overbrace{E(2c_A+2c_BS)}^{u(\partial v/\partial x)}}^{\partial(uv)/\partial x}=2c_A\alpha_A+2c_B\beta$$

Now bring the term $E(2c_A+2c_BS)$ to the right, divide both sides by $c_A^2+c_B^2+2c_Ac_BS$ and obtain

$$\frac{\partial E}{\partial c_A}=\frac{2c_A\alpha_A+2c_B\beta-E(2c_A+2c_BS)}{c_A^2+c_B^2+2c_Ac_BS}=\frac{2\{(\alpha_A-E)c_A+(\beta-SE)c_B\}}{c_A^2+c_B^2+2c_Ac_BS}$$

The expression for $\partial E/\partial c_B$ is found by exchanging the indices A and B. The two expressions for the derivatives are therefore

$$\frac{\partial E}{\partial c_A}=\frac{2\{\boxed{(\alpha_A-E)c_A+(\beta-SE)c_B}\}}{c_A^2+c_B^2+2c_Ac_BS}$$

$$\frac{\partial E}{\partial c_B}=\frac{2\{\boxed{(\alpha_B-E)c_B+(\beta-SE)c_A}\}}{c_A^2+c_B^2+2c_Ac_BS}$$

For the derivatives to be equal to 0, the numerators, and specifically the terms in boxes, of these expressions must vanish, leading to the **secular equations**:[2]

$$(\alpha_A-E)c_A+(\beta-SE)c_B=0 \qquad (9D.6a)$$
$$(\alpha_B-E)c_B+(\beta-SE)c_A=0 \qquad \text{Secular equations} \quad (9D.6b)$$

The quantities α_A, α_B, β, and S in the secular equations are

$$\alpha_A=\int\psi_A\hat{H}\psi_A\mathrm{d}\tau \quad \alpha_B=\int\psi_B\hat{H}\psi_B\mathrm{d}\tau \qquad \text{Coulomb integrals} \quad (9D.7a)$$

$$\beta=\int\psi_A\hat{H}\psi_B\mathrm{d}\tau=\int\psi_B\hat{H}\psi_A\mathrm{d}\tau \qquad \text{Resonance integral} \quad (9D.7b)$$

$$S=\int\psi_A\psi_B\mathrm{d}\tau \qquad \text{Overlap integral} \quad (9D.7c)$$

The parameter α is called a **Coulomb integral**. It is negative and can be interpreted as the energy of the electron when it occupies ψ_A (for α_A) or ψ_B (for α_B). In a homonuclear diatomic molecule, $\alpha_A=\alpha_B$. The parameter β is called a **resonance integral** (for classical reasons). It vanishes when the orbitals do not overlap, and at equilibrium bond lengths it is normally negative. The overlap integral S is discussed in Topic 9C.

In order to solve the secular equations for the coefficients c_A and c_B it is necessary to know the energy E and then use its value in eqn 9D.6. As for any set of simultaneous equations, the secular equations have a solution if the **secular determinant**, the determinant of the coefficients (*The chemist's toolkit* 9D.1), is zero. That is, if

$$\begin{vmatrix}\alpha_A-E & \beta-SE\\ \beta-SE & \alpha_B-E\end{vmatrix}=(\alpha_A-E)(\alpha_B-E)-(\beta-SE)^2$$
$$=(1-S^2)E^2+\{2\beta S-(\alpha_A+\alpha_B)\}E+(\alpha_A\alpha_B-\beta^2)=0 \qquad (9D.8)$$

This is a quadratic equation for E. A quadratic equation of the form $ax^2+bx+c=0$ has the solutions

$$x=\frac{-b\pm(b^2-4ac)^{1/2}}{2a}$$

In the present case, $a=1-S^2$, $b=2\beta S-(\alpha_A+\alpha_B)$, and $c=\alpha_A\alpha_B-\beta^2$, so the solutions (the energies) are

$$E_\pm=\frac{\alpha_A+\alpha_B-2\beta S\pm\{(2\beta S-(\alpha_A+\alpha_B))^2-4(1-S^2)(\alpha_A\alpha_B-\beta^2)\}^{1/2}}{2(1-S^2)} \qquad (9D.9a)$$

which, according to the variation principle, are the closest approximations to the true energy for a trial function of the form given in eqn 9D.1. They are the energies of the bonding and antibonding molecular orbitals formed from the two atomic orbitals.

Equation 9D.9a can be simplified. For a homonuclear diatomic, $\alpha_A=\alpha_B=\alpha$ and then

The chemist's toolkit 9D.1 Determinants

A 2 × 2 determinant is the entity

$$\begin{vmatrix}a & b\\ c & d\end{vmatrix}=ad-bc \qquad \text{2×2 Determinant}$$

A 3 × 3 determinant is evaluated by expanding it as a sum of 2 × 2 determinants:

$$\begin{vmatrix}a & b & c\\ d & e & f\\ g & h & i\end{vmatrix}=a\begin{vmatrix}e & f\\ h & i\end{vmatrix}-b\begin{vmatrix}d & f\\ g & i\end{vmatrix}+c\begin{vmatrix}d & e\\ g & h\end{vmatrix}$$
$$=a(ei-fh)-b(di-fg)+c(dh-eg) \qquad \text{3×3 Determinant}$$

Note the sign change in alternate columns (b occurs with a negative sign in the expansion). Determinants can be evaluated most easily by using mathematical software.

[2] The name 'secular' is derived from the Latin word for age or generation. The term comes from astronomy, where the same equations appear in connection with slowly accumulating modifications of planetary orbits.

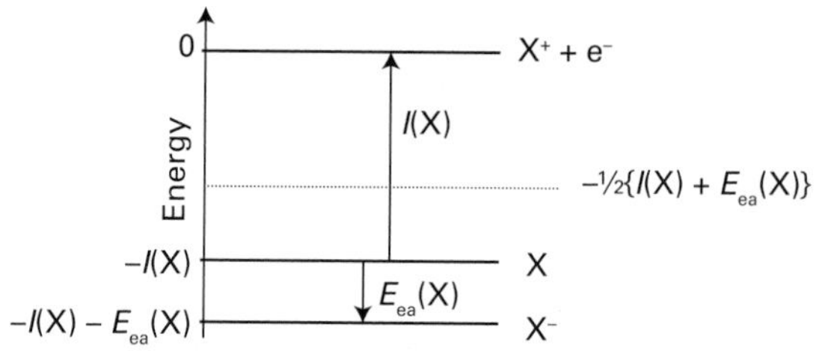

Figure 9D.2 The procedure for estimating the Coulomb integral in terms of the ionization energy and electron affinity.

$$E_{\pm} = \frac{2\alpha - 2\beta S \pm \{\overbrace{(2\beta S - 2\alpha)^2}^{(2\beta-2\alpha S)^2} - 4(1-S^2)(\alpha^2-\beta^2)\}^{1/2}}{2\underbrace{(1-S^2)}_{(1+S)(1-S)}}$$

$$= \frac{\alpha - \beta S \pm (\beta - \alpha S)}{(1+S)(1-S)} = \frac{(\alpha \pm \beta)(1 \mp S)}{(1+S)(1-S)}$$

That is,

$$E_+ = \frac{\alpha+\beta}{1+S} \quad E_- = \frac{\alpha-\beta}{1-S} \qquad \text{Homonuclear diatomics} \quad (9D.9b)$$

For $\beta < 0$, E_+ is the lower energy solution.

For heteronuclear diatomic molecules, making the approximation that $S = 0$ (simply to obtain a more transparent expression) turns eqn 9D.9a into

$$E_{\pm} = \tfrac{1}{2}(\alpha_A + \alpha_B) \pm \tfrac{1}{2}(\alpha_A - \alpha_B)\left\{1 + \left(\frac{2\beta}{\alpha_A - \alpha_B}\right)^2\right\}^{1/2} \qquad \text{Zero overlap approximation} \quad (9D.9c)$$

The values of the Coulomb integrals α_A and α_B may be estimated as follows. The extreme case of an atom X in a molecule is X^+ if it has lost control of the electron it supplied, X if it is sharing the electron pair equally with its bonded partner, and X^- if it has gained control of both electrons in the bond. If X^+ is taken as defining the energy 0, then X lies at $-I(X)$ and X^- lies at $-\{I(X) + E_{ea}(X)\}$, where I is the ionization energy and E_{ea} the electron affinity (Fig. 9D.2). The actual energy of the electron in the molecule lies at an intermediate value, and in the absence of further information, it is reasonable to estimate it as half-way down to the lower of these values, namely at $-\tfrac{1}{2}\{I(X) + E_{ea}(X)\}$. This quantity should be recognized (apart from its sign) as the Mulliken definition of electronegativity.

Brief illustration 9D.2

Consider HF. The general form of the molecular orbital is $\psi = c_H\psi_H + c_F\psi_F$, where ψ_H is an H1s orbital and ψ_F is an F2p$_z$ orbital (with z along the internuclear axis, the convention for linear molecules). The relevant data are as follows:

	I/eV	E_{ea}/eV	$-\tfrac{1}{2}\{I + E_{ea}\}$/eV
H	13.6	0.75	−7.2
F	17.4	3.34	−10.4

Therefore set $\alpha_A = \alpha_H = -7.2$ eV and $\alpha_B = \alpha_F = -10.4$ eV. With $\beta = -1.0$ eV, a typical value, and with $S = 0$ for simplicity, eqn 9D.9c becomes

$$E_{\pm}/\text{eV} = \tfrac{1}{2}(-7.2-10.4) \pm \tfrac{1}{2}(-7.2+10.4)\left\{1 + \left(\frac{-2.0}{-7.2+10.4}\right)^2\right\}^{1/2}$$
$$= -8.8 \pm 1.9 = -10.7 \text{ and } -6.9$$

These values, representing a bonding orbital at −10.7 eV and an antibonding orbital at −6.9 eV, are shown in Fig. 9D.3.

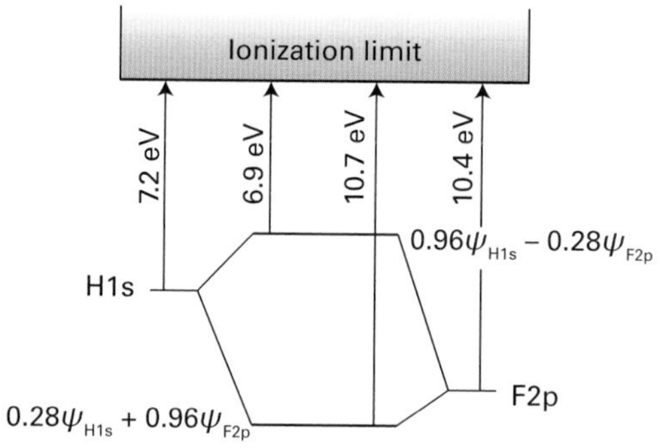

Figure 9D.3 The estimated energies of the Coulomb integrals α for H1s and F2p in HF and the resulting energies of the bonding and antibonding molecular orbitals.

9D.2(b) The features of the solutions

An important feature of eqn 9D.9c is that as the energy difference $|\alpha_A - \alpha_B|$ between the interacting atomic orbitals increases, the bonding and antibonding effects decrease (Fig. 9D.4).

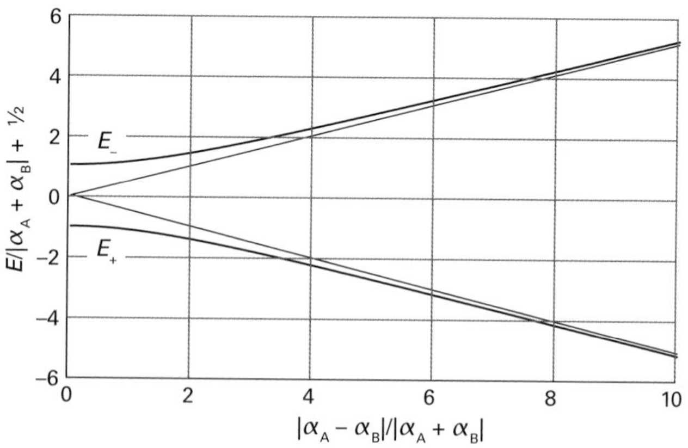

Figure 9D.4 The variation of the energies of the molecular orbitals (the thicker lines) as the energy difference of the contributing atomic orbitals is changed; the plots are for $\beta = -1$ (in the units adopted for E and α). The thinner lines are for the energies in the absence of mixing (i.e. $\beta = 0$) and correspond to energies α_A and α_B of the atomic orbitals.

When $|\alpha_B - \alpha_A| >> 2|\beta|$ it is possible to use the approximation $(1+x)^{1/2} \approx 1+\frac{1}{2}x$ (see the *Resource section*) to obtain

$$E_+ \approx \alpha_A + \frac{\beta^2}{\alpha_A - \alpha_B} \qquad E_- \approx \alpha_B - \frac{\beta^2}{\alpha_A - \alpha_B} \tag{9D.10}$$

As these expressions show, and as can be seen from the graph, when the energy difference $|\alpha_A - \alpha_B|$ is very large, the energies of the resulting molecular orbitals (the curved lines) differ only slightly from those of the atomic orbitals (the lines), which implies in turn that the bonding and antibonding effects are small. That is:

> The strongest bonding and antibonding effects are obtained when the two contributing orbitals have similar energies.

The large difference in energy between core and valence orbitals is the justification for neglecting the contribution of core orbitals to molecular orbitals constructed from valence atomic orbitals. Although the core orbitals of one atom have a similar energy to the core orbitals of the other atom, so might be expected to combine strongly, core–core interaction is largely negligible because the core orbitals are so contracted that the interaction between them, as measured by the value of $|\beta|$, is negligible. It is also a justification for treating the s and p_z contributions to σ orbital formation separately, an approximation used in Topic 9C in the discussion of homonuclear diatomic molecules.

The values of the coefficients in the linear combination in eqn 9D.1 are obtained by solving the secular equations after substituting the two energies obtained from the secular determinant. The lower energy, E_+, gives the coefficients for the bonding molecular orbital, the upper energy, E_-, the coefficients for the antibonding molecular orbital. The secular equations give expressions for the ratio of the coefficients. Thus, the first of the two secular equations in eqn 9D.6a, $(\alpha_A - E)c_A + (\beta - ES)c_B = 0$, gives

$$c_B = -\left(\frac{\alpha_A - E}{\beta - ES}\right)c_A \tag{9D.11}$$

The wavefunction should also be normalized. It has already been shown that $\int \psi^2 d\tau = c_A^2 + c_B^2 + 2c_Ac_BS$, so normalization requires that

$$c_A^2 + c_B^2 + 2c_Ac_BS = 1 \tag{9D.12}$$

When eqn 9D.11 is substituted into this expression, the result is

$$c_A = \frac{1}{\left\{1+\left(\frac{\alpha_A - E}{\beta - ES}\right)^2 - 2S\left(\frac{\alpha_A - E}{\beta - ES}\right)\right\}^{1/2}} \tag{9D.13}$$

which, together with eqn 9D.11, gives explicit expressions for the coefficients once the appropriate values of $E = E_\pm$ given in eqn 9D.9a are substituted.

As before, this expression becomes more transparent in two cases. First, for a homonuclear diatomic, with $\alpha_A = \alpha_B = \alpha$ and $E_\pm$ given in eqn 9D.9b, the results are

$$E_+ = \frac{\alpha+\beta}{1+S} \quad c_A = \frac{1}{\{2(1+S)\}^{1/2}} \quad c_B = c_A \qquad \text{Homonuclear diatomics} \tag{9D.14a}$$

$$E_- = \frac{\alpha-\beta}{1-S} \quad c_A = \frac{1}{\{2(1-S)\}^{1/2}} \quad c_B = -c_A \qquad \text{Homonuclear diatomics} \tag{9D.14b}$$

These values are the same as those asserted in Topic 9B on the basis of symmetry considerations.

For a heteronuclear diatomic with $S = 0$,

$$c_A = \frac{1}{\left\{1+\left(\frac{\alpha_A - E}{\beta}\right)^2\right\}^{1/2}} \qquad \text{Zero overlap approximation} \tag{9D.15}$$

with the appropriate values of $E = E_\pm$ taken from eqn 9D.9c. The coefficient c_B is then calculated from eqn 9D.11 (with $S = 0$).

Brief illustration 9D.3

Consider HF again. In *Brief illustration* 9D.2, with $\alpha_H = -7.2$ eV, $\alpha_F = -10.4$ eV, $\beta = -1.0$ eV, and $S = 0$, the two orbital energies were found to be $E_+ = -10.7$ eV and $E_- = -6.9$ eV. When these values are substituted into eqn 9D.15 the following coefficients are found:

$$E_+ = -10.7\ \text{eV} \quad \psi_+ = 0.28\psi_H + 0.96\psi_F$$
$$E_- = -6.9\ \text{eV} \quad \psi_- = 0.96\psi_H - 0.28\psi_F$$

Notice how the lower energy orbital (the one with energy −10.7 eV) has a composition that is more F2p orbital than H1s, and that the opposite is true of the higher energy, antibonding orbital.

Exercises

E9D.2 Using the data below estimate the orbital energies to use in a calculation of the molecular orbitals of HCl.

	I/(kJ mol^{-1})	E_{ea}/(kJ mol^{-1})
H	1312.0	72.8
Cl	1251.1	348.7

E9D.3 Use the values derived in E9D.2 to estimate the molecular orbital energies in HCl using (i) $S = 0$ and (ii) $S = 0.20$. Take $\beta = -1.0$ eV.

Checklist of concepts

- ☐ 1. A **polar bond** can be regarded as arising from a molecular orbital that is concentrated more on one atom than its partner.
- ☐ 2. The **electronegativity** of an element is a measure of the power of an atom to attract electrons to itself in a bond.
- ☐ 3. The electron pair in a bonding orbital is more likely to be found on the more electronegative atom; the opposite is true for electrons in an antibonding orbital.
- ☐ 4. The **variation principle** provides a criterion for optimizing a trial wavefunction.
- ☐ 5. A **basis set** is the set of atomic orbitals from which the molecular orbitals are constructed.
- ☐ 6. The bonding and antibonding effects are strongest when contributing atomic orbitals have similar energies.

Checklist of equations

Property	Equation	Comment	Equation number
Molecular orbital	$\psi = c_A\psi_A + c_B\psi_B$		9.D1
Pauling electronegativity	$\lvert\chi_A - \chi_B\rvert = \{hc\tilde{D}_0(\mathrm{AB})/\mathrm{eV} - \frac{1}{2}[hc\tilde{D}_0(\mathrm{AA})/\mathrm{eV} + hc\tilde{D}_0(\mathrm{BB})/\mathrm{eV}]\}^{1/2}$		9.D2
Mulliken electronegativity	$\chi = \frac{1}{2}(I + E_{ea})/\mathrm{eV}$		9.D3
Coulomb integral	$\alpha_A = \int \psi_A \hat{H} \psi_A \mathrm{d}\tau$	Definition	9D.7a
Resonance integral	$\beta = \int \psi_A \hat{H} \psi_B \mathrm{d}\tau$	Definition	9D.7b

TOPIC 9E Molecular orbital theory: polyatomic molecules

➤ **Why do you need to know this material?**

Molecular orbital theory is readily extendable to molecules considerably more complex than diatomic molecules. This material gives insight into how more sophisticated calculations work.

➤ **What is the key idea?**

Molecular orbitals can be expressed as linear combinations of all the atomic orbitals of the appropriate symmetry.

➤ **What do you need to know already?**

This Topic extends the approach used for heteronuclear diatomic molecules in Topic 9D, particularly the concepts of secular equations and secular determinants. The principal mathematical technique used is matrix algebra (*The chemist's toolkit* 9E.1). You should become familiar with the use of mathematical software to manipulate matrices numerically.

The principal difference between diatomic and polyatomic molecules lies in the greater range of shapes that are possible: a diatomic molecule is necessarily linear, but a triatomic molecule, for instance, may be either linear or angular (bent) with a characteristic bond angle. The shape of a polyatomic molecule—the specification of its bond lengths and its bond angles—can be predicted by calculating the total energy of the molecule for a variety of nuclear positions, and then identifying the conformation that corresponds to the lowest energy.

The molecular orbitals of polyatomic molecules are built in much the same way as in diatomic molecules (Topic 9D), the only difference being that more atomic orbitals are used to construct them. As for diatomic molecules, polyatomic molecular orbitals spread over the entire molecule and the electrons that occupy them spread like a web over the atoms to bind them all together. A molecular orbital has the general form

$$\psi = \sum_i c_i \psi_i \qquad \text{General form of LCAO-MO} \qquad (9E.1)$$

where ψ_i is an atomic orbital and the sum extends over all the valence orbitals of all the atoms in the molecule. As will be explained in detail as this Topic unfolds, the coefficients are found by setting up the secular equations, just as for diatomic molecules, and then solving them for the energies (Topic 9D). That step involves formulating the secular determinant and finding the values of the energy that ensure the determinant is equal to 0. Finally, these energies are used in the secular equations to find the coefficients of the atomic orbitals for each molecular orbital. Such calculations are best done using software, which is now widely available.

Symmetry considerations play a central role in the construction of molecular orbitals of polyatomic molecules, because only atomic orbitals of matching symmetry have non-zero overlap and contribute to a molecular orbital. To discuss these symmetry requirements fully requires the machinery developed in Focus 10, especially Topic 10C. There is one type of symmetry, however, that is intuitive: the planarity of conjugated hydrocarbons. That symmetry provides a distinction between the σ and π orbitals of the molecule, and in elementary approaches such molecules are commonly discussed in terms of the characteristics of their π orbitals, with the σ bonds providing a rigid framework that determines the general shape of the molecule.

9E.1 The Hückel approximation

The π molecular orbital energy level diagrams of conjugated molecules can be constructed by using a set of approximations suggested by Erich Hückel in 1931, and provide a simple introduction to this very technical subject. The method is introduced here by considering the π orbitals of ethene.

9E.1(a) An introduction to the method

The π orbitals are expressed as linear combinations of the C2p orbitals that lie perpendicular to the molecular plane. In ethene there are just two such orbitals, and so

$$\psi = c_A\psi_A + c_B\psi_B \qquad (9E.2)$$

where ψ_A and ψ_B are C2p orbitals on atoms A and B. Next, the optimum coefficients and energies are found by the variation principle as explained in Topic 9D. That is, the appropriate secular determinant is set up, equated to 0, and the equation solved for the energies. For ethene, the starting point is the pair of secular equations developed in Topic 9D:

$$\begin{aligned}(\alpha_A - E)c_A + (\beta - ES)c_B &= 0\\ (\beta - ES)c_A + (\alpha_B - E)c_B &= 0\end{aligned} \qquad (9E.3a)$$

The corresponding secular determinant is

$$\begin{vmatrix} \alpha_A - E & \beta - ES \\ \beta - ES & \alpha_B - E \end{vmatrix} \quad (9E.3b)$$

The determinant is much simpler to handle if the **Hückel approximations** are made. These approximations are

- All overlap integrals are set equal to zero.
- All resonance integrals between non-neighbours are set equal to zero.
- All Coulomb integrals for carbon atoms are set equal (to α).
- All remaining resonance integrals are set equal (to β).

Both α and β are negative. These approximations are obviously very severe and so the resulting molecular orbitals are only approximate. Nevertheless, the method is useful in that it gives, without too much computation, a general picture of the molecular orbitals and their energies.

When the Hückel approximations are used the secular determinant has the following structure:

- All diagonal elements: $\alpha - E$
- Off-diagonal elements between neighbouring atoms: β
- All other elements: 0

In the case of ethene the secular determinant (eqn 9E.3b) becomes

$$\begin{vmatrix} \alpha - E & \beta \\ \beta & \alpha - E \end{vmatrix}$$

The energies are found by setting the determinant equal to zero. The determinant is then expanded as explained in *The chemist's toolkit* 9D.1 to give

$$\begin{vmatrix} \alpha - E & \beta \\ \beta & \alpha - E \end{vmatrix} = (\alpha - E)^2 - \beta^2$$
$$= (\alpha - E + \beta)(\alpha - E - \beta) = 0 \quad (9E.4)$$

The second line follows from $a^2 - b^2 = (a-b)(a+b)$. The roots of the equation are $E = \alpha \pm \beta$. The + sign corresponds to the bonding combination (β is negative) and the − sign corresponds to the antibonding combination (Fig. 9E.1).

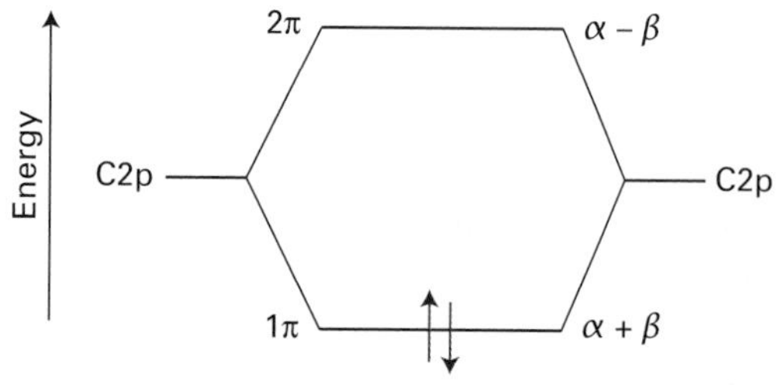

Figure 9E.1 The Hückel molecular orbital energy levels of ethene. Two electrons occupy the lower π orbital.

The building-up principle results in the configuration $1\pi^2$, because each carbon atom supplies one electron to the π system and both electrons can occupy the bonding orbital. The **highest occupied molecular orbital** in ethene, its HOMO, is the 1π orbital. The **lowest unoccupied molecular orbital**, its LUMO, is the 2π orbital. These two orbitals jointly form the **frontier orbitals** of the molecule. The frontier orbitals are important because they are largely responsible for many of the chemical and spectroscopic properties of this and analogous molecules.

Brief illustration 9E.1

The energy needed to excite a $\pi^* \leftarrow \pi$ transition is equal to the separation of the 1π and 2π orbitals, which in the Hückel framework is $2|\beta|$. This transition is known to occur at close to 40 000 cm^{-1}, corresponding to 5.0 eV. It follows that a plausible value of β is about −2.5 eV (−240 kJ mol^{-1}).

9E.1(b) The matrix formulation of the method

To make the Hückel theory readily applicable to bigger molecules, it is helpful to reformulate the secular equations in terms of matrices (see *The chemist's toolkit* 9E.1 for a short introduction to matrices; for more detail refer to the extended version of this toolkit available via the e-book).

To generalize the secular equations, eqn 9E.3a, the first step is to write $\alpha_J = H_{JJ}$ (with J = A or B), $\beta = H_{AB} = H_{BA}$, and then to invoke the Hückel approximation that $S = 0$:

$$(H_{AA} - E)c_A + H_{AB}c_B = 0$$
$$H_{BA}c_A + (H_{BB} - E)c_B = 0$$

The resonance integral β in the second equation has been written H_{BA} in order to show the symmetry of the two equations on interchanging A and B.

The next step is to note that these two simultaneous equations have two solutions. That is, they are satisfied when E has either of the two values E_1 and E_2, and for each energy there is a different wavefunction with its own set of coefficients. The coefficients in the wavefunction with energy E_1 are written $c_{1,A}$ and $c_{1,B}$, and those in the wavefunction with energy E_2 are $c_{2,A}$ and $c_{2,B}$. For the first solution the equations are therefore

$$(H_{AA} - E_1)c_{1,A} + H_{AB}c_{1,B} = 0$$
$$H_{BA}c_{1,A} + (H_{BB} - E_1)c_{1,B} = 0 \quad (9E.5a)$$

and for the second they are

$$(H_{AA} - E_2)c_{2,A} + H_{AB}c_{2,B} = 0$$
$$H_{BA}c_{2,A} + (H_{BB} - E_2)c_{2,B} = 0 \quad (9E.5b)$$

When the terms in E_1 are taken to the right in eqn 9E.5a the result is

$$H_{AA}c_{1,A} + H_{AB}c_{1,B} = E_1c_{1,A}$$
$$H_{BA}c_{1,A} + H_{BB}c_{1,B} = E_1c_{1,B}$$

These two equations can be written in matrix form as

$$\begin{pmatrix} H_{AA} & H_{AB} \\ H_{BA} & H_{BB} \end{pmatrix}\begin{pmatrix} c_{1,A} \\ c_{1,B} \end{pmatrix} = E_1\begin{pmatrix} c_{1,A} \\ c_{1,B} \end{pmatrix} \quad \text{(9E.6a)}$$

That eqn 9E.6a is true can be verified by multiplying the column vector by the matrix, as shown in *The chemist's toolkit* 9E.1. A similar matrix equation arises from eqn 9E.5b:

$$\begin{pmatrix} H_{AA} & H_{AB} \\ H_{BA} & H_{BB} \end{pmatrix}\begin{pmatrix} c_{2,A} \\ c_{2,B} \end{pmatrix} = E_2\begin{pmatrix} c_{2,A} \\ c_{2,B} \end{pmatrix} \quad \text{(9E.6b)}$$

Equations 9E.6a and 9E.6b can be expressed as a single equation by combining both sets of coefficients into a single matrix, with each pair of coefficients forming a column in the new matrix, and writing the energies as the elements of a diagonal matrix

$$\begin{pmatrix} H_{AA} & H_{AB} \\ H_{BA} & H_{BB} \end{pmatrix}\begin{pmatrix} c_{1,A} & c_{2,A} \\ c_{1,B} & c_{2,B} \end{pmatrix} = \begin{pmatrix} c_{1,A} & c_{2,A} \\ c_{1,B} & c_{2,B} \end{pmatrix}\begin{pmatrix} E_1 & 0 \\ 0 & E_2 \end{pmatrix} \quad \text{(9E.7)}$$

That eqn 9E.7 is true can be verified by multiplying the matrices on the left and right, and then comparing them element by element on the left and right. Multiplication gives

$$\begin{pmatrix} H_{AA}c_{1,A} + H_{AB}c_{1,B} & H_{AA}c_{2,A} + H_{AB}c_{2,B} \\ H_{BA}c_{1,A} + H_{BB}c_{1,B} & H_{BA}c_{2,A} + H_{BB}c_{2,B} \end{pmatrix} = \begin{pmatrix} E_1c_{1,A} & E_2c_{2,A} \\ E_1c_{1,B} & E_2c_{2,B} \end{pmatrix}$$

Then, for example, comparison of the top left-hand element on each side gives

$$H_{AA}c_{1,A} + H_{AB}c_{1,B} = E_1c_{1,A}$$

which is one of the original equations in eqn 9E.5a. Similarly comparison of the top right-hand element on each side gives

$$H_{AA}c_{2,A} + H_{AB}c_{2,B} = E_2c_{2,A}$$

which is one of the equations in eqn 9E.5b.

The matrix equation 9E.7 can be written

$$\boldsymbol{Hc} = \boldsymbol{cE} \quad \text{(9E.8)}$$

where

$$\boldsymbol{H} = \begin{pmatrix} H_{AA} & H_{AB} \\ H_{BA} & H_{BB} \end{pmatrix} \quad \boldsymbol{c} = \begin{pmatrix} c_{1,A} & c_{2,A} \\ c_{1,B} & c_{2,B} \end{pmatrix} \quad \boldsymbol{E} = \begin{pmatrix} E_1 & 0 \\ 0 & E_2 \end{pmatrix}$$

The first matrix, $\boldsymbol{H}$, is called the **hamiltonian matrix**. The final step is to multiply both sides of eqn 9E.8 from the left by $\boldsymbol{c}^{-1}$, the inverse of the matrix $\boldsymbol{c}$, to give

$$\boldsymbol{c}^{-1}\boldsymbol{Hc} = \overbrace{\boldsymbol{c}^{-1}\boldsymbol{c}}^{1}\boldsymbol{E}$$

By definition $\boldsymbol{c}^{-1}\boldsymbol{c} = \boldsymbol{1}$, and multiplication by the unit matrix $\boldsymbol{1}$ has no effect, Therefore

$$\boldsymbol{c}^{-1}\boldsymbol{Hc} = \boldsymbol{E} \quad \text{(9E.10)}$$

The matrix $\boldsymbol{E}$ is diagonal, with diagonal elements E_1 and E_2. The left-hand side must therefore also be a diagonal matrix. The interpretation of this equation is that the energies are calculated

The chemist's toolkit 9E.1 Matrices

A **matrix** is an array of numbers, as in the following two examples of 2 × 2 matrices:

$$\boldsymbol{M} = \begin{pmatrix} M_{11} & M_{12} \\ M_{21} & M_{22} \end{pmatrix} \quad \boldsymbol{N} = \begin{pmatrix} N_{11} & N_{12} \\ N_{21} & N_{22} \end{pmatrix}$$

A general element is denoted $M_{r[\text{ow}]c[\text{olumn}]}$. In a **diagonal matrix** the only non-zero elements are along the diagonal, that is elements for which the indices r and c are the same. A **unit matrix**, $\boldsymbol{1}$, is a diagonal matrix in which the elements along the diagonal are all 1. For example, a 2 × 2 unit matrix is

$$\boldsymbol{1} = \begin{pmatrix} 1 & 0 \\ 0 & 1 \end{pmatrix}$$

1 *Matrix addition and subtraction.* Add or subtract corresponding elements:

$$\boldsymbol{M} + \boldsymbol{N} = \begin{pmatrix} M_{11} + N_{11} & M_{12} + N_{12} \\ M_{21} + N_{21} & M_{22} + N_{22} \end{pmatrix}$$

$$\boldsymbol{M} - \boldsymbol{N} = \begin{pmatrix} M_{11} - N_{11} & M_{12} - N_{12} \\ M_{21} - N_{21} & M_{22} - N_{22} \end{pmatrix}$$

2 *Matrix multiplication:*

$$\boldsymbol{MN} = \begin{pmatrix} M_{11} & M_{12} \\ M_{21} & M_{22} \end{pmatrix}\begin{pmatrix} N_{11} & N_{12} \\ N_{21} & N_{22} \end{pmatrix}$$
$$= \begin{pmatrix} M_{11}N_{11} + M_{12}N_{21} & M_{11}N_{12} + M_{12}N_{22} \\ M_{21}N_{11} + M_{22}N_{21} & M_{21}N_{12} + M_{22}N_{22} \end{pmatrix}$$

Note that $\boldsymbol{MN}$ might differ from $\boldsymbol{NM}$. If $\boldsymbol{N}$ is a matrix with 2 rows and 1 column, a 'column vector', simply ignore all but the first column.

3 *Matrix division*

Just as x/y can be expressed as $y^{-1}x$, so matrix division is multiplication by the inverse of the matrix, the matrix $\boldsymbol{M}^{-1}$ such that $\boldsymbol{M}^{-1}\boldsymbol{M} = \boldsymbol{MM}^{-1} = \boldsymbol{1}$. Inverse matrices are best found by using mathematical software.

For more detail, see the extended version of this toolkit via the e-book.

by finding a matrix $\boldsymbol{c}$ which, by constructing the product $\boldsymbol{c}^{-1}\boldsymbol{Hc}$, results in a diagonal matrix. This procedure is called **matrix diagonalization** and is best carried out by using mathematical software (see the extended version of *The chemist's toolkit* 9E.1 available via the e-book). Once the software finds the matrix $\boldsymbol{c}$, the definitions following eqn 9E.8 show that its columns are the coefficients of the orbitals used as the basis set. Then, as implied by eqn 9E.10, the elements along the diagonal of the diagonalized matrix are the energies of the orbitals.

It follows that the molecular orbitals can be found as follows:

1. Construct the matrix $\boldsymbol{H}$
2. Find the matrix $\boldsymbol{c}$ that diagonalizes $\boldsymbol{H}$
3. The columns of $\boldsymbol{c}$ are the coefficients of the molecular orbitals
4. Their energies are given by the corresponding diagonal elements of the matrix $\boldsymbol{c}^{-1}\boldsymbol{Hc}$.

Although these equations have been developed for a basis of just two orbitals they apply equally well to larger basis sets. The matrix $\boldsymbol{H}$ is simply expanded accordingly, as is done in the following example.

Example 9E.1 Finding molecular orbitals by matrix diagonalization

Set up and solve the matrix equations within the Hückel approximation for the π orbitals of butadiene (**1**).

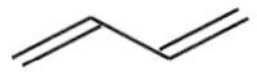

1 Butadiene

Collect your thoughts The matrices are four-dimensional (that is, 4 × 4) as each carbon atom in the π system contributes one out-of-plane p orbital. You need to follow the four steps set out in the text. Mathematical software operates with numerical matrix elements, so write $\boldsymbol{H}$, which is expressed in terms of the parameters α and β, as the sum of two matrices, one multiplied by α and the other multiplied by β. Then diagonalize the numerical matrix by using software.

The solution With the atoms labelled A, B, C, and D, the hamiltonian matrix $\boldsymbol{H}$ is

$$\boldsymbol{H}=\begin{pmatrix} H_{AA} & H_{AB} & H_{AC} & H_{AD} \\ H_{BA} & H_{BB} & H_{BC} & H_{BD} \\ H_{CA} & H_{CB} & H_{CC} & H_{CD} \\ H_{DA} & H_{DB} & H_{DC} & H_{DD} \end{pmatrix} \xrightarrow{\text{Hückel approximation}} \begin{pmatrix} \alpha & \beta & 0 & 0 \\ \beta & \alpha & \beta & 0 \\ 0 & \beta & \alpha & \beta \\ 0 & 0 & \beta & \alpha \end{pmatrix}$$

$$=\alpha\underbrace{\begin{pmatrix} 1 & 0 & 0 & 0 \\ 0 & 1 & 0 & 0 \\ 0 & 0 & 1 & 0 \\ 0 & 0 & 0 & 1 \end{pmatrix}}_{\boldsymbol{1}}+\beta\underbrace{\begin{pmatrix} 0 & 1 & 0 & 0 \\ 1 & 0 & 1 & 0 \\ 0 & 1 & 0 & 1 \\ 0 & 0 & 1 & 0 \end{pmatrix}}_{\boldsymbol{M}}=\alpha\boldsymbol{1}+\beta\boldsymbol{M}$$

The unit matrix is already diagonal. The matrix $\boldsymbol{c}$ that diagonalizes $\boldsymbol{M}$ is given by mathematical software as

$$\boldsymbol{c}=\begin{pmatrix} 0.372 & 0.602 & 0.602 & 0.372 \\ 0.602 & 0.372 & -0.372 & -0.602 \\ 0.602 & -0.372 & -0.372 & 0.602 \\ 0.372 & -0.602 & 0.602 & -0.372 \end{pmatrix}$$

and the diagonalized version of $\boldsymbol{H}$ is

$$\boldsymbol{E}=\begin{pmatrix} \alpha+1.62\beta & 0 & 0 & 0 \\ 0 & \alpha+0.62\beta & 0 & 0 \\ 0 & 0 & \alpha-0.62\beta & 0 \\ 0 & 0 & 0 & \alpha-1.62\beta \end{pmatrix}$$

The columns of the matrix $\boldsymbol{c}$ are the coefficients of the atomic orbitals for the corresponding molecular orbital. It follows that the energies and molecular orbitals are

$$\begin{aligned} E_1 &= \alpha+1.62\beta & \psi_1 &= 0.372\psi_A+0.602\psi_B+0.602\psi_C+0.372\psi_D \\ E_2 &= \alpha+0.62\beta & \psi_2 &= 0.602\psi_A+0.372\psi_B-0.372\psi_C-0.602\psi_D \\ E_3 &= \alpha-0.62\beta & \psi_3 &= 0.602\psi_A-0.372\psi_B-0.372\psi_C+0.602\psi_D \\ E_4 &= \alpha-1.62\beta & \psi_4 &= 0.372\psi_A-0.602\psi_B+0.602\psi_C-0.372\psi_D \end{aligned}$$

where the C2p atomic orbitals are denoted by $\psi_A,\ldots,\psi_D$. The molecular orbitals are mutually orthogonal and, with overlap neglected, normalized.

Comment. Note that $\psi_1,\ldots,\psi_4$ correspond to the 1π,…, 4π molecular orbitals of butadiene.

Self-test 9E.1 Repeat the exercise for the allyl radical, $\cdot CH_2–CH=CH_2$; assume that each carbon atom is sp^2 hybridized, and take as a basis one out-of-plane 2p orbital on each atom.

Answer: $E=\alpha+1.41\beta$, α, $\alpha-1.41\beta$; $\psi_1=0.500\psi_A+0.707\psi_B+0.500\psi_C$, $\psi_2=0.707\psi_A-0.707\psi_C$, $\psi_3=0.500\psi_A-0.707\psi_B+0.500\psi_C$

Exercise

E9E.1 Set up the secular determinants for (i) linear H_3, (ii) cyclic H_3 within the Hückel approximation.

9E.2 Applications

Although the Hückel method is very primitive, it can be used to account for some of the properties of conjugated polyenes.

9E.2(a) π-Electron binding energy

As seen in *Example* 9E.1, the energies of the four LCAO-MOs for butadiene are

$$E=\alpha\pm1.62\beta,\ \alpha\pm0.62\beta \tag{9E.11}$$

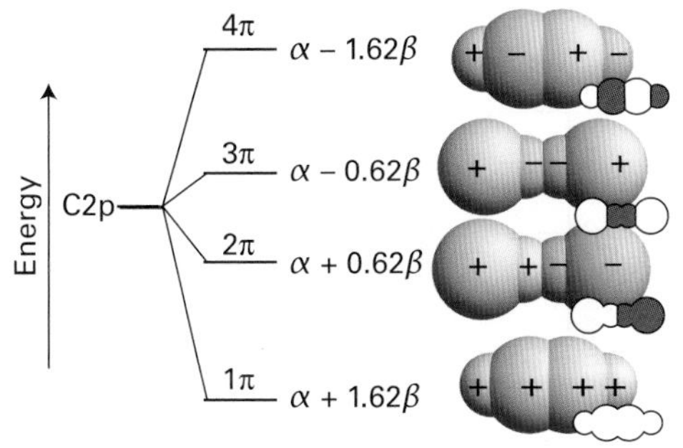

Figure 9E.2 The Hückel molecular orbital energy levels of butadiene and the top view of the corresponding π orbitals. The four p electrons (one supplied by each C) occupy the two lower π orbitals. Note that all the orbitals are delocalized.

These orbitals and their energies are drawn in Fig. 9E.2. Note that:

- The greater the number of internuclear nodes, the higher the energy of the orbital.
- There are four electrons to accommodate, so the ground-state configuration is $1\pi^2 2\pi^2$.
- The frontier orbitals of butadiene are the 2π orbital (the HOMO, which is largely bonding) and the 3π orbital (the LUMO, which is largely antibonding).

'Largely bonding' means that an orbital has both bonding and antibonding interactions between various neighbours, but the bonding effects dominate. 'Largely antibonding' indicates that the antibonding effects dominate.

An important point emerges by calculating the total **π-electron binding energy**, E_π, the sum of the energies of each π electron, and comparing it with the energy of a hypothetical molecule in which the π bonds are localized. For example, in butadiene π-electron binding energy is

$$E_\pi(\text{butadiene}) = 2(\alpha + 1.62\beta) + 2(\alpha + 0.62\beta) = 4\alpha + 4.48\beta$$

The π-electron binding energy of a localized π bond is simply that of the π bond in ethene. The two electrons are in the 1π orbital with energy $\alpha + \beta$, so the π-electron binding energy in ethene is

$$E_\pi(\text{ethene}) = 2(\alpha + \beta) = 2\alpha + 2\beta$$

The difference in π-electron binding energy between butadiene and a molecule with two localized π bonds is

$$E_\pi(\text{butadiene}) - 2 \times E_\pi(\text{ethene})$$
$$= (4\alpha + 4.48\beta) - 2(2\alpha + 2\beta) = 0.48\beta$$

Therefore, the energy of the delocalized butadiene molecule lies lower by 0.48β (about 115 kJ mol^{-1}) than of the hypothetical molecule with localized bonding (recall that β is negative). This extra stabilization of a conjugated system compared with a set of localized π bonds is called the **delocalization energy** of the molecule.

A closely related quantity is the **π-bond formation energy**, E_{bf}, the energy released when a π bond is formed. Because the contribution of α is the same in the molecule as in the atoms, the π-bond formation energy can be calculated from the π-electron binding energy by writing

$$E_{\text{bf}} = E_\pi - N_C\alpha \qquad \text{π-Bond formation energy [definition]} \qquad (9E.12)$$

where N_C is the number of carbon atoms in the molecule. The π-bond formation energy in butadiene, for instance, is 4.48β.

Example 9E.2 Estimating the delocalization energy

Use the Hückel approximation to find the energies of the π orbitals of cyclobutadiene, and estimate the delocalization energy.

Collect your thoughts Set up the hamiltonian matrix using the same basis as for butadiene, but note that atoms A and D are also now neighbours. Then diagonalize the matrix to find the energies. For the delocalization energy, subtract from the total π-bond energy the energy of two π-bonds.

The solution The hamiltonian matrix is

$$\boldsymbol{H} = \begin{pmatrix} \alpha & \beta & 0 & \beta \\ \beta & \alpha & \beta & 0 \\ 0 & \beta & \alpha & \beta \\ \beta & 0 & \beta & \alpha \end{pmatrix} = \alpha\mathbf{1} + \beta\overbrace{\begin{pmatrix} 0 & 1 & 0 & 1 \\ 1 & 0 & 1 & 0 \\ 0 & 1 & 0 & 1 \\ 1 & 0 & 1 & 0 \end{pmatrix}}^{M} = \alpha\mathbf{1} + \beta\boldsymbol{M}$$

Use mathematical software to diagonalize $\boldsymbol{M}$ by forming $\boldsymbol{c}^{-1}\boldsymbol{M}\boldsymbol{c}$:

$$\boldsymbol{c}^{-1}\boldsymbol{M}\boldsymbol{c} = \begin{pmatrix} 2 & 0 & 0 & 0 \\ 0 & 0 & 0 & 0 \\ 0 & 0 & 0 & 0 \\ 0 & 0 & 0 & -2 \end{pmatrix}$$

The diagonalized form of $\boldsymbol{H}$ (which is identified as $\boldsymbol{E}$) is therefore

$$\boldsymbol{E} = \begin{pmatrix} \alpha + 2\beta & 0 & 0 & 0 \\ 0 & \alpha & 0 & 0 \\ 0 & 0 & \alpha & 0 \\ 0 & 0 & 0 & \alpha - 2\beta \end{pmatrix}$$

and the energies of the orbitals are $\alpha + 2\beta, \alpha, \alpha, \alpha - 2\beta$.

Four electrons must be accommodated. Two occupy the lowest orbital (of energy $\alpha + 2\beta$), and two occupy the doubly degenerate orbitals (of energy α). The total energy is therefore $4\alpha + 4\beta$. Two isolated π bonds would have an energy $4\alpha + 4\beta$; therefore, in this case, the delocalization energy is zero.

Self-test 9E.2 Repeat the calculation for benzene (use software!).

Answer: See next subsection

9E.2(b) Aromatic stability

The most notable example of delocalization conferring extra stability is benzene and the aromatic molecules based on its structure. In elementary accounts, the structure of benzene, and other aromatic compounds, is often expressed in a mixture of valence-bond and molecular orbital terms, with typically valence-bond language (Topic 9A) used for its σ framework and molecular orbital language used to describe its π electrons.

First, the valence-bond component. The six C atoms are regarded as sp^2 hybridized, with a single unhybridized perpendicular 2p orbital. One H atom is bonded by (Csp^2,H1s) overlap to each C carbon, and the remaining hybrids overlap to give a regular hexagon of atoms (Fig. 9E.3). The internal angle of a regular hexagon is 120°, so sp^2 hybridization is ideally suited for forming σ bonds. The hexagonal shape of benzene permits strain-free σ bonding.

Now consider the molecular orbital component of the description. The six C2p orbitals overlap to give six π orbitals that spread all round the ring. Their energies are calculated within the Hückel approximation by diagonalizing the hamiltonian matrix

$$H=\begin{pmatrix}\alpha & \beta & 0 & 0 & 0 & \beta\\ \beta & \alpha & \beta & 0 & 0 & 0\\ 0 & \beta & \alpha & \beta & 0 & 0\\ 0 & 0 & \beta & \alpha & \beta & 0\\ 0 & 0 & 0 & \beta & \alpha & \beta\\ \beta & 0 & 0 & 0 & \beta & \alpha\end{pmatrix}$$

$$=\alpha\mathbf{1}+\beta\begin{pmatrix}0&1&0&0&0&1\\1&0&1&0&0&0\\0&1&0&1&0&0\\0&0&1&0&1&0\\0&0&0&1&0&1\\1&0&0&0&1&0\end{pmatrix}\xrightarrow{\text{Diagonalize}}\begin{pmatrix}2&0&0&0&0&0\\0&1&0&0&0&0\\0&0&1&0&0&0\\0&0&0&-1&0&0\\0&0&0&0&-1&0\\0&0&0&0&0&-2\end{pmatrix}$$

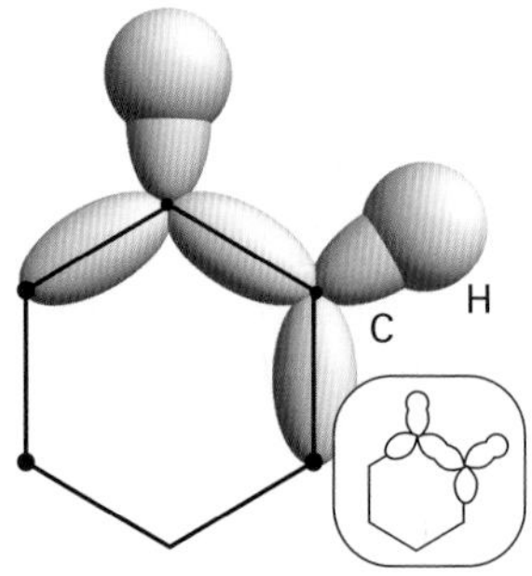

Figure 9E.3 The σ framework of benzene is formed by the overlap of Csp^2 hybrids, which fit without strain into a hexagonal arrangement.

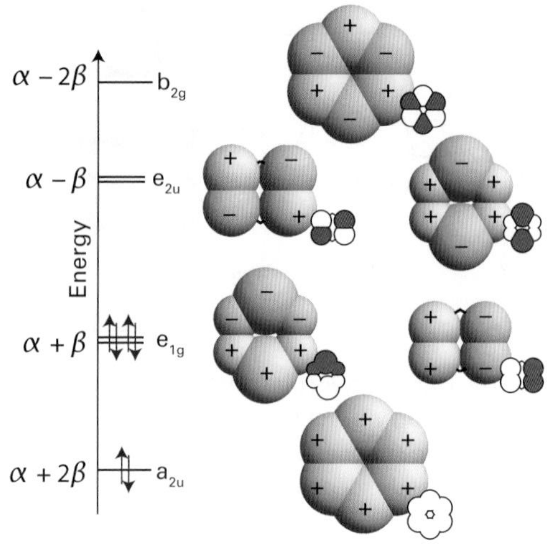

Figure 9E.4 The Hückel orbitals of benzene and the corresponding energy levels. The orbital labels are explained in Topic 10B. The bonding and antibonding character of the delocalized orbitals reflects the numbers of nodes between the atoms. In the ground state, only the bonding orbitals are occupied.

The MO energies are deduced from the diagonal elements of this matrix and are

$$E=\alpha+2\beta,\ \alpha+\beta,\ \alpha+\beta,\ \alpha-\beta,\ \alpha-\beta,\ \alpha-2\beta \qquad (9E.13)$$

as shown in Fig. 9E.4. The orbitals there have been given symmetry labels that are explained in Topic 10B. Note that the lowest energy orbital is bonding between all neighbouring atoms, the highest energy orbital is antibonding between each pair of neighbours, and the intermediate orbitals are a mixture of bonding, nonbonding, and antibonding character between adjacent atoms. There are two orbitals with energy $\alpha+\beta$, that is there is a degenerate pair of MOs, and likewise a second degenerate pair with energy $\alpha-\beta$.

Now apply the building-up principle to the π system. There are six electrons to accommodate (one from each C atom), so the three lowest orbitals (a_{2u} and the doubly degenerate pair e_{1g}) are fully occupied, giving the ground-state configuration $a_{2u}^2e_{1g}^4$. A significant point is that the only molecular orbitals occupied are those with net bonding character (the analogy with the strongly bonded N_2 molecule, Topic 9B, should be noted).

The π-electron binding energy of benzene is

$$E_\pi=2(\alpha+2\beta)+4(\alpha+\beta)=6\alpha+8\beta$$

If delocalization is ignored and the molecule is thought of as having three isolated π bonds, it would be ascribed a π-electron energy of only $3(2\alpha+2\beta)=6\alpha+6\beta$. The delocalization energy is therefore $2\beta\approx-480\ \mathrm{kJ\,mol^{-1}}$, which is considerably more than for butadiene. The π-bond formation energy in benzene is 8β.

This discussion suggests that aromatic stability can be traced to two main contributions. First, the shape of the regular hexagon is ideal for the formation of strong σ bonds: the σ framework is relaxed and without strain. Second, the π orbitals are such as to be able to accommodate all the electrons in bonding orbitals, and the delocalization energy is large.

Brief illustration 9E.2

In *Example* 9E.2 the delocalization energy of cyclobutadiene is found to be zero. The cyclobutadiene dication, $C_4H_4^{2+}$, has two π electrons to accommodate so the total π-electron binding energy is $2(\alpha + 2\beta) = 2\alpha + 4\beta$. The energy of a single localized π-bond is $2(\alpha + \beta)$, so the delocalization energy is 2β. Benzene, with six π electrons has a delocalization energy of 8β. Cyclic conjugated molecules with two or six π electrons are regarded as aromatic and benefit from extra stabilization as a result of delocalization. A molecule with four π electrons has no such stabilization and is not regarded as aromatic.

Exercises

E9E.2 Predict the electron configurations of (i) the benzene anion, (ii) the benzene cation. Estimate the π-electron binding energy in each case.

E9E.3 What is the delocalization energy and π-bond formation energy of (i) the benzene anion, (ii) the benzene cation?

9E.3 Computational chemistry

The power and speed of modern computers makes it possible to calculate molecular orbitals without resorting to the severe assumptions of the Hückel method. With readily available software packages it is possible to make quite sophisticated calculations on small to medium-sized molecules, and with the aid of supercomputers it is possible to tackle much larger molecules and obtain results which stand up well in comparison to experiment. These software packages go well beyond simply calculating molecular orbitals. They also offer the possibility of predicting the equilibrium geometries of molecules, as well as molecular properties such as vibrational frequencies and NMR chemical shifts, and even thermodynamic quantities such as enthalpies of formation. It is also possible to study chemical reactivity by exploring how the energy and electronic structures of reacting molecules change as they encounter one another. The development and exploitation of methods for the computation of molecular electronic structure has received an enormous amount of attention and has become a keystone of modern chemical research.

The underlying theory behind these calculations and their practical implementation as computer programs are both very complex matters and cannot be explored in detail here. This section simply aims to provide a brief introduction to the key ideas and language used in such calculations. An introduction to the background theory is given in Topic 9F.

Regardless of the precise computational method, all electronic structure calculations involve finding the molecular orbitals iteratively and self-consistently, just as for the self-consistent field (SCF) approach to atoms (Topic 8B). First, the molecular orbitals for the electrons present in the molecule are formulated (for example, as LCAOs). One molecular orbital is then selected and all the others are used to set up an expression for the potential energy of the electron in the chosen orbital. The form of the chosen orbital is then optimized for this potential, for example by adjusting the orbital coefficients or other parameters. The procedure is repeated for all the molecular orbitals and used to calculate the total energy of the molecule. Once one cycle of optimization is completed the orbitals will all have changed and so will the potential that each electron experiences. The process of optimizing each orbital is repeated until the computed orbitals and energy are constant to within some tolerance.

9E.3(a) Basis functions and basis sets

Although it is useful to think of a molecular orbital as a linear combination of atomic orbitals, the theory does not require that the MOs are formed from actual AOs. In fact, any convenient set of basis functions can be used. Numerical calculations of molecular orbitals require the evaluation of a large number of integrals (such as those occurring in the definitions of H_{JJ} and H_{IJ}) in which these basis functions appear. Therefore, it is important to choose a set of functions that facilitates such calculations.

It turns out that calculations using hydrogenic AOs are computationally time-consuming, so are not convenient basis functions. **Gaussian-type orbitals** (GTOs) are a computationally more convenient choice and are used widely. A GTO is centred on each nucleus and has the form

$$\psi = Nx^i y^j z^k \mathrm{e}^{-\zeta r^2} \qquad \text{Gaussian-type orbitals}$$

where *x*, *y*, and *z* are the Cartesian coordinates of the electron at a distance *r* from the nucleus, ζ is a positive adjustable parameter, and *i*, *j*, and *k* are non-negative integers. An **s-type** GTO has $i = j = k = 0$, a **p-type** GTO has one of the (i, j, k) set to 1 and the others are zero (for example $Nz\mathrm{e}^{-\zeta r^2}$), and a **d-type** GTO has $i + j + k = 2$ (for example $Nyz\mathrm{e}^{-\zeta r^2}$ or $Nx^2\mathrm{e}^{-\zeta r^2}$). There are three p-type and six d-type GTOs.

Important considerations when selecting a basis are the form of the GTOs and their number. There needs to be enough functions to construct the MOs to the required level of precision, but larger basis sets lead to more integrals and longer computational times. A great deal of effort has been put into devising relatively small basis sets that allow rapid calculations with reliable results.

The notation used to specify a basis set can seem daunting at first sight, but becomes clear when interpreted in stages. Carbon provides an example. Its ground-state electron configuration is $1s^2 2s^2 2p^2$, so at least five GTOs are needed, one for each type of occupied orbital (one 1s, one 2s, and three 2p orbitals). Such a basis is called a **single zeta** (SZ) basis, the name deriving from the single parameter ζ in each of the five GTOs. Such a small basis lacks the flexibility needed to describe adequately the contribution the carbon atom makes to MOs, so the number of basis functions may be doubled (giving a **double zeta**, DZ, basis) or tripled (giving a **triple zeta**, TZ, basis). In such bases, two or three GTOs with different values of ζ are used to describe a 1s orbital, and similarly for the other orbitals.

The core and valence orbitals are commonly treated separately in the basis set on the grounds that the core orbitals are likely to be so compact that they contribute little to the bonding, and so do not need to be represented very elaborately. Such an approach is described as a **split basis set**. One choice is to use a SZ basis for the core electrons and either a DZ or TZ basis for the valence electrons. For carbon, a split basis set with SZ for the core and TZ for the valence electrons would involve four s-type GTOs (one for the core and three for the valence shell) and a total of nine p-type GTOs (three sets of three), giving a total of 13 GTOs.

A single GTO is a rather poor approximation to an orbital wavefunction close to the nucleus. Therefore, the basis set is formed from linear combinations of GTOs called **contracted GTOs**. Each term in a contracted GTO has a different value of ζ. The coefficients of the linear combination and the ζ values are optimized for use in a particular basis set and are unchanged during the calculation.

The next step in nomenclature involves the specification of the basis. It is here that expressions such as **6-31G** are encountered. In this case the hyphen separates the core and valence orbitals. The number 6 in the 'core group' indicates that a contracted GTO formed from six GTOs is used to describe the core orbitals (for example, the single 1s orbital of carbon). The numbers after the hyphen specify the GTOs used to describe the valence orbitals. The 3 indicates that each type of valence orbital is represented by a contracted GTO formed from three GTOs (the *inner functions*) and the 1 indicates the use of one further (non-contracted) GTO (the *outer function*). In total 22 GTOs are used to describe carbon with this basis set: six for the core, twelve for the inner functions (three sets of four), and four for the outer functions.

The 6-31G basis is sufficient for simple organic molecules, but to cope with more unusual types of bonding, such as that found in small rings or so-called 'bent bonds', it is necessary to add more GTOs. These additional functions are called **polarization functions**, and correspond to d-type orbitals (for second-row elements) and p-type orbitals (for hydrogen). These functions allow for more flexibility in the form of the MOs than can be achieved simply with s- and p-type functions. The use of polarization functions is indicated by adding asterisks to the specification of the basis. If polarization functions are used for all but hydrogen, the basis is denoted 6-31G*, and if polarization functions are also used for hydrogen the basis is denoted 6-31G**. In an alternative notation, parentheses denote the number and type of polarization functions. For example, 6-31G(2df,2p) indicates that two d-type and one f-type GTO are used for non-hydrogen atoms, and that two p-type GTOs are used for hydrogen.

If the molecule is an anion or contains many lone pairs, it helps to add GTOs that decay more slowly (have smaller values of ζ) than those considered so far. These additional functions are called **diffuse functions**, and their use is indicated by adding a plus sign and more characters to the name of the basis set. For example, 6-31G+G(d) indicates the use of diffuse functions of the same number and type as used for the valence functions.

Here, in summary, is the anatomy of the notation 6-31G and its extensions:

Symbol	Significance
6	Number of GTOs in the contracted GTO used for the core orbitals
-	Separates core from valence shell description
3	Number of GTOs in the contracted GTO used for each valence orbital (inner functions)
1	Number of additional GTOs used for each valence orbital (outer functions)
G	Specifies a Gaussian basis
*	Polarization functions included for all atoms except H
**	Polarization functions included for all atoms
+	Diffuse functions included
G(d)	Diffuse functions of the same type as used for the valence functions included

9E.3(b) Electron correlation

The SCF approach treats the effects of electron–electron interactions by computing each MO on the basis that the electron experiences a potential that is the average of the interactions with all the other electrons. This approximation ignores **electron correlation**, the details of the interaction between the electrons. The failure to take into account electron correlation commonly gives calculated quantities that are in poor agreement with the experimental data.

Several methods have been developed that attempt to take into account the effects of electron correlation: they all involve additional calculations, and the most precise results require a significant increase in the amount of computation. Two widely used procedures are the **Møller–Plesset** (MP) and **coupled cluster singles and doubles** (CCSD) methods. For these highly technical emendations of the basic calculations you need to consult specialist texts.

9E.3(c) Density functional theory

A technique that has gained considerable ground in recent years to become one of the most widely used approaches for the calculation of molecular structure is **density functional theory** (DFT). It has proved to be computationally more efficient and often to give better results than the methods described so far.

The central focus of DFT is the electron density, ρ, rather than the wavefunction, ψ. The 'functional' part of the name comes from the fact that the energy of the molecule is a function of the electron density, written $E[\rho]$, and the electron density is itself a function of position, $\rho(\boldsymbol{r})$: in mathematics a function of a function is called a 'functional'. Topic 9F outlines how this procedure is used to calculate molecular orbitals. The method involves choosing a basis set and solving a set of equations iteratively and self-consistently. An important feature of DFT is that, in principle, the effects of electron correlation are included in the calculation.

An enormous amount of effort has gone into developing functionals that are computationally efficient and produce results that are in accord with experiment. Some functionals have been developed to give the best results for certain kinds of molecule or certain molecular properties. Artificial intelligence (AI) has also been used to identify reliable functionals. It is not uncommon for the functionals to be adjusted empirically to fit experimental data.

A very commonly used functional is **B3LYP**, developed by Lee, Yang, and Parr (hence LYP). The 'B' indicates the use of an 'exchange' term in the functional (Topic 9F) and the '3' indicates the presence of three parameters that can be adjusted to give the best fit to experimental data. There are variants on this functional; for example, the **B3LYP-D** functional has additional terms included designed to deal with long-range (dispersion) interactions.

9E.3(d) Practical calculations

How a calculation is approached depends on the computing power available, the type of molecule being considered, the properties to be calculated, and the precision required. A typical starting point is to use DFT with 6-31G* basis set and one of the widely used functionals such as B3LYP. This approach might be sufficient to obtain the general form of the MOs or the equilibrium geometry.

If more precise results are needed it is appropriate to refer to the literature to find an approach that has been shown to be effective for the intended computation. The calculation of vibrational frequencies, for instance, needs a different approach from the calculation of NMR chemical shifts.

The most precise calculations often involve using different methods in sequence. For example, the geometry of a molecule might first be optimized by using DFT, and then the energies of the orbitals are refined by taking into account electron correlation.

9E.3(e) Graphical representations

One of the most significant developments in computational chemistry has been the introduction of graphical representations of molecular orbitals and electron densities. The raw output of a molecular structure calculation is a list of the coefficients of the atomic orbitals in each molecular orbital and the energies of these orbitals. The graphical representation of a molecular orbital uses stylized shapes to represent the basis set, and then scales their size to indicate the coefficient in the linear combination. Different signs of the wavefunctions are represented by different colours.

Once the coefficients are known, it is possible to construct a representation of the electron density in the molecule by noting which orbitals are occupied and then forming the squares of those orbitals. The total electron density at any point is then the sum of the squares of the wavefunctions evaluated at that point. The outcome is commonly represented by an **isodensity surface**, a surface of constant total electron density (Fig. 9E.5). As shown in the illustration, there are several styles of representing an isodensity surface, as a solid form, as a transparent form with a ball-and-stick representation of the molecule within, or as a mesh. A related representation is a **solvent-accessible surface** in which the shape represents the shape of the molecule by imagining a sphere representing a solvent molecule rolling across the surface and plotting the locations of the centre of that sphere.

One of the most important aspects of a molecule other than its geometrical shape is the distribution of charge over its surface, which is commonly depicted as an **electrostatic potential surface** (an 'elpot surface'). The potential energy, E_p, of an imaginary positive charge Q at a point is calculated by taking into account its interaction with the nuclei and the electron density throughout the molecule. Then, because $E_p = Q\phi$, where ϕ is the electric potential, the potential energy can be interpreted as a potential and depicted as an appropriate colour (Fig. 9E.6). Electron-rich regions usually have negative potentials and electron-poor regions usually have positive potentials.

Representations such as those illustrated here are of critical importance in a number of fields. For instance, they may be used to identify an electron-poor region of a molecule that

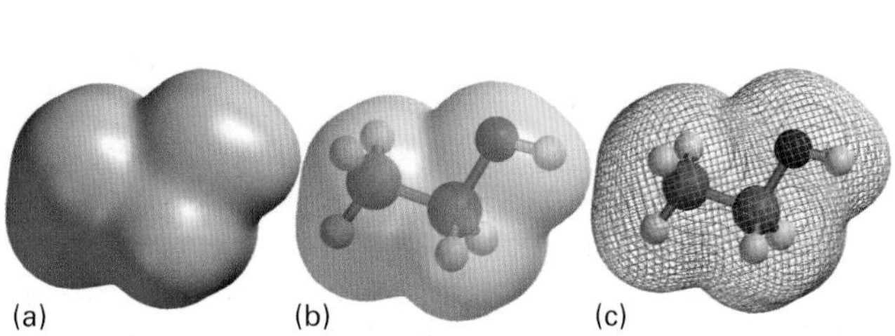

Figure 9E.5 Various representations of an isodensity surface of ethanol: (a) solid surface, (b) transparent surface, and (c) mesh surface.

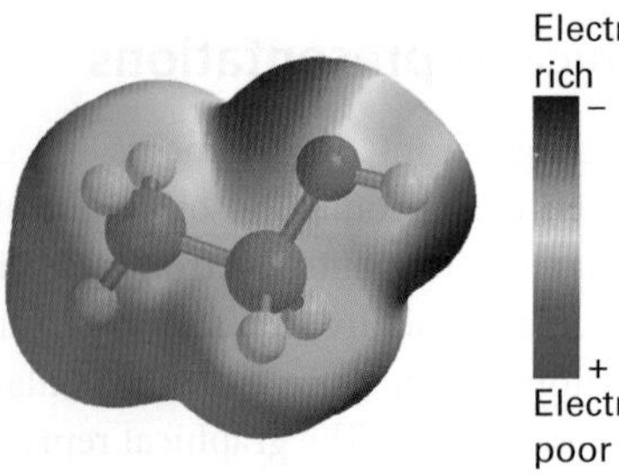

Figure 9E.6 An elpot diagram of ethanol; the molecule has the same orientation as in Fig. 9E.5. Red denotes regions of negative electrostatic potential and blue regions of positive potential (as in $^{\delta-}$O–H$^{\delta+}$).

is susceptible to association with or chemical attack by an electron-rich region of another molecule. Such considerations are important for assessing the pharmacological activity of potential drugs.

Exercises

E9E.4 Write down the form of the three p-type and six d-type GTOs.

E9E.5 How many GTOs are there in a single zeta basis set for (i) a carbon atom, (ii) a nitrogen atom?

E9E.6 Which GTOs are included in the 3-21G basis set?

Checklist of concepts

☐ 1. The **Hückel method** neglects overlap and interactions between orbitals on atoms that are not neighbours.

☐ 2. The highest occupied molecular orbital (HOMO) and the lowest unoccupied molecular orbital (LUMO) are the **frontier orbitals** of a molecule.

☐ 3. The Hückel method may be expressed in a compact manner by introducing matrices.

☐ 4. The **π-bond formation energy** is the energy released when a π bond is formed.

☐ 5. The **π-electron binding energy** is the sum of the energies of each π electron.

☐ 6. The **delocalization energy** is the difference between the π-electron binding energy and the energy of the same molecule with localized π bonds.

☐ 7. The stability of benzene arises from the geometry of the ring and the high delocalization energy.

☐ 8. **Density functional theories** develop equations based on the electron density rather than the wavefunction itself.

☐ 9. Graphical techniques are used to plot a variety of surfaces based on electronic structure calculations.

Checklist of equations

Property	Equation	Comment	Equation number
LCAO-MO	$\psi = \sum_i c_i \psi_i$	ψ_i are atomic orbitals	9E.1
Hückel equations	$\boldsymbol{Hc} = \boldsymbol{cE}$	Hückel approximations: $H_{AB} = 0$ except between neighbours; overlap neglected	9E.8
Diagonalization	$\boldsymbol{c}^{-1}\boldsymbol{Hc} = \boldsymbol{E}$		9E.10
π-Electron binding energy	E_π = sum of energies of π electrons	Definition	
π-Bond formation energy	$E_{bf} = E_\pi - N_C\alpha$	Definition; N_C is the number of carbon atoms	9E.12
π-Delocalization energy	$E_{deloc} = E_\pi - N_C(\alpha + \beta)$		

TOPIC 9F Computational chemistry

➤ Why do you need to know this material?

Modern computational procedures for predicting molecular structure and reactivity are now widely available and used throughout chemistry. Like any tool, it is important to know its foundations.

➤ What is the key idea?

Numerical procedures in computational chemistry typically proceed by solving equations until they do not change on successive iterations.

➤ What do you need to know already?

This Topic develops the approach introduced in Topic 9E and makes extensive use of matrix manipulations (see the extended version of *The chemist's toolkit* 9E.1 available on the website). It is based on the variation principle (Topic 9A).

The field of **computational chemistry**, the use of computers to predict molecular structure and reactivity, has grown in the past few decades due to the tremendous advances in computer hardware and to the development and wide availability of efficient software packages. The latter are now applied routinely to compute molecular properties in a wide variety of chemical applications including drug search and design, atmospheric and environmental chemistry, nanotechnology, and materials science. As mentioned in Topic 9E, most software packages have sophisticated graphical interfaces that permit the visualization of results.

9F.1 The central challenge

The central aim of electronic structure calculations in computational chemistry is the solution of the electronic Schrödinger equation, $\hat{H}\Psi = E\Psi$, where E is the electronic energy and Ψ is the many-electron wavefunction, a function of the coordinates of all the electrons and the nuclei. To make progress, it is necessary to invoke the Born–Oppenheimer approximation and the separation of electronic and nuclear motion (see the *Prologue* to this Focus).

To avoid a sequence of overly complicated equations, the computational procedures described in this Topic will be illustrated by using from the outset the simplest possible many-electron molecule, dihydrogen (H_2), with two electrons labelled 1 and 2 and two nuclei labelled A and B. Some of the techniques introduced do not need to be applied to this simple molecule, but they serve to illustrate them in a simple manner. Molecules with more electrons are then relatively straightforward elaborations of the equations for H_2: these more general equations are given in *A deeper look* 9F.1, available to read in the e-book accompanying this text.

Topic 9B introduces the electronic Hamiltonian for the one-electron system H_2^+. The corresponding hamiltonian for H_2 has additional terms to describe the kinetic energy of the second electron, its interaction with both nuclei, and the interaction between the two electrons:

$$\hat{H} = \overbrace{-\frac{\hbar^2}{2m_e}\left(\nabla_1^2+\nabla_2^2\right)}^{\text{kinetic energy of the electrons}} \overbrace{-\frac{e^2}{4\pi\varepsilon_0 r_{1A}}-\frac{e^2}{4\pi\varepsilon_0 r_{2A}}-\frac{e^2}{4\pi\varepsilon_0 r_{1B}}-\frac{e^2}{4\pi\varepsilon_0 r_{2B}}}^{\text{potential energy of attraction to the nuclei}} \overbrace{+\frac{e^2}{4\pi\varepsilon_0 r_{12}}}^{\text{electron–electron repulsion}} \qquad (9F.1)$$

In this expression r_{iJ} is the distance from electron i to nucleus J, and r_{12} is the distance between the electrons. As in Topic 9B the term $e^2/4\pi\varepsilon_0$ is written j_0,[1] and with this substitution the hamiltonian takes the form

$$\hat{H} = \hat{H}_1 + \hat{H}_2 + \hat{H}_{12} \qquad (9F.2)$$

with

$$\hat{H}_1 = -\frac{\hbar^2}{2m_e}\nabla_1^2 - j_0\left(\frac{1}{r_{1A}}+\frac{1}{r_{1B}}\right)$$

$$\hat{H}_2 = -\frac{\hbar^2}{2m_e}\nabla_2^2 - j_0\left(\frac{1}{r_{2A}}+\frac{1}{r_{2B}}\right)$$

$$\hat{H}_{12} = \frac{j_0}{r_{12}}$$

[1] In some accounts j_0 is replaced by 1, signalling that *atomic units* are being used.

It is hopeless to expect to find analytical eigenfunctions (wavefunctions) and eigenvalues (energies) for a hamiltonian of this complexity. The whole thrust of computational chemistry is to formulate and implement numerical procedures that give ever more reliable approximations to them.

Exercises

E9F.1 Write down the form of the hamiltonian for the three-electron system H_2^-.

E9F.2 Write the electronic hamiltonian for HeH^+.

9F.2 The Hartree–Fock formalism

The orbital approximation is introduced in Topic 8B in the context of many-electron atoms and is used here too. In this approximation it is supposed that each electron is described by its 'own' wavefunction, called an orbital, and that the overall wavefunction can be written as a product of these orbitals. For example, if electron 1 occupies an orbital with wavefunction ψ_a and electron 2 occupies an orbital orbital with wavefunction ψ_b, the overall wavefunction is written $\psi_a(\boldsymbol{r}_1)\psi_b(\boldsymbol{r}_2)$. This approximation is severe and loses many of the details of the dependence of the wavefunction on the relative locations of the electrons. To simplify the appearance of the expressions the wavefunction is written $\psi_a(1)\psi_b(2)$.

Now suppose that electron 1 is in orbital *a* with spin α, and electron 2 occupies the same orbital with spin β. The many-electron wavefunction is written as the product $\Psi=\psi_a^\alpha(1)\psi_a^\beta(2)$. The combination of an orbital and a spin function, such as $\psi_a^\alpha(1)$, is called a **spinorbital**. For example, the spinorbital ψ_a^α is interpreted as the product of the spatial wavefunction ψ_a and the spin wavefunction α:

$$\overbrace{\psi_a^\alpha(1)}^{\text{spinorbital}} = \overbrace{\psi_a(1)}^{\text{spatial wavefunction}} \times \overbrace{\alpha(1)}^{\text{spin wavefunction}}$$

and likewise for the other spinorbitals.

A simple product wavefunction such as $\psi_a^\alpha(1)\psi_a^\beta(2)$ does not satisfy the Pauli principle, which requires the function to change sign under the interchange of any pair of electrons (Topic 8B). To ensure that the wavefunction does satisfy the principle, it is expressed as a sum of all possible permutations of the occupation of spinorbitals by the electrons, using alternating plus and minus signs:

$$\Psi=\psi_a^\alpha(1)\psi_a^\beta(2)-\psi_a^\beta(1)\psi_a^\alpha(2)$$

This sum can be represented by a determinant called a **Slater determinant**, the expansion of which (see *The Chemists Toolkit* 9D.1) generates the same alternating sum of terms:

$$\Psi=\frac{1}{2^{1/2}}\begin{vmatrix}\psi_a^\alpha(1) & \psi_a^\beta(1)\\ \psi_a^\alpha(2) & \psi_a^\beta(2)\end{vmatrix} \qquad \text{Slater determinant} \quad (9F.3a)$$

The factor $1/2^{1/2}$ (and in general $1/N_e!^{1/2}$, where N_e is the number of electrons) ensures that the wavefunction is normalized if the component spinorbitals are normalized. To save the tedium of writing out large determinants when there are many electrons in a molecule, the wavefunction is normally written showing only the spinorbitals along the principal diagonal of the determinant:

$$\Psi=(1/2^{1/2})\left|\psi_a^\alpha(1)\psi_a^\beta(2)\right| \quad (9F.3b)$$

According to the variation principle (Topic 9A) the form of Ψ which is the best approximation to the actual wavefunction is the one that corresponds to the lowest achievable energy as the ψ are varied. That is, the wavefunctions ψ must be found that minimize the expectation value, $E=\int\Psi^*\hat{H}\Psi\mathrm{d}\tau$, of the hamiltonian.

D.R. Hartree and V. Fock showed that the optimum orbital wavefunctions each satisfy what is, at first sight, a very simple set of equations:

$$f_1\psi_a(1)=\varepsilon_a\psi_a(1) \qquad \text{Hartree–Fock equations} \quad (9F.4)$$

where f_1 is the **Fock operator** for electron 1. This is the equation to solve to find ψ_a; there are analogous equations for any other orbitals that are occupied. The explicit effect of the Fock operator on $\psi_a(1)$ is as follows

$$f_1\psi_a(1)=\overbrace{\hat{H}_1\psi_a(1)}^{\text{kinetic energy and attraction to the nuclei}}+\overbrace{2\hat{J}_a(1)\psi_a(1)}^{\text{repulsion of electrons}}-\overbrace{\hat{K}_a(1)\psi_a(1)}^{\text{electron exchange correction}} \quad (9F.5a)$$

where $\hat{H}_1$ is specified in eqn 9F.2, $\hat{J}_a$ is called the **coulomb operator**, and $\hat{K}_a$ is called the **exchange operator**. In general (when more than one orbital is occupied)

$$\hat{J}_b(1)\psi_a(1)=j_0\int\psi_a(1)\frac{1}{r_{12}}\psi_b^*(2)\psi_b(2)\,\mathrm{d}\tau_2 \qquad \text{Coulomb operator [definition]} \quad (9F.5b)$$

$$\hat{K}_b(1)\psi_a(1)=j_0\int\psi_b(1)\frac{1}{r_{12}}\psi_b^*(2)\psi_a(2)\,\mathrm{d}\tau_2 \qquad \text{Exchange operator [definition]}$$

For H_2 both electrons are in ψ_a and so there is no distinction between the two operators and

$$\hat{J}_a(1)\psi_a(1)=\hat{K}_a(1)\psi_a(1)=j_0\int\psi_a(1)\frac{1}{r_{12}}\psi_a^*(2)\psi_a(2)\mathrm{d}\tau_2 \quad (9F.5c)$$

Equation 9F.5a now becomes

$$-\frac{\hbar^2}{2m_e}\nabla_1^2\psi_a(1)-j_0\left(\frac{1}{r_{1A}}+\frac{1}{r_{1B}}\right)\psi_a(1)+j_0\int\psi_a(1)\frac{1}{r_{12}}\psi_a^*(2)\psi_a(2)\mathrm{d}\tau_2$$
$$=\varepsilon_a\psi_a(1) \quad (9F.5d)$$

Before exploring how this very complicated equation can be solved, note that it reveals a second principal approximation of the Hartree–Fock formalism (the first being its dependence on the orbital approximation). Instead of electron 1 (or any other electron) responding to the instantaneous positions of the other electrons in the molecule through terms of the form $1/r_{12}$, it responds to an *averaged* location of the other electrons through integrals of the kind that appear in eqn 9F.5d:

$$j_0 \int \psi_a(1)\frac{1}{r_{12}}\psi_a^*(2)\psi_a(2)\mathrm{d}\tau_2 \xrightarrow{\text{Interpreted as}}$$

$$\underbrace{j_0\psi_a(1)\overbrace{\int\frac{\overbrace{\psi_a^*(2)\psi_a(2)}^{\text{probability of electron 2 being in }\mathrm{d}\tau_2\text{ at }r_2}}{r_{12}}\mathrm{d}\tau_2}^{\text{average value of }1/r_{12}}}_{\text{average repulsion energy for an electron at }r_1}$$

That is, the Hartree–Fock method ignores **electron correlation**, the tendency of electrons to avoid one another to minimize repulsions. This failure to take into account electron correlation is a principal reason for inaccuracies in the calculations and leads to calculated energies that are higher than the 'true' (that is, experimental) values.

Although eqn 9F.5d is the equation to solve to find ψ_a, it is necessary to know the wavefunctions of all the occupied orbitals in order to set up the integrals J and K and hence to find ψ_a. Even when only one orbital is occupied, its form must be known in order to evaluate the integral that is then used to find that orbital. To make progress with this difficulty the calculation proceeds in the following way.

1. The initial form of ψ_a is guessed.
2. The assumed form of ψ_a is used to set up the integral in eqn 9F.5d.
3. The Hartree–Fock equations are solved to find a better version of ψ_a.
4. Steps 2 to 3 are repeated using this better wavefunction.

This process is continued until successive cycles of the calculation leaves the energies ε_a and orbital wavefunction ψ_a unchanged to within a chosen criterion. This is the origin of the term **self-consistent field** (SCF) for this type of procedure in general and of **Hartree–Fock self-consistent field** (HF-SCF) for the approach based on the orbital approximation. When more than one orbital is occupied, this procedure is repeated for all of them.

Exercise

E9F.3 What arrangement of electrons does the following Slater determinant represent?

$$\Psi = \frac{1}{2^{1/2}}\begin{vmatrix} \psi_a^\alpha(1) & \psi_a^\beta(1) \\ \psi_b^\alpha(2) & \psi_b^\beta(2) \end{vmatrix}$$

9F.3 The Roothaan equations

The difficulty with the HF-SCF procedure lies in the numerical solution of the Hartree–Fock equations, a challenging task even for powerful computers. As a result, a modification of the technique was needed before the procedure could be of use to chemists. In 1951 C.C.J. Roothaan and G.G. Hall independently found a way to convert the Hartree–Fock equations for the molecular orbitals into equations for the coefficients that appear in the LCAO used to construct the molecular orbital. Their approach (see *A deeper look* 9F.2, available to read in the e-book accompanying this text) leads to the **Roothaan equations**.

The atomic orbitals used to construct the LCAO are referred to as the **basis set** for the calculation, and the choice of basis is an important part of the procedure (see below). Note that if the basis set consists of N_o atomic orbitals, then N_o linearly independent linear combinations (molecular orbitals) can be constructed from them.

Once again, seeing how this procedure works for H_2 gives a hint of the general form of the Roothaan equations. As this account unfolds you will see many similarities between it and the matrix formulation of Hückel theory in Topic 9E, with equations having a similar structure but interpreted in a more sophisticated way.

First a molecular orbital is constructed by the linear combination of two basis orbitals χ_A and χ_B (think of these as hydrogen 1s orbitals, but they could be other types). This molecular orbital is written $\psi_m = c_{Am}\chi_A + c_{Bm}\chi_B$, with $m = a$ for what will become the occupied orbital and b for the unoccupied orbital (a 'virtual' molecular orbital in the jargon). Now write the coefficients c_{Am} and so on as the matrix $\boldsymbol{c}$:

$$\boldsymbol{c} = \begin{pmatrix} c_{Aa} & c_{Ab} \\ c_{Ba} & c_{Bb} \end{pmatrix}$$

The overlap matrix $\boldsymbol{S}$ of the individually normalized atomic orbitals is written

$$\boldsymbol{S} = \begin{pmatrix} 1 & S \\ S & 1 \end{pmatrix} \qquad S = \int \chi_A\chi_B \mathrm{d}\tau$$

Next, set up the **Fock matrix** using the Fock operator f that appears in eqn 9F.4:

$$\boldsymbol{F} = \begin{pmatrix} F_{AA} & F_{AB} \\ F_{BA} & F_{BB} \end{pmatrix} \qquad F_{XY} = \int \chi_X(1) f_1 \chi_Y(1)\mathrm{d}\tau_1$$

For now, regard the F_{XY} as variable quantities; keep in mind that they are simply energies arising from the kinetic energy of the electrons, their attractions to the nuclei, and repulsions between the electrons. Finally, define the matrix $\boldsymbol{\varepsilon}$, which is just a diagonal matrix of the energies ε_a and ε_b:

$$\boldsymbol{\varepsilon} = \begin{pmatrix} \varepsilon_a & 0 \\ 0 & \varepsilon_b \end{pmatrix}$$

The **Roothaan equations** are written in terms of these matrices

$$\overbrace{\begin{pmatrix} F_{AA} & F_{AB} \\ F_{BA} & F_{BB} \end{pmatrix}}^{F}\overbrace{\begin{pmatrix} c_{Aa} & c_{Ab} \\ c_{Ba} & c_{Bb} \end{pmatrix}}^{c}=\overbrace{\begin{pmatrix} 1 & S \\ S & 1 \end{pmatrix}}^{S}\overbrace{\begin{pmatrix} c_{Aa} & c_{Ab} \\ c_{Ba} & c_{Bb} \end{pmatrix}}^{c}\overbrace{\begin{pmatrix} \varepsilon_a & 0 \\ 0 & \varepsilon_b \end{pmatrix}}^{\varepsilon}$$

Roothaan equations for H_2 (9F.6)

As the annotation indicates, the general form of the Roothaan equations is $\boldsymbol{Fc} = \boldsymbol{Sc\varepsilon}$. Note their similarity to the Hückel treatment equations developed as eqn 9E.7 in Topic 9E:

$$\begin{pmatrix} H_{AA} & H_{AB} \\ H_{BA} & H_{BB} \end{pmatrix}\begin{pmatrix} c_{1,A} & c_{2,A} \\ c_{1,B} & c_{2,B} \end{pmatrix}=\begin{pmatrix} c_{1,A} & c_{2,A} \\ c_{1,B} & c_{2,B} \end{pmatrix}\begin{pmatrix} E_1 & 0 \\ 0 & E_2 \end{pmatrix}$$

After multiplying out the matrices in eqn 9F.6 and matching the elements corresponding to the occupied orbital a, they become

$$F_{AA}c_{Aa}+F_{AB}c_{Ba}=\varepsilon_a c_{Aa}+\varepsilon_a S c_{Ba}$$
$$F_{BA}c_{Aa}+F_{BB}c_{Ba}=\varepsilon_a c_{Ba}+\varepsilon_a S c_{Aa}$$

Then, by rearranging and collecting terms,

$$(F_{AA}-\varepsilon_a)c_{Aa}+(F_{AB}-\varepsilon_a S)c_{Ba}=0$$
$$(F_{BA}-\varepsilon_a S)c_{Aa}+(F_{BB}-\varepsilon_a)c_{Ba}=0$$

which are nothing other than a pair of simultaneous equations for the coefficients, as in Hückel theory, where they were

$$(H_{AA}-E_1)c_{1,A}+H_{AB}c_{1,B}=0$$
$$H_{BA}c_{1,A}+(H_{BB}-E_1)c_{1,B}=0$$

A general feature of simultaneous equations (*The Chemist's Toolkit* 9D.1) is that their solution requires the determinant of the factors multiplying the two unknowns, in this case c_{Aa} and c_{Ba}, to vanish. That is, it is necessary that

$$\begin{vmatrix} F_{AA}-\varepsilon_a & F_{AB}-\varepsilon_a S \\ F_{BA}-\varepsilon_a S & F_{BB}-\varepsilon_a \end{vmatrix}=0$$

As in Topic 9E, this determinant is called a secular determinant (in general the criterion is $|F - \varepsilon S| = 0$). When expanded, the resulting expression is a quadratic equation in the energy ε_a, which can easily be solved.

Brief illustration 9F.1

As has been emphasized, these equations are very similar to those arising in the Hückel method. One of the Hückel approximations is to set all overlap integrals equal to zero ($S = 0$). Next, noting that the F_{XY} are simply a variety of energies, the diagonal Fock matrix elements are set equal to α ($F_{AA} = F_{BB} = \alpha$), and the off-diagonal elements are set equal to β ($F_{AB} = F_{BA} = \beta$). The secular determinant then becomes

$$\begin{vmatrix} \alpha-\varepsilon_a & \beta \\ \beta & \alpha-\varepsilon_a \end{vmatrix}=0$$

which expands into $\varepsilon_a^2 - 2\alpha\varepsilon_a + \alpha^2 - \beta^2 = 0$ and the roots are $\varepsilon_a = \alpha \pm \beta$. If $\beta < 0$ (as in the Hückel treatment), the lower energy orbital is the one with $\varepsilon_a = \alpha + \beta$.

Despite this apparent simplicity, there is a hidden problem that becomes apparent as soon as the calculation becomes more sophisticated than the Hückel procedure. In general (and even for H_2) the F_{XY} depend on the coefficients that we are trying to find! For instance, the explicit form of F_{AA} for H_2 is

$$F_{AA}=\int\chi_A(1)\hat{H}_1\chi_A(1)d\tau_1+2j_0\int\chi_A(1)\chi_A(1)\frac{1}{r_{12}}\psi_a(2)\psi_a(2)d\tau_1 d\tau_2$$
$$-j_0\int\chi_A(1)\psi_a(1)\frac{1}{r_{12}}\psi_a(2)\chi_A(2)d\tau_1 d\tau_2 \qquad (9F.7)$$

But when $\psi_a = c_{Aa}\chi_A + c_{Ba}\chi_B$ is substituted into the second and third integrals the result is an expression that involves the coefficients and having the form

$$F_{AA}=A+Bc_{Aa}^2+Cc_{Ba}^2+Dc_{Aa}c_{Ba}$$

Therefore, solving the equations for the coefficients and the energies is far from straightforward. Once again, a self-consistency procedure has to be adopted. First, the coefficients are estimated and used to evaluate the F_{XY}. The resulting equations are solved for the coefficients and energies and the new coefficients are used to calculate another round of the F_{XY}. The procedure is repeated until the coefficients and energies no longer change significantly in each new cycle.

9F.4 Evaluation and approximation of the integrals

It should be clear that the computation of energies and coefficients depends upon knowing the values of numerous integrals, all of which must be evaluated numerically. How that is done is central to the variety of software packages currently available. The integrals that are appearing are becoming seriously cumbersome to write and so from now the following notation is used:

$$(AB|CD)=j_0\int\chi_A(1)\chi_B(1)\frac{1}{r_{12}}\chi_C(2)\chi_D(2)d\tau_1 d\tau_2$$

Integral notation (9F.8)

Integrals like this are fixed throughout the calculation (for a particular nuclear geometry) because they depend only on the choice of basis, so they can be tabulated once and for all and then used whenever required. In many cases the integrals have a clear physical significance, which can be used as a guide to their approximation if that is necessary.

Brief illustration 9F.2

When the molecular orbitals $\psi_a = c_{Aa}\chi_A + c_{Ba}\chi_B$ in eqn 9F.7 are expanded in terms of the basis orbitals, terms like the following are obtained:

$$j_0 \int \chi_A(1)\chi_A(1)\frac{1}{r_{12}}\psi_a(2)\psi_a(2)\mathrm{d}\tau_1\mathrm{d}\tau_2$$
$$= c_{Aa}^2(\mathrm{AA|AA}) + 2c_{Aa}c_{Ba}(\mathrm{AA|BA}) + c_{Ba}^2(\mathrm{AA|BB})$$

The last of the three integrals on the right is

$$(\mathrm{AA|BB}) = j_0 \int \chi_A^2(1)\frac{1}{r_{12}}\chi_B^2(2)\mathrm{d}\tau_1\mathrm{d}\tau_2$$

and represents the Coulombic interaction of one electron in an orbital on A with another electron in an orbital on B.

One approach to the evaluation of integrals is brute force: more formally, an ***a priori* method**. Each integral is evaluated numerically after making a particular choice of basis. In some cases it is possible to evaluate an integral analytically. For instance, if the basis consists of two H1s orbitals at a distance R apart, then the integral in *Brief illustration* 9F.2 is

$$(\mathrm{AA|BB}) = \frac{j_0}{R} - \frac{j_0}{2a_0}\left(\frac{2a_0}{R} + \frac{11}{4} + \frac{3R}{2a_0} + \frac{R^2}{2a_0^2}\right)e^{-2R/a_0} \qquad (9F.9)$$

where a_0 is the Bohr radius. All the integrals that occur for H_2 with the H1s basis set can be evaluated similarly. However, as soon as the molecules being considered become more complicated, this approach fails.

In all but the very simplest molecules some of the integrals that arise have atomic orbitals centred on four different atomic nuclei, and (AB|CD) is in general a so-called 'four-centre, two-electron integral'. The number of such integrals increases as the fourth power of the number of basis functions, N_o^4, so for a basis consisting of several dozen functions there will be tens of thousands of integrals of this form to evaluate.

For example, for CH_4 the simplest basis set would consist of four H1s orbitals, one C2s orbital, and three C2p orbitals, giving a total of eight functions. The number of integrals to evaluate would be $8^4 = 4096$. The efficient calculation of such integrals poses the greatest challenge in an HF-SCF calculation but can be alleviated by a clever choice of basis functions.

One of the earliest choices for basis set functions was that of **Slater-type orbitals** (STO) centred on each of the atomic nuclei in the molecule and of the form

$$\chi = Nr^a e^{-br} Y_{l,m_l}(\theta,\phi) \qquad \text{Slater-type orbitals} \quad (9F.10a)$$

where N is a normalization constant, a and b are (non-negative) parameters, Y_{l,m_l} is a spherical harmonic (Table 8A.1), and (r,θ,ϕ) are the spherical polar coordinates describing the location of the electron relative to the atomic nucleus. Several such basis functions are typically centred on each atom, with each basis function characterized by a unique set of values of a, b, l, and m_l.

The introduction of **Gaussian-type orbitals** (GTO), as mentioned in Topic 9E, largely overcame the problem of having to calculate huge numbers of integrals. Cartesian Gaussian functions centred on atomic nuclei have the form

$$\chi = Nx^i y^j z^k e^{-\zeta r^2} \qquad \text{Gaussian-type orbitals} \quad (9F.10b)$$

where (x, y, z) are the Cartesian coordinates of the electron at a distance r from the nucleus, (i, j, k) is a set of non-negative integers, and ζ is a positive constant. An **s-type** Gaussian has $i = j = k = 0$; a **p-type** Gaussian has $i + j + k = 1$; a **d-type** Gaussian has $i + j + k = 2$, and so on.

The advantage of GTOs is that the product of two Gaussian functions on different centres is equivalent to a single Gaussian function located at a point between the two centres. Therefore, two-electron integrals on three and four different atomic centres can be reduced to integrals over two different centres, which are much easier to evaluate numerically.

An alternative approach is to estimate the integrals, perhaps by varying them until certain computed molecular properties (such as enthalpies of formation) match experimentally determined values. An extreme version of these **semi-empirical procedures** is the Hückel procedure where many integrals are simply ignored. More sophisticated procedures discard integrals based on various criteria. For instance, in the procedure called **complete neglect of differential overlap** (CNDO), all (AB|CD) integrals are set to zero unless χ_A and χ_B are the same, and likewise for χ_C and χ_D. That is, only integrals of the form

$$(\mathrm{AA|CC}) = j_0 \int \chi_A^2(1)\frac{1}{r_{12}}\chi_C^2(2)\mathrm{d}\tau_1\mathrm{d}\tau_2$$

survive and are then set equal to values that result in agreement with experiment on certain classes of molecules.

The origin of the term 'differential overlap' is that what is normally taken to be a measure of 'overlap' is the integral $\int \chi_A \chi_B \mathrm{d}\tau$. The differential of an integral of a function is the function itself, so in this sense the 'differential' overlap is the product $\chi_A\chi_B$. The implication of the CNDO approach is to note which orbitals, χ_A and χ_B, occur in the integral. If they are the same, the integral is evaluated; if they are different, the integral is ignored. More recent semi-empirical methods make less severe decisions about which integrals are to be ignored, but they are all descendants of the early CNDO technique. For instance, in one modification, (AB|AB) is also allowed to survive if A and B are on the same atom (for instance, if A is a C2s orbital and B is a C2p orbital in methane).

Brief illustration 9F.3

The integral in eqn 9F.9 is expressed in terms of the three integrals (AA|AA), (AA|BA), and (AA|BB). Only the first and third integrals survive under CNDO, so in this approximation

$$j_0\int \chi_A(1)\chi_A(1)\frac{1}{r_{12}}\psi_a(2)\psi_a(2)\mathrm{d}\tau_1\mathrm{d}\tau_2 \approx c_{Aa}^2(\mathrm{AA|AA}) + c_{Ba}^2(\mathrm{AA|BB})$$

The integral (AA|AA) is the energy of interaction between the two electrons both localized on A, and the integral (AA|BB) is the energy of interaction when one electron is on nucleus A and the other is on B.

Exercises

E9F.4 Write out in full the integral represented by (AA|AA). What interaction does this integral represent?

E9F.5 Which values of a, l, and m_l in a Slater-type orbital would you choose to best represent (i) a 1s atomic orbital, (ii) a 2p atomic orbital, (iii) a 2s atomic orbital. Are there any features of the atomic orbital which the Slater-type orbital does not represent?

9F.5 Density functional theory

As remarked in Topic 9E, a technique that has gained considerable ground in recent years to become one of the most widely used procedures for the calculation of molecular structure is **density functional theory** (DFT). As mentioned there, the central focus of DFT is the electron probability density, ρ. The 'functional' part of the name comes from the fact that the energy of the molecule is a function of the electron density $\rho(\boldsymbol{r})$ and the electron density is itself a function of the location in the molecule. In mathematics a function of a function is called a *functional*, and in this specific case the energy is written as the **functional** $E[\rho]$. Functionals have been used elsewhere in this text but without using the terminology. For example the expectation value of the hamiltonian is the energy expressed as a functional of the wavefunction because a single value of the energy, $E[\psi]$, is associated with each function ψ.

As discussed earlier in this Topic, the energy of a molecule is a sum of the kinetic energy, the electron–nuclear interactions, and the electron–electron interaction. The first two contributions depend on the electron density distribution. The electron–electron interaction is likely to depend on the same quantity, but it is modified by a quantum mechanical exchange term (the contribution which in Hartree–Fock theory is expressed by the operator $\hat{K}$). That the exchange contribution can be expressed in terms of the electron density is not at all obvious, but in 1964 P. Hohenberg and W. Kohn were able to prove that the exact ground-state energy of a molecule is uniquely determined by the electron probability density. They showed that it is possible to write

$$E[\rho] = E_{\text{Classical}}[\rho] + E_{\text{XC}}[\rho] \qquad \text{Energy functional} \qquad (9\text{F}.11)$$

where $E_{\text{Classical}}[\rho]$ is the sum of the contributions of kinetic energy, electron–nucleus interactions, and the classical electron–electron potential energy, and $E_{\text{XC}}[\rho]$ is the **exchange–correlation energy**. This term takes into account all the non-classical electron–electron effects due to spin and applies small corrections to the kinetic energy part of $E_{\text{Classical}}$ that arise from electron–electron interactions. The crucial point is that these quantum mechanical effects can be expressed in terms of the classically significant electron density, not the quantum mechanical wavefunction itself. The **Hohenberg–Kohn theorem** guarantees the existence of $E_{\text{XC}}[\rho]$ but—like so many existence theorems in mathematics—gives no clue about how it should be calculated.

The first step in the implementation of this approach is to calculate the electron density. The relevant equations were deduced by W. Kohn and L.J. Sham in 1965, who showed that ρ can be expressed as a contribution from each electron present in the molecule: for H_2 it is simply $\rho(\boldsymbol{r}) = 2\psi_a^2(\boldsymbol{r})$. The function ψ_i is called a **Kohn–Sham orbital** and is a solution of the **Kohn–Sham equation**, which closely resembles the form of the Schrödinger equation (on which it is based). For a two-electron system the Kohn–Sham equation is

$$\hat{H}_1\psi_a(1) + j_0\int\frac{\rho(2)}{r_{12}}\mathrm{d}\tau_2\ \psi_a(1) + V_{\text{XC}}(1)\psi_a(1) = \varepsilon_a\psi_a(1)$$

Kohn-Sham equation (two electrons) (9F.12)

where, as usual, 1 stands for the location $\boldsymbol{r}_1$ of electron 1 and 2 for the location $\boldsymbol{r}_2$ of electron 2. The ε_a is the **Kohn–Sham orbital energy**. The first term on the left is familiar from the earlier discussion and accounts for the kinetic energy of electron 1 and its interactions with the nuclei. The second term is the classical interaction between electron 1 and electron 2. The third term, the one involving the **exchange–correlation potential**, V_{XC}, takes exchange effects into account. To understand its significance, note that if the electron density is modified everywhere (or just somewhere), the exchange correlation energy will also change by a certain amount. Specifically, V_{XC} is simply a function that enables the change in the exchange correlation energy to be related to any change in the electron density by evaluating the following integral:

$$\overbrace{\delta E_{\text{XC}}[\rho]}^{\text{change in } E_{\text{XC}}[\rho] \text{ arising from...}} = \int V_{\text{XC}}(\boldsymbol{r})\ \overbrace{\delta\rho(\boldsymbol{r})}^{\text{...a change in } \rho(\boldsymbol{r}) \text{ everywhere}}\ \mathrm{d}\tau \qquad \text{Exchange-correlation potential [definition]} \qquad (9\text{F}.13)$$

The greatest challenge in density functional theory is to find an accurate expression for the exchange–correlation energy and through that identify the exchange–correlation potential that relates it to changes in the electron density distribution in the molecule.

One widely used but approximate form for $E_{XC}[\rho]$ is based on the model of a uniform electron gas, a hypothetical electrically neutral system in which electrons move in a space of continuous and uniform distribution of positive charge. For a uniform electron gas

$$E_{XC}[\rho] = A\int \rho(\mathbf{r})^{4/3}\,d\tau \quad A = -\tfrac{9}{8}j_0\left(\frac{3}{\pi}\right)^{1/2}$$

Uniform electron gas (9F.14a)

It is now possible to identify the exchange-correlation energy.

How is that done? 9F.1 Identifying an exchange-correlation energy

To identify the exchange-correlation energy in eqn 9F.13 you need to know the definition of a 'functional derivative' of a functional $F[f]$ of a function f, $\delta F[f]$:

$$\delta F[f] = \int \frac{\delta F[f]}{\delta f}\,\delta f\,d\tau$$

For the functional $E_{XC}[\rho]$, substitute the expression in eqn 9F.14a:

$$\delta E_{XC}[\rho] = \int \underbrace{\frac{\delta E_{XC}[\rho]}{\delta \rho}}_{V_{XC}}\,\delta\rho\,d\tau = \int \frac{\delta A\rho^{4/3}}{\delta\rho}\,\delta\rho\,d\tau$$

$$= \int \tfrac{4}{3}A\rho^{1/3}\,\delta\rho\,d\tau$$

You can conclude that

$$V_{XC}(\mathbf{r}) = \tfrac{4}{3}A\rho(\mathbf{r})^{1/3}$$

(9F.15b) Uniform electron gas

There is now enough equipment assembled to understand the strategy of a DFT calculation, bearing in mind that the Kohn–Sham equations must be solved iteratively and self-consistently. First, the electron density is estimated; it is common to use a superposition of atomic electron probability densities. Second, the exchange–correlation potential is calculated by assuming an approximate form, such as that for the free electron gas. Next, the Kohn–Sham equations are solved to obtain an initial set of Kohn–Sham orbitals. This set of orbitals is used to obtain a better approximation to the electron probability density and the process is repeated until the density remains constant to within some specified tolerance.

To apply such a process to H_2, begin by assuming that the electron density is a sum of atomic electron densities arising from the presence of electrons in the atomic orbitals χ_A and χ_B (which may be STOs or GTOs) and write $\rho(i) = |\chi_A(i)|^2 + |\chi_B(i)|^2$ for electron $i = 1$ and 2. With the exchange–correlation potential in eqn 9F.14b, the Kohn–Sham orbital for the molecule is a solution of

$$\hat{H}_1\psi_a(1) + j_0\int\frac{\rho(2)}{r_{12}}d\tau_2\;\psi_a(1) + \tfrac{4}{3}\rho(1)^{1/3}\psi_a(1) = \varepsilon_a\psi_a(1)$$

Now insert the $\rho(1)$ and $\rho(2)$, and solve this equation numerically for ψ_a. Once that orbital has been obtained, replace the original guess at the electron density by $\rho(i) = |\psi_a(i)|^2$, $i = 1$ and 2. This density is then substituted back into the Kohn–Sham equation to obtain an improved function ψ_a and exchange–correlation energy, and hence the exchange–correlation potential. The process is repeated until the density and exchange–correlation energy are unchanged to within a specified tolerance on successive iterations.

When convergence of the iterations has been achieved, the electronic energy is calculated from

$$E[\rho] = 2\int\psi_a(1)\hat{H}_1\psi_a(1)d\tau_1 + j_0\int\frac{\rho(1)\rho(2)}{r_{12}}d\tau_1 d\tau_2 - A\int\rho(1)^{4/3}d\tau_1 \qquad (9F.16)$$

where the first term is the sum of the energies of the two electrons in the field of the two nuclei, the second term is the electron–electron repulsion, and the final term includes the correction due to nonclassical electron–electron effects.

Checklist of concepts

☐ 1. A **spinorbital** is the combination of a spatial orbital and a spin function.

☐ 2. A **Slater determinant** ensures that a many-electron wavefunction satisfies the Pauli principle.

☐ 3. The **Hartree–Fock self-consistent field** (HF-SCF) procedure is a systematic solution of the Schrödinger equation for a many-electron system.

☐ 4. The Hartree–Fock method ignores **electron correlation**, the tendency of electrons to avoid one another to minimize repulsions.

☐ 5. The **Roothaan equations** are used to express the Hartree–Fock procedure in terms of linear combinations of a basis set of atomic orbitals.

☐ 6. In the ***a priori* method**, each molecular integral is evaluated numerically.

☐ 7. In a **semi-empirical procedure** many integrals are either ignored or estimated on the basis of experimental data.

☐ 8. In **density functional theory** (DFT), the focus is on the electron density itself rather than the wavefunction.

☐ 9. The **Hohenberg–Kohn theorem** guarantees that the energy of a molecule is determined solely by the electron density.

☐ 10. A **Kohn–Sham orbital** is a solution of the **Kohn–Sham equation**.

Checklist of equations

Property	Equation	Comment	Equation number
Spinorbital	$\psi_a^\alpha(i) = \psi_a(i)\alpha(i)$	Electron i in spatial orbital ψ_a with spin α	
Slater determinant	$\Psi = \frac{1}{2^{1/2}}\begin{vmatrix}\psi_a^\alpha(1) & \psi_a^\beta(1)\\ \psi_a^\alpha(2) & \psi_a^\beta(2)\end{vmatrix}$	Satisfies Pauli principle	9F.3a
Hartree–Fock equations	$f_1\psi_a(1) = \varepsilon_a\psi_a(1)$		9F.4
Integral notation	$(AB\|CD) = j_0 \int \chi_A(1)\chi_B(1)\frac{1}{r_{12}}\chi_C(2)\chi_D(2)\mathrm{d}\tau_1\mathrm{d}\tau_2$		9F.8
Slater-type orbital	$\chi = Nr^a \mathrm{e}^{-br} Y_{l,m_l}(\theta,\phi)$		9F.10a
Gaussian-type orbital	$\chi = Nx^i y^j z^k \mathrm{e}^{-\zeta r^2}$		9F.10b

FOCUS 9 Molecular structure

To test your understanding of this material, work through the *Exercises, Additional exercises, Discussion questions,* and *Problems* found throughout this Focus.

Selected solutions can be found at the end of this Focus in the e-book. Solutions to even-numbered questions are available online only to lecturers.

TOPIC 9A Valence-bond theory

Discussion questions

D9A.1 Discuss the role of the Born–Oppenheimer approximation in the valence-bond calculation of a molecular potential energy curve or surface.

D9A.2 Why are promotion and hybridization invoked in valence-bond theory?

D9A.3 Describe the various types of hybrid orbitals and how they are used to describe the bonding in alkanes, alkenes, and alkynes. How does hybridization explain that in allene, $CH_2{=}C{=}CH_2$, the two CH_2 groups lie in perpendicular planes?

D9A.4 Why is spin-pairing so common a features of bond formation (in the context of valence-bond theory)?

D9A.5 What are the consequences of resonance?

Additional exercises

E9A.6 Write the valence-bond wavefunction for the triple bond in N_2.

E9A.7 Write the valence-bond wavefunction for the resonance hybrid $N_2 \leftrightarrow N^+N^- \leftrightarrow N^{2-}N^{2+} \leftrightarrow$ structures of similar energy.

E9A.8 Describe the structures of SO_2 and SO_3 in terms of valence-bond theory.

E9A.9 Account for the ability of phosphorus to form five bonds, as in PF_5.

E9A.10 Describe the bonding in 1,3-pentadiene using hybrid orbitals.

E9A.11 Describe the bonding in methylamine, CH_3NH_2, using hybrid orbitals.

E9A.12 Describe the bonding in pyridine, C_5H_5N, using hybrid orbitals.

E9A.13 Show that the linear combinations $h_1 = s + p_x + p_y + p_z$ and $h_2 = s - p_x - p_y + p_z$ are mutually orthogonal.

E9A.14 Show that the linear combinations $h_1 = (\sin\zeta)s + (\cos\zeta)p$ and $h_2 = (\cos\zeta)s - (\sin\zeta)p$ are mutually orthogonal for all values of the angle ζ (zeta).

E9A.15 Normalize to 1 the sp^2 hybrid orbital $h = s + 2^{1/2}p$ given that the s and p orbitals are each normalized to 1.

E9A.16 Normalize to 1 the linear combinations $h_1 = (\sin\zeta)s + (\cos\zeta)p$ and $h_2 = (\cos\zeta)s - (\sin\zeta)p$, given that the s and p orbitals are each normalized to 1.

Problems

P9A.1 Use the wavefunction for a H1s orbital to write a valence-bond wavefunction of the form $\Psi(1,2) = A(1)B(2) + A(2)B(1)$ in terms of the Cartesian coordinates of each electron, given that the internuclear separation (along the z-axis) is R.

P9A.2 An sp^2 hybrid orbital that lies in the xy-plane and makes an angle of 120° to the x-axis has the form

$$\psi = \frac{1}{3^{1/2}}\left(s - \frac{1}{2^{1/2}}p_x + \frac{3^{1/2}}{2^{1/2}}p_y\right)$$

Use a graphical argument to show that this function points in the specified direction. (*Hint:* Consider the p_x and p_y orbitals as being represented by unit vectors along x and y.)

P9A.3 Confirm that the hybrid orbitals in eqn 9A.7 make angles of 120° to each other. (*Hint:* Consider the p_x and p_y orbitals as being represented by unit vectors along x and y.)

P9A.4 Show that if two equivalent hybrid orbitals of the form sp^λ make an angle θ to each other, then $\lambda = \pm(-1/\cos\theta)^{1/2}$. Plot a graph of λ against θ and confirm that $\theta = 180°$ when no s orbital is included and $\theta = 120°$ when $\lambda = 2$.

TOPIC 9B Molecular orbital theory: the hydrogen molecule-ion

Discussion questions

D9B.1 Discuss the role of the Born–Oppenheimer approximation in the molecular-orbital calculation of a molecular potential energy curve or surface.

D9B.2 What feature of molecular orbital theory is responsible for bond formation?

D9B.3 Why is spin-pairing so common a feature of bond formation (in the context of molecular orbital theory)?

Additional exercises

E9B.4 Normalize to 1 the molecular orbital $\psi = \psi_A + \lambda\psi_B + \lambda'\psi'_B$ in terms of the parameters λ and λ' and the appropriate overlap integrals, where ψ_B and ψ'_B are mutually orthogonal and normalized orbitals on atom B.

E9B.5 Suppose that a molecular orbital has the (unnormalized) form $0.145A + 0.844B$. Find a linear combination of the orbitals A and B that is orthogonal to this combination and determine the normalization constants of both combinations using $S = 0.250$.

E9B.6 Suppose that a molecular orbital has the (unnormalized) form $0.727A + 0.144B$. Find a linear combination of the orbitals A and B that is orthogonal to this combination and determine the normalization constants of both combinations using $S = 0.117$.

E9B.7 The energy of H_2^+ with internuclear separation R is given by eqn 9B.4. The values of the contributions are given below. Plot the molecular potential energy curve and find the bond dissociation energy (in electronvolts) and the equilibrium bond length.

R/a_0	0	1	2	3	4
j/j_0	1.000	0.729	0.472	0.330	0.250
k/j_0	1.000	0.736	0.406	0.199	0.092
S	1.000	0.858	0.587	0.349	0.189

where $j_0 = 27.2$ eV, $a_0 = 52.9$ pm, and $E_{H1s} = -\frac{1}{2}j_0$.

E9B.8 The energy of H_2^+ with internuclear separation R is given by eqn 9B.4. The values of the contributions are given below.

R/a_0	0	1	2	3	4
j/E_h	1.000	0.729	0.472	0.330	0.250
k/E_h	1.000	0.736	0.406	0.199	0.092
S	1.000	0.858	0.587	0.349	0.189

where $E_h = 27.2$ eV, $a_0 = 52.9$ pm, and $E_{H1s} = -\frac{1}{2}E_h$. Plot the molecular potential energy curve for the antibonding orbital, which is given by eqn 9B.7.

E9B.9 Identify the g or u character of bonding and antibonding δ orbitals formed by face-to-face overlap of d atomic orbitals.

Problems

P9B.1 Calculate the (molar) energy of electrostatic repulsion between two hydrogen nuclei at the separation in H_2 (74.1 pm). The result is the energy that must be overcome by the attraction from the electrons that form the bond. Does the gravitational attraction between the nuclei play any significant role? *Hint:* The gravitational potential energy of two masses is equal to $-Gm_1m_2/r$; the gravitational constant G is listed on page i.

P9B.2 Imagine a small electron-sensitive probe of volume 1.00 pm^3 inserted into an H_2^+ molecule-ion in its ground state. Calculate the probability that it will register the presence of an electron at the following positions: (a) at nucleus A, (b) at nucleus B, (c) half way between A and B, (d) at a point 20 pm along the bond from A and 10 pm perpendicularly. Do the same for the molecule-ion the instant after the electron has been excited into the antibonding LCAO-MO. Take $R = 2.00a_0$.

P9B.3 Examine whether occupation of the bonding orbital in the H_2^+ molecule-ion by one electron has a greater or lesser bonding effect than occupation of the antibonding orbital by one electron. Is your conclusion true at all internuclear separations?

P9B.4 Use mathematical software or a spreadsheet to plot the amplitude of the σ and σ* wavefunctions (given in eqn 9B.2) along the z-axis for different values of the internuclear distance. Identify the features of the σ orbital that lead to bonding, and of the σ* orbital that lead to antibonding. The forms of the two H1s orbitals are $\psi_A = (\pi a_0^3)^{-1/2}e^{-r_{A1}/a_0}$ and $\psi_B = (\pi a_0^3)^{-1/2}e^{-r_{B1}/a_0}$ where r_{A1} is the distance from the electron to nucleus A, and r_{B1} is the distance from the electron to nucleus B.

P9B.5 (a) Calculate the total amplitude of the normalized bonding and antibonding LCAO-MOs that may be formed from two H1s orbitals at a separation of $2a_0 = 106$ pm. Plot the two amplitudes for positions along the molecular axis both inside and outside the internuclear region. (b) Plot the probability densities of the two orbitals. Then form the *difference density*, the difference between ψ^2 and $\frac{1}{2}(\psi_A^2 + \psi_B^2)$.

TOPIC 9C Molecular orbital theory: homonuclear diatomic molecules

Discussion questions

D9C.1 Draw diagrams to show the various orientations in which a p orbital and a d orbital on adjacent atoms may form bonding and antibonding molecular orbitals.

D9C.2 Outline the rules of the building-up principle for homonuclear diatomic molecules.

D9C.3 What is the justification for treating s and p atomic orbital contributions to molecular orbitals separately?

D9C.4 To what extent can orbital overlap be related to bond strength? To what extent might that be a correlation rather than an explanation?

Additional exercises

E9C.6 Give the ground-state electron configurations and bond orders of (i) F_2^-, (ii) N_2, and (iii) O_2^{2-}.

E9C.7 From the ground-state electron configurations of Li_2 and Be_2, predict which molecule should have the greater dissociation energy.

E9C.8 Arrange the species O_2^+, O_2, O_2^-, O_2^{2-} in order of increasing bond length.

E9C.9 Evaluate the bond order of each Period 2 homonuclear diatomic molecule.

E9C.10 Evaluate the bond order of each Period 2 homonuclear diatomic cation, X_2^+, and anion, X_2^-.

E9C.11 For each Period 2 homonuclear diatomic cation, X_2^+, and anion, X_2^-, specify which molecular orbital is the HOMO (the highest energy occupied orbital).

E9C.12 For each Period 2 homonuclear diatomic cation, X_2^+, and anion, X_2^-, specify which molecular orbital is the LUMO (the lowest energy unoccupied orbital).

E9C.13 What is the speed of a photoelectron ejected from a molecule with radiation of energy 21 eV and known to come from an orbital of ionization energy 12 eV?

Problems

P9C.1 Familiarity with the magnitudes of overlap integrals is useful when considering bonding abilities of atoms, and hydrogenic orbitals give an indication of their values. (a) The overlap integral between two hydrogenic 2s orbitals is

$$S(2s,2s)=\left\{1+\frac{ZR}{2a_0}+\frac{1}{12}\left(\frac{ZR}{a_0}\right)^2+\frac{1}{240}\left(\frac{ZR}{a_0}\right)^4\right\}e^{-ZR/2a_0}$$

Plot this expression. (b) For what internuclear distance is $S(2s,2s)=0.50$? (c) The side-by-side overlap of two 2p orbitals of atoms of atomic number Z is

$$S(2p,2p)=\left\{1+\frac{ZR}{2a_0}+\frac{1}{10}\left(\frac{ZR}{a_0}\right)^2+\frac{1}{120}\left(\frac{ZR}{a_0}\right)^3\right\}e^{-ZR/2a_0}$$

Plot this expression. (d) Evaluate $S(2s,2p)$ at the internuclear distance you calculated in part (b).

P9C.2 Before doing a calculation, sketch how the overlap between a 1s orbital and a 2p orbital directed towards it can be expected to depend on their separation. The overlap integral between an H1s orbital and an H2p orbital directed towards it on nuclei separated by a distance R is $S=(R/a_0)\{1+(R/a_0)+\frac{1}{3}(R/a_0)^2\}e^{-R/a_0}$. Plot this function, and find the separation for which the overlap is a maximum.

P9C.3‡ Use the $2p_x$ and $2p_z$ hydrogenic atomic orbitals to construct simple LCAO descriptions of $2p\sigma$ and $2p\pi$ molecular orbitals. (a) Make a probability density plot, and both surface and contour plots of the xz-plane amplitudes of the $2p_z\sigma$ and $2p_z\sigma$ molecular orbitals. (b) Plot the amplitude of the $2p_x\pi$ and $2p_x\pi$ molecular orbital wavefunctions in the xz-plane. Include plots for both an internuclear distance, R, of $10a_0$ and $3a_0$, where $a_0=52.9$ pm. Interpret the graphs, and explain why this graphical information is useful.

P9C.4 In a photoelectron spectrum using 21.21 eV photons, electrons were ejected with kinetic energies of 11.01 eV, 8.23 eV, and 15.22 eV. Sketch the molecular orbital energy level diagram for the species, showing the ionization energies of the three identifiable orbitals.

TOPIC 9D Molecular orbital theory: heteronuclear diatomic molecules

Discussion questions

D9D.1 Describe the Pauling and Mulliken electronegativity scales. Why should they be approximately in step?

D9D.2 Why do both ionization energy and electron affinity play a role in estimating the energy of an atomic orbital to use in a molecular orbital calculation?

D9D.3 Discuss the steps involved in the calculation of the energy of a system by using the variation principle. Are any assumptions involved?

D9D.4 What is the physical significance of the Coulomb and resonance integrals?

Additional exercises

E9D.4 Give the ground-state electron configurations of (i) CO, (ii) NO, and (iii) CN^-.

E9D.5 Give the ground-state electron configurations of (i) XeF, (ii) PN, and (iii) SO^-.

E9D.6 Sketch the molecular orbital energy level diagram for XeF and deduce its ground-state electron configuration. Is XeF likely to have a shorter bond length than XeF^+?

E9D.7 Sketch the molecular orbital energy level diagram for IF and deduce its ground-state electron configuration. Is IF likely to have a shorter bond length than IF^- or IF^+?

E9D.8 Use the electron configurations of NO^- and NO^+ to predict which is likely to have the shorter bond length.

E9D.9 Use the electron configurations of SO^- and SO^+ to predict which is likely to have the shorter bond length.

E9D.10 A reasonably reliable conversion between the Mulliken and Pauling electronegativity scales is given by eqn 9D.4. Use Table 9D.1 in the *Resource section* to assess how good the conversion formula is for Period 2 elements.

E9D.11 A reasonably reliable conversion between the Mulliken and Pauling electronegativity scales is given by eqn 9D.4. Use Table 9D.1 in the *Resource section* to assess how good the conversion formula is for Period 3 elements.

E9D.12 Estimate the orbital energies to use in a calculation of the molecular orbitals of HBr. For data, see Tables 8B.4 and 8B.5. Take $\beta=-1.00$ eV.

E9D.13 Estimate the molecular orbital energies in HBr; use $S=0$. *Hint:* Estimate the atomic orbital energies by using the data in Tables 8B.4 and 8B.5; take $\beta=-1.00$ eV.

E9D.14 Estimate the molecular orbital energies in HBr; use $S=0.20$. *Hint:* Estimate the atomic orbital energies by using the data in Tables 8B.4 and 8B.5; take $\beta=-1.00$ eV.

‡ These problems were supplied by Charles Trapp and Carmen Giunta.

Problems

P9D.1 Show, if overlap is ignored, (a) that if a molecular orbital is expressed as a linear combination of two atomic orbitals in the form $\psi = \psi_A \cos\theta + \psi_B \sin\theta$, where θ is a parameter that varies between 0 and π, with ψ_A and ψ_B are orthogonal and normalized to 1, then ψ is also normalized to 1. (b) To what values of θ do the bonding and antibonding orbitals in a homonuclear diatomic molecule correspond?

P9D.2 (a) Suppose that a molecular orbital of a heteronuclear diatomic molecule is built from the orbital basis A, B, and C, where B and C are both on one atom. Set up the secular equations for the values of the coefficients and the corresponding secular determinant. (b) Now let $\alpha_A = -7.2$ eV, $\alpha_B = -10.4$ eV, $\alpha_C = -8.4$ eV, $\beta_{AB} = -1.0$ eV, $\beta_{AC} = -0.8$ eV, and calculate the orbital energies and coefficients with both S_{AB} and S_{AC} equal to (i) 0, (ii) 0.2 (note that $S_{BC} = 0$ for orbitals on the same atom).

P9D.3 (a) Suppose that a molecular orbital of a heteronuclear diatomic molecule is built from the orbital basis A, B, and C, where B and C are both on one atom. Set up the secular equations for the values of the coefficients and the corresponding secular determinant. (b) Now let $\alpha_A = -7.2$ eV, $\alpha_B = -10.4$ eV, $\alpha_C = -8.4$ eV, $\beta_{AB} = -1.0$ eV, $\beta_{AC} = -0.8$ eV, and calculate the orbital energies and coefficients with both S_{AB} and S_{AC} equal to 0 (note that $S_{BC} = 0$ for orbitals on the same atom). (c) Explore the consequences of increasing the energy separation of the ψ_B and ψ_C orbitals. Are you justified in ignoring orbital ψ_C at any stage?

TOPIC 9E Molecular orbital theory: polyatomic molecules

Discussion questions

D9E.1 Discuss the scope, consequences, and limitations of the approximations on which the Hückel method is based.

D9E.2 Distinguish between delocalization energy, π-electron binding energy, and π-bond formation energy. Explain how each concept is employed.

D9E.3 Outline the computational steps used in the self-consistent field approach to electronic structure calculations.

D9E.4 Explain why the use of Gaussian-type orbitals is generally preferred over the use of hydrogenic orbitals in basis sets.

Additional exercises

E9E.7 Set up the secular determinants for (i) linear H_4, (ii) cyclic H_4 within the Hückel approximation.

E9E.8 Predict the electron configurations of (i) the allyl radical, $\bullet CH_2CHCH_2$, (ii) the cyclobutadiene cation $C_4H_4^+$. Estimate the π-electron binding energy in each case.

E9E.9 What is the delocalization energy and π-bond formation energy of (i) the allyl radical, (ii) the cyclobutadiene cation?

E9E.10 Set up the secular determinants for (i) anthracene (**1**), (ii) phenanthrene (**2**) within the Hückel approximation and using the out-of-plane C2p orbitals as the basis set.

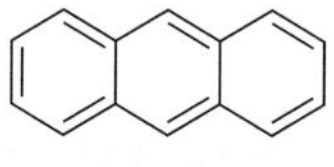

1 Anthracene

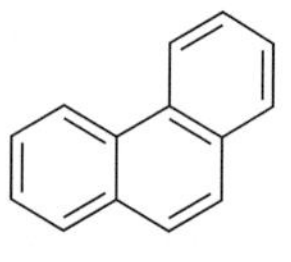

2 Phenanthrene

E9E.11 Set up the secular determinants for (i) azulene (**3**), (ii) acenaphthylene (**4**) within the Hückel approximation and using the out-of-plane C2p orbitals as the basis set.

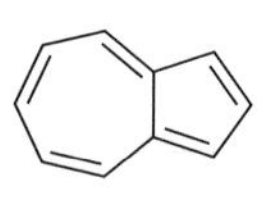

3 Azulene

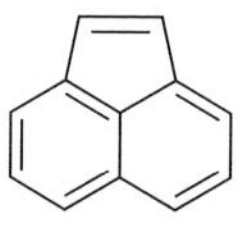

4 Acenaphthalene

E9E.12 Use mathematical software to estimate the π-electron binding energy of (i) anthracene (**1**), (ii) phenanthrene (**2**) within the Hückel approximation.

E9E.13 Use mathematical software to estimate the π-electron binding energy of (i) azulene (**3**), (ii) acenaphthylene (**4**) within the Hückel approximation.

E9E.14 Write the electronic hamiltonian for LiH^{2+}.

Problems

P9E.1 Set up and solve the Hückel secular equations for the π electrons of the triangular, planar CO_3^{2-} ion. Express the energies in terms of the Coulomb integrals α_O and α_C and the resonance integral β. Estimate the delocalization energy of the ion.

P9E.2 For monocyclic conjugated polyenes (such as cyclobutadiene and benzene) with each of N carbon atoms contributing an electron in a 2p orbital, simple Hückel theory gives the following expression for the energies E_k of the resulting π molecular orbitals (all are doubly degenerate except the lowest and highest values of k):

$$E_k = \alpha + 2\beta\cos\frac{2k\pi}{N} \quad k = 0, 1, \ldots, N/2 \text{ for } N \text{ even}$$

$$k = 0, 1, \ldots, (N-1)/2 \text{ for } N \text{ odd}$$

(a) Calculate the energies of the π molecular orbitals of benzene and cyclooctatetraene (**5**). Comment on the presence or absence of degenerate energy levels. (b) Calculate and compare the delocalization energies of benzene (using the expression above) and hexatriene. What do you conclude from your results? You will need the following expression for the energies of the π molecular orbitals of a linear polyene with N carbon atoms:

$E_k = \alpha + 2\beta\cos\{k\pi/(N+1)\}$, $k = 1,2,\ldots,N$. (c) Calculate and compare the delocalization energies of cyclooctatetraene and octatetraene. Are your conclusions for this pair of molecules the same as for the pair of molecules investigated in part (b)?

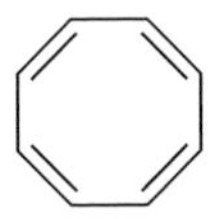

5 Cyclooctatetraene

P9E.3 Suppose that a molecular orbital of a heteronuclear diatomic molecule is built from the orbital basis ψ_A, ψ_B, and ψ_C, where ψ_B and ψ_C are both on one atom (they can be envisaged as F2s and F2p in HF, for instance). Set up the secular equations for the optimum values of the coefficients and set up the corresponding secular determinant.

P9E.4 Set up the secular determinants for the homologous series consisting of ethene, butadiene, hexatriene, and octatetraene and diagonalize them by using mathematical software. Use your results to show that the π molecular orbitals of linear polyenes obey the following rules:

- The π molecular orbital with lowest energy is delocalized over all carbon atoms in the chain.
- The number of nodal planes between C2p orbitals increases with the energy of the π molecular orbital.

P9E.5 Set up the secular determinants for cyclobutadiene, benzene, and cyclooctatetraene and diagonalize them by using mathematical software. Use your results to show that the π molecular orbitals of monocyclic polyenes with an even number of carbon atoms follow a pattern in which:

- The π molecular orbitals of lowest and highest energy are non-degenerate.
- The remaining π molecular orbitals exist as degenerate pairs.

P9E.6 Electronic excitation of a molecule may weaken or strengthen some bonds because bonding and antibonding characteristics differ between the HOMO and the LUMO. For example, a carbon–carbon bond in a linear polyene may have bonding character in the HOMO and antibonding character in the LUMO. Therefore, promotion of an electron from the HOMO to the LUMO weakens this carbon–carbon bond in the excited electronic state, relative to the ground electronic state. Consult Figs. 9E.2 and 9E.4 and discuss in detail any changes in bond order that accompany the $\pi^*\leftarrow\pi$ ultraviolet absorptions in butadiene and benzene.

P9E.7‡ In *Exercise* E9E.1 you are invited to set up the Hückel secular determinant for linear and cyclic H_3. The same secular determinant applies to the molecular ions H_3^+ and D_3^+. The molecular ion H_3^+ was discovered as long ago as 1912 by J.J. Thomson but the equilateral triangular structure was confirmed by M.J. Gaillard et al. (*Phys. Rev.* **A17**, 1797 (1978)) much more recently. The molecular ion H_3^+ is the simplest polyatomic species with a confirmed existence and plays an important role in chemical reactions occurring in interstellar clouds that may lead to the formation of water, carbon monoxide, and ethanol. The H_3^+ ion has also been found in the atmospheres of Jupiter, Saturn, and Uranus. (a) Solve the Hückel secular equations for the energies of the H_3 system in terms of the parameters α and β, draw an energy level diagram for the orbitals, and determine the binding energies of H_3^+, H_3, and H_3^-. (b) Accurate quantum mechanical calculations by G.D. Carney and R.N. Porter (*J. Chem. Phys.* **65**, 3547 (1976)) give the dissociation energy for the process $H_3^+ \rightarrow H + H + H^+$ as 849 kJ mol^{-1}. From this information and data in Table 9C.3, calculate the enthalpy of the reaction $H^+(g) + H_2(g) \rightarrow H_3^+(g)$. (c) From your equations and the information given, calculate a value for the resonance integral β in H_3^+. Then go on to calculate the binding energies of the other H_3 species in (a).

P9E.8‡ There is some indication that other hydrogen ring compounds and ions in addition to H_3 and D_3 species may play a role in interstellar chemistry. According to J.S. Wright and G.A. DiLabio (*J. Phys. Chem.* **96**, 10793 (1992)), H_5^-, H_6, and H_7^+ are particularly stable whereas H_4 and H_5^+ are not. Confirm these statements by Hückel calculations.

P9E.9 Use appropriate electronic structure software and basis sets of your or your instructor's choosing, perform self-consistent field calculations for the ground electronic states of H_2 and F_2. Determine ground-state energies and equilibrium geometries. Compare computed equilibrium bond lengths to experimental values.

P9E.10 Use an appropriate semi-empirical method to compute the equilibrium bond lengths and standard enthalpies of formation of (a) ethanol, (b) 1,4-dichlorobenzene. Compare to experimental values and suggest reasons for any discrepancies.

P9E.11 (a) For a linear conjugated polyene with each of N carbon atoms contributing an electron in a 2p orbital, the energies E_k of the resulting π molecular orbitals are given by:

$$E_k = \alpha + 2\beta\cos\frac{k\pi}{N+1} \qquad k = 1,2,\ldots,N$$

Use this expression to make a reasonable empirical estimate of the resonance integral β for the homologous series consisting of ethene, butadiene, hexatriene, and octatetraene given that $\pi\leftarrow\pi$ ultraviolet absorptions from the HOMO to the LUMO occur at 61 500, 46 080, 39 750, and 32 900 cm^{-1}, respectively. (b) Calculate the π-electron delocalization energy, $E_{\text{deloc}} = E_\pi - n(\alpha+\beta)$, of octatetraene, where E_π is the total π-electron binding energy and n is the total number of π-electrons. (c) In the context of this Hückel model, the π molecular orbitals are written as linear combinations of the carbon 2p orbitals. The coefficient of the jth atomic orbital in the kth molecular orbital is given by:

$$c_{kj} = \left(\frac{2}{N+1}\right)^{1/2} \sin\frac{jk\pi}{N+1} \qquad j = 1,2,\ldots,N$$

Evaluate the coefficients of each of the six 2p orbitals in each of the six π molecular orbitals of hexatriene. Match each set of coefficients (that is, each molecular orbital) with a value of the energy calculated with the expression given in part (a) of the molecular orbital. Comment on trends that relate the energy of a molecular orbital with its 'shape', which can be inferred from the magnitudes and signs of the coefficients in the linear combination that describes the molecular orbital.

FOCUS 9 Molecular structure

Integrated activities

I9.1 The languages of valence-bond theory and molecular orbital theory are commonly combined when discussing unsaturated organic compounds. Construct the molecular orbital energy level diagrams of ethene on the basis that the molecule is formed from the appropriately hybridized CH_2 or CH fragments.

I9.2 Here a molecular orbital theory treatment of the peptide group (**6**) is developed, a group that links amino acids in proteins, and establish the features that stabilize its planar conformation. (a) It will be familiar from introductory chemistry that valence-bond theory explains the planar conformation by invoking delocalization of the π bond over the oxygen, carbon, and nitrogen atoms by resonance:

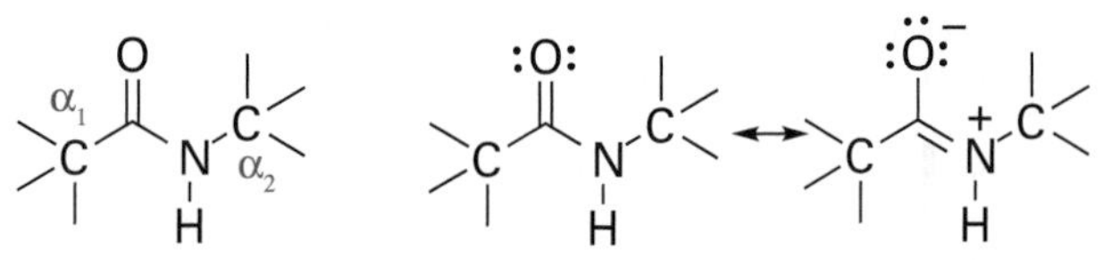

6 Peptide group

It follows that the peptide group can be modelled by using molecular orbital theory by constructing LCAO-MOs from 2p orbitals perpendicular to the plane defined by the O, C, and N atoms. The three combinations have the form:

$$\psi_1 = a\psi_O + b\psi_C + c\psi_N \quad \psi_2 = d\psi_O - e\psi_N \quad \psi_3 = f\psi_O - g\psi_C + h\psi_N$$

where the coefficients a to h are all positive. Sketch the orbitals ψ_1, ψ_2, and ψ_3 and characterize them as bonding, nonbonding, or antibonding. In a nonbonding molecular orbital, a pair of electrons resides in an orbital confined largely to one atom and not appreciably involved in bond formation. (b) Show that this treatment is consistent only with a planar conformation of the peptide link. (c) Draw a diagram showing the relative energies of these molecular orbitals and identify the occupancy of the orbitals. *Hint:* Convince yourself that there are four electrons to be distributed among the molecular orbitals. (d) Now consider a nonplanar conformation of the peptide link, in which the O2p and C2p orbitals are perpendicular to the plane defined by the O, C, and N atoms, but the N2p orbital lies on that plane. The LCAO-MOs are given by

$$\psi_4 = a\psi_O + b\psi_C \quad \psi_5 = e\psi_N \quad \psi_6 = f\psi_O - g\psi_C$$

Just as before, sketch these molecular orbitals and characterize them as bonding, nonbonding, or antibonding. Also, draw an energy level diagram and identify the occupancy of the orbitals. (e) Why is this arrangement of atomic orbitals consistent with a nonplanar conformation for the peptide link? (f) Does the bonding MO associated with the planar conformation have the same energy as the bonding MO associated with the nonplanar conformation? If not, which bonding MO is lower in energy? Repeat the analysis for the nonbonding and antibonding molecular orbitals. (g) Use your results from parts (a)–(f) to construct arguments that support the planar model for the peptide link.

I9.3 Molecular electronic structure methods may be used to estimate the standard enthalpy of formation of molecules in the gas phase. (a) Use a semi-empirical method of your or your instructor's choice to calculate the standard enthalpy of formation of ethene, butadiene, hexatriene, and octatetraene in the gas phase. (b) Consult a database of thermochemical data, and, for each molecule in part (a), calculate the difference between the calculated and experimental values of the standard enthalpy of formation. (c) A good thermochemical database will also report the uncertainty in the experimental value of the standard enthalpy of formation. Compare experimental uncertainties with the relative errors calculated in part (b) and discuss the reliability of your chosen semi-empirical method for the estimation of thermochemical properties of linear polyenes.

I9.4 The standard potential of a redox couple is a measure of the thermodynamic tendency of an atom, ion, or molecule to accept an electron (Topic 6D). Studies indicate that there is a correlation between the LUMO energy and the standard potential of aromatic hydrocarbons. Do you expect the standard potential to increase or decrease as the LUMO energy decreases? Explain your answer.

I9.5 Molecular orbital calculations may be used to predict trends in the standard potentials of conjugated molecules, such as the quinones and flavins, that are involved in biological electron transfer reactions. It is commonly assumed that decreasing the energy of the LUMO enhances the ability of a molecule to accept an electron into the LUMO, with an accompanying increase in the value of the molecule's standard potential. Furthermore, a number of studies indicate that there is a linear correlation between the LUMO energy and the reduction potential of aromatic hydrocarbons.

(a) The standard potentials at pH = 7 for the one-electron reduction of methyl-substituted 1,4-benzoquinones (**7**) to their respective semiquinone radical anions are:

R_2	R_3	R_5	R_6	$E^{\ominus}$/V
H	H	H	H	0.078
CH_3	H	H	H	0.023
CH_3	H	CH_3	H	−0.067
CH_3	CH_3	CH_3	H	−0.165
CH_3	CH_3	CH_3	CH_3	−0.260

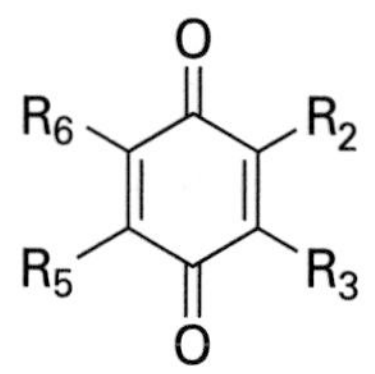

7

Use the computational method of your or your instructor's choice (semi-empirical, *ab initio*, or density functional theory methods) to calculate E_{LUMO}, the energy of the LUMO of each substituted 1,4-benzoquinone, and plot E_{LUMO} against $E^{\ominus}$. Do your calculations support a linear relation between E_{LUMO} and $E^{\ominus}$? (b) The 1,4-benzoquinone for which $R_2 = R_3 = CH_3$ and $R_5 = R_6 = OCH_3$ is a suitable model of ubiquinone, a component of the respiratory electron transport chain. Determine E_{LUMO} of this quinone and then use your results from part (a) to estimate its standard potential. (c) The 1,4-benzoquinone for which $R_2 = R_3 = R_5 = CH_3$ and $R_6 = H$ is a suitable model of plastoquinone, an electron carrier in photosynthesis. Determine E_{LUMO} of this quinone and then use your results from part (a) to estimate its standard potential. Is plastoquinone expected to be a better or worse oxidizing agent than ubiquinone?

I9.6 Molecular orbital calculations based on semi-empirical, *ab initio*, and DFT methods describe the spectroscopic properties of conjugated molecules better than simple Hückel theory. (a) Use the computational method of your or your instructor's choice (semi-empirical, *ab initio*, or density functional methods) to calculate the energy separation between the HOMO and LUMO of ethene, butadiene, hexatriene, and octatetraene. (b) Plot the HOMO–LUMO energy separations against the experimental frequencies for $\pi^* \leftarrow \pi$ ultraviolet absorptions for these molecules (61 500, 46 080, 39 750, and 32 900 cm^{-1}, respectively). Use mathematical software to find the polynomial equation that best fits the data. (b) Use your polynomial fit from part (b) to estimate the wavenumber and wavelength of the $\pi^* \leftarrow \pi$ ultraviolet absorption of decapentaene from the calculated HOMO–LUMO energy separation. (c) Discuss why the calibration procedure of part (b) is necessary.

I9.7 The variation principle can be used to formulate the wavefunctions of electrons in atoms as well as in molecules. Suppose that the function $\psi_{trial} = N(\alpha)e^{-ar^2}$ with $N(\alpha)$ the normalization constant and α an adjustable parameter, is used as a trial wavefunction for the 1s orbital of the hydrogen atom. Show that

$$E(\alpha) = \frac{3\hbar^2\alpha}{2\mu} - \frac{e^2}{2^{1/2}\pi^{3/2}\varepsilon_0}\alpha^{1/2}$$

where e is the fundamental charge, and μ is the reduced mass for the atom. What is the minimum energy associated with this trial wavefunction?

FOCUS 10

Molecular symmetry

In this Focus the concept of 'shape' is sharpened into a precise definition of 'symmetry'. As a result, symmetry and its consequences can be discussed systematically, thereby providing a powerful tool for the prediction and analysis of molecular structure and properties.

10A Shape and symmetry

This Topic shows how to classify any molecule according to its symmetry. Two immediate applications of this classification are the identification of whether or not a molecule can have an electric dipole moment (and so be polar) and whether or not it can be chiral (and so be optically active).

10B Group theory

The systematic treatment of symmetry is an application of 'group theory'. This theory represents the outcome of symmetry operations (such as rotations and reflections) by matrices. This step is important, because once symmetry operations are expressed numerically they can be manipulated quantitatively. The Topic introduces 'character tables' which are key to the application of group theory to chemical problems.

10C Applications of symmetry

Group theory provides simple criteria for deciding whether certain integrals necessarily vanish. One application is to decide whether the overlap integral between two atomic orbitals is necessarily zero and therefore to decide which atomic orbitals can contribute to molecular orbitals. Group theory also provides a way of constructing linear combinations of atomic orbitals that have the correct symmetry to overlap with one another and hence form molecular orbitals. By considering the symmetry properties of integrals, it is also possible to derive the selection rules that govern spectroscopic transitions.

➤ Go to the e-book for videos that feature the derivation and interpretation of equations, and applications of this material.

TOPIC 10A Shape and symmetry

> ➤ **Why do you need to know this material?**
>
> Symmetry arguments can be used to make immediate assessments of the properties of molecules; the initial step is to identify the symmetry a molecule possesses and then to classify it accordingly.
>
> ➤ **What is the key idea?**
>
> Molecules can be classified into groups according to their symmetry elements.
>
> ➤ **What do you need to know already?**
>
> This Topic does not draw on others directly, but it will be useful to be aware of the shapes of a variety of simple molecules and ions encountered in introductory chemistry courses.

Some objects are 'more symmetrical' than others. A sphere is more symmetrical than a cube because it looks the same after it has been rotated through any angle about any axis passing through the centre. A cube looks the same only if it is rotated through certain angles about specific axes (Fig. 10A.1). For example it looks the same after rotation by a multiple of 90° about an axis passing through the centres of opposite faces (a fourfold axis), or after rotation by a multiple of 120° about an axis passing through opposite corners (a threefold axis). Similarly, an NH_3 molecule is 'more symmetrical' than an H_2O molecule because NH_3 looks the same after rotations of 120° or 240° about the axis shown in Fig. 10A.2, whereas H_2O looks the same only after a rotation of 180°.

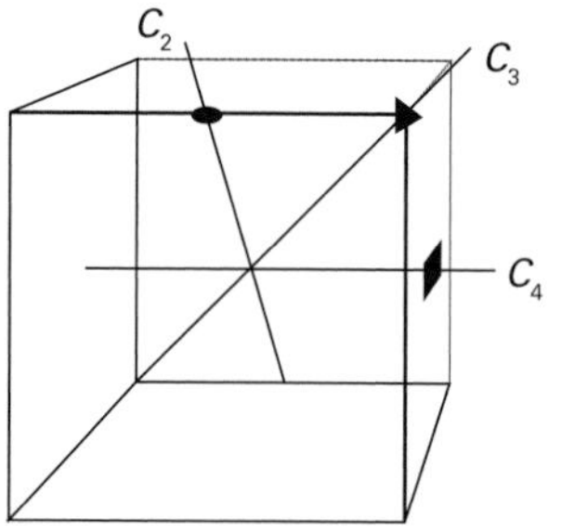

Figure 10A.1 Some of the symmetry elements of a cube. The twofold, threefold, and fourfold axes are labelled with the conventional symbols.

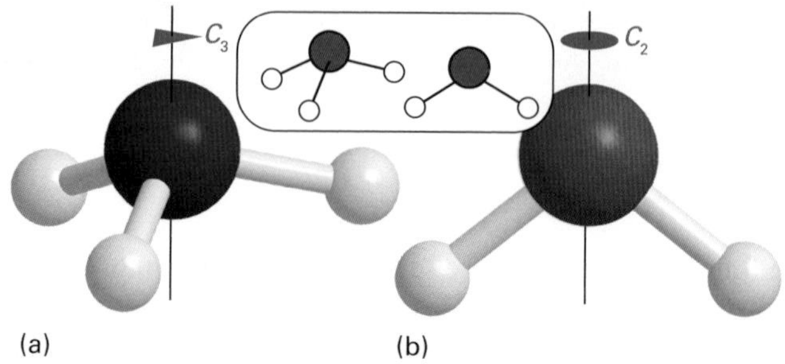

Figure 10A.2 (a) An NH_3 molecule has a threefold (C_3) axis and (b) an H_2O molecule has a twofold (C_2) axis. Both have other symmetry elements too.

This Topic puts these intuitive notions on a more formal foundation. It will be seen that molecules can be grouped together according to their symmetry, with the tetrahedral species CH_4 and SO_4^{2-} in one group and the pyramidal species NH_3 and SO_3^{2-} in another. It turns out that molecules in the same group share certain physical properties, so powerful predictions can be made about whole series of molecules once the group to which they belong has been identified.

10A.1 Symmetry operations and symmetry elements

An action that leaves an object looking the same after it has been carried out is called a **symmetry operation**. Typical symmetry operations include rotations, reflections, and inversions. There is a corresponding **symmetry element** for each symmetry operation, which is the point, line, or plane with respect to which the symmetry operation is performed. For instance, a rotation (a symmetry operation) is carried out around an axis (the corresponding symmetry element). Molecules can be classified by identifying all their symmetry elements, and then grouping together molecules that possess the same set of symmetry elements. This procedure, for example, puts the trigonal planar species BF_3 and CO_3^{2-} into one group and the species H_2O (bent) and ClF_3 (T-shaped) into another group.

An ***n*-fold rotation** (the operation) about an ***n*-fold rotation axis**, C_n (the corresponding element) is a rotation through $360°/n$. An H_2O molecule has one twofold rotation axis, C_2. An NH_3 molecule has one threefold rotation axis, C_3, with which is associated two symmetry operations, one being a rotation by 120° in a clockwise sense, and the other a rotation by 120° in an anticlockwise sense. There is only one twofold rotation associated with a C_2 axis because clockwise and anticlockwise 180° rotations are identical.

A pentagon has a C_5 rotation axis, which has a total of four rotations associated with it. The first two are a clockwise and an anticlockwise rotation through 72°. The second two are denoted C_5^2 and correspond to two successive C_5 rotations—clockwise through 144°, and anticlockwise through 144°.

A cube has three C_4 rotation axes, four C_3 rotation axes, and six C_2 rotation axes. However, even this high symmetry is exceeded by that of a sphere, which possesses an infinite number of rotation axes (along any axis passing through the centre) of all possible integral values of n.

If a molecule possesses several rotation axes, then the one with the highest value of n is called the **principal axis**. The principal axis of a benzene molecule is the sixfold rotation axis perpendicular to the hexagonal ring (**1**). If a molecule has more than one rotation axis with this highest value of n, and it is wished to designate one of them as the principal axis, then it is common to choose the axis that passes through the greatest number of atoms or, in the case of a planar molecule (such as naphthalene, **2**, which has three C_2 rotation axes competing for the title), to choose the axis perpendicular to the plane.

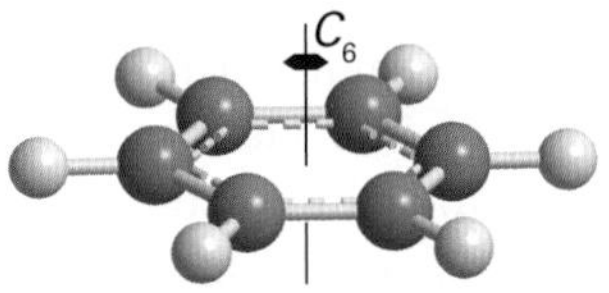

1 Benzene, C_6H_6

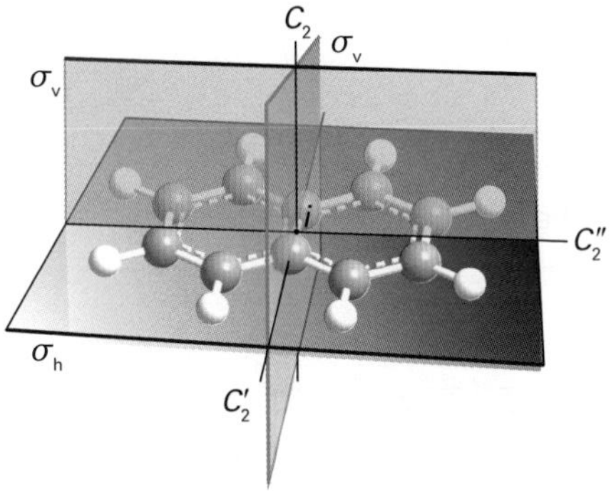

2 Naphthalene, $C_{10}H_8$

A **reflection** is the operation corresponding to a **mirror plane**, σ (the element). If the plane contains the principal axis, it is called 'vertical' and denoted σ_v. An H_2O molecule has two vertical mirror planes (Fig. 10A.3) and an NH_3 molecule has three. A vertical mirror plane that bisects the angle between two C_2 axes is called a 'dihedral plane' and is denoted σ_d (Fig. 10A.4). When the mirror plane is perpendicular to the principal axis it is called 'horizontal' and denoted σ_h. The benzene molecule has such a horizontal mirror plane, perpendicular to the C_6 (principal) axis.

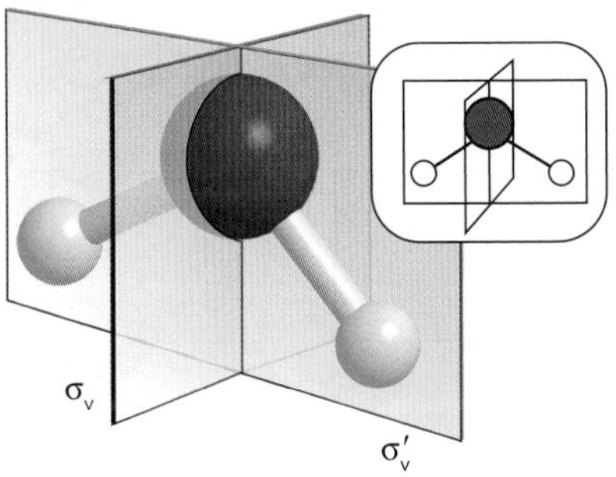

Figure 10A.3 An H_2O molecule has two mirror planes. They are both vertical (i.e. contain the principal axis), so are denoted σ_v and σ_v'.

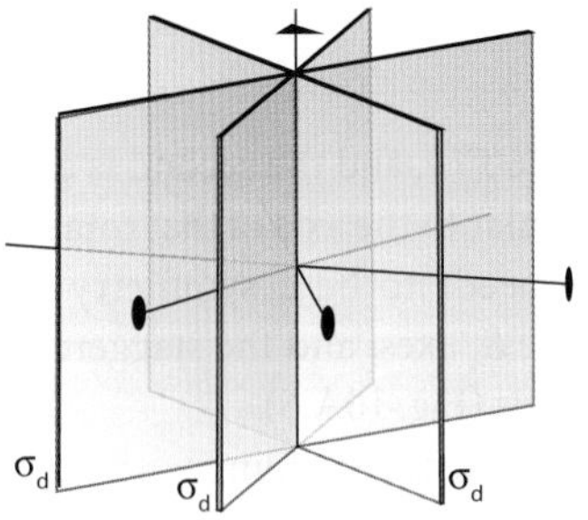

Figure 10A.4 Dihedral mirror planes (σ_d) bisect the C_2 axes perpendicular to the principal axis.

In an **inversion** (the operation) through a **centre of inversion**, i (the element), each point in a molecule is imagined as being moved in a straight line to the centre of inversion and then out the same distance on the other side; that is, the point (x, y, z) is taken into the point $(-x, -y, -z)$. Neither an H_2O molecule nor an NH_3 molecule has a centre of inversion, but a sphere and a cube do have one. A benzene molecule has a centre of inversion, as does a regular octahedron (Fig. 10A.5); a regular tetrahedron and a CH_4 molecule do not.

An ***n*-fold improper rotation** (the operation) about an ***n*-fold improper rotation axis**, S_n (the symmetry element) is composed of two successive transformations. The first is a

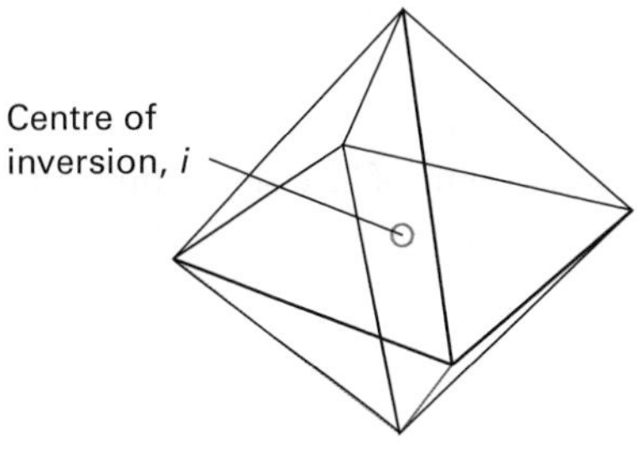

Figure 10A.5 A regular octahedron has a centre of inversion (i).

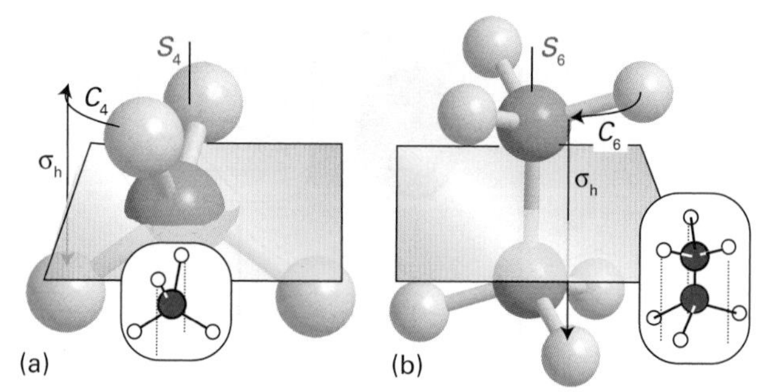

Figure 10A.6 (a) A CH_4 molecule has a fourfold improper rotation axis (S_4): the molecule is indistinguishable after a 90° rotation followed by a reflection across the horizontal plane, but neither operation alone is a symmetry operation. (b) The staggered form of ethane has an S_6 axis composed of a 60° rotation followed by a reflection.

rotation through 360°/n, and the second is a reflection through a plane perpendicular to the axis of that rotation. Neither transformation alone needs to be a symmetry operation. A CH_4 molecule has three S_4 axes, and the staggered conformation of ethane has an S_6 axis (Fig. 10A.6).

The **identity**, E, consists of doing nothing; the corresponding symmetry element is the entire object. Because every molecule is indistinguishable from itself if nothing is done to it, every object possesses at least the identity element. One reason for including the identity is that some molecules have only this symmetry element (**3**).

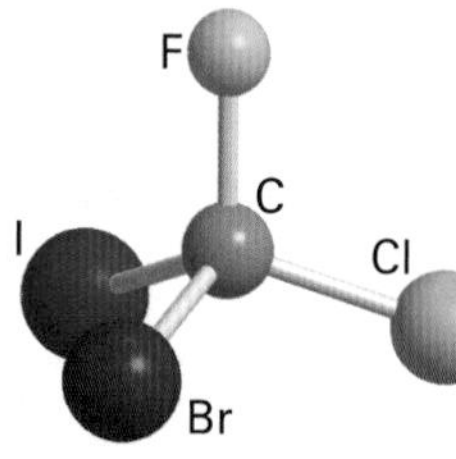

3 CBrClFI

Brief illustration 10A.1

To identify the symmetry elements of a naphthalene molecule (**2**), note that:

- Like all molecules, it has the identity element, E.
- There are three twofold rotation axes, C_2: one perpendicular to the plane of the molecule, and two others lying in the plane.
- With the C_2 axis perpendicular to the plane of the molecule chosen as the principal axis, there is a σ_h plane perpendicular to the principal axis, and two σ_v planes which contain the principal axis.
- There is also a centre of inversion, i, through the midpoint of the molecule, which is mid-way along the C–C bond at the ring junction.

Exercises

E10A.1 List the symmetry elements of CH_3Cl and locate them in a drawing of the molecule.

E10A.2 Locate the symmetry elements of anthracene and locate them in a drawing of the molecule.

E10A.3 Identify all of the symmetry elements possessed by the square planar anion $PtCl_4^{2-}$. Locate them in a drawing of the molecule.

10A.2 The symmetry classification of molecules

Objects are classified into groups according to the symmetry elements they possess. **Point groups** arise when objects are classified according to symmetry elements that correspond to operations leaving at least one common point unchanged. The five kinds of symmetry element identified so far are of this kind. When crystals are considered (Topic 15A), symmetries arising from translation through space also need to be taken into account, and the classification according to these elements gives rise to the more extensive **space groups**.

All molecules with the same set of symmetry elements belong to the same point group, and the name of the group is determined by this set of symmetry elements. There are two systems of notation (Table 10A.1). The **Schoenflies system** (in which a name looks like C_{4v}) is more common for the discussion of individual molecules, and the **Hermann–Mauguin system**, or **International system** (in which a name looks like 4*mm*), is used almost exclusively in the discussion of crystal symmetry. The identification of the point group to which a molecule belongs (in the Schoenflies system) is simplified by referring to the flow diagram in Fig. 10A.7 and to the shapes shown in Fig. 10A.8.

Table 10A.1 The notations for point groups*

C_i	$\bar{1}$								
C_s	m								
C_1	1	C_2	2	C_3	3	C_4	4	C_6	6
		C_{2v}	$2mm$	C_{3v}	$3m$	C_{4v}	$4mm$	C_{6v}	$6mm$
		C_{2h}	$2/m$	C_{3h}	$\bar{6}$	C_{4h}	$4/m$	C_{6h}	$6/m$
		D_2	222	D_3	32	D_4	422	D_6	622
		D_{2h}	mmm	D_{3h}	$\bar{6}2m$	D_{4h}	$4/mmm$	D_{6h}	$6/mmm$
		D_{2d}	$\bar{4}2m$	D_{3d}	$\bar{3}m$	S_4	$\bar{4}$	S_6	$\bar{3}$
T	23	T_d	$\bar{4}3m$	T_h	$m3$				
O	432	O_h	$m3m$						

* Schoenflies notation unboxed, Hermann–Mauguin (International system) boxed. In the Hermann–Mauguin system, a number n denotes the presence of an n-fold axis and m denotes a mirror plane. A slash (/) indicates that the mirror plane is perpendicular to the symmetry axis. It is important to distinguish symmetry elements of the same type but of different classes, as in 4/mmm, in which there are three classes of mirror plane. A bar over a number indicates that the element is combined with an inversion. The only groups listed here are the so-called 'crystallographic point groups'.

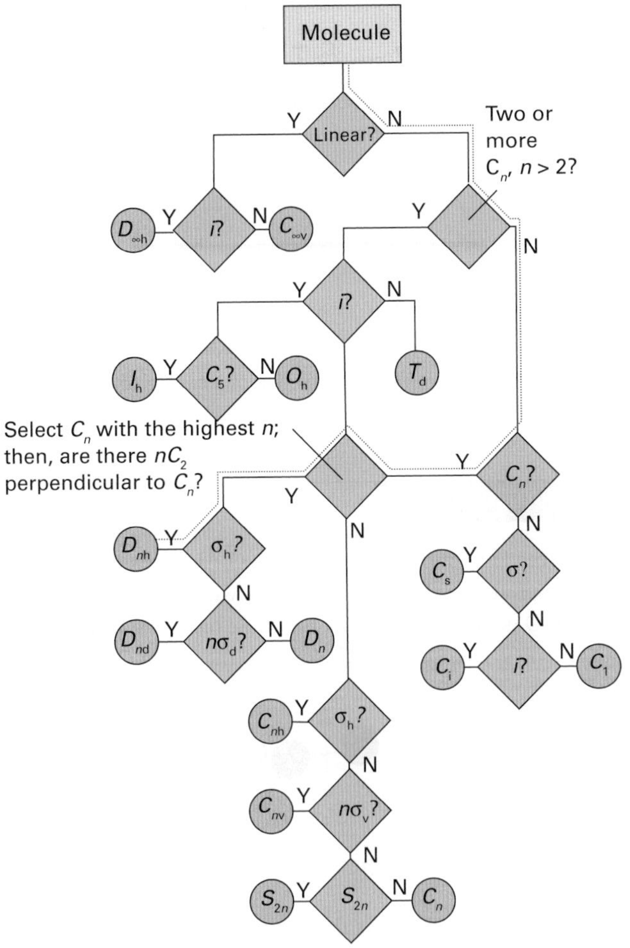

Figure 10A.7 A flow diagram for determining the point group of a molecule. Start at the top and answer the question posed in each diamond (Y = yes, N = no). The dotted line refers to the path taken in *Brief illustration* 10A.2.

Brief illustration 10A.2

To identify the point group to which a ruthenocene molecule (4) belongs, first identify the symmetry elements present and the use the flow diagram in Fig. 10A.7. Note that:

- The molecule has a fivefold rotation axis, and five twofold rotation axes which pass through the Ru atom and are perpendicular to the C_5 axis.
- There is a mirror plane, σ_h, perpendicular to the C_5 axis and passing through the Ru atom.
- There are five σ_v planes containing the principal axis: each passes through one carbon atom in a ring and the mid-point of the C–C bond on the opposite side. Each one of these planes contains one of the twofold rotation axes.

The path to trace in Fig. 10A.7 is shown by a dotted line; it ends at D_{nh}, and because the molecule has a fivefold axis, it belongs to the point group D_{5h}.

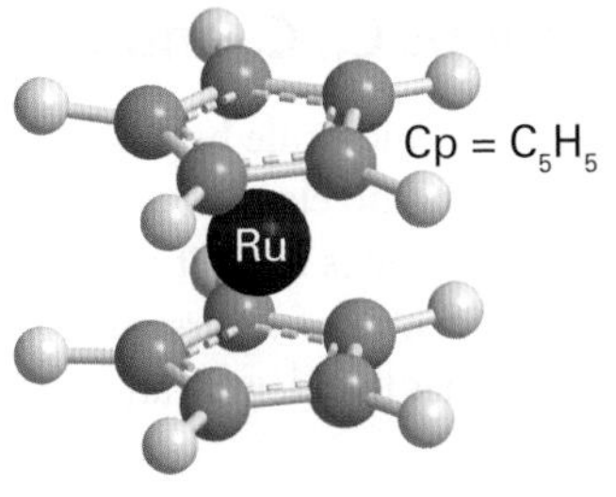

4 Ruthenocene, $Ru(Cp)_2$

If the rings are staggered, as they are in an excited state of ferrocene (5), the σ_h plane is absent. The other mirror planes are still present, but now they bisect the angles between twofold axes and so are described as σ_d. Tracing the appropriate path in Fig. 10A.7 gives the point group as D_{5d}.

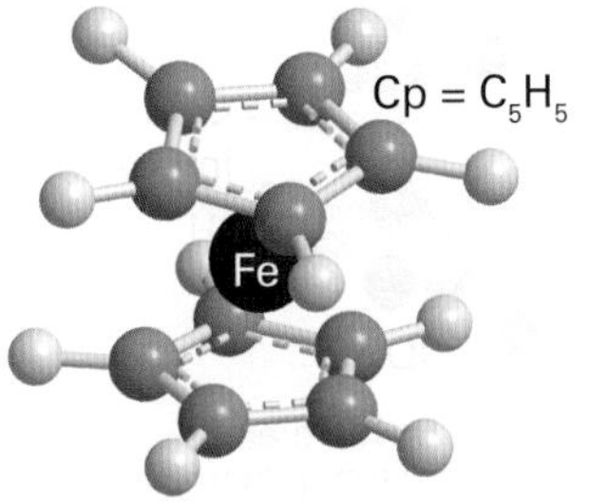

5 Ferrocene, $Fe(Cp)_2$ (excited state)

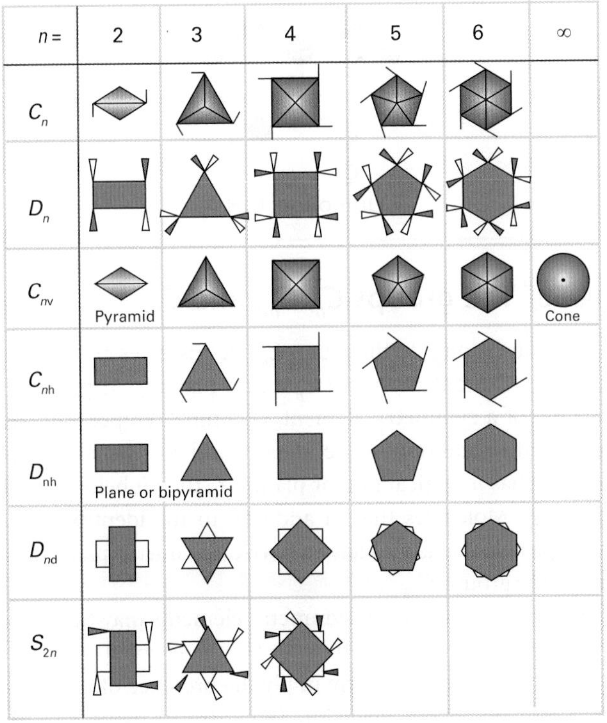

Figure 10A.8 A summary of the shapes corresponding to different point groups. The group to which a molecule belongs can often be identified from this diagram without going through the formal procedure in Fig. 10A.7.

10A.2(a) The groups C_1, C_i, and C_s

Name	Elements
C_1	E
C_i	E, i
C_s	E, σ

A molecule belongs to the group C_1 if it has no element other than the identity. It belongs to C_i if it has the identity and a centre of inversion alone, and to C_s if it has the identity and a mirror plane alone.

Brief illustration 10A.3

- The CBrClFI molecule (**3**) has only the identity element, and so belongs to the group C_1.
- *Meso*-tartaric acid (**6**) has the identity and a centre of inversion, and so belongs to the group C_i.

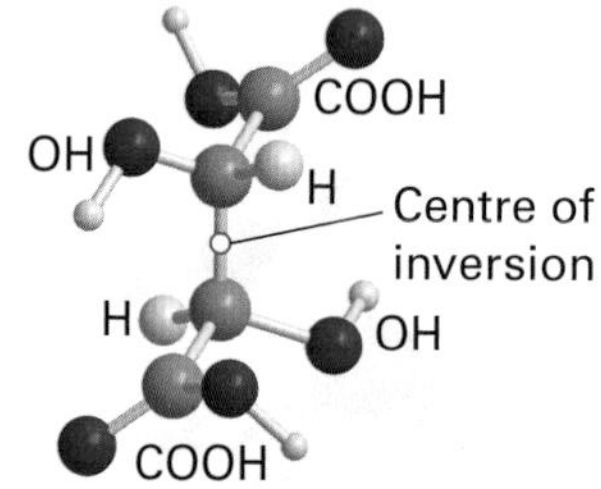

6 Meso-tartaric acid, HOOCCH(OH)CH(OH)COOH

- Quinoline (**7**) has the elements (E,σ), and so belongs to the group C_s.

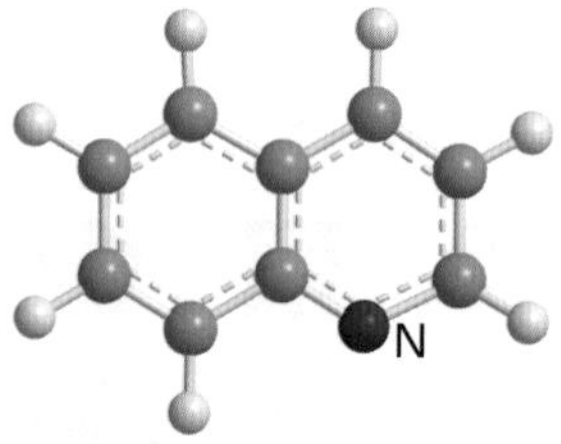

7 Quinoline, C_9H_7N

10A.2(b) The groups C_n, C_{nv}, and C_{nh}

A molecule belongs to the group C_n if it possesses an n-fold rotation axis. Note that symbol C_n is now playing a triple role: as the label of a symmetry element, a symmetry operation, and a group. If in addition to the identity and a C_n rotation axis a molecule has n vertical mirror planes σ_v, then it belongs to the group C_{nv}. Molecules that in addition to the identity and an n-fold principal axis also have a horizontal mirror plane σ_h belong to the groups C_{nh}.

The presence of certain symmetry elements may be implied by the presence of others: thus, in C_{2h} the elements C_2 and σ_h jointly imply the presence of a centre of inversion (Fig. 10A.9). Note also that the tables specify the *elements*, not the *operations*: for instance, there are two operations associated with a single C_3 axis (rotations by +120° and −120°).

Name	Elements
C_n	E, C_n
C_{nv}	E, C_n, $n\sigma_v$
C_{nh}	E, C_n, σ_h

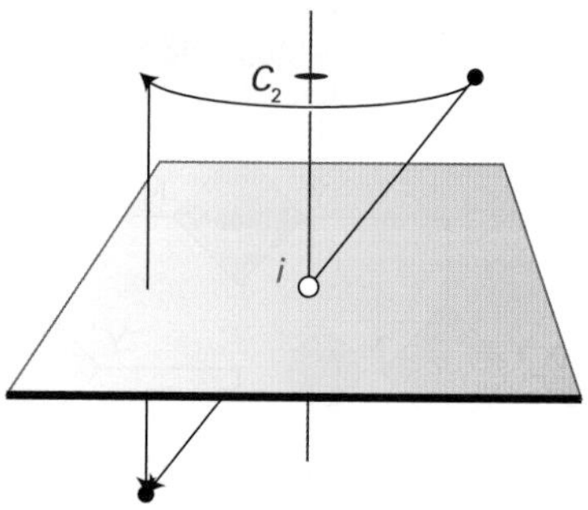

Figure 10A.9 The presence of a twofold axis and a horizontal mirror plane jointly imply the presence of a centre of inversion in the molecule.

Brief illustration 10A.4

- In the H_2O_2 molecule (**8**) the two O–H bonds make an angle of about 115° to one another when viewed down the O–O bond direction. The molecule has the symmetry elements E and C_2, and so belongs to the group C_2.

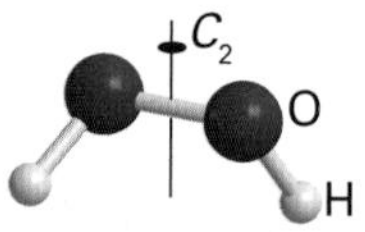

8 Hydrogen peroxide, H_2O_2

- An H_2O molecule has the symmetry elements E, C_2, and $2\sigma_v$, so it belongs to the group C_{2v}.
- An NH_3 molecule has the elements E, C_3, and $3\sigma_v$, so it belongs to the group C_{3v}.
- A heteronuclear diatomic molecule such as HCl belongs to the group $C_{\infty v}$ because rotations around the internuclear axis by any angle and reflections in any of the infinite number of planes that contain this axis are symmetry operations. Other members of the group $C_{\infty v}$ include the linear OCS molecule and a cone.
- The molecule *trans*-CHCl=CHCl (**9**) has the elements E, C_2, and σ_h, so belongs to the group C_{2h}.

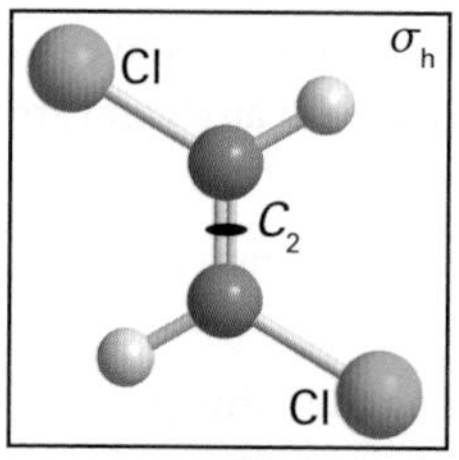

9 *trans*-CHCl=CHCl

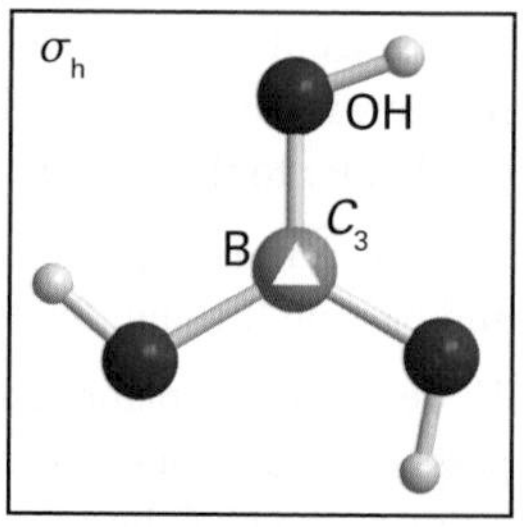

10 $B(OH)_3$

- The molecule $B(OH)_3$, in the planar conformation shown in (**10**), has a C_3 axis and a σ_h plane, and so belongs to the point group C_{3h}.

10A.2(c) The groups D_n, D_{nh}, and D_{nd}

Figure 10A.7 shows that a molecule that has an n-fold principal axis and n twofold rotation axes perpendicular to C_n belongs to the group D_n. A molecule belongs to D_{nh} if it also possesses a horizontal mirror plane. The linear molecules OCO and HCCH, and a uniform cylinder, all belong to the group $D_{\infty h}$. A molecule belongs to the group D_{nd} if in addition to the elements of D_n it possesses n dihedral mirror planes σ_d.

Name	Elements
D_n	E, C_n, nC_2'
D_{nh}	E, C_n, nC_2', σ_h
D_{nd}	$E, C_n, nC_2', n\sigma_d$

Brief illustration 10A.5

- The trigonal planar BF_3 molecule (**11**) has the elements E, C_3, $3C_2$ (with one C_2 axis along each B–F bond), and σ_h, so belongs to D_{3h}.

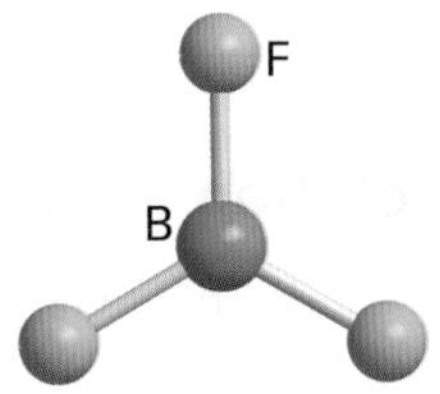

11 Boron trifluoride, BF_3

- The C_6H_6 molecule (**1**) has the elements E, C_6, $3C_2$, $3C_2'$, and σ_h together with some others that these elements imply, so it belongs to D_{6h}. Three of the C_2 axes bisect C–C bonds on opposite sides of the hexagonal ring formed by the carbon atoms, and the other three pass through vertices on opposite sides of the ring. The prime on $3C_2'$ indicates that these axes are different from the other three C_2 axes.
- All homonuclear diatomic molecules, such as N_2, belong to the group $D_{\infty h}$ because all rotations around the internuclear axis are symmetry operations, as are end-over-end rotations by 180°.
- PCl_5 (**12**) is another example of a D_{3h} species.

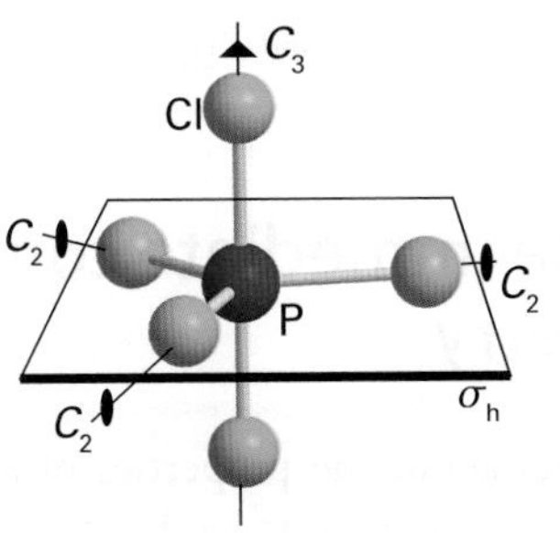

12 Phosphorus pentachloride, PCl_5 (D_{3h})

- Propadiene (an allene, **13**), in which the two CH_2 groups lie in perpendicular planes, belongs to the point group D_{2d}.

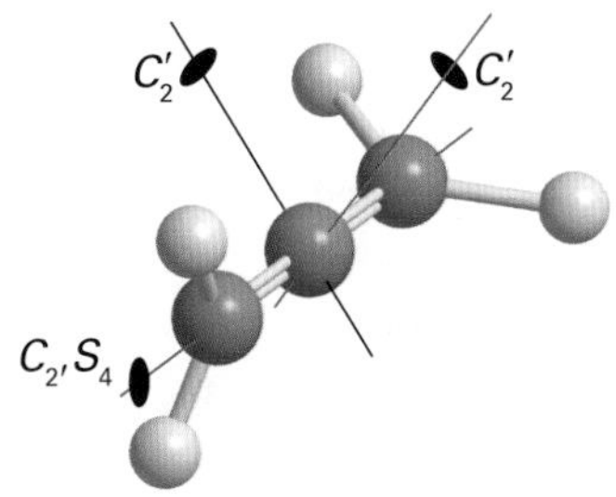

13 Propadiene, C_3H_4 (D_{2d})

10A.2(d) The groups S_n

Molecules that have not been classified into one of the groups mentioned so far, but which possess one S_n axis, belong to the groups S_n. Note that the group S_2 is the same as C_i, so such a molecule will already have been classified as C_i. Tetraphenylmethane (**14**) belongs to the point group S_4; molecules belonging to S_n with $n > 4$ are rare.

Name	Elements
S_n	E, S_n and not previously classified

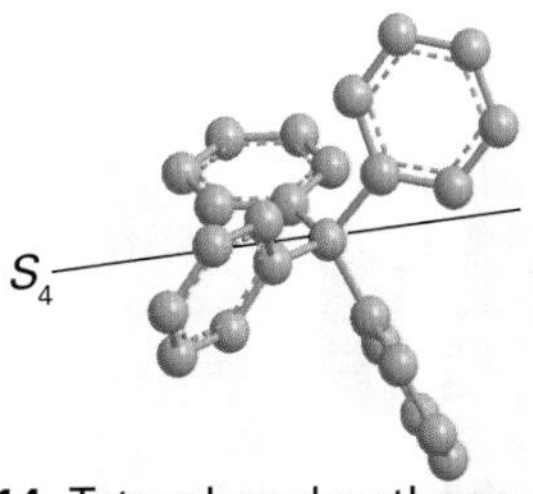

14 Tetraphenylmethane, $C(C_6H_5)_4$ (S_4)

10A.2(e) The cubic groups

A number of very important molecules possess more than one principal axis. Most belong to the **cubic groups**, and in particular to the **tetrahedral groups** T, T_d, and T_h (Fig. 10A.10a) or to the **octahedral groups** O and O_h (Fig. 10A.10b). A few icosahedral (20-faced) molecules belonging to the **icosahedral group**, I (Fig. 10A.10c), are also known.

The groups T_d and O_h are the groups of the regular tetrahedron and the regular octahedron, respectively. If the object possesses the rotational symmetry of the tetrahedron or the octahedron, but none of their planes of reflection, then it belongs to the simpler groups T or O (Fig. 10A.11). The group T_h is based on T but also contains a centre of inversion (Fig. 10A.12).

Name	Elements
T	$E, 4C_3, 3C_2$
T_d	$E, 3C_2, 4C_3, 3S_4, 6\sigma_d$
T_h	$E, 3C_2, 4C_3, i, 4S_6, 3\sigma_h$
O	$E, 3C_4, 4C_3, 6C_2$
O_h	$E, 3S_4, 3C_4, 6C_2, 4S_6, 4C_3, 3\sigma_h, 6\sigma_d, i$
I	$E, 6C_5, 10C_3, 15C_2$
I_h	$E, 6S_{10}, 10S_6, 6C_5, 10C_3, 15C_2, 15\sigma, i$

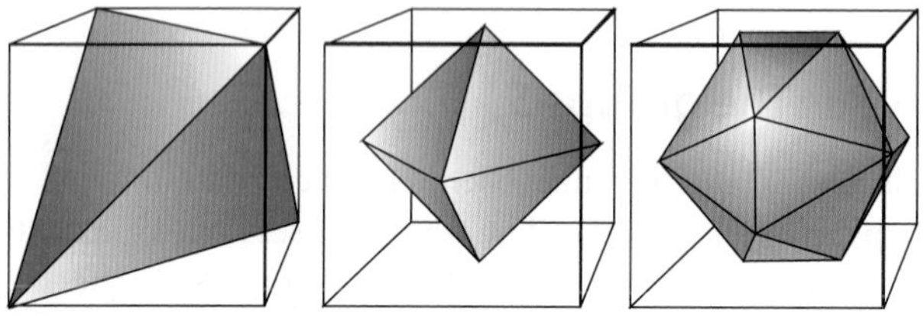

Figure 10A.10 (a) Tetrahedral, (b) octahedral, and (c) icosahedral shapes drawn to show their relation to a cube: they belong to the cubic groups T_d, O_h, and I_h, respectively.

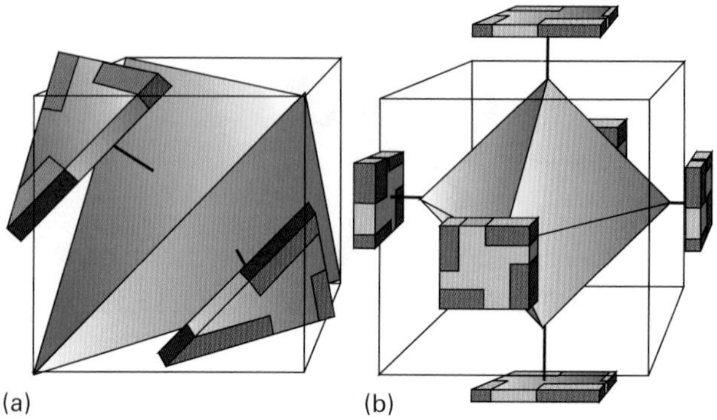

Figure 10A.11 Shapes corresponding to the point groups (a) T and (b) O. The presence of the decorated slabs reduces the symmetry of the object from T_d and O_h, respectively.

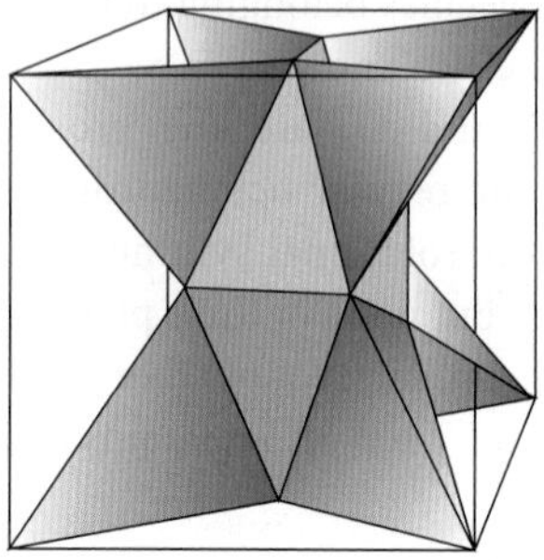

Figure 10A.12 The shape of an object belonging to the group T_h.

Brief illustration 10A.6

- The molecules CH_4 and SF_6 belong, respectively, to the groups T_d and O_h.
- Molecules belonging to the icosahedral group I include some of the boranes and buckminsterfullerene, C_{60} (**15**).

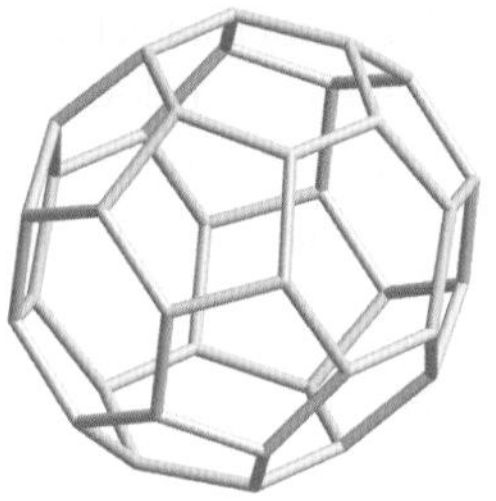

15 Buckminsterfullerene, C_{60} (I)

- The objects shown in Fig. 10A.11 belong to the groups T and O, respectively.

10A.2(f) The full rotation group

Name	Elements
R_3	$E, \infty C_2, \infty C_3, \ldots$

The full rotation group, R_3 (the 3 refers to rotation in three dimensions), consists of an infinite number of rotation axes with all possible values of n. A sphere and an atom belong to R_3, but no molecule does. Exploring the consequences of R_3 is a very important way of applying symmetry arguments to atoms, and is an alternative approach to the theory of orbital angular momentum.

Exercises

E10A.4 Identify the point groups to which the following objects belong: (i) a sphere, (ii) an isosceles triangle, (iii) an equilateral triangle, (iv) an unsharpened cylindrical pencil.

E10A.5 List the symmetry elements of the following molecules and name the point groups to which they belong: (i) NO_2, (ii) PF_5, (iii) $CHCl_3$, (iv) 1,4-difluorobenzene.

E10A.6 Assign (i) *cis*-dichloroethene and (ii) *trans*-dichloroethene to point groups.

10A.3 Some immediate consequences of symmetry

Some statements about the properties of a molecule can be made as soon as its point group has been identified.

10A.3(a) Polarity

A **polar molecule** is one with a permanent electric dipole moment (HCl, O_3, and NH_3 are examples). A dipole moment is a property of the molecule, so it follows that the dipole moment (which is represented by a vector) must be unaffected by any symmetry operation of the molecule because, by definition, such an operation leaves the molecule apparently unchanged.

If a molecule possesses a C_n axis ($n > 1$) then it is not possible for there to be a dipole moment perpendicular to this axis because such a dipole moment would change its orientation on rotation about the axis. It is, however, possible for there to be a dipole parallel to the axis, because it would not be affected by the rotation. For example, in H_2O the dipole lies in the plane of the molecule, pointing along the bisector of the HOH bond, which is the direction of the C_2 axis.

Similarly, if a molecule possesses a mirror plane there can be no dipole moment perpendicular to this plane, because reflection in the plane would reverse its direction. A molecule that possesses a centre of symmetry cannot have a dipole moment in any direction because the inversion operation would reverse it.

These considerations lead to the conclusion that

> Only molecules belonging to the groups C_n, C_{nv}, and C_s may have a permanent electric dipole moment.

For C_n and C_{nv}, the dipole moment must lie along the principal axis.

Brief illustration 10A.7

- Ozone, O_3, which has an angular structure, belongs to the group C_{2v} and is polar.
- Carbon dioxide, CO_2, which is linear and belongs to the group $D_{\infty h}$, is not polar.
- Tetraphenylmethane (**14**) belongs to the point group S_4 and so is not polar.

10A.3(b) Chirality

A **chiral molecule** (from the Greek word for 'hand') is a molecule that cannot be superimposed on its mirror image. An **achiral molecule** is a molecule that can be superimposed on its mirror image. Chiral molecules are **optically active** in the sense that they rotate the plane of polarized light. A chiral molecule and its mirror-image partner constitute an **enantiomeric pair** (from the Greek word for 'both') of isomers and rotate the plane of polarization by equal amounts but opposite directions.

> A molecule may be chiral, and therefore optically active, only if it does not possess an axis of improper rotation, S_n.

An S_n improper rotation axis may be present under a different name, and be implied by other symmetry elements that are present. For example, molecules belonging to the groups C_{nh} possess an S_n axis implicitly because they possess both C_n and σ_h, which are the two components of an improper rotation axis. A centre of inversion, i, is in fact the same as S_2 because the two corresponding operations achieve exactly the same result

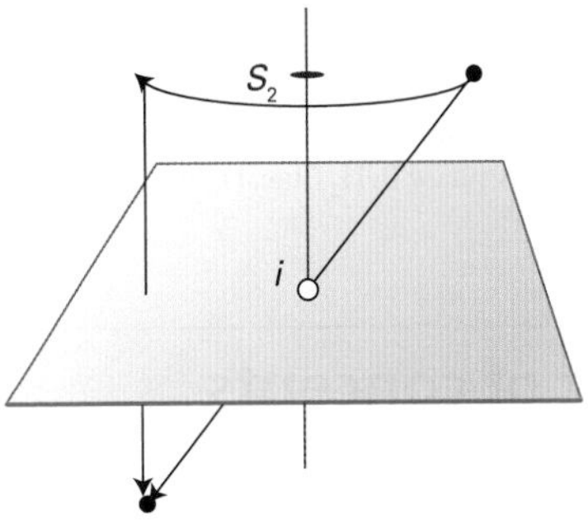

Figure 10A.13 The operations i and S_2 are equivalent in the sense that they achieve exactly the same outcome when applied to a point in the object.

(Fig. 10A.13). Furthermore, a mirror plane is the same as S_1 (rotation through 360° followed by reflection). Therefore molecules possessing a mirror plane or a centre of inversion effectively possess an improper rotation axis and so, by the above rule, are achiral.

Brief illustration 10A.8

- The amino acid alanine (**16**) does not possess a centre of inversion nor does it have any mirror planes: it is therefore chiral.

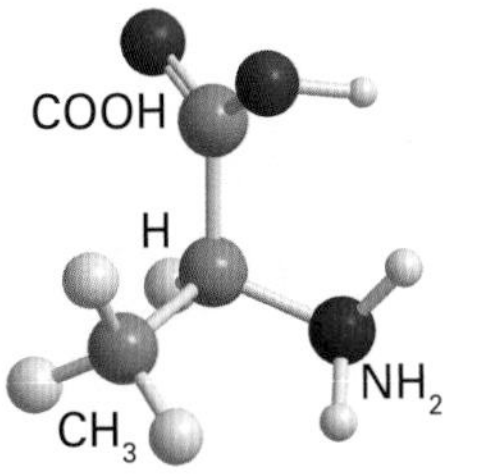

16 L-Alanine, $NH_2CH(CH_3)COOH$

- In contrast, glycine (**17**) has a mirror plane and so is achiral.

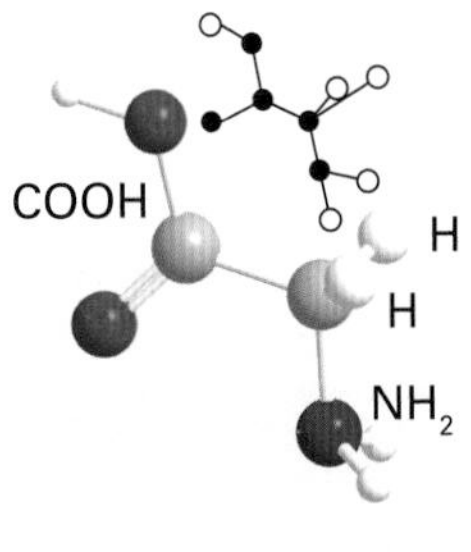

17 Glycine, NH_2CH_2COOH

- Tetraphenylmethane (**14**) belongs to the point group S_4; it does not possess a centre of inversion nor does it possess any mirror planes, but it is still achiral since it possesses an improper rotation axis (S_4).

Exercises

E10A.7 Which of the following molecules may be polar? (i) pyridine, (ii) nitroethane, (iii) gas-phase BeH_2 (linear), (iv) B_2H_6.

E10A.8 Can molecules belonging to the point groups D_{2h} or C_{3h} be chiral? Explain your answer.

Checklist of concepts

☐ 1. A **symmetry operation** is an action that leaves an object looking the same after it has been carried out.

☐ 2. A **symmetry element** is a point, line, or plane with respect to which a symmetry operation is performed.

☐ 3. The **notation** for point groups commonly used for molecules and solids is summarized in Table 10A.1.

☐ 4. To be **polar**, a molecule must belong to C_n, C_{nv}, or C_s (and have no higher symmetry).

☐ 5. A molecule will be **chiral** only if it does not possess an axis of improper rotation, S_n.

Checklist of symmetry operations and elements

Symmetry operation	Symbol	Symmetry element
n-Fold rotation	C_n	*n*-Fold rotation axis
Reflection	σ	Mirror plane
Inversion	i	Centre of inversion
n-Fold improper rotation	S_n	*n*-Fold improper rotation axis
Identity	E	Entire object

TOPIC 10B Group theory

> ➤ Why do you need to know this material?
>
> Group theory expresses qualitative ideas about symmetry mathematically and can be applied systematically to a wide variety of problems. The theory is also the origin of the labelling of atomic and molecular orbitals that is used throughout chemistry.
>
> ➤ What is the key idea?
>
> Symmetry operations may be represented by the effect of matrices on a basis.
>
> ➤ What do you need to know already?
>
> You need to know about the types of symmetry operation and element introduced in Topic 10A. This discussion uses matrix algebra, and especially matrix multiplication, as set out in *The chemist's toolkit* 9E.1 and in more detail in Part 1 of the *Resource section*.

The systematic discussion of symmetry is called **group theory**. Much of group theory is a summary of common sense about the symmetries of objects. However, because group theory is systematic, its rules can be applied in a straightforward, mechanical way. In most cases the theory gives a simple, direct method for arriving at useful conclusions with the minimum of calculation, and this is the aspect that is stressed here.

10B.1 The elements of group theory

A **group** in mathematics is a collection of transformations that satisfy four criteria. If the transformations are written as $R, R', \ldots$ (which might be the reflections, rotations, and so on introduced in Topic 10A), then they form a group if:

1. One of the transformations is the identity (that is, 'do nothing').
2. For every transformation R, the inverse transformation R^{-1} is included in the collection so that the combination RR^{-1} (the transformation R^{-1} followed by R) is equivalent to the identity.
3. The combination RR' (the transformation R' followed by R) is equivalent to a single member of the collection of transformations.
4. The combination $R(R'R'')$, the transformation $(R'R'')$ followed by R, is equivalent to $(RR')R''$, the transformation R'' followed by (RR').

Example 10B.1 Showing that the symmetry operations of a molecule form a group

The point group C_{2v} consists of the elements $\{E, C_2, \sigma_v, \sigma_v'\}$ and correspond to the operations $\{E, C_2, \sigma_v, \sigma_v'\}$. Show that this set of operations is a group in the mathematical sense.

Collect your thoughts You need to show that combinations of the operations match the criteria set out above. The operations are specified in Topic 10A, and illustrated in Figs. 10A.2 and 10A.3 for H_2O, which belongs to this group.

The solution

- Criterion 1 is fulfilled because the collection of symmetry operations includes the identity E.
- Criterion 2 is fulfilled because for each member of this collection of symmetry operations the inverse of the operation is the operation itself. For example, applying two successive twofold rotations brings the object back to its starting position. The effect of these two rotations is to do nothing, which is the effect of the identity operation. It follows that $C_2C_2 = E$. However, the inverse is defined as having the property $C_2C_2^{-1} = E$. Comparing these two equations shows that $C_2 = C_2^{-1}$. A similar argument shows that the inverse of a reflection and the identity are the operations themselves.
- Criterion 3 is fulfilled, because in each case one operation followed by another is the same as one of the four symmetry operations. For instance, a twofold rotation C_2 followed by the reflection σ_v is the same as the single reflection σ_v' (Fig. 10B.1); thus, $\sigma_v C_2 = \sigma_v'$. A 'group multiplication table' can be constructed in a similar way for all possible products of symmetry operations RR'; as required, each product is equivalent to another symmetry operation.

$R\downarrow\ R'\rightarrow$	E	C_2	σ_v	σ_v'
E	E	C_2	σ_v	σ_v'
C_2	C_2	E	σ_v'	σ_v
σ_v	σ_v	σ_v'	E	C_2
σ_v'	σ_v'	σ_v	C_2	E

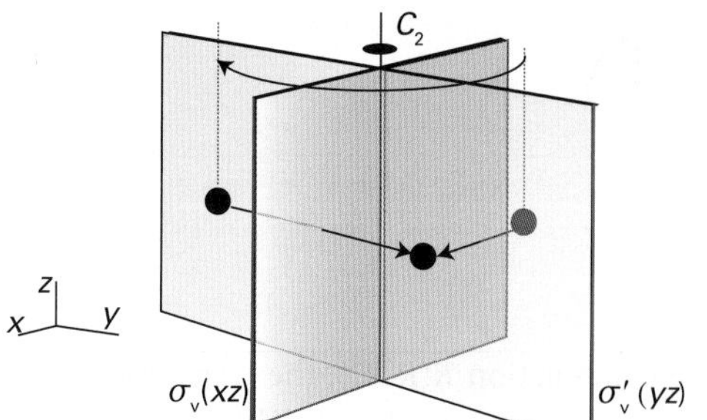

Figure 10B.1 A twofold rotation C_2 followed by the reflection σ_v gives the same result as the reflection σ'_v.

- Criterion 4 is fulfilled, as it is immaterial how the operations are grouped together. Thus $(\sigma_v\sigma'_v)C_2 = C_2C_2 = E$ and $\sigma_v(\sigma'_vC_2) = \sigma_v\sigma_v = E$, and likewise for all other combinations.

Self-test 10B.1 Confirm that C_{2h}, which has the elements $\{E,C_2,i,\sigma_h\}$ and hence the corresponding operations $\{E,C_2,i,\sigma_h\}$, is a group by constructing a group multiplication table.

Answer: Criteria are fulfilled

One potentially confusing point needs to be clarified at the outset. The entities that make up a group are its 'elements'. For applications in chemistry, these elements are almost always symmetry operations. However, as explained in Topic 10A, 'symmetry operations' are distinct from 'symmetry elements', the latter being the points, axes, and planes with respect to which the operations are carried out. A third use of the word 'element' is to denote the number lying in a particular location in a matrix. Be very careful to distinguish *element* (of a group), *symmetry element*, and *matrix element*.

Symmetry operations fall into the same **class** if they are of the same type (for example, rotations) and can be transformed into one another by a symmetry operation of the group. The two threefold rotations in C_{3v} belong to the same class because one can be converted into the other by a reflection (Fig. 10B.2); the three reflections all belong to the same class because each can be rotated into another by a threefold rotation. The formal definition of a class is that two operations R and R' belong to the same class if there is a member S of the group such that

$$R' = S^{-1}RS \qquad \text{Membership of a class} \qquad (10B.1)$$

where S^{-1} is the inverse of S.

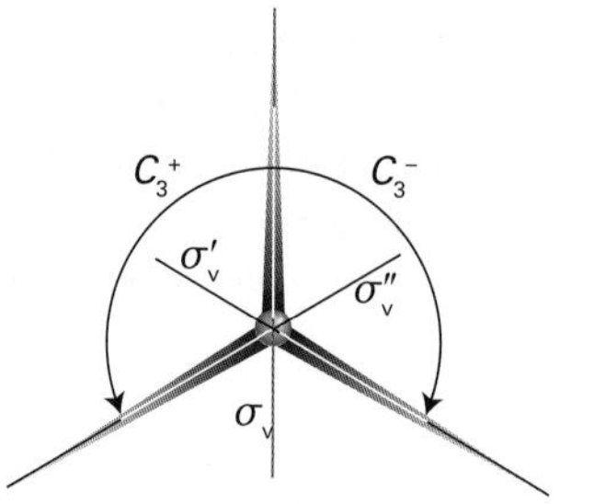

Figure 10B.2 Symmetry operations in the same class are related to one another by the symmetry operations of the group. Thus, the three mirror planes shown here are related by threefold rotations, and the two rotations shown here are related by reflection in σ_v.

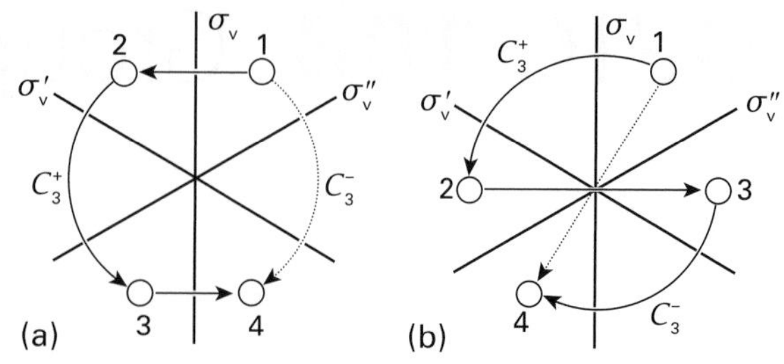

Figure 10B.3 (a) The sequence of operations $\sigma_v^{-1}C_3^+\sigma_v$ when applied to the point 1 takes it through the sequence $1 \rightarrow 2 \rightarrow 3 \rightarrow 4$ (σ_v^{-1} is the operation σ_v). The single operation C_3^- (dotted curve) takes point $1 \rightarrow 4$, so C_3^+ and C_3^- are in the same class. (b) The sequence of operations $(C_3^+)^{-1}\sigma_vC_3^+$ takes point $1 \rightarrow 4$, $((C_3^+)^{-1}$ is the operation C_3^-). The same transformation can be achieved with the single operation σ'_v (dotted line), so σ_v and σ'_v are in the same class.

Figure 10B.3a shows how eqn 10B.1 can be used to confirm that C_3^+ and C_3^- belong to the same class in the group C_{3v}. The transformation of interest is $\sigma_v^{-1}C_3^+\sigma_v$ and the figure shows how an arbitrary point initially at 1 behaves under this sequence of operations. The operation σ_v moves the point from 1 to 2, and then C_3^+ moves the point to 3. The inverse of a reflection is itself, $\sigma_v^{-1} = \sigma_v$, so the effect of σ_v^{-1} is to move the point to 4. From the diagram it can be seen that point 4 can be reached by applying C_3^- to point 1. It is therefore shown that $\sigma_v^{-1}C_3^+\sigma_v = C_3^-$, and hence that C_3^+ and C_3^- do indeed belong in the same class.

Brief illustration 10B.1

To show that σ_v and σ'_v are in the same class in the group C_{3v}, consider the transformation $(C_3^+)^{-1}\sigma_vC_3^+$. Because C_3^- is the inverse of C_3^+, this transformation is the same as $C_3^-\sigma_vC_3^+$. The effect of this sequence of operations on an arbitrary point 1 is shown in Fig. 10B.3b. The final position, 4, can also be reached from 1 by applying the operation σ'_v, thus showing that $C_3^-\sigma_vC_3^+ = \sigma'_v$ and hence that σ_v and σ'_v are in the same class.

Exercises

E10B.1 Show that all three C_2 operations in the group D_{3h} belong to the same class.

E10B.2 In the point group D_{3h}, determine the inverse of each of the following operations: (i) E, (ii) σ_v, (iii) C_3, (iv) S_3.

10B.2 Matrix representations

Group theory takes on great power when the notional ideas presented so far are expressed in terms of collections of numbers in the form of matrices. For basic information about how to handle matrices see *The chemist's toolkit* 9E.1 and a more extensive account in Part 1 of the *Resource section*.

10B.2(a) Representatives of operations

Consider the set of five p orbitals shown on the C_{2v} SO_2 molecule in Fig. 10B.4 and how they are affected by the reflection operation σ_v. The corresponding symmetry element is the mirror plane perpendicular to the plane of the molecule and passing through the S atom (with the labelling used here, σ_v is the xz plane). The effect of this reflection is to leave p_x and p_z unaffected, to change the sign of p_y, and to exchange p_A and p_B. Its effect can be written $(p_x\ -p_y\ p_z\ p_B\ p_A) \leftarrow (p_x\ p_y\ p_z\ p_A\ p_B)$. This transformation can be expressed by using matrix multiplication:

$$(p_x\ -p_y\ p_z\ p_B\ p_A) = (p_x\ p_y\ p_z\ p_A\ p_B)\overbrace{\begin{pmatrix} 1 & 0 & 0 & 0 & 0 \\ 0 & -1 & 0 & 0 & 0 \\ 0 & 0 & 1 & 0 & 0 \\ 0 & 0 & 0 & 0 & 1 \\ 0 & 0 & 0 & 1 & 0 \end{pmatrix}}^{D(\sigma_v)}$$
$$= (p_x\ p_y\ p_z\ p_A\ p_B)D(\sigma_v) \qquad (10B.2a)$$

The matrix $D(\sigma_v)$ is called a **representative** of the operation σ_v. Representatives take different forms according to the **basis**, the set of orbitals that has been adopted. In this case, the basis is the row vector $(p_x\, p_y\, p_z\, p_A\, p_B)$. Note that the matrix D appears to the right of the basis functions on which it acts.

The same technique can be used to find matrices that reproduce the other symmetry operations. For instance, C_2 has the effect $(-p_x\ -p_y\ p_z\ -p_B\ -p_A) \leftarrow (p_x\ p_y\ p_z\ p_A\ p_B)$, and its representative is

$$D(C_2) = \begin{pmatrix} -1 & 0 & 0 & 0 & 0 \\ 0 & -1 & 0 & 0 & 0 \\ 0 & 0 & 1 & 0 & 0 \\ 0 & 0 & 0 & 0 & -1 \\ 0 & 0 & 0 & -1 & 0 \end{pmatrix} \qquad (10B.2b)$$

The effect of σ_v' (reflection in the plane of the molecule, the yz plane as labelled here) is $(-p_x\ p_y\ p_z\ -p_A\ -p_B) \leftarrow (p_x\ p_y\ p_z\ p_A\ p_B)$; the oxygen orbitals remain in the same places, but change sign. The representative of this operation is

$$D(\sigma_v') = \begin{pmatrix} -1 & 0 & 0 & 0 & 0 \\ 0 & 1 & 0 & 0 & 0 \\ 0 & 0 & 1 & 0 & 0 \\ 0 & 0 & 0 & -1 & 0 \\ 0 & 0 & 0 & 0 & -1 \end{pmatrix} \qquad (10B.2c)$$

The identity operation has no effect on the basis, so its representative is the 5×5 unit matrix:

$$D(E) = \begin{pmatrix} 1 & 0 & 0 & 0 & 0 \\ 0 & 1 & 0 & 0 & 0 \\ 0 & 0 & 1 & 0 & 0 \\ 0 & 0 & 0 & 1 & 0 \\ 0 & 0 & 0 & 0 & 1 \end{pmatrix} \qquad (10B.2d)$$

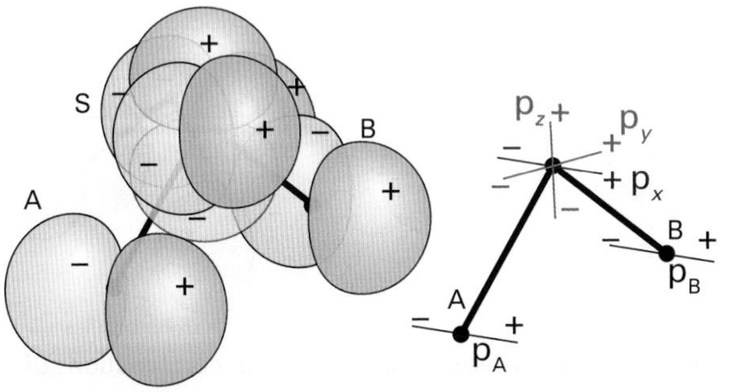

Figure 10B.4 The five p orbitals (three on the sulfur and one on each oxygen) that are used to illustrate the construction of a matrix representation in a C_{2v} molecule (SO_2). The σ_v' plane is the plane of the molecule (the yz plane), and the σ_v plane is perpendicular to the plane of the molecule, passing through the S atom (the xz plane).

10B.2(b) The representation of a group

The set of matrices that represents *all* the operations of the group is called a **matrix representation**, Γ (uppercase gamma), of the group in the basis that has been chosen. In the current example, there are five members of the basis and the representation is five-dimensional in the sense that the matrices are all 5×5 arrays. The matrices of a representation multiply together in the same way as the operations they represent. Thus, if for any two operations R and R', $RR' = R''$, then $D(R)D(R') = D(R'')$ for a given basis.

Brief illustration 10B.2

In the group C_{2v}, a twofold rotation followed by a reflection in a mirror plane is equivalent to a reflection in the second mirror plane: specifically, $\sigma_v' C_2 = \sigma_v$. Multiplying out the representatives specified in eqn 10B.2 gives

$$D(\sigma_v')D(C_2) = \begin{pmatrix} -1 & 0 & 0 & 0 & 0 \\ 0 & 1 & 0 & 0 & 0 \\ 0 & 0 & 1 & 0 & 0 \\ 0 & 0 & 0 & -1 & 0 \\ 0 & 0 & 0 & 0 & -1 \end{pmatrix}\begin{pmatrix} -1 & 0 & 0 & 0 & 0 \\ 0 & -1 & 0 & 0 & 0 \\ 0 & 0 & 1 & 0 & 0 \\ 0 & 0 & 0 & 0 & -1 \\ 0 & 0 & 0 & -1 & 0 \end{pmatrix}$$

$$= \begin{pmatrix} 1 & 0 & 0 & 0 & 0 \\ 0 & -1 & 0 & 0 & 0 \\ 0 & 0 & 1 & 0 & 0 \\ 0 & 0 & 0 & 0 & 1 \\ 0 & 0 & 0 & 1 & 0 \end{pmatrix} = D(\sigma_v)$$

As expected, this multiplication reproduces the same result as the group multiplication table. The same is true for multiplication of any two representatives, so the four matrices form a representation of the group.

The discovery of a matrix representation of the group means that a link has been established between symbolic manipulations of operations and algebraic manipulations of numbers. This link is the basis of the power of group theory in chemistry.

10B.2(c) Irreducible representations

Inspection of the representatives found above shows that they are all of **block-diagonal form**:

$$D = \begin{pmatrix} \blacksquare & 0 & 0 & 0 & 0 \\ 0 & \blacksquare & 0 & 0 & 0 \\ 0 & 0 & \blacksquare & 0 & 0 \\ 0 & 0 & 0 & \blacksquare & \blacksquare \\ 0 & 0 & 0 & \blacksquare & \blacksquare \end{pmatrix} \qquad \text{Block-diagonal form} \qquad (10B.3)$$

The block-diagonal form of the representatives implies that the symmetry operations of C_{2v} never mix p_x, p_y, and p_z together, nor do they mix these three orbitals with p_A and p_B, but p_A and p_B are mixed together by the operations of the group. Consequently, the basis can be cut into four parts: three for the individual p orbitals on S and the fourth for the two oxygen orbitals $(p_A\, p_B)$. The representations in the three one-dimensional bases are

For p_x: $D(E)=1$ $\quad D(C_2)=-1$ $\quad D(\sigma_v)=1$ $\quad D(\sigma_v')=-1$

For p_y: $D(E)=1$ $\quad D(C_2)=-1$ $\quad D(\sigma_v)=-1$ $\quad D(\sigma_v')=1$

For p_z: $D(E)=1$ $\quad D(C_2)=1$ $\quad D(\sigma_v)=1$ $\quad D(\sigma_v')=1$

These representations will be called $\Gamma^{(1)}$, $\Gamma^{(2)}$, and $\Gamma^{(3)}$, respectively. The remaining two functions $(p_A\, p_B)$ are a basis for a two-dimensional representation denoted Γ':

$$D(E) = \begin{pmatrix} 1 & 0 \\ 0 & 1 \end{pmatrix} \quad D(C_2) = \begin{pmatrix} 0 & -1 \\ -1 & 0 \end{pmatrix}$$

$$D(\sigma_v) = \begin{pmatrix} 0 & 1 \\ 1 & 0 \end{pmatrix} \quad D(\sigma_v') = \begin{pmatrix} -1 & 0 \\ 0 & -1 \end{pmatrix}$$

The original five-dimensional representation has been **reduced** to the 'direct sum' of three one-dimensional representations 'spanned' by each of the p orbitals on S, and a two-dimensional representation spanned by $(p_A\, p_B)$. The reduction is represented symbolically by writing[1]

$$\Gamma = \Gamma^{(1)} + \Gamma^{(2)} + \Gamma^{(3)} + \Gamma' \qquad \text{Direct sum} \qquad (10B.4)$$

The representations $\Gamma^{(1)}$, $\Gamma^{(2)}$, and $\Gamma^{(3)}$ cannot be reduced any further, and each one is called an **irreducible representation** of the group (an 'irrep').

For this basis and in this group the two-dimensional representation Γ' is reducible to a sum of two one-dimensional representations. This reduction can be demonstrated by forming linear combinations of the basis functions p_A and p_B, specifically the linear combinations $p_1 = p_A + p_B$ and $p_2 = p_A - p_B$ (illustrated in Fig. 10B.5).

The effect of the operation σ_v is to exchange p_A and p_B: $(p_B\, p_A) \leftarrow (p_A\, p_B)$, therefore the effect on the combination $p_A + p_B$ is $(p_B + p_A) \leftarrow (p_A + p_B)$. This is the same as $(p_1) \leftarrow (p_1)$. Similarly, $(p_B - p_A) \leftarrow (p_A - p_B)$, corresponding to $(-p_2) \leftarrow (p_2)$. It follows from these results, and similar ones for the other operations, that the representation in the basis $(p_1\, p_2)$ is

$$D(E) = \begin{pmatrix} 1 & 0 \\ 0 & 1 \end{pmatrix} \quad D(C_2) = \begin{pmatrix} -1 & 0 \\ 0 & 1 \end{pmatrix}$$

$$D(\sigma_v) = \begin{pmatrix} 1 & 0 \\ 0 & -1 \end{pmatrix} \quad D(\sigma_v') = \begin{pmatrix} -1 & 0 \\ 0 & -1 \end{pmatrix}$$

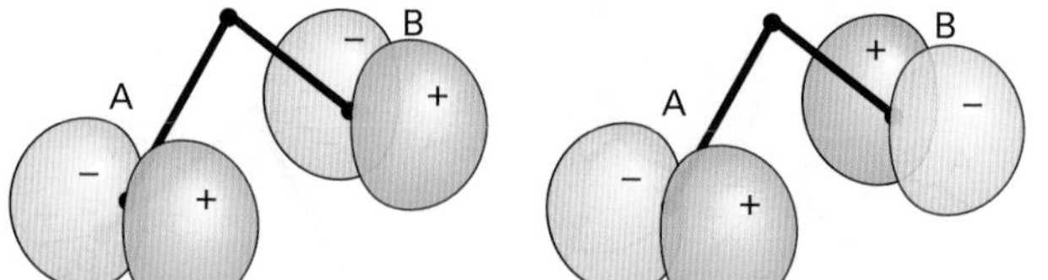

Figure 10B.5 Two symmetry-adapted linear combinations of the oxygen basis orbitals shown in Fig. 10B.4: on the left $p_1 = p_A + p_B$, on the right $p_2 = p_A - p_B$. The two combinations each span a one-dimensional irreducible representation, and their symmetry species are different.

[1] The symbol $\oplus$ is sometimes used to denote a direct sum to distinguish it from an ordinary sum, in which case eqn 10B.4 would be written $\Gamma = \Gamma^{(1)} \oplus \Gamma^{(2)} \oplus \Gamma^{(3)} \oplus \Gamma'$.

The new representatives are all in block-diagonal form, in this case in the form $\begin{pmatrix} \blacksquare & 0 \\ 0 & \blacksquare \end{pmatrix}$, and the two combinations are not mixed with each other by any operation of the group. The representation Γ′ has therefore been reduced to the sum of two one-dimensional representations. Thus, p_1 spans the one-dimensional representation

$$D(E)=1 \quad D(C_2)=-1 \quad D(\sigma_v)=1 \quad D(\sigma_v')=-1$$

which is the same as the representation $\Gamma^{(1)}$ spanned by p_x. The combination p_2 spans

$$D(E)=1 \quad D(C_2)=1 \quad D(\sigma_v)=-1 \quad D(\sigma_v')=-1$$

which is a new one-dimensional representation and denoted $\Gamma^{(4)}$. At this stage the original representation has been reduced into five one-dimensional representations as follows, with one representation, $\Gamma^{(1)}$, appearing twice:

$$\Gamma = 2\Gamma^{(1)} + \Gamma^{(2)} + \Gamma^{(3)} + \Gamma^{(4)}$$

10B.2(d) Characters

The **character**, χ (chi), of an operation in a particular matrix representation is the sum of the diagonal elements of the representative of that operation. Thus, in the original basis ($p_x\ p_y\ p_z\ p_A\ p_B$) the characters of the representatives are

R	E	C_2
$D(R)$	$\begin{pmatrix}1&0&0&0&0\\0&1&0&0&0\\0&0&1&0&0\\0&0&0&1&0\\0&0&0&0&1\end{pmatrix}$	$\begin{pmatrix}-1&0&0&0&0\\0&-1&0&0&0\\0&0&1&0&0\\0&0&0&0&-1\\0&0&0&-1&0\end{pmatrix}$
$\chi(R)$	5	−1

R	σ_v	σ_v'
$D(R)$	$\begin{pmatrix}1&0&0&0&0\\0&-1&0&0&0\\0&0&1&0&0\\0&0&0&0&1\\0&0&0&1&0\end{pmatrix}$	$\begin{pmatrix}-1&0&0&0&0\\0&1&0&0&0\\0&0&1&0&0\\0&0&0&-1&0\\0&0&0&0&-1\end{pmatrix}$
$\chi(R)$	1	−1

The characters of one-dimensional representatives are just the representatives themselves. For each operation, the sum of the characters of the reduced representations is the same as the character of the original representation (allowing for the appearance of $\Gamma^{(1)}$ twice in the reduction $\Gamma = 2\Gamma^{(1)} + \Gamma^{(2)} + \Gamma^{(3)} + \Gamma^{(4)}$):

R	E	C_2	σ_v	σ_v'
$\chi(R)$ for $\Gamma^{(1)}$	1	−1	1	−1
$\chi(R)$ for $\Gamma^{(1)}$	1	−1	1	−1
$\chi(R)$ for $\Gamma^{(2)}$	1	−1	−1	1
$\chi(R)$ for $\Gamma^{(3)}$	1	1	1	1
$\chi(R)$ for $\Gamma^{(4)}$	1	1	−1	−1
Sum for Γ:	5	−1	1	−1

At this point, four irreducible representations of the group C_{2v} have been found. Are these the only irreducible representations of the group C_{2v}? There are in fact no more irreducible representations in this group, a fact that can be deduced from a surprising theorem of group theory, which states that

$$\text{Number of irreducible representations} = \text{number of classes}$$

Number of irreducible representations (10B.5)

In C_{2v} there are four classes of operations (the four columns in the table), so there must be four irreducible representations. The ones already found are the only ones for this group.

The **order** of the group, h, is the total number of symmetry operations. Another powerful result from group theory relates the order to the sum of the squares of the dimensions, d_i, of all the irreducible representations $\Gamma^{(i)}$

$$\sum_{\substack{\text{irreducible}\\ \text{representations},\, i}} d_i^2 = h$$

Dimensionality and order (10B.6)

This result applies to all groups other than the pure rotation groups C_n with $n > 2$. In the group C_{2v} there are four irreducible representations, all of which are one-dimensional ($d_i = 1$) so

$$\sum_{\substack{\text{irreducible}\\ \text{representations},\, i}} d_i^2 = 1^2+1^2+1^2+1^2 = 4$$

There are indeed four symmetry operations of the group, $h = 4$, in accord with the theorem.

Brief illustration 10B.3

The operations of the group C_{3v} fall into three classes $\{E, 2C_3, 3\sigma_v\}$, so there are three irreducible representations. The order of the group is $h = 1 + 2 + 3 = 6$. Suppose that it is already known that two of the irreducible representations are one-dimensional. Equation 10B.6 is then used to find the dimension d_3 of the remaining irreducible representation: $1^2 + 1^2 + d_3^2 = 6$. It follows that $d_3 = 2$, meaning that the third irreducible representation is two-dimensional.

Exercises

E10B.3 Use as a basis the $2p_z$ orbitals on each atom in BF_3 to find the representative of the operation σ_h. Take z as perpendicular to the molecular plane.

E10B.4 Use the matrix representatives of the operations σ_h and C_3 in a basis of $2p_z$ orbitals on each atom in BF_3 to find the operation and its representative resulting from $\sigma_h C_3$. Take z as perpendicular to the molecular plane.

10B.3 Character tables

Tables showing all the characters of the operations of a group are called **character tables** and from now on they move to the centre of the discussion. The columns of a character table are labelled with the symmetry operations of the group. Although the notation $\Gamma^{(i)}$ is used to label general irreducible representations, in chemical applications and for displaying character tables it is more common to distinguish different irreducible representations by the use of the labels A, B, E, and T to denote the **symmetry species** of each representation:

A: one-dimensional representation, character +1 under the principal rotation

B: one-dimensional representation, character −1 under the principal rotation

E: two-dimensional irreducible representation

T: three-dimensional irreducible representation

Subscripts are used to distinguish the irreducible representations if there is more than one of the same type: A_1 is reserved for the representation with character 1 for all operations (called the **totally symmetric irreducible representation**); A_2 has 1 for the principal rotation but −1 for reflections. There appears to be no systematic way of attaching subscripts to B symmetry species, so care must be used when referring to character tables from different sources.

Table 10B.1 shows the character table for the group C_{2v}, with its four symmetry species (irreducible representations) and its four columns of symmetry operations. Table 10B.2 shows the table for the group C_{3v}. The columns are headed E, $2C_3$, and $3\sigma_v$: the numbers multiplying each operation are the number of members of each class. As inferred in *Brief illustration* 10B.3, there are three symmetry species, with one of them two-dimensional (E). Operations in the same class have the same character, which is why all three σ_v operations can be grouped together in one column.

Table 10B.1 The C_{2v} character table*

C_{2v}, $2mm$	E	C_2	$\sigma_v(xz)$	$\sigma_v'(yz)$	$h=4$	
A_1	1	1	1	1	z	z^2, y^2, x^2
A_2	1	1	−1	−1		xy
B_1	1	−1	1	−1	x	zx
B_2	1	−1	−1	1	y	yz

* More character tables are given in the *Resource section*.

Table 10B.2 The C_{3v} character table*

C_{3v}, $3m$	E	$2C_3$	$3\sigma_v$	$h=6$	
A_1	1	1	1	z	z^2, x^2+y^2
A_2	1	1	−1		
E	2	−1	0	(x, y)	$(xy, x^2-y^2), (yz, zx)$

* More character tables are given in the *Resource section*.

Character tables, and some of the data contained in them, are constructed on the assumption that the axis system is arranged in a particular way. This arrangement is specified in the character table when there is ambiguity. There is ambiguity in C_{2v} (and certain other groups), and so a more detailed specification of the symmetry operations is then necessary. The principal axis (a unique C_n axis with the greatest value of n), is taken to be the z-direction. If the molecule is planar, it is taken to lie in the yz-plane (referring to Fig. 10B.6). Then σ_v' is a reflection in the yz-plane and henceforth will be denoted $\sigma_v'(yz)$, and σ_v is a reflection in the xz-plane, and henceforth is denoted $\sigma_v(xz)$.

The irreducible representations are mutually orthogonal in the sense that if the set of characters is regarded as forming a row vector, the scalar product (as defined in *The chemist's toolkit* 8C.1) of the vectors corresponding to different irreducible representations is zero. In other words the vectors are mutually perpendicular.[2] Formally, this orthogonality is expressed as

$$\frac{1}{h}\sum_C N(C)\chi^{\Gamma^{(i)}}(C)\chi^{\Gamma^{(j)}}(C)=\begin{cases}0 & \text{for } i\neq j\\ 1 & \text{for } i=j\end{cases} \quad (10B.7)$$

Orthonormality of irreducible representations

where the sum is over the classes of the group, $N(C)$ is the number of operations in class C, and h is the number of operations in the group (its order). The division by h results in the scalar product of an irreducible representation with itself being 1, which means that the vectors, if divided by $h^{1/2}$, are normalized to 1. Vectors that are both orthogonal and normalized are said to be 'orthonormal'.

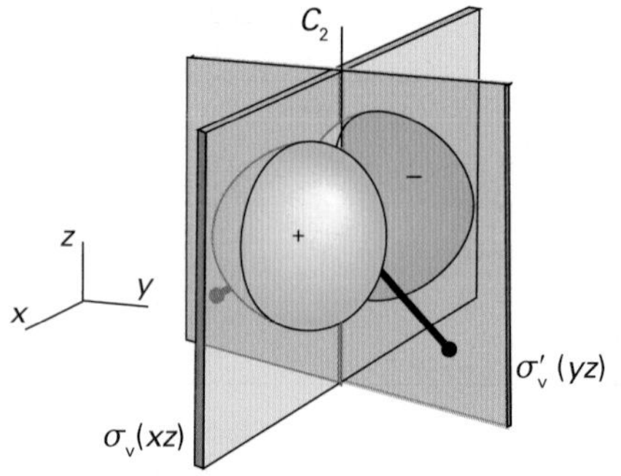

Figure 10B.6 A p_x orbital on the central atom of a C_{2v} molecule and the symmetry elements of the group.

[2] This result is a consequence of the 'great orthogonality theorem' of group theory; see our *Molecular quantum mechanics* (2011). In this Topic, the characters are taken to be real.

Brief illustration 10B.4

The point group C_{3v} has operations $\{E,2C_3,3\sigma_v\}$ which fall into three classes; the order h is 6. For the two irreducible representations with labels A_2 (with characters $\{1,1,-1\}$) and E (with characters $\{2,-1,0\}$), eqn 10B.7 is

$$\tfrac{1}{6}\{1\times1\times2+2\times1\times(-1)+3\times(-1)\times0\}=0$$

If the two irreducible representations are both E, the sum in eqn 10B.7 is

$$\tfrac{1}{6}\{1\times2\times2+2\times(-1)\times(-1)+3\times0\times0\}=1$$

The sum is also 1 if both irreducible representations are A_2:

$$\tfrac{1}{6}\{1\times1\times1+2\times1\times1+3\times(-1)\times(-1)\}=1$$

10B.3(a) The symmetry species of atomic orbitals

The characters of a one-dimensional irreducible representation indicate the behaviour of an orbital under the operation of the group to which each character corresponds. A character of 1 indicates that the orbital is unchanged, and a character of −1 indicates that it changes sign. It follows that the symmetry label of the orbital can be identified by comparing the changes that occur to an orbital under each operation, and then comparing the resulting 1 or −1 with the entries in a row of the character table for the relevant point group. By convention, the orbitals are labelled with the lower case equivalent of the symmetry species label (so an orbital of symmetry species A_1 is called an a_1 orbital).

Brief illustration 10B.5

Consider an H_2O molecule, point group C_{2v}, shown in Fig. 10B.6. The effect of C_2 on the oxygen $2p_x$ orbital is to cause it to change sign, so the character is −1; $\sigma_v'(yz)$ has the same effect and so has character −1. In contrast, $\sigma_v(xz)$ leaves the orbital unaffected and so has character 1, and of course the same is true of the identity operation. The characters of the operations $\{E,C_2,\sigma_v,\sigma_v'\}$ are therefore $\{1,-1,1,-1\}$. Reference to the C_{2v} character table (Table 10B.1) shows that $\{1,-1,1,-1\}$ are the characters for the symmetry species B_1; the orbital is therefore labelled b_1. A similar procedure gives the characters for oxygen $2p_y$ as $\{1,-1,-1,1\}$, which corresponds to B_2: the orbital is labelled b_2, therefore. Both the oxygen $2p_z$ and 2s are a_1.

The characters for irreducible representations of dimensionality greater than 1 (for example the E and T symmetry species) are the sums of the characters for the behaviour of the individual orbitals in the basis. Thus, if one member of a pair remains unchanged under a symmetry operation but the other changes sign (Fig. 10B.7), then the entry is reported as $\chi = 1 - 1 = 0$.

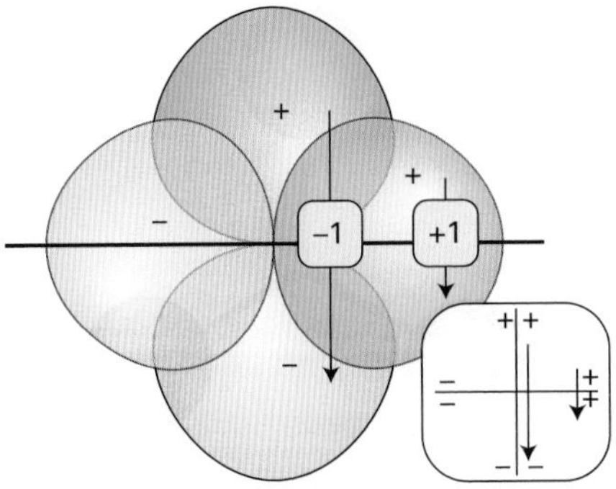

Figure 10B.7 The two orbitals shown here have different properties under reflection through the mirror plane: one changes sign (character −1), the other does not (character +1).

The symmetry species of s, p, and d orbitals on a central atom is easily read from the character table. The wavefunction of the p_z orbital is of the form $zf(r)$ and so it transforms in the same way as the cartesian function z; likewise the p_x and p_y orbitals transform as the cartesian functions y and z respectively. In the character table these cartesian functions are indicated on the right-hand side, in the row corresponding to their symmetry species.

For example, the character table for C_{3v} indicates that z has the symmetry species A_1, so the p_z orbital has this same symmetry species. The cartesian functions x and y are jointly of E symmetry. In technical terms, it is said that x and y, and hence p_x and p_y, jointly **span** an irreducible representation of symmetry species E. An s orbital on the central atom always spans the totally symmetric irreducible representation of a group as it is unchanged under all symmetry operations; in C_{3v} it has symmetry species A_1.

The five d orbitals of a shell are represented by the appropriate cartesian functions, for example xy for the d_{xy} orbital. These cartesian functions are also listed on the right of the character table. It can be seen at a glance that in C_{3v} d_{xy} and $d_{x^2-y^2}$ on a central atom jointly span E.

10B.3(b) The symmetry species of linear combinations of orbitals

The same technique may be applied to identify the symmetry species of linear combinations of orbitals, such as the combination $\psi_1 = s_A + s_B + s_C$ of the three H1s orbitals in the C_{3v} molecule NH_3 (Fig. 10B.8). This combination remains unchanged under a C_3 rotation and under any of the three vertical reflections of the group, so its characters are

$$\chi(E)=1 \quad \chi(C_3)=1 \quad \chi(\sigma_v)=1$$

Comparison with the C_{3v} character table shows that ψ_1 is of symmetry species A_1, and therefore has the label a_1.

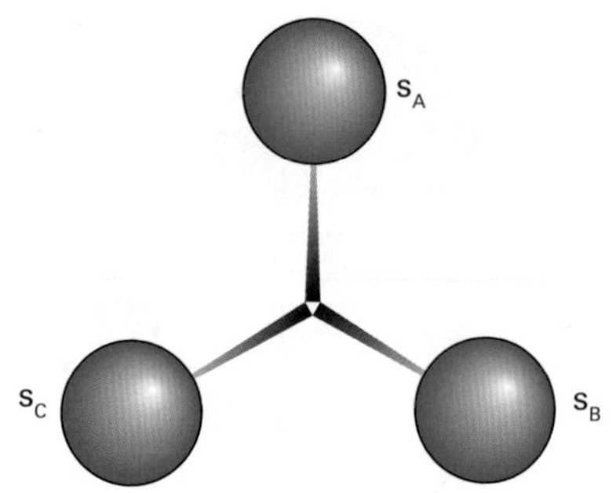

Figure 10B.8 The three H1s orbitals used to construct symmetry-adapted linear combinations in a C_{3v} molecule such as NH_3.

Example 10B.2 Identifying the symmetry species of orbitals

Identify the symmetry species of the orbital $\psi = \psi_A - \psi_B$ in a C_{2v} NO_2 molecule, where ψ_A is a $2p_x$ orbital on one O atom and ψ_B is a $2p_x$ orbital on the other O atom. As in Fig. 10B.6 take x to be perpendicular to the plane of the molecule.

Collect your thoughts The negative sign in ψ indicates that the sign of ψ_B is opposite to that of ψ_A. You need to consider how the combination changes under each operation of the group, and then write the character as 1, −1, or 0 as specified above. Then compare the resulting characters with each row in the character table for the point group, and hence identify the symmetry species.

The solution The combination is shown in Fig. 10B.9. Under C_2, ψ changes into itself, implying a character of 1. Under the reflections $\sigma_v(xz)$ and $\sigma'_v(yz)$ the wavefunction changes sign overall, $\psi \to -\psi$, implying a character of −1. The characters are therefore

$$\chi(E)=1 \quad \chi(C_2)=1 \quad \chi(\sigma_v(xz))=-1 \quad \chi(\sigma'_v(yz))=-1$$

These values match the characters of the A_2 symmetry species, so ψ is labelled a_2.

Self-test 10B.2 Consider $PtCl_4^{2-}$, in which the Cl^- ligands form a square planar array and the ion belongs to the point group D_{4h} (**1**). Identify the symmetry species of the combination of Cl s orbitals $\psi_A - \psi_B + \psi_C - \psi_D$. Note that in this group the C_2 axes coincide with the x and y axes, and σ_v planes coincide with the xz- and yz-planes; choose the x- and y-axes to pass through the corners of the square.

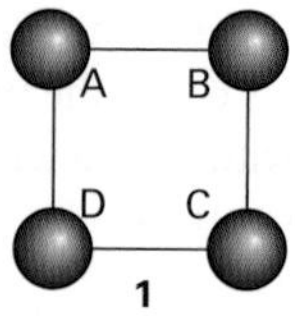

1

Answer: B_{1g}

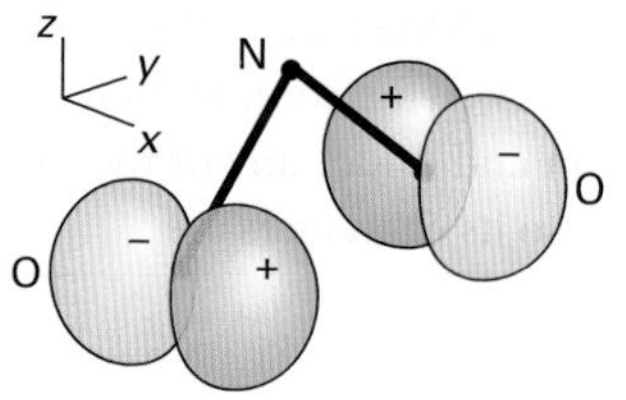

Figure 10B.9 One symmetry-adapted linear combination of $O2p_x$ orbitals in the C_{2v} NO_2 molecule.

10B.3(c) Character tables and degeneracy

In Topic 7D it is pointed out that degeneracy, which is when different wavefunctions have the same energy, is always related to symmetry. An energy level is degenerate if the wavefunctions corresponding to that energy can be transformed into each other by a symmetry operation (such as rotating a square well through 90°). Clearly, group theory should have a role in the identification of degeneracy.

A geometrically square well belongs to the group C_4 (Fig. 10B.10 and Table 10B.3), with the C_4 rotations (through 90°) converting x into y and vice versa.[3] As explained in Topic 7D, the two wavefunctions $\psi_{1,2} = (2/L)\sin(\pi x/L)\sin(2\pi y/L)$ and $\psi_{2,1} = (2/L)\sin(2\pi x/L)\sin(\pi y/L)$ both correspond to the energy $5h^2/8mL^2$, so that level is doubly degenerate. Under the operations of the group, these two functions transform as follows:

$$E:(\psi_{1,2}\,\psi_{2,1})\to(\psi_{1,2}\,\psi_{2,1}) \qquad C_4^+:(\psi_{1,2}\,\psi_{2,1})\to(\psi_{2,1}\,-\psi_{1,2})$$
$$C_4^-:(\psi_{1,2}\,\psi_{2,1})\to(-\psi_{2,1}\,\psi_{1,2}) \quad C_2:(\psi_{1,2}\,\psi_{2,1})\to(-\psi_{1,2}\,-\psi_{2,1})$$

The corresponding matrix representatives are

$$\boldsymbol{D}(E)=\begin{pmatrix}1 & 0\\ 0 & 1\end{pmatrix} \quad \boldsymbol{D}(C_4^+)=\begin{pmatrix}0 & 1\\ -1 & 0\end{pmatrix}$$
$$\boldsymbol{D}(C_4^-)=\begin{pmatrix}0 & -1\\ 1 & 0\end{pmatrix} \quad \boldsymbol{D}(C_2)=\begin{pmatrix}-1 & 0\\ 0 & -1\end{pmatrix}$$

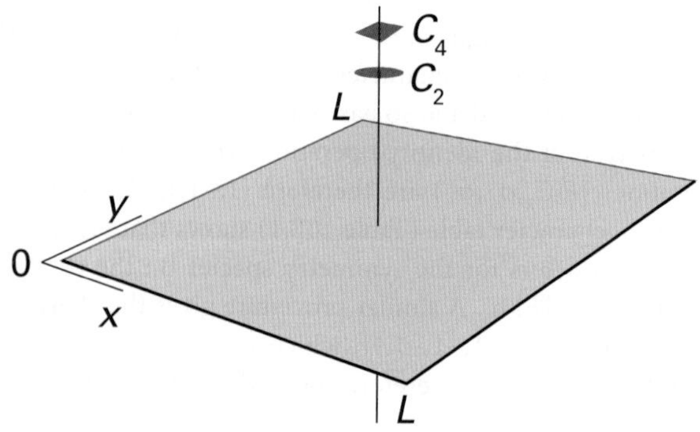

Figure 10B.10 A geometrically square well can be treated as belonging to the group C_4 (with elements $\{E,2C_4,C_2\}$).

[3] More complicated groups could be used, such as C_{4v} or D_{4h}, but C_4 captures the symmetry sufficiently.

Table 10B.3 The C_4 character table*

C_4, 4	E	$2C_4$	C_2	$h=4$	
A	1	1	1	z	z^2, x^2+y^2
B	1	−1	1		xy, x^2-y^2
E	2	0	−2	(x, y)	(yz, zx)

* More character tables are given in the *Resource section*.

and their characters are

$$\chi(E)=2 \quad \chi(C_4^+)=0 \quad \chi(C_4^-)=0 \quad \chi(C_2)=-2$$

A glance at the character table in Table 10B.3 (noting that the rotations C_4^+ and C_4^- belong to the same class and appear in the column labelled $2C_4$) shows that the basis spans the irreducible representation of symmetry species E. The same is true of all the doubly degenerate energy levels. There are no triply degenerate (or higher) energy levels in the system. Notice too that in the group C_4 there are no irreducible representations of dimension 3 or higher. These two observations illustrate the general principle that:

> The highest dimensionality of irreducible representation in a group is the maximum degree of degeneracy in the group.

Thus, if there is an E irreducible representation in a group, 2 is the highest degree of degeneracy; if there is a T irreducible representation in a group, then 3 is the highest degree of degeneracy. Some groups have irreducible representations of higher dimension, and therefore allow higher degrees of degeneracy. Furthermore, because the character of the identity operation is always equal to the dimensionality of the representation, the maximum degeneracy can be identified by noting the maximum value of $\chi(E)$ in the relevant character table.

Brief illustration 10B.6

- A trigonal planar molecule such as BF_3 cannot have triply degenerate orbitals because its point group is D_{3h} and the character table for this group (in the *Resource section*) does not have a T symmetry species.
- A methane molecule belongs to the tetrahedral point group T_d and because that group has irreducible representations of T symmetry, it can have triply degenerate orbitals. The same is true of tetrahedral P_4, which, with just four atoms, is the simplest kind of molecule with triply degenerate orbitals.
- A buckminsterfullerene molecule, C_{60}, belongs to the icosahedral point group (I_h) and its character table (in the *Resource section*) shows that the maximum dimensionality of its irreducible representations is 5, so it can have five-fold degenerate orbitals.

Exercises

E10B.5 For the point group C_{2h}, confirm that all the irreducible representations are orthonormal according to the property defined in eqn 10B.7. The character table will be found in the *Resource Section*.

E10B.6 By inspection of the character table for D_{3h}, state the symmetry species of the 3p and 3d orbitals located on the central Al atom in AlF_3.

E10B.7 What is the maximum degeneracy of the wavefunctions of a particle confined to the interior of an octahedral hole in a crystal?

Checklist of concepts

- ☐ 1. A **group** is a collection of transformations that satisfy the four criteria set out at the start of the Topic.
- ☐ 2. The **order** of a group is the number of its symmetry operations.
- ☐ 3. A **matrix representative** is a matrix that represents the effect of an operation on a basis.
- ☐ 4. The **character** is the sum of the diagonal elements of a matrix representative of an operation.
- ☐ 5. A **matrix representation** is the collection of matrix representatives for the operations in the group.
- ☐ 6. A **character table** consists of entries showing the characters of all the irreducible representations of a group.
- ☐ 7. A **symmetry species** is a label for an irreducible representation of a group.
- ☐ 8. The highest dimensionality of irreducible representation in a group is the maximum degree of degeneracy in the group.

Checklist of equations

Property	Equation	Comment	Equation number
Class membership	$R' = S^{-1}RS$	All elements members of the group	10B.1
Number of irreducible representations	Number of irreducible representations = number of classes		10B.5
Dimensionality and order	$\sum_{\text{irreps } i} d_i^2 = h$	For groups other than pure rotation groups with $n > 2$	10B.6
Orthonormality of irreducible representations	$\frac{1}{h}\sum_C N(C)\chi^{\Gamma^{(i)}}(C)\chi^{\Gamma^{(j)}}(C) = \begin{cases} 0 \text{ for } i \neq j \\ 1 \text{ for } i = j \end{cases}$	Sum over classes	10B.7

TOPIC 10C Applications of symmetry

> **Why do you need to know this material?**
> Group theory is a key tool for constructing molecular orbitals and formulating spectroscopic selection rules.
>
> **What is the key idea?**
> An integral can be non-zero only if the integrand is invariant under the symmetry operations of a molecule.
>
> **What do you need to know already?**
> This Topic develops the material in Topic 10A, where the classification of molecules on the basis of their symmetry elements is introduced, and draws heavily on the properties of characters and character tables described in Topic 10B. You need to be aware that many quantum-mechanical properties, including transition dipole moments (Topic 8C), depend on integrals involving products of wavefunctions.

Group theory shows its power when brought to bear on a variety of problems in chemistry, among them the construction of molecular orbitals and the formulation of spectroscopic selection rules.

10C.1 Vanishing integrals

Any integral, I, of a function $f(x)$ over a symmetric range around $x = 0$ is zero if the function is antisymmetric (odd), meaning that $f(-x) = -f(x)$:

$$I = \int_{-a}^{+a} f(x)\,\mathrm{d}x = 0 \qquad \text{if } f(-x) = -f(x)$$

For an analogous integral in two dimensions over a symmetrical range in x and y the area of integration is a square (Fig. 10C.1). The square has various symmetry operations including a C_4 rotation perpendicular to the plane. The integral of the integrand $f(x,y)$ has contributions from regions that are related by this C_4 rotation and if $f(x,y)$ changes sign under this operation, the contribution of the first region is cancelled by that from the symmetry-related region and the integral is zero (Fig. 10C.1b).

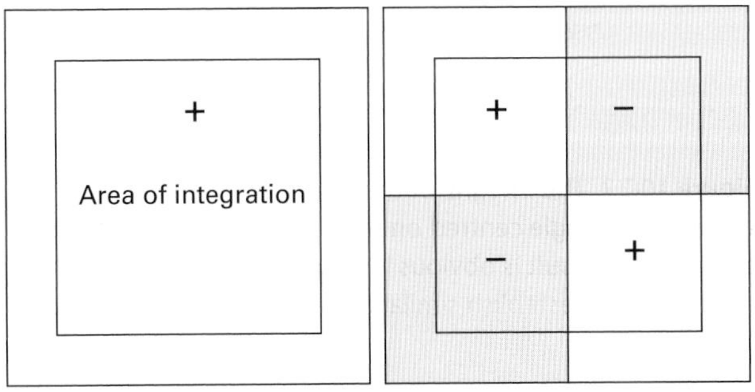

Figure 10C.1 (a) Only if the integrand is unchanged under each symmetry operation of the group (here C_4) can its integral over the region indicated be non-zero. (b) If the integrand changes sign under any operation, its integral is necessarily zero.

The integral may be non-zero only if the integrand is invariant under the operation (Fig. 10C.1a).

The same argument applies to each of the symmetry operations of the area of integration. In addition, it is possible that the integrand can be expressed as a sum of contributions, one of which is invariant under these symmetry operations. In that case, the integral may be non-zero.

In summary the integral may be non-zero only if the integrand, or a contribution to it, is invariant under each symmetry operation of the group that reflects the shape of the area of integration or, in three dimensions, the volume of integration. In group-theoretical terms:

> An integral over a region of space can be non-zero only if the integrand (or a contribution to it) spans the totally symmetric irreducible representation of the point group of the region.

The totally symmetric irreducible representation has all characters equal to 1, and is typically the symmetry species denoted A_1.

Brief illustration 10C.1

To decide whether the integral of the function $f = xy$ may be non-zero when evaluated over a region the shape of an equilateral triangle centred on the origin (Fig. 10C.2), recognize that the triangle belongs to the point group C_3. Reference to the character table of the group shows that xy is a member of a basis that spans the irreducible representation E. Therefore, its

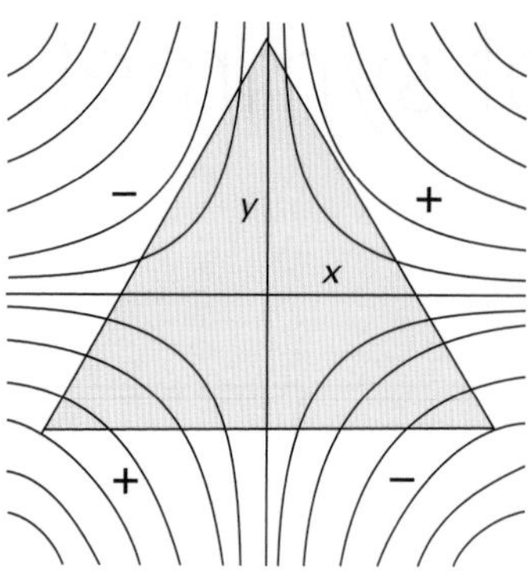

Figure 10C.2 The integral of the function $f = xy$ over an equilateral triangle centred on the origin (shaded) is zero. In this case, the result is obvious by inspection, but group theory can be used to establish similar results in less obvious cases.

integral must be zero, because the integrand has no component that spans the totally symmetric irreducible representation (A in this point group).

10C.1(a) Integrals of the product of functions

Suppose the integral of interest is of a product of two functions, f_1 and f_2, taken over all space and over all relevant variables (represented, as is usual in quantum mechanics, by $d\tau$):

$$I = \int f_1 f_2 d\tau \qquad (10C.1)$$

For example, f_1 and f_2 might be atomic orbitals on different atoms, in which case I would be their overlap integral. The implication of such an integral being zero is that a molecular orbital does not result from the overlap of these two orbitals. It follows from the general point made above that the integral may be non-zero only if the integrand itself, the product $f_1 f_2$, is unchanged by any symmetry operation of the molecular point group and so spans the totally symmetric irreducible representation (typically the symmetry species with label A_1).

To decide whether the product $f_1 f_2$ does indeed span the totally symmetric irreducible representation, it is necessary to form the **direct product** of the symmetry species spanned by f_1 and f_2 separately. The procedure is as follows:

- Write down a table with columns headed by the symmetry operations, R, of the group.
- In the first row write down the characters of the symmetry species spanned by f_1; in the second row write down the characters of the symmetry species spanned by f_2.
- Multiply the numbers in the two rows together, column by column. The resulting set of numbers are the characters of the representation spanned by $f_1 f_2$.

Brief illustration 10C.2

Suppose that in the point group C_{2v} f_1 has the symmetry species A_2, and f_2 has the symmetry species B_1. From the character table the characters for these species are 1,1,−1,−1 and 1,−1,1,−1, respectively. The direct product of these two species is found by setting up the following table

	E	C_2	$\sigma_v(xz)$	$\sigma_v'(yz)$
A_2	1	1	−1	−1
B_1	1	−1	1	−1
product	1	−1	−1	1

Now recognize that the characters in the final row are those of the symmetry species B_2. It follows that the symmetry species of the product $f_1 f_2$ is B_2. Because the direct product does not contain the totally symmetric irreducible representation (A_1), the integral of $f_1 f_2$ over all space must be zero.

The direct product of two irreducible representations $\Gamma^{(i)}$ and $\Gamma^{(j)}$ is written $\Gamma^{(i)} \times \Gamma^{(j)}$ or sometimes $\Gamma^{(i)} \otimes \Gamma^{(j)}$. Direct products have some simplifying features.

- The direct product of the totally symmetric irreducible representation with any other representation is the latter irreducible representation itself: $A_1 \times \Gamma^{(i)} = \Gamma^{(i)}$.

All the characters of A_1 are 1, so multiplication by them leaves the characters of $\Gamma^{(i)}$ unchanged. It follows that if one of the functions in eqn 10C.1 transforms as A_1, then the integral will vanish if the other function does not also transform as A_1.

- The direct product of two irreducible representations is A_1 only if the two irreducible representations are identical: $\Gamma^{(i)} \times \Gamma^{(j)}$ contains A_1 only if $i = j$.

For one-dimensional irreducible representations the characters are either 1 or −1, and the character 1 is obtained only if the characters of $\Gamma^{(i)}$ and $\Gamma^{(j)}$ are the same (both 1 or both −1). For example, in C_{2v}, $A_1 \times A_1$, $A_2 \times A_2$, $B_1 \times B_1$, and $B_2 \times B_2$, but no other combination, all give A_1. This requirement is also true for higher-dimensional representations, as is demonstrated at the end of the following section.

It follows that if f_1 and f_2 transform as different symmetry species, then the product cannot transform as the totally symmetric irreducible representation. In this case the integral of $f_1 f_2$ is necessarily zero. If, on the other hand, both functions transform as the same symmetry species, then the product transforms as the totally symmetric irreducible representation (and possibly has contributions from other symmetry species too). Now the integral is not necessarily zero.

An important point is that group theory is specific about when an integral must be zero, but integrals that it allows to be non-zero may be zero for reasons unrelated to symmetry. For example, in ammonia the 1s orbital on N and the combination $(s_1+s_2+s_3)$, where s_1 is a 1s orbital on H_1 and so on, both transform as A_1. Group theory therefore predicts that the orbitals have non-zero overlap, but in practice the 1s orbital on N is so contracted that its overlap with the H1s orbitals is negligible at the N—H distance found in this molecule.

Integrals of the form

$$I=\int f_1 f_2 f_3 \mathrm{d}\tau \qquad (10C.2)$$

are also common in quantum mechanics, and it is important to know when they are necessarily zero. For example, they appear in the calculation of transition dipole moments (Topic 8C). As for integrals over two functions, for I to be non-zero, *the product $f_1 f_2 f_3$ must span the totally symmetric irreducible representation or contain a component that spans that representation.* To test whether this is so, the characters of all three irreducible representations are multiplied together in the same way as in the rules set out above.

Example 10C.1 Deciding if an integral must be zero

Does the integral $\int(d_{z^2})x(d_{xy})\mathrm{d}\tau$ vanish in a C_{2v} molecule, where d_{z^2} and d_{xy} are the wavefunctions of d orbitals indicated by the subscripts?

Collect your thoughts Use the C_{2v} character table to find the characters of the irreducible representations spanned by $3z^2-r^2$ (the form of the d_{z^2} orbital), x, and xy. Then set up a table to work out the triple direct product and identify whether the symmetry species it spans includes A_1.

The solution The C_{2v} character table shows that the function xy, and hence the orbital d_{xy}, transforms as A_2, that z^2 transforms as A_1, and that x transforms as B_1. The table is therefore

	E	C_2	$\sigma_v(xz)$	$\sigma_v'(yz)$	
A_2	1	1	−1	−1	$f_3=d_{xy}$
B_1	1	−1	1	−1	$f_2=x$
A_1	1	1	1	1	$f_1=d_{z^2}$
	1	−1	−1	1	product

The characters in the bottom row are those of B_2, not of A_1. Therefore, the integral is necessarily zero.

A quicker solution involves noting that A_1 (for f_1) has no effect on the outcome of the triple direct product (by the first feature mentioned above), and therefore, by the second feature, the symmetry species of the two functions f_2 and f_3 must be the same for their direct product to be A_1; but in this example they are not the same.

Self-test 10C.1 Does the integral $\int(d_{xz})x(p_z)\mathrm{d}\tau$ necessarily vanish in a C_{2v} molecule?

Answer: No

10C.1(b) Decomposition of a representation

In some cases, it turns out that the direct product is a sum of irreducible representations (symmetry species), not just a single species. For instance, in C_{3v} the characters of the direct product $E\times E$ are {4,1,0}, which can be decomposed as A_1, A_2, and E:

	E	$2C_3$	$3\sigma_v$
A_1	1	1	1
A_2	1	1	−1
E	2	−1	0
sum	4	1	0

This decomposition is written symbolically $E\times E=A_1+A_2+E$.[1]

In simple cases the decomposition can be done by inspection. Group theory, however, provides a systematic way of using the characters of the representation to find the irreducible representations of which it is composed. The formal recipe for finding the number of times, $n(\Gamma^{(i)})$, that irreducible representation $\Gamma^{(i)}$ occurs is:[2]

$$n(\Gamma^{(i)})=\frac{1}{h}\sum_C N(C)\chi^{(\Gamma^{(i)})}(C)\chi(C) \qquad \text{Decomposition of a representation} \qquad (10C.3a)$$

Note that the sum is over the classes of operations in the group. In this expression h is the order of the group, $\chi^{(\Gamma^{(i)})}(C)$ is the character for operations in class C for the irreducible representation $\Gamma^{(i)}$, and $\chi(C)$ is the corresponding character of the representation being decomposed. In the character table the number of operations in each class, $N(C)$, is indicated in the header of the columns.

If all that is of interest is the occurrence of the totally symmetric irreducible representation in the decomposition the calculation is somewhat simpler. All the characters of the totally symmetric irreducible representation (symmetry species A_1) are 1, so setting $\Gamma^{(i)}=A_1$ and $\chi^{(A_1)}(C)=1$ for all C in eqn 10C.3a gives

$$n(A_1)=\frac{1}{h}\sum_C N(C)\chi(C) \qquad \text{Occurrence of } A_1 \qquad (10C.3b)$$

Brief illustration 10C.3

In the character table for C_{3v}, the columns are headed E, $2C_3$, and $3\sigma_v$, indicating that the numbers in each class are 1, 2, and 3, respectively and $h=1+2+3=6$. To decide whether A_1

[1] As mentioned in Topic 10B, a direct sum is sometimes denoted ⊕. The analogous symbol for a direct product is ⊗. The symbolic expression is then written $E\otimes E=A_1\oplus A_2\oplus E$.

[2] This result arises from the 'great orthogonality theorem': see our *Molecular quantum mechanics* (2011). In this Topic, the characters are taken to be real.

occurs in the representation with characters {4,1,0} in C_{3v} use eqn 10C.3b to give

$$n(A_1) = \frac{1}{h}\sum_C N(C)\chi(C)$$
$$= \tfrac{1}{6}\{1\times\chi(E) + 2\times\chi(C_3) + 3\times\chi(\sigma_v)\}$$
$$= \tfrac{1}{6}\{1\times 4 + 2\times 1 + 3\times 0\} = 1$$

A_1 therefore occurs once in the decomposition.

It is asserted in the preceding section that the direct product of two irreducible representations is A_1 only if the two irreducible representations are identical. That this is so can now be shown with the aid of eqn 10C.3b.

How is that done? 10C.1 Confirming the criterion for a direct product to contain the totally symmetric irreducible representation

Start by considering the characters of the direct product between irreducible representations $\Gamma^{(i)}$ and $\Gamma^{(j)}$. For irreducible representation $\Gamma^{(i)}$ the character of an operation in class C is $\chi^{(\Gamma^{(i)})}(C)$, and likewise $\chi^{(\Gamma^{(j)})}(C)$ for irreducible representation $\Gamma^{(j)}$. It follows that the characters in the direct product are $\chi(C) = \chi^{(\Gamma^{(i)})}(C)\chi^{(\Gamma^{(j)})}(C)$. The number of times that the totally symmetric irreducible representation (A_1) occurs in this direct-product representation is given by eqn 10C.3b as

$$n(A_1) = \frac{1}{h}\sum_C N(C)\chi^{(\Gamma^{(i)})}(C)\chi^{(\Gamma^{(j)})}(C)$$

Irreducible representations are orthonormal in the sense that (eqn 10B.7)

$$\frac{1}{h}\sum_C N(C)\chi^{(\Gamma^{(i)})}(C)\chi^{(\Gamma^{(j)})}(C) = \begin{cases} 0 \text{ if } i \neq j \\ 1 \text{ if } i = j \end{cases}$$

It follows that

$$n(A_1) = \begin{cases} 0 \text{ if } i \neq j \\ 1 \text{ if } i = j \end{cases}$$

In other words, the direct product of two irreducible representations has a component that spans A_1 only if the two irreducible representations belong to the same symmetry species. This result is independent of the dimensionality of the irreducible representations.

Exercises

E10C.1 Use symmetry properties to determine whether or not the integral $\int p_x z p_z \mathrm{d}\tau$ is necessarily zero in a molecule with symmetry C_{2v}.

E10C.2 Show that the function xy has symmetry species B_{1g} in the group D_{2h}.

E10C.3 A set of basis functions is found to span a reducible representation of the group C_{4v} with characters 5,1,1,3,1 (in the order of operations in the character table in the *Resource section*). What irreducible representations does it span?

10C.2 Applications to molecular orbital theory

The rules outlined so far can be used to decide which atomic orbitals may have non-zero overlap in a molecule. Group theory also provides procedures for constructing linear combinations of atomic orbitals of a specified symmetry.

10C.2(a) Orbital overlap

The overlap integral, S, between orbitals ψ_1 and ψ_2 is

$$S = \int \psi_2^* \psi_1 \mathrm{d}\tau \qquad \text{Overlap integral} \qquad (10C.4)$$

It follows from the discussion of eqn 10C.1 that this integral can be non-zero only if the two orbitals span the same symmetry species. In other words,

> Only orbitals of the same symmetry species may have non-zero overlap ($S \neq 0$) and hence go on to form bonding and antibonding combinations.

The selection of atomic orbitals with non-zero overlap is the central and initial step in the construction of molecular orbitals as LCAOs.

Example 10C.2 Identifying which orbitals can contribute to bonding

The four H1s orbitals of methane span $A_1 + T_2$. With which of the C2s and C2p atomic orbitals can they overlap? What additional overlap would be possible if d orbitals on the C atom were also considered?

Collect your thoughts Refer to the T_d character table (in the *Resource section*) and look for s, p, and d orbitals spanning A_1 or T_2. As explained in Topic 10B, the symmetry species can be identified by looking for the appropriate cartesian functions listed on the right of the table.

The solution A C2s orbital spans A_1 in the group T_d, so it may have non-zero overlap with the A_1 combination of H1s orbitals. From the table (x,y,z) jointly span T_2, so the three C2p orbitals together transform as T_2; they may have non-zero overlap with the T_2 combination of H1s orbitals.

The combinations (xy,yz,zx) span T_2, therefore the d_{xy}, d_{yz}, and d_{zx} orbitals do the same and so they may overlap with the T_2 combination of H1s orbitals. The other two d orbitals span E and so they cannot overlap with the A_1 or T_2 H1s orbitals.

It follows that in methane there are a_1 orbitals arising from (C2s,H1s)-overlap and t_2 orbitals arising from (C2p,H1s)-overlap. The C3d orbitals might contribute to the latter. The lowest energy configuration is probably $a_1^2t_2^6$, with all bonding orbitals occupied.

Self-test 10C.2 Consider the octahedral SF_6 molecule, with the bonding arising from overlap of orbitals on S and a 2p orbital on each fluorine directed towards the central sulfur atom. The latter span $A_{1g} + E_g + T_{1u}$. Which sulfur orbitals have non-zero overlap with these F orbitals? Suggest what the ground-state configuration is likely to be.

Answer: 3s(A_{1g}),3p(T_{1u}),($3d_{x^2-y^2}$,$3d_{z^2}$)(E_g); $a_{1g}^2t_{1u}^6e_g^4$

10C.2(b) Symmetry-adapted linear combinations

Topic 10B introduces the idea of generating a combination of atomic orbitals designed to transform as a particular symmetry species or irreducible representation. Such a combination is an example of a **symmetry-adapted linear combination** (SALC), which is a combination of orbitals constructed from equivalent atoms and having a specified symmetry. SALCs are very useful in constructing molecular orbitals because a given SALC has non-zero overlap only with other orbitals of the same symmetry.

The technique for building SALCs is derived by using the full power of group theory and involves the use of a **projection operator**, $P^{(\Gamma^{(i)})}$, an operator that takes one of the basis orbitals and generates from it—projects from it—a SALC of the symmetry species $\Gamma^{(i)}$:

$$P^{(\Gamma^{(i)})} = \frac{1}{h}\sum_R \chi^{(\Gamma^{(i)})}(R)R \qquad \psi^{(\Gamma^{(i)})} = P^{(\Gamma^{(i)})}\psi_n \qquad \text{Projection operator} \qquad (10C.5)$$

Here ψ_n is one of the basis orbitals and $\psi^{(\Gamma^{(i)})}$ is a SALC (there might be more than one) that transforms as the symmetry species $\Gamma^{(i)}$; the sum is over the operations (not the classes) of the group of order h. To implement this rule, do the following:

- Construct a table with the columns headed by each symmetry operation R of the group; include a column for each operation, not just for each class.
- Select a basis function and work out the effect that each operation has on it. Enter the resulting function beneath each operation.
- On the next row enter the characters of the symmetry species of interest, $\chi^{(\Gamma^{(i)})}(R)$.
- Multiply the entries in the previous two rows, operation by operation.
- Sum the result, and divide it by the order of the group, h.

Brief illustration 10C.4

To construct the B_1 SALC from the two O2p_x orbitals in NO_2, point group C_{2v} (Fig. 10C.3), choose p_A as the basis function and draw up the following table

	E	C_2	$\sigma_v(xz)$	$\sigma_v'(yz)$
Effect on p_A	p_A	$-p_B$	p_B	$-p_A$
Characters for B_1	1	−1	1	−1
Product of rows 1 and 2	p_A	p_B	p_B	p_A

The sum of the final row, divided by the order of the group ($h = 4$), gives $\psi^{(B_1)} = \frac{1}{2}(p_A + p_B)$.

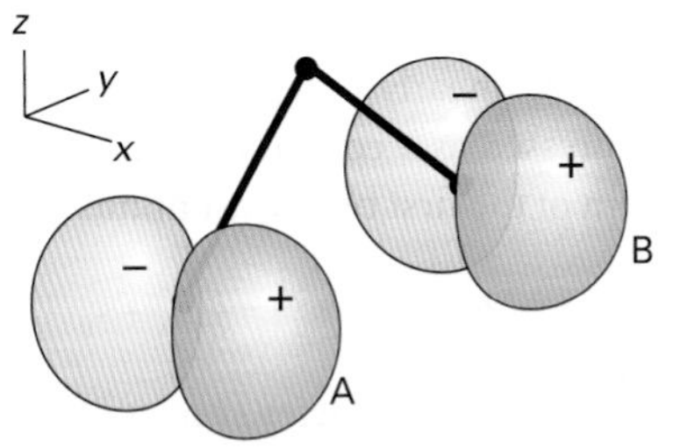

Figure 10C.3 The two O2p_x atomic orbitals in NO_2 (point group C_{2v}) can be used as a basis for forming SALCs.

If an attempt is made to generate a SALC with symmetry that is not spanned by the basis functions, the result is zero. For example, if in *Brief illustration* 10C.4 an attempt is made to project an A_1 symmetry orbital, all the characters in the second row of the table will be 1, so when the product of rows 1 and 2 is formed the result is $p_A - p_B + p_B - p_A = 0$.

A difficulty is encountered when aiming to generate an SALC of symmetry species of dimension higher than 1, because then the rules generate sums of SALCs. Consider, for instance, the generation of SALCs from the three H1s atomic orbitals in NH_3 (point group C_{3v}). The molecule and the orbitals are shown in Fig. 10C.4. The table below shows the effect of applying the projection operator to s_A, s_B, and s_C in turn to give the SALC of symmetry species E.

Row		E	C_3^+	C_3^-	σ_v	σ_v'	σ_v''
1	effect on s_A	s_A	s_C	s_B	s_A	s_C	s_B
2	characters for E	2	−1	−1	0	0	0
3	product of rows 1 and 2	$2s_A$	$-s_C$	$-s_B$			
4	effect on s_B	s_B	s_A	s_C	s_C	s_B	s_A
5	product of rows 4 and 2	$2s_B$	$-s_A$	$-s_C$			
6	effect on s_C	s_C	s_B	s_A	s_B	s_A	s_C
7	product of rows 6 and 2	$2s_C$	$-s_B$	$-s_A$			

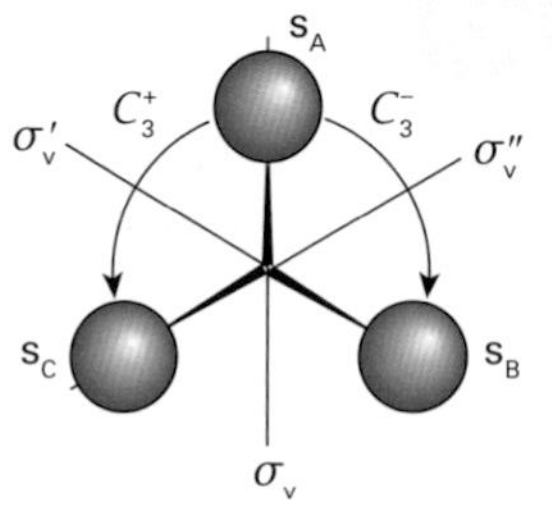

Figure 10C.4 The three H1s atomic orbitals in NH_3 (point group C_{3v}) can be used as a basis for forming SALCs.

Application of the projection operator to a different basis function gives a different SALC in each case (rows 3, 5, and 7).

$$\tfrac{1}{6}(2s_A - s_B - s_C) \quad \tfrac{1}{6}(2s_B - s_A - s_C) \quad \tfrac{1}{6}(2s_C - s_A - s_B)$$

These three are not 'linearly independent' because any one of them can be expressed as a combination of the other two. For example, the sum of the first two is just minus the third:

$$\tfrac{1}{6}(2s_A - s_B - s_C) + \tfrac{1}{6}(2s_B - s_A - s_C) = \tfrac{1}{6}(s_A + s_B - 2s_C)$$

Furthermore, because the symmetry species of the SALCs being projected is two-dimensional, it is expected that there will be just two SALCs.

It is possible to find two combinations of the SALCs from the table which are linearly independent; they turn out to be $(s_A - s_B)$ and $(2s_C - s_A - s_B)$. These SALCs are used in the construction of the molecular orbitals. Note that this choice is not unique, and that other pairs of SALCs formed by cyclic permutation of the indices A, B, and C are equally valid.

According to the discussion in Topic 9E concerning the construction of molecular orbitals of polyatomic molecules, only orbitals with the same symmetry can overlap to give a molecular orbital. In the language introduced here, only SALCs of the same symmetry species have non-zero overlap and contribute to a molecular orbital. In NH_3, for instance, the molecular orbitals will have the form

$$\psi(a_1) = c_{a_1} s_N + c_{a_2}(s_A + s_B + s_C)$$
$$\psi(e_x) = c_{e_1} p_{Nx} + c_{e_2}(2s_C - s_A - s_B)$$
$$\psi(e_y) = c'_{e_1} p_{Ny} + c'_{e_2}(s_A - s_B)$$

Group theory is silent on the values of the coefficients: they have to be determined by one of the methods outlined in Topic 9E.

Exercises

E10C.4 Consider the C_{2v} molecule OF_2; take the molecule to lie in the yz-plane, with z directed along the C_2 axis; the mirror plane σ'_v is the yz-plane, and σ_v is the xz-plane. The combination $p_z(A) + p_z(B)$ of the two F atoms spans A_1, and the combination $p_z(A) - p_z(B)$ of the two F atoms spans B_2. Are there any valence orbitals of the central O atom that can have a non-zero overlap with these combinations of F orbitals? How would the situation be different in SF_2, where 3d orbitals might be available?

E10C.5 Consider the C_{2v} molecule NO_2. The combination $p_x(A) - p_x(B)$ of the two O atoms (with x perpendicular to the plane) spans A_2. Is there any valence orbital of the central N atom that can have a non-zero overlap with that combination of O orbitals? What would be the case in SO_2, where 3d orbitals might be available?

10C.3 Selection rules

The intensity of a spectral line arising from a transition between some initial state with wavefunction ψ_i and a final state with wavefunction ψ_f depends on the (electric) transition dipole moment, μ_{fi} (Topic 8C). The q-component, where q is x, y, or z, of this vector, is defined through

$$\mu_{q,fi} = -e\int \psi_f^* q \psi_i \, d\tau \qquad \text{Transition dipole moment [definition]} \qquad (10C.6)$$

where $-e$ is the charge of the electron. The transition moment has the form of the integral over $f_1 f_2 f_3$ (eqn 10C.2). Therefore, once the symmetry species of the wavefunctions and q are known, group theory can be used to formulate the selection rules for the transitions.

Example 10C.3 Deducing a selection rule

Is $p_y \rightarrow p_z$ an allowed electric dipole transition in a molecule with C_{2v} symmetry?

Collect your thoughts You need to decide whether the product $p_z q p_y$, with $q = x$, y, or z, spans A_1. The symmetry species for p_y, p_z, and q can be read off from the right-hand side of the character table.

The solution The p_y orbital spans B_2 and p_z spans A_1, so the required direct product is $A_1 \times \Gamma^{(q)} \times B_2$, where $\Gamma^{(q)}$ is the symmetry species of x, y, or z. It does not matter in which order the direct products are calculated, so noting that $A_1 \times B_2 = B_2$ implies that $A_1 \times \Gamma^{(q)} \times B_2 = \Gamma^{(q)} \times B_2$. This direct product can be equal to A_1 only if $\Gamma^{(q)}$ is B_2, which is the symmetry species of y. Therefore, provided $q = y$ the integral may be non-zero and the transition allowed.

Comment. The analysis implies that the electromagnetic radiation involved in the transition has a component of its electric vector in the y-direction.

Self-test 10C.3 Are (a) $p_x \rightarrow p_y$ and (b) $p_x \rightarrow p_z$ allowed electric dipole transitions in a molecule with C_{2v} symmetry?

Answer: (a) No; (b) yes, with $q = x$

Exercises

E10C.6 Is the transition $A_1 \rightarrow A_2$ forbidden for electric dipole transitions in a C_{3v} molecule?

E10C.7 The electronic ground state of NO_2 is A_1 in the group C_{2v}. To what excited states may it be excited by electric dipole transitions, and what polarization of light is it necessary to use?

E10C.8 What states of (i) benzene, (ii) naphthalene may be reached by electric dipole transitions from their (totally symmetrical) ground states?

Checklist of concepts

☐ 1. Character tables can be used to decide whether an integral is necessarily zero.

☐ 2. For an integral to be non-zero, the integrand must include a component that transforms as the totally symmetric irreducible representation (A_1).

☐ 3. Only orbitals of the same symmetry species may have non-zero overlap.

☐ 4. A **symmetry-adapted linear combination** (SALC) is a linear combination of atomic orbitals constructed from equivalent atoms and having a specified symmetry.

Checklist of equations

Property	Equation	Comment	Equation number
Decomposition of a representation	$n(\Gamma^{(i)}) = \frac{1}{h}\sum_C N(C)\chi^{(\Gamma^{(i)})}(C)\chi(C)$	Real characters*	10C.3a
Presence of A_1 in a decomposition	$n(A_1) = \frac{1}{h}\sum_C N(C)\chi(C)$	Real characters*	10C.3b
Overlap integral	$S = \int \psi_2^* \psi_1 \mathrm{d}\tau$	Definition	10C.4
Projection operator	$P^{(\Gamma^{(i)})} = \frac{1}{h}\sum_R \chi^{(\Gamma^{(i)})}(R)R$	To generate $\psi^{(\Gamma^{(i)})} = P^{(\Gamma^{(i)})}\psi_n$	10C.5
Transition dipole moment	$\mu_{q,\mathrm{fi}} = -e\int \psi_\mathrm{f}^* q \psi_\mathrm{i} \mathrm{d}\tau$	q-Component, $q = x, y$, or z	10C.6

* In general, characters may have complex values; throughout this text only real values are encountered.

FOCUS 10 Molecular symmetry

To test your understanding of this material, work through the *Exercises, Additional exercises, Discussion questions*, and *Problems* found throughout this Focus.

Selected solutions can be found at the end of this Focus in the e-book. Solutions to even-numbered questions are available online only to lecturers.

TOPIC 10A Shape and symmetry

Discussion questions

D10A.1 Explain how a molecule is assigned to a point group.

D10A.2 List the symmetry operations and the corresponding symmetry elements that occur in point groups.

D10A.3 State and explain the symmetry criteria that allow a molecule to be polar.

D10A.4 State the symmetry criterion that allows a molecule to be optically active.

Additional exercises

E10A.9 The BF_3 molecule belongs to the point group D_{3h}. List the symmetry elements of the group and locate them in a drawing of the molecule.

E10A.10 Identify the group to which the *trans*-difluoroethene molecule belongs and locate the symmetry elements in a drawing of the molecule.

E10A.11 Identify the point groups to which the following objects belong: (i) a sharpened cylindrical pencil, (ii) a box with a rectangular cross-section, (iii) a coffee mug with a handle, (iv) a three-bladed propeller (assume the sector-like blades are flat), (v) a three-bladed propeller (assume the blades are twisted out of the plane, all by the same amount).

E10A.12 List the symmetry elements of the following molecules and name the point groups to which they belong: (i) furan (**1**), (ii) γ-pyran (**2**), (iii) 1,2,5-trichlorobenzene.

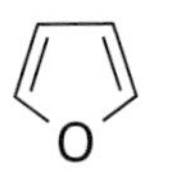

1 Furan

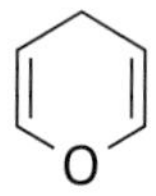

2 γ-Pyran

E10A.13 Assign the following molecules to point groups: (i) HF, (ii) IF_7 (pentagonal bipyramid), (iii) ClF_3 (T-shaped), (iv) $Fe_2(CO)_9$ (**3**), (v) cubane, C_8H_8, (vi) tetrafluorocubane, $C_8H_4F_4$ (**4**).

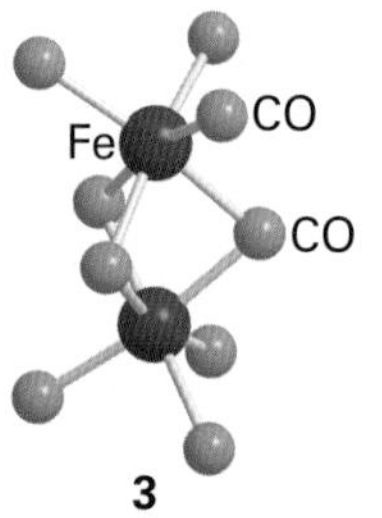

3

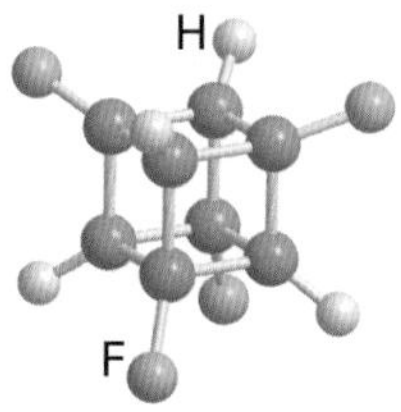

4

E10A.14 Which of the following molecules may be polar? (i) CF_3H, (ii) PCl_5, (iii) *trans*-difluoroethene, (iv) 1,2,4-trinitrobenzene.

E10A.15 Identify the point group to which each of the possible isomers of dichloronaphthalene belong.

E10A.16 Identify the point group to which each of the possible isomers of dichloroanthracene belong.

E10A.17 Can molecules belonging to the point groups T_h or T_d be chiral? Explain your answer.

Problems

P10A.1 List the symmetry elements of the following molecules and name the point groups to which they belong: (a) staggered CH_3CH_3, (b) chair and boat cyclohexane, (c) B_2H_6, (d) $[Co(en)_3]^{3+}$, where en is 1,2-diaminoethane (ignore its detailed structure), (e) crown-shaped S_8. Which of these molecules can be (i) polar, (ii) chiral?

P10A.2 Consider the series of molecules SF_6, SF_5Cl, SF_4Cl_2, SF_3Cl_3. Assign each to the relevant point group and state whether or not the molecule is expected to be polar. If isomers are possible for any of these molecules, consider all possible structures.

P10A.3 (a) Identify the symmetry elements in ethene and in allene, and assign each molecule to a point group. (b) Consider the biphenyl molecule, Ph–Ph, in which different conformations are possible according to the value of the dihedral angle between the planes of the two benzene rings: if this angle is 0°, the molecule is planar, if it is 90°, the two rings are perpendicular to one another. For each of the following dihedral angles, identify the symmetry elements present and hence assign the point group: (i) 0°, (ii) 90°, (iii) 45°, (iv) 60°.

P10A.4 Find the point groups of all the possible geometrical isomers for the complex $MA_2B_2C_2$ in which there is 'octahedral' coordination around the central atom M and where the ligands A, B, and C are treated as structureless points. Which of the isomers are chiral?

P10A.5‡ In the square-planar complex anion $[trans\text{-}Ag(CF_3)_2(CN)_2]^-$, the Ag–CN groups are collinear. (a) Assume free rotation of the CF_3 groups (i.e. disregarding the AgCF and AgCN angles) and identify the point group of this complex ion. (b) Now suppose the CF_3 groups cannot rotate freely (because the ion was in a solid, for example). Structure (**5**) shows a plane

‡ These problems were provided by Charles Trapp and Carmen Giunta.

which bisects the NC–Ag–CN axis and is perpendicular to it. Identify the point group of the complex if each CF_3 group has a CF bond in that plane (so the CF_3 groups do not point to either CN group preferentially) and the CF_3 groups are (i) staggered, (ii) eclipsed.

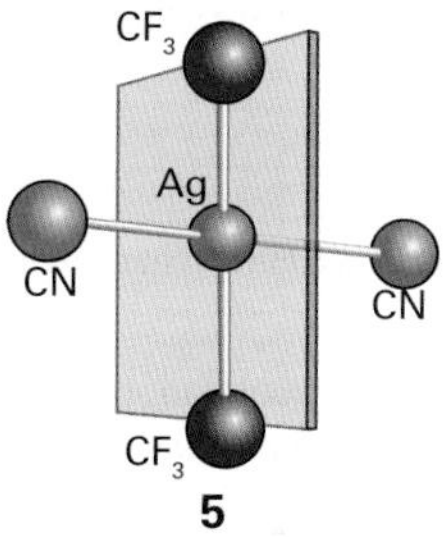

5

P10A.6‡ B.A. Bovenzi and G.A. Pearse, Jr (*J. Chem. Soc. Dalton Trans.*, 2763 (1997)) synthesized coordination compounds of the tridentate ligand pyridine-2,6-diamidoxime ($C_7H_9N_5O_2$, **6**). Reaction with $NiSO_4$ produced a complex in which two of the essentially planar ligands are bonded at right angles to a single Ni atom. Identify the point group and the symmetry operations of the resulting $[Ni(C_7H_9N_5O_2)_2]^{2+}$ complex cation.

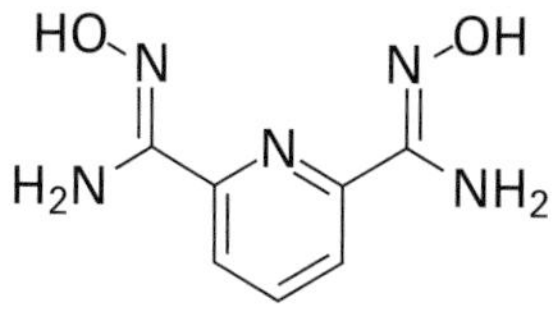

6

TOPIC 10B Group theory

Discussion questions

D10B.1 Explain what is meant by a 'group'.

D10B.2 Explain what is meant by (a) a representative and (b) a representation in the context of group theory.

D10B.3 Explain the construction and content of a character table.

D10B.4 Explain what is meant by the reduction of a representation to a direct sum of irreducible representations.

D10B.5 Discuss the significance of the letters and subscripts used to denote the symmetry species of an irreducible representation.

Additional exercises

E10B.8 Use as a basis the $2p_z$ orbitals on each atom in BF_3 to find the representative of the operation C_3. Take z as perpendicular to the molecular plane.

E10B.9 Use the matrix representatives of the operations σ_h and C_3 in a basis of $2p_z$ orbitals on each atom in BF_3 to find the operation and its representative resulting from $C_3\sigma_h$. Take z as perpendicular to the molecular plane.

E10B.10 Show that all three σ_v operations in the group D_{3h} belong to the same class.

E10B.11 For the point group D_{3h}, confirm that the irreducible representation E′ is orthogonal (in the sense defined by eqn 10B.7) to the irreducible representations A_1', A_2', and E″.

E10B.12 By inspection of the character table for D_{4h}, state the symmetry species of the 4s, 4p, and 3d orbitals located on the central Ni atom in $Ni(CN)_4^{2-}$.

E10B.13 What is the maximum degeneracy of the wavefunctions of a particle confined to the interior of an icosahedral nanoparticle?

E10B.14 What is the maximum possible degree of degeneracy of the orbitals in benzene?

E10B.15 What is the maximum possible degree of degeneracy of the orbitals in 1,4-dichlorobenzene?

Problems

P10B.1 The group C_{2h} consists of the elements E, C_2, σ_h, i. Construct the group multiplication table. Give an example of a molecule that belongs to the group.

P10B.2 The group D_{2h} has a C_2 axis perpendicular to the principal axis and a horizontal mirror plane. Show that the group must therefore have a centre of inversion.

P10B.3 Consider the H_2O molecule, which belongs to the group C_{2v}. Take the molecule to lie in the yz-plane, with z directed along the C_2 axis; the mirror plane σ_v' is the yz-plane, and σ_v is the xz-plane. Take as a basis the two H1s orbitals and the four valence orbitals of the O atom and set up the 6 × 6 matrices that represent the group in this basis. (a) Confirm, by explicit matrix multiplication, that $C_2\sigma_v = \sigma_v'$ and $\sigma_v\sigma_v' = C_2$. (b) Show that the representation is reducible and spans $3A_1 + B_1 + 2B_2$.

P10B.4 Find the representatives of the operations of the group T_d in a basis of four H1s orbitals, one at each apex of a regular tetrahedron (as in CH_4). You need give the representative for only *one* member of each class.

P10B.5 Find the representatives of the operations of the group D_{2h} in a basis of the four H1s orbitals of ethene. Take the molecule as lying in the xy-plane, with x directed along the C–C bond.

P10B.6 Find the representatives of the operations of the group D_{2h} in a basis of the four H1s orbitals of ethene. Then confirm that the representatives reproduce the group multiplications $C_2^zC_2^y = C_2^x$, $\sigma^{xz}C_2^z = C_2^y$ and $iC_2^y = \sigma^{xz}$. Take the molecule as lying in the xy-plane, with x directed along the C–C bond.

P10B.7 The (one-dimensional) matrices $D(C_3) = 1$ and $D(C_2) = 1$, and $D(C_3) = 1$ and $D(C_2) = -1$ both represent the group multiplication $C_3C_2 = C_6$ in the group C_{6v} with $D(C_6) = +1$ and -1, respectively. Use the character table to confirm these remarks. What are the representatives of σ_v and σ_d in each case?

P10B.8 Construct the multiplication table of the Pauli spin matrices, σ, and the 2 × 2 unit matrix:

$$\sigma_x = \begin{pmatrix} 0 & 1 \\ 1 & 0 \end{pmatrix} \quad \sigma_y = \begin{pmatrix} 0 & -i \\ i & 0 \end{pmatrix} \quad \sigma_z = \begin{pmatrix} 1 & 0 \\ 0 & -1 \end{pmatrix} \quad \sigma_0 = \begin{pmatrix} 1 & 0 \\ 0 & 1 \end{pmatrix}$$

Do the four matrices form a group under multiplication?

P10B.9 The algebraic forms of the f orbitals are a radial function multiplied by one of the factors (a) $z(5z^2 - 3r^2)$, (b) $y(5y^2 - 3r^2)$, (c) $x(5x^2 - 3r^2)$, (d) $z(x^2 - y^2)$, (e) $y(x^2 - z^2)$, (f) $x(z^2 - y^2)$, (g) xyz. Identify the irreducible representations spanned by these orbitals in the point group C_{2v}. (*Hint:* Because r is the radius, r^2 is invariant to any operation.)

P10B.10 Using the same approach as in Section 10B.3(c) find the representatives using as a basis two wavefunctions $\psi_{2,3} = (2/L)\sin(2\pi x/L) \times \sin(3\pi y/L)$ and $\psi_{3,2} = (2/L)\sin(3\pi x/L)\sin(2\pi y/L)$ in the point group C_4, and hence show that these functions span a degenerate irreducible representation.

TOPIC 10C Applications of symmetry

Discussion questions

D10C.1 Identify and list four applications of character tables.

D10C.2 Explain how symmetry arguments are used to construct molecular orbitals.

Additional exercises

E10C.9 Use symmetry properties to determine whether or not the integral $\int p_x z p_z d\tau$ is necessarily zero in a molecule with symmetry D_{3h}.

E10C.10 Is the transition $A_{1g} \rightarrow E_{2u}$ forbidden for electric dipole transitions in a D_{6h} molecule?

E10C.11 Show that the function xyz has symmetry species A_u in the group D_{2h}.

E10C.12 Consider the C_{2v} molecule OF_2; take the molecule to lie in the yz-plane, with z directed along the C_2 axis; the mirror plane σ_v' is the yz-plane, and σ_v is the xz-plane. Find the irreducible representations spanned by the combinations $p_y(A) + p_y(B)$ and $p_y(A) - p_y(B)$. Are there any valence orbitals of the central O atom that can have a non-zero overlap with these combinations of F orbitals?

E10C.13 Consider BF_3 (point group D_{3h}). There are SALCs from the F valence orbitals which transform as A_2'' and E''. Are there any valence orbitals of the central B atom that can have a non-zero overlap with these SALCs? How would your conclusion differ for AlF_3, where 3d orbitals might be available?

E10C.14 The ClO_2 molecule (which belongs to the group C_{2v}) was trapped in a solid. Its ground state is known to be B_1. Light polarized parallel to the y-axis (parallel to the OO separation) excited the molecule to an upper state. What is the symmetry species of that state?

E10C.15 A set of basis functions is found to span a reducible representation of the group D_2 with characters 6,−2,0,0 (in the order of operations in the character table in the *Resource section*). What irreducible representations does it span?

E10C.16 A set of basis functions is found to span a reducible representation of the group D_{4h} with characters 4,0,0,2,0,0,0,4,2,0 (in the order of operations in the character table in the *Resource section*). What irreducible representations does it span?

E10C.17 A set of basis functions is found to span a reducible representation of the group O_h with characters 6,0,0,2,2,0,0,0,4,2 (in the order of operations in the character table in the *Resource section*). What irreducible representations does it span?

E10C.18 What states of (i) anthracene, (ii) coronene (**7**) may be reached by electric dipole transitions from their (totally symmetrical) ground states?

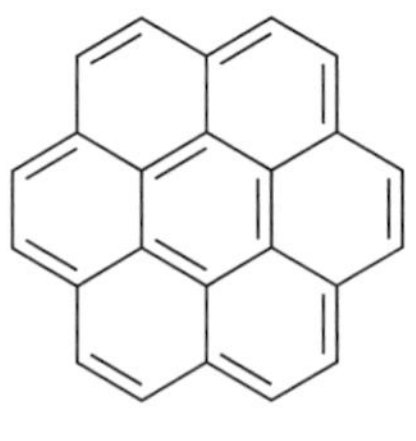

7 Coronene

Problems

P10C.1 What irreducible representations do the four H1s orbitals of CH_4 span? Are there s and p orbitals of the central C atom that may form molecular orbitals with them? In SiH_4, where 3d orbitals might be available, could these orbitals play a role in forming molecular orbitals by overlapping with the H1s orbitals?

P10C.2 Suppose that a methane molecule became distorted to (a) C_{3v} symmetry by the lengthening of one bond, (b) C_{2v} symmetry, by a kind of scissors action in which one bond angle opened and another closed slightly. Would more d orbitals on the carbon become available for bonding?

P10C.3 Does the integral of the function $3x^2 - 1$ necessarily vanish when integrated over a symmetrical range in (a) a cube, (b) a tetrahedron, (c) a hexagonal prism, each centred on the origin?

P10C.4‡ In a spectroscopic study of C_{60}, Negri et al. (*J. Phys. Chem.* **100**, 10849 (1996)) assigned peaks in the fluorescence spectrum. The molecule has icosahedral symmetry (I_h). The ground electronic state is A_{1g}, and the lowest-lying excited states are T_{1g} and G_g. (a) Are photon-induced transitions allowed from the ground state to either of these excited states? Explain your answer. (b) What if the molecule is distorted slightly so as to remove its centre of inversion?

P10C.5 In the square planar XeF_4 molecule, consider the symmetry-adapted linear combination $p_1 = p_A - p_B + p_C - p_D$, where p_A, p_B, p_C, and p_D are $2p_z$ atomic orbitals on the fluorine atoms (clockwise labelling of the F atoms). Decide which of the various s, p, and d atomic orbitals on the central Xe atom can form molecular orbitals with p_1.

P10C.6 The chlorophylls that participate in photosynthesis and the haem (heme) groups of cytochromes are derived from the porphine dianion group (**8**), which belongs to the D_{4h} point group. The ground electronic state is A_{1g} and the lowest-lying excited state is E_u. Is a photon-induced transition allowed from the ground state to the excited state? Explain your answer.

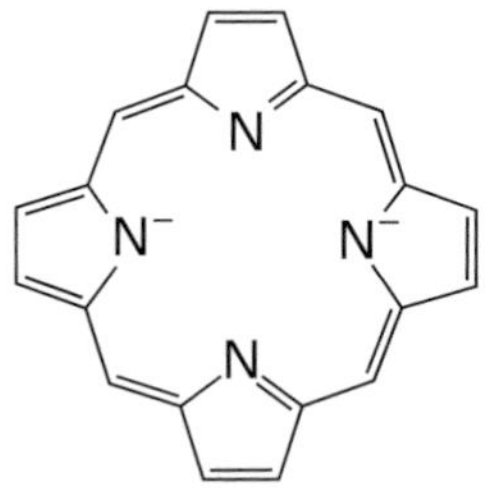

8

P10C.7 Consider the ethene molecule (point group D_{2h}), and take it as lying in the xy-plane, with x directed along the C–C bond. By applying the projection formula to one of the hydrogen 1s orbitals generate SALCs which have symmetry A_g, B_{2u}, B_{3u}, and B_{1g}. What happens when you try to project out a SALC with symmetry B_{1u}?

P10C.8 Consider the molecule $F_2C{=}CF_2$ (point group D_{2h}), and take it as lying in the xy-plane, with x directed along the C–C bond. (a) Consider a basis formed from the four $2p_z$ orbitals from the fluorine atoms: show that the basis spans B_{1u}, B_{2g}, B_{3g}, and A_u. (b) By applying the projection formula to one of the $2p_z$ orbitals, generate the SALCs with the indicated symmetries. (c) Repeat the process for a basis formed from four $2p_x$ orbitals (the symmetry species will be different from those for $2p_z$).

FOCUS 11

Molecular spectroscopy

The origin of spectral lines in molecular spectroscopy is the absorption, emission, or scattering of a photon accompanied by a change in the energy of a molecule. The difference from atomic spectroscopy (Topic 8C) is that the energy of a molecule can undergo not only electronic transitions but also changes of rotational and vibrational state. Molecular spectra are therefore more complex than atomic spectra. However, they contain information relating to more properties, and their analysis leads to values of bond strengths, lengths, and angles. They also provide a way of determining a variety of molecular properties, such as dissociation energies and dipole moments.

11A General features of molecular spectroscopy

This Topic begins with a discussion of the theory of absorption and emission of radiation, leading to the factors that determine the intensities and widths of spectral lines. The features of the instrumentation used to monitor the absorption, emission, and scattering of radiation spanning a wide range of frequencies are also described.

11B Rotational spectroscopy

This Topic shows how expressions for the values of the rotational energy levels of diatomic and polyatomic molecules are derived. The most direct procedure, which is used here, is to identify the expressions for the energy and angular momentum obtained in classical physics, and then to transform these expressions into their quantum mechanical counterparts. The Topic then focuses on the interpretation of pure rotational and rotational Raman spectra, in which only the rotational state of a molecule changes. The observation that not all molecules can occupy all rotational states is shown to arise from symmetry constraints resulting from the presence of nuclear spin.

11C Vibrational spectroscopy of diatomic molecules

The harmonic oscillator (Topic 7E) is a good starting point for modelling the vibrations of diatomic molecules, but it is shown that the description of real molecules requires deviations from harmonic behaviour to be taken into account. The vibrational spectra of gaseous samples show features due to the rotational transitions that accompany the excitation of vibrations.

11D Vibrational spectroscopy of polyatomic molecules

The vibrational spectra of polyatomic molecules can be discussed as though they consist of a set of independent harmonic oscillators. Their spectra can then be understood in much the same way as those of diatomic molecules.

11E Symmetry analysis of vibrational spectra

The atomic displacements involved in the vibrations of polyatomic molecules can be classified according to the symmetry possessed by the molecule. This classification makes it possible to decide which vibrations can be studied spectroscopically.

11E.1 Classification of normal modes according to symmetry; 11E.2 Symmetry of vibrational wavefunctions

11F Electronic spectra

This Topic introduces the key idea that electronic transitions occur within a stationary nuclear framework. The electronic spectra of diatomic molecules are considered first, and it is seen that in the gas phase it is possible to observe simultaneous vibrational and rotational transitions that accompany the electronic transition. The general features of the electronic spectra of polyatomic molecules are also described.

11F.1 Diatomic molecules; 11F.2 Polyatomic molecules

11G Decay of excited states

This Topic begins with an account of spontaneous emission by molecules, including the phenomena of 'fluorescence' and 'phosphorescence'. It also explains how non-radiative decay of excited states can result in the transfer of energy as heat to the surroundings or result in molecular dissociation. The stimulated radiative decay of excited states is the key process responsible for the action of lasers.

11G.1 Fluorescence and phosphorescence; 11G.2 Dissociation and predissociation; 11G.3 Lasers

What is an application of this material?

Molecular spectroscopy is also useful to astrophysicists and environmental scientists. *Impact 16*, accessed via the e-book, discusses how the identities of molecules found in interstellar space can be inferred from their rotational and vibrational spectra. *Impact 17*, accessed via the e-book, focuses back on Earth and shows how the vibrational properties of its atmospheric constituents can affect its climate.

➤ **Go to the e-book for videos that feature the derivation and interpretation of equations, and applications of this material.**

TOPIC 11A General features of molecular spectroscopy

➤ **Why do you need to know this material?**

To interpret data from the wide range of varieties of molecular spectroscopy you need to understand the experimental and theoretical features shared by them all.

➤ **What is the key idea?**

A transition from a low energy state to one of higher energy can be stimulated by absorption of radiation; a transition from a higher to a lower state, resulting in emission of a photon, may be either spontaneous or stimulated by radiation.

➤ **What do you need to know already?**

You need to be familiar with the fact that molecular energy is quantized (Topics 7E and 7F) and be aware of the concept of selection rules (Topic 8C).

In **emission spectroscopy** the electromagnetic radiation that arises from molecules undergoing a transition from a higher energy state to a lower energy state is detected and its frequency analysed. In **absorption spectroscopy**, the net absorption of radiation passing through a sample is monitored over a range of frequencies. It is necessary to specify the *net* absorption because not only can radiation be absorbed but it can also stimulate the emission of radiation, so the net absorption is detected. In **Raman spectroscopy**, the frequencies of radiation scattered by molecules is analysed to determine the changes in molecular states that accompany the scattering process. Throughout this discussion it is important to be able to express the characteristics of radiation variously as a frequency, ν, a wavenumber, $\tilde{\nu} = \nu/c$, or a wavelength, $\lambda = c/\nu$, as set out in *The chemist's toolkit* 7A.1.

In each case, the emission, absorption, or scattering of radiation can be interpreted in terms of individual photons. When a molecule undergoes a transition between a lower state with energy E_l and an upper state with energy E_u a photon with energy $h\nu$ is emitted or absorbed. The energy of the photon depends on the difference in energy between the two states according to the Bohr frequency condition (eqn 7A.10 of Topic 7A), $h\nu = E_u - E_l$, where ν is the frequency of the radiation emitted or absorbed. Emission and absorption spectroscopy give the same information about electronic, vibrational, or rotational energy level separations, but practical considerations generally determine which technique is employed.

In Raman spectroscopy the sample is exposed to monochromatic (single frequency) radiation and therefore a stream of photons all of the same energy. When the photons encounter the molecules, most are scattered elastically (without change in their energy): this process is called **Rayleigh scattering**. About 1 in 10^7 of the photons are scattered inelastically (with different energy). In **Stokes scattering** the photons lose energy to the molecules and the emerging radiation has a lower frequency. In **anti-Stokes scattering**, a photon gains energy from a molecule and the emerging radiation has a higher frequency (Fig. 11A.1). By analysing the frequencies of the scattered radiation it is possible to gather information about the energy levels of the molecules. Raman spectroscopy is used to study molecular vibrations and rotations.

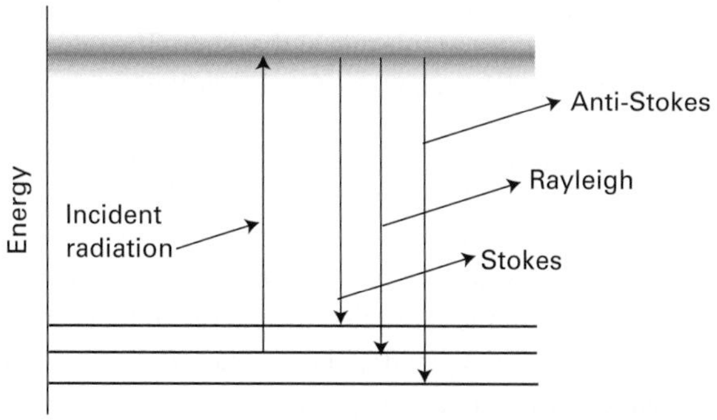

Figure 11A.1 In Raman spectroscopy, incident photons are scattered from a molecule. Most photons are scattered elastically and so have the same energy as the incident photons (Rayleigh scattering). Some photons lose energy to the molecule and so emerge as Stokes radiation; others gain energy and so emerge as anti-Stokes radiation. The scattering can be regarded as taking place by an excitation of the molecule from its initial state to a series of excited states (represented by the shaded band), and the subsequent return to a final state. Any net energy change is either supplied from, or carried away by, the photon.

11A.1 The absorption and emission of radiation

The separation of rotational energy levels (in small molecules, $\Delta E \approx 0.01$ zJ, corresponding to about $0.01\,\text{kJ mol}^{-1}$) is smaller than that of vibrational energy levels ($\Delta E \approx 10$ zJ, corresponding to $10\,\text{kJ mol}^{-1}$), which itself is smaller than that of electronic energy levels ($\Delta E \approx 0.1$–1 aJ, corresponding to about 10^2–$10^3\,\text{kJ mol}^{-1}$). From the Bohr frequency condition in the form $\nu = \Delta E/h$, the corresponding frequencies of the photons involved in these different kinds of transitions are about 10^{10} Hz for rotation, 10^{13} Hz for vibration, and in the range 10^{14}–10^{15} Hz for electronic transitions. It follows that rotational, vibrational, and electronic transitions result from the absorption or emission of microwave, infrared, and ultraviolet/visible radiation, respectively.

11A.1(a) Stimulated and spontaneous radiative processes

Albert Einstein identified three processes by which radiation could be either generated or absorbed by matter as a result of transitions between states. In **stimulated absorption** a transition from a lower energy state l to a higher energy state u is driven by an electromagnetic field oscillating at the frequency ν corresponding to the energy separation of the two states: $h\nu = E_u - E_l$. The rate of such transitions is proportional to the intensity of the incident radiation at the transition frequency: the more intense the radiation, the greater the number of photons impinging on the molecules and the greater the probability that a photon will be absorbed. The rate is also proportional to the number of molecules in the lower state, N_l, because the greater the population of that state the more likely it is that a photon will encounter a molecule in that state. The rate of stimulated absorption, $W_{u\leftarrow l}$, can therefore be written

$$W_{u\leftarrow l} = B_{u,l}N_l\rho(\nu) \quad \text{Rate of stimulated absorption} \quad (11A.1a)$$

In this expression, $\rho(\nu)$ is the energy spectral density, such that $\rho(\nu)\mathrm{d}\nu$ is the energy density of radiation in the frequency range from ν to $\nu + \mathrm{d}\nu$, in the sense that the energy in that range in a region of volume V is $V\rho(\nu)\mathrm{d}\nu$. The constant $B_{u,l}$ is the **Einstein coefficient of stimulated absorption**.

Einstein also supposed that the radiation could induce the molecule in an upper state to undergo a transition to a lower state and thereby generate a photon of frequency ν. This process is called **stimulated emission**, and its rate depends on the number of molecules in the upper level N_u and the intensity of the radiation at the transition frequency. By analogy with eqn 11A.1a, the rate of stimulated emission can be written as

$$W^{\text{stimulated}}_{u\rightarrow l} = B_{l,u}N_u\rho(\nu) \quad \text{Rate of stimulated emission} \quad (11A.1b)$$

In this expression $B_{l,u}$ is the **Einstein coefficient of stimulated emission.**

Einstein went on to suppose that a molecule could lose energy by **spontaneous emission** in which the molecule makes a transition to a lower state without it being driven by the presence of radiation. The rate of spontaneous emission is written as

$$W^{\text{spontaneous}}_{u\rightarrow l} = A_{l,u}N_u \quad \text{Rate of spontaneous emission} \quad (11A.1c)$$

where $A_{l,u}$ is the **Einstein coefficient of spontaneous emission.** When both stimulated and spontaneous emission are taken into account, the total rate of emission is

$$W_{u\rightarrow l} = B_{l,u}N_u\rho(\nu) + A_{l,u}N_u \quad \text{Total rate of emission} \quad (11A.1d)$$

When the molecules and radiation are in equilibrium, the rates given in eqns 11A.1a and 11A.1d must be equal, and the populations then have their equilibrium values N_l^{eq} and N_u^{eq}. Therefore

$$B_{u,l}N_l^{\text{eq}}\rho(\nu) = B_{l,u}N_u^{\text{eq}}\rho(\nu) + A_{l,u}N_u^{\text{eq}} \quad (11A.2a)$$

and so

$$\rho(\nu) = \frac{A_{l,u}/B_{u,l}}{N_l^{\text{eq}}/N_u^{\text{eq}} - B_{l,u}/B_{u,l}} \quad (11A.2b)$$

However, the ratio of the equilibrium populations must be in accord with the Boltzmann distribution (as specified in *Energy: A first look* and Topic 13A):

$$\frac{N_u^{\text{eq}}}{N_l^{\text{eq}}} = \mathrm{e}^{-(E_u - E_l)/kT} = \mathrm{e}^{-h\nu/kT} \quad (11A.3)$$

and therefore, at equilibrium,

$$\rho(\nu) = \frac{A_{l,u}/B_{u,l}}{\mathrm{e}^{h\nu/kT} - B_{l,u}/B_{u,l}} \quad (11A.4)$$

Moreover, at equilibrium the radiation density is given by the Planck distribution of radiation in equilibrium with a black body (eqn 7A.6b, Topic 7A):

$$\rho(\nu) = \frac{8\pi h\nu^3/c^3}{\mathrm{e}^{h\nu/kT} - 1} \quad \text{Planck distribution} \quad (11A.5)$$

By comparing eqns 11A.4 and 11A.5 it follows that

$$A_{l,u} = \frac{8\pi h\nu^3}{c^3}B_{l,u} \quad \text{and} \quad B_{l,u} = B_{u,l} \quad (11A.6a)$$

Although these relations have been derived on the assumption that the molecules and radiation are in equilibrium, they are properties of the molecules themselves and are independent of the spectral distribution of the radiation (that is, whether it is black-body or not) and can be used in eqn 11A.1 for any energy densities.

The ratio of the rate of spontaneous to stimulated emission can be found by combining eqns 11A.1b, 11A.1c, and 11A.6a to give

$$\frac{W_{u\to l}^{\text{spontaneous}}}{W_{u\to l}^{\text{stimulated}}}=\frac{A_{l,u}}{B_{l,u}\rho(\nu)}=\frac{8\pi h\nu^3}{c^3\rho(\nu)} \quad (11A.6b)$$

This relation shows that:

- For a given spectral density, the relative importance of spontaneous emission increases as the cube of the transition frequency and therefore that spontaneous emission is most likely to be of importance at high frequencies.
- Spontaneous emission can be ignored at low frequencies, in which case the intensities of such transitions can be discussed in terms of stimulated emission and absorption alone.

Brief illustration 11A.1

On going from infrared to visible radiation, the frequency increases by a factor of about 100, so the ratio of the rates of spontaneous to stimulated emission increases by a factor of 10^6 for the same spectral density. This strong increase accounts for the observation that whereas electronic transitions are often monitored by emission spectroscopy, vibrational spectroscopy is an absorption technique and spontaneous (but not stimulated) emission is negligible.

11A.1(b) Selection rules and transition moments

A 'selection rule' is a statement about whether a transition is forbidden or allowed (Topic 8C). The underlying idea is that, for the molecule to be able to interact with the electromagnetic field and absorb or create a photon of frequency ν, it must possess, at least transiently, an electric dipole oscillating at that frequency. This transient dipole is expressed quantum mechanically in terms of the **transition dipole moment**, μ_{fi}, between the initial and final states with wavefunctions ψ_i and ψ_f:[1]

$$\mu_{fi}=\int\psi_f^*\hat{\mu}\psi_i\,d\tau \quad \text{Transition dipole moment [definition]} \quad (11A.7)$$

where $\hat{\mu}$ is the electric dipole moment operator. The magnitude of the transition dipole moment can be regarded as a measure of the charge redistribution that accompanies a transition. Moreover, a transition is active (and generates or absorbs a photon) only if the accompanying charge redistribution is dipolar (Fig. 11A.2). It follows that, to identify the selection rules, the conditions for which $\mu_{fi}\neq 0$ must be established.

[1] Equation 11A.7 is derived in *A deeper look* 11A.1 , available to read in the e-book accompanying this text.

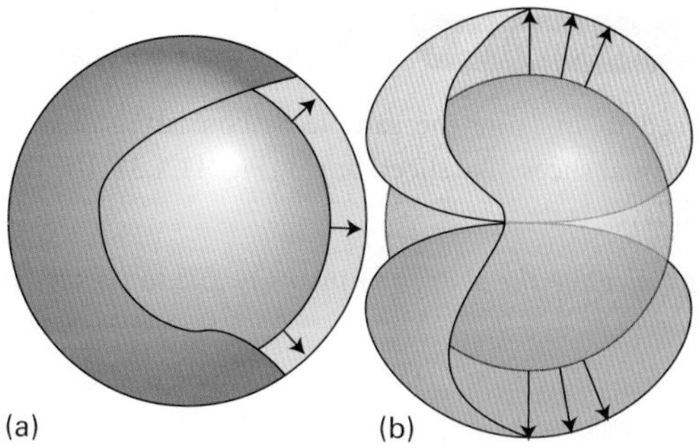

Figure 11A.2 (a) When an electron undergoes a transition from a 1s orbital to a 2s orbital, there is a spherical migration of charge. There is no dipole moment associated with this migration of charge, so this transition is electric-dipole forbidden. (b) In contrast, when an electron moves from 1s to 2p, there is a dipole associated with the charge migration; this transition is electric-dipole allowed.

A **gross selection rule** specifies the general features that a molecule must have if it is to have a spectrum of a given kind. For instance, in Topic 11B it is shown that a molecule gives a rotational spectrum only if it has a permanent electric dipole moment. This rule, and others like it for other types of transition, is explained in the relevant Topic. A detailed study of the transition moment leads to the **specific selection rules**, which express the allowed transitions in terms of the changes in quantum numbers or various symmetry features of the molecule.

11A.1(c) The Beer–Lambert law

It is found empirically that when electromagnetic radiation passes through a sample of length L and molar concentration [J] of the absorbing species J, the incident and transmitted intensities, I_0 and I, are related by the **Beer–Lambert law**:

$$I=I_0 10^{-\varepsilon[\mathrm{J}]L} \quad \text{Beer–Lambert law} \quad (11A.8)$$

The quantity ε (epsilon) is called the **molar absorption coefficient** (formerly, and still widely, the 'extinction coefficient'); it depends on the frequency (or wavenumber and wavelength) of the incident radiation and is greatest where the absorption is most intense.

The dimensions of ε are 1/(concentration × length), and it is normally convenient to express it in cubic decimetres per mole per centimetre ($dm^3\,mol^{-1}\,cm^{-1}$); in SI base units it is expressed in metres squared per mole ($m^2\,mol^{-1}$). The latter units imply that ε may be regarded as a (molar) cross-section for absorption, and that the greater the cross-sectional area of the molecule for absorption, the greater is its ability to block the passage of the incident radiation at a given frequency.

The Beer–Lambert law is an empirical result. However, its form can be derived on the basis of a simple model.

How is that done? 11A.1 Justifying the Beer–Lambert law

You need to imagine the sample as consisting of a stack of infinitesimal slices, like sliced bread (Fig. 11A.3). The thickness of each layer is dx.

Step 1 *Calculate the change in intensity due to passage through one slice*

The change in intensity, dI, that occurs when electromagnetic radiation passes through one particular slice is proportional to the thickness of the slice, the molar concentration of the absorber J, and (because the absorption is stimulated) the intensity of the incident radiation at that slice of the sample, so $\mathrm{d}I \propto [\mathrm{J}]I\mathrm{d}x$. The intensity is reduced by absorption, which means that dI is negative and can therefore be written

$$\mathrm{d}I = -\kappa[\mathrm{J}]I\mathrm{d}x$$

where κ (kappa) is the proportionality coefficient. Division of both sides by I gives

$$\frac{\mathrm{d}I}{I} = -\kappa[\mathrm{J}]\mathrm{d}x$$

This expression applies to each successive slice.

Step 2 *Evaluate the total change in intensity due to passage through successive slices*

To obtain the intensity that emerges from a sample of thickness L when the intensity incident on one face of the sample is I_0, you need the sum of all the successive changes. Assume that the molar concentration of the absorbing species is uniform and may be treated as a constant. Because a sum over infinitesimally small increments is an integral, it follows that:

$$\overbrace{\int_{I_0}^{I}\frac{\mathrm{d}I}{I}}^{\text{Integral A.2}} = -\kappa\int_0^L[\mathrm{J}]\mathrm{d}x \overset{[\mathrm{J}]\text{ a constant}}{=} -\kappa[\mathrm{J}]\overbrace{\int_0^L \mathrm{d}x}^{\text{Integral A.1}}$$

Therefore

$$\ln\frac{I}{I_0} = -\kappa[\mathrm{J}]L$$

Now express the natural logarithm as a common logarithm (to base 10) by using $\ln x = (\ln 10)\log x$, and a new constant ε defined as $\varepsilon = \kappa/(\ln 10)$ to give

$$\log\frac{I}{I_0} = -\varepsilon[\mathrm{J}]L$$

Take common antilogarithms of each side (that is, form 10^x of each side) and obtain the Beer–Lambert law (eqn 11A.8).

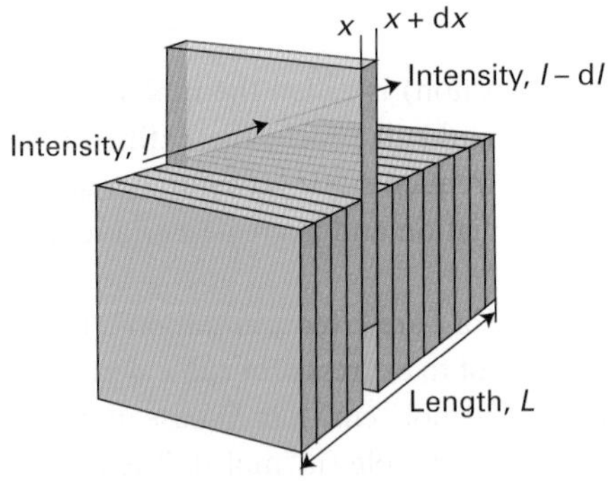

Figure 11A.3 To establish the Beer–Lambert law, the sample is supposed to be divided into a large number of thin slices. The reduction in intensity caused by one slice is proportional to the intensity incident on it (after passing through the preceding slices), the thickness of the slice, and the concentration of the absorbing species.

The spectral characteristics of a sample are commonly reported as the **transmittance**, T, of the sample at a given frequency:

$$T = \frac{I}{I_0} \qquad \text{Transmittance [definition]} \qquad (11\mathrm{A}.9\mathrm{a})$$

or its **absorbance**, A:

$$A = \log\frac{I_0}{I} \qquad \text{Absorbance [definition]} \qquad (11\mathrm{A}.9\mathrm{b})$$

The two quantities are related by $A = -\log T$ (note the common logarithm). In terms of the absorbance the Beer–Lambert law becomes

$$A = \varepsilon[\mathrm{J}]L \qquad (11\mathrm{A}.9\mathrm{c})$$

The product $\varepsilon[\mathrm{J}]L$ was known formerly as the *optical density* of the sample.

Example 11A.1 Determining a molar absorption coefficient

Radiation of wavelength 280 nm passed through 1.0 mm of a solution that contained an aqueous solution of the amino acid tryptophan at 0.50 mmol dm^{-3}. The intensity is reduced to 54 per cent of its initial value (so $T = 0.54$). Calculate the absorbance and the molar absorption coefficient of tryptophan at 280 nm. What would be the transmittance through a cell of thickness 2.0 mm?

Collect your thoughts From $A = -\log T = \varepsilon[\mathrm{J}]L$, it follows that $\varepsilon = A/[\mathrm{J}]L$. For the transmittance through the thicker cell, you need to calculate the absorbance by using $A = -\log T = \varepsilon[\mathrm{J}]L$ and the computed value of ε; the transmittance is $T = 10^{-A}$.

The solution The absorbance is $A = -\log 0.54 = 0.27$, and so the molar absorption coefficient is

$$\varepsilon = \frac{-\log 0.54}{(5.0\times10^{-4}\ \mathrm{mol\,dm^{-3}})\times(1.0\ \mathrm{mm})} = 5.4\times10^{2}\ \mathrm{dm^3\,mol^{-1}\,mm^{-1}}$$

These units are convenient for the rest of the calculation (but the outcome could be reported as $5.4 \times 10^3\,dm^3\,mol^{-1}\,cm^{-1}$ if desired or even as $5.4 \times 10^2\,m^2\,mol^{-1}$). The absorbance of a sample of length 2.0 mm is

$$A = (5.4\times10^2\ dm^3\ mol^{-1}\ mm^{-1})\times(5.0\times10^{-4}\ mol\,dm^{-3})\times(2.0\ mm) = 0.54$$

The transmittance is now $T = 10^{-A} = 10^{-0.54} = 0.29$.

Self-test 11A.1 The transmittance of an aqueous solution containing the amino acid tyrosine at a molar concentration of $0.10\ mmol\,dm^{-3}$ was measured as 0.14 at 240 nm in a cell of length 5.0 mm. Calculate the absorbance of the solution and the molar absorption coefficient of tyrosine at that wavelength. What would be the transmittance through a cell of length 1.0 mm?

Answer: $A = 0.85$, $1.7 \times 10^4\ dm^3\,mol^{-1}\,cm^{-1}$, $T = 0.67$

The maximum value of the molar absorption coefficient, ε_{max}, is an indication of the intensity of a transition. However, because absorption bands generally spread over a range of wavenumbers, quoting the absorption coefficient at a single wavenumber might not give a true indication of the intensity of a transition. The **integrated absorption coefficient**, $\mathcal{A}$, is the sum of the absorption coefficients over the entire band (Fig. 11A.4), and corresponds to the area under the plot of the molar absorption coefficient against wavenumber:

$$\mathcal{A} = \int_{band} \varepsilon(\tilde{\nu})\,d\tilde{\nu} \qquad \text{Integrated absorption coefficient [definition]} \qquad (11A.10)$$

For bands of similar widths, the integrated absorption coefficients are proportional to the heights of the bands. Equation 11A.10 also applies to the individual lines that contribute to a band: a *spectroscopic* line is not a geometrically thin line, but has a width.

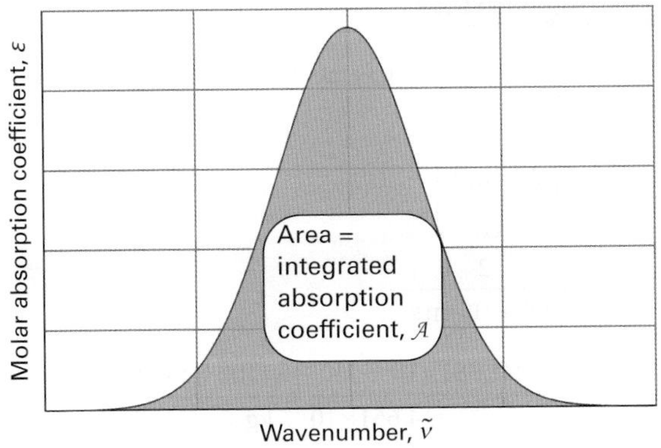

Figure 11A.4 The integrated absorption coefficient of a transition is the area under a plot of the molar absorption coefficient against the wavenumber of the incident radiation.

Exercises

E11A.1 Calculate the ratio A/B for transitions with the following characteristics: (i) 70.8 pm X-rays, (ii) 500 nm visible light, (iii) $3000\ cm^{-1}$ infrared radiation.

E11A.2 The molar absorption coefficient of a substance dissolved in hexane is known to be $723\ dm^3\,mol^{-1}\,cm^{-1}$ at 260 nm. Calculate the percentage reduction in intensity when ultraviolet radiation of that wavelength passes through 2.50 mm of a solution of concentration $4.25\ mmol\,dm^{-3}$.

E11A.3 A solution of a certain component of a biological sample when placed in an absorption cell of path length 1.00 cm transmits 18.1 per cent of ultraviolet radiation of wavelength 320 nm incident upon it. If the concentration of the component is $0.139\ mmol\,dm^{-3}$, what is the molar absorption coefficient?

11A.2 Spectral linewidths

Several effects contribute to the widths of spectroscopic lines. The design of the spectrometer itself affects the linewidth, and there are other contributions that arise from physical processes in the sample. Some of the latter can be minimized by altering the conditions whereas others are intrinsic to the molecules and cannot be altered.

11A.2(a) Doppler broadening

One important broadening process in gaseous samples is the **Doppler effect**, in which radiation is shifted in frequency when its source is moving towards or away from the observer. When a molecule emitting electromagnetic radiation of frequency ν_0 moves with a speed s relative to an observer, the observer detects radiation of frequency

$$\nu_{receding} = \left(\frac{1-s/c}{1+s/c}\right)^{1/2}\nu_0 \qquad \nu_{approaching} = \left(\frac{1+s/c}{1-s/c}\right)^{1/2}\nu_0 \qquad \text{Doppler shifts} \qquad (11A.11a)$$

where c is the speed of light. For nonrelativistic speeds ($s << c$), these expressions simplify to

$$\nu_{receding} \approx (1-s/c)\nu_0 \qquad \nu_{approaching} \approx (1+s/c)\nu_0 \qquad (11A.11b)$$

Atoms and molecules reach high speeds in all directions in a gas, and a stationary observer detects the corresponding Doppler-shifted range of frequencies. Some molecules approach the observer, some move away; some move quickly, others slowly. The detected spectral 'line' is the absorption or emission profile arising from all the resulting Doppler shifts. The challenge is to relate the observed linewidth to the spread of speeds in the gas, and in turn to see how that spread depends on the temperature.

How is that done? 11A.2 Deriving an expression for Doppler broadening

You need to relate the spread of Doppler shifts to the distribution of molecular kinetic energy as expressed by the Boltzmann distribution.

Step 1 *Establish the relation between the observed frequency and the molecular speed*

It follows from the Boltzmann distribution (see *Energy: A first look*) that the probability that an atom or molecule of mass m and speed s in a gas phase sample at a temperature T has kinetic energy $E_k = \frac{1}{2}ms^2$ is proportional to $e^{-ms^2/2kT}$. When $s << c$, the Doppler shifts for receding and approaching molecules are given by the expressions in eqn 11A.11b. It follows that the shift between the observed frequency and the true frequency is $\nu_{obs} - \nu_0 \approx \pm\nu_0 s/c$. This expression can be rearranged to give

$$s = \pm c(\nu_{obs} - \nu_0)/\nu_0$$

Step 2 *Evaluate the distribution of frequencies arising from a distribution of speeds*

The intensity I of a transition at ν_{obs} is proportional to the probability of there being a molecule that emits or absorbs at ν_{obs}. Such a molecule would have a speed given by the expression just derived, $s = \pm c(\nu_{obs} - \nu_0)/\nu_0$, so it follows from the Boltzmann distribution that

$$I(\nu_{obs}) \propto e^{-ms^2/2kT} = e^{-mc^2(\nu_{obs}-\nu_0)^2/2\nu_0^2 kT}$$

which has the form of a Gaussian function. The width of the absorption line at half its maximum height, $\delta\nu_{obs}$, can be inferred directly from the general form of such a function (as specified in *The chemist's toolkit* 11A.1):

$$\delta\nu_{obs} = \frac{2\nu_0}{c}\left(\frac{2kT\ln 2}{m}\right)^{1/2} \quad (11A.12a)$$

Doppler broadening

Doppler broadening increases with temperature (Fig. 11A.5) because the molecules then acquire a wider range of speeds. Conversely, reducing the temperature results in narrower lines.

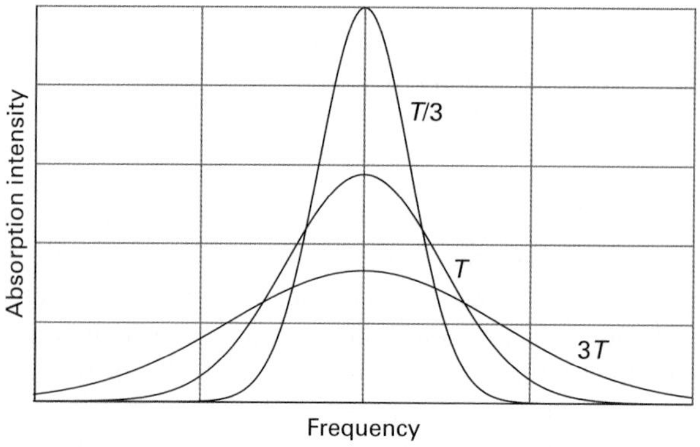

Figure 11A.5 The Gaussian shape of a Doppler-broadened spectral line reflects the Boltzmann distribution of translational kinetic energies in the sample at the temperature of the experiment. The line broadens as the temperature is increased.

The chemist's toolkit 11A.1 Exponential and Gaussian functions

An **exponential function** is a function of the form

$$f(x) = ae^{-bx}$$

Exponential function

This function has the value a at $x = 0$ and decays toward zero as $x \to \infty$. This decay is faster when b is large than when it is small. The function rises rapidly to infinity as $x \to -\infty$. See Sketch 1.

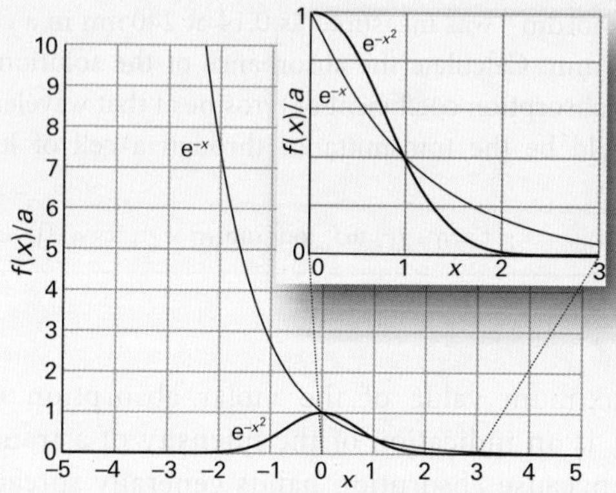

Sketch 1

The general form of a **Gaussian function** is

$$f(x) = ae^{-(x-b)^2/2\sigma^2}$$

Gaussian function

The graph of this function is a symmetrical bell-shaped curve centred on $x = b$; the function has its maximum values of a at its centre. The width of the function, measured at half its height, is $\delta x = 2\sigma(2\ln 2)^{1/2}$; the greater σ, the greater is the width at half-height. Sketch 1 also shows a Gaussian function with $b = 0$.

Note that the Doppler linewidth is proportional to the frequency, so Doppler broadening becomes more important as higher frequencies are observed.

Brief illustration 11A.2

For a molecule such as CO at T = 300 K, and noting that $1\,J = 1\,kg\,m^2\,s^{-2}$,

$$\frac{\delta\nu_{obs}}{\nu_0} = \frac{2}{c}\left(\frac{2kT\ln 2}{m_{CO}}\right)^{1/2}$$
$$= \frac{2}{2.998\times10^8\ m\,s^{-1}} \times \left(\frac{2\times(1.381\times10^{-23}\ J\,K^{-1})\times(300\ K)\times\ln 2}{4.651\times10^{-26}\ kg}\right)^{1/2}$$
$$= 2.34\times10^{-6}$$

For a transition wavenumber of 2150 cm^{-1} from the infrared spectrum of CO, corresponding to a frequency of 64.4 THz (1 THz = 10^{12} Hz), the linewidth is 151 MHz or 5.0×10^{-3} cm^{-1}.

The linewidth due to Doppler broadening can also be expressed in terms of wavelength as

$$\delta\lambda_{\text{obs}} = \frac{2\lambda_0}{c}\left(\frac{2kT\ln 2}{m}\right)^{1/2} \quad \text{Doppler broadening} \quad (11\text{A}.12\text{b})$$

11A.2(b) Lifetime broadening

At any instant a molecule exists in a specific state but it does not remain in that state indefinitely. For example, the molecule might collide with another and in the process change its state. Alternatively, the molecule may fall to a lower level and emit a photon by the process of spontaneous emission. How long the state persists depends on the rates of these processes and is characterized by a lifetime, τ. When the Schrödinger equation is analysed for a state that persists for a time τ, it turns out that the energy is uncertain to an extent $\delta E \approx \hbar/\tau$. Therefore, spectroscopic transitions that involve this state have a linewidth of the order of

$$\frac{\delta E}{h} = \frac{1}{2\pi\tau} \quad \text{Lifetime broadening} \quad (11\text{A}.13)$$

It follows that the shorter the lifetime, the broader is the line. This process is called **lifetime broadening**. When the lifetime is limited by the process of spontaneous emission rather than external causes such as collisions, the resulting linewidth is called the **natural linewidth**.

Brief illustration 11A.3

Excited electronic states of molecules often have short lifetimes due to the high rate of spontaneous emission. A typical lifetime might be 10 ns, which would lead to a natural linewidth of $1/(2\pi \times 10 \times 10^{-9}\,\text{s}) = 16\,\text{MHz}$ or $5.3 \times 10^{-4}\,\text{cm}^{-1}$. As can be inferred from *Brief illustration* 11 A.2, the Doppler linewidth is typically much greater than the natural linewidth.

Collisions between molecules are generally efficient at changing their rotational or vibrational energies, so a good estimate of the resulting **collisional lifetime**, τ_{col}, is to equate it to $1/z$, where z is the collision frequency (Topic 1B). If it is assumed that each collision results in a change of rotational or vibrational state, the lifetime of a state can be taken as τ_{col}, and hence the resulting broadening is $\delta E/h = 1/2\pi\tau_{\text{col}} = z/2\pi$; this contribution to the linewidth is often referred to as **collisional line broadening**. The collision frequency for two molecules with masses m_A and m_B is given by eqn 1B.12b as

$$z = \frac{\sigma v_{\text{rel}} p}{kT} \quad \text{with} \quad v_{\text{rel}} = \left(\frac{8kT}{\pi\mu}\right)^{1/2} \quad \text{and} \quad \mu = \frac{m_A m_B}{m_A + m_B}$$

Note that the collision frequency, and hence the linewidth, is proportional to the pressure, which is why the broadening due to collisions is sometimes referred to as **pressure broadening**. This contribution to the linewidth can be minimized by lowering the pressure as much as possible, although doing so decreases the intensity of the absorption because there are fewer molecules to absorb the radiation. In contrast to Doppler broadening, pressure broadening is independent of the transition frequency.

Brief illustration 11A.4

The linewidth due to pressure broadening in methane gas at 1 bar and 300 K can be estimated by using the expressions just quoted; the collision cross-section σ is $0.46\,\text{nm}^2$. Taking m_A and m_B both to be the mass of a methane molecule, v_{rel} is $888\,\text{m s}^{-1}$. Hence

$$\begin{aligned} z &= \frac{\sigma v_{\text{rel}} p}{kT} \\ &= \frac{(0.46\times10^{-18}\,\text{m}^2)\times(888\,\text{m s}^{-1})\times(1\times10^5\,\text{N m}^{-2})}{(1.381\times10^{-23}\,\text{J K}^{-1})\times(300\,\text{K})} \\ &= 9.9\times10^9\,\text{s}^{-1} \end{aligned}$$

The linewidth is therefore $z/2\pi = 1.6\,\text{GHz}$ or $0.053\,\text{cm}^{-1}$. In *Brief illustration* 11A.2 the Doppler linewidth for a transition in the infrared is estimated as 150 MHz, which is much less than the pressure broadening estimated for the present set of conditions. As the frequency is raised, the Doppler broadening increases in proportion, but the pressure broadening remains unchanged, so Doppler broadening might become dominant.

Exercises

E11A.4 What is the Doppler-broadened linewidth of the electronic transition at 821 nm in atomic hydrogen at 300 K?

E11A.5 Estimate the lifetime of a state that gives rise to a line of width (i) $0.20\,\text{cm}^{-1}$, (ii) $2.0\,\text{cm}^{-1}$.

E11A.6 A molecule in a liquid undergoes about 1.0×10^{13} collisions in each second. Suppose that (i) every collision is effective in deactivating the molecule vibrationally and (ii) that one collision in 100 is effective. Calculate the width (in cm^{-1}) of vibrational transitions in the molecule.

11A.3 Experimental techniques

Common to all spectroscopic techniques is a *spectrometer*, an instrument used to detect the characteristics of radiation scattered, emitted, or absorbed by atoms and molecules. Figure 11A.6 shows the general layout of an absorption spectrometer. Radiation from an appropriate source is directed towards a sample and the radiation transmitted strikes a device that separates it into different frequencies or wavelengths. The intensity of radiation at each frequency is then analysed by a suitable detector. It is usual to record one spectrum with the

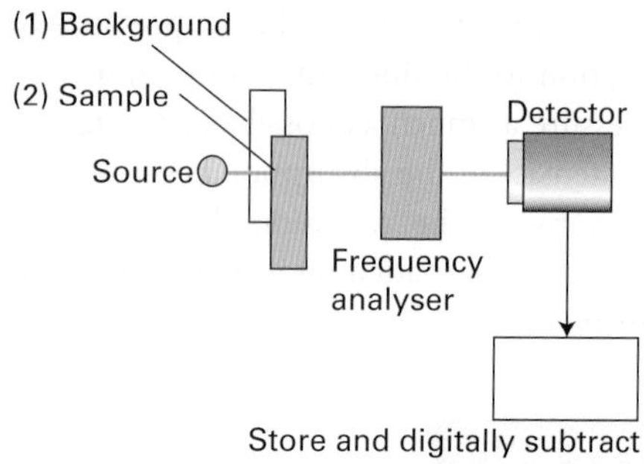

Figure 11A.6 The layout of a typical absorption spectrometer. Radiation from the source passes through the background material (1), typically an empty container or a container containing the solvent only, or a sample (2) held in a container that matches the characteristics of that used as background. In each case the radiation is then dispersed according to frequency and the intensity at each frequency is measured with a detector. The absorption spectrum of the sample is then constructed by digitally subtracting the background signal from the sample signal.

sample in place, and one with the sample removed (the 'background spectrum'): the difference between these two spectra eliminates any absorption not due to the sample itself.

11A.3(a) Sources of radiation

Sources of radiation are either *monochromatic*, those spanning a very narrow range of frequencies around a central value, or *polychromatic*, those spanning a wide range of frequencies. In the microwave region *frequency synthesizers* and various solid state devices can be used to generate monochromatic radiation that can be tuned over a wide range of frequencies. Certain kinds of lasers and light-emitting diodes are often used to provide monochromatic radiation from the infrared to the ultraviolet region. Polychromatic black-body radiation from hot materials (Topic 7A) can be used over the same range. Examples include mercury arcs inside a quartz envelope (usable in the range 35–200 cm^{-1}), *Nernst filaments* and *globars* (200–4000 cm^{-1}), and *quartz–tungsten–halogen lamps* (320–2500 nm).

A *gas discharge lamp* is a common source of ultraviolet and visible radiation. In a *xenon discharge lamp*, an electrical discharge excites xenon atoms to excited states, which then emit ultraviolet radiation. In a *deuterium lamp*, excited D_2 molecules dissociate into electronically excited D atoms that emit intense radiation in the range 200–400 nm.

For certain applications, radiation is generated in a *synchrotron storage ring*, which consists of an electron beam travelling in a circular path with circumferences of up to several hundred metres. As electrons travelling in a circle are constantly accelerated by the forces that constrain them to their path, they generate radiation (Fig. 11A.7). This 'synchrotron radiation' spans

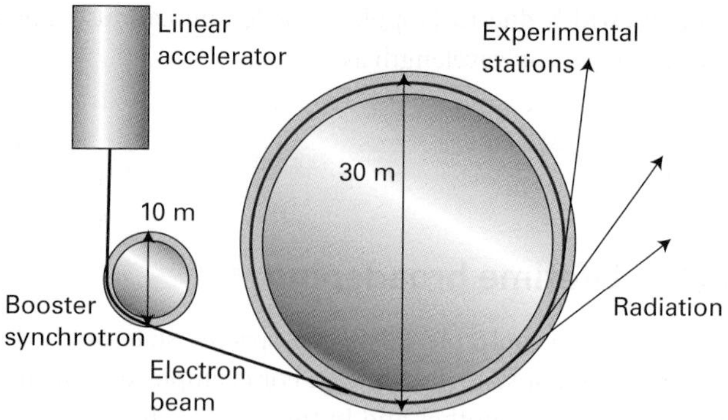

Figure 11A.7 A simple synchrotron storage ring. The electrons injected into the ring from the linear accelerator and booster synchrotron are accelerated to high speed in the main ring. An electron in a curved path is subject to constant acceleration, and an accelerated charge radiates electromagnetic energy.

a wide range of frequencies, including infrared radiation and X-rays. Except in the microwave region, synchrotron radiation is much more intense than can be obtained by most conventional sources.

11A.3(b) Spectral analysis

A common device for the analysis of the frequencies, wavenumbers, or wavelengths in a beam of radiation is a *diffraction grating*, which consists of a glass or ceramic plate into which fine grooves have been cut and covered with a reflective aluminium coating. For work in the visible region of the spectrum, the grooves are cut about 1000 nm apart (a spacing comparable to the wavelength of visible light). The grating causes interference between waves reflected from its surface, and constructive interference occurs at specific angles that depend on the wavelength of the radiation being used. Thus, each wavelength

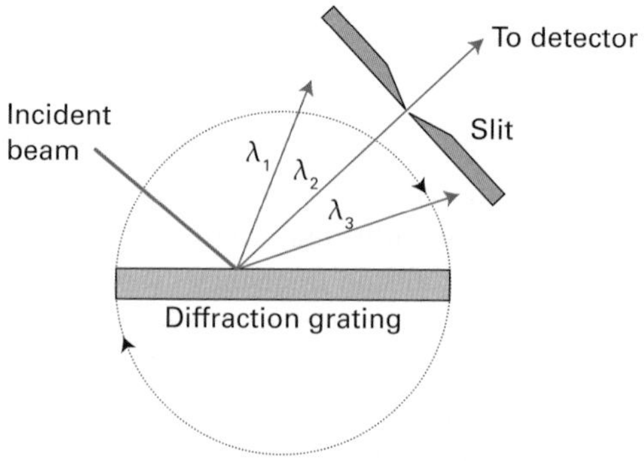

Figure 11A.8 A polychromatic beam is dispersed by a diffraction grating into three component wavelengths λ_1, λ_2, and λ_3. In the configuration shown, only radiation with λ_2 passes through a narrow slit and reaches the detector. Rotation of the diffraction grating (as shown by the arrows on the dotted circle) allows λ_1 or λ_3 to reach the detector.

of light is diffracted into a specific direction (Fig. 11A.8). In a *monochromator*, a narrow exit slit allows only a narrow range of wavelengths to reach the detector. Turning the grating on an axis perpendicular to the incident and diffracted beams allows different wavelengths to be analysed; in this way, the absorption spectrum is built up one narrow wavelength range at a time. In a *polychromator*, there is no slit and a broad range of wavelengths can be analysed simultaneously by *array detectors*, such as those discussed below.

Currently, almost all spectrometers operating in the infrared and near-infrared use 'Fourier transform' techniques for spectral detection and analysis. (Fourier transforms are discussed, but in more detail than needed here, in *The chemist's toolkit* 12C.1.) The heart of a Fourier transform spectrometer is a *Michelson interferometer*, a device for analysing the wavelengths present in a composite signal. The Michelson interferometer works by splitting the beam after it has passed through the sample into two and arranging for them to take different routes through the instrument before eventually recombining at the detector (Fig. 11A.9). One beam is reflected from mirror M_1 and one from mirror M_2; by moving M_1 it is therefore possible to introduce a difference in the length of the path traversed by the two beams.

Consider first the simplest case in which a beam of monochromatic light of wavelength λ is passed into the interferometer. If the path length difference p is 0, the two beams interfere constructively; the same is true if p is an integer number of wavelengths: λ, 2λ, 3λ,.... If p is one half of a wavelength, the two beams interfere destructively and cancel; the same is true if p is an odd multiple of half-wavelengths: $\lambda/2$, $3\lambda/2$, $5\lambda/2$,.... Therefore, as the mirror M_1 is moved the detected signal goes through a series of peaks and troughs depending on whether the two beams interfere constructively or destructively, and the net signal varies as $1 + \cos(2\pi p/\lambda)$, or $1+\cos(2\pi p\tilde{\nu})$ (Fig. 11A.10).

In a spectroscopic observation a mixture of radiation of different wavelengths and intensities is passed into the

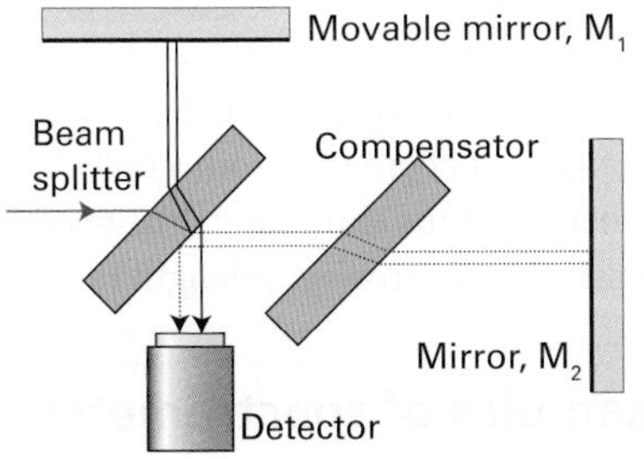

Figure 11A.9 A Michelson interferometer. The beam-splitting element divides the incident beam into two beams with a path difference that depends on the location of the movable mirror M_1. The compensator ensures that both beams pass through the same thickness of material.

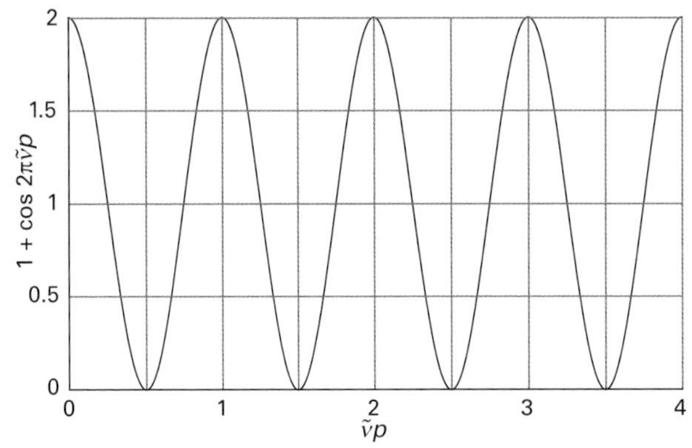

Figure 11A.10 An interferogram produced as the path length difference p is changed in the interferometer shown in Fig. 11A.9. Only a single wavelength component is present in the signal, so the graph is a plot of $1+\cos(2\pi\tilde{\nu}p)$.

spectrometer. Each component gives rise an interference pattern proportional to $1+\cos(2\pi p\tilde{\nu})$, and the signal recorded by the detector is their sum. Thus, if the intensity of the radiation entering the spectrometer consists of a mixture of wavenumber $\tilde{\nu}_i$ with intensities $I(\tilde{\nu}_i)$, the signal measured at the detector is given by the sum

$$\tilde{I}(p) = \sum_i I(\tilde{\nu}_i)\{1+\cos(2\pi\tilde{\nu}_i p)\} \qquad (11A.14)$$

A plot of $\tilde{I}(p)$ against p, which is detected by the system and recorded, is called an **interferogram**. The problem is to find $I(\tilde{\nu})$, the variation of the intensity with wavenumber, which is the spectrum, from the measured $\tilde{I}(p)$. This conversion can be carried out by using a standard mathematical technique, called the *Fourier transform*, which involves evaluating the integral

$$I(\tilde{\nu}) = 4\int_0^\infty \{\tilde{I}(p) - \tfrac{1}{2}\tilde{I}(0)\}\cos(2\pi\tilde{\nu}p)\mathrm{d}p \qquad \text{Fourier transform} \qquad (11A.15)$$

In practice, the measured values of $\tilde{I}(p)$ are digitized, stored in a computer attached to the spectrometer, and then the Fourier transform is computed numerically.

Example 11A.2 Relating a spectrum to an interferogram

Suppose the light entering the interferometer consists of three components with the following characteristics:

$\tilde{\nu}_i/\mathrm{cm}^{-1}$	150	250	450
$I(\tilde{\nu}_i)$	1	3	6

where the intensities are relative to the first value listed. Plot the interferogram associated with this signal. Then calculate and plot the Fourier transform of the interferogram.

Collect your thoughts For a signal consisting of just these three component beams, you can use eqn 11A.14 directly. Although in this case (where $\tilde{I}(p)$ is simply the sum of trigonometric functions) the Fourier transform $I(\tilde{\nu})$ can be carried out exactly, in general it is best done numerically by using mathematical software.

The solution From the data, the interferogram is

$$\begin{aligned}\tilde{I}(p) &= (1+\cos 2\pi\tilde{\nu}_1 p) + 3\times(1+\cos 2\pi\tilde{\nu}_2 p) + 6\times(1+\cos 2\pi\tilde{\nu}_3 p)\\ &= 10+\cos 2\pi\tilde{\nu}_1 p + 3\cos 2\pi\tilde{\nu}_2 p + 6\cos 2\pi\tilde{\nu}_3 p\end{aligned}$$

This function is plotted in Fig. 11A.11. The result of evaluating the Fourier transform numerically is shown in Fig. 11A.12.

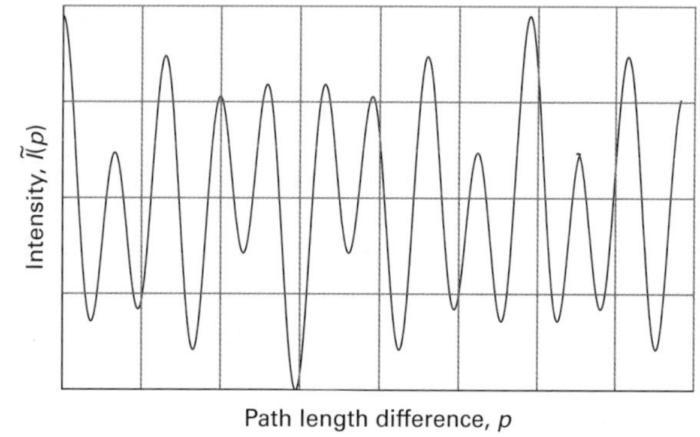

Figure 11A.11 The interferogram calculated from data in *Example* 11A.2.

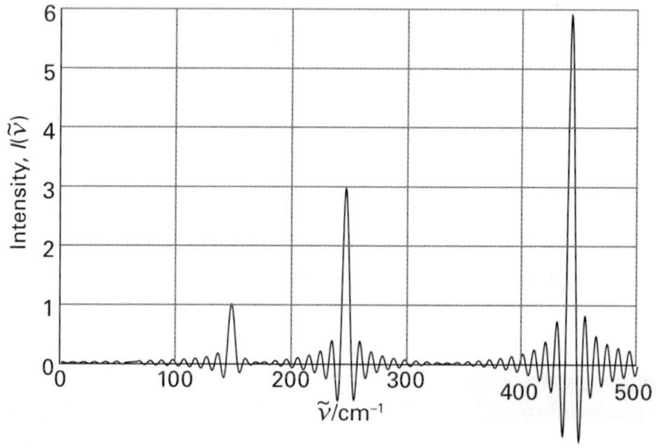

Figure 11A.12 The Fourier transform of the interferogram shown in Fig. 11A.11. The oscillations arise from the way that the signal in Fig. 11A.11 is sampled. As the sampling is extended to greater path-length differences, the oscillations disappear, and the peaks become sharper. *Interact with the dynamic version of this graph in the e-book.*

Self-test 11A.2 Explore the effect of varying the wavenumbers of the three components of the radiation on the shape of the interferogram by changing the value of $\tilde{\nu}_3$ to 550 cm^{-1}.

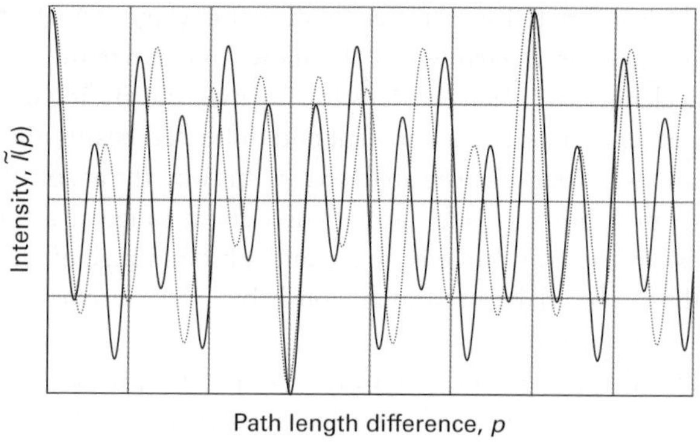

Figure 11A.13 The interferogram calculated from the data in Self-test 11A.2 superimposed on the interferogram obtained in the Example itself.

Answer: Fig. 11A.13

11A.3(c) Detectors

A **detector** is a device that converts radiation into an electric signal for processing and display. Detectors may consist of a single radiation sensing element or of several small elements arranged in one- or two-dimensional arrays.

A microwave detector is typically a *crystal diode* consisting of a tungsten tip in contact with a semiconductor. The most common detectors found in commercial infrared spectrometers are sensitive in the mid-infrared region. In a *photovoltaic device* the potential difference changes upon exposure to radiation. In a *pyroelectric device* the capacitance is sensitive to temperature and hence to the presence of infrared radiation.

A common detector for work in the ultraviolet and visible ranges is a *photomultiplier tube* (PMT), in which the photoelectric effect (Topic 7A) is used to generate an electrical signal proportional to the intensity of light that strikes the detector. A common, but less sensitive, alternative to the PMT is a *photodiode*, a solid-state device that conducts electricity when struck by photons because light-induced electron transfer reactions in the detector material create mobile charge carriers (negatively charged electrons and positively charged 'holes').

A *charge-coupled device* (CCD) is a two-dimensional array of several million small photodiode detectors. With a CCD, a wide range of wavelengths that emerge from a polychromator are detected simultaneously, thus eliminating the need to measure the radiation intensity one narrow wavelength range at a time.

11A.3(d) Examples of spectrometers

With an appropriate choice of spectrometer, absorption spectroscopy can be used to probe electronic, vibrational, and rotational transitions in molecules. It is often necessary to modify the general design of Fig. 11A.6 in order to detect weak signals. For example, to detect rotational transitions with a microwave spectrometer it is useful to modulate the transmitted intensity by varying the energy levels with an oscillating electric field. In this *Stark modulation*, an electric field of about $10^5\,\mathrm{V\,m^{-1}}$ ($1\,\mathrm{kV\,cm^{-1}}$) and a frequency of between 10 and 100 kHz is applied to the sample.

In a typical Raman spectroscopy experiment, a monochromatic incident laser beam is scattered from the front face of the sample and monitored (Fig. 11A.14). Lasers are used as the source of the incident radiation because the scattered beam is then more intense. The monochromatic character of laser radiation makes possible the observation of Stokes and anti-Stokes lines with frequencies that differ only slightly from that of the incident radiation. Such high resolution is particularly useful for observing rotational transitions by Raman spectroscopy.

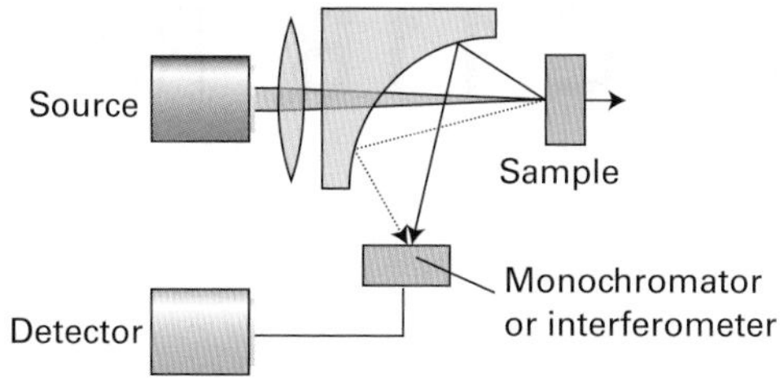

Figure 11A.14 A common arrangement adopted in Raman spectroscopy. A laser beam passes through a lens and then through a small hole in a mirror with a curved reflecting surface. The focused beam strikes the sample and scattered light is both deflected and focused by the mirror. The spectrum is analysed by a monochromator or an interferometer.

Checklist of concepts

- ☐ 1. In **emission spectroscopy** the electromagnetic radiation that arises from molecules undergoing a transition from a higher energy state to a lower energy state is detected.
- ☐ 2. In **absorption spectroscopy**, the net absorption of radiation passing through a sample is monitored.
- ☐ 3. In **Raman spectroscopy**, changes in molecular state are explored by examining the energies (frequencies) of the photons scattered by molecules.
- ☐ 4. Photons that are scattered elastically give rise to **Rayleigh scattering**.
- ☐ 5. In **Stokes scattering** a photon gives up some of its energy to a molecule; in **anti-Stokes scattering** the photon gains energy from the molecule.
- ☐ 6. **Stimulated absorption** is a process in which a transition from a low energy state to one of higher energy is driven by an electromagnetic field oscillating at the transition frequency; its rate is determined in part by the **Einstein coefficient of stimulated absorption**.
- ☐ 7. **Stimulated emission** is a process in which a transition from a high energy state to one of lower energy is driven by an electromagnetic field oscillating at the transition frequency; its rate is determined in part by the **Einstein coefficient of stimulated emission**.
- ☐ 8. **Spontaneous emission** is the transition from a high energy state to a lower energy state at a rate independent of any radiation also present. The relative importance of spontaneous emission increases as the cube of the transition frequency.
- ☐ 9. A **gross selection rule** specifies the general features a molecule must have if it is to have a spectrum of a given kind; a **specific selection rule** expresses the allowed transitions in terms of the changes in quantum numbers.
- ☐ 10. **Collisional line broadening** arises from the shortened lifetime due to collisions. The broadening is proportional to the pressure, and is often termed **pressure broadening**.

Checklist of equations

Property	Equation	Comment	Equation number
Ratio of Einstein coefficients of spontaneous and stimulated emission	$A_{\mathrm{l,u}} = (8\pi h\nu^3/c^3)B_{\mathrm{l,u}}$	$B_{\mathrm{u,l}} = B_{\mathrm{l,u}}$	11A.6a
Transition dipole moment	$\boldsymbol{\mu}_{\mathrm{fi}} = \int \psi_{\mathrm{f}}^* \hat{\boldsymbol{\mu}} \psi_{\mathrm{i}}\,\mathrm{d}\tau$	Electric dipole transitions	11A.7
Beer–Lambert law	$I = I_0 10^{-\varepsilon[\mathrm{J}]L}$	Uniform sample	11A.8
Absorbance and transmittance	$A = \log(I_0/I) = -\log T$	Definition	11A.9b
Integrated absorption coefficient	$\mathcal{A} = \int_{\mathrm{band}} \varepsilon(\tilde{\nu})\,\mathrm{d}\tilde{\nu}$	Definition	11A.10
Doppler broadening	$\delta\nu_{\mathrm{obs}} = (2\nu_0/c)(2kT\ln 2/m)^{1/2}$		11A.12a
Lifetime broadening	$\delta E/h = 1/2\pi\tau$	τ is the lifetime of the state	11A.13

TOPIC 11B Rotational spectroscopy

> **➤ Why do you need to know this material?**
>
> Rotational spectroscopy provides very precise values of bond lengths, bond angles, and dipole moments of molecules in the gas phase.
>
> **➤ What is the key idea?**
>
> The spacing of the lines in rotational spectra is used to determine the rotational constants of molecules and, through them, values of their bond lengths and angles.
>
> **➤ What do you need to know already?**
>
> You need to be familiar with the classical description of rotational motion (*Energy: A first look*), the quantization of angular momentum (Topic 7F), the general principles of molecular spectroscopy (Topic 11A), and the Pauli principle (Topic 8B). You also need to be familiar with the concepts of dipole moment and polar molecules, as described in introductory chemistry courses.

Pure rotational spectra, in which only the rotational state of a molecule changes, can be observed only in the gas phase. In spite of this limitation, rotational spectroscopy can provide a wealth of information about molecules, including precise bond lengths, bond angles, and dipole moments.

11B.1 Rotational energy levels

The classical expression for the energy of a body rotating about an axis q is

$$E_q = \tfrac{1}{2}I_q\omega_q^2 \qquad (11B.1)$$

where ω_q is the angular velocity about the axis $q = x, y, z$ and I_q is the corresponding moment of inertia. The **moment of inertia**, I, of a molecule about an axis passing through the centre of mass is defined as (Fig. 11B.1)

$$I = \sum_i m_i x_i^2 \qquad \text{Moment of inertia [definition]} \qquad (11B.2)$$

where m_i is the mass of the atom i treated as a point and x_i is its perpendicular distance from the axis of rotation. In general, the rotational properties of any molecule can be expressed in terms of its three **principal moments of inertia** I_q about three mutually

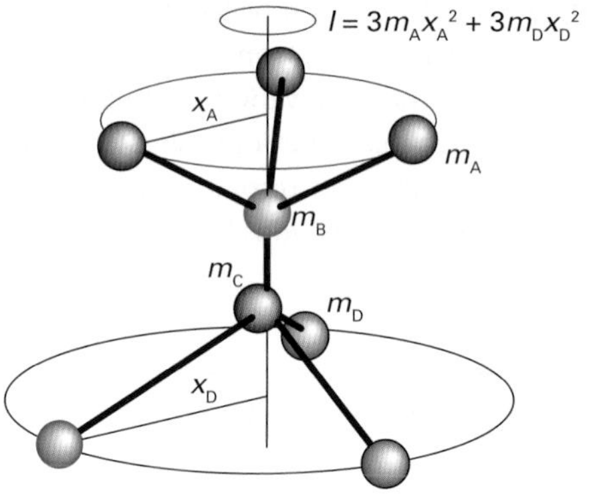

Figure 11B.1 The definition of moment of inertia. In this molecule there are three atoms with mass m_A attached to the B atom and three atoms with mass m_D attached to the C atom. The moment of inertia about the axis passing through the B and C atoms depends on the perpendicular distance x_A from this axis to the A atoms, and the perpendicular distance x_D to the D atoms.

perpendicular axes, $q = x, y, z$. For linear molecules, the moment of inertia around the internuclear axis is zero (because $x_i = 0$ for all the atoms) and the two remaining moments of inertia, which are equal, are denoted simply I. Explicit expressions for the moments of inertia of some symmetrical molecules are given in Table 11B.1. The principal moments of inertia are also commonly recorded as I_a, I_b, and I_c, with $I_c \geq I_b \geq I_a$.

The energy of a body free to rotate about three axes is

$$E = \tfrac{1}{2}I_x\omega_x^2 + \tfrac{1}{2}I_y\omega_y^2 + \tfrac{1}{2}I_z\omega_z^2 \qquad (11B.3)$$

The classical angular momentum about the axis q is $J_q = I_q\omega_q$. It follows that

$$E = \frac{J_x^2}{2I_x} + \frac{J_y^2}{2I_y} + \frac{J_z^2}{2I_z} \qquad \text{Rotational energy: classical expression} \qquad (11B.4)$$

Table 11B.1 Moments of inertia*

1. Diatomic molecules	
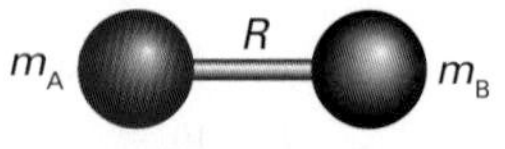	$I = \mu R^2 \quad \mu = \dfrac{m_A m_B}{m}$
2. Triatomic linear rotors	
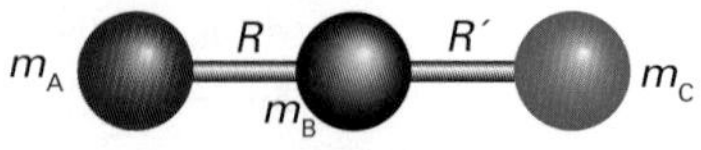	$I = m_A R^2 + m_C R'^2 - \dfrac{(m_A R - m_C R')^2}{m}$
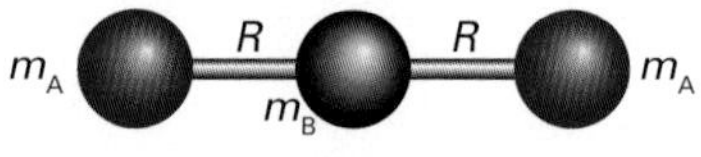	$I = 2m_A R^2$

3. Symmetric rotors

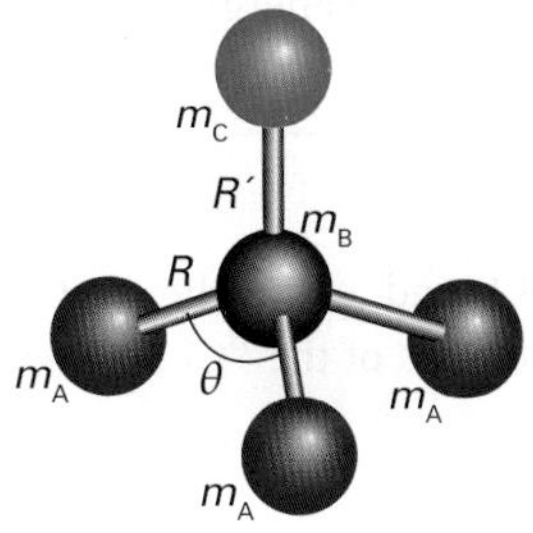

$$I_{\parallel} = 2m_A f_1(\theta)R^2$$

$$I_{\perp} = m_A f_1(\theta)R^2 + \frac{m_A}{m}(m_B + m_C)f_2(\theta)R^2 + \frac{m_C}{m}\{(3m_A + m_B)R' + 6m_A R\left[\tfrac{1}{3}f_2(\theta)\right]^{1/2}\}R'$$

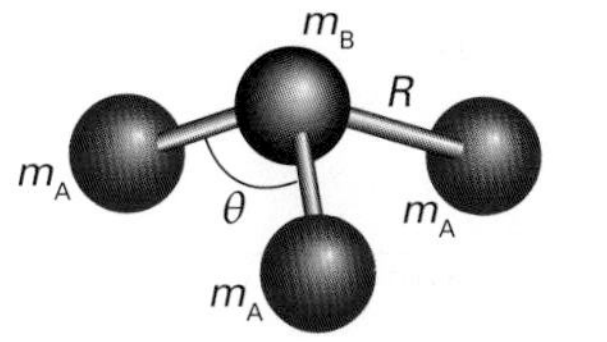

$$I_{\parallel} = 2m_A f_1(\theta)R^2$$

$$I_{\perp} = m_A f_1(\theta)R^2 + \frac{m_A m_B}{m} f_2(\theta)R^2$$

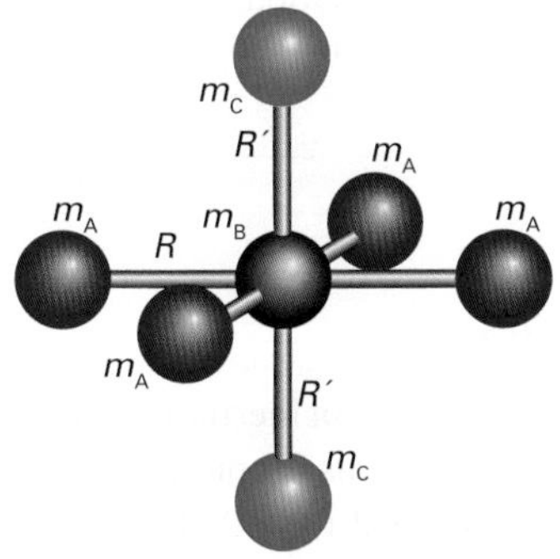

$$I_{\parallel} = 4m_A R^2$$

$$I_{\perp} = 2m_A R^2 + 2m_C R'^2$$

4. Spherical rotors

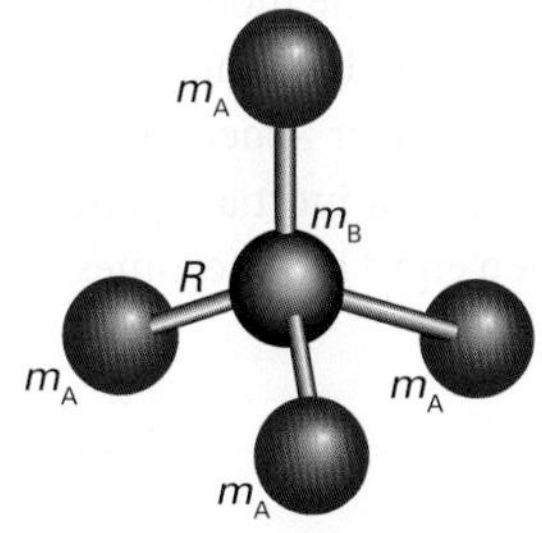

$$I = \tfrac{8}{3}m_A R^2$$

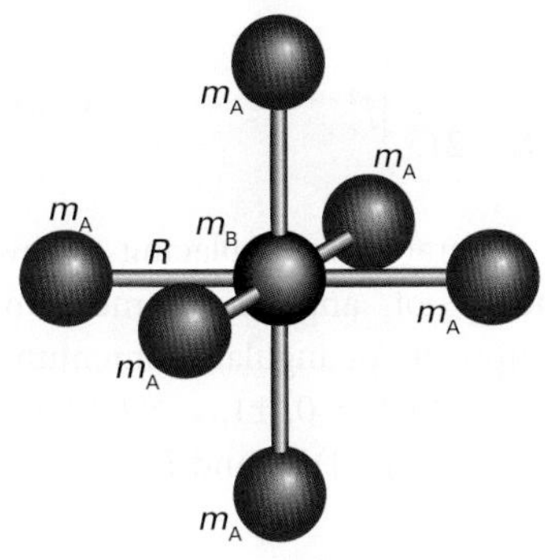

$$I = 4m_A R^2$$

* $f_1(\theta) = 1 - \cos\theta$, $f_2(\theta) = 1 + 2\cos\theta$; in each case, m is the total mass of the molecule.

Example 11B.1 Calculating the moment of inertia of a molecule

Calculate the moment of inertia of a ^1H_2O molecule around the axis defined by the bisector of the HOH angle (**1**). The HOH bond angle is 104.5° and the OH bond length is 95.7 pm.

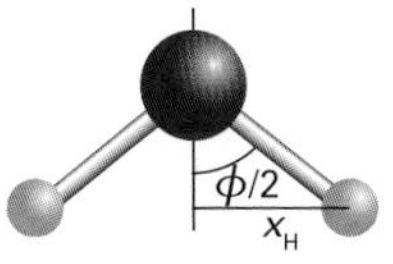

1

Collect your thoughts You can compute the moment of inertia by using eqn 11B.2. The mass to use is the actual atomic mass (expressed in terms of the atomic mass constant m_u), not the element's molar mass. The x_i are the perpendicular distances from each atom to the axis of rotation, and you will be able to calculate these distances by using trigonometry and the bond angle and bond length.

The solution From eqn 11B.2,

$$I = \sum_i m_i x_i^2 = m_H x_H^2 + 0 + m_H x_H^2 = 2m_H x_H^2$$

In this case, m_H is the mass of a ^{1}H atom, so $m_H = 1.0078m_u$. If the bond angle of the molecule is ϕ and the OH bond length is R, trigonometry gives $x_H = R\sin\frac{1}{2}\phi$. It follows that

$$I = 2m_H R^2 \sin^2\tfrac{1}{2}\phi$$

Substitution of the data gives

$$\begin{aligned} I &= 2 \times (1.0078 \times 1.6605 \times 10^{-27}\ \text{kg}) \times (9.57 \times 10^{-11}\ \text{m})^2 \\ &\quad \times \sin^2\left(\tfrac{1}{2} \times 104.5°\right) \\ &= 1.92 \times 10^{-47}\ \text{kg m}^2 \end{aligned}$$

Note that the mass of the O atom makes no contribution to the moment of inertia because the axis passes through this atom.

Self-test 11B.1 Calculate the moment of inertia of a $CH^{35}Cl_3$ molecule around a rotational axis along the C–H bond. The C–Cl bond length is 177 pm and the HCCl angle is 107°; $m(^{35}Cl) = 34.97m_u$.

Answer: 4.99×10^{-45} kg m^2

11B.1(a) Spherical rotors

Spherical rotors have all three moments of inertia equal, as in CH_4 and SF_6. If these moments of inertia have the value I, the classical expression for the energy is

$$E = \frac{J_x^2 + J_y^2 + J_z^2}{2I} = \frac{J^2}{2I} \tag{11B.5}$$

where J^2 is the square of the magnitude of the angular momentum. The corresponding quantum expression is generated by making the replacement

$$J^2 \rightarrow J(J+1)\hbar^2 \quad J=0,1,2,\ldots$$

where J is the angular momentum quantum number. Therefore, the energy of a spherical rotor is confined to the values

$$E_J = J(J+1)\frac{\hbar^2}{2I} \quad J=0,1,2,\ldots \qquad \text{Energy levels of a spherical rotor} \quad (11B.6)$$

The resulting ladder of energy levels is illustrated in Fig. 11B.2. The energy is normally expressed in terms of the **rotational constant**, $\tilde{B}$ (a wavenumber), of the molecule, where

$$hc\tilde{B} = \frac{\hbar^2}{2I} \quad \text{so} \quad \tilde{B} = \frac{\hbar}{4\pi cI} \qquad \text{Rotational constant [definition]} \quad (11B.7)$$

The expression for the energy is then

$$E_J = hc\tilde{B}J(J+1) \quad J=0,1,2,\ldots \qquad \text{Energy levels of a spherical rotor} \quad (11B.8)$$

It is also common to express the rotational constant as a frequency and to denote it B. Then $B = \hbar/4\pi I$ and the energy is $E_J = hBJ(J+1)$. The two quantities are related by $B = c\tilde{B}$.

The energy of a rotational state is normally reported as the **rotational term**, $\tilde{F}(J)$, a wavenumber, by division of both sides of eqn 11B.8 by hc:

$$\tilde{F}(J) = \tilde{B}J(J+1) \qquad \text{Rotational terms of spherical rotor} \quad (11B.9)$$

To express the rotational term as a frequency, use $F = c\tilde{F}$. The separation of successive terms is

$$\tilde{F}(J+1)-\tilde{F}(J)=\tilde{B}(J+1)(J+2)-\tilde{B}J(J+1)=2\tilde{B}(J+1) \quad (11B.10)$$

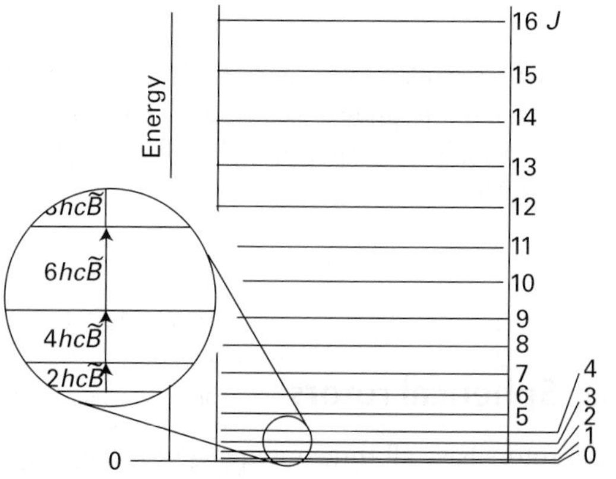

Figure 11B.2 The rotational energy levels of a linear or spherical rotor. Note that the energy separation between successive levels increases as J increases.

Because the rotational constant is inversely proportional to I, large molecules have closely spaced rotational energy levels.

Brief illustration 11B.1

Consider $^{12}C^{35}Cl_4$. From Table 11B.1 and given the C–Cl bond length ($R_{C-Cl} = 177$ pm) and the mass of the ^{35}Cl nuclide ($m(^{35}Cl) = 34.97m_u$), the moment of inertia is

$$I = \tfrac{8}{3}m(^{35}\text{Cl})R_{\text{C-Cl}}^2 = \tfrac{8}{3}\times\overbrace{(5.807\times10^{-26}\ \text{kg})}^{34.97m_u}\times(1.77\times10^{-10}\ \text{m})^2$$
$$= 4.85\ldots\times10^{-45}\ \text{kg m}^2$$

From eqn 11B.7 the rotational constant is

$$\tilde{B} = \frac{1.05457\times10^{-34}\ \overbrace{\text{J}}^{\text{kg m}^2\text{s}^{-2}}\ \text{s}}{4\pi\times(2.998\times10^8\ \text{m s}^{-1})\times(4.85\ldots\times10^{45}\ \text{kg m}^2)}$$
$$= 5.77\ \text{m}^{-1} = 0.0577\ \text{cm}^{-1}$$

It follows from eqn 11B.10 that the separation between the $J = 0$ and $J = 1$ terms is $\tilde{F}(1)-\tilde{F}(0)=2\tilde{B}=0.1154\ \text{cm}^{-1}$, corresponding to 3.46 GHz.

11B.1(b) Symmetric rotors

Symmetric rotors have two equal moments of inertia and a third that is non-zero. In group theoretical terms (Topic 10A), such rotors have an n-fold axis of rotation, with $n > 2$. The unique axis of a symmetric rotor (such as CH_3Cl, NH_3, and C_6H_6) is its **principal axis** (or *figure axis*). If the moment of inertia about the principal axis is larger than the other two, the rotor is classified as **oblate** (like a pancake, and C_6H_6). If the moment of inertia around the principal axis is smaller than the other two, the rotor is classified as **prolate** (like a rugby ball or American football, and CH_3Cl). The two equal moments of inertia (I_x and I_y) are denoted $I_\perp$ and I_z is denoted $I_\parallel$. Then eqn 11B.4 becomes

$$E = \frac{J_x^2+J_y^2}{2I_\perp}+\frac{J_z^2}{2I_\parallel} \quad (11B.11)$$

This expression can be written in terms of $J^2 = J_x^2+J_y^2+J_z^2$:

$$E = \frac{J^2-J_z^2}{2I_\perp}+\frac{J_z^2}{2I_\parallel}=\frac{J^2}{2I_\perp}+\left(\frac{1}{2I_\parallel}-\frac{1}{2I_\perp}\right)J_z^2 \quad (11B.12)$$

The quantum expression is generated by replacing J^2 by $J(J+1)\hbar^2$. The quantum theory of angular momentum (Topic 7F) also restricts the component of angular momentum about any axis to the values $K\hbar$, with $K = 0, \pm1,\ldots, \pm J$. Then, after making the replacements $J^2 \rightarrow J(J+1)\hbar^2$ and $J_z^2 \rightarrow K^2\hbar^2$ the rotational terms are

$$\tilde{F}(J,K) = \tilde{B}J(J+1) + (\tilde{A} - \tilde{B})K^2 \qquad J = 0,1,2,\ldots \quad K = 0,\pm 1,\ldots,\pm J$$ Rotational terms of a symmetric rotor (11B.13a)

with

$$\tilde{A} = \frac{\hbar}{4\pi c I_{\parallel}} \qquad \tilde{B} = \frac{\hbar}{4\pi c I_{\perp}} \qquad (11\text{B}.13\text{b})$$

Equation 11B.13a matches what is expected for the dependence of the energy levels on the two distinct moments of inertia of the molecule:[1]

- When $K = 0$, there is no component of angular momentum about the principal axis, and the energy levels depend only on $I_{\perp}$ (Fig. 11B.3a).
- When $K = \pm J$, almost all the angular momentum arises from rotation around the principal axis, and the energy levels are determined largely by $I_{\parallel}$ (Fig. 11B.3b).
- The sign of K does not affect the energy because opposite values of K correspond to opposite senses of rotation, and the energy does not depend on the sense of rotation.

Example 11B.2 Calculating the rotational energy levels of a symmetric rotor

A $^{14}N^{1}H_3$ molecule is a symmetric rotor with bond length 101.2 pm and HNH bond angle 106.7°. Calculate its rotational terms.

Collect your thoughts Begin by calculating the moments of inertia by using the expressions given in Table 11B.1. Then use eqn 11B.13a to find the rotational terms. The rotational constants are found by using eqn 11B.13b.

The solution Substitution of $m_A = 1.0078m_u$, $m_B = 14.0031m_u$, $R = 101.2$ pm, and $\theta = 106.7°$ into the second of the symmetric rotor expressions in Table 11B.1 gives $I_{\parallel} = 4.4128 \times 10^{-47}$ kg m^2 and $I_{\perp} = 2.8059 \times 10^{-47}$ kg m^2. The expressions in eqn. 11B.13b give $\tilde{A} = 6.344$ cm^{-1} and $\tilde{B} = 9.977$ cm^{-1}. It follows from eqn 11B.13a that

$$\tilde{F}(J,K)/\text{cm}^{-1} = 9.977J(J+1) - 3.633K^2$$

Multiplication by c converts $\tilde{F}(J,K)$ to a frequency, denoted $F(J,K)$:

$$F(J,K)/\text{GHz} = 299.1J(J+1) - 108.9K^2$$

If $J = 1$, $F(1,1) = 16.32$ cm^{-1} (489.3 GHz) for $K = \pm 1$, and $F(1,0) = 19.95$ cm^{-1} (598.1 GHz) for $K = 0$.

Self-test 11B.2 A $^{12}C^{1}H_3^{35}Cl$ molecule has a C–Cl bond length of 178 pm, a C–H bond length of 111 pm, and an HCH angle of 110.5°. Identify whether the molecule is oblate or prolate, and calculate its rotational energy terms.

Answer: $I_{\perp} = 6.262 \times 10^{-46}$ kg m^2, $I_{\parallel} = 5.568 \times 10^{-47}$ kg m^2; prolate; $\tilde{A} = 5.0275$ cm^{-1} and $\tilde{B} = 0.4470$ cm^{-1}; $\tilde{F}(J,K)/\text{cm}^{-1} = 0.447J(J+1) + 4.58K^2$

[1] The quantum number K is used to signify the component on the principal axis, as distinct from the quantum number M_J, which is used to signify the component on an externally defined axis.

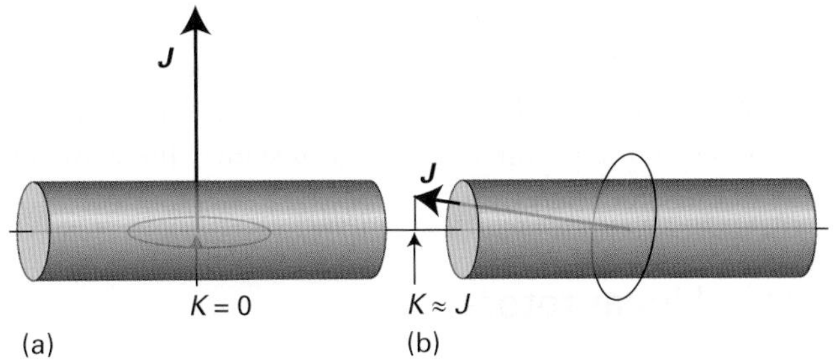

Figure 11B.3 The significance of the quantum number K. (a) When $K = 0$ the molecule has no angular momentum about its principal axis: it is undergoing end-over-end rotation. (b) When $|K|$ is close to its maximum value, J, most of the molecular rotation is around the principal axis.

The energy of a symmetric rotor depends on J and K. Because the states with K and $-K$ have the same energy, each level, except those with $K = 0$, is doubly degenerate. In addition, the angular momentum of the molecule has a component on an external, laboratory-fixed axis. This component is quantized, and its permitted values are $M_J\hbar$, with $M_J = 0, \pm 1, \ldots, \pm J$, giving $2J + 1$ values in all (Fig. 11B.4). The quantum number M_J does not appear in the expression for the energy, but it is necessary for a complete specification of the state of the rotor. Consequently, all $2J + 1$ orientations of the rotating molecule have the same energy. It follows that a symmetric rotor level is $2(2J + 1)$-fold degenerate for $K \neq 0$ and $(2J + 1)$-fold degenerate for $K = 0$.

A spherical rotor can be regarded as a version of a symmetric rotor in which $I_{\perp} = I_{\parallel}$ and therefore $\tilde{A} = \tilde{B}$. The quantum number K still takes any one of $2J + 1$ values, but the energy is independent of which value it takes. Therefore, as well as having a $(2J + 1)$-fold degeneracy arising from its orientation in space, the rotor also has a $(2J + 1)$-fold degeneracy arising from its

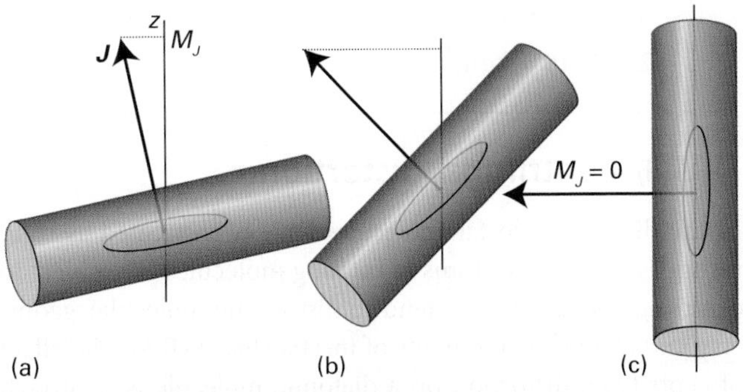

Figure 11B.4 The significance of the quantum number M_J. (a) When M_J is close to its maximum value, J, most of the molecular rotation is around the laboratory axis (taken as the z-axis). (b) An intermediate value of M_J. (c) When $M_J = 0$ the molecule has no angular momentum about the z-axis. All three diagrams correspond to a state with $K = 0$; there are corresponding diagrams for different values of K, in which the angular momentum makes a different angle to the principal axis of the molecule.

orientation with respect to an arbitrary axis in the molecule. The overall degeneracy of a symmetric rotor energy level with quantum number J is therefore $(2J + 1)^2$. This degeneracy increases very rapidly: when $J = 10$, for instance, there are 441 states of the same energy.

11B.1(c) Linear rotors

For a linear rotor (such as CO_2, HCl, and C_2H_2), in which the atoms are regarded as mass points, the rotation occurs only about an axis perpendicular to the internuclear axis and there is no rotation around that axis. Therefore the component of angular momentum around the internuclear axis of a linear rotor is identically zero, and $K \equiv 0$ in eqn 11B.13a. The rotational terms of a linear molecule are therefore

$$\tilde{F}(J) = \tilde{B}J(J+1) \quad J = 0, 1, 2, \ldots \qquad \text{Rotational terms of linear rotor} \quad (11B.14)$$

This expression is the same as eqn 11B.9 but arrived at it in a significantly different way: here $K \equiv 0$, but for a spherical rotor $\tilde{A} = \tilde{B}$ and K has a range of values. The angular momentum of a linear rotor has $2J + 1$ components on an external axis, so its degeneracy is just $2J + 1$ rather than the $(2J + 1)^2$-fold degeneracy of a spherical rotor.

Brief illustration 11B.2

Equation 11B.10 for the energy separation of successive levels of a spherical rotor also applies to linear rotors, so $\tilde{F}(3) - \tilde{F}(2) = 6\tilde{B}$. Spectroscopic measurements on $^1H^{35}Cl$ give $\tilde{F}(3) - \tilde{F}(2) = 63.56\ \text{cm}^{-1}$, so it follows that $6\tilde{B} = 63.56\ \text{cm}^{-1}$, $\tilde{B} = 10.59\ \text{cm}^{-1}$, and therefore

$$I = \frac{\hbar}{4\pi c\tilde{B}} = \frac{1.05457\times10^{-34}\ \text{J s}}{4\pi\times(2.998\times10^{10}\ \text{cm s}^{-1})\times(10.59\ \text{cm}^{-1})}$$
$$= 2.643\times10^{-47}\ \text{kg m}^2$$

11B.1(d) Centrifugal distortion

In the discussion so far molecules have been treated as rigid rotors. However, the atoms of rotating molecules are subject to centrifugal forces, which tend to distort the molecular geometry and change its moments of inertia (Fig. 11B.5). The effect of centrifugal distortion on a diatomic molecule is to stretch the bond and hence to increase the moment of inertia. As a result, the rotational constant is reduced and the energy levels are slightly closer together than the rigid-rotor expressions predict. The effect is usually taken into account by including in the energy expression a negative term that becomes more important as J increases:

$$\tilde{F}(J) = \tilde{B}J(J+1) - \tilde{D}_J J^2(J+1)^2 \qquad \text{Rotational terms affected by centrifugal distortion} \quad (11B.15)$$

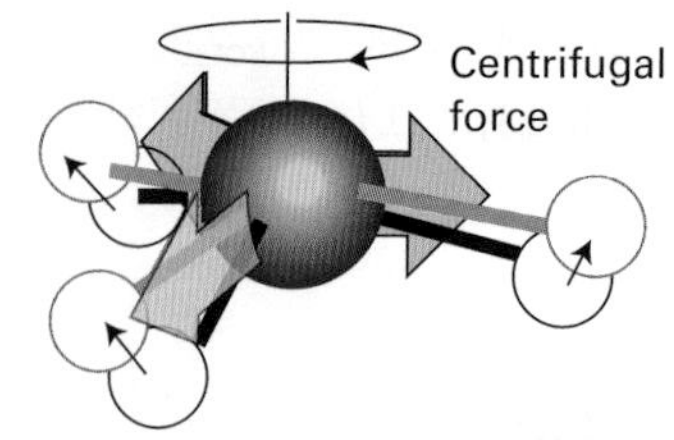

Figure 11B.5 The effect of rotation on a molecule. The centrifugal force arising from rotation distorts the molecule, opening out bond angles and stretching bonds slightly. The effect is to increase the moment of inertia of the molecule and hence to decrease its rotational constant.

The parameter $\tilde{D}_J$ is the **centrifugal distortion constant**. The centrifugal distortion constant of a diatomic molecule is related to the vibrational wavenumber of the bond, $\tilde{\nu}$ (which, as seen in Topic 11C, is a measure of its stiffness), through the approximate relation

$$\tilde{D}_J = \frac{4\tilde{B}^3}{\tilde{\nu}^2} \qquad \text{Centrifugal distortion constant} \quad (11B.16)$$

As expected, a bond that is easily stretched, and therefore has a low vibrational wavenumber, has a high centrifugal distortion constant.

Brief illustration 11B.3

For $^{12}C^{16}O$, $\tilde{B} = 1.931\ \text{cm}^{-1}$ and $\tilde{\nu} = 2170\ \text{cm}^{-1}$. It follows that

$$\tilde{D}_J = \frac{4\times(1.931\ \text{cm}^{-1})^3}{(2170\ \text{cm}^{-1})^2} = 6.116\times10^{-6}\ \text{cm}^{-1}$$

Because $\tilde{D}_J << \tilde{B}$, centrifugal distortion has a very small effect on the energy levels until J is large. For $J = 20$, $\tilde{D}_J J^2(J+1)^2 = 1.08\ \text{cm}^{-1}$ (corresponding to 32 GHz).

Exercises

E11B.1 Calculate the moment of inertia around the bisector of the OOO angle and the corresponding rotational constant of an $^{16}O_3$ molecule (bond angle 117°; OO bond length 128 pm).

E11B.2 Classify the following rotors: (i) O_3, (ii) CH_3CH_3, (iii) XeO_4, (iv) $FeCp_2$ (Cp denotes the cyclopentadienyl group, C_5H_5).

E11B.3 Calculate the HC and CN bond lengths in HCN from the rotational constants $B(^1H^{12}C^{14}N) = 44.316$ GHz and $B(^2H^{12}C^{14}N) = 36.208$ GHz.

E11B.4 Estimate the centrifugal distortion constant for $^1H^{127}I$, for which $\tilde{B} = 6.511\ \text{cm}^{-1}$ and $\tilde{\nu} = 2308\ \text{cm}^{-1}$. By what factor would the constant change when ^2H is substituted for ^1H?

11B.2 Microwave spectroscopy

Typical values of the rotational constant $\tilde{B}$ for small molecules are in the region of 0.1–10 cm^{-1}; two examples are 0.356 cm^{-1} for NF_3 and 10.59 cm^{-1} for HCl. It follows that rotational

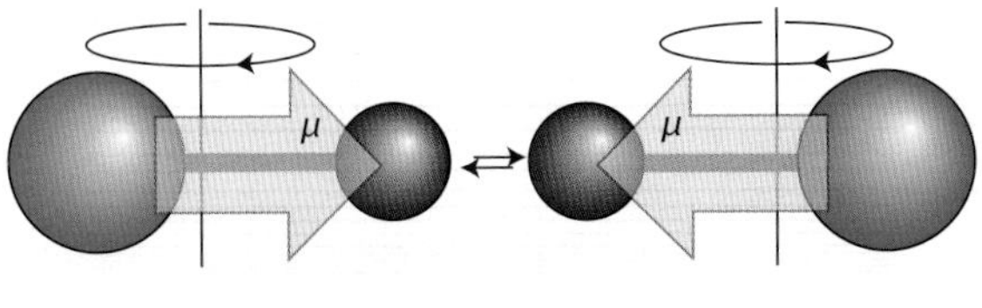

Figure 11B.6 To a stationary observer, a rotating polar molecule looks like an oscillating dipole which will generate an oscillating electromagnetic wave (or, in the case of absorption, interact with such a wave). This picture is the classical origin of the gross selection rule for rotational transitions.

transitions can be studied with **microwave spectroscopy**, a technique that monitors the absorption of radiation in the microwave region of the spectrum.

11B.2(a) Selection rules

As usual in spectroscopy, the selection rules can be established by considering the relevant transition dipole moment. The details of the calculation are shown in *A deeper look* 11B.1, available to read in the e-book accompanying this text. The conclusion is that

> The gross selection rule for the observation of a pure rotational transition is that a molecule must have a permanent electric dipole moment.

The classical basis of this rule is that a polar molecule appears to possess a fluctuating dipole when rotating, but a nonpolar molecule does not (Fig. 11B.6). The permanent dipole can be regarded as a handle with which the molecule stirs the electromagnetic field into oscillation (and vice versa for absorption).

Brief illustration 11B.4

Homonuclear diatomic molecules and nonpolar polyatomic molecules such as CO_2, $CH_2{=}CH_2$, and C_6H_6 do not give rise to microwave spectra. On the other hand, OCS and H_2O are polar and have microwave spectra. Spherical rotors cannot have electric dipole moments unless they become distorted by rotation, so they are rotationally inactive except in special cases. An example of a spherical rotor that does become sufficiently distorted for it to acquire a dipole moment is SiH_4, which has a dipole moment of about 8.3 μD by virtue of its rotation when $J \approx 10$ (for comparison, HCl has a permanent dipole moment of 1.1 D; molecular dipole moments and their units are discussed in Topic 14A).

The analysis also shows that, for a linear molecule, the transition moment vanishes unless the following conditions are fulfilled:

$$\Delta J = \pm 1 \quad \Delta M_J = 0, \pm 1 \qquad \text{Rotational selection rules: linear rotors} \tag{11B.17}$$

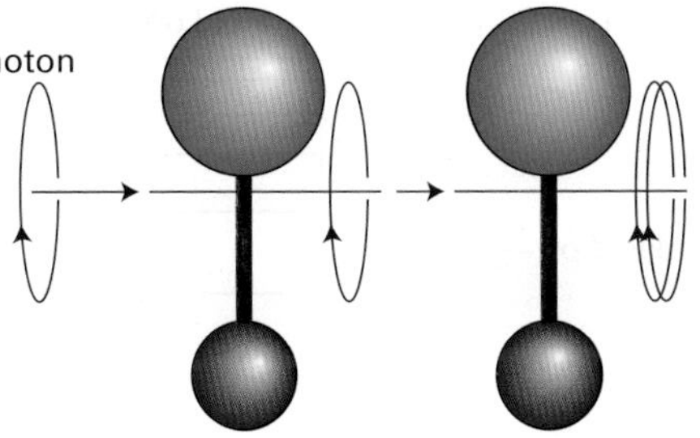

Figure 11B.7 When a photon is absorbed by a molecule, the angular momentum of the combined system is conserved. If the molecule is rotating in the same sense as the spin of the incoming photon, then J increases by 1.

The transition $\Delta J = +1$ corresponds to absorption and the transition $\Delta J = -1$ corresponds to emission.

- The allowed change in J arises from the conservation of angular momentum when a photon, a spin-1 particle, is emitted or absorbed (Fig. 11B.7).
- The allowed change in M_J also arises from the conservation of angular momentum when a photon is emitted or absorbed in a specific direction.

When the transition moment is evaluated for all possible relative orientations of the molecule to the line of flight of the photon, it is found that the total $J+1 \leftrightarrow J$ transition intensity is proportional to

$$|\mu_{J+1,J}|^2 = \frac{J+1}{2J+1}\mu_0^2 \tag{11B.18}$$

where μ_0 is the permanent electric dipole moment of the molecule. The intensity is proportional to the square of μ_0, so strongly polar molecules give rise to much more intense rotational lines than less polar molecules.

Rotation of a symmetric rotor about its principal (figure) axis does not lead to any change in the orientation of the dipole; there is no fluctuating dipole to interact with the radiation, and therefore no change in K is possible. For symmetric rotors the selection rules are therefore:

$$\Delta J = \pm 1 \quad \Delta M_J = 0, \pm 1 \quad \Delta K = 0 \qquad \text{Rotational selection rules: symmetric rotors} \tag{11B.19}$$

The degeneracy associated with the quantum number M_J (the orientation of the rotation in space) is partly removed when an electric field is applied to a polar molecule (Fig. 11B.8). The splitting of states by an electric field is called the **Stark effect**. The energy shift depends on the permanent electric dipole moment, μ_0, so the observation of the Stark effect in a rotational spectrum can be used to measure the magnitudes of electric dipole moments.

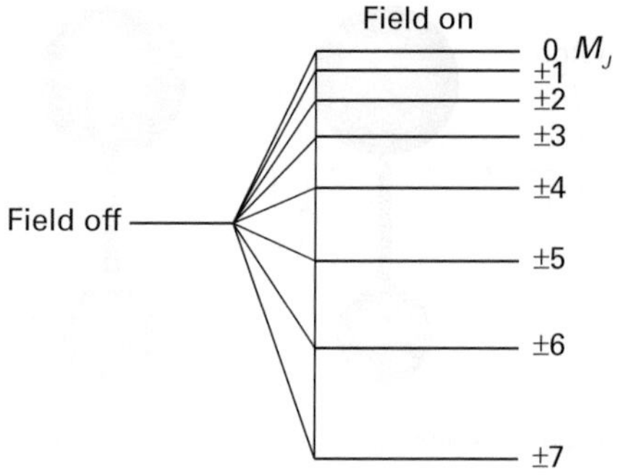

Figure 11B.8 The effect of an electric field on the energy level with $J = 7$ of a polar linear rotor. All levels are doubly degenerate except that with $M_J = 0$.

11B.2(b) The appearance of microwave spectra

When the selection rules are applied to the expressions for the energy levels of a linear rigid rotor (eqn 11B.14), it follows that the wavenumbers of the allowed $J + 1 \leftarrow J$ absorptions are

$$\tilde{\nu}(J+1 \leftarrow J) = \tilde{F}(J+1) - \tilde{F}(J) = 2\tilde{B}(J+1) \quad J = 0, 1, 2, \ldots$$

Wavenumbers of rotational transitions: linear rotor (11B.20a)

When centrifugal distortion is taken into account, the corresponding expression obtained from eqn 11B.15 is

$$\tilde{\nu}(J+1 \leftarrow J) = 2\tilde{B}(J+1) - 4\tilde{D}_J(J+1)^3 \qquad (11\text{B}.20\text{b})$$

However, because the second term is typically very small compared with the first, the appearance of the spectrum closely resembles that predicted from eqn 11B.20a.

Example 11B.3 Predicting the appearance of a rotational spectrum

Predict the form of the rotational spectrum of $^{14}NH_3$, which is an oblate symmetric rotor with $\tilde{B} = 9.977\ \text{cm}^{-1}$.

Collect your thoughts The rotational terms are given by eqn 11B.13a. Because $\Delta J = \pm 1$ and $\Delta K = 0$, the expression for the wavenumbers of the rotational transitions is identical to eqn 11B.20a and depends only on $\tilde{B}$.

The solution The following table can be drawn up for the $J + 1 \leftarrow J$ transitions.

J	0	1	2	3	…
$\tilde{\nu}/\text{cm}^{-1}$	19.95	39.91	59.86	79.82	…
ν/GHz	598.1	1196	1795	2393	…

The line spacing is $19.95\ \text{cm}^{-1}$ (598.1 GHz).

Self-test 11B.3 Repeat the problem for $CH_3^{35}Cl$, a prolate symmetric rotor for which $\tilde{B} = 0.444\ \text{cm}^{-1}$.

Answer: Lines of separation $0.888\ \text{cm}^{-1}$ (26.6 GHz)

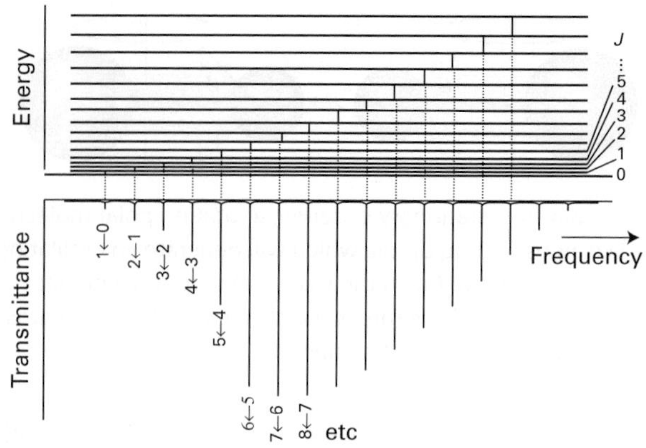

Figure 11B.9 The rotational energy levels of a linear rotor, the transitions allowed by the selection rule $\Delta J = +1$, and a typical pure rotational absorption spectrum (displayed here in terms of the radiation transmitted through the sample). The intensities reflect the populations of the initial level in each case and the strengths of the transition dipole moments.

The form of the spectrum predicted by eqn 11B.20a is shown in Fig. 11B.9. The most significant feature is that it consists of a series of lines with wavenumbers $2\tilde{B}, 4\tilde{B}, 6\tilde{B}, \ldots$ and of separation $2\tilde{B}$. The measurement of the line spacing therefore gives $\tilde{B}$, and hence the moment of inertia $I_\perp$ perpendicular to the principal axis of the molecule. Because the masses of the atoms are known, it is a simple matter to deduce the bond length of a diatomic molecule. However, in the case of a polyatomic molecule such as OCS or NH_3, a knowledge of one moment of inertia is insufficient data from which to infer, for example, the two bond lengths in OCS, or the bond length and bond angle in NH_3.

This difficulty can be overcome by measuring the spectra of **isotopologues**, isotopically substituted molecules. The spectrum from each isotopologue gives a separate moment of inertia and, if it is assumed that the bond lengths and angles are unaffected by isotopic substitution, the extra data make it possible to extract values of the bond lengths and angles. A good example of this procedure is the study of OCS; the actual calculation is worked through in Problem P11B.7. The assumption that bond lengths are unchanged in isotopologues is only an approximation, but it is a good one in most cases. Nuclear spin (Topic 12A), which differs from one isotope to another, also affects the appearance of high-resolution rotational spectra because spin is a source of angular momentum and can couple with the rotation of the molecule itself and hence affect the rotational energy levels.

The intensities of spectral lines increase with increasing J and pass through a maximum before tailing off as J becomes large. The most important reason for this behaviour is the existence of a maximum in the population of rotational levels. The Boltzmann distribution (see *Energy: A first look* and Topic 13A) implies that the population of a state decreases exponentially as its energy increases. However, the population of a *level* is also proportional to its degeneracy, and in the case of rotational levels this degeneracy increases with J. These two opposite trends result in the population of the energy levels (as distinct from the individual states) passing through a maximum. Specifically, the population N_J of a rotational energy level J is given by the Boltzmann expression

$$N_J \propto N g_J e^{-E_J/kT}$$

where N is the total number of molecules in the sample and g_J is the degeneracy of the level J. The value of J corresponding to a maximum of this expression is found by treating J as a continuous variable, differentiating with respect to J, and then setting the result equal to zero. The result for a linear rotor, for which $g_J = 2J+1$, (see Problem P11B.11) is

$$J_{\max} \approx \left(\frac{kT}{2hc\tilde{B}}\right)^{1/2} - \tfrac{1}{2} \quad \text{Rotational level with largest population: linear rotor} \quad (11B.21)$$

For a typical molecule (e.g. OCS, with $\tilde{B} = 0.2\ \mathrm{cm^{-1}}$) $kT/2hc\tilde{B} \approx 500$ at room temperature, so $J_{\max} \approx 22$. However, the transition dipole moment depends on the value of J (eqn 11B.18) and, because the radiation can also cause stimulated emission (Topic 11A), the intensity also depends on the population *difference* between the two states involved in the transition. Hence the value of J corresponding to the most intense line is not quite the same as the value of J for the most highly populated level.

Exercises

E11B.5 Which of the following molecules may show a pure rotational microwave absorption spectrum: (i) H_2, (ii) HCl, (iii) CH_4, (iv) CH_3Cl, (v) CH_2Cl_2?

E11B.6 Calculate the frequency and wavenumber of the $J = 3 \leftarrow 2$ transition in the pure rotational spectrum of $^{14}N^{16}O$. The equilibrium bond length is 115 pm. Would the frequency increase or decrease if centrifugal distortion is considered?

E11B.7 The spacing of lines in the microwave spectrum of $^{27}Al^{1}H$ is 12.604 $\mathrm{cm^{-1}}$; calculate the moment of inertia and bond length of the molecule.

E11B.8 What is the most highly populated rotational level of Cl_2 at (i) 25 °C, (ii) 100 °C? Take $\tilde{B} = 0.244\ \mathrm{cm^{-1}}$.

11B.3 Rotational Raman spectroscopy

Raman scattering (Topic 11A) can also arise as a result of rotational transitions. The gross selection rule for rotational Raman transitions is that the molecule must be anisotropically polarizable. To understand this criterion it is necessary to know that the distortion of the electron density of a molecule in an electric field is determined by its polarizability, α (Topic 14A). More precisely, if the strength of the field is $\mathcal{E}$, then the molecule acquires an induced dipole moment of magnitude

$$\mu = \alpha \mathcal{E} \quad (11B.22)$$

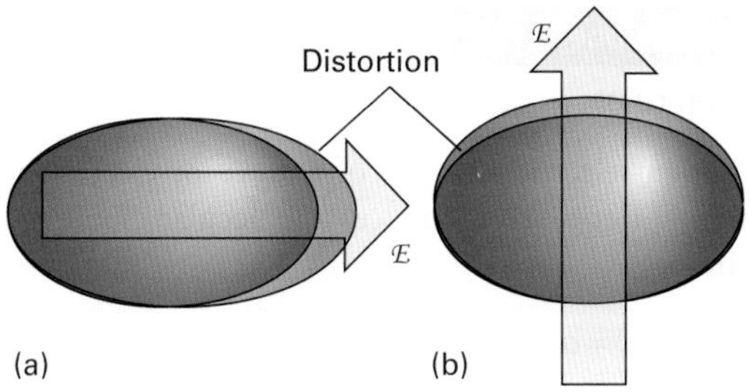

Figure 11B.10 An electric field $\mathcal{E}$ applied to a molecule results in a distortion of its electron density, and the molecule acquires a contribution to its dipole moment (even if it is nonpolar initially). The polarizability may be different when the field is applied (a) parallel or (b) perpendicular to the molecular axis (or, in general, in different directions relative to the molecule); if that is so, then the molecule has an anisotropic polarizability.

in addition to any permanent dipole moment it might have. An atom is isotropically polarizable: that is, the same distortion is induced whatever the direction of the applied field. The polarizability of a spherical rotor is also isotropic. However, non-spherical rotors have polarizabilities that do depend on the direction of the field relative to the molecule, so these molecules are anisotropically polarizable (Fig. 11B.10). The electron distribution in H_2, for example, is more distorted when the field is applied parallel to the bond than when it is applied perpendicular to it, and so $\alpha_{\parallel} > \alpha_{\perp}$.

All linear molecules, including both heteronuclear and homonuclear diatomics, have anisotropic polarizabilities and so are rotationally Raman active. This activity is one reason for the importance of rotational Raman spectroscopy, because the technique can be used to study many of the molecules that are inaccessible to microwave spectroscopy. Spherical rotors such as CH_4 and SF_6, however, are rotationally Raman inactive as well as microwave inactive. This inactivity does not mean that such molecules are never found in rotationally excited states. Molecular collisions do not have to obey such restrictive selection rules, and hence collisions between molecules can result in the population of any rotational state.

As usual, to establish the selection rules, it is necessary to consider the transition dipole moment. The full calculation can be found in *A deeper look* 11B.1, available to read in the e-book accompanying this text, and leads to the conclusion that the specific rotational Raman selection rules are

$$\begin{aligned} &\text{Linear rotors:} && \Delta J = 0, \pm 2 \\ &\text{Symmetric rotors:} && \Delta J = 0, \pm 1, \pm 2 \\ & && \Delta K = 0 \end{aligned} \quad \text{Rotational Raman selection rules} \quad (11B.23)$$

The $\Delta J = 0$ transitions do not lead to a shift in frequency of the scattered photon and therefore contribute to the unshifted radiation (Rayleigh scattering, Topic 11A). A classical argument can be used to give physical insight into the quantum mechanical calculation.

How is that done? 11B.1 Justifying the rotational Raman selection rules

The incident electric field, $\mathcal{E}$, of a wave of electromagnetic radiation of frequency ω_i induces a molecular dipole moment given by

$$\mu_{ind} = \alpha\mathcal{E}(t) = \alpha\mathcal{E}\cos\omega_i t$$

If the molecule is rotating at an angular frequency ω_R, it appears to an external observer that the polarizability is also time dependent (if it is anisotropic). This dependence can be written

$$\alpha = \alpha_0 + \Delta\alpha\cos 2\omega_R t$$

where $\Delta\alpha = \alpha_{\parallel} - \alpha_{\perp}$ and α ranges from $\alpha_0 + \Delta\alpha$ to $\alpha_0 - \Delta\alpha$ as the molecule rotates. The $2\omega_R$ appears because the polarizability returns to its initial value twice each revolution (Fig. 11B.11). Combining these expressions gives

$$\begin{aligned}\mu_{ind} &= (\alpha_0 + \Delta\alpha\cos 2\omega_R t) \times (\mathcal{E}\cos\omega_i t)\\ &= \alpha_0\mathcal{E}\cos\omega_i t + \mathcal{E}\,\Delta\alpha\cos\omega_i t\cos 2\omega_R t\end{aligned}$$

$\cos x\cos y = \tfrac{1}{2}\{\cos(x+y)+\cos(x-y)\}$

$$= \alpha_0\mathcal{E}\cos\omega_i t + \tfrac{1}{2}\mathcal{E}\,\Delta\alpha\{\cos(\omega_i + 2\omega_R)t + \cos(\omega_i - 2\omega_R)t\}$$

This calculation shows that the induced dipole has a component oscillating at the incident frequency (which results in Rayleigh scattering), and that it also has components at $\omega_i \pm 2\omega_R$, which give rise to the shifted Raman lines. These lines appear only if $\Delta\alpha \neq 0$; hence the polarizability must be anisotropic for there to be Raman lines. This is the gross selection rule for rotational Raman spectroscopy.

The distortion induced in the molecule by the incident electric field returns to its initial value after a rotation of 180° (i.e. twice a revolution). This is the classical origin of the specific selection rule $\Delta J = \pm 2$.

To predict the form of the Raman spectrum of a linear rotor the selection rule $\Delta J = \pm 2$ is applied to the rotational energy levels (Fig. 11B.12). For Stokes lines, $\Delta J = +2$ and the scattered radiation is at a lower wavenumber than the incident radiation at $\tilde{\nu}_i$, the shift being the difference $\tilde{F}(J+2) - \tilde{F}(J)$

$$\begin{aligned}\tilde{\nu}(J+2 \leftarrow J) &= \tilde{\nu}_i - \{\tilde{F}(J+2) - \tilde{F}(J)\}\\ &= \tilde{\nu}_i - 2\tilde{B}(2J+3)\end{aligned} \qquad \text{Wavenumbers of Stokes lines: linear rotor} \qquad (11B.24a)$$

For anti-Stokes lines $\Delta J = -2$ and the scattered radiation is at a higher wavenumber, the shift being the difference $\tilde{F}(J) - \tilde{F}(J-2)$:

$$\begin{aligned}\tilde{\nu}(J-2 \leftarrow J) &= \tilde{\nu}_i + \{\tilde{F}(J) - \tilde{F}(J-2)\}\\ &= \tilde{\nu}_i + 2\tilde{B}(2J-1)\end{aligned} \qquad \text{Wavenumbers of anti-Stokes lines: linear rotor} \qquad (11B.24b)$$

The Stokes lines appear to low frequency of the incident radiation and at displacements $6\tilde{B}, 10\tilde{B}, 14\tilde{B},\ldots$ from $\tilde{\nu}_i$ for $J = 0, 1, 2, \ldots$. The anti-Stokes lines appear to high frequency of the incident radiation at displacements of $6\tilde{B}, 10\tilde{B}, 14\tilde{B},\ldots$ from $\tilde{\nu}_i$ for $J = 2, 3, 4, \ldots$. Note that $J = 2$ is the lowest state from which

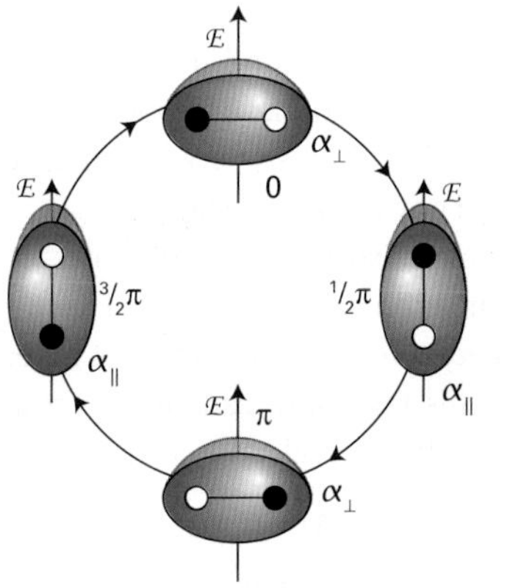

Figure 11B.11 The distortion induced in the electron density of a molecule by an applied electric field returns the polarizability to its initial value after a rotation of only 180° (i.e. twice a revolution). This is the origin of the $\Delta J = \pm 2$ selection rule in rotational Raman spectroscopy.

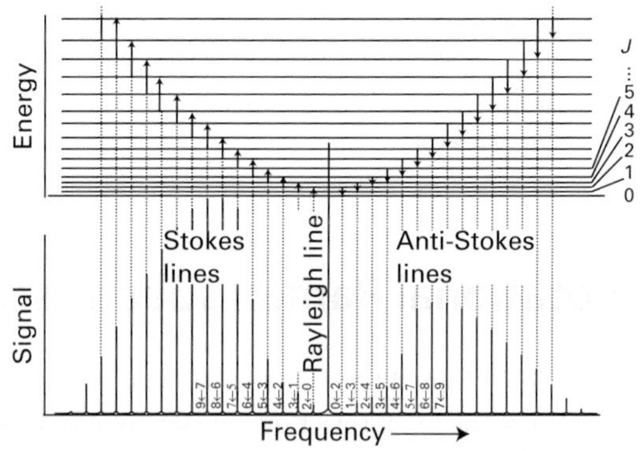

Figure 11B.12 The rotational energy levels of a linear rotor and the transitions allowed by the $\Delta J = \pm 2$ Raman selection rules. The form of a typical rotational Raman spectrum is also shown. In practice the Rayleigh line is much stronger than depicted in the figure.

a transition obeying the selection rule $\Delta J = -2$ can take place. The separation of adjacent lines in both the Stokes and the anti-Stokes regions is $4\tilde{B}$, so from the spacing $I_\perp$ can be determined and then used to find the bond length exactly as in the case of microwave spectroscopy.

Example 11B.4 Predicting the form of a Raman spectrum

Predict the form of the rotational Raman spectrum of $^{14}N_2$, for which $\tilde{B} = 1.99\ cm^{-1}$, when it is exposed to 336.732 nm laser radiation.

Collect your thoughts The molecule is rotationally Raman active because end-over-end rotation modulates its polarizability as viewed by a stationary observer. The wavenumbers of the Stokes and anti-Stokes lines are given by eqn 11B.24.

The solution The incident radiation with wavelength 336.732 nm corresponds to a wavenumber of $\tilde{\nu}_i = 29697.2\ cm^{-1}$; eqns 11B.24a and 11B.24b give the following line positions:

J	0	1	2	3
Stokes lines				
$\tilde{\nu}/cm^{-1}$	29 685.3	29 677.3	29 669.3	29 661.4

J	0	1	2	3
Anti-Stokes lines				
$\tilde{\nu}/cm^{-1}$			29 709.1	29 717.1

There will be a strong central line at 336.732 nm accompanied on either side by lines of increasing and then decreasing intensity (as a result of transition moment and population effects). The spread of the entire spectrum is very small, so the incident light must be highly monochromatic.

Self-test 11B.4 Repeat the calculation for the rotational Raman spectrum of $^{35}Cl_2$ ($\tilde{B} = 0.9752\ cm^{-1}$).

Answer: Stokes lines at 29 691.3, 29 687.4, 29 683.5, 29 679.6 cm^{-1}, anti-Stokes lines at 29 703.1, 29 707.0 cm^{-1}

Exercises

E11B.9 Which of the following molecules may show a pure rotational Raman spectrum: (i) H_2, (ii) HCl, (iii) CH_4, (iv) CH_3Cl?

E11B.10 The wavenumber of the incident radiation in a Raman spectrometer is 20 487 cm^{-1}. What is the wavenumber of the scattered Stokes radiation for the $J = 2 \leftarrow 0$ transition of $^{14}N_2$? Take $\tilde{B} = 1.9987\ cm^{-1}$.

E11B.11 The rotational Raman spectrum of $^{35}Cl_2$ shows a series of Stokes lines separated by 0.9752 cm^{-1} and a similar series of anti-Stokes lines. Calculate the bond length of the molecule.

11B.4 Nuclear statistics and rotational states

If eqn 11B.24 is used to analyse the rotational Raman spectrum of $C^{16}O_2$, the rotational constant derived from the spacing of the lines is inconsistent with other measurements of C–O bond lengths. The results are consistent if it is supposed that the molecule can exist only in states with even values of J, so the observed Stokes lines are $2 \leftarrow 0$, $4 \leftarrow 2, \ldots$; the lines $3 \leftarrow 1$, $5 \leftarrow 3, \ldots$ are missing.

The explanation of the missing lines lies in the Pauli principle (Topic 8B) and the fact that ^{16}O nuclei are spin-0 bosons: just as the Pauli principle excludes certain electronic states, so too does it exclude certain molecular rotational states. The Pauli principle states that, when two identical bosons are exchanged, the overall wavefunction must remain unchanged. When a $C^{16}O_2$ molecule rotates through 180°, two identical ^{16}O nuclei are interchanged, so the overall wavefunction of the molecule must remain unchanged. However, inspection of the form of the rotational wavefunctions (which have the same angular dependence as the s, p, etc. orbitals of atoms) shows that they change sign by $(-1)^J$ under such a rotation (Fig. 11B.13). Therefore, only even values of J are permissible for $C^{16}O_2$, and hence the Raman spectrum shows only alternate lines.

The selective existence of rotational states that stems from the Pauli principle is termed **nuclear statistics**. Nuclear statistics must be taken into account whenever a rotation interchanges equivalent nuclei. However, the consequences are not always as simple as for $C^{16}O_2$ because there are complicating features when the nuclei have non-zero spin: it is found that there are several different relative nuclear spin orientations consistent with even values of J and a different number of spin orientations consistent with odd values of J. For $^1H_2$ and $^{19}F_2$, which have two identical spin-$\frac{1}{2}$ nuclei, by using the Pauli principle it can be shown that there are three times as many ways of achieving a state with odd J than with even J, and there is a corresponding 3:1 alternation in intensity in their rotational Raman spectra (Fig. 11B.14).

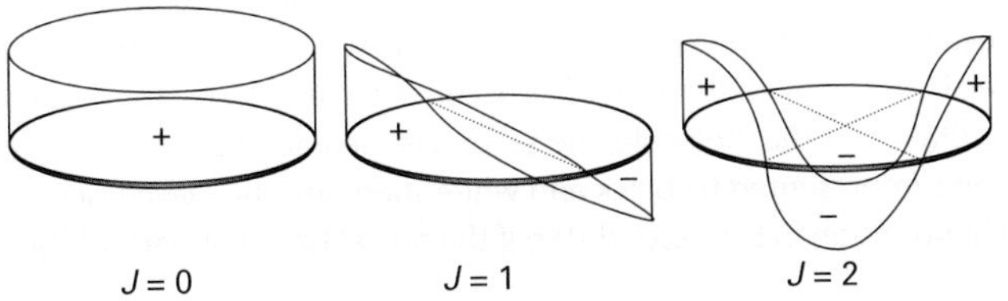

Figure 11B.13 The symmetries of rotational wavefunctions (shown here, for simplicity as a two-dimensional rotor) under a rotation through 180° depend on the value of J. Wavefunctions with J even do not change sign; those with J odd do change sign.

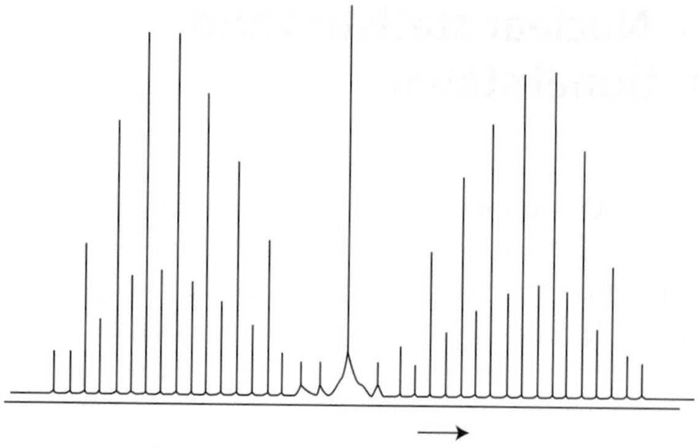

Figure 11B.14 The rotational Raman spectrum of a homonuclear diatomic molecule with two identical spin-$\frac{1}{2}$ nuclei shows an alternation in intensity as a result of nuclear statistics. In practice the Rayleigh line is much stronger than depicted in the figure.

How is that done? 11B.2 Identifying the effect of nuclear statistics

Because ^{1}H nuclei have $I=\frac{1}{2}$, like electrons, they are fermions and the Pauli principle requires the overall wavefunction to change sign under particle interchange. However, the rotation of a $^1H_2$ molecule through 180° has a more complicated effect than simply relabelling the nuclei (Fig. 11B.15).

There are four nuclear spin wavefunctions: three correspond to a total nuclear spin $I_{total}=1$ (parallel spins, ↑↑); and one with $I_{total}=0$ (paired spins, ↑↓). The three wavefunctions with $I_{total}=1$ are $\alpha(A)\alpha(B)$, $\alpha(A)\beta(B)+\alpha(B)\beta(A)$, and $\beta(A)\beta(B)$ with $M_I=+1$, 0, and −1, respectively. Rotation of the molecule through 180° interchanges the labels A and B, but overall these three wavefunctions are unchanged. Therefore, to achieve an overall change of sign, the rotational wavefunction must change sign, and so only odd values of J are allowed.

The fourth wavefunction, with $I_{total}=0$ and $M_I=0$, is $\alpha(A)\beta(B)-\alpha(B)\beta(A)$. When the labels A and B are interchanged the nuclear spin wavefunction changes sign: $\alpha(A)\beta(B)-\alpha(B)\beta(A)\rightarrow\alpha(B)\beta(A)-\alpha(A)\beta(B)\equiv-\{\alpha(A)\beta(B)-\alpha(B)\beta(A)\}$. Therefore, in this case for the overall wavefunction to change sign requires that the rotational wavefunction not change sign. Hence, only even values of J are allowed.

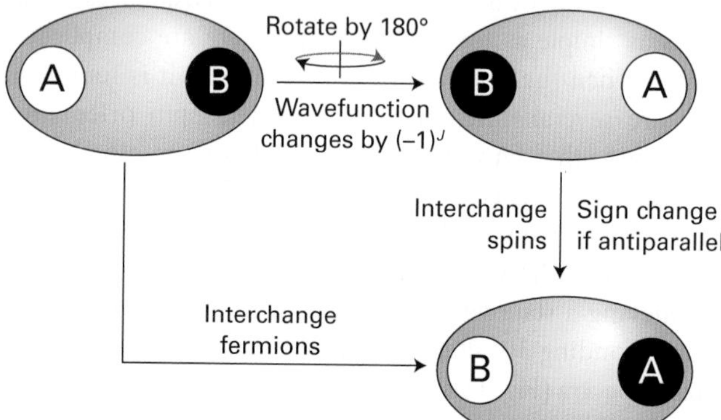

Figure 11B.15 The interchange of two identical fermion nuclei results in the change in sign of the overall wavefunction. The relabelling can be thought of as occurring in two steps: the first is a rotation of the molecule; the second is the interchange of unlike spins (represented by black and white shading). The wavefunction changes sign in the second step if the nuclei have antiparallel spins.

The analysis leads to the conclusion that there are three nuclear spin wavefunctions that can be combined with odd values of J, and one wavefunction that can be combined with even values of J. Therefore, the ratio of the number of ways of achieving odd J to even J is 3:1. In general, for a homonuclear diatomic molecule with nuclei of spin I, the numbers of ways of achieving states of odd and even J are in the ratio

$$\frac{\text{Number of ways of achieving odd } J}{\text{Number of ways of achieving even } J}=\begin{cases}(I+1)/I & \text{for half-integral spin nuclei}\\ I/(I+1) & \text{for integral spin nuclei}\end{cases}$$

Nuclear statistics: homonuclear diatomics (11B.25)

For $^1H_2$, $I=\frac{1}{2}$ and the ratio is 3:1. For $^{14}N_2$, with $I=1$ the ratio is 1:2. Additional complications arise when the electronic state of the molecule is not totally symmetric (as for O_2, Topic 11F).

Nuclear statistics have consequences outside spectroscopy. Different relative nuclear spin orientations change into one another only very slowly, so a $^1H_2$ molecule with parallel nuclear spins remains distinct from one with paired nuclear spins for long periods. The form with parallel nuclear spins is called ***ortho*-hydrogen** and the form with paired nuclear spins is called ***para*-hydrogen**. Because *ortho*-hydrogen cannot exist in a state with $J=0$, it continues to rotate at very low temperatures and has an effective rotational zero-point energy.

Exercise

E11B.12 What is the ratio of weights of populations due to the effects of nuclear statistics for $^{35}Cl_2$?

Checklist of concepts

☐ 1. A **rigid rotor** is a body that does not distort under the stress of rotation.

☐ 2. Rigid rotors are classified as **spherical, symmetric, linear**, or **asymmetric** by noting the number of equal principal moments of inertia (or their symmetry).

☐ 3. Symmetric rotors are classified as **prolate** or **oblate**.

☐ 4. **Centrifugal distortion** arises from forces that change the geometry of a molecule.

☐ 5. The **gross selection rule** for a molecule to give a pure rotational spectrum is that it must be polar.

☐ 6. The **specific selection rules** for microwave spectroscopy are $\Delta J = \pm 1$, $\Delta M_J = 0, \pm 1$; for symmetric rotors the additional rule $\Delta K = 0$ also applies.

☐ 7. A molecule must be **anisotropically polarizable** for it to give rise to rotational Raman scattering.

☐ 8. The **specific selection rules** for rotational Raman spectroscopy are: (i) linear rotors, $\Delta J = 0, \pm 2$; (ii) symmetric rotors, $\Delta J = 0, \pm 1, \pm 2$; $\Delta K = 0$.

☐ 9. The appearance of rotational spectra is affected by **nuclear statistics**, the selective occupation of rotational states that stems from the Pauli principle.

Checklist of equations

Property	Equation	Comment	Equation number
Moment of inertia	$I = \sum_i m_i x_i^2$	x_i is the perpendicular distance of atom i from the axis of rotation	11B.2
Rotational terms of a spherical or linear rotor	$\tilde{F}(J) = \tilde{B}J(J+1)$	$J = 0, 1, 2, \ldots$ $\tilde{B} = \hbar/4\pi c I$	11B.9, 11B.14
Rotational terms of a symmetric rotor	$\tilde{F}(J,K) = \tilde{B}J(J+1) + (\tilde{A} - \tilde{B})K^2$	$J = 0, 1, 2, \ldots$ $K = 0, \pm 1, \ldots, \pm J$	11B.13a
		$\tilde{A} = \hbar/4\pi c I_\parallel$ $\tilde{B} = \hbar/4\pi c I_\perp$	11B.13b
Centrifugal distortion	$\tilde{F}(J) = \tilde{B}J(J+1) - \tilde{D}_J J^2(J+1)^2$	Spherical or linear rotor	11B.15
Centrifugal distortion constant	$\tilde{D}_J = 4\tilde{B}^3/\tilde{\nu}^2$		11B.16
Wavenumbers of rotational transitions	$\tilde{\nu}(J+1 \leftarrow J) = 2\tilde{B}(J+1)$	$J = 0, 1, 2, \ldots$; linear rigid rotors	11B.20a
Rotational state with largest population	$J_{\max} \approx (kT/2hc\tilde{B})^{1/2} - \frac{1}{2}$	Linear rotors	11B.21
Wavenumbers of (i) Stokes and (ii) anti-Stokes lines in the rotational Raman spectrum of linear rotors	(i) $\tilde{\nu}(J+2 \leftarrow J) = \tilde{\nu}_i - 2\tilde{B}(2J+3)$ (ii) $\tilde{\nu}(J-2 \leftarrow J) = \tilde{\nu}_i + 2\tilde{B}(2J-1)$	(i) $J = 0, 1, 2, \ldots$ (ii) $J = 2, 3, 4, \ldots$	11B.24a 11B.24b

TOPIC 11C Vibrational spectroscopy of diatomic molecules

➤ Why do you need to know this material?

The observation of vibrational transition frequencies is used to determine the strengths and rigidities of bonds. Measurements in the gas phase can also be used to calculate the bond lengths of diatomic molecules.

➤ What is the key idea?

The vibrational spectrum of a diatomic molecule can be interpreted by using the harmonic oscillator model, with modifications that account for bond dissociation and the coupling of rotational and vibrational motion.

➤ What do you need to know already?

You need to be familiar with the harmonic oscillator (Topic 7E) and rigid rotor (Topic 11B) models of molecular motion and the general principles of spectroscopy (Topic 11A).

One internal mode of motion of a diatomic molecule is its vibration, in which the internuclear separation increases and decreases periodically. This motion, and the transitions between the allowed quantum states, can be treated initially as an example of harmonic motion like that described in Topic 7E.

11C.1 Vibrational motion

Figure 11C.1 shows a typical potential energy curve of a diatomic molecule (it is essentially a reproduction of Fig. 7E.1 of Topic 7E). The potential energy $V(x)$, where $x = R - R_e$ (the displacement from equilibrium), can be expanded around its minimum by using a Taylor series (see Part 1 of the *Resource section*):

$$V(x) = V(0) + \left(\frac{dV}{dx}\right)_0 x + \frac{1}{2}\left(\frac{d^2V}{dx^2}\right)_0 x^2 + \cdots \qquad (11C.1a)$$

The notation $(\ldots)_0$ means that the derivative is evaluated at $x = 0$. The term $V(0)$ can be set arbitrarily to zero, and the first derivative of V is zero at the minimum. Therefore, the first surviving term is proportional to the square of the displacement. For small displacements all the higher terms can be ignored so the potential energy can be written

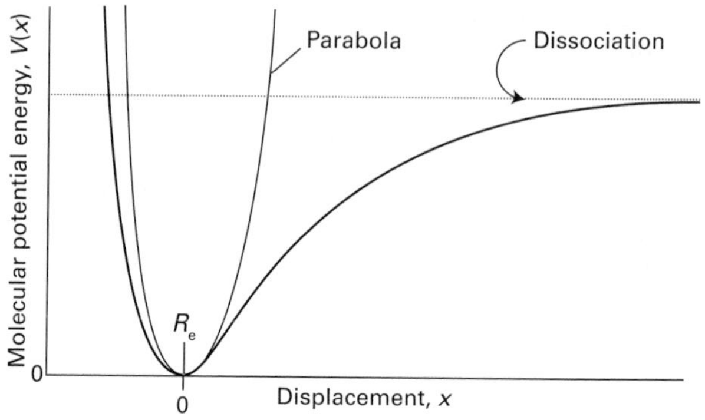

Figure 11C.1 A molecular potential energy curve can be approximated by a parabola near the bottom of the well. The parabolic potential energy results in harmonic oscillations. At high excitation energies the parabolic approximation is poor (the true potential energy is less confining), and is totally wrong near the dissociation limit.

$$V(x) \approx \frac{1}{2}\left(\frac{d^2V}{dx^2}\right)_0 x^2 \qquad (11C.1b)$$

Therefore, the first approximation to a molecular potential energy curve is a parabolic potential of the form

$$V(x) = \frac{1}{2}k_f x^2 \quad x = R - R_e \qquad \text{Parabolic potential energy} \qquad (11C.2a)$$

where k_f is the **force constant** of the bond, a measure of its stiffness:

$$k_f = \left(\frac{d^2V}{dx^2}\right)_0 \qquad \text{Force constant [definition]} \qquad (11C.2b)$$

If the potential energy curve is sharply curved close to its minimum, then k_f will be large and the bond stiff. Conversely, if the potential energy curve is wide and shallow, then k_f will be small and the bond easily stretched or compressed (Fig. 11C.2).

The Schrödinger equation for the relative motion of two atoms of masses m_1 and m_2 with a parabolic potential energy is

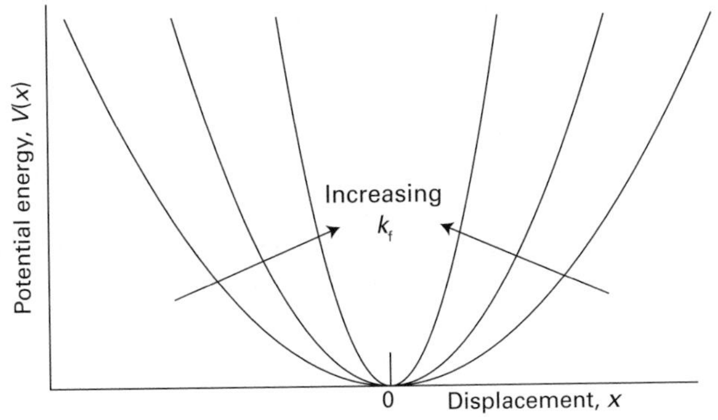

Figure 11C.2 The force constant is a measure of the curvature of the potential energy close to the equilibrium extension of the bond. A strongly confining well (one with steep sides, a stiff bond) corresponds to high values of k_f.

$$-\frac{\hbar^2}{2m_{\text{eff}}}\frac{d^2\psi}{dx^2}+\tfrac{1}{2}k_f x^2\psi = E\psi \qquad (11C.3a)$$

where m_{eff} is the **effective mass**:[1]

$$m_{\text{eff}} = \frac{m_1 m_2}{m_1 + m_2} \qquad \text{Effective mass [definition]} \quad (11C.3b)$$

These equations are derived by using the separation of variables procedure (see *A deeper look* 8A.1, available to read in the e-book accompanying this text) to separate the relative motion of the atoms from the motion of the molecule as a whole.

Apart from the appearance of the effective mass, the Schrödinger equation in eqn 11C.3a is the same as eqn 7E.2 for a particle of mass m undergoing harmonic motion. Therefore, the results from that Topic can be used to write the permitted vibrational energy levels:

$$E_v = \left(v+\tfrac{1}{2}\right)\hbar\omega, \quad \omega = \left(\frac{k_f}{m_{\text{eff}}}\right)^{1/2} \quad v = 0,1,2,\ldots$$

Vibrational energy levels [diatomic molecule] (11C.4a)

The **vibrational terms** of a molecule, the energies of its vibrational states expressed as wavenumbers, are denoted $\tilde{G}(v)$, with $E_v = hc\tilde{G}(v)$. Therefore, with $\omega = 2\pi\nu$ and $\tilde{\nu} = \nu/c = \omega/2\pi c$:

$$\tilde{G}(v) = \left(v+\tfrac{1}{2}\right)\tilde{\nu}, \quad \tilde{\nu} = \frac{1}{2\pi c}\left(\frac{k_f}{m_{\text{eff}}}\right)^{1/2}$$

Vibrational terms [diatomic molecule] (11C.4b)

[1] Distinguish *effective mass* from *reduced mass*. The former is a measure of the mass that is moved during a vibration. The latter is the quantity that emerges from the separation of relative internal and overall translational motion. For a diatomic molecule the two are the same, but that is not true in general for vibrations of polyatomic molecules. Many, however, do not make this distinction and refer to both quantities as the 'reduced mass'.

The vibrational wavefunctions are the same as those discussed in Topic 7E for a harmonic oscillator.

The vibrational terms depend on the *effective* mass of the molecule, not directly on its total mass. This dependence is physically reasonable, because if atom 1 is very much heavier than atom 2, then the effective mass is close to m_2, the mass of the lighter atom. The vibration would then be that of the light atom relative to an essentially stationary heavy atom. For a homonuclear diatomic molecule $m_1 = m_2$, and the effective mass is half the total mass: $m_{\text{eff}} = \tfrac{1}{2}m$.

Brief illustration 11C.1

The force constant of the bond in HCl is 516 N m^{-1}, a reasonably typical value for a single bond. The effective mass of $^1H^{35}Cl$ is 1.63×10^{-27} kg (note that this mass is very close to the mass of the hydrogen atom, 1.67×10^{-27} kg, implying that the H atom is essentially vibrating against a stationary Cl atom). The vibrational frequency is therefore

$$\omega = \left(\frac{516\ \overbrace{\text{N}}^{\text{kg m s}^{-2}}\text{m}^{-1}}{1.63\times10^{-27}\ \text{kg}}\right)^{1/2} = 5.63\times10^{14}\ \text{s}^{-1}$$

and the corresponding wavenumber is

$$\tilde{\nu} = \frac{\omega}{2\pi c} = \frac{5.63\times10^{14}\ \text{s}^{-1}}{2\pi\times(2.998\times10^{10}\ \text{cm s}^{-1})} = 2.99\times10^{3}\ \text{cm}^{-1}$$

Exercises

E11C.1 Calculate the percentage difference in the fundamental vibrational wavenumbers of $^{23}Na^{35}Cl$ and $^{23}Na^{37}Cl$ on the assumption that their force constants are the same. The mass of ^{23}Na is $22.9898m_u$.

E11C.2 The wavenumber of the fundamental vibrational transition of $^{35}Cl_2$ is 564.9 cm^{-1}. Calculate the force constant of the bond.

11C.2 Infrared spectroscopy

The gross and specific selection rules for vibrational transitions are established, as usual, by considering the properties of the electric transition dipole moment. The detailed calculation is shown in *A deeper look* 11C.1, available to read in the e-book accompanying this text. The conclusion is that

> The gross selection rule for a change in vibrational state brought about by absorption or emission of radiation is that the electric dipole moment of the molecule must change when the atoms are displaced relative to one another.

Such vibrations are said to be **infrared active**. The classical basis of this rule is that an oscillating electric dipole generates an electromagnetic wave. Similarly, the oscillating electric field of an incoming electromagnetic wave induces an oscillating dipole in the molecule.

Note that the molecule need not have a permanent dipole moment: the rule requires only a *change* in dipole moment. Some vibrations do not affect the dipole moment of the molecule (for instance, the stretching motion of a homonuclear diatomic molecule), so they neither absorb nor generate radiation: such vibrations are said to be **infrared inactive**. Weak infrared transitions can be observed from homonuclear diatomic molecules trapped within various nanomaterials. For instance, when incorporated into solid C_{60}, H_2 molecules interact through van der Waals forces with the surrounding C_{60} molecules and acquire dipole moments, with the result that they have observable infrared spectra.

The calculation also shows that the specific selection rule is

$$\Delta v = \pm 1 \qquad \text{Specific selection rule [harmonic oscillator]} \qquad (11C.5)$$

Transitions for which $\Delta v = +1$ correspond to absorption and those with $\Delta v = -1$ correspond to emission. It follows that the wavenumbers of allowed vibrational transitions, which are denoted $\Delta\tilde{G}_{v+\frac{1}{2}}$ for the transition $v+1 \leftarrow v$, are

$$\Delta\tilde{G}_{v+\frac{1}{2}} = \tilde{G}(v+1) - \tilde{G}(v) = \tilde{\nu} \qquad \text{Vibrational wavenumbers [harmonic oscillator]} \qquad (11C.6)$$

The wavenumbers of vibrational transitions correspond to radiation in the infrared region of the electromagnetic spectrum, so vibrational transitions absorb and generate infrared radiation.

At room temperature $kT/hc \approx 200\,\text{cm}^{-1}$, and because most vibrational wavenumbers are significantly greater than $200\,\text{cm}^{-1}$ it follows from the Boltzmann distribution that almost all the molecules are in their vibrational ground states. Hence, the dominant spectral transition is the **fundamental transition**, $1 \leftarrow 0$. As a result, the spectrum is expected to consist of a single absorption line.

If the molecules are formed in a vibrationally excited state, such as when vibrationally excited HF molecules are formed in the reaction $H_2 + F_2 \rightarrow 2\,HF^{\star}$, where the star indicates a vibrationally 'hot' molecule, the transitions $5\rightarrow4$, $4\rightarrow3$,… may also appear (in emission). In the harmonic approximation, all these lines lie at the same frequency, and the spectrum is also a single line. However, the breakdown of the harmonic approximation causes the transitions to lie at slightly different frequencies, so several lines are observed.

11C.3 Anharmonicity

The vibrational terms in eqn 11C.4b are only approximate because they are based on a parabolic approximation to the actual potential energy curve. A parabola cannot be correct at all extensions because it does not allow a bond to dissociate. At high vibrational excitations the separation of the atoms (more precisely, the spread of the vibrational wavefunction) allows the molecule to explore regions of the potential energy curve where the parabolic approximation is poor and additional terms in the Taylor expansion of V (eqn 11C.1a) must be retained. The motion then becomes **anharmonic**, in the sense that the restoring force is no longer proportional to the displacement. Because the actual curve is less confining than a parabola, it can be anticipated that the energy levels become more closely spaced at high excitations.

11C.3(a) The convergence of energy levels

One approach to the calculation of the energy levels in the presence of anharmonicity is to use a function that resembles the true potential energy more closely. One example is the **Morse potential energy** function:

$$V(x) = hc\tilde{D}_e(1-e^{-ax})^2 \qquad a = \left(\frac{m_{\text{eff}}\omega^2}{2hc\tilde{D}_e}\right)^{1/2} \qquad \text{Morse potential energy} \qquad (11C.7)$$

At $x = 0$, $V(0) = 0$, and at large positive displacements, $V(x)$ approaches $hc\tilde{D}_e$. Therefore $hc\tilde{D}_e$ can be identified as the depth of the potential energy well (Fig. 11C.3). Near the well minimum the variation of V with displacement resembles a parabola (as can be checked by expanding the exponential and retaining the first two terms). The Schrödinger equation can be solved for the Morse potential energy and the permitted levels are

$$\tilde{G}(v) = \left(v+\tfrac{1}{2}\right)\tilde{\nu} - \left(v+\tfrac{1}{2}\right)^2 x_e\tilde{\nu} \qquad v = 0,1,2,\ldots,v_{\max} \qquad x_e = \frac{a^2\hbar}{2m_{\text{eff}}\omega} = \frac{\tilde{\nu}}{4\tilde{D}_e} \qquad \text{Vibrational terms [Morse potential energy]} \qquad (11C.8)$$

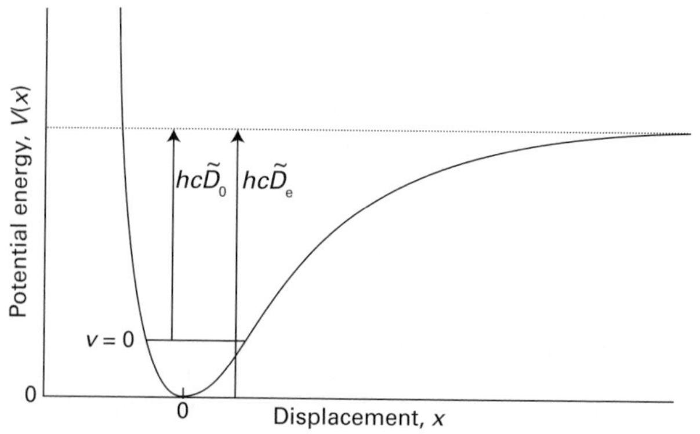

Figure 11C.3 The dissociation energy of a molecule, $hc\tilde{D}_0$, differs from the depth of the potential well, $hc\tilde{D}_e$, on account of the zero-point energy of the vibration of the bond.

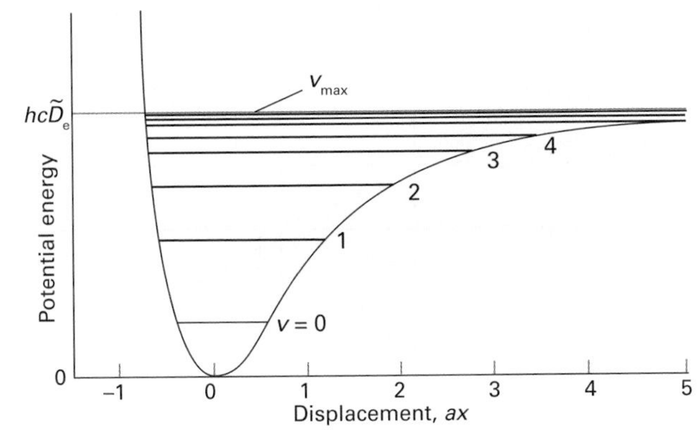

Figure 11C.4 The Morse potential energy curve reproduces the general shape of a molecular potential energy curve. The corresponding Schrödinger equation can be solved, and the values of the energies obtained. The number of bound levels is finite. *Interact with the dynamic version of this graph in the e-book.*

The positive dimensionless parameter x_e is called the **anharmonicity constant**. The number of vibrational levels of a Morse oscillator is finite, as shown in Fig. 11C.4. The second term in the expression for $\tilde{G}$ subtracts from the first with increasing effect as v increases, and hence gives rise to the convergence of the levels at high quantum numbers. The **dissociation energy** $hc\tilde{D}_0$ is the energy difference between the lowest vibrational state ($v = 0$) and the infinitely separated atoms. As can be seen in Fig. 11C.3, the dissociation energy and the well depth are related by $\tilde{D}_e = \tilde{D}_0 + \tilde{G}(0)$.

Although the Morse oscillator is quite useful theoretically, in practice the more general expression

$$\tilde{G}(v) = \left(v+\tfrac{1}{2}\right)\tilde{\nu} - \left(v+\tfrac{1}{2}\right)^2 x_e\tilde{\nu} + \left(v+\tfrac{1}{2}\right)^3 y_e\tilde{\nu} + \cdots \qquad (11C.9a)$$

where x_e, y_e,... are empirical dimensionless constants characteristic of the molecule, is used to fit the experimental data and to determine the dissociation energy of the molecule. In this case the wavenumbers of transitions with $\Delta v = +1$ are

$$\Delta\tilde{G}_{v+\frac{1}{2}} = \tilde{G}(v+1) - \tilde{G}(v) = \tilde{\nu} - 2(v+1)x_e\tilde{\nu} + \cdots \qquad (11C.9b)$$

Equation 11C.9b shows that, because $x_e > 0$, the transitions move to lower wavenumbers as v increases.

In addition to the strong fundamental transition 1←0, a set of weaker absorption lines are also seen and correspond to the transitions 2←0, 3←0,.... These transitions are forbidden for a harmonic oscillator, but become weakly allowed as a result of anharmonicity. The transition 2←0 is known as the **first overtone**, 3←0 is the **second overtone**, and so on. The wavenumber of the overtone transition v←0 is given by

$$\tilde{G}(v) - \tilde{G}(0) = v\tilde{\nu} - v(v+1)x_e\tilde{\nu} + \cdots \qquad (11C.10)$$

The reason for the appearance of overtones is that the selection rule is derived from the properties of harmonic oscillator wavefunctions, which are only approximately valid when anharmonicity is present. Therefore, the selection rule is also only an approximation. For an anharmonic oscillator, all values of Δv are allowed, but transitions with $\Delta v > 1$ are allowed only weakly if the anharmonicity is slight. Typically, the first overtone is only about one-tenth as intense as the fundamental.

Example 11C.1 Estimating an anharmonicity constant

Estimate the anharmonicity constant x_e for $^{35}Cl^{19}F$ given that the wavenumbers of the fundamental and first overtones are found to be 773.8 and 1535.3 cm^{-1}, respectively.

Collect your thoughts You can find an expression for the wavenumber of the fundamental transition 1←0, by using eqn 11C.9b with $v = 0$, and for the wavenumber of the first overtone 2←0 by using eqn 11C.10 with $v = 2$. You then need to solve the two equations to give values for $\tilde{\nu}$ and $x_e\tilde{\nu}$, and hence find x_e itself.

The solution From eqn 11C.9b the expression for the wavenumber of the fundamental is $\tilde{\nu} - 2x_e\tilde{\nu}$, and eqn 11C.10 gives the expression for the wavenumber of the first overtone as $2\tilde{\nu} - 6x_e\tilde{\nu}$. From the data it follows that $773.8\ \text{cm}^{-1} = \tilde{\nu} - 2x_e\tilde{\nu}$ and $1535.3\ \text{cm}^{-1} = 2\tilde{\nu} - 6x_e\tilde{\nu}$. The terms in $\tilde{\nu}$ alone are eliminated by noting that $(\tilde{\nu} - 2x_e\tilde{\nu}) - \tfrac{1}{2}(2\tilde{\nu} - 6x_e\tilde{\nu}) = x_e\tilde{\nu}$ to give

$$x_e\tilde{\nu} = 773.8\ \text{cm}^{-1} - \tfrac{1}{2}\times 1535.3\ \text{cm}^{-1} = 6.15\ \text{cm}^{-1}$$

This value for $x_e\tilde{\nu}$ can then be substituted into $773.8\ \text{cm}^{-1} = \tilde{\nu} - 2x_e\tilde{\nu}$ to give

$$\tilde{\nu} = 773.8\ \text{cm}^{-1} + 2x_e\tilde{\nu} = 773.8\ \text{cm}^{-1} + 2\times 6.15\ \text{cm}^{-1} = 786.1\ \text{cm}^{-1}$$

It follows that

$$x_e = \frac{x_e\tilde{\nu}}{\tilde{\nu}} = \frac{6.15\ \text{cm}^{-1}}{786.1\ \text{cm}^{-1}} = 7.82\times 10^{-3}$$

Self-test 11C.1 Predict the wavenumber of the second overtone for this molecule.

Answer: 2284.5 cm^{-1}

11C.3(b) The Birge–Sponer plot

When several vibrational transitions are detectable, a graphical technique called a **Birge–Sponer plot** can be used to determine the dissociation energy of the bond. The basis of the plot is that the sum of the successive intervals $\Delta\tilde{G}_{v+\frac{1}{2}}$ (eqn 11C.9b) from $v = 0$ to the dissociation limit is the dissociation wavenumber $\tilde{D}_0$:

$$\tilde{D}_0 = \Delta\tilde{G}_{1/2} + \Delta\tilde{G}_{3/2} + \cdots = \sum_v \Delta\tilde{G}_{v+\frac{1}{2}} \qquad (11C.11)$$

just as the height of a ladder is the sum of the separations of its rungs (Fig. 11C.5). The construction in Fig. 11C.6 shows

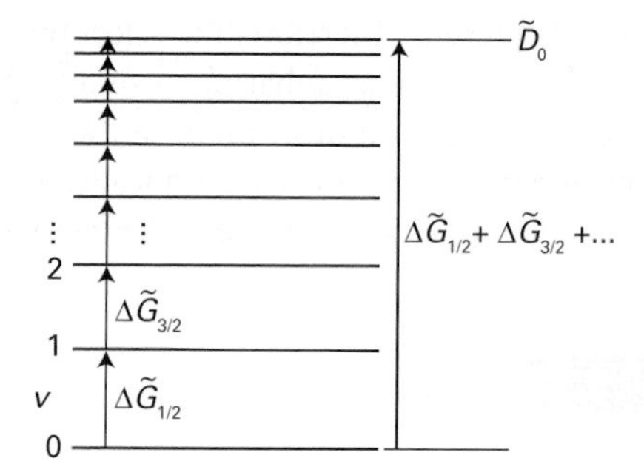

Figure 11C.5 The dissociation wavenumber is the sum of the separations $\Delta\tilde{G}_{v+\frac{1}{2}}$ of the vibrational terms up to the dissociation limit.

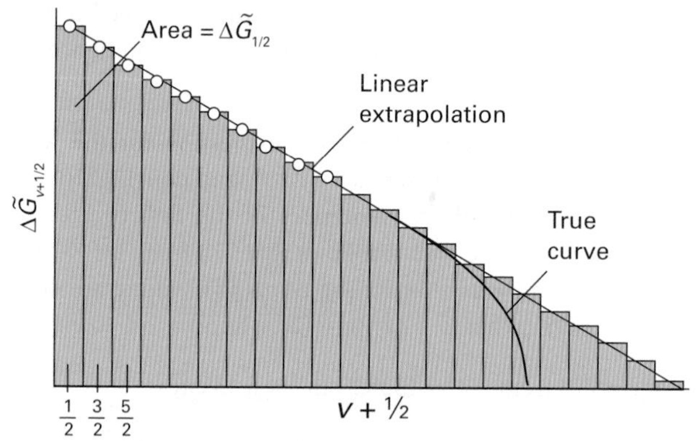

Figure 11C.6 The area under a plot of $\Delta\tilde{G}_{v+\frac{1}{2}}$ against vibrational quantum number is equal to the dissociation wavenumber of the molecule. The assumption that the differences approach zero linearly is the basis of the Birge–Sponer extrapolation.

that the area under the plot of $\Delta\tilde{G}_{v+\frac{1}{2}}$ against $v+\frac{1}{2}$ is equal to the sum, and therefore to $\tilde{D}_0$. The successive terms decrease linearly when only the x_e anharmonicity constant is taken into account. If this is the case, any missing data for high vibrational quantum numbers can be estimated by linear extrapolation. Most actual plots differ from the linear plot as shown in Fig. 11C.6, so the value of $\tilde{D}_0$ obtained in this way is usually an overestimate of the true value.

Example 11C.2 Using a Birge–Sponer plot

The observed vibrational intervals of H_2^+ lie at the following values for 1←0, 2←1,..., respectively (in cm^{-1}): 2191, 2064, 1941, 1821, 1705, 1591, 1479, 1368, 1257, 1145, 1033, 918, 800, 677, 548, 411. Determine the dissociation energy of the molecule.

Collect your thoughts Plot the separations against $v+\frac{1}{2}$, extrapolate linearly to the point cutting the horizontal axis, and then measure the area under the curve.

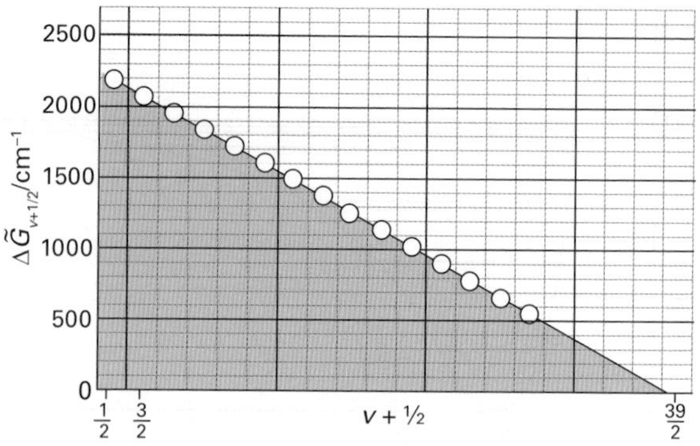

Figure 11C.7 The Birge–Sponer plot used in *Example* 11C.2. The area is obtained simply by counting the squares beneath the line or using the formula for the area of a triangle (area = $\frac{1}{2}$ × base × height).

The solution The points are plotted in Fig. 11C.7, and a linear extrapolation. The area under the curve (use the formula for the area of a triangle or count the squares) is 216. Each square corresponds to 100 cm^{-1} (refer to the scale of the vertical axis); hence the dissociation energy, expressed as a wavenumber, is 21 600 cm^{-1} (corresponding to 258 kJ mol^{-1}).

Self-test 11C.2 The vibrational levels of HgH converge rapidly, and successive intervals are 1203.7 (which corresponds to the 1←0 transition), 965.6, 632.4, and 172 cm^{-1}. Estimate the molar dissociation energy.

Answer: 35.6 kJ mol^{-1}

Exercises

E11C.3 For $^{16}O_2$, $\Delta\tilde{G}$ values for the transitions $v = 1 \leftarrow 0$, $2 \leftarrow 0$, and $3 \leftarrow 0$ are, respectively, 1556.22, 3088.28, and 4596.21 cm^{-1}. Calculate $\tilde{\nu}$ and x_e. Assume y_e to be zero.

E11C.4 The first five vibrational energy levels of HCl are at 1481.86, 4367.50, 7149.04, 9826.48, and 12 399.8 cm^{-1}. Calculate the dissociation energy of the molecule in reciprocal centimetres and electronvolts.

11C.4 Vibration–rotation spectra

Each line of the high resolution vibrational spectrum of a gas-phase heteronuclear diatomic molecule is found to consist of a large number of closely spaced components (Fig. 11C.8). Hence, molecular spectra are often called **band spectra**. The separation between the components is less than 10 cm^{-1}, which suggests that the structure is due to rotational transitions accompanying the vibrational transition. A rotational change should be expected because classically a vibrational transition can be thought of as leading to a sudden increase or decrease in the instantaneous bond length. Just as ice-skaters rotate more rapidly when they bring their arms in, and more slowly when

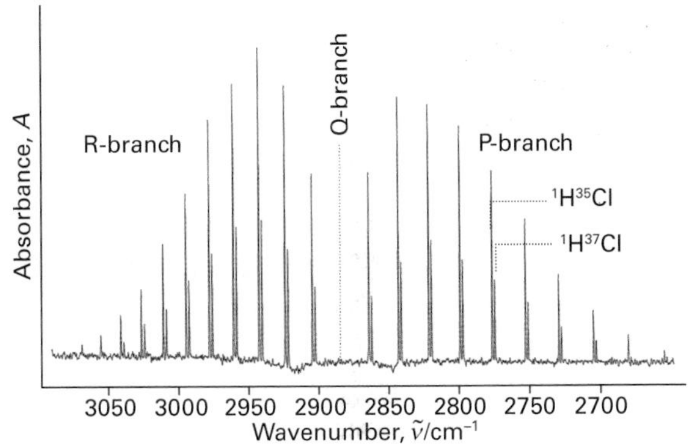

Figure 11C.8 A high-resolution vibration–rotation spectrum of HCl. The lines appear in pairs because $H^{35}Cl$ and $H^{37}Cl$ both contribute (their abundance ratio is 3:1). There is no Q branch (see below), because $\Delta J = 0$ is forbidden for this molecule.

they throw them out, so the molecular rotation is either accelerated or retarded by a vibrational transition.

11C.4(a) Spectral branches

A detailed analysis of the quantum mechanics of simultaneous vibrational and rotational changes shows that the rotational quantum number J changes by ±1 during the vibrational transition of a diatomic molecule. If the molecule also possesses angular momentum about its axis, as in the case of the electronic orbital angular momentum of the molecule NO with its configuration $\ldots\pi^1$, then the selection rules also allow $\Delta J = 0$.

The appearance of the vibration–rotation spectrum of a diatomic molecule can be discussed by using the combined vibration–rotation terms, $\tilde{S}$:

$$\tilde{S}(v,J) = \tilde{G}(v) + \tilde{F}(J) \qquad (11C.12a)$$

If anharmonicity and centrifugal distortion are ignored, $\tilde{G}(v)$ can be replaced by the expression in eqn 11C.4b, and $\tilde{F}(J)$ can be replaced by the expression in eqn 11B.9, $\tilde{F}(J) = \tilde{B}J(J+1)$, to give

$$\tilde{S}(v,J) = (v+\tfrac{1}{2})\tilde{\nu} + \tilde{B}J(J+1) \qquad (11C.12b)$$

In a more detailed treatment, $\tilde{B}$ is allowed to depend on the vibrational state and written $\tilde{B}_v$.

In the vibrational transition $v+1 \leftarrow v$, J changes by ±1 and in some cases by 0 (when $\Delta J = 0$ is allowed). The absorptions then fall into three groups called **branches** of the spectrum. The **P branch** consists of all transitions with $\Delta J = -1$:

$$\tilde{\nu}_P(J) = \tilde{S}(v+1,J-1) - \tilde{S}(v,J) = \tilde{\nu} - 2\tilde{B}J$$
$$J = 1, 2, 3\ldots \qquad \text{P branch transitions} \qquad (11C.13a)$$

This branch consists of lines extending to the low wavenumber side of $\tilde{\nu}$ at $\tilde{\nu}-2\tilde{B}$, $\tilde{\nu}-4\tilde{B}$,... with an intensity distribution reflecting both the populations of the rotational levels and the magnitude of the $J-1 \leftarrow J$ transition moment (Fig. 11C.9). The

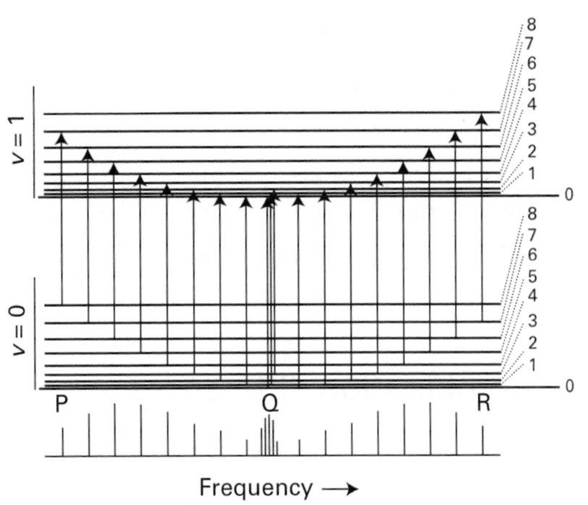

Figure 11C.9 The formation of P, Q, and R branches in a vibration–rotation spectrum. The intensities reflect the populations of the initial rotational levels and magnitudes of the transition moments. If the value of the rotational constant $\tilde{B}$ is independent of the vibrational state, all the Q branch transitions occur at the same frequency, leading to a single spectral line. In most cases the values of $\tilde{B}$ for the $v = 0$ and $v = 1$ states do differ slightly, so the Q branch transitions occur at slightly different frequencies.

Q branch consists of all transitions with $\Delta J = 0$, and its wavenumbers are the same for all values of J:

$$\tilde{\nu}_Q(J) = \tilde{S}(v+1,J) - \tilde{S}(v,J) = \tilde{\nu} \qquad \text{Q branch transitions} \qquad (11C.13b)$$

This branch, when it is allowed (as in NO), appears at the vibrational transition wavenumber $\tilde{\nu}$. In Fig. 11C.8 there is a gap at the expected location of the Q branch because it is forbidden in HCl, which has zero electronic angular momentum around its internuclear axis. The **R branch** consists of lines with $\Delta J = +1$:

$$\tilde{\nu}_R(J) = \tilde{S}(v+1,J+1) - \tilde{S}(v,J) = \tilde{\nu} + 2\tilde{B}(J+1)$$
$$J = 0, 1, 2, \ldots \qquad \text{R branch transitions} \qquad (11C.13c)$$

This branch consists of lines extending to the high-wavenumber side of $\tilde{\nu}$ at $\tilde{\nu}+2\tilde{B}$, $\tilde{\nu}+4\tilde{B}$,....

The separation between the lines in the P and R branches of a vibrational transition gives the value of $\tilde{B}$. Therefore, the bond length can be deduced in the same way as from microwave spectra (Topic 11B). However, the latter technique gives more precise bond lengths because microwave frequencies can be measured with greater precision than infrared frequencies.

Brief illustration 11C.2

The infrared absorption spectrum of $^1H^{81}Br$ contains a band arising from $v = 0$. It follows from eqn 11C.13c and the data in Table 11C.1 that the wavenumber of the line in the R branch originating from the rotational state with $J = 2$ is

$$\tilde{\nu}_R(2) = \tilde{\nu} + 6\tilde{B} = 2648.98\ \text{cm}^{-1} + 6\times(8.465\ \text{cm}^{-1})$$
$$= 2699.77\ \text{cm}^{-1}$$

Table 11C.1 Properties of diatomic molecules*

	$\tilde{\nu}/\text{cm}^{-1}$	R_e/pm	$\tilde{B}_e/\text{cm}^{-1}$	$k_f/(\text{N m}^{-1})$	$\tilde{D}_0/(10^4\ \text{cm}^{-1})$
$^1H_2$	4400	74	60.86	575	3.61
$^1H^{35}Cl$	2991	127	10.59	516	3.58
$^1H^{127}I$	2308	161	6.51	314	2.46
$^{35}Cl_2$	560	199	0.244	323	2.00

* More values are given in the *Resource section.*

11C.4(b) Combination differences

A more detailed analysis of the rotational fine structure shows that the rotational constant decreases as the vibrational quantum number v increases. The origin of this effect is that the average value of $1/R^2$ decreases because the asymmetry of the potential well results in the average bond length increasing with vibrational energy. A harmonic oscillator also shows this effect because although the average value of R is unchanged with increasing v, the average value of $1/R^2$ does change (see Problem P11C.13). Typically, $\tilde{B}_1$ is 1–2 per cent smaller than $\tilde{B}_0$.

The result of $\tilde{B}_1$ being smaller than $\tilde{B}_0$ is that the Q branch (if it is present) consists of a series of closely spaced lines; the lines of the R branch converge slightly as J increases, and those of the P branch diverge. It follows from eqn 11C.12b with $\tilde{B}_v$ in place of $\tilde{B}$

$$\begin{aligned}\tilde{\nu}_P(J) &= \tilde{\nu} - (\tilde{B}_1 + \tilde{B}_0)J + (\tilde{B}_1 - \tilde{B}_0)J^2 \\ \tilde{\nu}_Q(J) &= \tilde{\nu} + (\tilde{B}_1 - \tilde{B}_0)J(J+1) \\ \tilde{\nu}_R(J) &= \tilde{\nu} + (\tilde{B}_1 + \tilde{B}_0)(J+1) + (\tilde{B}_1 - \tilde{B}_0)(J+1)^2\end{aligned} \quad (11C.14)$$

The method of **combination differences** is used to determine the two rotational constants individually. The method involves setting up expressions for the difference in the wavenumbers of transitions to a common state. The resulting expression then depends solely on properties of the other states.

As can be seen from Fig. 11C.10, the transitions $\tilde{\nu}_R(J-1)$ and $\tilde{\nu}_P(J+1)$ have a common upper state, and so the difference between these transitions can be anticipated to depend on $\tilde{B}_0$. From the diagram it can be seen that $\tilde{\nu}_R(J-1) - \tilde{\nu}_P(J+1) = \tilde{S}(0,J+1) - \tilde{S}(0,J-1)$. The right-hand side is evaluated by using the expression for $\tilde{S}(v,J)$ in eqn 11C.12b (with $\tilde{B}_0$ in place of $\tilde{B}$) to give

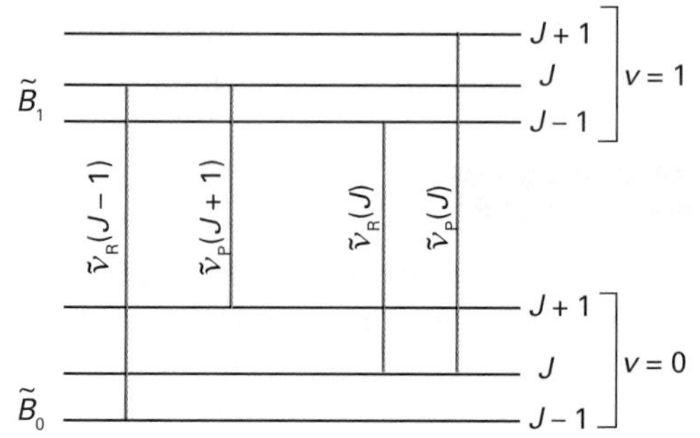

Figure 11C.10 The method of combination differences makes use of the fact that certain pairs of transitions share a common state.

$$\tilde{\nu}_R(J-1) - \tilde{\nu}_P(J+1) = 4\tilde{B}_0\left(J+\tfrac{1}{2}\right) \quad (11C.15a)$$

Therefore, a plot of the combination difference $\tilde{\nu}_R(J-1) - \tilde{\nu}_P(J+1)$ against $J+\frac{1}{2}$ should be a straight line of slope $4\tilde{B}_0$ and intercept (with the vertical axis) zero; the value of $\tilde{B}_0$ can therefore be determined from the slope. The presence of centrifugal distortion results in the intercept deviating from zero, but has little effect on the quality of the straight line.

The two lines $\tilde{\nu}_R(J)$ and $\tilde{\nu}_P(J)$ have a common lower state, and hence their combination difference depends on $\tilde{B}_1$. As before, from Fig. 11C.10 it can be seen that $\tilde{\nu}_R(J) - \tilde{\nu}_P(J) = \tilde{S}(1,J+1) - \tilde{S}(1,J-1)$ which is

$$\tilde{\nu}_R(J) - \tilde{\nu}_P(J) = 4\tilde{B}_1\left(J+\tfrac{1}{2}\right) \quad (11C.15b)$$

Brief illustration 11C.3

The rotational constants of $\tilde{B}_0$ and $\tilde{B}_1$ can be estimated from a calculation involving only a few transitions. For $^1H^{35}Cl$, $\tilde{\nu}_R(0) - \tilde{\nu}_P(2) = 62.6\ \text{cm}^{-1}$, and it follows from eqn 11C.15a, with J = 1, that $\tilde{B}_0 = 62.6/4\left(1+\frac{1}{2}\right)\text{cm}^{-1} = 10.4\ \text{cm}^{-1}$. Similarly, $\tilde{\nu}_R(1) - \tilde{\nu}_P(1) = 60.8\ \text{cm}^{-1}$, and it follows from eqn 11C.15b, again with J = 1 that $\tilde{B}_1 = 60.8/4\left(1+\frac{1}{2}\right)\text{cm}^{-1} = 10.1\ \text{cm}^{-1}$. If more lines are used to make combination difference plots, the values $\tilde{B}_0 = 10.440\ \text{cm}^{-1}$ and $\tilde{B}_1 = 10.136\ \text{cm}^{-1}$ are found. The two rotational constants differ by about 3 per cent of $\tilde{B}_0$.

It is common to assume that the rotational constants vary linearly with the vibrational quantum number v according to

$$\tilde{B}_v = \tilde{B}_e - \tilde{a}\left(v+\tfrac{1}{2}\right) \quad (11C.16)$$

where $\tilde{B}_e$ is the rotational constant for a molecule in which the atoms are separated by the equilibrium bond length R_e.

Exercise

E11C.5 Infrared absorption by $^1H^{127}I$ gives rise to an R branch from $v = 0$. What is the wavenumber of the line originating from the rotational state with $J = 2$? *Hint:* Use data from Table 11C.1.

11C.5 Vibrational Raman spectra

The gross and specific selection rules for vibrational Raman transitions are established, as usual, by considering the appropriate transition moment. The details are set out in *A deeper look* 11C.1, available to read in the e-book accompanying this text. The conclusion is that

> The gross selection rule for vibrational Raman transitions is that the polarizability must change as the molecule vibrates.

The polarizability plays a role in vibrational Raman spectroscopy because the molecule must be squeezed and stretched by the incident radiation in order for vibrational excitation to occur during the inelastic photon–molecule collision. Both homonuclear and heteronuclear diatomic molecules swell and contract during a vibration, the control of the nuclei over the electrons varies, and hence the molecular polarizability changes. Both types of diatomic molecule are therefore vibrationally Raman active. The analysis also shows that the specific selection rule for vibrational Raman transitions in the harmonic approximation is $\Delta v = \pm 1$, just as for infrared transitions.

The lines to high frequency of the incident radiation, in the language introduced in Topic 11A the 'anti-Stokes lines', are those for which $\Delta v = -1$. The lines to low frequency, the 'Stokes lines', correspond to $\Delta v = +1$. The intensities of the anti-Stokes and Stokes lines are governed largely by the Boltzmann populations of the vibrational states involved in the transition. It follows that anti-Stokes lines are usually weak because the populations of the excited vibrational states are very small.

In gas-phase spectra, the Stokes and anti-Stokes lines have a branch structure arising from the simultaneous rotational transitions that accompany the vibrational excitation (Fig. 11C.11). The selection rules are $\Delta J = 0, \pm 2$ (as in pure rotational Raman spectroscopy), and give rise to the **O branch** ($\Delta J = -2$), the **Q branch** ($\Delta J = 0$), and the **S branch** ($\Delta J = +2$):

$$\tilde{\nu}_O(J) = \tilde{\nu}_i - \tilde{\nu} + 4\tilde{B}\left(J - \tfrac{1}{2}\right) \quad J = 2, 3, 4, \ldots \qquad \text{O branch transitions} \quad (11C.17a)$$

$$\tilde{\nu}_Q(J) = \tilde{\nu}_i - \tilde{\nu} \qquad \text{Q branch transitions} \quad (11C.17b)$$

$$\tilde{\nu}_S(J) = \tilde{\nu}_i - \tilde{\nu} - 4\tilde{B}\left(J + \tfrac{3}{2}\right) \quad J = 0, 1, 2, \ldots \qquad \text{S branch transitions} \quad (11C.17c)$$

where $\tilde{\nu}_i$ is the wavenumber of the incident radiation. Note that, unlike in infrared spectroscopy, a Q branch is obtained for all linear molecules. The spectrum of CO, for instance, is shown in Fig. 11C.12: rather than being a single line, as implied by eqn 11C.17, the Q branch appears as a broad feature. Its breadth arises from the presence of several overlapping lines arising from the difference in rotational constants of the upper and lower vibrational states.

The information available from vibrational Raman spectra adds to that from infrared spectroscopy because homonuclear diatomics can also be studied. The spectra can be interpreted in terms of the force constants, dissociation energies, and bond lengths, and some of the information obtained is included in Table 11C.1.

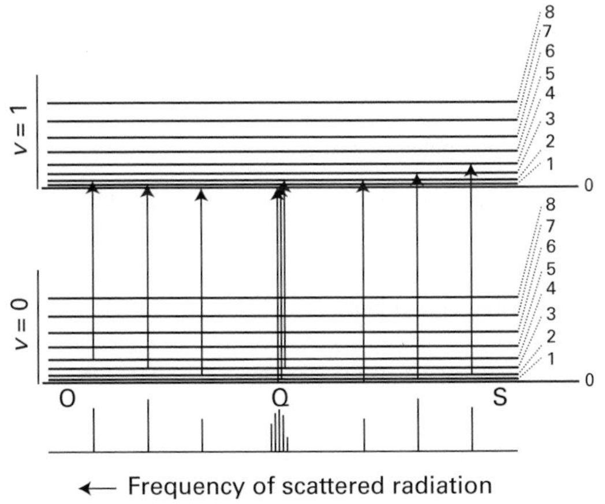

Figure 11C.11 The formation of O, Q, and S branches in a vibration–rotation Raman spectrum of a diatomic molecule (its Stokes lines). Note that the frequency scale runs in the opposite direction to that in Fig. 11C.9, because the higher energy transitions (on the right) extract more energy from the incident beam, resulting in scattered photons with lower frequencies.

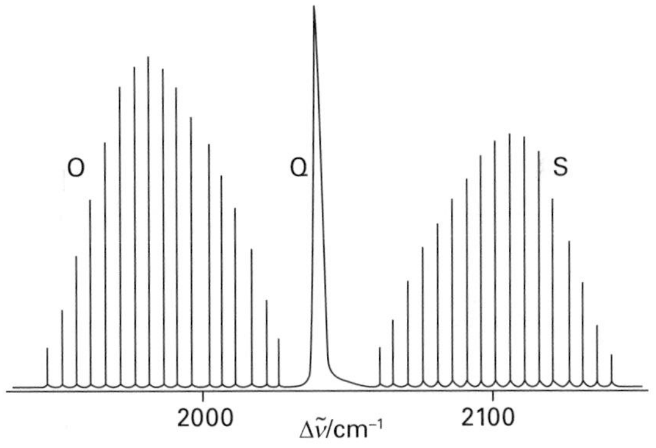

Figure 11C.12 The structure of a vibrational line in the vibrational Raman spectrum (the Stokes lines) of carbon monoxide, showing the O, Q, and S branches. The horizontal axis represents the wavenumber difference between the incident and scattered radiation. For these Stokes lines the wavenumber of the scattered radiation (as distinct from the difference, which represents the energy deposited in the molecule) increases to the left, as in Fig. 11C.11.

Checklist of concepts

☐ 1. The vibrational energy levels of a diatomic molecule modelled as a harmonic oscillator depend on the **force constant** k_f (a measure of the stiffness of the bond) and the **effective mass** of the vibration.

☐ 2. The **gross selection rule** for infrared spectra is that the electric dipole moment of the molecule must depend on the bond length.

☐ 3. The **specific selection rule** for infrared spectra (within the harmonic approximation) is $\Delta v = \pm 1$.

☐ 4. The **Morse potential energy** can be used to model anharmonic vibration.

☐ 5. The strongest infrared transitions are the **fundamental transitions** ($v = 1 \leftarrow v = 0$).

☐ 6. Anharmonicity gives rise to weaker **overtone transitions** ($v = 2 \leftarrow v = 0$, $v = 3 \leftarrow v = 0,\ldots$).

☐ 7. A **Birge–Sponer** plot may be used to determine the dissociation energy of a diatomic molecule.

☐ 8. In the gas phase, vibrational transitions have a **P, R branch structure** due to simultaneous rotational transitions; some molecules also have a **Q branch**.

☐ 9. For a vibration to be **Raman active**, the polarizability must change as the molecule vibrates.

☐ 10. The **specific selection rule** for vibrational Raman spectra (within the harmonic approximation) is $\Delta v = \pm 1$.

☐ 11. In gas-phase spectra, the Stokes and anti-Stokes lines in a Raman spectrum have an **O, Q, S branch structure.**

Checklist of equations

Property	Equation	Comment	Equation number
Vibrational terms	$\tilde{G}(v) = (v+\tfrac{1}{2})\tilde{\nu},\ \tilde{\nu} = (1/2\pi c)(k_f/m_{\text{eff}})^{1/2}$ $m_{\text{eff}} = m_1 m_2/(m_1+m_2)$	Diatomic molecules; harmonic approximation	11C.4b
Infrared spectra (vibrational)	$\Delta\tilde{G}_{v+\frac{1}{2}} = \tilde{\nu}$	Diatomic molecules; harmonic approximation	11C.6
Morse potential energy	$V(x) = hc\tilde{D}_e(1-\mathrm{e}^{-ax})^2$ $a = (m_{\text{eff}}\omega^2/2hc\tilde{D}_e)^{1/2}$		11C.7
Vibrational terms (diatomic molecules)	$\tilde{G}(v) = (v+\tfrac{1}{2})\tilde{\nu} - (v+\tfrac{1}{2})^2 x_e\tilde{\nu},\ x_e = \tilde{\nu}/4\tilde{D}_e$	Morse potential energy	11C.8
Infrared spectra (vibrational)	$\Delta\tilde{G}_{v+\frac{1}{2}} = \tilde{\nu} - 2(v+1)x_e\tilde{\nu} + \cdots$	Anharmonic oscillator	11C.9b
	$\tilde{G}(v) - \tilde{G}(0) = v\tilde{\nu} - v(v+1)x_e\tilde{\nu} + \cdots$	Overtones	11C.10
Dissociation wavenumber	$\tilde{D}_0 = \Delta\tilde{G}_{1/2} + \Delta\tilde{G}_{3/2} + \cdots = \sum_v \Delta\tilde{G}_{v+\frac{1}{2}}$	Birge–Sponer plot	11C.11
Vibration–rotation terms (diatomic molecules)	$\tilde{S}(v,J) = (v+\tfrac{1}{2})\tilde{\nu} + \tilde{B}J(J+1)$	Rotation coupled to vibration	11C.12b
Infrared spectra (vibration–rotation)	$\tilde{\nu}_P(J) = \tilde{S}(v+1,J-1) - \tilde{S}(v,J) = \tilde{\nu} - 2\tilde{B}J$ $J = 1,2,3\ldots$	P branch ($\Delta J = -1$)	11C.13a
	$\tilde{\nu}_Q(J) = \tilde{S}(v+1,J) - \tilde{S}(v,J) = \tilde{\nu}$	Q branch ($\Delta J = 0$)	11C.13b
	$\tilde{\nu}_R(J) = \tilde{S}(v+1,J+1) - \tilde{S}(v,J) = \tilde{\nu} + 2\tilde{B}(J+1)$ $J = 0,1,2,\ldots$	R branch ($\Delta J = +1$)	11C.13c
	$\tilde{\nu}_R(J-1) - \tilde{\nu}_P(J+1) = 4\tilde{B}_0(J+\tfrac{1}{2})$ $\tilde{\nu}_R(J) - \tilde{\nu}_P(J) = 4\tilde{B}_1(J+\tfrac{1}{2})$	Combination differences	11C.15
Raman spectra (vibration–rotation)	$\tilde{\nu}_O(J) = \tilde{\nu}_i - \tilde{\nu} + 4\tilde{B}(J-\tfrac{1}{2})\quad J = 2, 3, 4, \ldots$	O branch ($\Delta J = -2$)	11C.17a
	$\tilde{\nu}_Q(J) = \tilde{\nu}_i - \tilde{\nu}$	Q branch ($\Delta J = 0$)	11C.17b
	$\tilde{\nu}_S(J) = \tilde{\nu}_i - \tilde{\nu} - 4\tilde{B}(J+\tfrac{3}{2})\quad J = 0, 1, 2, \ldots$	S branch ($\Delta J = +2$)	11C.17c

TOPIC 11D Vibrational spectroscopy of polyatomic molecules

➤ Why do you need to know this material?

The analysis of vibrational spectra is a widely used analytical technique that provides information about the identity and shapes of polyatomic molecules in the gas and condensed phases.

➤ What is the key idea?

The vibrational spectrum of a polyatomic molecule can be interpreted in terms of its normal modes.

➤ What do you need to know already?

You need to be familiar with the harmonic oscillator (Topic 7E), the general principles of spectroscopy (Topic 11A), and the selection rules for vibrational infrared and Raman spectroscopy (Topic 11C).

There is only one mode of vibration for a diatomic molecule: the periodic stretching and compression of the bond. In polyatomic molecules there are many bond lengths and angles that can change, and as a result the vibrational motion of the molecule is very complex. Some order can be brought to this complexity by introducing the concept of 'normal modes'.

11D.1 Normal modes

The first step in the analysis of the vibrations of a polyatomic molecule is to calculate the total number of vibrational modes.

How is that done? 11D.1 Counting the number of vibrational modes

The total number of coordinates needed to specify the locations of N atoms is $3N$. Each atom may change its location by varying each of its three coordinates (x, y, and z), so the total number of displacements available is $3N$. These displacements can be grouped together in a physically sensible way. For example, three coordinates are needed to specify the location of the centre of mass of the molecule, so three of the $3N$ displacements correspond to the translational motion of the molecule as a whole. The remaining $3N - 3$ displacements are 'internal' modes of the molecule.

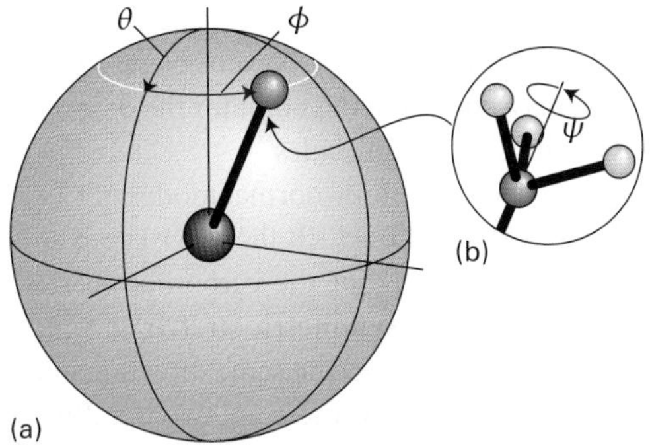

Figure 11D.1 (a) The orientation of a linear molecule requires the specification of two angles. (b) The orientation of a nonlinear molecule requires the specification of three angles.

Two angles are needed to specify the orientation of a linear molecule in space: in effect, only the latitude and longitude of the direction in which the molecular axis is pointing need be specified (Fig. 11D.1a). However, three angles are needed for a nonlinear molecule because the orientation of the molecule around the direction defined by the latitude and longitude also needs to be specified (Fig. 11D.1b). Therefore, for linear molecules two of the $3N - 3$ internal displacements are rotational, whereas for nonlinear molecules three of the displacements are rotational. That leaves $3N - 5$ (linear) or $3N - 6$ (nonlinear) non-rotational internal displacements of the atoms: these are the vibrational modes. It follows that the number of modes of vibration is:

$$\text{Linear molecule:} \quad N_{\text{vib}} = 3N - 5$$
$$\text{Nonlinear molecule:} \quad N_{\text{vib}} = 3N - 6 \qquad (11\text{D}.1)$$

Numbers of vibrational modes

Brief illustration 11D.1

Water, H_2O, is a nonlinear triatomic molecule with $N = 3$, and so has $3N - 6 = 3$ modes of vibration; CO_2 is a linear triatomic molecule, and has $3N - 5 = 4$ modes of vibration. Methylbenzene has 15 atoms and 39 modes of vibration.

The greatest simplification of the description of vibrational motion is obtained by analysing it in terms of 'normal modes'. A **normal mode** is a vibration of the molecule in which the centre of mass remains fixed, the orientation is unchanged, and the atoms move synchronously. When a normal mode is excited, the energy remains in that mode and does not migrate into other normal modes of the molecule.

A normal mode analysis is possible only if it is assumed that the potential energy is parabolic (as in a harmonic oscillator, Topic 11C). In reality, the potential energy is not parabolic, the vibrations are anharmonic (Topic 11C), and the normal modes are not completely independent. Nevertheless, a normal mode analysis remains a good starting point for the description of the vibrations of polyatomic molecules.

Figure 11D.2 shows the four normal modes of CO_2. Mode ν_1 is the **symmetric stretch** in which the two oxygen atoms move in and out synchronously but the carbon atom remains stationary. Mode ν_2, the **antisymmetric stretch**, in which the two oxygen atoms always move in the same direction and opposite to that of the carbon. Finally, there are two **bending modes** ν_3 in which the oxygen atoms move perpendicular to the internuclear axis in one direction and the carbon atom moves in the opposite direction: this bending motion can take place in either of two perpendicular planes. In all these modes, the position of the centre of mass and orientation of the molecule are unchanged by the vibration.

In the harmonic approximation, each normal mode, q, behaves like an independent harmonic oscillator and has an energy characterized by the quantum number v_q. Expressed as a wavenumber, these terms are

$$\tilde{G}_q(v) = \left(v_q + \tfrac{1}{2}\right)\tilde{\nu}_q, \quad v_q = 0,1,2,\ldots \quad \tilde{\nu}_q = \frac{1}{2\pi c}\left(\frac{k_{\mathrm{f},q}}{m_q}\right)^{1/2}$$

Vibrational terms of normal modes [harmonic oscillator] (11D.2)

where $\tilde{\nu}_q$ is the wavenumber of mode q; this quantity depends on the force constant $k_{\mathrm{f},q}$ for the mode and on the effective mass m_q of the mode: stiff bonds and low effective masses correspond to high wavenumber and high frequency vibrations. The effective mass of the mode is a measure of the mass that moves in the vibration and in general is a combination of the masses of the atoms. For example, in the symmetric stretch of CO_2, the carbon atom is stationary, and the effective mass depends on the masses of only the oxygen atoms. In the antisymmetric stretch and in the bends all three atoms move, so the masses of all three atoms contribute (but to different extents) to the effective mass of each mode.

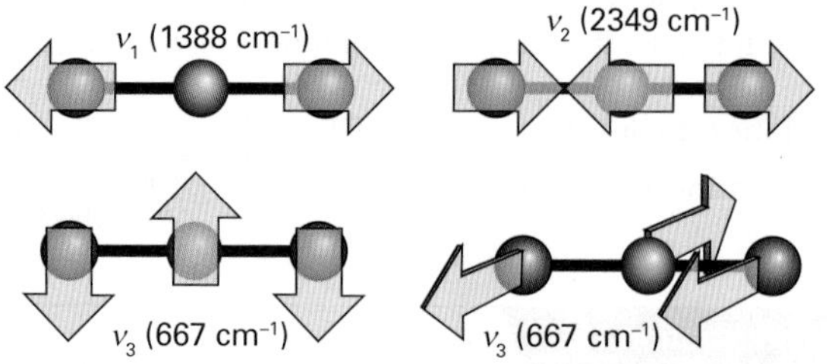

Figure 11D.2 The four normal modes of CO_2. The two bending motions (ν_3) have the same vibrational frequency. In this representation, an arrow indicates displacement in one direction, but the atoms oscillate back and forth about their equilibrium positions.

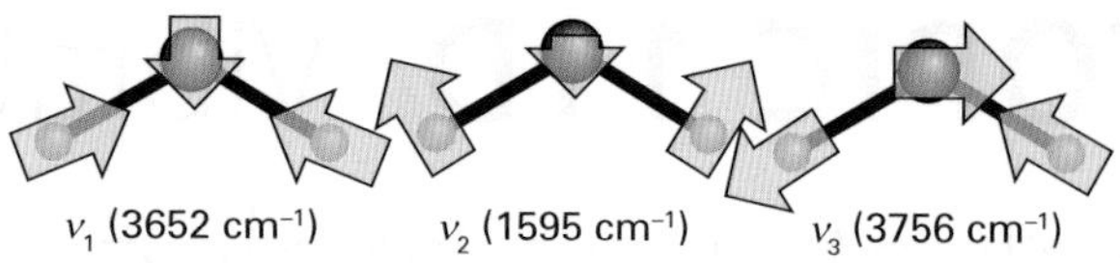

Figure 11D.3 The three normal modes of H_2O. The mode ν_2 is predominantly bending, and occurs at lower wavenumber than the other two.

The three normal modes of H_2O are shown in Fig. 11D.3: note that the predominantly bending mode (ν_2) has a lower frequency (and wavenumber) than the others, which are predominantly stretching modes. It is generally the case that the frequencies of bending motions are lower than those of stretching modes. Only in special cases (such as the CO_2 molecule) are the normal modes purely stretches or purely bends. In general, a normal mode is a composite motion involving simultaneous stretching of bonds and changes to bond angles. In a given normal mode, heavy atoms generally move less than light atoms.

The vibrational state of a polyatomic molecule is specified by the vibrational quantum number v_q for each of the normal modes. For example, for H_2O with three normal modes, the vibrational state is designated (v_1,v_2,v_3). The vibrational ground state of an H_2O molecule is therefore (0,0,0); the state (0,1,0) implies that modes 1 and 3 are in their ground states, and mode 2 is in the first excited state.

Exercises

E11D.1 How many normal modes of vibration are there for the following molecules: (i) H_2O, (ii) H_2O_2, (iii) C_2H_4?

E11D.2 Write an expression for the vibrational term for the ground vibrational state of H_2O in terms of the wavenumbers of the normal modes. Neglect anharmonicities, as in eqn 11D.2.

11D.2 Infrared absorption spectra

The gross selection rule for infrared activity is a straightforward generalization of the rule for diatomic molecules (Topic 11C):[1]

> The motion corresponding to a normal mode must be accompanied by a change of electric dipole moment.

[1] Topic 11E describes a systematic procedure based on group theory for deciding whether a vibrational mode is infrared active.

Simple inspection of atomic motions is sometimes all that is needed in order to assess whether a normal mode is infrared active. For example, the symmetric stretch of CO_2 leaves the dipole moment unchanged (at zero, see Fig. 11D.2), so this mode is infrared inactive. The antisymmetric stretch, however, changes the dipole moment because the molecule becomes unsymmetrical as it vibrates, so this mode is infrared active. Because the dipole moment change is parallel to the principal axis, the transitions arising from this mode are classified as **parallel bands** in the spectrum. Both bending modes are infrared active: they are accompanied by a changing dipole perpendicular to the principal axis, so transitions involving them lead to a **perpendicular band** in the spectrum.

Example 11D.1 Using the gross selection rule for infrared spectroscopy

State which of the following molecules are infrared active: N_2O, OCS, H_2O, $CH_2{=}CH_2$.

Collect your thoughts Molecules that are infrared active have a normal mode (or modes) in which there is a change in dipole moment during the course of the motion. Therefore, to decide if a molecule is infrared active you need to decide whether there is any distortion of the molecule that results in a change in its electric dipole moment (including changes from zero).

The solution The linear molecules N_2O and OCS both have permanent electric dipole moments that change as a result of stretching any of the bonds; in addition, bending perpendicular to the internuclear axis results in a dipole moment in that direction: both molecules are therefore infrared active. An H_2O molecule also has a permanent dipole moment which changes either by stretching the bonds or by altering the bond angle: the molecule is infrared active. A $CH_2{=}CH_2$ molecule does not have a permanent dipole moment (it possesses a centre of inversion) but there are vibrations in which the symmetry is reduced and a dipole moment forms, for example, by stretching the two C–H bonds on one carbon atom and simultaneously compressing the two C–H bonds on the other carbon atom.

Self-test 11D.1 Identify an infrared inactive normal mode of $CH_2{=}CH_2$.

Answer: A 'breathing' mode in which all the C–H bonds contract and stretch synchronously

The specific selection rule in the harmonic approximation is $\Delta v_q = \pm 1$. In this approximation the quantum number of only one active mode can change in the interaction of a molecule with a photon. A **fundamental transition** is a transition from the ground state of the molecule to the next higher energy level of the specified mode. For example, in H_2O there are three such fundamentals corresponding to the excitation of each of the three normal modes: $(1,0,0) \leftarrow (0,0,0)$, $(0,1,0) \leftarrow (0,0,0)$, and $(0,0,1) \leftarrow (0,0,0)$.

Anharmonicity also allows transitions in which more than one quantum of excitation takes place: such transitions are referred to as **overtones**. A transition such as $(0,0,2) \leftarrow (0,0,0)$ in H_2O is described as a **first overtone**, and a transition such as $(0,0,3) \leftarrow (0,0,0)$ is a **second overtone** of the mode ν_3. **Combination bands** (or *combination lines*) corresponding to the simultaneous excitation of more than one normal mode in the transition, as in $(1,1,0) \leftarrow (0,0,0)$, are also possible in the presence of anharmonicity.

As for diatomic molecules (Topic 11C), transitions between vibrational levels can be accompanied by simultaneous changes in rotational state, so giving rise to band spectra rather than the single absorption line of a pure vibrational transition. The spectra of linear polyatomic molecules show branches similar to those of diatomic molecules. For nonlinear molecules, the rotational fine structure is considerably more complex and difficult to analyse: even in moderately complex molecules the presence of several normal modes gives rise to several fundamental transitions, many overtones, and many combination lines, each with associated rotational fine structure, and results in infrared spectra of considerable complexity.

These complications are eliminated (or at least concealed) by recording infrared spectra of samples in the condensed phase (liquids, solutions, or solids). Molecules in liquids do not rotate freely but collide with each other after only a small change of orientation. As a result, the lifetimes of rotational states in liquids are very short, which results in a broadening of the associated energies (Topic 11A). Collisions occur at a rate of about $10^{13}\,s^{-1}$ and, even allowing for only a 10 per cent success rate in changing the molecule into another rotational state, a lifetime broadening of more than $1\,cm^{-1}$ can easily result. The rotational structure of the vibrational spectrum is blurred by this effect, so the infrared spectra of molecules in condensed phases usually consist of bands without any resolved branch structure.

Infrared spectroscopy is commonly used in routine chemical analysis, most usually on samples in solution, prepared as a fine dispersion (a 'mull'), or compressed as solids into a very thin layer. The resulting spectra show many absorption bands, even for moderately complex molecules. There is no chance of analysing such complex spectra in terms of the normal modes. However, they are of great utility for it turns out that certain groups within a molecule (such as a carbonyl group or an $-NH_2$ group) give rise to absorption bands in a particular range of wavenumbers. The spectra of a very large number of molecules have been recorded and these data have been used to draw up charts and tables of the expected range of the wavenumbers of absorptions from different groups. Comparison of the features in the spectrum of an unknown molecule or the product of a chemical reaction with entries in these data tables is a common first step towards identifying the molecule.

Exercises

E11D.3 Which of the vibrations of an AB_2 molecule are infrared active when it is (i) angular, (ii) linear?

E11D.4 Consider the vibrational mode that corresponds to the uniform expansion of the benzene ring. Is it infrared active?

11D.3 Vibrational Raman spectra

As for diatomic molecules, the normal modes of vibration of molecules are Raman active if they are accompanied by a changing polarizability. A closer analysis of infrared and Raman activity of normal modes based on considerations of symmetry leads to the **exclusion rule**:

> If the molecule has a centre of inversion, then no mode can be both infrared and Raman active.

(A mode may be inactive in both.) Because it is often possible to judge intuitively if a mode changes the molecular dipole moment, this rule can be used to identify modes that are not Raman active.

Brief illustration 11D.2

The symmetric stretch of CO_2 alternately swells and contracts the molecule: this motion changes the size and hence the polarizability of the molecule, so the mode is Raman active. Because CO_2 has a centre of inversion the exclusion rule applies, so the stretching mode cannot be infrared active. The antisymmetric stretch and the two bends are infrared active, and so cannot be Raman active.

The assignment of Raman lines to particular vibrational modes is aided by noting the state of polarization of the scattered light. The **depolarization ratio**, ρ, of a line is the ratio of the intensities of the scattered light with polarization perpendicular and parallel, $I_\perp$ and $I_\parallel$, to the plane of the incident radiation (Fig. 11D.4):

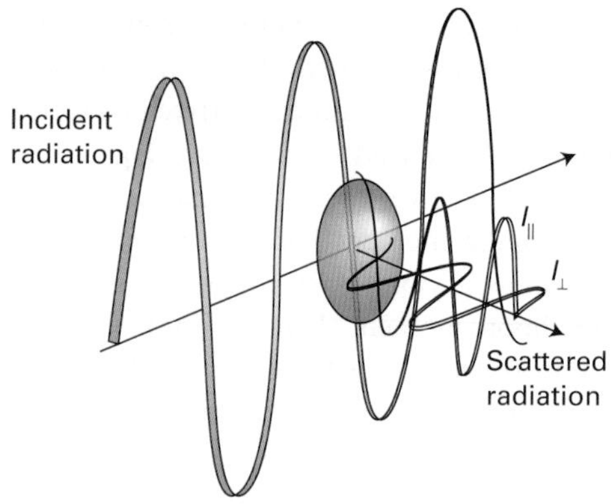

Figure 11D.4 The definition of the planes used for the specification of the depolarization ratio, ρ, in Raman scattering.

$$\rho = \frac{I_\perp}{I_\parallel} \qquad \text{Depolarization ratio [definition]} \qquad (11D.3)$$

To measure ρ, the intensity of a Raman line is measured with a polarizing filter (a 'half-wave plate') first parallel and then perpendicular to the polarization of the incident beam. If the emergent light is not polarized, then both intensities are the same and ρ is close to 1; if the light retains its initial polarization, then $I_\perp = 0$, so $\rho = 0$. A line is classified as **depolarized** if it has ρ close to or greater than 0.75 and as **polarized** if $\rho < 0.75$. Only totally symmetrical vibrations give rise to polarized lines in which the incident polarization is largely preserved. Vibrations that are not totally symmetrical give rise to depolarized lines because the incident radiation can give rise to radiation in the perpendicular direction too.

Exercises

E11D.5 Which of the vibrations of an AB_2 molecule are Raman active when it is (i) angular, (ii) linear?

E11D.6 Consider the vibrational mode that corresponds to the uniform expansion of the benzene ring. Is it Raman active?

E11D.7 Does the exclusion rule apply to H_2O?

Checklist of concepts

☐ 1. A **normal mode** is a synchronous displacement of the atoms in which the centre of mass and orientation of the molecule remains fixed. In the harmonic approximation, normal modes are mutually independent.

☐ 2. A normal mode is **infrared active** if it is accompanied by a change of electric dipole moment; in the harmonic approximation the specific selection rule is $\Delta v_q = \pm 1$.

☐ 3. A normal mode is **Raman active** if it is accompanied by a change in polarizability; in the harmonic approximation the specific selection rule is $\Delta v_q = \pm 1$.

☐ 4. The **exclusion rule** states that, if the molecule has a centre of inversion, then no mode can be both infrared and Raman active.

☐ 5. **Polarized lines** preserve the polarization of the incident radiation in the Raman spectrum and arise from totally symmetrical vibrations.

Checklist of equations

Property	Equation	Comment	Equation number
Number of normal modes	$N_{vib} = 3N - 5$ (linear) $N_{vib} = 3N - 6$ (nonlinear)	Independent if harmonic; N is the number of atoms	11D.1
Vibrational terms of normal modes	$\tilde{G}_q(v_q) = \left(v_q + \frac{1}{2}\right)\tilde{\nu}_q$, $\tilde{\nu}_q = (1/2\pi c)(k_{f,q}/m_q)^{1/2}$	Harmonic approximation	11D.2
Depolarization ratio	$\rho = I_\perp / I_\parallel$	Depolarized lines: ρ close to or greater than 0.75; polarized lines: $\rho < 0.75$	11D.3

TOPIC 11E Symmetry analysis of vibrational spectra

➤ **Why do you need to know this material?**

The analysis of vibrational spectra is aided by understanding the relationship between the symmetry of the molecule, its normal modes, and the selection rules that govern the transitions.

➤ **What is the key idea?**

The vibrational modes of a molecule can be classified according to the symmetry of the molecule.

➤ **What do you need to know already?**

You need to be familiar with the vibrational spectra of polyatomic molecules (Topic 11D) and the treatment of symmetry in Focus 10.

The classification of the normal modes of vibration of a polyatomic molecule according to their symmetry makes it possible to predict in a very straightforward way which are infrared or Raman active.

11E.1 Classification of normal modes according to symmetry

Each normal mode can be classified as belonging to one of the symmetry species of the irreducible representations of the molecular point group. The classification proceeds as follows:

1. The basis functions are the three displacement vectors (x, y, and z) on each atom: there are $3N$ such basis functions for a molecule with N atoms.
2. The character, $\chi(C)$, for each class, C, of operations in the group is found by considering the effect of one operation of the class and counting 1 for each basis function that is unchanged by the operation, −1 if the basis function changes sign, and 0 if it changes into some other displacement.
3. The resulting representation is decomposed into its component irreducible representations, denoted Γ with characters $\chi^{(\Gamma)}(C)$, by using the relevant character table in conjunction with eqn 10C.3a in the form:

$$n(\Gamma)=\frac{1}{h}\sum_{C} N(C)\chi^{(\Gamma)}(C)\chi(C)$$

where h is the order of the group and $N(C)$ the number of operations in the class C.

4. The symmetry species corresponding to x, y, and z (corresponding to translations) and those corresponding to the rotations about x, y, and z (denoted R_x, R_y, and R_z) are removed. Their symmetry species are listed in the character tables.
5. The remaining symmetry species correspond to the normal modes.

Example 11E.1 Identifying the symmetry species of the normal modes of H_2O

Identify the symmetry species of the normal modes of H_2O, which belongs to the point group C_{2v}.

Collect your thoughts You need to identify the axes in the molecule and then refer to the character table (in the *Resource section*) for the symmetry operations and their characters. You need consider only one symmetry operation of each class, because all members of the same class have the same character. (In C_{2v}, there is only one member of each class anyway.) Then follow the five steps outlined in the text. Note from the character table the symmetry species of the translations and rotations, which are given in the right-hand column.

The solution The molecule lies in the yz-plane, with the z-axis bisecting the HOH bond angle. The three displacement vectors of each atom are shown in Fig. 11E.1. From the character table for C_{2v}, the symmetry operations are E, C_2, $\sigma_v(xz)$, and $\sigma'_v(yz)$. It is seen that:

- None of the nine displacement vectors is affected by the operation E, so $\chi(E) = 9$.
- The C_2 operation moves all the displacement vectors on the H atoms to other positions, so these count 0;

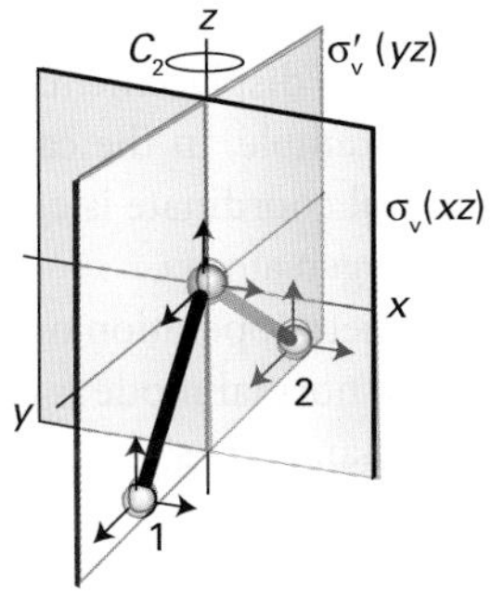

Figure 11E.1 The atomic displacements of H_2O and the symmetry elements used to calculate the characters.

the x and y displacement vectors on the O atom change sign, giving a count of -1 each, whereas the z displacement vector is unaffected, giving a count of $+1$. Hence $\chi(C_2) = -1 - 1 + 1 = -1$.

- The operation $\sigma_v(xz)$ moves all the displacement vectors on the H atoms to other positions, changes the sign of the y displacement vector on the O atom, and leaves the x and z vectors unaffected: hence $\chi(\sigma_v) = -1 + 1 + 1 = 1$.
- The operation $\sigma'_v(yz)$ changes the sign of the x displacement vectors on both H atoms and leaves the sign of the y and z displacement vectors unaffected. For the O atom, the x vector changes sign and the y and z displacement vectors are unaffected: hence $\chi(\sigma'_v) = -1-1+1+1+1+1-1+1+1 = 3$.
- The characters of the representation are therefore 9, -1, 1, 3.

Decomposition by using eqn 10C.3a shows that the reducible representation spans the symmetry species $3A_1 + A_2 + 2B_1 + 3B_2$. The translations have symmetry species B_1, B_2, and A_1 and the rotations have symmetry species B_1, B_2, and A_2; their removal leaves $2A_1 + B_2$ as the symmetry species of the normal modes. As expected, there are three such modes.

Comment. In Fig. 11D.3 (Topic 11D) depicting the normal modes of H_2O, ν_1 and ν_2 have symmetry A_1, and ν_3 has symmetry B_2. This assignment is evident from the fact that the combination of displacements for both ν_1 and ν_2 are unchanged by any of the operations of the group, so the characters are all 1 as required for A_1. In contrast, for ν_3 the displacements shown change sign under C_2 and σ_v giving the characters 1, -1, -1, 1, which correspond to B_2.

Self-test 11E.1 Identify the symmetry species of the normal modes of methanal, $H_2C{=}O$, point group C_{2v} (orientate the molecule in the same way as H_2O, with the CH_2 group in the yz-plane).

Answer: $3A_1 + B_1 + 2B_2$

All the normal modes of H_2O have symmetry species A or B and therefore are non-degenerate. There are no two- or higher-dimensional irreducible representations in C_{2v} molecules, so vibrational degeneracy never arises. Degeneracy can arise in molecules with higher symmetry, as illustrated in the following example.

Example 11E.2 Identifying the symmetry species of the normal modes of BF_3

Identify the symmetry species of the normal modes of vibration of BF_3, which is trigonal planar and belongs to the point group D_{3h}.

Collect your thoughts The overall procedure is the same as in *Example* 11E.1. However, because the molecule is D_{3h}, which has two-dimensional irreducible representations (E′ and E″), you need to be alert for the possibility that there are doubly degenerate pairs of normal modes. You can treat the displacement vectors on the B atom separately from those on the F atoms because no symmetry operation interconverts these two sets: this separation simplifies the calculations. Because the molecule is nonlinear with 4 atoms, there are 6 normal modes.

The solution The C_3 axis is the principal axis and defines the z-direction; the molecule lies in the xy-plane. The three C_2 axes pass along the B–F bonds, and the three σ_v planes contain the B–F bonds and are perpendicular to the plane of the molecule. The σ_h plane lies in the plane of the molecule, and the S_3 axis is coincident with the C_3 axis.

First, consider the displacement vectors on the B atom. Because this atom lies on the principal axis, the z displacement vector must transform as the function z, which from the character table has the symmetry species A''_2. Similarly, the x and y displacement vectors together transform as E′.

Next, consider the nine displacement vectors on the F atoms:

- The identity operation has no effect, so $\chi(E) = 9$.
- The C_3 operation moves all these vectors, so $\chi(C_3) = 0$.
- A C_2 operation about a particular B–F bond has no effect on the displacement vector that points along the bond, but the other two vectors change sign; the displacement vectors on the other F atoms are moved, hence $\chi(C_2) = 1 - 1 - 1 = -1$.
- Under σ_h the z displacement vector on each F changes sign, but the x and y vectors do not. The character is therefore $\chi(\sigma_h) = 3 \times (-1 + 1 + 1) = 3$.
- The character for S_3 is the same as for C_3, $\chi(S_3) = 0$.
- A σ_v reflection in a plane containing a particular B–F bond has no effect on the displacement vector that points along the bond, nor on the z displacement vector; however, the other vector changes sign. The displacement vectors on the other atoms are moved, hence $\chi(\sigma_v) = 1 + 1 - 1 = 1$.

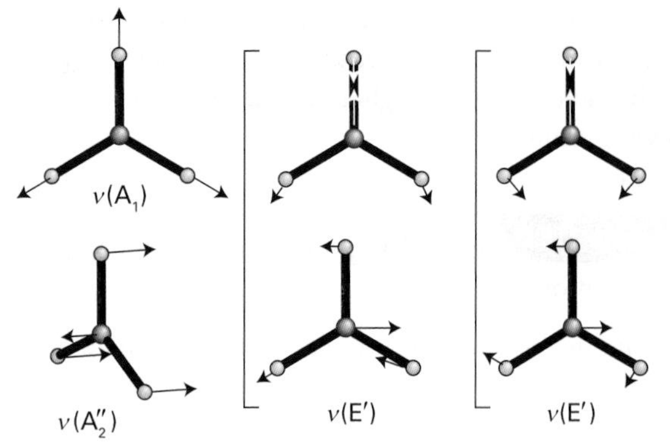

Figure 11E.2 The normal modes of vibration of BF_3.

The characters of the reducible representation are therefore 9, 0, −1, 3, 0, 1; this set can be decomposed into the symmetry species $A_1' + A_2' + 2E' + A_1'' + E''$ for the displacement vectors on the F atoms. The displacement vectors on the B atom transform as $A_2'' + E'$, so the complete set of symmetry species is $A_1' + A_2' + 3E' + 2A_2'' + E''$.

The character table shows that z transforms as A_2'', and x and y together span E′. The rotation about z, R_z, transforms as A_2' and rotations about x and y together (R_x,R_y) transform as E″. Removing these symmetry species from the complete set leaves $A_1' + 2E' + A_2''$ as the symmetry species of the vibrational modes. Figure 11E.2 shows these normal modes.

Comment. Because the E′ symmetry species is two-dimensional, the corresponding normal mode is doubly degenerate. The above analysis shows that there are two E′ symmetry species present, which correspond to two different doubly degenerate normal modes. The total number of normal modes represented by $A_1' + 2E' + A_2''$ is therefore $1 + 2 \times 2 + 1 = 6$.

Self-test 11E.2 Identify the symmetry species of the normal modes of ammonia NH_3, point group C_{3v}.

Answer: $2A_1 + 2E$

Exercise

E11E.1 The molecule CH_2Cl_2 belongs to the point group C_{2v}. The displacements of the atoms span $5A_1 + 2A_2 + 4B_1 + 4B_2$. What are the symmetry species of the normal modes of vibration?

11E.2 Symmetry of vibrational wavefunctions

For a one-dimensional harmonic oscillator the ground-state wavefunction (with $v = 0$) is proportional to $e^{-x^2/2\alpha^2}$, where x is the displacement from the equilibrium position and α is a constant (Topic 7E). For the first excited state, with $v = 1$, the wavefunction is proportional to $xe^{-x^2/2\alpha^2}$. The same wavefunctions apply to a normal mode q of a more complex molecule provided that x is replaced by the **normal coordinate**, Q_q, which is the combination of displacements that corresponds to the normal mode. For example, in the case of the symmetric stretch of CO_2 the normal coordinate is $z_{O,1} - z_{O,2}$, where $z_{O,i}$ is the z-displacement of oxygen atom i.

The effect of any symmetry operation on the normal coordinate of a non-degenerate normal mode is either to leave it unchanged or at most to change its sign. In other words, all the characters are either 1 or −1. The ground-state wavefunction is a function of the square of the normal coordinate, so regardless of whether $Q_q \rightarrow +Q_q$, or $Q_q \rightarrow -Q_q$ as a result of any symmetry operation, the effect on Q_q^2 is to leave it unaffected. Therefore, all the characters for Q_q^2 are 1, so the ground-state wavefunction transforms as the totally symmetric irreducible representation (typically A_1).

The first excited state wavefunction is a product of a part that depends on Q_q^2 (the exponential term) and a factor proportional to Q_q. As has already been seen, Q_q^2 transforms as the totally symmetric irreducible representation and Q_q has the same symmetry species as the normal mode. The direct product of the totally symmetric irreducible representation with any symmetry species leaves the latter unaffected, hence it follows that the symmetry of the first excited state wavefunction is the same as that of the normal mode.

11E.2(a) Infrared activity of normal modes

Once the symmetry of a particular normal mode is known it is a simple matter to determine from the appropriate character table whether or not the fundamental transition of that mode is allowed and therefore if the mode is infrared active.

How is that done? 11E.1 Determining the infrared activity of a normal mode

You need to note that the fundamental transition of a particular normal mode is the transition from the ground state, $v_q = 0$, to the first excited state, $v_q = 1$. You already know that the state with $v_q = 0$ transforms as the totally symmetric irreducible representation, and the state with $v_q = 1$ has the same symmetry as the corresponding normal mode.

Step 1 *Formulate the integral used to identify the selection rule*

Whether or not the transition between $v_q = 0$ and $v_q = 1$ is allowed is assessed by evaluating the transition dipole between ψ_0 and ψ_1, $\mu_{10} = \int \psi_1^* \hat{\mu} \psi_0 \mathrm{d}\tau$ (Topic 11A); the dipole moment operator transforms as x, y, or z (Topic 10C). As is shown in Topic 10C, this integral may be non-zero only if the integrand, that is the product $\psi_1^* \hat{\mu} \psi_0$, spans the totally symmetric irreducible representation.

Step 2 *Identify the symmetry species spanned by the integrand*

You can find the symmetry species of $\psi_1^* \hat{\mu} \psi_0$ by taking the direct product of the symmetry species spanned by each of

ψ_1, $\hat{\mu}$, and ψ_0 separately (Topic 10C). Because ψ_0 transforms as the totally symmetric irreducible representation it has no effect on the symmetry of $\psi_1^*\hat{\mu}\psi_0$, and so you need consider only the symmetry of the product $\psi_1^*\hat{\mu}$. As is shown in Topic 10C, the only way for this product to span the totally symmetric irreducible representation is for ψ_1^* and $\hat{\mu}$ to span the same symmetry species. In other words, the integral can be non-zero only if ψ_1^*, and hence the normal mode, has the same symmetry species as x, y, or z.

The result of the analysis can be summarized by the following rule:

A mode is infrared active only if its symmetry species is the same as the symmetry species of any of x, y, or z.

Brief illustration 11E.1

The normal modes of BF_3 (point group D_{3h}) have symmetry species $A_1' + 2E' + A_2''$ (*Example* 11E.2). From the character table it can be seen that z transforms as A_2'' and (x,y) together transform as E'. The A_2'' normal mode and the two doubly degenerate E' normal modes are therefore infrared active. The A_1' mode is not.

11E.2(b) Raman activity of normal modes

Symmetry arguments also provide a systematic way of deciding whether or not the fundamental of a normal mode gives rise to Raman scattering; that is, whether or not the mode is Raman active. The argument is similar to that for assessing infrared activity except that it is based on the symmetry of the polarizability operator rather than the dipole moment operator. That operator transforms in the same way as the quadratic forms (x^2, xy, and so on, which are listed in the character tables), and leads to the following rule:

A mode is Raman active only if its symmetry species is the same as the symmetry species of a quadratic form.

Brief illustration 11E.2

The normal modes of BF_3 (point group D_{3h}) have symmetry species $A_1' + 2E' + A_2''$ (*Example* 11E.2). From the character table it can be seen that x^2, y^2, and z^2 all transform as A_1', and that (x^2-y^2, xy) together transform as E'. The A_1' normal mode and the two doubly degenerate E' normal modes are therefore Raman active. The A_2'' mode is not active because no quadratic form has this symmetry species. The E' modes are both infrared and Raman active: the exclusion rule does not apply because BF_3 does not have a centre of symmetry. The A_1' normal mode is the highly symmetrical breathing mode in which all the B–F bonds stretch together. The corresponding Raman line is expected to be polarized. The E' modes are expected to give depolarized lines.

11E.2(c) The symmetry basis of the exclusion rule

The exclusion rule, Topic 11D, can be derived by using a symmetry argument. If a molecule has a centre of inversion, then all the symmetry species of its displacements are either g or u according to their behaviour under inversion. If the character for this operation is positive, indicating that the displacement or displacements are unchanged by the operation, the label is g, whereas if the character is negative, indicating that the sign of the displacement or displacements are changed, the label is u.

The functions x, y, and z (which occur in the transition dipole moment) all change sign under inversion, so they must correspond to symmetry species with a label u. In contrast, the quadratic forms (which govern the Raman activity) are all unchanged by inversion and so have the label g. For example, the effect of the inversion on xz is to transform it into $(-x)(-z) = xz$.

Any normal mode in a molecule with a centre of inversion corresponds to a symmetry species that is either g or u. If the normal mode has the same symmetry species as x, y, or z, it is infrared active; such a mode must be u. If the normal mode has the same symmetry species as a quadratic form, it is Raman active; such a mode must be g. Because a normal mode cannot be both g and u, no mode can be both infrared and Raman active.

Exercises

E11E.2 Which of the normal modes of CH_2Cl_2 (found in *Exercise* E11E.1) are infrared active? Which are Raman active?

E11E.3 Which of the normal modes of (i) H_2O, (ii) H_2CO are infrared active?

Checklist of concepts

☐ 1. A normal mode is infrared active if its symmetry species is the same as the symmetry species of x, y, or z.

☐ 2. A normal mode is Raman active if its symmetry species is the same as the symmetry species of a quadratic form.

TOPIC 11F Electronic spectra

> ➤ Why do you need to know this material?
>
> The study of spectroscopic transitions between different electronic states of molecules gives access to data on the electronic structure of molecules, and hence insight into bonding, as well as vibrational frequencies and bond lengths.
>
> ➤ What is the key idea?
>
> Electronic transitions occur within a stationary nuclear framework.
>
> ➤ What do you need to know already?
>
> You need to be familiar with the general features of spectroscopy (Topic 11A), the quantum mechanical origins of selection rules (Topics 8C, 11B, and 11C), and vibration–rotation spectra (Topic 11C). It would be helpful to be aware of atomic term symbols (Topic 8C).

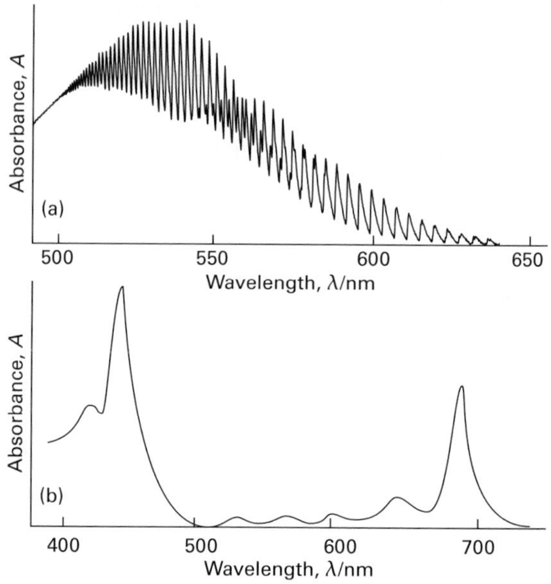

Figure 11F.1 Electronic absorption spectra recorded in the visible region. (a) The spectrum of I_2 in the gas phase shows resolved vibrational structure. (b) The spectrum of chlorophyll recorded in solution, shows only broad bands with no resolved structure. (Absorbance, *A*, is defined in Topic 11A.)

Electronic spectra arise from transitions between the electronic energy levels of molecules. These transitions may also be accompanied by simultaneous changes in vibrational energy; for small molecules in the gas phase the resulting spectral features can be resolved (Fig. 11F.1a), but in a liquid or solid the individual lines usually merge together and result in a broad, almost featureless band (Fig. 11F.1b).

The energies needed to change the electron distributions of molecules are of the order of several electronvolts (1 eV is equivalent to about 8000 cm^{-1} or 100 kJ mol^{-1}). Consequently, the photons emitted or absorbed when such changes occur lie in the visible and ultraviolet regions of the spectrum (Table 11F.1).

Table 11F.1 Colour, frequency, and energy of light*

Colour	λ/nm	$\nu/(10^{14}$ Hz)	E/(kJ mol^{-1})
Infrared	>1000	<3.0	<120
Red	700	4.3	170
Yellow	580	5.2	210
Blue	470	6.4	250
Ultraviolet	<400	>7.5	>300

* More values are given in the *Resource section*.

11F.1 Diatomic molecules

Topic 8C explains how the states of atoms are described by using term symbols. The electronic states of diatomic molecules are also specified by using term symbols, the key difference being that the full spherical symmetry of atoms is replaced by the cylindrical symmetry defined by the axis of the molecule.

11F.1(a) Term symbols

In a diatomic molecule only the component of the total orbital angular momentum along the internuclear axis can be specified: the quantum number for this component is Λ (uppercase lambda). Its value is found by adding together the component of the orbital angular momentum along the internuclear axis, λ_i, for each electron present:

$$\Lambda = \lambda_1 + \lambda_2 + \cdots \qquad (11F.1)$$

For an electron in a σ molecular orbital (which is cylindrically symmetric), $\lambda = 0$; for an electron in one of the degenerate pair of π orbitals $\lambda = \pm 1$. In the molecular term symbol the value of Λ is represented by an uppercase Greek letter in the following way

$\lvert\Lambda\rvert$	0	1	2	…
	Σ	Π	Δ	…

These labels are the analogues of S, P, D,… used for atomic states with $L = 0, 1, 2, \ldots$. The total spin, S, of a linear molecule is specified in the same way as for an atom. As in an atomic term symbol, the value of $2S + 1$ is shown as a left superscript and denotes the multiplicity of the term.

Configurations such σ^2 and π^4 with all electrons paired have $S = 0$. Such configurations do not contribute to the total orbital angular momentum, either because both electrons have $\lambda = 0$ or because there are equal numbers of electrons with $\lambda = +1$ and -1. The term symbol for the ground state of H_2, configuration $1\sigma_g^2$, is therefore $^1\Sigma$ (read as 'singlet sigma'), and the same is true of the ground state of N_2, configuration $1\sigma_g^2 1\sigma_u^2 2\sigma_g^2 2\sigma_u^2 1\pi_u^4 3\sigma_g^2$.

Brief illustration 11F.1

The ground-state configuration of H_2^+ is $1\sigma_g^1$. The single σ electron has $\lambda = 0$, so $\Lambda = 0$; for a single electron $S = \frac{1}{2}$, so $2S + 1 = 2$. The term symbol is therefore $^2\Sigma$ (read as 'doublet sigma'). The ground-state configuration of NO is $\ldots 1\pi^1$, where … indicates completed orbitals that make no contribution to either S or Λ. The single electron can occupy either of the degenerate π orbitals, so $\Lambda = +1$ or -1; for a single electron, $S = \frac{1}{2}$. The term symbol is therefore $^2\Pi$ (read as 'doublet pi'). The ground-state configuration of O_2 is $\ldots 1\pi_g^2$. If the two electrons occupy the same π orbital, $\Lambda = (+1) + (+1) = +2$ (or $\Lambda = (-1) + (-1) = -2$); in this arrangement the electron spins must be paired, so $S = 0$. The resulting term is $^1\Delta$ ('singlet delta'); a $^3\Delta$ term is not possible as it would require two electrons with parallel spins to occupy one of the π orbitals. If the electrons occupy different π orbitals, $\Lambda = (+1) + (-1) = 0$; in this arrangement the spins can be paired or parallel, so $S = 0$ or $S = 1$. Two further terms therefore arise: $^1\Sigma$ and $^3\Sigma$ ('triplet sigma'); the latter turns out to be the lowest in energy of all the three terms.

As explained in Topic 9B, homonuclear diatomic molecules (but not heteronuclear diatomic molecules) possess a centre of inversion and their orbitals are labelled g or u according to their parity (the behaviour under inversion). Orbitals that are unchanged upon inversion are g and orbitals that change sign are u. Parity labels also apply to centrosymmetric polyatomic linear molecules, such as CO_2 and HC≡CH. The overall parity of a configuration of a many-electron homonuclear diatomic molecule is found by noting the parity of each occupied orbital and using

$$\mathrm{g}\times\mathrm{g}=\mathrm{g} \quad \mathrm{u}\times\mathrm{u}=\mathrm{g} \quad \mathrm{u}\times\mathrm{g}=\mathrm{u} \qquad (11\mathrm{F}.2)$$

for each electron. (These rules are generated by interpreting g as +1 and u as −1.) The resulting parity label g or u is added as a right-subscript to the term symbol. Any molecule in which the occupied orbitals are full (in the sense of being occupied by a pair of electrons) must have overall parity g because there is an even number of electrons. The term symbol for such a homonuclear diatomic molecule is therefore $^1\Sigma_g$ ('singlet sigma g').

Brief illustration 11F.2

Dinitrogen, N_2, has the configuration $1\sigma_g^2 1\sigma_u^2 2\sigma_g^2 2\sigma_u^2 1\pi_u^4 3\sigma_g^2$ in which all the occupied orbitals are full; the same is true of H_2 and F_2: all three therefore have the term symbol $^1\Sigma_g$. The configuration of He_2^+ is $1\sigma_g^2 1\sigma_u^1$. There is one electron outside the doubly occupied bonding orbital, and the parity of the orbital it occupies is u. Because $S = \frac{1}{2}$ and $\Lambda = 0$, its term symbol is $^2\Sigma_u$ ('doublet sigma u'). The ground-state configuration of O_2 is $\ldots 1\pi_g^2$. Although the π_g orbitals may be both singly occupied, both electrons are in orbitals with g parity, so the overall parity is $\mathrm{g} \times \mathrm{g} = \mathrm{g}$. The three terms arising from this configuration are therefore $^1\Sigma_g$, $^3\Sigma_g$, and $^1\Delta_g$ (see *Brief illustration* 11F.1).

Diatomic molecules (and all linear molecules) possess a mirror plane containing the internuclear axis. All σ orbitals (both bonding and antibonding) are symmetric with respect to reflection in this plane. The overall symmetry of a configuration is found by assigning +1 to an electron in a symmetric orbital and −1 to an electron in an orbital that changes sign under reflection and then multiplying the numbers for all the electrons. For example, for the ground state of H_2, in which both electrons are in σ orbitals, the overall symmetry is $(+1) \times (+1) = +1$. A + sign is added as a right-superscript to the term symbol: $^1\Sigma_g^+$ ('singlet sigma g plus'). Any configuration consisting of electrons solely in σ orbitals necessarily has + overall reflection symmetry; for example, the ground state of He_2^+ (*Brief illustration* 11F.2) is $^2\Sigma_u^+$.

The behaviour of the degenerate pair of π molecular orbitals under reflection is more complex: as is shown in Fig. 11F.2, one

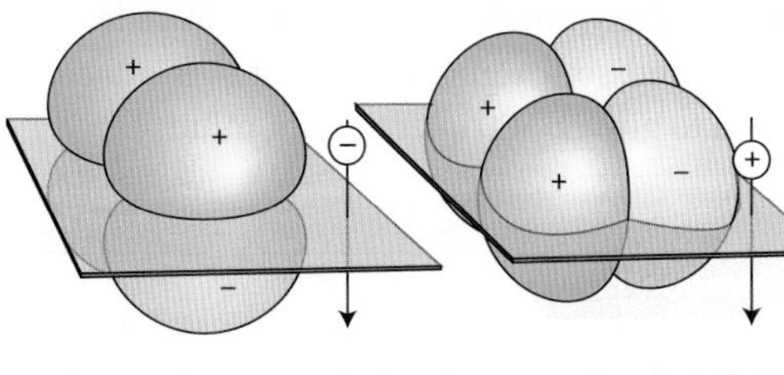

Figure 11F.2 A molecular orbital can be classified as symmetric (+) or antisymmetric (–) according to whether it changes sign under reflection in a plane containing the internuclear axis. Shown here are the two degenerate π MOs: one is symmetric and one is antisymmetric under reflection in the plane shown.

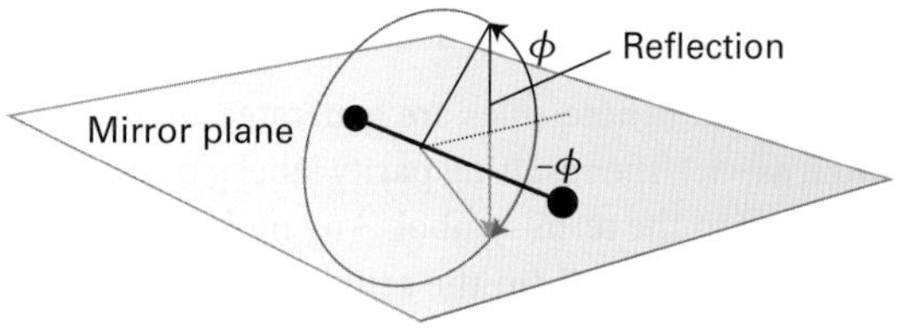

Figure 11F.3 In a linear molecule, the molecular orbital depends on the azimuthal angle ϕ. Reflection in the mirror plane is equivalent to reversing the sign of ϕ.

of the orbitals changes sign but the other does not. The consequences of this observation can be explored by considering the mathematical form of the π orbitals and how they depend on the angle ϕ shown in Fig. 11F.3. The orbital π_x is proportional to cos ϕ and therefore has a nodal plane at $\phi = \pi/2$ (the yz-plane) with positive and negative lobes on either side of this plane; it is unchanged by reflection in the xz-plane. The orbital π_y is proportional to sin ϕ, so the xz-plane at $\phi = 0$ is a nodal plane; the orbital changes sign on reflection in the xz-plane. These two wavefunctions are degenerate, so any linear combination of them is also an acceptable wavefunction. For the present discussion the combinations $\pi_+ = \cos\phi + \mathrm{i}\sin\phi = \mathrm{e}^{+\mathrm{i}\phi}$ and $\pi_- = \cos\phi - \mathrm{i}\sin\phi = \mathrm{e}^{-\mathrm{i}\phi}$ are convenient for they correspond to $\lambda = +1$, and $\lambda = -1$, respectively. As shown in Fig. 11F.3, reflection in the xz-plane changes the sign of ϕ. Because $\cos(-\phi) = \cos\phi$, the term cos ϕ is unchanged by this reflection. However, sin ϕ changes sign because $\sin(-\phi) = -\sin\phi$. It follows that this reflection interconverts π_+ and π_-.

Now consider O_2, which has the configuration $\ldots 1\pi_g^2$. The triplet state ($S = 1$), in which the two electrons have parallel spins and necessarily occupy different orbitals, is the state of lowest energy. The triplet spin wavefunction is symmetric with respect to the interchange of the two electrons (it is of the form $\alpha(1)\alpha(2)$, and so on), so it follows from the Pauli principle (Topic 8B) that the spatial part of the wavefunction must be antisymmetric with respect to interchange. Such a wavefunction, in which one electron occupies the π_+ orbital and the other occupies π_-, is $\Psi_-(1,2) = \pi_+(r_1)\pi_-(r_2) - \pi_+(r_2)\pi_-(r_1)$. Reflection in the mirror plane, which interconverts π_+ and π_-, gives $\pi_-(r_1)\pi_+(r_2) - \pi_-(r_2)\pi_+(r_1) = -\Psi_-(1,2)$. That is, the spatial wavefunction of the triplet state is antisymmetric with respect to reflection in the mirror plane, and so a right-superscript − is attached to the term symbol, to give ${}^3\Sigma_g^-$ ('triplet sigma g minus').

Brief illustration 11F.3

An alternative, higher energy configuration of O_2 has the outermost two electrons in separate π orbitals but with their spins paired (a ${}^1\Sigma_g$ term). The spin wavefunction for a singlet (which is proportional to $\alpha(1)\beta(2) - \alpha(2)\beta(1)$) is antisymmetric with respect to the interchange of the electrons. The spatial function therefore must be symmetric. A suitable wavefunction is $\Psi_+(1,2) = \pi_+(r_1)\pi_-(r_2) + \pi_+(r_2)\pi_-(r_1)$. Reflection changes this function to $\pi_-(r_1)\pi_+(r_2) + \pi_-(r_2)\pi_+(r_1)$ which is $+\Psi_+(1,2)$. The state is symmetric with respect to this reflection, and a superscript + is added to the term symbol, to give ${}^1\Sigma_g^+$ ('singlet sigma g plus').

As for atoms, it is sometimes necessary to specify the total electronic angular momentum, the sum of the orbital and spin contributions, and hence the different 'levels' of a term. In a linear molecule, only the component of the total electronic angular momentum along the internuclear axis is well defined, and is specified by the quantum number Ω (uppercase omega). For light molecules, where the spin–orbit coupling is weak, Ω is obtained by adding together the components of the orbital angular momentum along the axis (the value of Λ) and the component of the electron spin on that axis (Fig. 11F.4). The latter is denoted Σ, where $\Sigma = S, S-1, S-2, \ldots, -S$.[1] Then

$$\Omega = \Lambda + \Sigma \qquad (11F.3)$$

The value of $|\Omega|$ is then attached to the term symbol as a right-subscript (just like J is used in atoms) to denote the different levels. These levels differ in energy, as in atoms, as a result of spin–orbit coupling.

Brief illustration 11F.4

The ground-state configuration of NO is $\ldots 1\pi^1$, so it is a ${}^2\Pi$ term with $\Lambda = \pm 1$ and $S = \frac{1}{2}$; from the latter it follows that $\Sigma = \pm\frac{1}{2}$. There are two levels of the term, one with $\Omega = \pm\frac{1}{2}$ and the other with $\pm\frac{3}{2}$, denoted ${}^2\Pi_{1/2}$ and ${}^2\Pi_{3/2}$, respectively. Each level is doubly degenerate (corresponding to the opposite signs of Ω). It turns out that, in NO, ${}^2\Pi_{1/2}$ ('doublet pi one half') lies slightly lower in energy than ${}^2\Pi_{3/2}$.

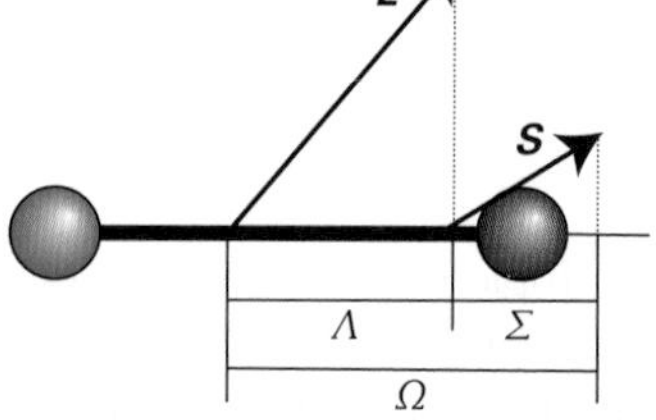

Figure 11F.4 The coupling of spin and orbital angular momenta in a linear molecule: only their components along the internuclear axis (Σ and Λ) are well defined.

[1] It is important to distinguish between the upright term symbol Σ and the sloping quantum number Σ.

11F.1(b) Selection rules

A number of selection rules govern which transitions can be observed in the electronic spectrum of a molecule. The selection rules concerned with changes in angular momentum in a linear molecule are

$$\Delta\Lambda = 0, \pm 1 \quad \Delta S = 0 \quad \Delta\Sigma = 0 \quad \Delta\Omega = 0, \pm 1$$

Selection rules for electronic spectra of linear molecules (11F.4)

As in atoms (Topic 8C), the origins of these rules are conservation of angular momentum during a transition and the fact that a photon has a spin of 1.

Two selection rules can be deduced on the basis of symmetry.

How is that done? 11F.1 Establishing symmetry-based selection rules

As usual when establishing selection rules, you need to consider properties of the electric-dipole transition moment introduced in Topic 8C, $\mu_{fi} = \int \psi_f^* \hat{\mu} \psi_i d\tau$, and to note that it vanishes unless the integrand is invariant under all symmetry operations of the molecule (Topic 10C).

The z-component (the component parallel to the axis of the molecules) of the electric dipole moment operator is responsible for $\Sigma \leftrightarrow \Sigma$ transitions (the other components of μ perpendicular to the axis have Π symmetry and cannot make a contribution to this transition). The z-component of μ has (+) symmetry with respect to reflection in a plane containing the internuclear axis. Therefore, for a $(+) \leftrightarrow (-)$ transition, the overall symmetry of the integrand is $(+) \times (+) \times (-) = (-)$, so the integral must be zero and hence $\Sigma^+ \leftrightarrow \Sigma^-$ transitions are not allowed. The integrands for $\Sigma^+ \leftrightarrow \Sigma^+$ and $\Sigma^- \leftrightarrow \Sigma^-$ transitions transform as $(+) \times (+) \times (+) = (+)$ and $(-) \times (+) \times (-) = (+)$, respectively. The integrals are therefore not necessarily zero and so both transitions are allowed.

The three components of the dipole moment operator transform like x, y, and z, and in a centrosymmetric molecule are all u. Therefore, for a $g \rightarrow g$ transition, the overall parity of the integrand is $g \times u \times g = u$, so the integral must be zero. Likewise, for a $u \rightarrow u$ transition, the overall parity is $u \times u \times u = u$, so the integral is again zero. Hence, transitions without a change of parity are forbidden. For a $g \leftrightarrow u$ transition the integrand transforms as $g \times u \times u = g$, so the transition is allowed.

The first part of this analysis can be summarized as follows:

For Σ terms, only $\Sigma^+ \leftrightarrow \Sigma^+$ and $\Sigma^- \leftrightarrow \Sigma^-$ are allowed.

The second part is in fact the **Laporte selection rule** for centrosymmetric molecules (those with a centre of inversion, not only linear molecules) which states that *the only transitions allowed are accompanied by a change of parity*. That is,

For centrosymmetric molecules, only $u \rightarrow g$ and $g \rightarrow u$ transitions are allowed.

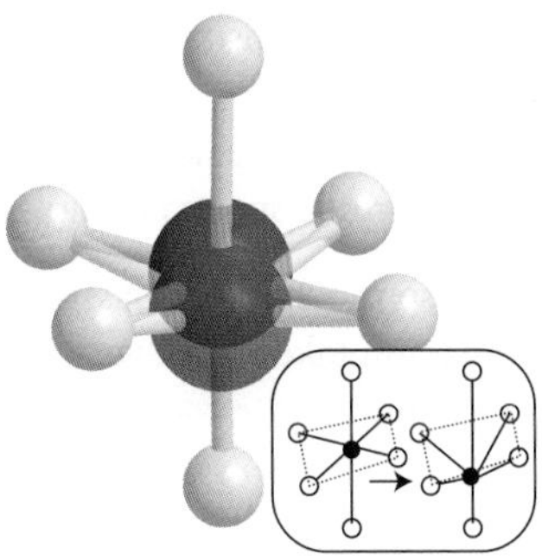

Figure 11F.5 A d–d transition is parity-forbidden because it corresponds to a g–g transition. However, a vibration of the molecule can destroy the inversion symmetry of the molecule and the g,u classification no longer applies. The removal of the centre of inversion gives rise to a vibronically allowed transition.

A forbidden $g \rightarrow g$ transition can become allowed if the centre of inversion is eliminated by an asymmetrical vibration, such as the one shown in Fig. 11F.5. When the centre of inversion is lost, $g \rightarrow g$ and $u \rightarrow u$ transitions are no longer parity-forbidden and become weakly allowed. A transition that derives its intensity from an asymmetrical vibration of a molecule is called a **vibronic transition**.

Brief illustration 11F.5

Three possible transitions in the electronic spectrum of O_2, ${}^3\Sigma_g^- \leftarrow {}^3\Sigma_u^-$, ${}^3\Sigma_g^- \leftarrow {}^1\Delta_g$, ${}^3\Sigma_g^- \leftarrow {}^3\Sigma_u^+$, can be considered in the light of the selection rules in eqn 11F.4 to see which are allowed. A table can be drawn up, in which forbidden values are shown in boxes.

	ΔS	$\Delta\Lambda$	$\Sigma^{\pm} \leftarrow \Sigma^{\pm}$	Change of parity	
${}^3\Sigma_g^- \leftarrow {}^3\Sigma_u^-$	0	0	$\Sigma^- \leftarrow \Sigma^-$	$g \leftarrow u$	Allowed
${}^3\Sigma_g^- \leftarrow {}^1\Delta_g$	$\boxed{+1}$	$\boxed{-2}$	Not applicable	$\boxed{g \leftarrow g}$	Forbidden
${}^3\Sigma_g^- \leftarrow {}^3\Sigma_u^+$	0	0	$\boxed{\Sigma^- \leftarrow \Sigma^+}$	$g \leftarrow u$	Forbidden

11F.1(c) Vibrational fine structure

An electronic transition may be accompanied by a simultaneous change in the vibrational state of a molecule, giving rise to **vibrational fine structure** in the spectrum. In the case of absorption spectra, the transitions are from the ground electronic state, and typically it is only the ground vibrational level, $v'' = 0$, of this state that is occupied significantly. In some cases, the transition from $v'' = 0$ in the electronic ground state to $v' = 0$ in the upper electronic state is found to be the strongest, with a sharp decline in intensity as v' increases. In other cases, transitions with significant intensity to a range of v' levels are seen (as in Fig. 11F.1a).

The **Franck–Condon principle** accounts for the vibrational fine structure in the electronic spectra of molecules:

> Because nuclei are so much more massive than electrons, an electronic transition takes place very much faster than the nuclei can respond.

The physical basis of this principle is as follows. As a result of the electronic transition, electron density is built up rapidly in new regions of the molecule and removed from others. In classical terms, the initially stationary nuclei suddenly experience a new force field, to which they respond by beginning to vibrate and (in classical terms) swing backwards and forwards from their original separation which was maintained during the rapid electronic excitation. The stationary equilibrium separation of the nuclei in the initial electronic state therefore becomes a stationary turning point in the final electronic state. The transition can be thought of as taking place up the vertical line in Fig. 11F.6. This interpretation is the origin of the expression **vertical transition**, which denotes an electronic transition that occurs without change of nuclear geometry and, in classical terms, with the nuclei remaining stationary.

Now consider the two potential energy curves shown in Fig. 11F.7a in which the equilibrium bond lengths are the same and initially the molecule is not vibrating. The vertical transition takes place from the minimum of the lower curve, the nuclei remain at the same separation, and ends at the minimum of the upper curve.

Next consider the case shown in Fig. 11F.7b, in which the equilibrium bond length in the upper state is greater than that in the ground electronic state, and the molecule is initially not vibrating. Preservation of the nuclear separation during the transition takes the molecule up the vertical line. The nuclei are not moving initially, and do not start to move during the transition, so the transition terminates at the turning point of the upper electronic state where the nuclei are still stationary.

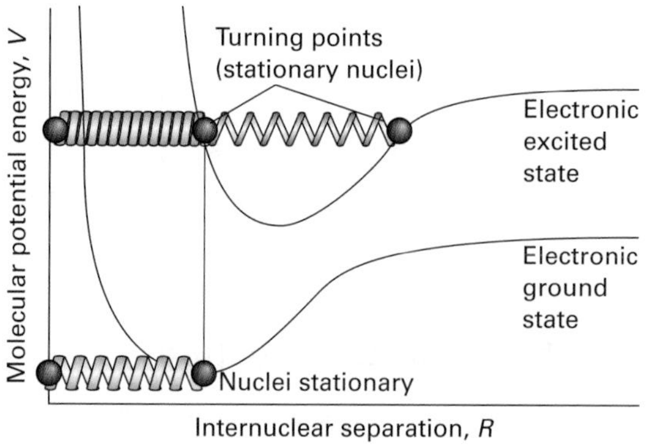

Figure 11F.6 According to the Franck–Condon principle, the most intense vibronic transition is from the ground vibrational state to the vibrational state lying vertically above it. As a result of the vertical transition, the nuclei suddenly experience a new force field, to which they respond through their vibrational motion. The equilibrium separation of the nuclei in the initial electronic state therefore becomes a turning point in the final electronic state. Transitions to other vibrational levels also occur, but with lower intensity.

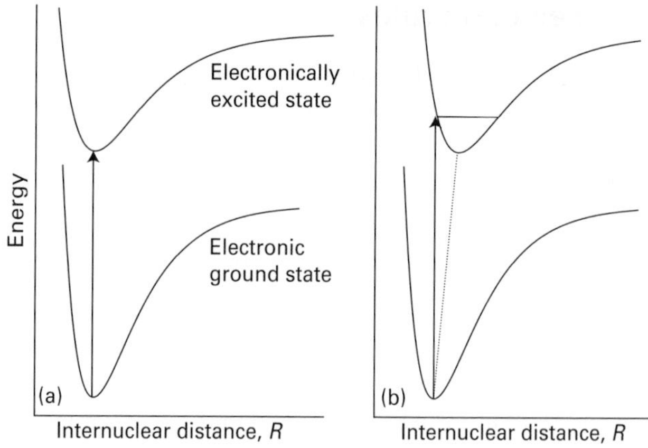

Figure 11F.7 (a) If the equilibrium bond lengths of the ground and excited electronic states are the same, a vertical transition leaves the vibrational state of the molecule unexcited. (b) If the equilibrium bond length is greater in the upper electronic state, the vertical transition ends at a compressed state of the bond and results in vibrational excitation.

The quantum mechanical version of the Franck–Condon principle refines this picture. Instead of saying that the nuclei stay at the same locations and are stationary during the transition, it replaces that statement by the assertion that *the nuclei retain their initial dynamical state*. In quantum mechanics, the dynamical state is expressed by the wavefunction, so an equivalent statement is that the vibrational wavefunction does not change during the electronic transition. Initially the molecule is in the lowest vibrational state of its ground electronic state with a bell-shaped wavefunction centred on the equilibrium bond length (Fig. 11F.8). To find the nuclear state to which the transition takes place, it is necessary to look for the vibrational wavefunction of the upper electronic state that most closely resembles this initial wavefunction, for that corresponds to the nuclear dynamical state that is least changed in the transition. That final wavefunction is the one with a large peak close to the position of the initial bell-shaped function. As explained in Topic 7E, provided the vibrational quantum number is not zero, the biggest peaks of vibrational wavefunctions occur close to the edges of the confining potential, so the transition can be expected to occur to those vibrational states, in accord with the classical description. However, several vibrational states have their major peaks in similar positions, so transitions can be expected to occur to a range of vibrational states, to give rise to a **vibrational progression**, a series of transitions to different vibrational states of the upper electronic state. In a typical progression, the vertical transition is the most intense.

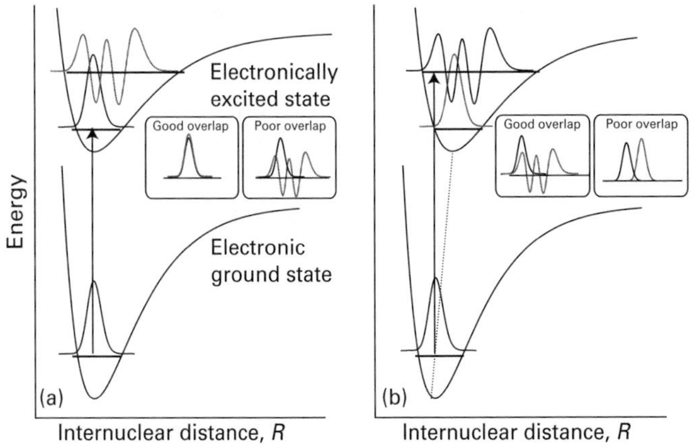

Figure 11F.8 (a) If the equilibrium bond lengths of the ground and excited electronic states are the same, the wavefunctions for $v''=0$ and $v'=0$ are similar and the most probable transition leaves the molecule vibrationally unexcited. (b) If the equilibrium bond length of the upper state is greater than that of the ground state, the wavefunction that most resembles the ground state vibrational wavefunction is that of an excited state. Other transitions also occur with lower intensity.

The quantitative version of the Franck–Condon principle involves considering how the transition dipole moment for a given electronic transition varies with the vibrational levels in the two electronic states.

How is that done? 11F.2 Expressing the Franck–Condon principle quantitatively

Once again, you need to consider the properties of the electric-dipole transition moment. First, note that the electric dipole moment operator is a sum over all nuclei and electrons in the molecule:

$$\hat{\mu} = -e\sum_i \boldsymbol{r}_i + e\sum_N Z_N \boldsymbol{R}_N$$

where the origin of the vectors $\boldsymbol{r}$ (for locating electrons) and $\boldsymbol{R}$ (for locating nuclei) is the centre of charge of the molecule, i labels the electrons, and N labels the nuclei. Within the Born–Oppenheimer approximation (the separation of electronic and vibrational motion, the *Prologue* to Focus 9), the overall state of the molecule consists of an electronic contribution, labelled ε, and a vibrational contribution, labelled v. Therefore, the transition dipole moment factorizes as follows:

$$\mu_{\rm fi} = \int \psi^*_{\varepsilon,\rm f}\psi^*_{v,\rm f}\Big\{-e\sum_i \boldsymbol{r}_i + e\sum_N Z_N\boldsymbol{R}_N\Big\}\psi_{\varepsilon,\rm i}\psi_{v,\rm i}\,{\rm d}\tau$$

$$= -e\sum_i \int \psi^*_{\varepsilon,\rm f}\boldsymbol{r}_i\psi_{\varepsilon,\rm i}\,{\rm d}\tau_{\rm e}\int \psi^*_{v,\rm f}\psi_{v,\rm i}\,{\rm d}\tau_{\rm N}$$

$$+ e\sum_N Z_N \overbrace{\boxed{\int \psi^*_{\varepsilon,\rm f}\psi_{\varepsilon,\rm i}\,{\rm d}\tau_{\rm e}}}^{0}\int \psi^*_{v,\rm f}\boldsymbol{R}_N\psi_{v,\rm i}\,{\rm d}\tau_{\rm N}$$

where ${\rm d}\tau_{\rm e}$ indicates integration over the electronic coordinates, and ${\rm d}\tau_{\rm N}$ integration over the nuclear coordinates. Because the two different electronic states are orthogonal (they are eigenstates of the same hamiltonian but correspond to different eigenvalues) the integral in the box is zero, which leaves

$$\mu_{\rm fi} = \overbrace{-e\sum_i \int \psi^*_{\varepsilon,\rm f}\boldsymbol{r}_i\psi_{\varepsilon,\rm i}\,{\rm d}\tau_{\rm e}}^{\mu_{\varepsilon,\rm fi}}\overbrace{\int \psi^*_{v,\rm f}\psi_{v,\rm i}\,{\rm d}\tau_{\rm N}}^{S(v_{\rm f},v_{\rm i})} = \mu_{\varepsilon,\rm fi}S(v_{\rm f},v_{\rm i})$$

The quantity $\mu_{\varepsilon,\rm fi}$ is the electric-dipole transition moment arising from the change in the electronic wavefunction: this term describes the interaction of the electrons with the electromagnetic field. The factor $S(v_{\rm f},v_{\rm i})$, is the overlap integral between the vibrational level with quantum number $v_{\rm i}$ in the initial electronic state of the molecule, and the vibrational level with quantum number $v_{\rm f}$ in the final electronic state of the molecule.

The transition intensity is proportional to the square of the magnitude of the transition dipole moment, so is proportional to the square of $S(v_{\rm f},v_{\rm i})$, which is known as the **Franck–Condon factor**

$$|S(v_{\rm f},v_{\rm i})|^2 = \left(\int \psi^*_{v,\rm f}\psi_{v,\rm i}\,{\rm d}\tau_{\rm N}\right)^2 \quad \text{Franck–Condon factor} \quad (11\text{F}.5)$$

The integral on the right-hand side of eqn 11F.5 is the overlap between the two vibrational wavefunctions: the greater this overlap (physically, the greater the resemblance of the vibrational wavefunctions), the greater is the intensity of the transition.

Example 11F.1 Calculating a Franck–Condon factor

Consider the transition from one electronic state to another, their equilibrium bond lengths being $R_{\rm e}$ and $R'_{\rm e}$; the force constants of the two states are assumed to be equal. Calculate the Franck–Condon factor for the $v''=0$ to $v'=0$ transition (the 0–0 transition) and show that the transition is most intense when the bond lengths are equal.

Collect your thoughts You need to calculate $S(0,0)$, the overlap integral of the two ground-state vibrational wavefunctions, and then take its square. The difference between harmonic and anharmonic vibrational wavefunctions is negligible for $v=0$, so it is safe for you to use the harmonic oscillator wavefunctions.

The solution The (real) wavefunctions are (Topic 7E)

$$\psi_0 = \left(\frac{1}{\alpha\pi^{1/2}}\right)^{1/2} {\rm e}^{-x^2/2\alpha^2} \qquad \psi'_0 = \left(\frac{1}{\alpha\pi^{1/2}}\right)^{1/2} {\rm e}^{-x'^2/2\alpha^2}$$

where $x = R - R_{\rm e}$ and $x' = R - R'_{\rm e}$, with $\alpha = (\hbar^2/\mu k_{\rm f})^{1/4}$. It is convenient to introduce $\delta R = R_{\rm e} - R'_{\rm e}$ so that $x' = x + \delta R$. The overlap integral is

$$S(0,0) = \int_{-\infty}^{\infty}\psi'_0\psi_0\,{\rm d}R = \frac{1}{\alpha\pi^{1/2}}\int_{-\infty}^{\infty}{\rm e}^{-(x^2+x'^2)/2\alpha^2}\,{\rm d}x$$

$$= \frac{1}{\alpha\pi^{1/2}}\int_{-\infty}^{\infty}{\rm e}^{-\{x^2+(x+\delta R)^2\}/2\alpha^2}\,{\rm d}x$$

The integral can most easily be evaluated by using mathematical software (to find the integral by hand use the substitution $x^2 + (x + \delta R)^2 = 2\left(x + \frac{1}{2}\delta R\right)^2 + \frac{1}{2}\delta R^2$. The result is

$$S(0,0) = e^{-\delta R^2/4\alpha^2}$$

and hence the Franck–Condon factor is

$$S(0,0)^2 = e^{-(R_e - R_e')^2/2\alpha^2}$$

This factor is equal to 1 when $R_e' = R_e$ and decreases as the equilibrium bond lengths diverge from each other (Fig. 11F.9).

For $^{79}Br_2$, R_e = 228 pm and there is an upper state with $R_e' = 266$ pm. With the vibrational wavenumber taken as 250 cm^{-1} it follows that $\alpha^2 = 3.42 \times 10^{-23}$ m^2 and hence $S(0,0)^2 = 6.7 \times 10^{-10}$. That is, the intensity of the 0–0 transition is only 6.7×10^{-10} times what it would have been if the potential curves had been directly above each other.

Self-test 11F.1 Suppose the normalized vibrational wavefunctions can be approximated by rectangular functions of width W and W', centred on the equilibrium bond lengths (Fig. 11F.10). Find the corresponding Franck–Condon factors when the centres are coincident and $W' < W$.

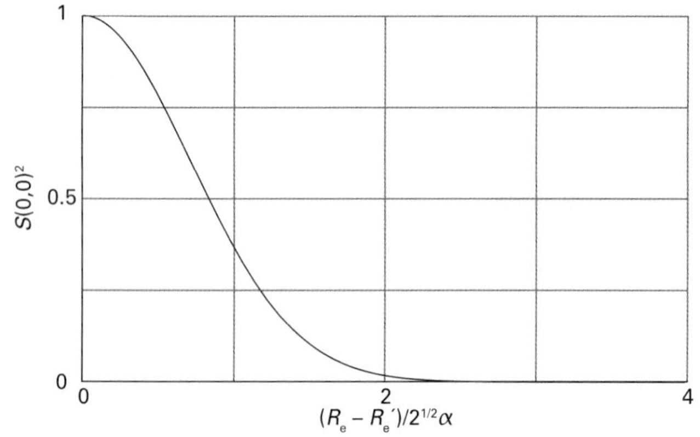

Figure 11F.9 The Franck–Condon factor for the arrangement discussed in *Example* 11F.1.

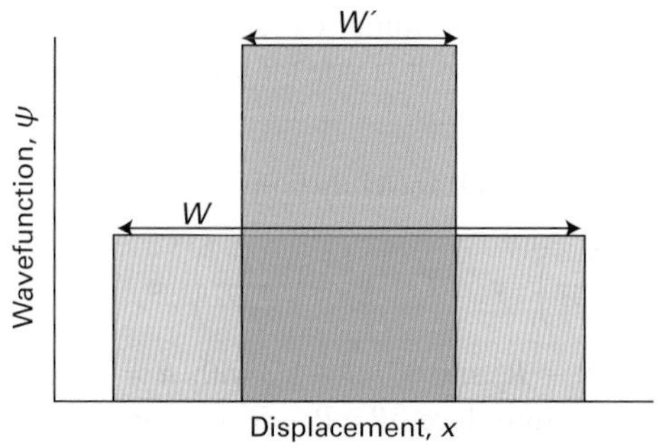

Figure 11F.10 The model wavefunctions used in *Self-test* 11F.1.

Answer: $S^2 = W'/W$

11F.1(d) Rotational fine structure

Electronic transitions may be accompanied by simultaneous changes in both vibrational and rotational energy. Therefore, when the lines dues to vibrational fine structure are inspected at higher resolution they are found to have **rotational fine structure** and to consist of P, Q, and R branches of the type discussed in Topic 11C. Because electronic excitation can result in much larger changes in bond length than vibrational excitation causes alone, the rotational branches have a more complex structure than in vibration–rotation spectra.

The rotational constants of the electronic ground and excited states are denoted $\tilde{B}$ and $\tilde{B}'$, respectively. The rotational terms of the initial and final states are

$$\tilde{F}(J) = \tilde{B}J(J+1) \text{ and } \tilde{F}(J') = \tilde{B}'J'(J'+1) \qquad (11F.6)$$

When a transition occurs with $\Delta J = -1$ the wavenumber of the vibrational contribution to the electronic transition changes from $\tilde{\nu}$ to

$$\tilde{\nu} + \tilde{B}'(J-1)J - \tilde{B}J(J+1) = \tilde{\nu} - (\tilde{B}' + \tilde{B})J + (\tilde{B}' - \tilde{B})J^2$$

This transition is a line in the P branch (just as in Topic 11C). There are corresponding lines in the Q and R branches with wavenumbers that are calculated similarly. All three branches are:

$$\text{P branch } (\Delta J = -1){:}\quad \tilde{\nu}_P(J) = \tilde{\nu} - (\tilde{B}' + \tilde{B})J + (\tilde{B}' - \tilde{B})J^2 \qquad (11F.7a)$$

$$\text{Q branch } (\Delta J = 0){:}\quad \tilde{\nu}_Q(J) = \tilde{\nu} + (\tilde{B}' - \tilde{B})J(J+1) \qquad (11F.7b)$$

$$\text{R branch } (\Delta J = +1){:}\quad \tilde{\nu}_R(J) = \tilde{\nu} + (\tilde{B}' + \tilde{B})(J+1) + (\tilde{B}' - \tilde{B})(J+1)^2 \qquad (11F.7c)$$

Spectroscopic data may be analysed by the method of 'combination differences' introduced in Topic 11C, as shown in the following *Example*.

Example 11F.2 Estimating rotational constants from electronic spectra

The following rotational transitions were observed in the 0–0 band of the $^1\Sigma^+ \leftarrow {}^1\Sigma^+$ electronic transition of $^{63}Cu^2H$: $\tilde{\nu}_R(3) =$ 23 347.69 cm^{-1}, $\tilde{\nu}_P(3) = 23298.85$ cm^{-1}, and $\tilde{\nu}_P(5) = 23275.77$ cm^{-1}. Estimate the values of $\tilde{B}'$ and $\tilde{B}$.

Collect your thoughts According to the method of combination differences, form the differences $\tilde{\nu}_R(J) - \tilde{\nu}_P(J)$ and $\tilde{\nu}_R(J-1) - \tilde{\nu}_P(J+1)$ from eqns 11F.7a and 11F.7c, then use the resulting expressions to calculate the rotational constants $\tilde{B}'$ and $\tilde{B}$ from the data provided.

The solution From eqns 11F.7a and 11F.7c it follows that the differences are

$$\begin{aligned}\tilde{\nu}_R(J) - \tilde{\nu}_P(J) &= (\tilde{B}' + \tilde{B})(J+1) + (\tilde{B}' - \tilde{B})(J+1)^2 \\ &\quad -\{-(\tilde{B}' + \tilde{B})J + (\tilde{B}' - \tilde{B})J^2\} = 4\tilde{B}'\left(J + \tfrac{1}{2}\right)\end{aligned}$$

$$\tilde{\nu}_R(J-1)-\tilde{\nu}_P(J+1)=(\tilde{B}'+\tilde{B})J+(\tilde{B}'-\tilde{B})J^2$$
$$-\{-(\tilde{B}'+\tilde{B})(J+1)+(\tilde{B}'-\tilde{B})(J+1)^2\}$$
$$=4\tilde{B}\left(J+\tfrac{1}{2}\right)$$

(These equations are analogous to eqn 11C.14.) Then insert the data:

$$\text{For } J=3: \quad \tilde{\nu}_R(3)-\tilde{\nu}_P(3)=\overbrace{48.84}^{23347.69-23298.85}\ \text{cm}^{-1}=14\tilde{B}'$$

$$\text{For } J=4: \quad \tilde{\nu}_R(3)-\tilde{\nu}_P(5)=\overbrace{71.92}^{23347.69-23275.77}\ \text{cm}^{-1}=18\tilde{B}$$

Therefore, $\tilde{B}'=3.489\ \text{cm}^{-1}$ and $\tilde{B}=3.996\ \text{cm}^{-1}$.

Self-test 11F.2 The following rotational transitions were observed in the ${}^1\Sigma^+ \leftarrow {}^1\Sigma^+$ electronic transition of RhN: $\tilde{\nu}_R(5)=22\,387.06\ \text{cm}^{-1}$, $\tilde{\nu}_P(5)=22\,376.87\ \text{cm}^{-1}$, and $\tilde{\nu}_P(7)=22\,373.95\ \text{cm}^{-1}$. Estimate the values of $\tilde{B}'$ and $\tilde{B}$.

Answer: $\tilde{B}'=0.4632\ \text{cm}^{-1}$, $\tilde{B}=0.5042\ \text{cm}^{-1}$

If the bond length in the electronically excited state is greater than that in the ground state, then $\tilde{B}'<\tilde{B}$ and therefore $\tilde{B}'-\tilde{B}<0$. In this case the lines of the R branch converge with increasing J. At sufficiently high values of J, the negative term in $(J+1)^2$ in eqn 11F.7c will dominate the positive term in $(J+1)$ and the lines will start to appear at successively decreasing wavenumbers. That is, the R branch has a **band head** (Fig. 11F.11a) corresponding to the line of highest frequency achieved in the branch.

The value of J at which the band head appears can be identified by finding the maximum wavenumber of the lines in the R branch; that is, by identifying an integral value of J close to where $d\tilde{\nu}_R(J)/dJ=0$. This maximum occurs at $J_{max}\approx(\tilde{B}-3\tilde{B}')/2(\tilde{B}'-\tilde{B})$. When the bond is shorter in the excited state than in the ground state, $\tilde{B}'>\tilde{B}$ and $\tilde{B}'-\tilde{B}>0$. In this case, the lines of the P branch begin to converge and form a band head when $J_{max}\approx(\tilde{B}'+\tilde{B})/2(\tilde{B}'-\tilde{B})$, as shown in Fig. 11F.11b.

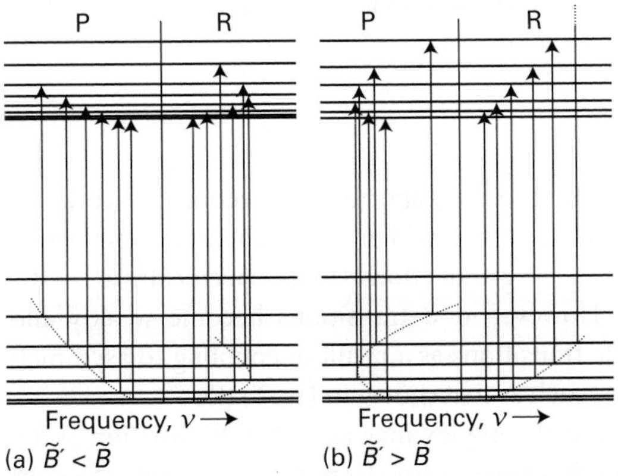

Figure 11F.11 (a) The formation of a head in the R branch when $\tilde{B}'<\tilde{B}$; (b) the formation of a head in the P branch when $\tilde{B}'>\tilde{B}$. The dotted curve shows the wavenumbers of the lines in the two branches as they spread away from the centre of the band.

Brief illustration 11F.6

For the transition described in *Example* 11F.2, $\tilde{B}'<\tilde{B}$, so a band head is expected in the R branch. The approximate value of J at which this occurs is given by

$$J_{max}\approx\frac{\tilde{B}-3\tilde{B}'}{2(\tilde{B}'-\tilde{B})}=\frac{(3.996\ \text{cm}^{-1})-3\times(3.489\ \text{cm}^{-1})}{2\times\{(3.489\ \text{cm}^{-1})-(3.996\ \text{cm}^{-1})\}}=6.38$$

The closest integer is $J=6$.

Exercises

E11F.1 What is the full term symbol of the ground electronic state of Li_2^+?

E11F.2 One of the excited states of the C_2 molecule has the valence electron configuration $1\sigma_g^2 1\sigma_u^2 1\pi_u^3 1\pi_g^1$. Give the multiplicity and parity of the term.

E11F.3 Which of the following transitions are electric-dipole allowed? (i) ${}^2\Pi \leftrightarrow {}^2\Pi$, (ii) ${}^1\Sigma \leftrightarrow {}^1\Sigma$, (iii) $\Sigma \leftrightarrow \Delta$, (iv) $\Sigma^+ \leftrightarrow \Sigma^-$, (v) $\Sigma^+ \leftrightarrow \Sigma^+$.

E11F.4 The R-branch of the ${}^1\Pi_u \leftarrow {}^1\Sigma_g^+$ transition of H_2 shows a band head at the very low value of $J=1$. The rotational constant of the ground state is $60.80\ \text{cm}^{-1}$. What is the rotational constant of the upper state? Has the bond length increased or decreased in the transition?

11F.2 Polyatomic molecules

The absorption of a photon can often be traced to the excitation of electrons that belong to a small group of atoms in a polyatomic molecule. For example, when a carbonyl group (C=O) is present, an absorption at about 290 nm is normally observed. Groups with characteristic optical absorption bands are called **chromophores** (from the Greek for 'colour bringer'), and their presence often accounts for the colours of substances (Table 11F.2).

Table 11F.2 Absorption characteristics of some groups and molecules*

Group	$\tilde{\nu}$/cm^{-1}	λ_{max}/nm	ε_{max}/(dm^3 mol^{-1} cm^{-1})
C=C ($\pi^* \leftarrow \pi$)	61 000	163	15 000
C=O ($\pi^* \leftarrow$ n)	35 000–37 000	270–290	10–20
H_2O	60 000	167	7000

* More values are given in the *Resource section*; ε_{max} is the molar absorption coefficient (see Topic 11A). The wavenumbers and wavelengths are the values for maximum absorption.

11F.2(a) d-Metal complexes

In a free atom, all five d orbitals of a given shell are degenerate. In a d-metal complex, where the immediate environment of the atom is no longer spherical, the d orbitals are not all degenerate, and the complex can absorb photons. The result is promotion of electrons from orbitals of lower energy to orbitals of higher energy.

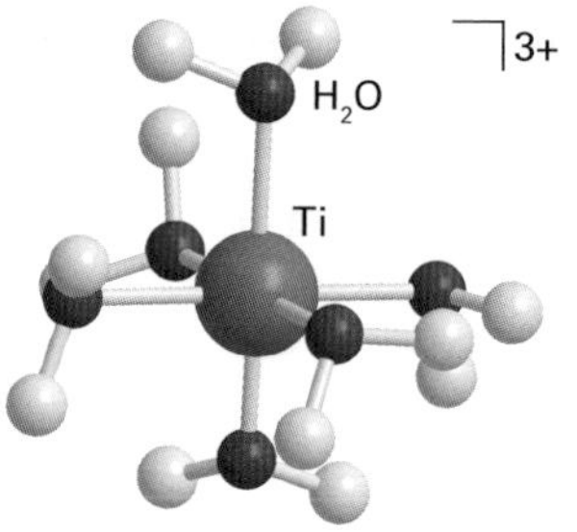

1

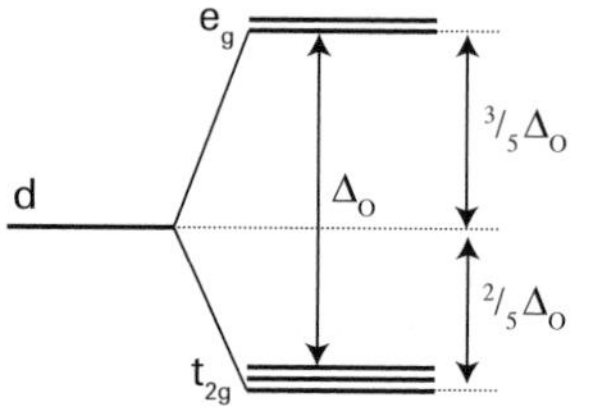

2 $[Ti(OH_2)_6]^{3+}$

To see the origin of this splitting in an octahedral complex such as $[Ti(OH_2)_6]^{3+}$ (**1**), the six ligands can be regarded as point negative charges that repel the d electrons of the central ion (Fig. 11F.12). As a result, the orbitals fall into two groups, with $d_{x^2-y^2}$ and d_{z^2} pointing directly towards the ligand positions, and d_{xy}, d_{yz}, and d_{zx} pointing between them. An electron occupying an orbital of the former group has a less favourable potential energy than when it occupies any of the three orbitals of the other group, and so the d orbitals split into the two sets shown in (**2**): a triply degenerate set comprising the d_{xy}, d_{yz}, and d_{zx} orbitals and labelled t_{2g}, and a doubly degenerate set comprising the $d_{x^2-y^2}$ and d_{z^2} orbitals and labelled e_g (these symmetry labels are explained in Topic 10B). The t_{2g} orbitals lie below the e_g orbitals in energy; the difference in energy Δ_O is called the **ligand field splitting parameter** (the o denotes octahedral symmetry). The ligand field splitting is typically about 10 per cent of the overall energy of interaction between the ligands and the central metal ion, which is largely responsible for the existence of the complex. The d orbitals also divide into two sets in a tetrahedral complex, but in this case the two e orbitals lie below the three t_2 orbitals (no g or u label is given because a tetrahedral complex has no centre of inversion); the separation of these groups of orbitals is written Δ_T.

The values of Δ_O and Δ_T are such that transitions between the two sets of orbitals typically occur in the visible region of the spectrum. The transitions are responsible for many of the colours that are so characteristic of d-metal complexes.

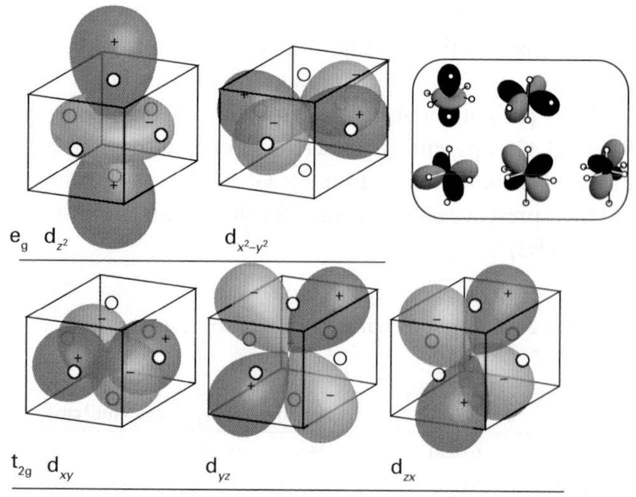

Figure 11F.12 The classification of d orbitals in an octahedral environment. The open circles represent the positions of the six (point-charge) ligands.

Brief illustration 11F.7

The spectrum of $[Ti(OH_2)_6]^{3+}$ near 24 000 cm^{-1} (500 nm) is shown in Fig. 11F.13, and can be ascribed to the promotion of its single d electron from a t_{2g} orbital to an e_g orbital. The wavenumber of the absorption maximum suggests that $\Delta_O \approx 24\,000\ \mathrm{cm^{-1}}$ for this complex, which corresponds to about 3.0 eV.

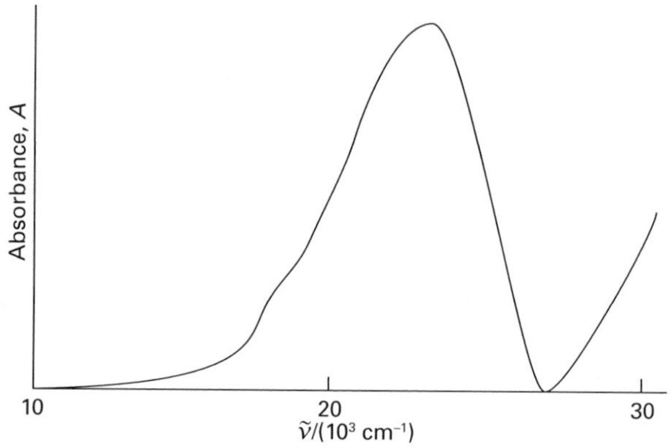

Figure 11F.13 The electronic absorption spectrum of $[Ti(OH_2)_6]^{3+}$ in aqueous solution.

According to the Laporte rule (Section 11F.1b), d–d transitions are parity-forbidden in octahedral complexes because they are g → g transitions (more specifically, $e_g \leftarrow t_{2g}$ transitions). However, d–d transitions become weakly allowed as vibronic transitions as a result of coupling to asymmetrical vibrations such as that shown in Fig. 11F.5.

A d-metal complex may also absorb radiation as a result of the transfer of an electron from the ligands into the d orbitals of the central atom, or vice versa. In such **charge-transfer transitions** the electron moves through a considerable distance, which means that the transition dipole moment may be large and the absorption correspondingly intense. In the permanganate ion,

MnO_4^-, the charge redistribution that accompanies the migration of an electron from the O atoms to the central Mn atom results in a strong transition in the range 475–575 nm that accounts for the intense purple colour of the ion.

An electronic migration from the ligands to the metal corresponds to a **ligand-to-metal charge-transfer transition** (LMCT). The reverse migration, a **metal-to-ligand charge-transfer transition** (MLCT), can also occur. An example is the migration of a d electron onto the antibonding π orbitals of an aromatic ligand. The resulting excited state may have a very long lifetime if the electron is extensively delocalized over several aromatic rings.

In common with other transitions, the intensities of charge-transfer transitions are proportional to the square of the transition dipole moment. The transition moment can be thought of as a measure of the distance moved by the electron as it migrates between metal and ligand, with a large distance of migration corresponding to a large transition dipole moment and therefore a high intensity of absorption. However, the integrand in the expression for the transition moment is proportional to the product of the initial and final wavefunctions; this product is zero unless the two wavefunctions have non-zero values in the same region of space. Therefore, although large distances of migration favour high intensities, the diminished overlap of the initial and final wavefunctions for large separations of metal and ligands favours low intensities (see Problem P11F.9).

11F.2(b) $\pi^* \leftarrow \pi$ and $\pi^* \leftarrow$ n transitions

Absorption by a C=C double bond results in the excitation of a π electron into an antibonding π^* orbital (Fig. 11F.14). The chromophore activity is therefore due to a **$\pi^* \leftarrow \pi$ transition** (a 'π to π-star transition'). Its energy is about 6.9 eV for an unconjugated double bond, which corresponds to an absorption at 180 nm (in the ultraviolet). When the double bond is part of a conjugated chain, the energies of the molecular orbitals lie closer together and the $\pi^* \leftarrow \pi$ transition moves to a longer wavelength; it may even lie in the visible region if the conjugated system is long enough.

One of the transitions responsible for absorption in carbonyl compounds can be traced to the lone pairs of electrons on the O atom. The Lewis concept of a 'lone pair' of electrons is represented in molecular orbital theory by a pair of electrons in an orbital confined largely to one atom and not appreciably involved in bond formation. One of these electrons may be excited into an empty π^* orbital of the carbonyl group (Fig. 11F.15), which gives rise to an **$\pi^* \leftarrow$ n transition** (an 'n to π-star transition'). Typical absorption energies are about 4.3 eV (290 nm). Because $\pi^* \leftarrow$ n transitions in carbonyls are symmetry forbidden, the absorptions are weak. By contrast, the $\pi^* \leftarrow \pi$ transition in a carbonyl, which corresponds to excitation of a π electron of the C=O double bond, is allowed by symmetry and results in relatively strong absorption.

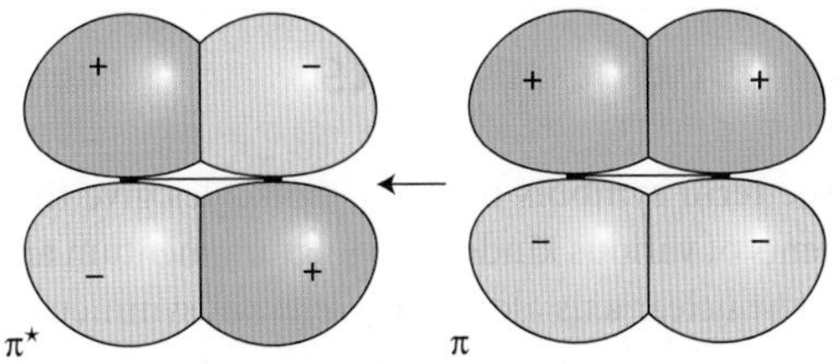

Figure 11F.14 A C=C double bond acts as a chromophore. In the $\pi^* \leftarrow \pi$ transition illustrated here, an electron is promoted from a π orbital to the corresponding antibonding orbital.

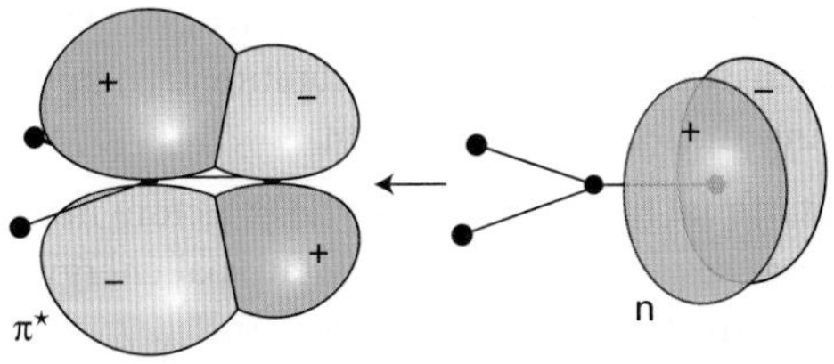

Figure 11F.15 A carbonyl group (C=O) acts as a chromophore partly on account of the excitation of a nonbonding O lone-pair electron to an antibonding CO π^* orbital; this transition is denoted $\pi^* \leftarrow$ n.

Brief illustration 11F.8

The compound $CH_3CH{=}CHCHO$ has a strong absorption in the ultraviolet at 46 950 cm^{-1} (213 nm) and a weak absorption at 30 000 cm^{-1} (330 nm). The former is a $\pi^* \leftarrow \pi$ transition associated with the delocalized π system C=C–C=O. Delocalization extends the range of the C=O $\pi^* \leftarrow \pi$ transition to lower wavenumbers (longer wavelengths). The latter is an $\pi^* \leftarrow$ n transition associated with the carbonyl chromophore.

Exercises

E11F.5 The complex ion $[Fe(OH_2)_6]^{3+}$ has an electronic absorption spectrum with a maximum at 700 nm. Estimate a value of Δ_O for the complex.

E11F.6 The two compounds 2,3-dimethyl-2-butene and 2,5-dimethyl-2,4-hexadiene are to be distinguished by their ultraviolet absorption spectra. The maximum absorption in one compound occurs at 192 nm and in the other at 243 nm. Match the maxima to the compounds and justify the assignment.

Checklist of concepts

☐ 1. The **term symbols** of linear molecules give the components of various kinds of angular momentum around the internuclear axis along with relevant symmetry labels.

☐ 2. The **Laporte selection rule** states that, for centrosymmetric molecules, only u → g and g → u transitions are allowed.

☐ 3. The **Franck–Condon principle** asserts that electronic transitions occur within an unchanging nuclear framework.

☐ 4. **Vibrational fine structure** is the structure in a spectrum that arises from changes in vibrational energy accompanying an electronic transition.

☐ 5. **Rotational fine structure** is the structure in a spectrum that arises from changes in rotational energy accompanying an electronic transition.

☐ 6. In gas phase samples, rotational fine structure can be resolved and under some circumstances **band heads** are formed.

☐ 7. In d-metal complexes, the presence of ligands removes the degeneracy of d orbitals and vibrationally allowed **d–d transitions** can occur between them.

☐ 8. **Charge-transfer transitions** typically involve the migration of electrons between the ligands and the central metal atom.

☐ 9. A **chromophore** is a group with a characteristic optical absorption band.

Checklist of equations

Property	Equation	Comment	Equation number
Selection rules (angular momentum)	$\Delta\Lambda = 0, \pm 1; \Delta S = 0; \Delta\Sigma = 0; \Delta\Omega = 0, \pm 1$	Linear molecules	11F.4
Franck–Condon factor	$\lvert S(v_f, v_i)\rvert^2 = \left(\int \psi^*_{v,f}\psi_{v,i}\,d\tau_N\right)^2$		11F.5
Rotational structure of electronic spectra (diatomic molecules)	$\tilde{\nu}_P(J) = \tilde{\nu} - (\tilde{B}' + \tilde{B})J + (\tilde{B}' - \tilde{B})J^2$	P branch ($\Delta J = -1$)	11F.7a
	$\tilde{\nu}_Q(J) = \tilde{\nu} + (\tilde{B}' - \tilde{B})J(J+1)$	Q branch ($\Delta J = 0$)	11F.7b
	$\tilde{\nu}_R(J) = \tilde{\nu} + (\tilde{B}' + \tilde{B})(J+1) + (\tilde{B}' - \tilde{B})(J+1)^2$	R branch ($\Delta J = +1$)	11F.7c

TOPIC 11G Decay of excited states

➤ Why do you need to know this material?

Information about the electronic structure of a molecule can be obtained by observing the radiative decay of excited electronic states back to the ground state. Such decay is also used in lasers, which are of exceptional technological importance.

➤ What is the key idea?

Molecules in excited electronic states discard their excess energy by emission of electromagnetic radiation, transfer as heat to the surroundings, or fragmentation.

➤ What do you need to know already?

You need to be familiar with electronic transitions in molecules (Topic 11F), the difference between spontaneous and stimulated emission of radiation (Topic 11A), and the general features of spectroscopy (Topic 11A). You need to be aware of the difference between singlet and triplet states (Topic 8C) and of the Franck–Condon principle (Topic 11F).

Radiative decay is a process in which a molecule discards its excitation energy as a photon (Topic 11A); depending on the nature of the excited state, this process is classified as either 'fluorescence' or 'phosphorescence'. A more common fate of an electronically excited molecule is **non-radiative decay**, in which the excess energy is transferred into the vibration, rotation, and translation of the surrounding molecules. This thermal degradation converts the excitation energy into thermal motion of the environment (i.e. into 'heat'). An excited molecule may also dissociate or take part in a chemical reaction (Topic 17G).

11G.1 Fluorescence and phosphorescence

In **fluorescence**, spontaneous emission of radiation occurs while the sample is being irradiated and ceases as soon as the exciting radiation is extinguished (Fig. 11G.1). In **phosphorescence**, the spontaneous emission may persist for long periods (even hours, but more commonly seconds or fractions of seconds). The difference suggests that fluorescence is a fast conversion of absorbed radiation into re-emitted energy, and that phosphorescence involves the storage of energy in a reservoir from which it slowly leaks.

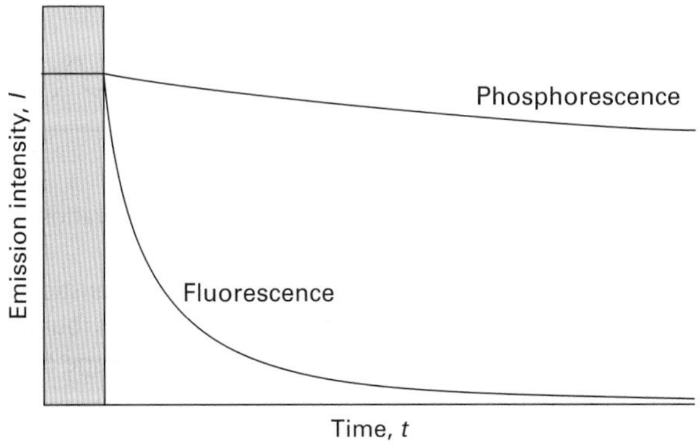

Figure 11G.1 The empirical (observation-based) distinction between fluorescence and phosphorescence is that the former is extinguished very quickly after the exciting radiation (indicated by the tinted area on the left) is removed, whereas the latter continues with relatively slowly diminishing intensity.

Figure 11G.2 shows the sequence of steps involved in fluorescence from molecules in solution. The initial stimulated absorption takes the molecule to an excited electronic state; if the absorption spectrum were monitored it would look like the one shown in blue in Fig. 11G.3. The excited molecule is subjected to collisions with the surrounding molecules. As it

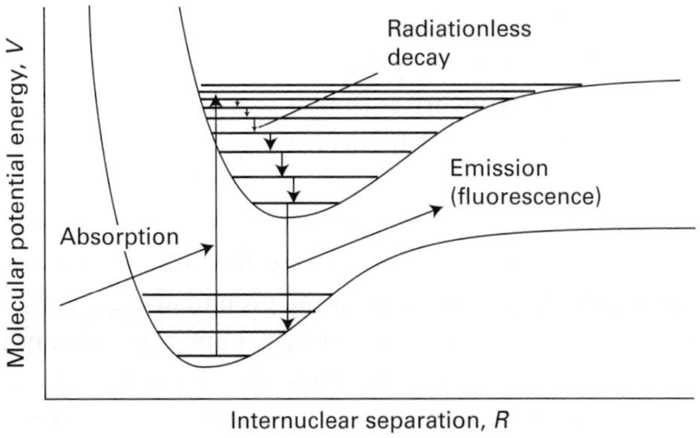

Figure 11G.2 The sequence of steps leading to fluorescence by molecules in solution. After the initial absorption, the upper vibrational states undergo radiationless decay by giving up energy to the surrounding molecules. A radiative transition then occurs from the vibrational ground state of the upper electronic state.

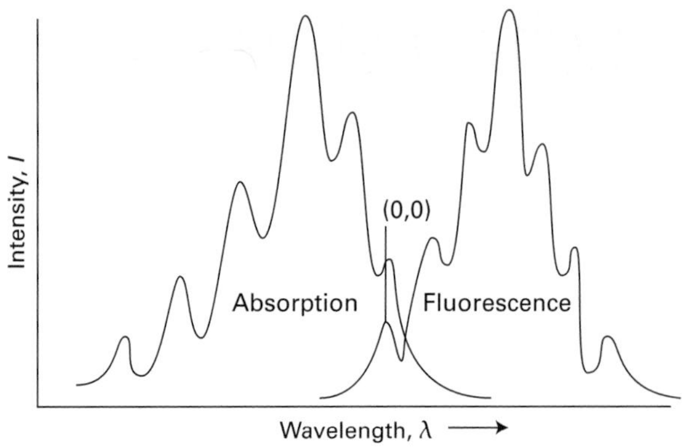

Figure 11G.3 An absorption spectrum (left) shows vibrational structure characteristic of the upper electronic state. A fluorescence spectrum (right) shows structure characteristic of the lower state; it is also displaced to lower frequencies (but the 0–0 transitions are coincident) and is often a mirror image of the absorption spectrum.

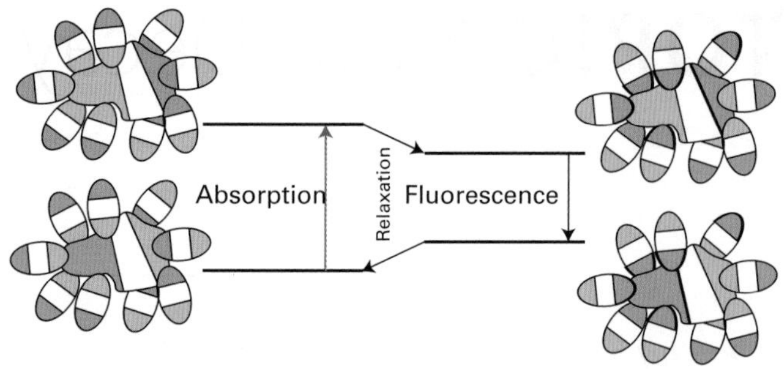

Figure 11G.4 Solvent–solute interactions can result in a shift in the fluorescence spectrum relative to the absorption spectrum. On the left absorption occurs with the solvent (depicted by the ellipsoids) in the arrangement characteristic of the ground electronic state of the molecule (the central blob). However, before fluorescence occurs, the solvent molecules relax into a new arrangement, and that arrangement is preserved during the subsequent radiative transition, the fluorescence.

gives up energy to them non-radiatively it steps down (typically within picoseconds) the ladder of vibrational levels to the lowest vibrational level of the excited electronic state. The surrounding molecules, however, might now be unable to accept the larger energy difference needed to lower the molecule to the ground electronic state. The excited electronic state might therefore survive long enough to undergo spontaneous emission and emit the remaining excess energy as radiation. The downward electronic transition is vertical, in accord with the Franck–Condon principle (Topic 11F), and the fluorescence spectrum has vibrational structure characteristic of the *lower* electronic state (the spectrum shown on the right in Fig. 11G.3b).

Provided they can be seen, the 0–0 absorption and fluorescence transitions (where the numbers are the values of v_f and v_i, the vibrational quantum numbers for the final and initial states) can be expected to be coincident. The absorption spectrum arises from $0 \leftarrow 0$, $1 \leftarrow 0$, $2 \leftarrow 0, \ldots$ transitions, which occur at progressively higher wavenumbers (shorter wavelengths) and with intensities governed by the Franck–Condon principle. The fluorescence spectrum arises from $0 \rightarrow 0$, $0 \rightarrow 1, \ldots$ downward transitions, which occur with decreasing wavenumbers (longer wavelengths).

The 0–0 absorption and fluorescence peaks are not always exactly coincident, however, because the solvent may interact differently with the solute in the ground and excited electronic states (for instance, the hydrogen bonding pattern might differ). Because the solvent molecules do not have time to rearrange during the transition, the absorption occurs in an environment characteristic of the solvated ground state; however, the fluorescence occurs in an environment characteristic of the solvated excited state (Fig. 11G.4).

Fluorescence occurs at lower frequencies than the incident radiation because the emissive transition occurs after some vibrational energy has been discarded into the surroundings. The vivid oranges and greens of fluorescent dyes are an everyday manifestation of this effect: they absorb in the ultraviolet and blue, and fluoresce in the visible. The mechanism also suggests that the intensity of the fluorescence ought to depend on the ability of the solvent molecules to accept the electronic and vibrational quanta. It is indeed found that a solvent composed of molecules with widely spaced vibrational levels (such as water) can in some cases accept the large quantum of electronic energy and so extinguish, or 'quench', the fluorescence. The rate at which fluorescence is quenched by other molecules also gives valuable kinetic information (Topic 17G).

Figure 11G.5 shows the sequence of events leading to phosphorescence for a molecule with a singlet ground state (denoted S_0). The first steps are the same as in fluorescence, but the

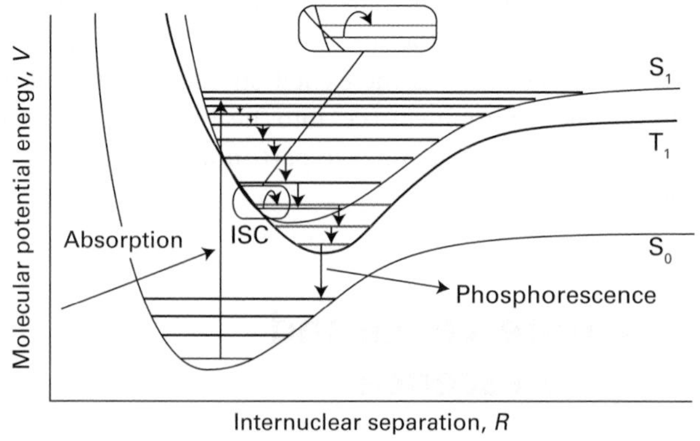

Figure 11G.5 The sequence of steps leading to phosphorescence. The important step is the intersystem crossing (ISC), the switch from a singlet state (S_1) to a triplet state (T_1) brought about by spin–orbit coupling. The triplet state acts as a slowly radiating reservoir because the return to the ground state is spin-forbidden.

presence of a triplet excited state (T_1) at an energy close to that of the singlet excited state (S_1) plays a decisive role. The singlet and triplet excited states share a common geometry at the point where their potential energy curves intersect. Hence, if there is a mechanism for unpairing two electron spins (and achieving the conversion of ↑↓ to ↑↑), the molecule may undergo **intersystem crossing**, a non-radiative transition between states of different multiplicity, and become a triplet state. As in the discussion of atomic spectra (Topic 8C), singlet–triplet transitions may occur in the presence of spin–orbit coupling. Intersystem crossing is expected to be important when a molecule contains a moderately heavy atom (such as sulfur), because then the spin–orbit coupling is large.

Once an excited molecule has crossed into a triplet state, it continues to discard energy into the surroundings. However, it is now stepping down the triplet's vibrational ladder and ends in its lowest vibrational level. The triplet state is lower in energy than the corresponding singlet state (Hund's rule, Topic 8B). The solvent cannot absorb the final, large quantum of electronic excitation energy, and the molecule cannot radiate its energy because return to the ground state is spin-forbidden. The radiative transition, however, is not totally forbidden because the spin–orbit coupling that was responsible for the intersystem crossing also weakens the selection rule. The molecules are therefore able to emit weakly, and the emission may continue long after the original excited state was formed.

This mechanism accounts for the observation that the excitation energy seems to get trapped in a slowly leaking reservoir. It also suggests (as is confirmed experimentally) that phosphorescence should be most intense from solid samples: energy transfer is then less efficient and intersystem crossing has time to occur as the singlet excited state steps slowly past the intersection point. The mechanism also suggests that the phosphorescence efficiency should depend on the presence of a moderately heavy atom (with significant spin–orbit coupling), which is in fact the case.

The various types of non-radiative and radiative transitions that can occur in molecules are often represented on a schematic **Jablonski diagram** of the type shown in Fig. 11G.6.

Brief illustration 11G.1

Fluorescence efficiency decreases, and the phosphorescence efficiency increases, in the series of compounds: naphthalene, 1-chloronaphthalene, 1-bromonaphthalene, 1-iodonaphthalene. The replacement of an H atom by successively heavier atoms enhances intersystem crossing from S_1 into T_1, thereby decreasing the efficiency of fluorescence. The rate of the radiative transition from T_1 to S_0 is also enhanced by the presence of heavier atoms, thereby increasing the efficiency of phosphorescence.

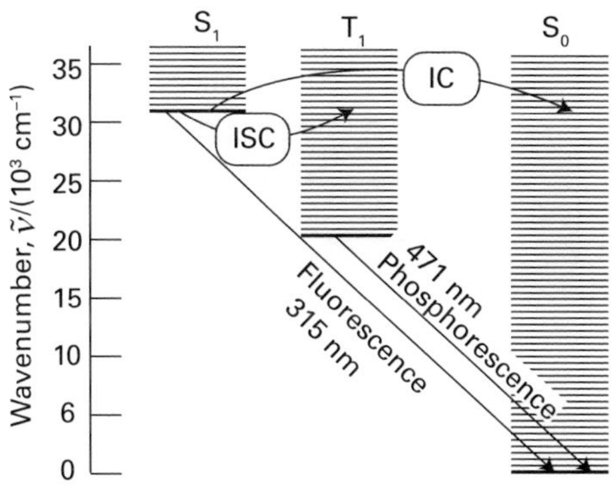

Figure 11G.6 A Jablonski diagram (here, for naphthalene) is a simplified portrayal of the relative positions of the electronic energy levels of a molecule. Vibrational levels of states of a given electronic state lie above each other, but the relative horizontal locations of the columns bear no relation to the nuclear separations in the states. The ground vibrational states of each electronic state are correctly located vertically but the other vibrational states are shown only schematically. (IC: internal conversion; ISC: intersystem crossing.)

11G.2 Dissociation and predissociation

A chemically important fate for an electronically excited molecule is **dissociation**, the breaking of bonds (Fig. 11G.7). The onset of dissociation can be detected in an absorption spectrum by noting that the vibrational fine structure of a band terminates at a certain frequency. Absorption occurs in a continuous band above this **dissociation limit** because the final state is an unquantized translational motion of the fragments. Locating the dissociation limit is a valuable way of determining the bond dissociation energy.

In some cases, the vibrational structure disappears but resumes at higher frequencies of the incident radiation. This

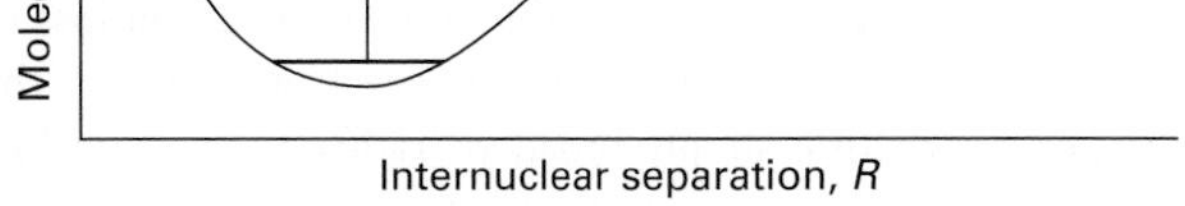

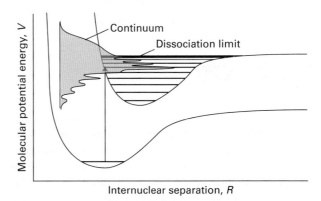

Figure 11G.7 When absorption occurs to unbound states of the upper electronic state, the molecule dissociates and the absorption spectrum is a continuum. Below the dissociation limit the electronic spectrum shows a normal vibrational structure.

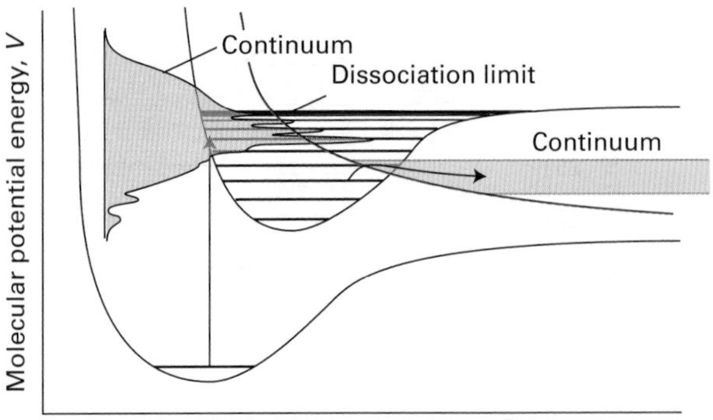

Figure 11G.8 When a dissociative state crosses a bound state, as in the upper part of the illustration, molecules excited to levels near the crossing may dissociate. This predissociation is detected in the spectrum as a loss of vibrational structure that resumes at higher frequencies.

effect provides evidence of **predissociation**, which can be interpreted in terms of the molecular potential energy curves shown in Fig. 11G.8. When a molecule is excited to a high vibrational level of the upper electronic state, its electrons may undergo a redistribution that results in it undergoing an **internal conversion**, a radiationless conversion to another electronic state of the same multiplicity. An internal conversion occurs most readily at the point of intersection of the two molecular potential energy curves, because there the nuclear geometries of the two electronic states are the same, as are their energies. The state into which the molecule converts may be dissociative, so the states near the intersection have a finite lifetime and hence their energies are imprecisely defined (as a result of lifetime broadening, Topic 11A). As a result, the absorption spectrum is blurred. When the incoming photon brings enough energy to excite the molecule to a vibrational level high above the intersection, the internal conversion does not occur (the nuclei are unlikely to have the same geometry). Consequently, the levels resume their well-defined, vibrational character with correspondingly well-defined energies, and the line structure resumes on the high-frequency side of the blurred region.

Brief illustration 11G.2

The O_2 molecule absorbs ultraviolet radiation in a transition from its ${}^3\Sigma_g^-$ ground electronic state to a ${}^3\Sigma_u^-$ excited state that is energetically close to a dissociative ${}^3\Pi_u$ state. In this case, the effect of predissociation is more subtle than the abrupt loss of vibrational–rotational structure in the spectrum; instead, the vibrational structure simply broadens rather than being lost completely. As before, the broadening is explained by the short lifetimes of the excited vibrational states near the intersection of the curves describing the bound and dissociative excited electronic states.

Exercise

E11G.1 An oxygen molecule absorbs ultraviolet radiation in a transition from its ${}^3\Sigma_g^-$ ground electronic state to an excited state that is energetically close to a dissociative ${}^5\Pi_u$ state. The absorption band has a relatively large experimental linewidth. Account for this observation.

11G.3 Lasers

An excited state can be driven to discard its excess energy by using radiation to induce stimulated emission. The word laser is an acronym formed from **l**ight **a**mplification by **s**timulated **e**mission of **r**adiation. In stimulated emission (Topic 11A), an excited state is stimulated to emit a photon by radiation of the same frequency: the more photons that are present, the greater is the probability of emission.

Laser radiation has a number of striking characteristics (Table 11G.1). Each of them (sometimes in combination with the others) opens up opportunities for applications in physical chemistry. Raman spectroscopy (Topics 11B–D) has flourished on account of the high intensity of monochromatic radiation available from lasers, and the ultrashort pulses that lasers can generate make possible the study of light-initiated reactions on timescales of femtoseconds and even attoseconds.

One requirement of laser action is the existence of a **metastable excited state**, an excited state with a long enough lifetime for it to undergo stimulated emission. For stimulated emission to dominate absorption, it is necessary for there to be a **population inversion** in which the population of the excited state is greater than that of the lower state. Figure 11G.9 illustrates one way to achieve population inversion indirectly

Table 11G.1 Characteristics of laser radiation and their chemical applications

Characteristic	Advantage	Application
High power	Multiphoton process	Spectroscopy
	Low detector noise	Improved sensitivity
	High scattering intensity	Raman spectroscopy (Topics 11B–11D)
Monochromatic	High resolution	Spectroscopy
	State selection	Photochemical studies (Topic 17G)
		Reaction dynamics (Topic 18D)
Collimated beam	Long path lengths	Improved sensitivity
	Forward-scattering observable	Raman spectroscopy (Topics 11B–11D)
Pulsed	Precise timing of excitation	Fast reactions (Topics 17G, 18C)
		Relaxation (Topic 17C)
		Energy transfer (Topic 17G)

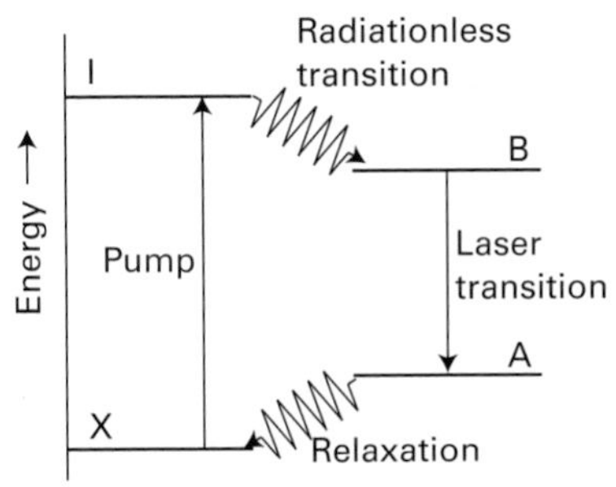

Figure 11G.9 The transitions involved in a four-level laser. Because the laser transition terminates in an excited state (A), the population inversion between A and B is much easier to achieve than when the lower state of the laser transition is the ground state, X.

through an intermediate state I. In this process, the molecule or atom is excited to I, which then gives up some of its energy non-radiatively (for example, by passing energy on to vibrations of the surroundings) and decays to a lower state B. Four levels are involved overall, so this arrangement is termed a **four-level laser**. One advantage of this arrangement is that the population inversion of the A and B levels is easier to achieve than one involving the heavily populated ground state. The transition from X to I is caused by irradiation with intense light (either continuously or as a flash) in the process called **pumping**. In some cases pumping is achieved with an electric discharge through xenon or with the radiation from another laser.

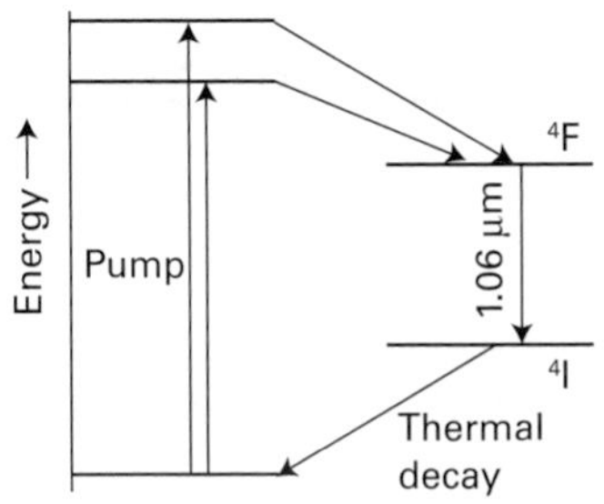

Figure 11G.10 The transitions involved in a neodymium laser.

Brief illustration 11G.3

The neodymium laser is an example of a four-level solid-state laser (Fig. 11G.10). In one form it consists of Nd^{3+} ions at low concentration in yttrium aluminium garnet (YAG, specifically $Y_3Al_5O_{12}$), and is then known as an Nd:YAG laser. A neodymium laser operates at a number of wavelengths in the infrared. The most common wavelength of operation is 1064 nm, which corresponds to the electronic transition from the ^4F to the ^4I state of the Nd^{3+} ion.

Many of the most important laser systems are solid-state devices; they are discussed in Topic 15G.

Checklist of concepts

☐ 1. **Fluorescence** is radiative decay between states of the same multiplicity; it ceases soon after the exciting radiation is removed.

☐ 2. **Phosphorescence** is radiative decay between states of different multiplicity; it persists after the exciting radiation is removed.

☐ 3. **Intersystem crossing** is the non-radiative conversion to an electronic state of different multiplicity.

☐ 4. A **Jablonski diagram** is a schematic diagram showing the types of non-radiative and radiative transitions that can occur in molecules.

☐ 5. An additional fate of an electronically excited species is **dissociation**.

☐ 6. **Internal conversion** is a non-radiative conversion to an electronic state of the same multiplicity.

☐ 7. **Predissociation** is the observation of the effects of dissociation before the dissociation limit is reached.

☐ 8. **Laser action** is the stimulated emission of coherent radiation between states between which there is a population inversion.

☐ 9. A **metastable excited state** is an excited state with a long enough lifetime for it to undergo stimulated emission.

☐ 10. A **population inversion** is a condition in which the population of an upper state is greater than that of a relevant lower state.

☐ 11. **Pumping**, the stimulation of an absorption with an external source of intense radiation, is a process by which a population inversion is created.

FOCUS 11 Molecular spectroscopy

To test your understanding of this material, work through the *Exercises, Additional exercises, Discussion questions*, and *Problems* found throughout this Focus.

Selected solutions can be found at the end of this Focus in the e-book. Solutions to even-numbered questions are available online only to lecturers.

Note: The masses of nuclides are listed in the *Resource section*.

TOPIC 11A General features of molecular spectroscopy

Discussion questions

D11A.1 What is the physical origin of a selection rule?

D11A.2 Describe the physical origins of linewidths in absorption and emission spectra. Do you expect the same contributions for species in condensed and gas phases?

D11A.3 Describe the basic experimental arrangements commonly used for absorption, emission, and Raman spectroscopy.

Additional exercises

E11A.7 Calculate the ratio A/B for transitions with the following characteristics: (i) 500 MHz radiofrequency radiation, (ii) 3.0 cm microwave radiation.

E11A.8 The molar absorption coefficient of a substance dissolved in hexane is known to be 227 $dm^3 mol^{-1} cm^{-1}$ at 290 nm. Calculate the percentage reduction in intensity when ultraviolet radiation of that wavelength passes through 2.00 mm of a solution of concentration 2.52 $mmol\,dm^{-3}$.

E11A.9 When ultraviolet radiation of wavelength 400 nm passes through 2.50 mm of a solution of an absorbing substance at a concentration 0.717 $mmol\,dm^{-3}$, the transmission is 61.5 per cent. Calculate the molar absorption coefficient of the solute at this wavelength. Express your answer in square centimetres per mole ($cm^2 mol^{-1}$).

E11A.10 The molar absorption coefficient of a solute at 540 nm is 386 $dm^3 mol^{-1} cm^{-1}$. When light of that wavelength passes through a 5.00 mm cell containing a solution of the solute, 38.5 per cent of the light was absorbed. What is the molar concentration of the solute?

E11A.11 The molar absorption coefficient of a solute at 440 nm is 423 $dm^3 mol^{-1} cm^{-1}$. When light of that wavelength passes through a 6.50 mm cell containing a solution of the solute, 48.3 per cent of the light was absorbed. What is the molar concentration of the solute?

E11A.12 The following data were obtained for the absorption at 450 nm by a dye in carbon tetrachloride when using a 2.0 mm cell. Calculate the molar absorption coefficient of the dye at the wavelength employed:

[dye]/(mol dm^{-3})	0.0010	0.0050	0.0100	0.0500
T/(per cent)	81.4	35.6	12.7	3.0×10^{-3}

E11A.13 The following data were obtained for the absorption at 600 nm by a dye dissolved in methylbenzene using a 2.50 mm cell. Calculate the molar absorption coefficient of the dye at the wavelength employed:

[dye]/(mol dm^{-3})	0.0010	0.0050	0.0100	0.0500
T/(per cent)	72	20	4.0	1.00×10^{-5}

E11A.14 A 2.0 mm cell was filled with a solution of benzene in a non-absorbing solvent. The concentration of the benzene was 0.010 $mol\,dm^{-3}$ and the wavelength of the radiation was 256 nm (where there is a maximum in the absorption). Calculate the molar absorption coefficient of benzene at this wavelength given that the transmission was 48 per cent. What will the transmittance be through a 4.0 mm cell at the same wavelength?

E11A.15 A 5.00 mm cell was filled with a solution of a dye. The concentration of the dye was 18.5 $mmol\,dm^{-3}$. Calculate the molar absorption coefficient of the dye at this wavelength given that the transmission was 29 per cent. What will the transmittance be through a 2.50 mm cell at the same wavelength?

E11A.16 A swimmer enters a gloomier world (in one sense) on diving to greater depths. Given that the mean molar absorption coefficient of sea water in the visible region is 6.2×10^{-5} $dm^3 mol^{-1} cm^{-1}$, calculate the depth at which a diver will experience (i) half the surface intensity of light, (ii) one tenth the surface intensity.

E11A.17 Given that the maximum molar absorption coefficient of a molecule containing a carbonyl group is 30 $dm^3 mol^{-1} cm^{-1}$ near 280 nm, calculate the thickness of a sample that will result in (i) half the initial intensity of radiation, (ii) one tenth the initial intensity. Take the absorber concentration to be 10 $mmol\,dm^{-3}$.

E11A.18 The absorption associated with a particular transition begins at 220 nm, peaks sharply at 270 nm, and ends at 300 nm. The maximum value of the molar absorption coefficient is 2.21×10^4 $dm^3 mol^{-1} cm^{-1}$. Estimate the integrated absorption coefficient of the transition assuming a triangular lineshape.

E11A.19 The absorption associated with a certain transition begins at 167 nm, peaks sharply at 200 nm, and ends at 250 nm. The maximum value of the molar absorption coefficient is $3.35 \times 10^4\,\mathrm{dm^3\,mol^{-1}\,cm^{-1}}$. Estimate the integrated absorption coefficient of the transition assuming an inverted parabolic lineshape (see figure below).

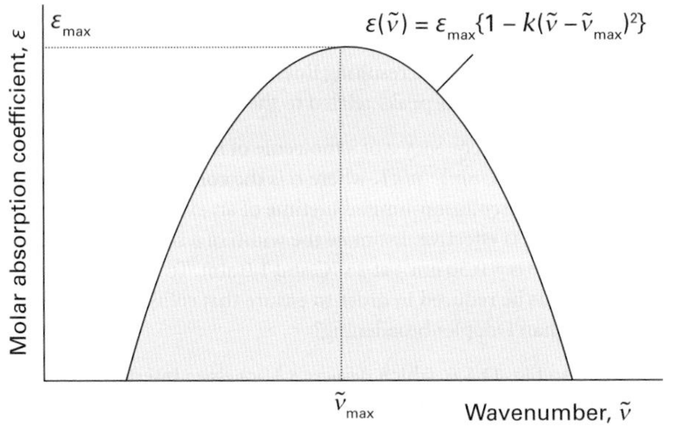

E11A.20 What is the Doppler-broadened linewidth of the vibrational transition at $2308\,\mathrm{cm^{-1}}$ in $^1H^{127}I$ at 400 K?

E11A.21 What is the Doppler-shifted wavelength of a red (680 nm) traffic light approached at $60\,\mathrm{km\,h^{-1}}$?

E11A.22 At what speed of approach would a red (680 nm) traffic light appear green (530 nm)?

E11A.23 Estimate the lifetime of a state that gives rise to a line of width (i) 200 MHz, (ii) $2.45\,\mathrm{cm^{-1}}$.

E11A.24 A molecule in a gas undergoes about 1.0×10^9 collisions in each second. Suppose that (i) every collision is effective in deactivating the molecule rotationally and (ii) that one collision in 10 is effective. Calculate the width (in hertz) of rotational transitions in the molecule.

Problems

P11A.1 The flux of visible photons reaching Earth from the North Star is about $4 \times 10^3\,\mathrm{mm^{-2}\,s^{-1}}$. Of these photons, 30 per cent are absorbed or scattered by the atmosphere and 25 per cent of the surviving photons are scattered by the surface of the cornea of the eye. A further 9 per cent are absorbed inside the cornea. The area of the pupil at night is about $40\,\mathrm{mm^2}$ and the response time of the eye is about 0.1 s. Of the photons passing through the pupil, about 43 per cent are absorbed in the ocular medium. How many photons from the North Star are focused onto the retina in 0.1 s? For a continuation of this story, see R.W. Rodieck, *The first steps in seeing*, Sinauer, Sunderland (1998).

P11A.2 The Beer–Lambert law is derived on the basis that the concentration of absorbing species is uniform. Suppose, instead, that the concentration falls exponentially as $[J] = [J]_0 e^{-x/x_0}$. Develop an expression for the variation of I with sample length; suppose that $L >> x_0$.

P11A.3 It is common to make measurements of absorbance at two wavelengths and use them to find the individual concentrations of two components A and B in a mixture. Show that the molar concentrations of A and B in a cell of length L are

$$[A] = \frac{\varepsilon_{B2}A_1 - \varepsilon_{B1}A_2}{(\varepsilon_{A1}\varepsilon_{B2} - \varepsilon_{A2}\varepsilon_{B1})L} \qquad [B] = \frac{\varepsilon_{A1}A_2 - \varepsilon_{A2}A_1}{(\varepsilon_{A1}\varepsilon_{B2} - \varepsilon_{A2}\varepsilon_{B1})L}$$

where A_1 and A_2 are absorbances of the mixture at wavelengths λ_1 and λ_2, and the molar extinction coefficients of A (and B) at these wavelengths are ε_{A1} and ε_{A2} (and ε_{B1} and ε_{B2}).

P11A.4 When pyridine is added to a solution of iodine in carbon tetrachloride the 520 nm band of absorption shifts toward 450 nm. However, the absorbance of the solution at 490 nm remains constant: this feature is called an *isosbestic point*. Show that an isosbestic point should occur when two absorbing species are in equilibrium. *Hint:* Assume that pyridine and iodine form a 1:1 complex, and that the absorption is due to pyridine and the complex only.

P11A.5‡ Ozone, uniquely among other abundant atmospheric constituents, absorbs ultraviolet radiation in a part of the electromagnetic spectrum energetic enough to disrupt DNA in biological organisms. This spectral range, which is denoted UV-B, spans from about 290 nm to 320 nm. The molar extinction coefficient of ozone over this range is given in the table below (DeMore et al., *Chemical kinetics and photochemical data for use in stratospheric modeling: Evaluation Number 11*, JPL Publication 94–26 (1994)).

λ/nm	292.0	296.3	300.8	305.4	310.1	315.0	320.0
$\varepsilon/(\mathrm{dm^3\,mol^{-1}\,cm^{-1}})$	1512	865	477	257	135.9	69.5	34.5

Evaluate the integrated absorption coefficient of ozone over the wavelength range 290–320 nm. *Hint:* $\varepsilon(\tilde{\nu})$ can be fitted to an exponential function quite well.

P11A.6 In many cases it is possible to assume that an absorption band has a Gaussian lineshape (one proportional to e^{-x^2}) centred on the band maximum. Assume such a lineshape, and show that $\mathcal{A} = \int \varepsilon(\tilde{\nu})\mathrm{d}\tilde{\nu} \approx 1.0645\varepsilon_{max}\Delta\tilde{\nu}_{1/2}$, where $\Delta\tilde{\nu}_{1/2}$ is the width at half-height. The absorption spectrum of azoethane ($CH_3CH_2N_2$) between $24\,000\,\mathrm{cm^{-1}}$ and $34\,000\,\mathrm{cm^{-1}}$ is shown in the figure below. First, estimate $\mathcal{A}$ for the band by assuming that it is Gaussian. Then use mathematical software to fit a polynomial (or a Gaussian) to the absorption band, and integrate the result analytically.

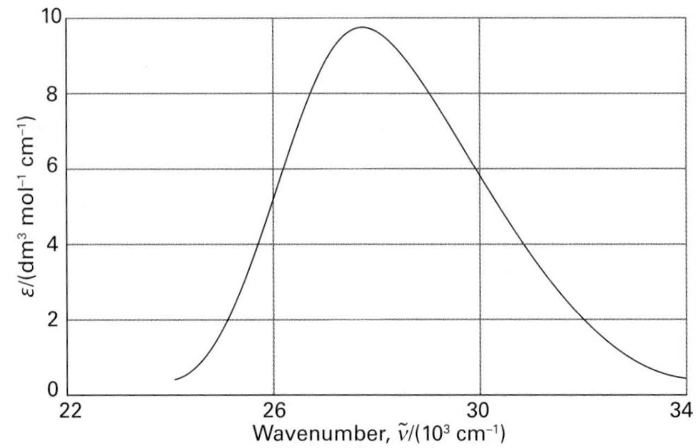

P11A.7‡ Wachewsky et al. (*J. Phys. Chem.* **100**, 11559 (1996)) examined the ultraviolet absorption spectrum of CH_3I, a species of interest in connection with stratospheric ozone chemistry. They found the integrated absorption coefficient to be dependent on temperature and pressure to an extent inconsistent with internal structural changes in isolated CH_3I molecules. They explained the changes as due to dimerization of a substantial fraction of the CH_3I, a process which would naturally be pressure and temperature

‡ These problems were supplied by Charles Trapp and Carmen Giunta.

dependent. (a) Compute the integrated absorption coefficient of CH_3I over a triangular lineshape in the range 31 250–34 483 cm^{-1} and a maximal molar absorption coefficient of 150 $dm^3\,mol^{-1}\,cm^{-1}$ at the mid-point of the range. (b) Suppose 1.0 per cent of the CH_3I units in a sample at 2.4 Torr and 373 K exists as dimers. Evaluate the absorbance expected at the mid-point of the absorption lineshape in a sample cell of length 12.0 cm. (c) Suppose 18 per cent of the CH_3I units in a sample at 100 Torr and 373 K exists as dimers. Calculate the absorbance expected at the wavenumber corresponding to the mid-point of the lineshape for a sample cell of length 12.0 cm; compute the molar absorption coefficient that would be inferred from this absorbance if dimerization is not considered.

P11A.8 When a star emitting electromagnetic radiation of frequency ν moves with a speed s relative to an observer, the observer detects radiation of frequency $\nu_{\text{receding}} = \nu f$ or $\nu_{\text{approaching}} = \nu/f$, where $f = \{(1 - s/c)/(1 + s/c)\}^{1/2}$ and c is the speed of light. (a) Three Fe I lines of the star HDE 271 182, which belongs to the Large Magellanic Cloud, occur at 438.882 nm, 441.000 nm, and 442.020 nm. The same lines occur at 438.392 nm, 440.510 nm, and 441.510 nm in the spectrum of an Earth-bound iron arc. Decide whether HDE 271 182 is receding from or approaching the Earth and estimate the star's radial speed with respect to the Earth. (b) What additional information would you need to calculate the radial velocity of HDE 271 182 with respect to the Sun?

P11A.9 When a star emitting electromagnetic radiation of frequency ν moves with a speed s relative to an observer, the observer detects radiation of frequency $\nu_{\text{receding}} = \nu f$ or $\nu_{\text{approaching}} = \nu/f$, where $f = \{(1 - s/c)/(1 + s/c)\}^{1/2}$ and c is the speed of light. A spectral line of $^{48}Ti^{8+}$ (of mass $47.95m_u$) in a distant star was found to be shifted from 654.2 nm to 706.5 nm and to be broadened to 61.8 pm. What is the speed of recession and the surface temperature of the star?

P11A.10 The Gaussian shape of a Doppler-broadened spectral line reflects the Maxwell distribution of speeds (see Topic 1B) in the sample at the temperature of the experiment. In a spectrometer that makes use of *phase-sensitive detection* the output signal is proportional to the first derivative of the signal intensity, $dI/d\nu$. Plot the resulting lineshape for various temperatures. How is the separation of the peaks related to the temperature?

P11A.11 The collision frequency z of a molecule of mass m in a gas at a pressure p is $z = 4\sigma(kT/\pi m)^{1/2}p/kT$, where σ is the collision cross-section. Find an expression for the collision-limited lifetime of an excited state assuming that every collision is effective. Estimate the width of a rotational transition at 63.56 cm^{-1} in HCl ($\sigma = 0.30\ nm^2$) at 25 °C and 1.0 atm. To what value must the pressure of the gas be reduced in order to ensure that collision broadening is less important than Doppler broadening?

P11A.12 Refer to Fig. 11A.9, which depicts a Michelson interferometer. The mirror M_1 moves in discrete distance increments, so the path difference p is also incremented in discrete steps. Explore the effect of increasing the step size on the shape of the interferogram for a monochromatic beam of wavenumber $\tilde{\nu}$ and intensity I_0. That is, draw plots of $I(p)/I_0$ against $\tilde{\nu}p$, each with a different number of data points spanning the same total distance path taken by the movable mirror M_1.

P11A.13 Use mathematical software to elaborate on the results of *Example* 11A.2 by: (a) exploring the effect of varying the wavenumbers and intensities of the three components of the radiation on the shape of the interferogram; and (b) calculating the Fourier transforms of the functions you generated in part (a).

TOPIC 11B Rotational spectroscopy

Discussion questions

D11B.1 Account for the rotational degeneracy of the various types of rigid rotor. Would the loss of rigidity affect your conclusions?

D11B.2 Does centrifugal distortion increase or decrease the separation between adjacent rotational energy levels?

D11B.3 Distinguish between an oblate and a prolate symmetric rotor and give several examples of each.

D11B.4 Describe the physical origins of the gross selection rule for microwave spectroscopy.

D11B.5 Describe the physical origins of the gross selection rule for rotational Raman spectroscopy.

D11B.6 Does $Be^{19}F_2$ exist in *ortho* and *para* forms? *Hints:* (a) Determine the geometry of BeF_2, then (b) decide whether fluorine nuclei are fermions or bosons.

D11B.7 Describe the role of nuclear statistics in the occupation of energy levels in $^1H^{12}C{\equiv}^{12}C^1H$, $^1H^{13}C{\equiv}^{13}C^1H$, and $^2H^{12}C{\equiv}^{12}C^2H$. For nuclear spin data, see Table 12A.2.

D11B.8 Account for the existence of a rotational zero-point energy in molecular hydrogen.

Additional exercises

E11B.13 Calculate the moment of inertia around the threefold symmetry axis and the corresponding rotational constant of a $^{31}P^1H_3$ molecule (HPH bond angle 93.5°; PH bond length 142 pm).

E11B.14 Plot the expressions for the two moments of inertia of a pyramidal symmetric top version of an AB_4 molecule (Table 11B.1) with equal bond lengths but with the angle θ increasing from 90° to the tetrahedral angle.

E11B.15 Plot the expressions for the two moments of inertia of a pyramidal symmetric top version of an AB_4 molecule (Table 11B.1) with θ equal to the tetrahedral angle but with one A–B bond varying. *Hint:* Write $\rho = R'_{AB}/R_{AB}$, and allow ρ to vary from 2 to 1.

E11B.16 Classify the following rotors: (i) $CH_2{=}CH_2$, (ii) SO_3, (iii) ClF_3, (iv) N_2O.

E11B.17 Calculate the CO and CS bond lengths in OCS from the rotational constants $B(^{16}O^{12}C^{32}S) = 6081.5$ MHz, $B(^{16}O^{12}C^{34}S) = 5932.8$ MHz.

E11B.18 Estimate the centrifugal distortion constant for $^{79}Br^{81}Br$, for which $\tilde{B} = 0.0809\ cm^{-1}$ and $\tilde{\nu} = 323.2\ cm^{-1}$. By what factor would the constant change when the ^{79}Br is replaced by ^{81}Br?

E11B.19 Which of the following molecules may show a pure rotational microwave absorption spectrum: (i) H_2O, (ii) H_2O_2, (iii) NH_3, (iv) N_2O?

E11B.20 Calculate the frequency and wavenumber of the $J = 2 \leftarrow 1$ transition in the pure rotational spectrum of $^{12}C^{16}O$. The equilibrium bond length is 112.81 pm. Would the frequency increase or decrease if centrifugal distortion is considered?

E11B.21 The wavenumber of the $J = 3 \leftarrow 2$ rotational transition of $^1H^{35}Cl$ considered as a rigid rotor is 63.56 cm^{-1}; what is the H—Cl bond length?

E11B.22 The wavenumber of the $J = 1 \leftarrow 0$ rotational transition of $^1H^{81}Br$ considered as a rigid rotor is 16.93 cm^{-1}; what is the H—Br bond length?

E11B.23 The spacing of lines in the microwave spectrum of $^{35}Cl^{19}F$ is 1.033 cm^{-1}; calculate the moment of inertia and bond length of the molecule.

E11B.24 What is the most highly populated rotational level of Br_2 at (i) 25 °C, (ii) 100 °C? Take $\tilde{B} = 0.0809$ cm^{-1}.

E11B.25 Which of the following molecules may show a pure rotational Raman spectrum: (i) CH_2Cl_2, (ii) CH_3CH_3, (iii) SF_6, (iv) N_2O?

E11B.26 The wavenumber of the incident radiation in a Raman spectrometer is 20 623 cm^{-1}. What is the wavenumber of the scattered Stokes radiation for the $J = 3 \leftarrow 1$ transition of $^{16}O_2$? Take $\tilde{B} = 1.4457$ cm^{-1}.

E11B.27 The rotational Raman spectrum of $^{35}Cl_2$ shows a series of Stokes lines separated by 0.9752 cm^{-1} and a similar series of anti-Stokes lines. Calculate the bond length of the molecule.

E11B.28 The rotational Raman spectrum of $^{19}F_2$ shows a series of Stokes lines separated by 3.5312 cm^{-1} and a similar series of anti-Stokes lines. Calculate the bond length of the molecule.

E11B.29 What is the ratio of weights of populations due to the effects of nuclear statistics for $^{12}C^{32}S_2$? What effect would be observed when ^{12}C is replaced by ^{13}C? For nuclear spin data, see Table 12A.2.

Problems

P11B.1 Show that the moment of inertia of a diatomic molecule composed of atoms of masses m_A and m_B and bond length R is equal to $m_{\text{eff}}R^2$, where $m_{\text{eff}} = m_A m_B/(m_A + m_B)$.

P11B.2 Confirm the expression given in Table 11B.1 for the moment of inertia of a linear ABC molecule. *Hint:* Begin by locating the centre of mass.

P11B.3 The rotational constant of NH_3 is 298 GHz. Calculate the separation of the pure rotational spectrum lines as a frequency (in GHz) and a wavenumber (in cm^{-1}), and show that the value of B is consistent with an N—H bond length of 101.4 pm and a bond angle of 106.78°.

P11B.4 Rotational absorption lines from $^1H^{35}Cl$ gas were found at the following wavenumbers (R.L. Hausler and R.A. Oetjen, *J. Chem. Phys.* **21**, 1340 (1953)): 83.32, 104.13, 124.73, 145.37, 165.89, 186.23, 206.60, 226.86 cm^{-1}. Calculate the moment of inertia and the bond length of the molecule. Predict the positions of the corresponding lines in $^2H^{35}Cl$.

P11B.5 Is the bond length in ^1HCl the same as that in ^2HCl? The wavenumbers of the $J = 1 \leftarrow 0$ rotational transitions for $^1H^{35}Cl$ and $^2H^{35}Cl$ are 20.8784 and 10.7840 cm^{-1}, respectively. Accurate atomic masses are 1.007 825m_u and 2.0140m_u for ^1H and ^2H, respectively. The mass of ^{35}Cl is 34.968 85m_u. Based on this information alone, can you conclude that the bond lengths are the same or different in the two molecules?

P11B.6 Thermodynamic considerations suggest that the copper monohalides CuX should exist mainly as polymers in the gas phase, and indeed it proved difficult to obtain the monomers in sufficient abundance to detect spectroscopically. This problem was overcome by flowing the halogen gas over copper heated to 1100 K (Manson et al., *J. Chem. Phys.* **63**, 2724 (1975)). For $^{63}Cu^{79}Br$ the $J = 14 \leftarrow 13$, $15 \leftarrow 14$, and $16 \leftarrow 15$ transitions occurred at 84 421.34, 90 449.25, and 96 476.72 MHz, respectively. Calculate the rotational constant and bond length of $^{63}Cu^{79}Br$. The mass of ^{63}Cu is 62.9296m_u.

P11B.7 The microwave spectrum of $^{16}O^{12}CS$ gave absorption lines (in GHz) as follows:

J	1	2	3	4
^{32}S	24.325 92	36.488 82	48.651 64	60.814 08
^{34}S	23.732 33		47.462 40	

Use the expressions for moments of inertia in Table 11B.1, assuming that the bond lengths are unchanged by substitution, to calculate the CO and CS bond lengths in OCS.

P11B.8 Equation 11B.20b may be rearranged into

$$\tilde{\nu}(J+1 \leftarrow J)/\{2(J+1)\} = \tilde{B} - 2\tilde{D}_J(J+1)^2$$

which is the equation of a straight line when the left-hand side is plotted against $(J + 1)^2$. The following wavenumbers of transitions (in cm^{-1}) were observed for $^{12}C^{16}O$:

J	0	1	2	3	4
$\tilde{\nu}$/cm^{-1}	3.845 033	7.689 919	11.534 510	15.378 662	19.222 223

Evaluate $\tilde{B}$ and $\tilde{D}_J$ for CO.

P11B.9‡ In a study of the rotational spectrum of the linear FeCO radical, Tanaka et al. (*J. Chem. Phys.* **106**, 6820 (1997)) report the following $J + 1 \leftarrow J$ transitions:

J	24	25	26	27	28	29
ν/MHz	214 777.7	223 379.0	231 981.2	240 584.4	249 188.5	257 793.5

Evaluate the rotational constant of the molecule. Also, estimate the value of J for the most highly populated rotational energy level at 298 K and at 100 K.

P11B.10 The rotational terms of a symmetric top, allowing for centrifugal distortion, are commonly written

$$\tilde{F}(J,K) = \tilde{B}J(J+1) + (\tilde{A} - \tilde{B})K^2 - \tilde{D}_J J^2(J+1)^2 - \tilde{D}_{JK}J(J+1)K^2 - \tilde{D}_K K^4$$

(a) Develop an expression for the wavenumbers of the allowed rotational transitions. (b) The following transition frequencies (in gigahertz, GHz) were observed for CH_3F:

51.0718 102.1403 102.1421 153.1987 153.2095

Evaluate as many constants in the expression for the rotational terms as these values permit.

P11B.11 Develop an expression for the value of J corresponding to the most highly populated rotational energy level of a diatomic rotor at a temperature T remembering that the degeneracy of each level is $2J + 1$. Evaluate the expression for ICl (for which $\tilde{B} = 0.1142$ cm^{-1}) at 25 °C. Repeat the problem for the most highly populated level of a spherical rotor, taking note of the fact that each level is $(2J + 1)^2$-fold degenerate. Evaluate the expression for CH_4 (for which $\tilde{B} = 5.24$ cm^{-1}) at 25 °C. *Hint:* To develop the expression, recall that the first derivative of a function is zero when the function reaches either a maximum or minimum value.

P11B.12 A. Dalgarno, in *Chemistry in the interstellar medium, Frontiers of Astrophysics*, ed. E.H. Avrett, Harvard University Press, Cambridge, MA (1976), notes that although both CH and CN spectra show up strongly in the interstellar medium in the constellation Ophiuchus, the CN spectrum has become the standard for the determination of the temperature of the cosmic microwave background radiation. Demonstrate through a calculation why CH would not be as useful for this purpose as CN. The rotational constants $\tilde{B}$ for CH and CN are 14.190 cm^{-1} and 1.891 cm^{-1}, respectively.

P11B.13 The space immediately surrounding stars, the *circumstellar space*, is significantly warmer because stars are very intense black-body emitters with temperatures of several thousand kelvin. Discuss how such factors as cloud temperature, particle density, and particle velocity may affect the rotational spectrum of CO in an interstellar cloud. What new features in the spectrum of CO can be observed in gas ejected from and still near a star with temperatures of about 1000 K, relative to gas in a cloud with temperature of about 10 K? Explain how these features may be used to distinguish between circumstellar and interstellar material on the basis of the rotational spectrum of CO.

P11B.14 Pure rotational Raman spectra of gaseous C_6H_6 and C_6D_6 yield the following rotational constants: $\tilde{B}(C_6H_6) = 0.189\,60\ cm^{-1}$, $\tilde{B}(C_6H_6) = 0.156\,81\ cm^{-1}$. The moments of inertia of the molecules about any axis perpendicular to the C_6 axis were calculated from these data as $I(C_6H_6) = 1.4759 \times 10^{-45}\ kg\,m^2$, $I(C_6D_6) = 1.7845 \times 10^{-45}\ kg\,m^2$. Calculate the CC and CH bond lengths.

TOPIC 11C Vibrational spectroscopy of diatomic molecules

Discussion questions

D11C.1 Discuss the strengths and limitations of the parabolic and Morse functions as approximations to the true potential energy curve of a diatomic molecule.

D11C.2 Describe the effect of vibrational excitation on the rotational constant of a diatomic molecule.

D11C.3 How is the method of combination differences used in rotation–vibration spectroscopy to determine rotational constants?

D11C.4 In what ways may the rotational and vibrational spectra of molecules change as a result of isotopic substitution?

Additional exercises

E11C.6 An object of mass 100 g suspended from the end of a rubber band has a vibrational frequency of 2.0 Hz. Calculate the force constant of the rubber band.

E11C.7 An object of mass 1.0 g suspended from the end of a spring has a vibrational frequency of 10.0 Hz. Calculate the force constant of the spring.

E11C.8 Calculate the percentage difference in the fundamental vibrational wavenumbers of $^{1}H^{35}Cl$ and $^{2}H^{37}Cl$ on the assumption that their force constants are the same.

E11C.9 The wavenumber of the fundamental vibrational transition of $^{79}Br^{81}Br$ is $323.2\ cm^{-1}$. Calculate the force constant of the bond.

E11C.10 The hydrogen halides have the following fundamental vibrational wavenumbers: $4141.3\ cm^{-1}$ ($^{1}H^{19}F$); $2988.9\ cm^{-1}$ ($^{1}H^{35}Cl$); $2649.7\ cm^{-1}$ ($^{1}H^{81}Br$); $2309.5\ cm^{-1}$ ($H^{127}I$). Calculate the force constants of the hydrogen–halogen bonds.

E11C.11 The hydrogen halides have the following fundamental vibrational wavenumbers: $4141.3\ cm^{-1}$ ($^{1}H^{19}F$); $2988.9\ cm^{-1}$ ($^{1}H^{35}Cl$); $2649.7\ cm^{-1}$ ($^{1}H^{81}Br$); $2309.5\ cm^{-1}$ ($H^{127}I$). Predict the fundamental vibrational wavenumbers of the corresponding deuterium halides.

E11C.12 Calculate the relative numbers of Cl_2 molecules ($\tilde{\nu} = 559.7\ cm^{-1}$) in the ground and first excited vibrational states at (i) 298 K, (ii) 500 K.

E11C.13 Calculate the relative numbers of Br_2 molecules ($\tilde{\nu} = 321\ cm^{-1}$) in the second and first excited vibrational states at (i) 298 K, (ii) 800 K.

E11C.14 For $^{14}N_2$, $\Delta\tilde{G}$ values for the transitions $v = 1 \leftarrow 0$, $2 \leftarrow 0$, and $3 \leftarrow 0$ are, respectively, 2329.91, 4631.20, and $6903.69\ cm^{-1}$. Calculate $\tilde{\nu}$ and x_e. Assume y_e to be zero.

E11C.15 The first five vibrational energy levels of HI are at 1144.83, 3374.90, 5525.51, 7596.66, and $9588.35\ cm^{-1}$. Calculate the dissociation energy of the molecule in reciprocal centimetres and electronvolts.

E11C.16 Infrared absorption by $^{1}H^{81}Br$ gives rise to a P branch from $v = 0$. What is the wavenumber of the line originating from the rotational state with $J = 2$? *Hint:* Use data from Table 11C.1.

Problems

P11C.1 Use molecular modelling software and the computational method of your choice to construct molecular potential energy curves like the one shown in Fig. 11C.1. Consider the hydrogen halides (HF, HCl, HBr, and HI): (a) plot the calculated energy of each molecule against the bond length, and (b) identify the order of force constants of the H–Hal bonds.

P11C.2 Derive an expression for the force constant of an oscillator that can be modelled by a Morse potential energy (eqn 11C.7).

P11C.3 Suppose a particle confined to a cavity in a microporous material has a potential energy of the form $V(x) = V_0(e^{-a^2/x^2} - 1)$. Sketch $V(x)$. What is the value of the force constant corresponding to this potential energy? Would the particle undergo simple harmonic motion? Sketch the likely form of the first two vibrational wavefunctions.

P11C.4 The vibrational levels of $^{23}Na^{127}I$ lie at the wavenumbers 142.81, 427.31, 710.31, and $991.81\ cm^{-1}$. Show that they fit the expression $(v+\frac{1}{2})\tilde{\nu} - (v+\frac{1}{2})^2 x_e\tilde{\nu}$, and deduce the force constant, zero-point energy, and dissociation energy of the molecule.

P11C.5 The $^{1}H^{35}Cl$ molecule is quite well described by the Morse potential energy with $hc\tilde{D}_e = 5.33\ eV$, $\tilde{\nu} = 2989.7\ cm^{-1}$, and $x_e\tilde{\nu} = 52.05\ cm^{-1}$. Assuming that the potential is unchanged on deuteration, predict the dissociation energies ($hc\tilde{D}_0$, in electronvolts) of (a) $^{1}H^{35}Cl$, (b) $^{2}H^{35}Cl$.

P11C.6 The Morse potential energy (eqn 11C.7) is very useful as a simple representation of the actual molecular potential energy. When $^{85}Rb^{1}H$ was studied, it was found that $\tilde{\nu} = 936.8\ cm^{-1}$ and $x_e\tilde{\nu} = 14.15\ cm^{-1}$. Plot the potential energy curve from 50 pm to 800 pm around $R_e = 236.7$ pm. Then go on to explore how the rotation of a molecule may weaken its bond by allowing for the kinetic energy of rotation of a molecule and plotting $V^* = V + hc\tilde{B}J(J+1)$ with $\tilde{B} = \hbar/4\pi c\mu R^2$. Plot these curves on the same diagram for $J = 40$, 80, and 100, and observe how the dissociation energy is affected by the rotation. *Hints:* Taking $\tilde{B} = 3.020\ cm^{-1}$ at the equilibrium bond length will greatly simplify the calculation. The mass of ^{85}Rb is $84.9118m_u$.

P11C.7‡ Luo et al. (*J. Chem. Phys.* **98**, 3564 (1993)) reported the observation of the He_2 complex, a species which had escaped detection for a long time. The fact that the observation required temperatures in the neighbourhood of 1 mK is consistent with computational studies which suggest that $hc\tilde{D}_e$ for He_2 is about 1.51×10^{-23} J, $hc\tilde{D}_0 \approx 2\times10^{-26}$ J, and R_e about 297 pm. (a) Estimate the fundamental vibrational wavenumber, force constant, moment of inertia, and rotational constant based on the harmonic oscillator and rigid-rotor approximations. (b) Such a weakly bound complex is hardly likely to be rigid. Estimate the vibrational wavenumber and anharmonicity constant based on the Morse potential energy.

P11C.8 Confirm that a Morse oscillator has a finite number of bound states, the states with $E_v < hc\tilde{D}_e$. Determine the value of v_{max} for the highest bound state.

P11C.9 Provided higher order terms are neglected, eqn 11C.9b for the vibrational wavenumbers of an anharmonic oscillator, $\Delta\tilde{G}_{v+1/2} = \tilde{\nu} - 2(v+1)x_e\tilde{\nu} + \cdots$, is the equation of a straight line when the left-hand side is plotted against $v + 1$. Use the following data on CO to determine the values of $\tilde{\nu}$ and $x_e\tilde{\nu}$ for CO:

v	0	1	2	3	4
$\Delta\tilde{G}_{v+1/2}/\text{cm}^{-1}$	2143.1	2116.1	2088.9	2061.3	2033.5

P11C.10 The rotational constant for CO is 1.9225 cm^{-1} and 1.9050 cm^{-1} in the ground and first excited vibrational states, respectively. By how much does the internuclear distance change as a result of this transition?

P11C.11 The average spacing between the rotational lines of the P and R branches of $^{12}C_2{}^{1}H_2$ and $^{12}C_2{}^{2}H_2$ is 2.352 cm^{-1} and 1.696 cm^{-1}, respectively. Estimate the CC and CH bond lengths.

P11C.12 Absorptions in the $v = 1 \leftarrow 0$ vibration–rotation spectrum of $^{1}H^{35}Cl$ were observed at the following wavenumbers (in cm^{-1}):

2998.05	2981.05	2963.35	2944.99	2925.92
2906.25	2865.14	2843.63	2821.59	2799.00

Assign the rotational quantum numbers and use the method of combination differences to calculate the rotational constants of the two vibrational levels.

P11C.13 Suppose that the internuclear distance may be written $R = R_e + x$ where R_e is the equilibrium bond length. Also suppose that the potential well is symmetrical and confines the oscillator to small displacements. Deduce expressions for $1/\langle R\rangle^2$, $1/\langle R^2\rangle$, and $\langle 1/R^2\rangle$ to the lowest non-zero power of $\langle x^2\rangle/R_e^2$ and confirm that the values are not the same.

P11C.14 Continue the development of Problem P11C.13 by using the virial theorem (Topic 7E) to relate $\langle x^2\rangle$ to the vibrational quantum number. Does your result imply that the rotational constant increases or decreases as the oscillator becomes excited to higher quantum states? What would be the effect of anharmonicity?

P11C.15 The rotational constant for a diatomic molecule in the vibrational state with quantum number v typically fits the expression $\tilde{B}_v = \tilde{B}_e - \tilde{a}\left(v + \frac{1}{2}\right)$, where $\tilde{B}_e$ is the rotational constant corresponding to the equilibrium bond length. For the interhalogen molecule IF it is found that $\tilde{B}_e = 0.27971\ \text{cm}^{-1}$ and $\tilde{a} = 0.187\ \text{m}^{-1}$ (note the change of units). Calculate $\tilde{B}_0$ and $\tilde{B}_1$ and use these values to calculate the wavenumbers of the transitions originating from $J = 3$ of the P and R branches. You will need the following additional information: $\tilde{\nu} = 610.258\ \text{cm}^{-1}$ and $x_e\tilde{\nu} = 3.141\ \text{cm}^{-1}$. Estimate the dissociation energy of the IF molecule.

P11C.16 Develop eqn 11B.16 ($\tilde{D}_J = 4\tilde{B}^3/\tilde{\nu}^2$) for the centrifugal distortion constant $\tilde{D}_J$ of a diatomic molecule of effective mass m_{eff}. Treat the bond as an elastic spring with force constant k_f and equilibrium length R_e that is subjected to a centrifugal distortion to a new length R_c. Begin the derivation by letting the particles experience a restoring force of magnitude $k_f(R_c - R_e)$ that is countered perfectly by a centrifugal force $m_{eff}\,\omega^2 R_c$, where ω is the angular velocity of the rotating molecule. Then introduce quantum mechanical effects by writing the angular momentum as $\{J(J+1)\}^{1/2}\hbar$. Finally, write an expression for the energy of the rotating molecule, compare it with eqn 11B.15, and infer an expression for $\tilde{D}_J$.

P11C.17 At low resolution, the strongest absorption band in the infrared absorption spectrum of $^{12}C^{16}O$ is centred at 2150 cm^{-1}. Upon closer examination at higher resolution, this band is observed to be split into two sets of closely spaced peaks, one on each side of the centre of the spectrum at 2143.26 cm^{-1}. The separation between the peaks immediately to the right and left of the centre is 7.655 cm^{-1}. Make the harmonic oscillator and rigid rotor approximations and calculate from these data: (a) the vibrational wavenumber of a CO molecule, (b) its molar zero-point vibrational energy, (c) the force constant of the CO bond, (d) the rotational constant $\tilde{B}$, and (e) the bond length of CO.

P11C.18 For $^{12}C^{16}O$, $\tilde{\nu}_R(0) = 2147.084\ \text{cm}^{-1}$, $\tilde{\nu}_R(1) = 2150.858\ \text{cm}^{-1}$, $\tilde{\nu}_P(1) = 2139.427\ \text{cm}^{-1}$ and $\tilde{\nu}_P(2) = 2135.548\ \text{cm}^{-1}$. Estimate the values of $\tilde{B}_0$ and $\tilde{B}_1$.

P11C.19 The analysis of combination differences summarized in the text considered the R and P branches. Extend the analysis to the O and S branches of a Raman spectururm.

TOPIC 11D Vibrational spectroscopy of polyatomic molecules

Discussion questions

D11D.1 Describe the physical origin of the gross selection rule for infrared spectroscopy.

D11D.2 Describe the physical origin of the gross selection rule for vibrational Raman spectroscopy.

D11D.3 Can a linear, nonpolar molecule like CO_2 have a Raman spectrum?

Additional exercises

E11D.8 Which of the following molecules may show infrared absorption spectra: (i) H_2, (ii) HCl, (iii) CO_2, (iv) H_2O?

E11D.9 Which of the following molecules may show infrared absorption spectra: (i) CH_3CH_3, (ii) CH_4, (iii) CH_3Cl, (iv) N_2?

E11D.10 How many normal modes of vibration are there for the following molecules: (i) C_6H_6, (ii) $C_6H_5CH_3$, (iii) HC≡C–C≡C–H?

E11D.11 How many vibrational modes are there for the molecule NC–(C≡C–C≡C–)$_{10}$CN detected in an interstellar cloud?

E11D.12 How many vibrational modes are there for the molecule NC–(C≡C–C≡C–)$_8$CN detected in an interstellar cloud?

E11D.13 Write an expression for the vibrational term for the ground vibrational state of SO_2 in terms of the wavenumbers of the normal modes. Neglect anharmonicities, as in eqn 11D.2.

E11D.14 Is the out-of-plane mode of a planar AB_3 molecule infrared or Raman active?

E11D.15 Consider the vibrational mode that corresponds to the boat-like bending of a benzene ring. Is it (i) Raman, (ii) infrared active?

E11D.16 Does the exclusion rule apply to C_2H_4?

Problems

P11D.1 Suppose that the out-of-plane distortion of an AB_3 planar molecule is described by a potential energy $V = V_0(1 - e^{-bh^4})$, where h is the distance by which the central atom A is displaced. Sketch this potential energy as a function of h (allow h to be both negative and positive). What could be said about (a) the force constant, (b) the vibrations? Sketch the form of the ground-state wavefunction.

P11D.2 Predict the shape of the nitronium ion, NO_2^+, from its Lewis structure and the VSEPR model. It has one Raman active vibrational mode at 1400 cm^{-1}, two strong IR active modes at 2360 and 540 cm^{-1}, and one weak absorption in the IR at 3735 cm^{-1}. Are these data consistent with the predicted shape of the molecule? Assign the vibrational wavenumbers to the modes from which they arise.

P11D.3 The computational methods discussed in Topic 9E can be used to simulate the vibrational spectrum of a molecule, and it is then possible to determine the correspondence between a vibrational frequency and the atomic displacements that give rise to a normal mode. (a) Using molecular modelling software and the computational method of your choice, calculate the fundamental vibrational wavenumbers and depict the vibrational normal modes of SO_2 in the gas phase graphically. (b) The experimental values of the fundamental vibrational wavenumbers of SO_2 in the gas phase are 525 cm^{-1}, 1151 cm^{-1}, and 1336 cm^{-1}. Compare the calculated and experimental values. Even if agreement is poor, is it possible to establish a correlation between an experimental value of the vibrational wavenumber with a specific vibrational normal mode?

TOPIC 11E Symmetry analysis of vibrational spectra

Discussion question

D11E.1 Suppose that you wish to characterize the normal modes of benzene in the gas phase. Why is it important to obtain both infrared absorption and Raman spectra of the molecule?

Additional exercises

E11E.4 A carbon disulfide molecule belongs to the point group $D_{\infty h}$. The nine displacements of the three atoms span $A_{1g} + 2A_{1u} + 2E_{1u} + E_{1g}$. What are the symmetry species of the normal modes of vibration?

E11E.5 The molecule BeH_2 is linear and centrosymmetric; its normal modes have symmetry species $A_{1g} + A_{1u} + E_{1u}$. Which of these normal modes are infrared active? Which are Raman active?

E11E.6 Which of the normal modes of (i) H_2O, (ii) H_2CO are Raman active?

Problems

P11E.1 Consider the molecule CH_3Cl. (a) To what point group does the molecule belong? (b) How many normal modes of vibration does the molecule have? (c) What are the symmetry species of the normal modes of vibration of this molecule? (d) Which of the vibrational modes of this molecule are infrared active? (e) Which of the vibrational modes of this molecule are Raman active?

P11E.2 Suppose that three conformations are proposed for the nonlinear molecule H_2O_2 (**1**, **2**, and **3**). The infrared absorption spectrum of gaseous H_2O_2 has bands at 870, 1370, 2869, and 3417 cm^{-1}. The Raman spectrum of the same sample has bands at 877, 1408, 1435, and 3407 cm^{-1}. All bands correspond to fundamental vibrational wavenumbers and you may assume that: (a) the 870 and 877 cm^{-1} bands arise from the same normal mode, and (b) the 3417 and 3407 cm^{-1} bands arise from the same normal mode. (i) If H_2O_2 were linear, how many normal modes of vibration would it have? (ii) Give the symmetry point group of each of the three proposed conformations of nonlinear H_2O_2. (iii) Determine which of the proposed conformations is inconsistent with the spectroscopic data. Explain your reasoning.

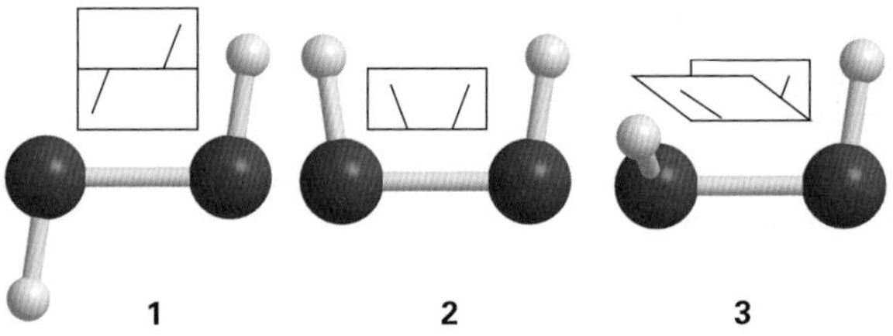

1 **2** **3**

TOPIC 11F Electronic spectra

Discussion questions

D11F.1 Explain the origin of the term symbol $^3\Sigma_g^-$ for the ground state of a dioxygen molecule.

D11F.2 Explain the basis of the Franck–Condon principle and how it leads to the formation of a vibrational progression.

D11F.3 How do the band heads in P and R branches arise? Could the Q branch show a head?

D11F.4 Explain how colour can arise from molecules.

D11F.5 Suppose that you are a colour chemist and had been asked to intensify the colour of a dye without changing the type of compound, and that the dye in question was a conjugated polyene. (a) Would you choose to lengthen or to shorten the chain? (b) Would the modification to the length shift the apparent colour of the dye towards the red or the blue?

D11F.6 Can a complex of the Zn^{2+} ion have a d–d electronic transition? Explain your answer.

Additional exercises

E11F.7 What is the value of S and the term symbol for the ground state of H_2?

E11F.8 The term symbol for one of the lowest excited states of H_2 is $^3\Pi_u$. To which excited-state configuration does this term symbol correspond?

E11F.9 What are the levels of the term for the ground electronic state of O_2^-?

E11F.10 Another of the excited states of the C_2 molecule has the valence electron configuration $1\sigma_g^2 1\sigma_u^2 1\pi_u^2 1\pi_g^2$. Give the multiplicity and parity of the term.

E11F.11 Which of the following transitions are electric-dipole allowed? (i) $^1\Sigma_g^+ \leftrightarrow {}^1\Sigma_u^+$, (ii) $^3\Sigma_g^+ \leftrightarrow {}^3\Sigma_u^+$, (ii) $\pi^\star \leftrightarrow$ n.

E11F.12 The ground-state wavefunction of a certain molecule is described by the vibrational wavefunction $\psi_0 = N_0 e^{-ax^2/2}$. Calculate the Franck–Condon factor for a transition to a vibrational state described by the wavefunction $\psi_0' = N_0 e^{-a(x-x_0)^2/2}$. The normalization constants are given by eqn 7E.10.

E11F.13 The ground-state wavefunction of a certain molecule is described by the vibrational wavefunction $\psi_0 = N_0 e^{-ax^2/2}$. Calculate the Franck–Condon factor for a transition to a vibrational state described by the wavefunction $\psi_1' = N_1(x - x_0)e^{-a(x-x_0)^2/2}$. The normalization constants are given by eqn 7E.10.

E11F.14 Suppose that the ground vibrational state of a molecule is modelled by using the particle-in-a-box wavefunction $\psi_0 = (2/L)^{1/2}\sin(\pi x/L)$ for $0 \le x \le L$ and 0 elsewhere. Calculate the Franck–Condon factor for a transition to a vibrational state described by the wavefunction $\psi' = (2/L)^{1/2}\sin\{\pi(x-L/4)/L\}$ for $L/4 \le x \le 5L/4$ and 0 elsewhere.

E11F.15 Suppose that the ground vibrational state of a molecule is modelled by using the particle-in-a-box wavefunction $\psi_0 = (2/L)^{1/2}\sin(\pi x/L)$ for $0 \le x \le L$ and 0 elsewhere. Calculate the Franck–Condon factor for a transition to a vibrational state described by the wavefunction $\psi' = (2/L)^{1/2}\sin\{\pi(x - L/2)/L\}$ for $L/2 \le x \le 3L/2$ and 0 elsewhere.

E11F.16 Use eqn 11F.7a to infer the value of J corresponding to the location of the band head of the P branch of a transition.

E11F.17 Use eqn 11F.7c to infer the value of J corresponding to the location of the band head of the R branch of a transition.

E11F.18 The following parameters describe the electronic ground state and an excited electronic state of SnO: $\tilde{B} = 0.3540\ \text{cm}^{-1}$, $\tilde{B}' = 0.3101\ \text{cm}^{-1}$. Which branch of the transition between them shows a head? At what value of J will it occur?

E11F.19 The following parameters describe the electronic ground state and an excited electronic state of BeH: $\tilde{B} = 10.308\ \text{cm}^{-1}$, $\tilde{B}' = 10.470\ \text{cm}^{-1}$. Which branch of the transition between them shows a head? At what value of J will it occur?

E11F.20 The P-branch of the $^2\Pi \leftarrow {}^2\Sigma^+$ transition of CdH shows a band head at $J = 25$. The rotational constant of the ground state is $5.437\ \text{cm}^{-1}$. What is the rotational constant of the upper state? Has the bond length increased or decreased in the transition?

E11F.21 The complex ion $[Fe(CN)_6]^{3-}$ has an electronic absorption spectrum with a maximum at 305 nm. Estimate a value of Δ_o for the complex.

E11F.22 Suppose that a charge-transfer transition in a one-dimensional system can be modelled as a process in which a rectangular wavefunction that is non-zero in the range $0 \le x \le a$ makes a transition to another rectangular wavefunction that is non-zero in the range $\frac{1}{2}a \le x \le b$. Evaluate the transition moment $\int \psi_f x \psi_i dx$. (Assume $a < b$.) *Hint:* Don't forget to normalize each wavefunction to 1.

E11F.23 Suppose that a charge-transfer transition in a one-dimensional system can be modelled as a process in which an electron described by a rectangular wavefunction that is non-zero in the range $0 \le x \le a$ makes a transition to another rectangular wavefunction that is non-zero in the range $ca \le x \le a$ where $0 \le c \le 1$. Evaluate the transition moment $\int \psi_f x \psi_i dx$ and explore its dependence on c. *Hint:* Don't forget to normalize each wavefunction to 1.

E11F.24 Suppose that a charge-transfer transition in a one-dimensional system can be modelled as a process in which a Gaussian wavefunction centred on $x = 0$ and width a makes a transition to another Gaussian wavefunction of the same width centred on $x = \frac{1}{2}a$. Evaluate the transition moment $\int \psi_f x \psi_i dx$. *Hint:* Don't forget to normalize each wavefunction to 1.

E11F.25 Suppose that a charge-transfer transition can be modelled in a one-dimensional system as a process in which an electron described by a Gaussian wavefunction centred on $x = 0$ and width a makes a transition to another Gaussian wavefunction of width $a/2$ and centred on $x = 0$. Evaluate the transition moment $\int \psi_f x \psi_i dx$. *Hint:* Don't forget to normalize each wavefunction to 1.

E11F.26 3-Buten-2-one (4) has a strong absorption at 213 nm and a weaker absorption at 320 nm. Assign the ultraviolet absorption transitions, giving your reasons.

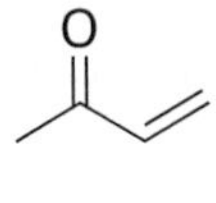

4 3-Buten-2-one

Problems

P11F.1 Which of the following electronic transitions are allowed in O_2: $^3\Sigma_g^- \leftrightarrow {}^1\Sigma_g^+$, and $^3\Sigma_g^- \leftrightarrow {}^3\Delta_u$?

P11F.2‡ J.G. Dojahn et al. (*J. Phys. Chem.* **100**, 9649 (1996)) characterized the potential energy curves of the ground and electronic states of homonuclear diatomic halogen anions. These anions have a $^2\Sigma_u^+$ ground state and $^2\Pi_g$, $^2\Pi_u$, and $^2\Sigma_g^+$ excited states. To which of the excited states are electric-dipole transitions allowed from the ground state? Explain your conclusion.

P11F.3 The vibrational wavenumber of the oxygen molecule in its electronic ground state is $1580\ \text{cm}^{-1}$, whereas that in the excited state (Σ_u^-), to which there is an allowed electronic transition, is $700\ \text{cm}^{-1}$. Given that the separation in energy between the minima in their respective potential energy curves of these two electronic states is 6.175 eV, what is the wavenumber of the lowest energy transition in the band of transitions originating from the $v = 0$ vibrational state of the electronic ground state to this excited state? Ignore any rotational structure or anharmonicity.

P11F.4 A transition of particular importance in O_2 gives rise to the Schumann–Runge band in the ultraviolet region. The wavenumbers (in cm^{-1}) of transitions from the ground state to the vibrational levels of the first excited state ($^3\Sigma_u^-$) are 50 062.6, 50 725.4, 51 369.0, 51 988.6, 52 579.0, 53 143.4, 53 679.6, 54 177.0, 54 641.8, 55 078.2, 55 460.0, 55 803.1, 56 107.3, 56 360.3, 56 570.6. What is the dissociation energy of the upper electronic state? (Use a Birge–Sponer plot, Topic 11C.) The same excited state is known to dissociate into one ground-state O atom and one excited-state atom with an energy 190 kJ mol^{-1} above the ground state. (This excited atom is responsible for a great deal of photochemical mischief in the atmosphere.) Ground-state O_2 dissociates into two ground-state atoms. Use this information to calculate the dissociation energy of ground-state O_2 from the Schumann–Runge data.

P11F.5 You are now ready to understand more deeply the features of photoelectron spectra (Topic 9B). The figure below shows the photoelectron spectrum of HBr. Disregarding for now the fine structure, the HBr lines fall into two main groups. The least tightly bound electrons (with the lowest ionization energies and hence highest kinetic energies when ejected) are those in the lone pairs of the Br atom. The next ionization energy lies at 15.2 eV, and corresponds to the removal of an electron from the HBr σ bond. (a) The spectrum shows that ejection of a σ electron is accompanied by a lot of vibrational excitation. Use the Franck–Condon principle to account for this observation. (b) Go on to explain why the lack of much vibrational structure in the other band is consistent with the nonbonding role of the Br4p_x and Br4p_y lone-pair electrons.

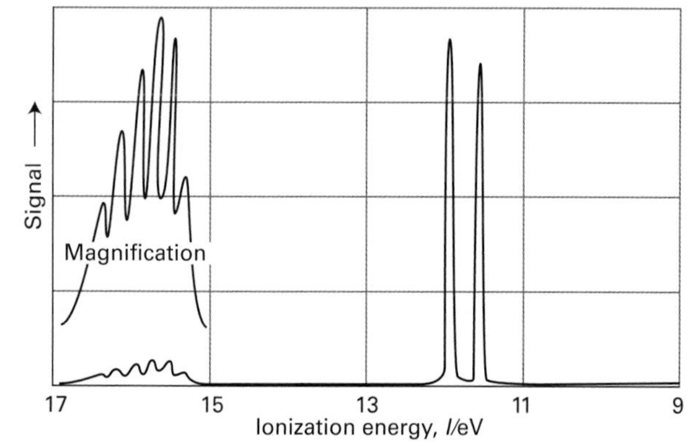

P11F.6 The highest kinetic energy electrons in the photoelectron spectrum of H_2O using 21.22 eV radiation are at about 9 eV and show a large vibrational spacing of 0.41 eV. The symmetric stretching mode of the neutral H_2O molecule lies at 3652 cm^{-1}. (a) What conclusions can be drawn from the nature of the orbital from which the electron is ejected? (b) In the same spectrum of H_2O, the band near 7.0 eV shows a long vibrational series with spacing 0.125 eV. The bending mode of H_2O lies at 1596 cm^{-1}. What conclusions can you draw about the characteristics of the orbital occupied by the photoelectron?

P11F.7 Assume that the states of the π electrons of a conjugated molecule can be approximated by the wavefunctions of a particle in a one-dimensional box, and that the magnitude of the dipole moment can be related to the displacement along this length by $\mu = -ex$. Show that the transition probability for the transition $n = 1 \rightarrow n = 2$ is non-zero, whereas that for $n = 1 \rightarrow n = 3$ is zero. *Hints:* The following relation will be useful: $\sin x \sin y = \frac{1}{2}\cos(x-y) - \frac{1}{2}\cos(x+y)$. Relevant integrals are given in the *Resource section*.

P11F.8 1,3,5-Hexatriene (a kind of 'linear' benzene) was converted into benzene itself. On the basis of a free-electron molecular orbital model (in which hexatriene is treated as a linear box and benzene as a ring), would you expect the lowest energy absorption to rise or fall in energy as a result of the conversion?

P11F.9 Estimate the magnitude of the transition dipole moment of a charge-transfer transition modelled as the migration of an electron from a H1s orbital on one atom to another H1s orbital on an atom a distance R away. Approximate the transition moment by $-eRS$ where S is the overlap integral of the two orbitals. Sketch the transition moment as a function of R using the expression for S given in Table 9C.1. Why does the intensity of a charge-transfer transition fall to zero as R approaches 0 and infinity?

P11F.10 The figure below shows the UV-visible absorption spectra of a selection of amino acids. Suggest reasons for their different appearances in terms of the structures of the molecules.

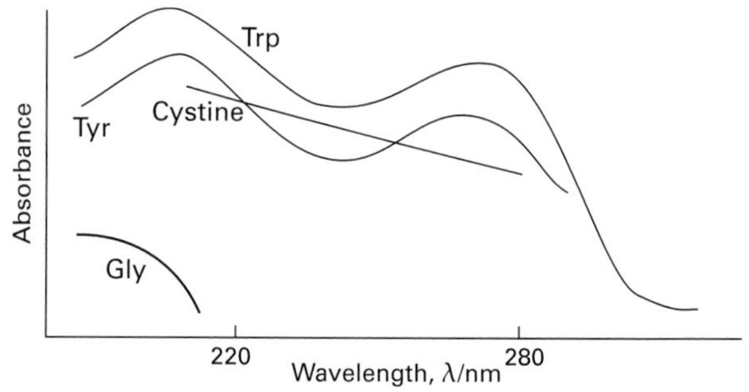

P11F.11 Propanone (acetone, $(CH_3)_2CO$) has a strong absorption at 189 nm and a weaker absorption at 280 nm. Identify the chromophore and assign the absorptions to $\pi^* \leftarrow n$ or $\pi^* \leftarrow \pi$ transitions.

P11F.12 Spin angular momentum is conserved when a molecule dissociates into atoms. What atom multiplicities are permitted when the ground state of an O_2 molecule, (b) an N_2 molecule dissociates into atoms?

TOPIC 11G Decay of excited states

Discussion questions

D11G.1 Describe the mechanism of fluorescence. In what respects is a fluorescence spectrum not the exact mirror image of the corresponding absorption spectrum?

D11G.2 What is the evidence for the usual explanation of the mechanism of (a) fluorescence, (b) phosphorescence?

D11G.3 Consider an aqueous solution of a chromophore that fluoresces strongly. Is the addition of iodide ion to the solution likely to increase or decrease the efficiency of phosphorescence of the chromophore?

D11G.4 What can be estimated from the wavenumber of the onset of predissociation?

D11G.5 Describe the principles of a four-level laser.

Additional exercises

E11G.2 The line marked A in the figure below is the fluorescence spectrum of benzophenone in solid solution in ethanol at low temperatures observed when the sample is illuminated with 360 nm ultraviolet radiation. (a) What can be said about the vibrational energy levels of the carbonyl group in (i) its ground electronic state and (ii) its excited electronic state? (b) When naphthalene is illuminated with 360 nm ultraviolet radiation it does not absorb, but the line marked B in the figure below is the phosphorescence spectrum of a frozen solution of a mixture of naphthalene and benzophenone in ethanol. Now a component of fluorescence from naphthalene can be detected. Account for this observation.

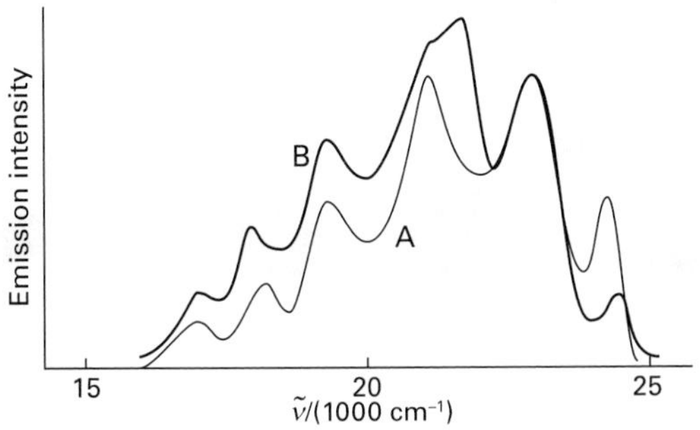

E11G.3 A hydrogen molecule absorbs ultraviolet radiation in a transition from its ${}^1\Sigma_g^+$ ground electronic state to an excited state that is energetically close to a dissociative ${}^1\Sigma_u^+$ state. The absorption band has a relatively large experimental linewidth. Account for this observation.

Problems

P11G.1 The fluorescence spectrum of anthracene vapour shows a series of peaks of increasing intensity with individual maxima at 440 nm, 410 nm, 390 nm, and 370 nm followed by a sharp cut-off at shorter wavelengths. The absorption spectrum rises sharply from zero to a maximum at 360 nm with a trail of peaks of lessening intensity at 345 nm, 330 nm, and 305 nm. Account for these observations.

P11G.2 The Beer–Lambert law states that the absorbance of a sample at a wavenumber $\tilde{\nu}$ is proportional to the molar concentration [J] of the absorbing species J and to the length L of the sample (eqn 11A.8). In this problem you are asked to show that the intensity of fluorescence emission from a sample of J is also proportional to [J] and L. Consider a sample of J that is illuminated with a beam of intensity $I_0(\tilde{\nu})$ at the wavenumber $\tilde{\nu}$. Before fluorescence can occur, a fraction of $I_0(\tilde{\nu})$ must be absorbed and an intensity $I(\tilde{\nu})$ will be transmitted. However, not all the absorbed intensity is re-emitted and the intensity of fluorescence depends on the *fluorescence quantum yield*, ϕ_F, the efficiency of photon emission. The fluorescence quantum yield ranges from 0 to 1 and is proportional to the ratio of the integral of the fluorescence spectrum over the integrated absorption coefficient. Because of a shift of magnitude $\Delta\tilde{\nu}$, fluorescence occurs at a wavenumber $\tilde{\nu}_f$, with $\tilde{\nu}_f + \Delta\tilde{\nu} = \tilde{\nu}$. It follows that the fluorescence intensity at $\tilde{\nu}_f$, $I_f(\tilde{\nu}_f)$, is proportional to ϕ_f and to the intensity of exciting radiation that is absorbed by J, $I_{abs}(\tilde{\nu}) = I_0(\tilde{\nu}) - I(\tilde{\nu})$. (a) Use the Beer–Lambert law to express $I_{abs}(\tilde{\nu})$ in terms of $I_0(\tilde{\nu})$, [J], L, and $\varepsilon(\tilde{\nu})$, the molar absorption coefficient of J at $\tilde{\nu}$. (b) Use your result from part (a) to show that $I_f(\tilde{\nu}_f) \propto I_0(\tilde{\nu})\varepsilon(\tilde{\nu})\phi_f[\mathrm{J}]L$.

P11G.3 A laser medium is confined to a cavity that ensures that only certain photons of a particular frequency, direction of travel, and state of polarization are generated abundantly. The cavity is essentially a region between two mirrors, which reflect the light back and forth. This arrangement can be regarded as a version of the particle in a box, with the particle now being a photon. As in the treatment of a particle in a box (Topic 7D), the only wavelengths that can be sustained satisfy $n \times \frac{1}{2}\lambda = L$, where n is an integer and L is the length of the cavity. That is, only an integral number of half-wavelengths fit into the cavity; all other waves undergo destructive interference with themselves. These wavelengths characterize the *resonant modes* of the laser. For a laser cavity of length 1.00 m, calculate (a) the allowed frequencies and (b) the frequency difference between successive resonant modes.

P11G.4 Laser radiation is spatially coherent in the sense that the electromagnetic waves are all in step across the cross-section of the beam emerging from the laser cavity (see Problem P11G.3). The *coherence length*, l_C, is the distance across the beam over which the waves remain coherent, and is related to the range of wavelengths, $\Delta\lambda$, present in the beam by $l_C = \lambda^2/2\Delta\lambda$. When many wavelengths are present, and $\Delta\lambda$ is large, the waves get out of step in a short distance and the coherence length is small. (a) How does the coherence length of a typical light bulb ($l_C = 400$ nm) compare with that of a He–Ne laser with $\lambda = 633$ nm and $\Delta\lambda = 2.0$ pm? (b) What is the condition that would lead to an infinite coherence length?

P11G.5 A *continuous-wave laser* emits a continuous beam of radiation, whereas a *pulsed laser* emits pulses of radiation. The peak power, P_{peak}, of a pulse is defined as the energy delivered in a pulse divided by its duration. The average power, $P_{average}$, is the total energy delivered by a large number of pulses divided by the duration of the time interval over which that total energy is measured. Suppose that a certain laser can generate radiation in 3.0 ns pulses, each of which delivers an energy of 0.10 J, at a pulse repetition frequency of 10 Hz. Calculate the peak power and the average power of this laser.

P11G.6 Light-induced degradation of molecules, also called *photobleaching*, is a serious problem in applications that require very high intensities. A molecule of a fluorescent dye commonly used to label biopolymers can withstand about 10^6 excitations by photons before light-induced reactions destroy its π system and the molecule no longer fluoresces. For how long will a single dye molecule fluoresce while being excited by 1.0 mW of 488 nm radiation from a continuous-wave laser? You may assume that the dye has an absorption spectrum that peaks at 488 nm and that every photon delivered by the laser is absorbed by the molecule.

FOCUS 11 Molecular spectroscopy

Integrated activities

I11.1 In the group theoretical language developed in Focus 10, a spherical rotor is a molecule that belongs to a cubic or icosahedral point group, a symmetric rotor is a molecule with at least a threefold axis of symmetry, and an asymmetric rotor is a molecule without a threefold (or higher) axis. Linear molecules are linear rotors. Classify each of the following molecules as a spherical, symmetric, linear, or asymmetric rotor and justify your answers with group theoretical arguments: (a) CH_4, (b) CH_3CN, (c) CO_2, (d) CH_3OH, (e) benzene, (f) pyridine.

I11.2‡ The H_3^+ ion has been found in the interstellar medium and in the atmospheres of Jupiter, Saturn, and Uranus. The rotational energy levels of H_3^+, an oblate symmetric rotor, are given by eqn 11B.13a, with $\tilde{C}$ replacing $\tilde{A}$, when centrifugal distortion and other complications are ignored. Experimental values for vibrational–rotational constants are $\tilde{\nu}(E') = 2521.6\ cm^{-1}$, $\tilde{B} = 43.55\ cm^{-1}$, and $\tilde{C} = 20.71\ cm^{-1}$. (a) Show that for a planar molecule (such as H_3^+) $I_{\parallel} = 2I_{\perp}$. The rather large discrepancy with the experimental values is due to factors ignored in eqn 11B.13. (b) Calculate an approximate value of the H–H bond length in H_3^+. (c) The value of R_e obtained from the best quantum mechanical calculations by J.B. Anderson (*J. Chem. Phys.* **96**, 3702 (1991)) is 87.32 pm. Use this result to calculate the values of the rotational constants $\tilde{B}$ and $\tilde{C}$. (d) Assuming that the geometry and force constants are the same in D_3^+ and H_3^+, calculate the spectroscopic constants of D_3^+. The molecular ion D_3^+ was first produced by Shy et al. (*Phys. Rev. Lett* **45**, 535 (1980)) who observed the $\nu_2(E')$ band in the infrared.

I11.3 Use appropriate electronic structure software to perform calculations on H_2O and CO_2 with basis sets of your or your instructor's choosing. (a) Compute ground-state energies, equilibrium geometries and vibrational frequencies for each molecule. (b) Compute the magnitude of the dipole moment of H_2O; the experimental value is 1.854 D. (c) Compare computed values to experiment and suggest reasons for any discrepancies.

I11.4 The protein haemerythrin is responsible for binding and carrying O_2 in some invertebrates. Each protein molecule has two Fe^{2+} ions that are in very close proximity and work together to bind one molecule of O_2. The Fe_2O_2 group of oxygenated haemerythrin is coloured and has an electronic absorption band at 500 nm. The Raman spectrum of oxygenated haemerythrin obtained with laser excitation at 500 nm has a band at $844\ cm^{-1}$ that has been attributed to the O–O stretching mode of bound $^{16}O_2$. (a) Proof that the $844\ cm^{-1}$ band arises from a bound O_2 species may be obtained by conducting experiments on samples of haemerythrin that have been mixed with $^{18}O_2$, instead of $^{16}O_2$. Predict the fundamental vibrational wavenumber of the ^{18}O–^{18}O stretching mode in a sample of haemerythrin that has been treated with $^{18}O_2$. (b) The fundamental vibrational wavenumbers for the O–O stretching modes of O_2, O_2^- (superoxide anion), and O_2^{2-} (peroxide anion) are 1555, 1107, and $878\ cm^{-1}$, respectively. Explain this trend in terms of the electronic structures of O_2, O_2^-, and O_2^{2-}. *Hint:* Review Topic 9C. What are the bond orders of O_2, O_2^-, and O_2^{2-}? (c) Based on the data given above, which of the following species best describes the Fe_2O_2 group of haemerythrin: $Fe^{2+}{}_2O_2$, $Fe^{2+}Fe^{3+}O_2^-$, or $Fe^{3+}{}_2O_2^{2-}$? Explain your reasoning. (d) The Raman spectrum of haemerythrin mixed with $^{16}O^{18}O$ has two bands that can be attributed to the O–O stretching mode of bound oxygen. Discuss how this observation may be used to exclude one or more of the four proposed schemes (5–8) for binding of O_2 to the Fe_2 site of haemerythrin.

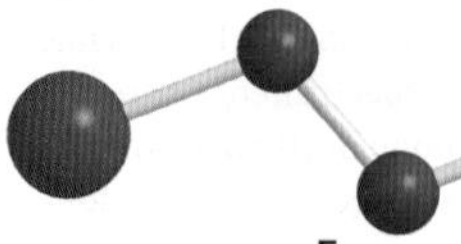
5

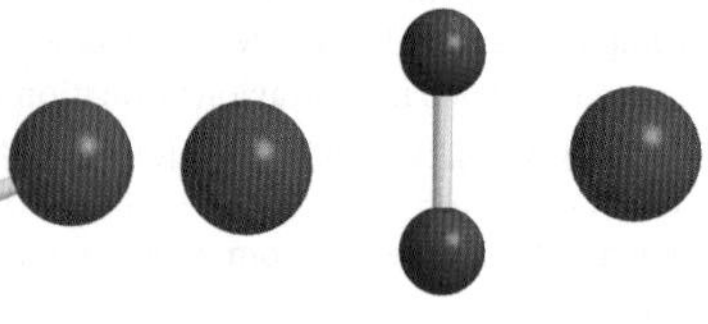
6

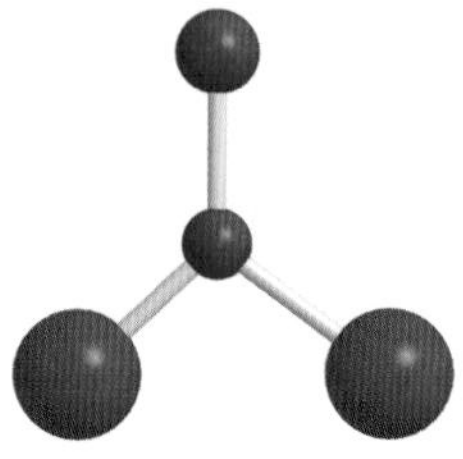
7

8

I11.5 The moments of inertia of the linear mercury(II) halides are very large, so the O and S branches of their vibrational Raman spectra show little rotational structure. Nevertheless, the position of greatest intensity in each branch can be identified and these data have been used to measure the rotational constants of the molecules (R.J.H. Clark and D.M. Rippon, *J. Chem. Soc. Faraday Soc. II* **69**, 1496 (1973)). Show, from a knowledge of the value of J corresponding to the intensity maximum, that the separation of the peaks of the O and S branches is given by the Placzek–Teller relation $\delta = (32\tilde{B}kT/hc)^{1/2}$. The following widths were obtained at the temperatures stated:

	$Hg^{35}Cl_2$	$Hg^{79}Br_2$	$Hg^{127}I_2$
$\theta/°C$	282	292	292
δ/cm^{-1}	23.8	15.2	11.4

Calculate the bond lengths in the three molecules.

I11.6‡ A mixture of carbon dioxide (2.1 per cent) and helium, at 1.00 bar and 298 K in a gas cell of length 10 cm has an infrared absorption band centred at $2349\ cm^{-1}$ with absorbances, $A(\tilde{\nu})$, described by:

$$A(\tilde{\nu}) = \frac{a_1}{1 + a_2(\tilde{\nu} - a_3)^2} + \frac{a_4}{1 + a_5(\tilde{\nu} - a_6)^2}$$

where the coefficients are $a_1 = 0.932$, $a_2 = 0.005050\ cm^2$, $a_3 = 2333\ cm^{-1}$, $a_4 = 1.504$, $a_5 = 0.01521\ cm^2$, $a_6 = 2362\ cm^{-1}$. (a) Draw graphs of $A(\tilde{\nu})$ and $\varepsilon(\tilde{\nu})$. What is the origin of both the band and the band width? What are the allowed and forbidden transitions of this band? (b) Calculate the transition wavenumbers and absorbances of the band with a simple harmonic oscillator–rigid rotor model and compare the result with the experimental spectra. The CO bond length is 116.2 pm. (c) Within what height, h, is basically all the infrared emission from the Earth in this band absorbed by atmospheric carbon dioxide? The mole fraction of CO_2 in the atmosphere is 3.3×10^{-4} and $T/K = 288 - 0.0065(h/m)$ below 10 km. Draw a surface plot of the atmospheric transmittance of the band as a function of both height and wavenumber.

I11.7‡ One of the principal methods for obtaining the electronic spectra of unstable radicals is to study the spectra of comets, which are almost entirely due to radicals. Many radical spectra have been detected in comets, including that due to CN. These radicals are produced in comets by the absorption of far-ultraviolet solar radiation by their parent compounds. Subsequently, their fluorescence is excited by sunlight of longer wavelength. The spectra of comet Hale–Bopp (C/1995 O1) have been the subject of many recent studies. One such study is that of the fluorescence spectrum of CN in the comet at large heliocentric distances by R.M. Wagner and D.G. Schleicher (*Science* **275**, 1918 (1997)), in which the authors determine the spatial distribution and rate of production of CN in the coma (the cloud constituting the major part of the head of the comet). The (0–0) vibrational band is centred on 387.6 nm and the weaker (1–1) band with relative intensity 0.1 is centred on 386.4 nm. The band heads for (0–0) and (0–1) are known to be 388.3 and 421.6 nm, respectively. From these data, calculate the energy of the excited S_1 state relative to the ground S_0 state, the vibrational wavenumbers and the difference in the vibrational wavenumbers of the two states, and the relative populations of the $v = 0$ and $v = 1$ vibrational levels of the S_1 state. Also estimate the effective temperature of the molecule in the excited S_1 state. Only eight rotational levels of the S_1 state are thought to be populated. Is that observation consistent with the effective temperature of the S_1 state?

I11.8 Use a group theoretical argument to decide which of the following transitions are electric-dipole allowed: (a) the $\pi^* \leftarrow \pi$ transition in ethene, (b) the $\pi^* \leftarrow n$ transition in a carbonyl group in a C_{2v} environment.

I11.9 Use molecule (**9**) as a model of the *trans* conformation of the chromophore found in rhodopsin. In this model, the methyl group bound to the nitrogen atom of the protonated Schiff's base replaces the protein. (a) Use molecular modelling software and the computational method of your instructor's choice, to calculate the energy separation between the HOMO and LUMO of (**9**). (b) Repeat the calculation for the 11-*cis* form of (**9**). (c) Based on your results from parts (a) and (b), do you expect the experimental frequency for the $\pi^* \leftarrow \pi$ visible absorption of the *trans* form of (**9**) to be higher or lower than that for the 1-*cis* form of (**9**)?

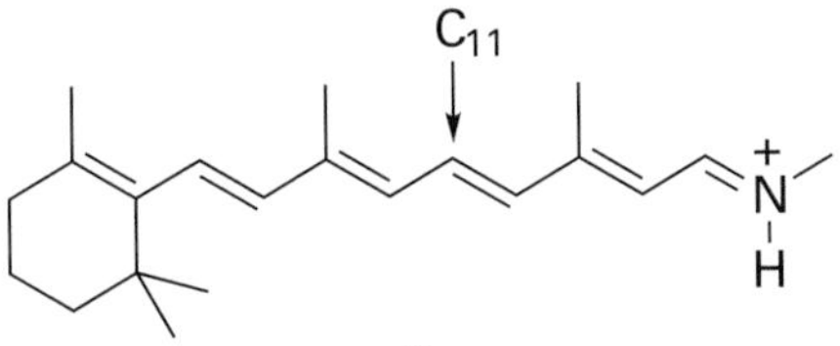

9

I11.10 Aromatic hydrocarbons and I_2 form complexes from which charge-transfer electronic transitions are observed. The hydrocarbon acts as an electron donor and I_2 as an electron acceptor. The energies $h\nu_{max}$ of the charge-transfer transitions for a number of hydrocarbon–I_2 complexes are given below:

Hydrocarbon	benzene	biphenyl	naphthalene	phenanthrene	pyrene	anthracene
$h\nu_{max}$/eV	4.184	3.654	3.452	3.288	2.989	2.890

Investigate the hypothesis that there is a correlation between the energy of the HOMO of the hydrocarbon (from which the electron comes in the charge-transfer transition) and $h\nu_{max}$. Use one of the computational methods discussed in Topic 9E to determine the energy of the HOMO of each hydrocarbon in the data set.

I11.11 A lot of information about the energy levels and wavefunctions of small inorganic molecules can be obtained from their ultraviolet spectra. An example of a spectrum with considerable vibrational structure, that of gaseous SO_2 at 25 °C, is shown in the accompanying figure. Estimate the integrated absorption coefficient for the band centred at 280 nm. What electronic states are accessible from the A_1 ground state of this C_{2v} molecule by electric-dipole transitions?

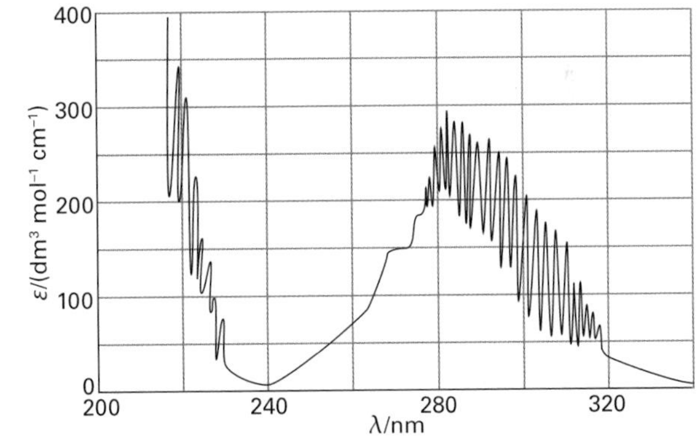

FOCUS 12

Magnetic resonance

The techniques of 'magnetic resonance' observe transitions between spin states of nuclei and electrons in molecules. 'Nuclear magnetic resonance' (NMR) spectroscopy observes nuclear spin transitions and is one of the most widely used spectroscopic techniques for the exploration of the structures and dynamics of molecules ranging from simple organic species to biopolymers. 'Electron paramagnetic resonance' (EPR) spectroscopy is a similar technique that observes electron spin transitions in species with unpaired electrons.

12A General principles

This Topic gives an account of the principles that govern the energies and spectroscopic transitions between spin states of nuclei and electrons in molecules in the presence of a magnetic field. It describes simple experimental arrangements for the detection of these transitions.

12B Features of NMR spectra

This Topic describes conventional NMR spectroscopy. It explains how the properties of a magnetic nucleus are affected by its electronic environment and the presence of magnetic nuclei in its vicinity. These concepts are used to explain how molecular structure governs the appearance of NMR spectra both in solution and in the solid state.

12C Pulse techniques in NMR

The modern implementation of NMR spectroscopy employs pulses of radiofrequency radiation followed by analysis of the resulting signal. This approach opens up many possibilities for the development of more sophisticated experiments. The Topic includes a discussion of spin relaxation in NMR and how it can be exploited for structural studies.

12D Electron paramagnetic resonance

The detailed form of an EPR spectrum reflects the molecular environment of the unpaired electron and the nuclei with which it interacts magnetically. From an analysis of the spectrum it is possible to infer the distribution of electron spin density.

What is an application of this material?

Magnetic resonance is ubiquitous in chemistry, as it is an enormously powerful analytical and structural technique, especially in organic chemistry and biochemistry. One of the most striking applications of nuclear magnetic resonance is in medicine. 'Magnetic resonance imaging' (MRI) is a portrayal of the distribution of protons in a solid object (*Impact* 18, accessed via the e-book). This technique is particularly useful for diagnosing disease. *Impact* 19, accessed via the e-book, highlights an application of electron paramagnetic resonance in materials science and biochemistry: the use of a 'spin probe', a radical that interacts with biopolymers or nanostructures, and has an EPR spectrum that is sensitive to the local structure and dynamics of its environment.

➤ **Go to the e-book for videos that feature the derivation and interpretation of equations, and applications of this material.**

TOPIC 12A General principles

> ➤ Why do you need to know this material?
>
> Nuclear magnetic resonance spectroscopy and electron paramagnetic resonance spectroscopy are used widely in chemistry to identify molecules and to determine their structures. To understand these techniques, you need to understand how a magnetic field affects the energies of the spin states of nuclei and electrons.
>
> ➤ What is the key idea?
>
> The spin states of nuclei and electrons have different energies in a magnetic field, and resonant transitions can take place between them when electromagnetic radiation of the appropriate frequency is applied.
>
> ➤ What do you need to know already?
>
> You need to be familiar with the quantum mechanical concept of spin (Topic 8B), the Boltzmann distribution (see *Energy: A first look* and Topic 13A), and the general features of spectroscopy (Topic 11A).

Electrons and many nuclei have the property called 'spin', an intrinsic angular momentum. This spin gives rise to a magnetic moment and results in them behaving like small bar magnets. The energies of these magnetic moments depend on their orientation with respect to an applied magnetic field.

Spectroscopic techniques that measure transitions between nuclear and electron spin energy levels rely on the phenomenon of **resonance**, the strong coupling of oscillators of the same frequency. In fact, all spectroscopy is a form of resonant coupling between the electromagnetic field and the molecules, but in magnetic resonance, at least in its original form, the energy levels are adjusted to match the electromagnetic field rather than vice versa.

12A.1 Nuclear magnetic resonance

The **nuclear spin quantum number**, I, is a fixed characteristic property of a nucleus[1] and, depending on the nuclide, it is either an integer (including zero) or a half-integer (Table 12A.1). The angular momentum associated with nuclear spin has the same properties as other kinds of angular momentum (Topic 7F):

- The magnitude of the angular momentum is $\{I(I+1)\}^{1/2}\hbar$.
- The component of the angular momentum along a specified axis (the 'z-axis') is $m_I\hbar$ where $m_I = I, I-1,\ldots,-I$.
- The orientation of the angular momentum, and hence of the magnetic moment, is determined by the value of m_I.

According to the second property, the angular momentum, and hence the magnetic moment, of the nucleus may lie in $2I+1$ different orientations relative to an axis. A ^{1}H nucleus has $I=\frac{1}{2}$ so its magnetic moment may adopt either of two orientations $\left(m_I=+\frac{1}{2},-\frac{1}{2}\right)$. The $m_I=+\frac{1}{2}$ state is commonly denoted α and the $m_I=-\frac{1}{2}$ state is commonly denoted β. A ^{14}N nucleus has $I=1$ so there are three orientations ($m_I=+1, 0, -1$). Examples of nuclei with $I=0$, and hence no magnetic moment, are ^{12}C and ^{16}O.

12A.1(a) The energies of nuclei in magnetic fields

The energy of a magnetic moment $\boldsymbol{\mu}$ in a magnetic field $\boldsymbol{\mathcal{B}}$ is equal to their scalar product:

$$E = -\boldsymbol{\mu}\cdot\boldsymbol{\mathcal{B}} \qquad (12A.1)$$

More formally, $\boldsymbol{\mathcal{B}}$ is the 'magnetic induction' and is measured in tesla, T; $1\,\mathrm{T} = 1\,\mathrm{kg\,s^{-2}\,A^{-1}}$. The (non-SI) unit gauss, G, is also occasionally used: $1\,\mathrm{T} = 10^4\,\mathrm{G}$. The corresponding expression for the hamiltonian is

$$\hat{H} = -\hat{\boldsymbol{\mu}}\cdot\boldsymbol{\mathcal{B}} \qquad (12A.2)$$

The magnetic moment operator of a nucleus is proportional to its spin angular momentum operator and is written

$$\hat{\boldsymbol{\mu}} = \gamma_{\mathrm{N}}\hat{\boldsymbol{I}} \qquad (12A.3a)$$

Table 12A.1 Nuclear constitution and the nuclear spin quantum number*

Number of protons	Number of neutrons	I
Even	Even	0
Odd	Odd	Integer (1, 2, 3,...)
Even	Odd	Half-integer $\left(\frac{1}{2},\frac{3}{2},\frac{5}{2},\ldots\right)$
Odd	Even	Half-integer $\left(\frac{1}{2},\frac{3}{2},\frac{5}{2},\ldots\right)$

* For nuclear ground states.

[1] Excited nuclear states, which are states in which the nucleons are arranged differently from the ground state, can have different spin from the ground state. Only nuclear ground states are considered here.

Table 12A.2 Nuclear spin properties*

Nucleus	Natural abundance/%	Spin, I	g-factor, g_I	Magnetogyric ratio, $\gamma_N/(10^7\,T^{-1}\,s^{-1})$	NMR frequency at 1 T, ν/MHz
^1H	99.98	$\frac{1}{2}$	5.586	26.75	42.576
^2H	0.02	1	0.857	4.11	6.536
^{13}C	1.11	$\frac{1}{2}$	1.405	6.73	10.708
^{11}B	80.4	$\frac{3}{2}$	1.792	8.58	13.663
^{14}N	99.64	1	0.404	1.93	3.078

* More values are given in the *Resource section*.

The constant of proportionality, γ_N, is the **nuclear magnetogyric ratio** (also called the 'gyromagnetic ratio'); its value depends on the identity of the nucleus and is determined empirically (Table 12A.2). If the magnetic field defines the z-direction and has magnitude $\mathcal{B}_0$, then eqn 12A.2 becomes

$$\hat{H} = -\hat{\mu}_z\mathcal{B}_0 = -\gamma_N\mathcal{B}_0\hat{I}_z \qquad (12A.3b)$$

The eigenvalues of the operator $\hat{I}_z$ for the z-component of the spin angular momentum are $m_I\hbar$. The eigenvalues of the hamiltonian in eqn 12A.3b, the allowed energy levels of the nucleus in a magnetic field, are therefore

$$E_{m_I} = -\gamma_N\hbar\mathcal{B}_0 m_I \qquad \text{Energies of a nuclear spin in a magnetic field} \qquad (12A.4a)$$

It is common to rewrite this expression in terms of the **nuclear magneton**, μ_N,

$$\mu_N = \frac{e\hbar}{2m_p} \qquad \text{Nuclear magneton [definition]} \qquad (12A.4b)$$

(where m_p is the mass of the proton) and an experimentally determined dimensionless constant called the **nuclear g-factor**, g_I,

$$g_I = \frac{\gamma_N\hbar}{\mu_N} \qquad \text{Nuclear } g\text{-factor [definition]} \qquad (12A.4c)$$

Equation 12A.4a then becomes

$$E_{m_I} = -g_I\mu_N\mathcal{B}_0 m_I \qquad \text{Energies of a nuclear spin in a magnetic field} \qquad (12A.4d)$$

The value of the nuclear magneton is $\mu_N = 5.051 \times 10^{-27}\,J\,T^{-1}$. Typical values of nuclear g-factors range between −6 and +6 (Table 12A.2). Positive values of g_I and γ_N denote a magnetic moment that lies in the same direction as the spin angular momentum; negative values indicate that the magnetic moment and spin lie in opposite directions.

When $\gamma_N > 0$, as is the case for the most commonly observed nuclei ^1H and ^{13}C, in a magnetic field the energies of states with $m_I > 0$ lie below states with $m_I < 0$. For a **spin-$\frac{1}{2}$ nucleus**, a nucleus for which $I = \frac{1}{2}$, the α state lies lower in energy than the β state, and the separation between them is

$$\Delta E = E_{-1/2} - E_{+1/2} = \tfrac{1}{2}\gamma_N\hbar\mathcal{B}_0 - \left(-\tfrac{1}{2}\gamma_N\hbar\mathcal{B}_0\right) = \gamma_N\hbar\mathcal{B}_0 \qquad (12A.5)$$

The corresponding frequency of electromagnetic radiation for a transition between these states is given by the Bohr frequency condition $\Delta E = h\nu$ (Fig. 12A.1). Therefore,

$$h\nu = \gamma_N\hbar\mathcal{B}_0 \quad \text{or} \quad \nu = \frac{\gamma_N\mathcal{B}_0}{2\pi} \qquad (12A.6)$$

This relation is called the **resonance condition**, and ν is called the **NMR frequency** for that nucleus. Although eqn 12A.6 has been derived for a spin-$\frac{1}{2}$ nucleus, the same expression applies for any nucleus with non-zero spin because the allowed transitions are between states that differ by ±1 in the value of m_I.

It is sometimes useful to compare the quantum mechanical treatment with the classical picture in which magnetic nuclei are pictured as tiny bar magnets. A bar magnet in a magnetic field undergoes the motion called **precession** as it twists round the direction of the field and sweeps out the surface of a cone (Fig. 12A.2). The rate of precession ν_L is called the **Larmor precession frequency**:

$$\nu_L = \frac{\gamma_N\mathcal{B}_0}{2\pi} \qquad \text{Larmor frequency of a nucleus [definition]} \qquad (12A.7)$$

The Larmor precession frequency is the same as the resonance frequency given by eqn 12A.6. In other words, the frequency of radiation that causes resonant transitions between the α and β states is the same as the Larmor precession frequency. The

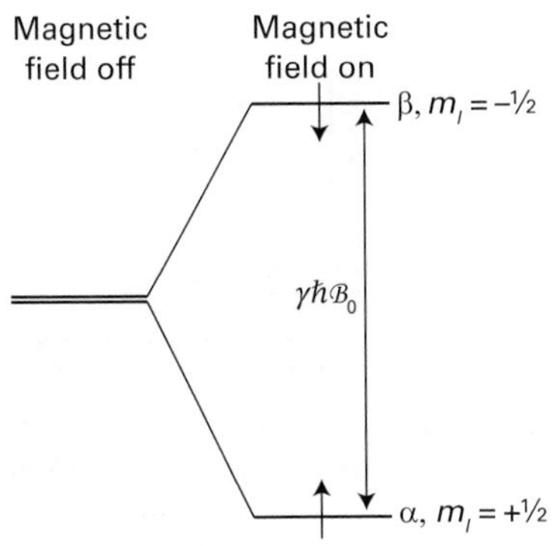

Figure 12A.1 The nuclear spin energy levels of a spin-$\frac{1}{2}$ nucleus with positive magnetogyric ratio (e.g. ^1H or ^{13}C) in a magnetic field. Resonant absorption of radiation occurs when the energy separation of the levels matches the energy of the photons.

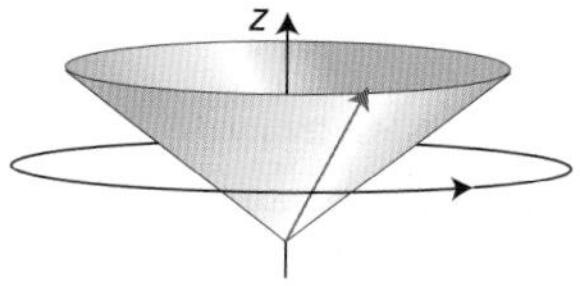

Figure 12A.2 The classical view of magnetic nuclei pictures them as behaving as tiny bar magnets. In an externally applied magnetic field the resulting magnetic moment, here represented as a vector, precesses round the direction of the field.

achievement of resonance absorption can therefore be pictured as changing the applied magnetic field until the bar magnet representing the nuclear magnetic moment precesses at the same frequency as the magnetic component of the electromagnetic field to which it is exposed.

Brief illustration 12A.1

The NMR frequency for ^{1}H nuclei ($I=\frac{1}{2}$) in a 12.0 T magnetic field can be found using eqn 12A.6, with the relevant value of γ_N taken from Table 12A.1:

$$\nu = \frac{\overbrace{(2.6752\times10^8\ \mathrm{T^{-1}\,s^{-1}})}^{\gamma_N}\times\overbrace{(12.0\ \mathrm{T})}^{\mathcal{B}_0}}{2\pi} = 5.11\times10^8\ \mathrm{s^{-1}} = 511\ \mathrm{MHz}$$

This radiation lies in the radiofrequency region of the electromagnetic spectrum, close to frequencies used for radio communication.

12A.1(b) The NMR spectrometer

The key component of an NMR spectrometer (Fig. 12A.3) is the magnet into which the sample is placed. Most modern spectrometers use superconducting magnets capable of producing fields of 12 T or more. Such magnets have the advantages that the field they produce is stable over time and no electrical power is needed to maintain the field. With the currently available magnets, all NMR frequencies fall in the radiofrequency range (see *Brief illustration* 12A.1). Therefore, a radiofrequency transmitter and receiver are needed to excite and detect the transitions taking place between nuclear spin states. The details of how the transitions are excited and detected are discussed in Topic 12C.

The sample being studied is most commonly in the form of a solution contained in a glass tube placed within the magnet. It is also possible to study solid samples by using more specialized techniques. Although the superconducting magnet itself has to be held close to the temperature of liquid helium (4 K), the magnet is designed so as to have a room-temperature clear space into which the sample can be placed.

The intensity of an NMR transition depends on a number of factors. They can be identified by considering the populations of the two spin states.

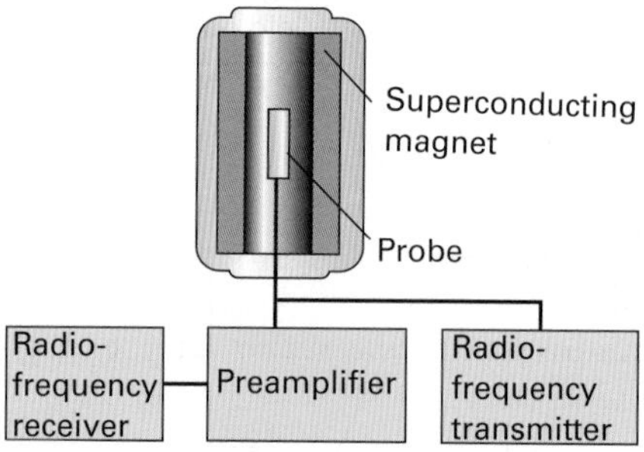

Figure 12A.3 The layout of a typical NMR spectrometer. The sample is held within the probe, which is placed at the centre of the magnetic field.

How is that done? 12A.1 Identifying the contributions to the absorption intensity

The rate of absorption of electromagnetic radiation is proportional to the population of the lower energy state (N_α in the case of a spin-$\frac{1}{2}$ nucleus with a positive magnetogyric ratio) and the rate of stimulated emission is proportional to the population of the upper state (N_β). At the low frequencies typical of magnetic resonance, spontaneous emission can be neglected as it is very slow. Therefore, the net rate of absorption is proportional to the difference in populations:

$$\text{Rate of absorption} \propto N_\alpha - N_\beta$$

Step 1 *Write an expression for the intensity of absorption in terms of the population difference*

The intensity of absorption, the rate at which energy is absorbed, is proportional to the product of the rate of absorption (the rate at which photons are absorbed) and the energy of each photon. The latter is proportional to the frequency ν of the incident radiation (through $E=h\nu$). At resonance, this frequency is proportional to the applied magnetic field (through $\nu=\gamma_N\mathcal{B}_0/2\pi$), so it follows that

$$\begin{aligned}\text{Intensity of absorption} &\propto \text{rate of absorption}\\ &\quad\times\text{energy of photon}\\ &\propto (N_\alpha - N_\beta)\times h\gamma_N\mathcal{B}_0/2\pi\end{aligned} \qquad (12A.8a)$$

Step 2 *Write an expression for the ratio of populations*

Now use the Boltzmann distribution (see the *Energy: A first look* and Topic 13A) to write an expression for the ratio of populations:

$$\frac{N_\beta}{N_\alpha} = e^{-\overbrace{\gamma_N\hbar\mathcal{B}_0}^{\Delta E}/kT} \overset{e^{-x}=1-x+\cdots}{\approx} 1-\frac{\gamma_N\hbar\mathcal{B}_0}{kT}$$

The expansion of the exponential term (see Part 1 of the *Resource section*) is appropriate for $\Delta E = \gamma_N \hbar \mathcal{B}_0 << kT$, a condition usually met for nuclear spins.

Step 3 *Use that ratio to write an expression for the population difference*

Consider the ratio of the population difference to the total number of spins, N: $(N_\alpha - N_\beta)/N$ and use the expression for the ratio of populations to write this ratio as

$$\underbrace{\frac{N_\alpha - N_\beta}{N_\alpha + N_\beta}}_{N} = \frac{N_\alpha(1-N_\beta/N_\alpha)}{N_\alpha(1+N_\beta/N_\alpha)} = \frac{1-\overbrace{N_\beta/N_\alpha}^{1-\gamma_N\hbar\mathcal{B}_0/kT}}{1+\underbrace{N_\beta/N_\alpha}_{1-\gamma_N\hbar\mathcal{B}_0/kT}}$$

$$\approx \frac{1-(1-\gamma_N\hbar\mathcal{B}_0/kT)}{\underbrace{1+(1-\gamma_N\hbar\mathcal{B}_0/kT)}_{\approx 1}} \approx \frac{\gamma_N\hbar\mathcal{B}_0/kT}{2}$$

Therefore

$$N_\alpha - N_\beta \approx \frac{N\gamma_N\hbar\mathcal{B}_0}{2kT} \quad \text{Population difference [spin-}\tfrac{1}{2}\text{ nuclei]} \quad (12A.8b)$$

The intensity is now obtained by substituting this expression into eqn 12A.8a which gives, after discarding constants that do not refer to the spins,

$$\text{Intensity} \propto \frac{N\gamma_N^2\mathcal{B}_0^2}{T} \quad (12A.8c)$$

Absorption intensity

The intensity of absorption is proportional to $\mathcal{B}_0^2$. It follows that the signal can be enhanced significantly by increasing the strength of the applied magnetic field. The use of high magnetic fields also simplifies the appearance of spectra (a point explained in Topic 12B) and so allows them to be interpreted more readily. The intensity is also proportional to γ_N^2, so nuclei with large magnetogyric ratios (^{1}H, for instance) give more intense signals than those with small magnetogyric ratios (^{13}C, for instance).

Brief illustration 12A.2

For ^{1}H nuclei $\gamma_N = 2.675 \times 10^8\ \text{T}^{-1}\,\text{s}^{-1}$. Therefore, for 1 000 000 of these nuclei in a field of 10 T at 20 °C,

$$N_\alpha - N_\beta \approx \frac{\overbrace{1000000}^{N} \times \overbrace{(2.675\times10^8\ \text{T}^{-1}\,\text{s}^{-1})}^{\gamma_N} \times \overbrace{(1.055\times10^{-34}\ \text{J s})}^{\hbar} \times \overbrace{(10\ \text{T})}^{\mathcal{B}_0}}{2\times\underbrace{(1.381\times10^{-23}\ \text{J K}^{-1})}_{k}\times\underbrace{(293\ \text{K})}_{T}}$$

$$\approx 35$$

Even in such a strong field there is only a tiny imbalance of population of about 35 in a million. This small population difference means that special techniques had to be developed before NMR became a viable technique.

Exercises

E12A.1 For a ^{1}H nucleus (a proton), what are the magnitude of the spin angular momentum and what are its allowed components along the z-axis? Express your answer in multiples of $\hbar$. What angles does the angular momentum make with the z-axis?

E12A.2 What is the NMR frequency of a ^{1}H nucleus (a proton) in a magnetic field of 13.5 T? Express your answer in megahertz.

E12A.3 Calculate the relative population differences $(N_\alpha - N_\beta)/N$ for ^{1}H nuclei in fields of (i) 0.30 T, (ii) 1.5 T, and (iii) 10 T at 25 °C.

12A.2 Electron paramagnetic resonance

The observation of resonant transitions between the energy levels of an electron in a magnetic field is the basis of **electron paramagnetic resonance** (EPR; or *electron spin resonance*, ESR). This kind of spectroscopy has several features in common with NMR.

12A.2(a) The energies of electrons in magnetic fields

The magnetic moment of an electron is proportional to its spin angular momentum. Its magnetic moment operator and the hamiltonian for its interaction with a magnetic field are

$$\hat{\boldsymbol{\mu}} = \gamma_e \hat{\boldsymbol{s}} \quad \text{and} \quad \hat{H} = -\gamma_e \hat{\boldsymbol{s}}\cdot\boldsymbol{\mathcal{B}} \quad (12A.9a)$$

where $\hat{\boldsymbol{s}}$ is the spin angular momentum operator and γ_e is the **magnetogyric ratio** of the electron:

$$\gamma_e = -\frac{g_e e}{2m_e} \quad \text{Magnetogyric ratio of electron} \quad (12A.9b)$$

with $g_e = 2.002\,319\ldots$ as the **g-value** of the free electron. (Note that the current convention is to include the g-value in the definition of the magnetogyric ratio.) Dirac's relativistic theory, his modification of the Schrödinger equation to make it consistent with Einstein's special relativity, gives $g_e = 2$; the additional $0.002\,319\ldots$ arises from interactions of the electron with the electromagnetic fluctuations of the vacuum that surrounds it. The negative sign of γ_e (arising from the sign of the charge on the electron) shows that the magnetic moment is opposite in direction to the vector representing its angular momentum.

The hamiltonian when the magnetic field lies in the z-direction and has magnitude $\mathcal{B}_0$ is

$$\hat{H} = -\gamma_e \mathcal{B}_0 \hat{s}_z \quad (12A.10)$$

where $\hat{s}_z$ is the operator for the z-component of the spin angular momentum. It follows that the energies of an electron spin in a magnetic field are

$$E_{m_s} = -\gamma_e \hbar \mathcal{B}_0 m_s \quad \text{Energies of an electron spin in a magnetic field} \quad (12A.11a)$$

with $m_s = \pm\frac{1}{2}$. It is common to write this expression in terms of the **Bohr magneton**, μ_B, defined as

$$\mu_B = \frac{e\hbar}{2m_e} \quad \text{Bohr magneton} \quad (12A.11b)$$

Its value is $9.274 \times 10^{-24}\,\mathrm{J\,T^{-1}}$. This positive quantity is often regarded as the fundamental quantum of magnetic moment. Note that the Bohr magneton is about 2000 times bigger than the nuclear magneton, so electron magnetic moments are that much bigger than nuclear magnetic moments. By using eqn 12A.9b the term $\gamma_e\hbar$ in eqn 12A.11a can be expressed as $-g_e e\hbar/2m_e$, which in turn can be written $-g_e\mu_B$ by introducing the definition of the Bohr magneton from eqn 12A.11b. It then follows that

$$E_{m_s} = g_e \mu_B \mathcal{B}_0 m_s \quad \text{Energies of an electron spin in a magnetic field} \quad (12A.11c)$$

The energy separation between the $m_s = +\frac{1}{2}$ (α) and $m_s = -\frac{1}{2}$ (β) states is therefore

$$\Delta E = E_{+1/2} - E_{-1/2} = \tfrac{1}{2} g_e \mu_B \mathcal{B}_0 - \left(-\tfrac{1}{2} g_e \mu_B \mathcal{B}_0\right) = g_e \mu_B \mathcal{B}_0 \quad (12A.12a)$$

with β the lower state. This energy separation comes into resonance with electromagnetic radiation of frequency ν when (Fig. 12A.4)

$$h\nu = g_e \mu_B \mathcal{B}_0 \quad \text{Resonance condition for EPR} \quad (12A.12b)$$

Brief illustration 12A.3

A typical commercial EPR spectrometer uses a magnetic field of about 0.33 T. The EPR resonance frequency is

$$\nu = \frac{\overbrace{(2.0023)}^{g_e} \times \overbrace{(9.274 \times 10^{-24}\,\mathrm{J\,T^{-1}})}^{\mu_B} \times \overbrace{(0.33\,\mathrm{T})}^{\mathcal{B}_0}}{\underbrace{6.626 \times 10^{-34}\,\mathrm{J\,s}}_{h}}$$
$$= 9.2 \times 10^9\,\mathrm{s^{-1}} = 9.2\,\mathrm{GHz}$$

This frequency corresponds to a wavelength of 3.2 cm, which is in the microwave region, and specifically in the 'X band' of frequencies.

12A.2(b) The EPR spectrometer

Most commercial EPR spectrometers use magnetic field strengths that result in EPR frequencies in the microwave region (see *Brief illustration* 12A.3). The layout of a typical EPR spectrometer is shown in Fig. 12A.5. It consists of a fixed-frequency microwave source (typically a Gunn oscillator, based on a solid-state device), a cavity into which the sample (held in a glass or quartz tube) is inserted, a microwave detector, and an electromagnet with a variable magnetic field. The sample in an EPR observation must have unpaired electrons, so is either a radical or a d-metal complex. For technical reasons related to the detection procedure, the spectrum shows the first derivative of the absorption line (Fig. 12A.6).

As in NMR, the intensities of spectral lines in EPR depend on the difference in populations between the ground and excited states. For an electron, the β state lies below the α state in energy and, by a similar argument that led to eqn 12A.8b for nuclei,

$$N_\beta - N_\alpha \approx \frac{N g_e \mu_B \mathcal{B}_0}{2kT} \quad \text{Population difference [electrons]} \quad (12A.13)$$

where N is the total number of electron spins.

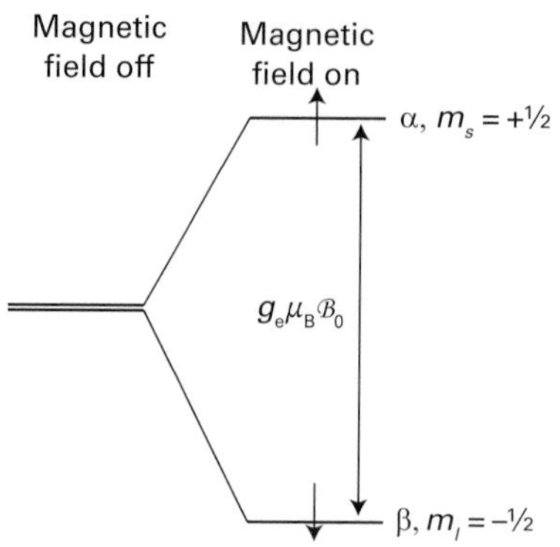

Figure 12A.4 Electron spin levels in a magnetic field. Note that the β state is lower in energy than the α state (because the magnetogyric ratio of an electron is negative). Resonant absorption occurs when the frequency of the incident radiation matches the frequency corresponding to the energy separation.

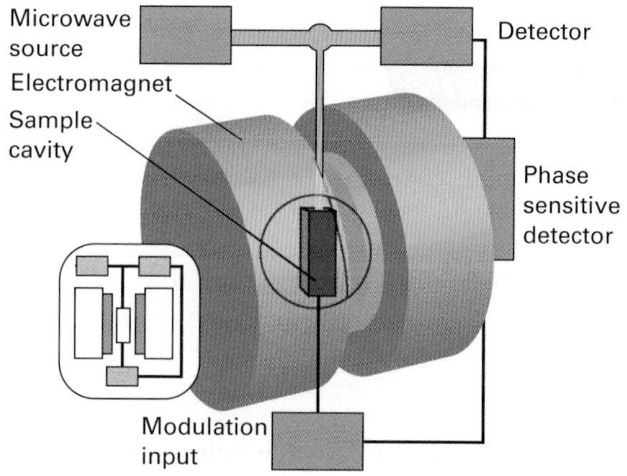

Figure 12A.5 The layout of a typical EPR spectrometer. A typical magnetic field is 0.3 T, which requires 9 GHz (3 cm) microwaves for resonance.

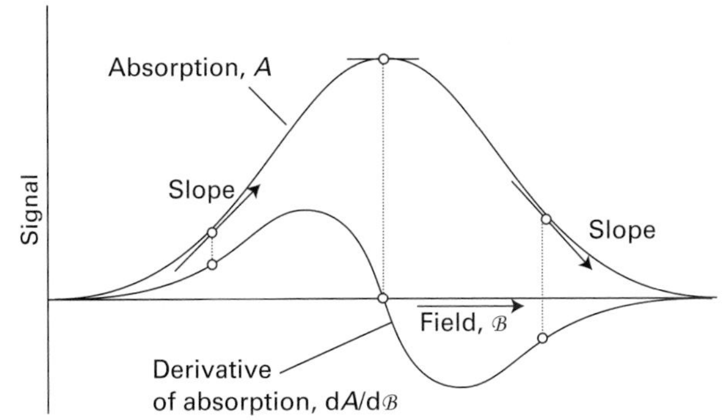

Figure 12A.6 When phase-sensitive detection is used, the signal is the first derivative of the absorption intensity. Note that the peak of the absorption corresponds to the point where the derivative passes through zero.

Brief illustration 12A.4

When 1000 electron spins experience a magnetic field of 1.0 T at 20 °C (293 K), the population difference is

$$N_\beta - N_\alpha \approx \frac{\overbrace{1000}^{N}\times\overbrace{2.0023}^{g_e}\times\overbrace{(9.274\times10^{-24}\ \mathrm{J\,T^{-1}})}^{\mu_B}\times\overbrace{(1.0\ \mathrm{T})}^{\mathcal{B}_0}}{2\times\underbrace{(1.381\times10^{-23}\ \mathrm{J\,K^{-1}})}_{k}\times\underbrace{(293\ \mathrm{K})}_{T}}$$

$$\approx 2.3$$

There is an imbalance of populations of only about two electrons in a thousand. However, the imbalance is much larger for electron spins than for nuclear spins (*Brief illustration* 12A.2) because the energy separation between the spin states of electrons is larger than that for nuclear spins even at the lower magnetic field strengths normally employed for EPR.

Exercises

E12A.4 In which of the following systems is the energy level separation larger? (i) A ^{14}N nucleus in a magnetic field that corresponds to an NMR frequency for ^{1}H of 600 MHz, (ii) an electron in a field of 0.300 T.

E12A.5 Some commercial EPR spectrometers use 8 mm microwave radiation (the 'Q band'). What magnetic field is needed to satisfy the resonance condition?

Checklist of concepts

- ☐ 1. The **nuclear spin quantum number**, I, of a nucleus is either a non-negative integer or half-integer; I can be zero.
- ☐ 2. In the presence of a magnetic field a nucleus has $2I + 1$ energy levels characterized by different values of m_I.
- ☐ 3. **Nuclear magnetic resonance** (NMR) is the observation of the absorption of radiofrequency electromagnetic radiation by nuclei in a magnetic field.
- ☐ 4. In NMR the absorption intensity increases with the strength of the applied magnetic field (as $\mathcal{B}_0^2$) and is also proportional to the square of the magnetogyric ratio of the nucleus.
- ☐ 5. In the presence of a magnetic field, an electron has two energy levels corresponding to the α and β spin states.
- ☐ 6. **Electron paramagnetic resonance** (EPR) is the observation of the resonant absorption of microwave electromagnetic radiation by unpaired electrons in a magnetic field.

Checklist of equations

Property	Equation	Comment	Equation number
Energies of a nuclear spin in a magnetic field	$E_{m_I} = -\gamma_N \hbar \mathcal{B}_0 m_I$		12A.4a
	$= -g_I \mu_N \mathcal{B}_0 m_I$		12A.4d
Nuclear magneton	$\mu_N = e\hbar/2m_p$	$\mu_N = 5.051\times10^{-27}\ \mathrm{J\,T^{-1}}$	12A.4b
Resonance condition (spin-$\frac{1}{2}$ nuclei)	$h\nu = \gamma_N \hbar \mathcal{B}_0$		12A.6
Larmor frequency	$\nu_L = \gamma_N \mathcal{B}_0/2\pi$		12A.7
Magnetogyric ratio (electron)	$\gamma_e = -g_e e/2m_e$	$g_e = 2.002\,319\ldots$	12A.9b
Energies of an electron spin in a magnetic field	$E_{m_s} = -\gamma_e \hbar \mathcal{B}_0 m_s$		12A.11a
	$= g_e \mu_B \mathcal{B}_0 m_s$		12A.11c
Bohr magneton	$\mu_B = e\hbar/2m_e$	$\mu_B = 9.274\times10^{-24}\ \mathrm{J\,T^{-1}}$	12A.11b
Resonance condition (electrons)	$h\nu = g_e \mu_B \mathcal{B}_0$		12A.12b

TOPIC 12B Features of NMR spectra

> ➤ **Why do you need to know this material?**
>
> To analyse NMR spectra and extract the wealth of information they contain you need to understand how the appearance of a spectrum arises from molecular structure.
>
> ➤ **What is the key idea?**
>
> The resonance frequency of a magnetic nucleus is affected by its electronic environment and the presence of nearby magnetic nuclei.
>
> ➤ **What do you need to know already?**
>
> You need to be familiar with the general principles of magnetic resonance (Topic 12A).

Nuclear magnetic moments interact with the *local* magnetic field, the field at the location of the nucleus in question. The local field differs from the applied field due to the effects of the electrons surrounding the nucleus and the presence of other magnetic nuclei in the molecule. The overall effect is that the NMR frequency of a given nucleus is sensitive to its molecular environment.

12B.1 The chemical shift

The applied magnetic field can be thought of as causing a circulation of electrons through the molecule. This circulation is analogous to an electric current and so gives rise to a magnetic field. The **local magnetic field**, $\mathcal{B}_{\text{loc}}$, the total field experienced by the nucleus, is the sum of the applied field $\mathcal{B}_0$ and the additional field $\delta\mathcal{B}$ due to the circulation of the electrons

$$\mathcal{B}_{\text{loc}} = \mathcal{B}_0 + \delta\mathcal{B} \qquad (12\text{B}.1)$$

The additional field is proportional to the applied field, and it is conventional to write

$$\delta\mathcal{B} = -\sigma\mathcal{B}_0 \qquad \text{Shielding constant [definition]} \qquad (12\text{B}.2)$$

where the dimensionless quantity σ is called the **shielding constant** of the nucleus. The ability of the applied field to induce an electronic current in the molecule, and hence affect the strength of the resulting local magnetic field, depends on the details of the electronic structure near the magnetic nucleus of interest. Therefore, nuclei in different chemical groups have different shielding constants. As a result, the Larmor frequency ν_L of the nucleus (and therefore its resonance frequency) changes from $\gamma_N\mathcal{B}_0/2\pi$ to

$$\nu_L = \frac{\gamma_N\mathcal{B}_{\text{loc}}}{2\pi} = \frac{\gamma_N(\mathcal{B}_0 + \delta\mathcal{B})}{2\pi} = \frac{\gamma_N(\mathcal{B}_0 - \sigma\mathcal{B}_0)}{2\pi} = \frac{\gamma_N\mathcal{B}_0}{2\pi}(1-\sigma) \qquad (12\text{B}.3)$$

The Larmor frequency is different for nuclei in different environments, even if those nuclei are of the same element.

The **chemical shift** of a nucleus is the difference between its resonance frequency and that of a reference standard. The standard for ^{1}H and ^{13}C is the resonance in tetramethylsilane, $Si(CH_3)_4$, commonly referred to as TMS. The frequency separation between the resonance from a particular nucleus and that from the standard increases with the strength of the applied magnetic field because the Larmor frequency (eqn 12B.3) is proportional to the applied field.

Chemical shifts are reported on the **δ scale**, which is defined as

$$\delta = \frac{\nu - \nu^\circ}{\nu^\circ} \times 10^6 \qquad \delta \text{ scale [definition]} \qquad (12\text{B}.4a)$$

where ν° is the resonance frequency (Larmor frequency) of the standard. Because ν° is very close to the operating frequency of the spectrometer ν_{spect}, which is typically chosen to be in the middle of the range of Larmor frequencies exhibited by the nucleus being studied, the ν° in the denominator of eqn 12B.4a can safely be replaced by ν_{spect} to give

$$\delta = \frac{\nu - \nu^\circ}{\nu_{\text{spect}}} \times 10^6 \qquad (12\text{B}.4b)$$

The advantage of the δ scale is that shifts reported with it are independent of the applied field (because both numerator and denominator in eqn 12B.4a are proportional to the applied field). The resonance frequencies themselves, however, do depend on the applied field through[1]

$$\nu = \nu^\circ + \left(\frac{\nu_{\text{spect}}}{10^6}\right)\delta \qquad (12\text{B}.5)$$

[1] In much of the literature, chemical shifts are reported in parts per million, ppm, in recognition of the factor of 10^6 in the definition; this is unnecessary. If you see '$\delta = 10$ ppm', interpret it, and use it in eqn 12B.5, as $\delta = 10$.

Brief illustration 12B.1

In an NMR spectrometer operating at 500.130 00 MHz the resonance from TMS is found to be at a frequency of 500.127 50 MHz. The chemical shift of a resonance at a frequency of 500.128 25 MHz is

$$\delta = \frac{\nu - \nu^\circ}{\nu_{\text{spect}}} \times 10^6 = \frac{(500.128\,25\ \text{MHz}) - (500.127\,50\ \text{MHz})}{500.130\,00\ \text{MHz}} \times 10^6$$
$$= 1.5$$

On the same spectrometer the frequency separation of two resonances with chemical shifts $\delta_1 = 1.25$ and $\delta_2 = 5.75$ is found using eqn 12B.5

$$\nu_2 - \nu_1 = \frac{\nu_{\text{spect}}}{10^6} \times (\delta_2 - \delta_1) = \frac{500.130\,00 \times 10^6\ \text{Hz}}{10^6} \times (5.75 - 1.25)$$
$$= 2250\ \text{Hz}$$

The relation between δ and σ is obtained by substituting eqn 12B.3 into eqn 12B.4a:

$$\delta = \frac{(1-\sigma)\mathcal{B}_0 - (1-\sigma^\circ)\mathcal{B}_0}{(1-\sigma^\circ)\mathcal{B}_0} \times 10^6$$
$$= \frac{\sigma^\circ - \sigma}{1 - \sigma^\circ} \times 10^6 \overset{|\sigma^\circ| \ll 1}{\approx} (\sigma^\circ - \sigma) \times 10^6 \qquad \text{Relation between } \delta \text{ and } \sigma \qquad (12B.6)$$

where σ° is the shielding constant of the reference standard. A decrease in σ (reduction in shielding) therefore leads to an increase in δ. Therefore, nuclei with large chemical shifts are said to be strongly **deshielded**. Some typical chemical shifts are given in Fig. 12B.1. As can be seen from the illustration, the nuclei of different elements have very different ranges of chemical shifts. The ranges exhibit the variety of electronic environments of the nuclei in molecules: the higher the atomic number of the element, the greater the number of electrons around the nucleus and hence the greater the range of the extent of shielding. By convention, NMR spectra are plotted with δ decreasing from left to right.

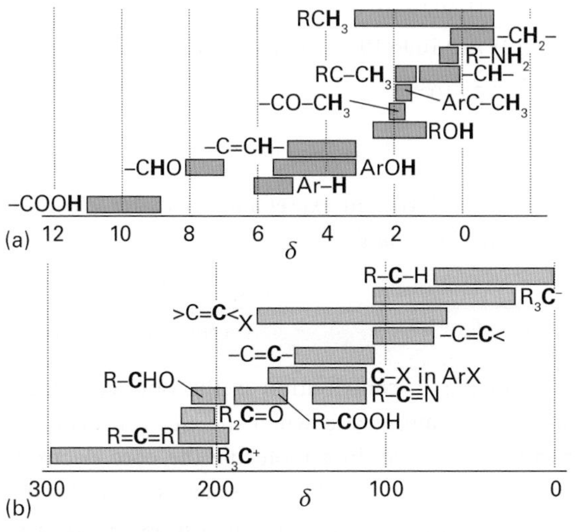

Figure 12B.1 The range of typical chemical shifts for (a) ^{1}H resonances and (b) ^{13}C resonances.

Example 12B.1 Interpreting a ^{1}H NMR spectrum

Figure 12B.2 shows the ^{1}H (proton) NMR spectrum of 1-methoxy-2-propanone, $CH_3OCH_2COCH_3$. Account for the observed chemical shifts.

Collect your thoughts You need to consider the effect of any electron-withdrawing atom: it deshields strongly the protons to which it is bound and has a diminishing effect on more distant protons. To identify which of the B and C resonances correspond to H atoms 2 and 3 you can take either of two approaches. One is to look at the large compilations of chemical shift data available. The second approach is to make use of the 'integral' of a line, the area under the resonance peak, which is proportional to the number of nuclei giving rise to the peak. These integrals are commonly shown by step-like curves superimposed on the spectrum, as is the case in Fig. 12B.2: the integral is proportional to the height of the step.

The solution The H atoms labelled 2 and 3 are all attached to a C atom that is attached to the strongly electron-withdrawing O atom, whereas the H atoms labelled 1 are further away from any O atoms. You can expect the deshielding of H atoms 2 and 3 to be greater than that of H atoms 1, and so the chemical shifts of 2 and 3 will be larger than that of 1. Resonance A at δ = 2.2 can therefore confidently be assigned to the H atoms at position 1. It is evident from the spectrum that the integral of peak B is greater than that of C (in fact they are in the ratio 3:2), immediately identifying peak B as corresponding to the H atoms at position 3, and peak C to those at position 2.

Self-test 12B.1 The ^{1}H (proton) NMR spectrum of ethanal (acetaldehyde) has lines at δ = 2.20 and δ = 9.80. Which line can be assigned to the H atom in the CHO group?

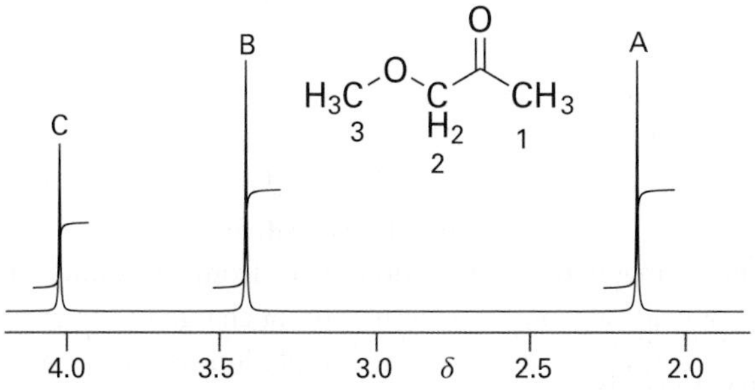

Figure 12B.2 The ^{1}H (proton) NMR spectrum of 1-methoxy-2-propanone. The step-like curve indicates the integral of the peak (the area under the peak) with the height of the step being proportional to the integral.

Answer: δ = 9.80

Exercises

E12B.1 The ^{1}H resonance from TMS is found to occur at 500.130 000 MHz. What is the chemical shift (on the δ scale) of a peak at 500.132 500 MHz?

E12B.2 What is the frequency separation, in hertz, between two peaks in a ^{1}H NMR spectrum with chemical shifts $\delta = 9.80$ and $\delta = 2.2$ in a spectrometer operating at 400.130 000 MHz for ^{1}H?

12B.2 The origin of shielding constants

The calculation of shielding constants (and hence chemical shifts) is difficult because it requires detailed knowledge of the distribution of electron density in the ground and excited states and the electronic excitation energies of the molecule. Nevertheless, it is helpful to understand the different contributions to chemical shifts so that patterns and trends can be identified.

A useful approach is to assume that the observed shielding constant is the sum of three contributions:

$$\sigma = \sigma(\text{local}) + \sigma(\text{neighbour}) + \sigma(\text{solvent}) \qquad (12\text{B}.7)$$

The **local contribution**, σ(local), is essentially the contribution of the electrons of the atom that contains the nucleus in question. The **neighbouring group contribution**, σ(neighbour), is the contribution from the groups of atoms that form the rest of the molecule. The **solvent contribution**, σ(solvent), is the contribution from the solvent molecules.

12B.2(a) The local contribution

It is convenient to regard the local contribution to the shielding constant as the sum of a **diamagnetic contribution**, σ_d, and a **paramagnetic contribution**, σ_p:

$$\sigma(\text{local}) = \sigma_d + \sigma_p \qquad \text{Local contribution to the shielding constant} \qquad (12\text{B}.8)$$

The diamagnetic contribution arises from additional fields that oppose the applied magnetic field and hence shield the nucleus; σ_d is therefore positive. The paramagnetic contribution arises from additional fields that reinforce the applied field and hence lead to deshielding; σ_p is therefore negative.

The diamagnetic contribution arises from the ability of the applied field to generate a circulation of charge in the ground-state electron distribution. The circulation generates a magnetic field which opposes the applied field and hence shields the nucleus. The magnitude of σ_d depends on the electron density close to the nucleus and for atoms it can be calculated from the **Lamb formula**:[2]

[2] For a derivation, see our *Molecular quantum mechanics* (2011).

$$\sigma_d = \frac{e^2\mu_0}{12\pi m_e}\langle 1/r\rangle \qquad \text{Lamb formula} \qquad (12\text{B}.9)$$

where μ_0 is the magnetic constant (vacuum permeability), r is the electron–nucleus distance, and the angle brackets $\langle\ldots\rangle$ indicate an expectation value.

Example 12B.2 Using the Lamb formula

Calculate the shielding constant for the nucleus in a free H atom.

Collect your thoughts To calculate σ_d from the Lamb formula, you need to calculate the expectation value of $1/r$ for a hydrogen 1s orbital. The radial part of the wavefunction can be found from Table 8A.1 and the angular part from Table 7F.1.

The solution The normalized wavefunction for a hydrogen 1s orbital is, in spherical polar coordinates (see *The chemist's toolkit* 7F.2),

$$\psi = \left(\frac{1}{\pi a_0^3}\right)^{1/2} e^{-r/a_0}$$

In this coordinate system the volume element is $d\tau = \sin\theta\, r^2 dr d\theta d\phi$, so the expectation value of $1/r$ is

$$\langle 1/r\rangle = \int \frac{\psi^*\psi}{r}d\tau = \frac{1}{\pi a_0^3}\overbrace{\int_0^{2\pi}d\phi}^{2\pi}\overbrace{\int_0^{\pi}\sin\theta\, d\theta}^{2}\overbrace{\int_0^{\infty}re^{-2r/a_0}dr}^{\text{Integral E.2}}$$
$$= \frac{1}{\pi a_0^3}\times 4\pi\times\frac{a_0^2}{4} = \frac{1}{a_0}$$

Therefore,

$$\sigma_d = \frac{e^2\mu_0}{12\pi m_e a_0} = \frac{(1.602\times10^{-19}\ \text{C})^2\times(4\pi\times10^{-7}\ \text{J s}^2\,\text{C}^{-2}\,\text{m}^{-1})}{12\pi\times(9.109\times10^{-31}\ \text{kg})\times(5.292\times10^{-11}\ \text{m})}$$
$$= \frac{(1.602\times10^{-19})^2\times(4\pi\times10^{-7})}{12\pi\times(9.109\times10^{-31})\times(5.292\times10^{-11})}\times\frac{\text{C}^2\,\text{J s}^2\,\text{C}^{-2}\,\text{m}^{-1}}{\text{kg m}}$$
$$= 1.775\times10^{-5}$$

where $1\ \text{J} = 1\ \text{kg m}^2\,\text{s}^{-2}$ has been used.

Self-test 12B.2 Derive an expression for σ_d for a hydrogenic atom with nuclear charge Z.

Answer: $\sigma_d = Ze^2\mu_0/12\pi m_e a_0$

The diamagnetic contribution is the only contribution in closed-shell free atoms and when the electron distribution is spherically symmetric. In a molecule the core electrons near to a particular nucleus are likely to have spherical symmetry, even if the valence electron distribution is highly distorted. Therefore, core electrons contribute only to the diamagnetic part of the shielding. The diamagnetic contribution is broadly

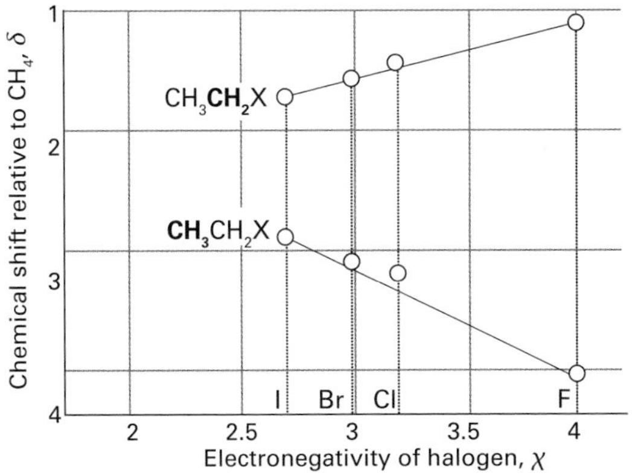

Figure 12B.3 The variation of chemical shielding with electronegativity. The shifts for the methyl protons follow the simple expectation that increasing the electronegativity of the halogen will increase the chemical shift. However, to emphasize that chemical shifts are subtle phenomena, notice that the trend for the methylene protons is opposite to that expected. For these protons another contribution (the magnetic anisotropy of C–H and C–X bonds) is dominant.

proportional to the electron density of the atom containing the nucleus of interest. It follows that the shielding is decreased if the electron density on the atom is reduced by the influence of an electronegative atom nearby. That reduction in shielding as the electronegativity of a neighbouring atom increases translates into an increase in the chemical shift δ (Fig. 12B.3).

The local paramagnetic contribution, σ_p, arises from the ability of the applied field to force electrons to circulate through the molecule by making use of orbitals that are unoccupied in the ground state. It is absent in free atoms and also in linear molecules (such as ethyne, HC≡CH) when the applied field lies along the symmetry axis; in this arrangement the electrons can circulate freely and the applied field is unable to force them into other orbitals. Large paramagnetic contributions can be expected for light atoms (because the valence electrons, and hence the induced currents, are close to the nucleus) and in molecules with low-lying excited states (because an applied field can then induce significant currents). In fact, the paramagnetic contribution is the dominant local contribution for atoms other than hydrogen.

12B.2(b) Neighbouring group contributions

The neighbouring group contribution arises from the currents induced in nearby groups of atoms. Consider the influence of the neighbouring group X on the hydrogen atom in a molecule such as H–X. The applied field generates currents in the electron distribution of X and gives rise to an induced magnetic moment (an induced magnetic dipole) proportional to the applied field; the constant of proportionality is the magnetic susceptibility, χ (chi), of the group X: $\mu_{\text{induced}} = \chi\mathcal{B}_0$. The susceptibility is negative for a diamagnetic group because the induced moment is opposite to the direction of the applied field.

The induced moment gives rise to a magnetic field which is experienced by neighbouring nuclei. As explained in *The chemist's toolkit* 12B.1, a nucleus at distance r and angle θ (defined in **1**) from the induced moment experiences a local field that has the form

$$\mathcal{B}_{\text{local}} \propto \frac{\mu_{\text{induced}}}{r^3}(1-3\cos^2\theta) \qquad \text{Local dipolar field} \qquad (12B.10a)$$

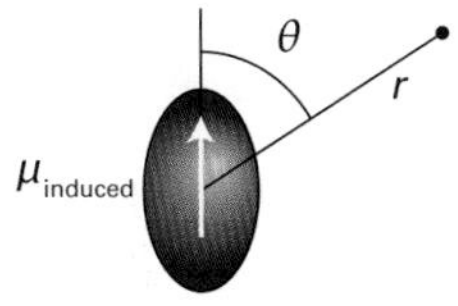

1

This local field is parallel to the applied field, and the angle θ is measured from the direction of the applied field. Note that the strength of the field is inversely proportional to the cube of the distance r between H and X. If the magnetic susceptibility is independent of the orientation of the molecule (that is, it is 'isotropic') then the induced dipole is also independent of orientation. It follows that the local field given in eqn 12B.10a averages to zero because, when averaged over a sphere, $1 - 3\cos^2\theta$ is zero.

However, if the magnetic susceptibility, and hence the induced dipole, varies with the orientation of the molecule with

The chemist's toolkit 12B.1 Dipolar magnetic fields

Standard electromagnetic theory gives the magnetic field at a point $\boldsymbol{r}$ from a point magnetic dipole $\boldsymbol{\mu}$ as

$$\mathcal{B} = -\frac{\mu_0}{4\pi r^3}\left(\boldsymbol{\mu} - \frac{3(\boldsymbol{\mu}\cdot\boldsymbol{r})\boldsymbol{r}}{r^2}\right)$$

where μ_0 is the magnetic constant (previously known as the vacuum permeability, a fundamental constant with the value $1.257 \times 10^{-6}\,\text{T}^2\,\text{J}^{-1}\,\text{m}^3$). The component of magnetic field in the z-direction is

$$\mathcal{B}_z = -\frac{\mu_0}{4\pi r^3}\left(\mu_z - \frac{3(\boldsymbol{\mu}\cdot\boldsymbol{r})z}{r^2}\right)$$

with $z = r\cos\theta$, the z-component of the distance vector $\boldsymbol{r}$. If the magnetic dipole is also parallel to the z-direction and has magnitude μ, it follows that

$$\mathcal{B}_z = -\frac{\mu_0}{4\pi r^3}\left(\overbrace{\mu}^{\mu_z} - \frac{3\overbrace{(\mu r\cos\theta)}^{\boldsymbol{\mu}\cdot\boldsymbol{r}}\overbrace{(r\cos\theta)}^{z}}{r^2}\right) = -\frac{\mu\mu_0}{4\pi r^3}(1-3\cos^2\theta)$$

respect to the magnetic field, the local field may average to a non-zero value. For instance, suppose that the neighbouring group has axial symmetry (as might be the case for a triple bond): when the applied field is parallel to the symmetry axis the susceptibility is $\chi_{\parallel}$, and when it is perpendicular the susceptibility is $\chi_{\perp}$. After averaging over all orientations of the molecule the contribution to the shielding constant of a nucleus at a distance R has the following form

$$\sigma(\text{neighbour}) \propto (\chi_{\parallel} - \chi_{\perp})\left(\frac{1-3\cos^2\Theta}{R^3}\right)$$

Neighbouring group contribution (12B.10b)

where Θ (uppercase theta) is the angle between the symmetry axis and the vector to the nucleus (**2**). Equation 12B.10b shows that the neighbouring group contribution may be positive or negative according to the relative magnitudes of the two magnetic susceptibilities and the direction given by Θ. If $54.7° < \Theta < 125.3°$, then $1 - 3\cos^2\Theta$ is positive, but it is negative otherwise (Figs. 12B.4 and 12B.5).

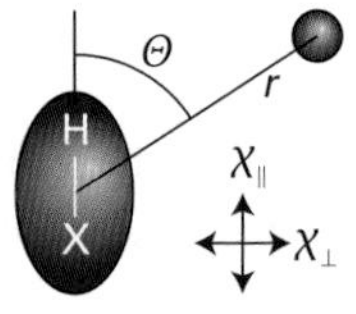

2

A special case of a neighbouring group effect is found in aromatic compounds. The strong anisotropy of the magnetic susceptibility of the benzene ring is ascribed to the ability of the field to induce a **ring current**, a circulation of electrons around the ring, when the field is applied perpendicular to the molecular plane. Protons in the plane are deshielded (Fig. 12B.6), but any that happen to lie above or below the plane (as members of substituents of the ring) are shielded.

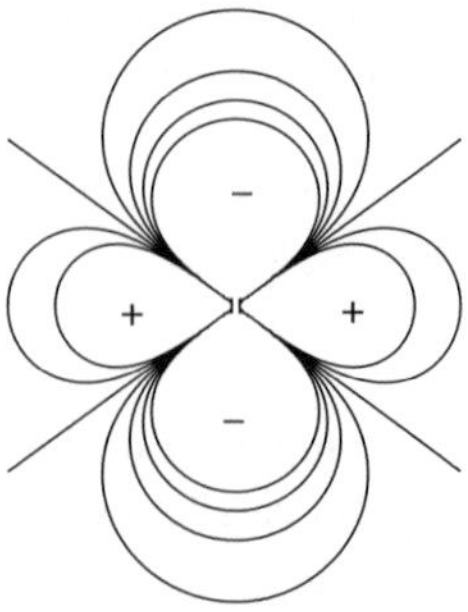

Figure 12B.4 A depiction of the field arising from a point magnetic dipole. The contours represent the strength of field declining with distance (as $1/R^3$). The field is zero on the straight lines.

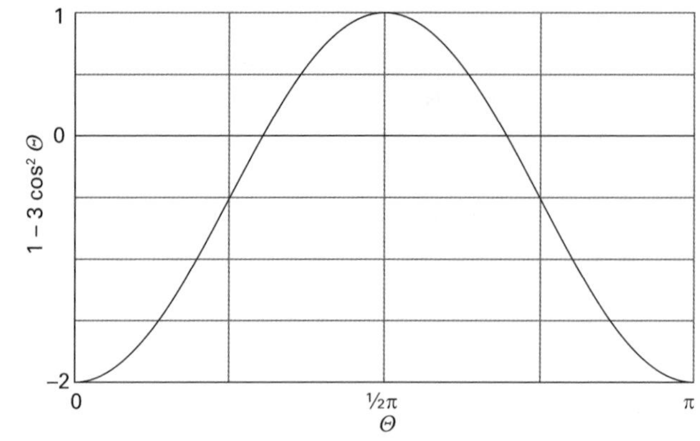

Figure 12B.5 The variation of the function $1 - 3\cos^2\Theta$ with the angle Θ.

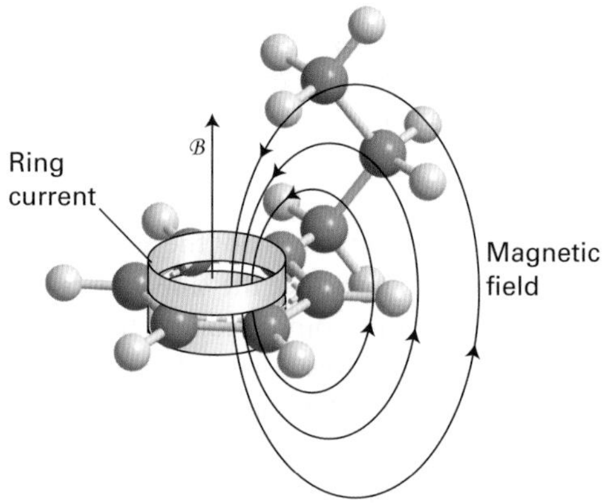

Figure 12B.6 The shielding and deshielding effects of the ring current induced in the benzene ring by the applied field. Protons attached to the ring are deshielded but a proton attached to a substituent that projects above the ring is shielded.

12B.2(c) The solvent contribution

A solvent can influence the local magnetic field experienced by a nucleus in a variety of ways. Some of these effects arise from specific interactions between the solute and the solvent (such as hydrogen bond formation and other forms of Lewis acid–base complex formation). The anisotropy of the magnetic susceptibility of the solvent molecules, especially if they are aromatic, can also be the source of a local magnetic field. Moreover, if there are steric interactions that result in a loose but specific interaction between a solute molecule and a solvent molecule, then protons in the solute molecule may experience shielding or deshielding effects according to their location relative to the solvent molecule. An aromatic solvent such as benzene can give rise to local currents that shield or deshield a proton in a solute molecule. The arrangement shown in Fig. 12B.7 leads to shielding of a proton on the solute molecule.

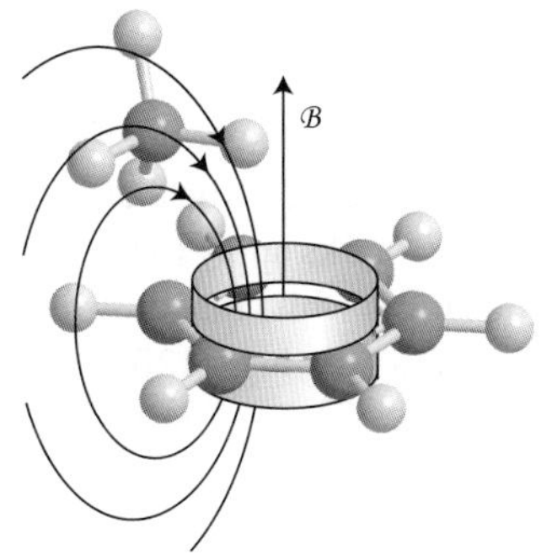

Figure 12B.7 An aromatic solvent (such as benzene shown here) can give rise to local currents that shield or deshield a proton in a solute molecule. In the orientation shown the proton on the solute molecule is shielded.

Exercise

E12B.3 The chemical shift of the CH_3 protons in ethanal (acetaldehyde) is $\delta = 2.20$ and that of the CHO proton is 9.80. What is the difference in local magnetic field between the two regions of the molecule when the applied field is (i) 1.5 T, (ii) 15 T?

12B.3 The fine structure

Figure 12B.8 shows the ^{1}H (proton) NMR spectrum of chloroethane. In this molecule there are two different types of ^{1}H, the methylene (CH_2) and the methyl (CH_3) protons, each with a characteristic chemical shift. In addition, the spectrum shows **fine structure**, the splitting of a resonance line into several components. The groups of lines are called **multiplets**.

This fine structure arises from **scalar coupling** in which the resonance frequency of one nucleus is affected by the spin state of another nucleus. Qualitatively, the effect of scalar coupling arises when the local magnetic field at one nucleus depends on the relative orientation of the other spin. If the spin is in one state (α, for instance) the local field is increased, whereas when the spin is in the other state (β in this case), the local field is decreased. Therefore, rather than there being one line in the spectrum, there are two because there are two possible values for the local field, corresponding to the second nucleus being either α or β.

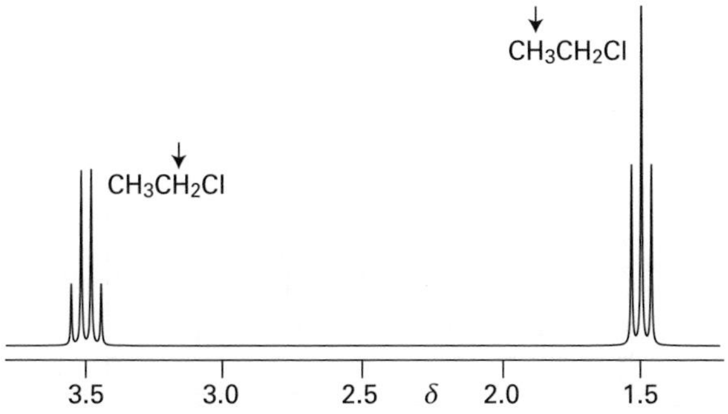

Figure 12B.8 The ^{1}H (proton) NMR spectrum of chloroethane. The arrows indicate the protons giving rise to each multiplet; the multiplets arise due to scalar coupling between the protons.

The scalar coupling interaction is represented by a term $(hJ/\hbar^2)\hat{\boldsymbol{I}}_1 \cdot \hat{\boldsymbol{I}}_2$ in the hamiltonian, where $\hat{\boldsymbol{I}}_N$, with $N = 1$ or 2, is the operator for the nuclear spin angular momentum of nucleus N. That the coupling term is a scalar product simply expresses the fact that the energy of interaction depends on the relative orientation of the spins of the two nuclei. The strength of the interaction is given by the value of the **scalar coupling constant**, J. The presence of $\hbar^2$ in $(hJ/\hbar^2)\hat{\boldsymbol{I}}_1 \cdot \hat{\boldsymbol{I}}_2$ cancels the $\hbar^2$ arising from the eigenvalues of the two angular momenta, leaving the energy as hJ, so J is a frequency (measured in hertz, Hz). The coupling constant can be positive or negative, and it is independent of the field strength.

If the Larmor frequencies of the two coupled nuclei are significantly different, they precess at very different rates and the x- and y-components of their magnetic moments are never in step. Only the z-components remain in alignment whatever the precession rates, and so the only surviving term in the scalar product is $(hJ/\hbar^2)\hat{I}_{1z}\hat{I}_{2z}$. The eigenvalues of each $\hat{I}_{Nz}$ are $m_{I_N}\hbar$, so it follows that the eigenvalues (the energies) of the coupling term are

$$E_{m_{I_1}m_{I_2}} = hJm_{I_1}m_{I_2} \qquad \text{Spin–spin coupling energy} \qquad (12B.11)$$

12B.3(a) The appearance of the spectrum

In NMR, letters far apart in the alphabet (typically A and X) are used to indicate nuclei with very different chemical shifts in the sense that the difference in chemical shift corresponds to a frequency that is large compared to J; letters close together (such as A and B) are used for nuclei with similar chemical shifts.

Consider first an AX system, a molecule that contains two spin-$\frac{1}{2}$ nuclei A and X with very different chemical shifts, so eqn 12B.11 can be used for the spin–spin coupling energy. Nucleus A has two spin states with $m_A = \pm\frac{1}{2}$ corresponding to the states denoted α_A and β_A. The X nucleus also has two spin states with $m_X = \pm\frac{1}{2}$ (α_X and β_X). In the AX system there are therefore four spin states: $\alpha_A\alpha_X$, $\alpha_A\beta_X$, $\beta_A\alpha_X$, and $\beta_A\beta_X$. The energies of these states, neglecting any scalar coupling, are therefore

$$\begin{aligned} E_{m_A m_X} &= -\gamma_N\hbar(1-\sigma_A)\mathcal{B}_0 m_A - \gamma_N\hbar(1-\sigma_X)\mathcal{B}_0 m_X \\ &= -h\nu_A m_A - h\nu_X m_X \end{aligned} \qquad (12B.12a)$$

where ν_A and ν_X are the Larmor frequencies of A and X (eqn 12B.3). This expression gives the four levels illustrated on the left of Fig. 12B.9. When spin–spin coupling is included (by using eqn 12B.11) the energy levels are

$$E_{m_A m_X} = -h\nu_A m_A - h\nu_X m_X + hJm_A m_X \quad (12B.12b)$$

The resulting energy level diagram (for $J > 0$) is shown on the right of Fig. 12B.9. The $\alpha_A\alpha_X$ and $\beta_A\beta_X$ states are both raised by $\frac{1}{4}hJ$ and the $\alpha_A\beta_X$ and $\beta_A\alpha_X$ states are both lowered by $\frac{1}{4}hJ$. For $J > 0$, the effect of the coupling term is to lower the energy of the $\alpha_A\beta_X$ and $\beta_A\alpha_X$ states, and raise the energy of the other two states. The opposite is the case for $J < 0$.

In a transition, only one nucleus changes its orientation, so the selection rule is that either m_A or m_X can change by ± 1, but not both. There are two transitions in which the spin state of A changes while that of X remains fixed: $\beta_A\alpha_X \leftarrow \alpha_A\alpha_X$ and $\beta_A\beta_X \leftarrow \alpha_A\beta_X$. They are shown in Fig. 12B.9 and in a slightly different form in Fig. 12B.10. The energies of the transitions are

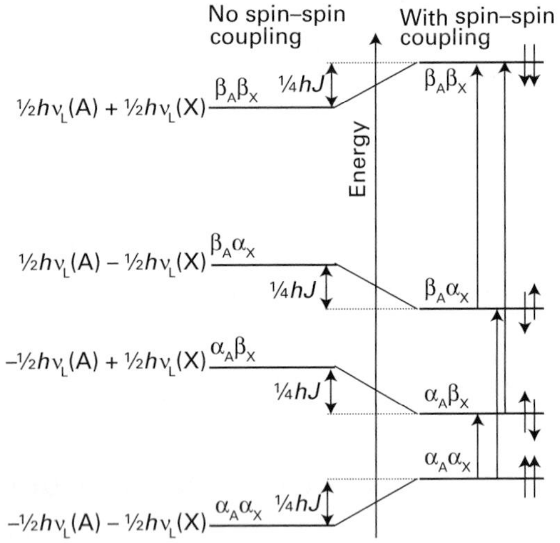

Figure 12B.9 The energy levels of an AX spin system. The four levels on the left are those in the case of no spin–spin coupling. The four levels on the right show how a positive spin–spin coupling constant affects the energies. The effect of the coupling on the energy levels has been exaggerated greatly for clarity; in practice, the change in energy caused by spin–spin coupling is much smaller than that caused by the applied field.

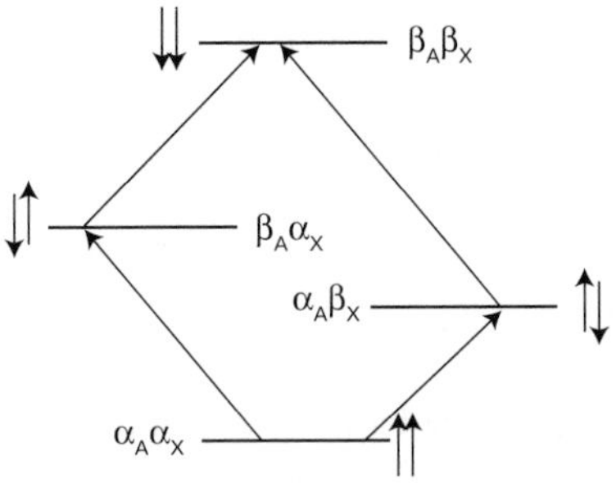

Figure 12B.10 An alternative depiction of the energy levels and transitions shown in Fig. 12B.9. Once again, the effect of spin–spin coupling has been exaggerated.

$$\Delta E = h\nu_A \pm \tfrac{1}{2}hJ \quad (12B.13a)$$

The spectrum due to A transitions therefore consists of a doublet of separation J centred on the Larmor frequency of A (Fig. 12B.11). Similar remarks apply to the transitions in which the spin state of X changes while that of A remains fixed. These transitions are also shown in Figs. 12B.9 and 12B.10, and the transition energies are

$$\Delta E = h\nu_X \pm \tfrac{1}{2}hJ \quad (12B.13b)$$

It follows that there is a second doublet with the same separation J as the first, but now centred on the Larmor frequency of X (as shown in Fig. 12B.11). Overall, the spectrum of an AX spin system consists of two doublets.

Suppose that there is another X nucleus in the molecule with the same chemical shift as the first X thus giving an AX_2 spin system; the two X nuclei are said to be *equivalent*. The X resonance is split into a doublet by A, just as for AX (Fig. 12B.12).

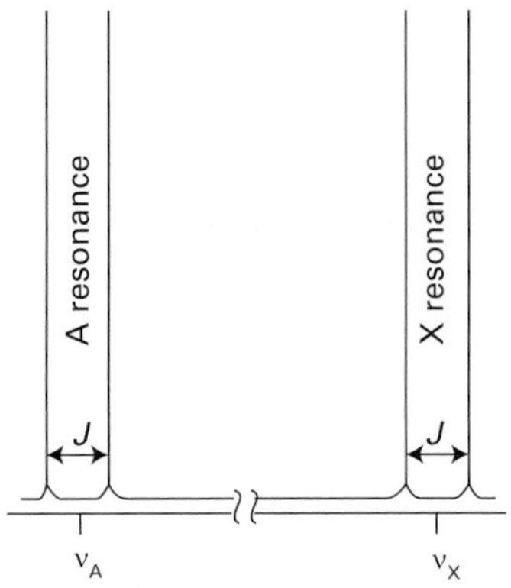

Figure 12B.11 The effect of spin–spin coupling on an AX spectrum. Each resonance is split into two lines, a doublet, separated by J. There is a doublet centred on the Larmor frequency (chemical shift) of A, and one centred on the Larmor frequency of X.

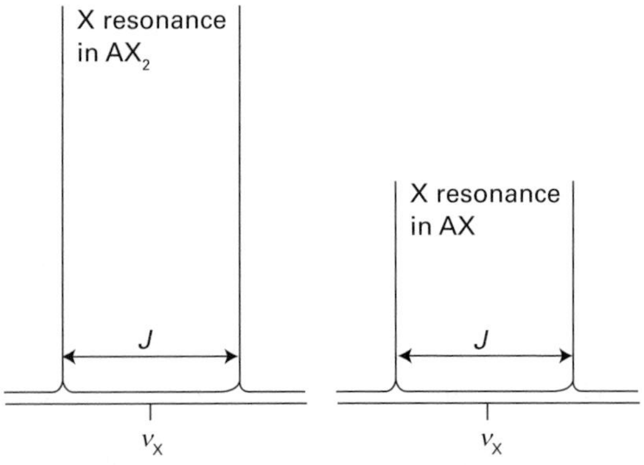

Figure 12B.12 The X resonance of an AX_2 spin system is a doublet because of the coupling to the A spin. However, the overall absorption is twice as intense as that of an AX spin system.

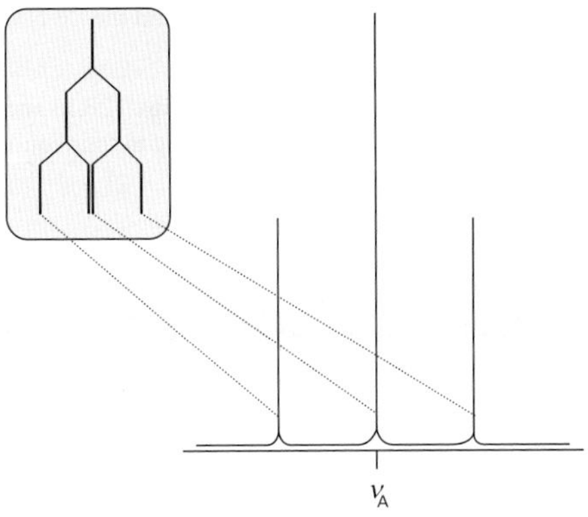

Figure 12B.13 The origin of the 1:2:1 triplet in the A resonance of an AX_2 spin system. The resonance of A is split into two by coupling with one X nucleus (as shown in the inset), and then each of those two lines is split into two by coupling to the second X nucleus. Because each X nucleus causes the same splitting, the two central transitions are coincident and give rise to an absorption line of double the intensity of the outer lines.

The resonance of A is split into a doublet by one X, and each line of the doublet is split again by the same amount by the second X (Fig. 12B.13). This splitting results in three lines in the intensity ratio 1:2:1 (because the central frequency can be obtained in two ways).

Three equivalent X nuclei (an AX_3 spin system) split the resonance of A into four lines of intensity ratio 1:3:3:1 (Fig. 12B.14). The X resonance remains a doublet as a result of the splitting caused by A. In general, N equivalent spin-$\frac{1}{2}$ nuclei split the resonance of a nearby spin or group of equivalent spins into $N+1$ lines with an intensity distribution given by Pascal's triangle (**3**).

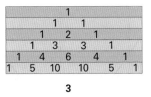

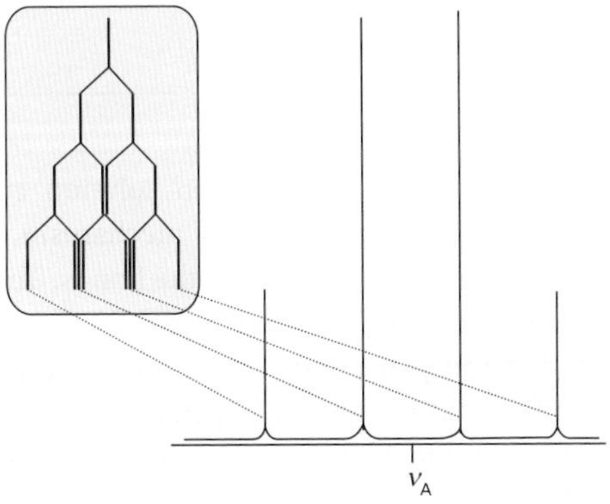

Figure 12B.14 The origin of the 1:3:3:1 quartet in the A resonance of an AX_3 spin system. The third X nucleus splits each of the lines shown in Fig. 12B.13 for an AX_2 spin system into a doublet, and the intensity distribution reflects the number of transitions that have the same energy.

Successive rows of this triangle are formed by adding together the two adjacent numbers in the line above.

1
1 1
1 2 1
1 3 3 1
1 4 6 4 1
1 5 10 10 5 1

3

Example 12B.3 Accounting for the fine structure in a spectrum

Account for the fine structure in the ¹H (proton) NMR spectrum of chloroethane shown in Fig. 12B.8.

Collect your thoughts You need to consider how each group of equivalent protons (for instance, the three methyl protons) splits the resonances of the other groups of protons. There is no splitting within groups of equivalent protons. You can identify the pattern of intensities within a multiplet by referring to Pascal's triangle.

The solution The three protons of the CH_3 group split the resonance of the CH_2 protons into a 1:3:3:1 quartet with a splitting J. Likewise, the two protons of the CH_2 group split the resonance of the CH_3 protons into a 1:2:1 triplet with the same splitting J.

Self-test 12B.3 What fine structure can be expected in the ¹H spectrum, and in the ¹⁵N spectrum, of $^{15}NH_4^+$? Nitrogen-15 is a spin-$\frac{1}{2}$ nucleus.

Answer: The ¹H spectrum is a 1:1 doublet and the ¹⁵N spectrum is a 1:4:6:4:1 quintet

12B.3(b) The magnitudes of coupling constants

The scalar coupling constant of two nuclei separated by N bonds is denoted ^NJ, with subscripts to indicate the types of nuclei involved. Thus, $^1J_{CH}$ is the coupling constant for a proton joined directly to a ^{13}C atom, and $^2J_{CH}$ is the coupling constant when the same two nuclei are separated by two bonds (as in ^{13}C–C–H). A typical value of $^1J_{CH}$ is in the range 120–250 Hz; $^2J_{CH}$ is between 10 and 20 Hz. Couplings over both three and four bonds (^3J and ^4J) can give detectable effects in a spectrum, but couplings over larger numbers of bonds can generally be ignored.

As remarked in the discussion following eqn 12B.12b, the sign of J_{XY} determines whether a particular energy level is raised or lowered as a result of the coupling interaction. If $J > 0$, the levels with antiparallel spins are lowered in energy, whereas if $J < 0$ the levels with parallel spins are lowered. Experimentally, it is found that $^1J_{CH}$ is invariably positive, $^2J_{HH}$ is often negative,

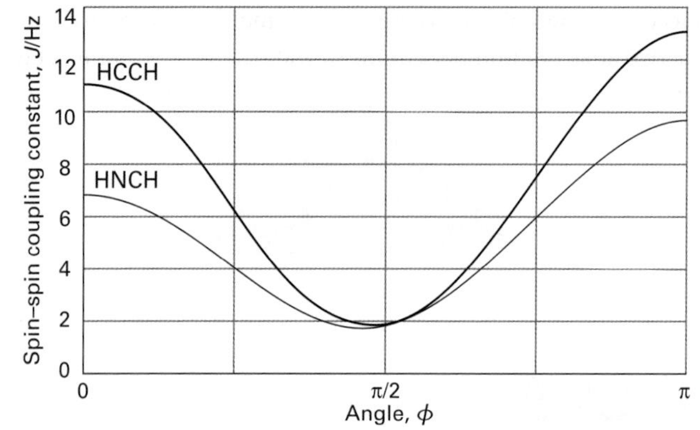

Figure 12B.15 The variation of the spin–spin coupling constant with dihedral angle predicted by the Karplus equation for an HCCH group and an HNCH group.

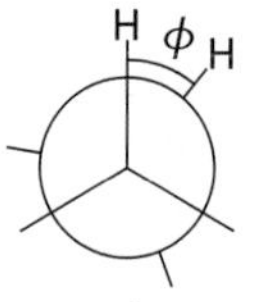

4

and $^3J_{HH}$ is often positive. An additional point is that J varies with the dihedral angle between the bonds (Fig. 12B.15). Thus, a $^3J_{HH}$ coupling constant is often found to depend on the dihedral angle ϕ (4) according to the **Karplus equation**:

$$^3J_{HH} = A + B\cos\phi + C\cos 2\phi \quad \text{Karplus equation} \quad (12B.14)$$

with A, B, and C empirical constants with values close to +7 Hz, −1 Hz, and +5 Hz, respectively, for an HCCH fragment. It follows that the measurement of $^3J_{HH}$ in a series of related compounds can be used to determine their conformations. The coupling constant $^1J_{CH}$ also depends on the hybridization of the C atom, as the following values indicate:

	sp	sp^2	sp^3
$^1J_{CH}$/Hz	250	160	125

Brief illustration 12B.2

The investigation of H–N–C–H couplings in polypeptides can help reveal their conformation. For $^3J_{HH}$ coupling in such a group, A = +5.1 Hz, B = −1.4 Hz, and C = +3.2 Hz. For a helical polymer, ϕ is close to 120°, which gives $^3J_{HH} \approx$ 4 Hz. For the sheet-like conformation, ϕ is close to 180°, which gives $^3J_{HH} \approx$ 10 Hz. Experimental measurements of the value of $^3J_{HH}$ should therefore make it possible to distinguish between the two possible structures.

Exercises

E12B.4 Make a sketch, roughly to scale, of the ^{1}H NMR spectrum expected for an AX spin system with δ_A = 1.00, δ_X = 2.00, and J_{AX} = 10 Hz recorded on a spectrometer operating at (i) 250 MHz, (ii) 800 MHz. The horizontal scale should be in hertz, taking the resonance from TMS as the origin.

E12B.5 Sketch the form of the ^{19}F NMR spectrum and the ^{10}B NMR spectrum of $^{10}BF_4^-$.

E12B.6 Use an approach similar to that shown in Figs. 12B.13 and 12B.14 to predict the multiplet you would expect for coupling to four equivalent spin-$\frac{1}{2}$ nuclei.

12B.4 The origin of spin–spin coupling

Some insight into the origin of coupling, if not its precise magnitude—or always reliably its sign—can be obtained by considering the magnetic interactions within molecules. A nucleus with the z-component of its spin angular momentum specified by the quantum number m_I gives rise to a magnetic field with z-component $\mathcal{B}_{nuc}$ at a distance R, where, to a good approximation,

$$\mathcal{B}_{nuc} = -\frac{\gamma_N \hbar \mu_0}{4\pi R^3}(1 - 3\cos^2\theta)m_I \quad (12B.15)$$

The angle θ is defined in (1); this expression is a version of eqn 12B.10a. However, in solution molecules tumble rapidly so it is necessary to average $\mathcal{B}_{nuc}$ over all values of θ. As has already been noted, the average of $1 - 3\cos^2\theta$ is zero, therefore the direct dipolar interaction between spins cannot account for the fine structure seen in the spectra of molecules in solution.

Brief illustration 12B.3

There can be a direct dipolar interaction between nuclei in solids, where the molecules do not rotate. The z-component of the magnetic field arising from a ^{1}H nucleus with $m_I = +\frac{1}{2}$, at R = 0.30 nm, and at an angle θ = 0 is[3]

$$\mathcal{B}_{nuc} = -\frac{\overbrace{(2.821\times10^{-26}\ \mathrm{J\,T^{-1}})}^{\gamma_N\hbar}\times\overbrace{(1.2566\times10^{-6}\ \mathrm{T^2\,J^{-1}\,m^3})}^{\mu_0}}{4\pi\times\underbrace{(3.0\times10^{-10}\ \mathrm{m})^3}_{R^3}}\times\overbrace{(-1)}^{(1-3\cos^2\theta)m_I}$$

$$= 1.0\times10^{-4}\ \mathrm{T} = 0.10\ \mathrm{mT}$$

Spin–spin coupling in molecules in solution can be explained in terms of the **polarization mechanism**, in which the interaction is transmitted through the bonds. The simplest case to consider is that of $^1J_{XY}$, where X and Y are spin-$\frac{1}{2}$ nuclei joined by an electron-pair bond. The coupling mechanism depends on the fact that the energy depends on the relative orientation of the bonding electrons and the nuclear spins. This electron–nucleus coupling is magnetic in origin, and may be

[3] As seen in the front matter of the book, $\mu_0 = 1.2566\times10^{-6}\ \mathrm{J\,s^2\,C^{-2}\,m^{-1}}$. But $1\ \mathrm{T} = 1\ \mathrm{J\,s\,C^{-1}\,m^{-2}}$, so it is also possible to write $\mu_0 = 1.2566\times10^{-6}\ \mathrm{T^2\,J^{-1}\,m^3}$, with these units proving more useful in this calculation.

either a dipolar interaction or a **Fermi contact interaction**. A pictorial description of the latter is as follows.

First, regard the magnetic moment of the nucleus as arising from the circulation of a current in a tiny loop with a radius similar to that of the nucleus (Fig. 12B.16). Far from the nucleus the field generated by this loop is indistinguishable from the field generated by a point magnetic dipole. Close to the loop, however, the field differs from that of a point dipole. The magnetic interaction between this non-dipolar field and the electron's magnetic moment is the contact interaction. The contact interaction—essentially the failure of the point-dipole approximation—depends on the very close approach of an electron to the nucleus and hence can occur only if the electron occupies an s orbital (which is the reason why $^1J_{CH}$ depends on the hybridization ratio).

Suppose that it is energetically favourable for an electron spin and a nuclear spin to be antiparallel (as is the case for a proton and an electron in a hydrogen atom). If the X nucleus is α, a β electron of the bonding pair will tend to be found nearby, because that is an energetically favourable arrangement (Fig. 12B.17). The second electron in the bond, which must have α spin if the other is β (by the Pauli principle; Topic 8B), will be found mainly at the far end of the bond because electrons tend to stay apart to reduce their mutual repulsion. Because it is energetically favourable for the spin of Y to be antiparallel to an electron spin, a Y nucleus with β spin has a lower energy than when it has α spin. The opposite is true when X is β, for now the α spin of Y has the lower energy. In other words, the antiparallel arrangement of nuclear spins lies lower in energy than the parallel arrangement as a result of their magnetic coupling with the bond electrons. That is, $^1J_{CH}$ is positive.

To account for the value of $^2J_{XY}$, as for $^2J_{HH}$ in H–C–H, a mechanism is needed that can transmit the spin alignments through the central C atom (which may be ^{12}C, with no nuclear spin of its own). In this case (Fig. 12B.18), an X nucleus with α spin polarizes the electrons in its bond, and the α electron is likely to be found closer to the C nucleus. The more favourable arrangement of two electrons on the same atom is with their spins parallel (Hund's rule, Topic 8B), so the more favourable arrangement is for the α electron of the neighbouring bond to be close to the C nucleus. Consequently, the β electron of that bond is more likely to be found close to the Y nucleus, and therefore that nucleus will have a lower energy if it is α. Hence, according to this mechanism, the lower energy will be obtained if the Y spin is parallel to that of X. That is, $^2J_{HH}$ is negative.

The coupling of nuclear spin to electron spin by the Fermi contact interaction is most important for proton spins, but it is not necessarily the most important mechanism for other nuclei. These nuclei may also interact by a dipolar mechanism with the electron magnetic moments and with their orbital motion, and there is no simple way of specifying whether J will be positive or negative. The dipolar interaction does not average to zero as the molecule tumbles if it accounts for the interaction of *both* nuclei with their surrounding electrons because then $1 - 3\cos^2\theta$ appears as its square and therefore with a non-negative value at all orientations, and its average value is no longer zero.

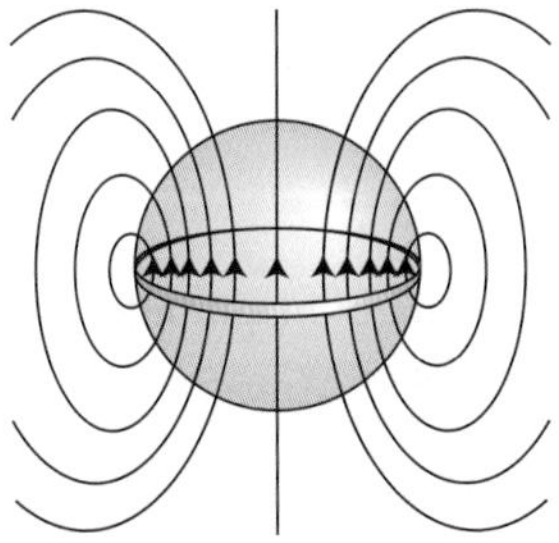

Figure 12B.16 The origin of the Fermi contact interaction. From far away, the magnetic field pattern arising from a ring of current (representing the rotating charge of the nucleus, the pale grey sphere) is that of a point dipole. However, if an electron can sample the field close to the region indicated by the sphere, the field distribution differs significantly from that of a point dipole. For example, if the electron can penetrate the sphere, then the spherical average of the field it experiences is not zero.

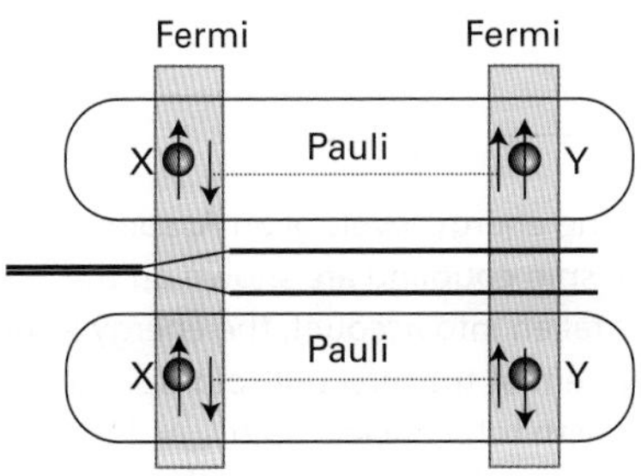

Figure 12B.17 The polarization mechanism for spin–spin coupling ($^1J_{CH}$). The two arrangements have slightly different energies. In this case, J is positive, corresponding to a lower energy when the nuclear spins are antiparallel.

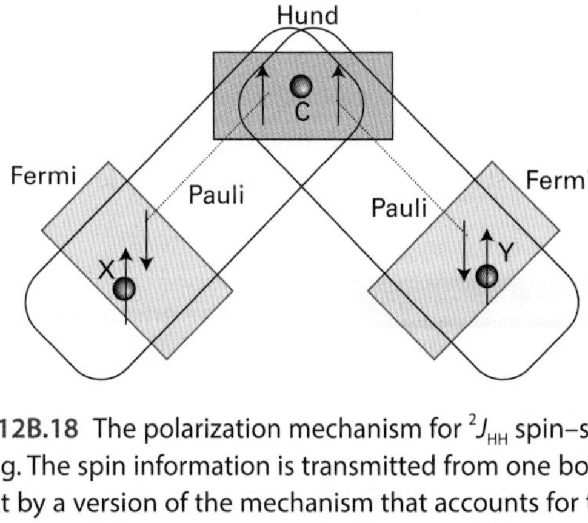

Figure 12B.18 The polarization mechanism for $^2J_{HH}$ spin–spin coupling. The spin information is transmitted from one bond to the next by a version of the mechanism that accounts for the lower energy of electrons with parallel spins in different atomic orbitals (Hund's rule of maximum multiplicity). In this case, $J < 0$, corresponding to a lower energy when the nuclear spins are parallel.

12B.4(a) Equivalent nuclei

A group of identical nuclei are **chemically equivalent** if they are related by a symmetry operation of the molecule and have the same chemical shifts. Chemically equivalent nuclei are from atoms that would be regarded as 'equivalent' according to ordinary chemical criteria. Nuclei are **magnetically equivalent** if, as well as being chemically equivalent, they also have identical spin–spin interactions with any other magnetic nuclei in the molecule.

Brief illustration 12B.4

The difference between chemical and magnetic equivalence is illustrated by CH_2F_2 and $H_2C{=}CF_2$ (recall that ^{19}F is a spin-$\frac{1}{2}$ nucleus). In each of these molecules the ^1H nuclei (protons) are chemically equivalent because they are related by symmetry. The protons in CH_2F_2 are magnetically equivalent, but those in $CH_2{=}CF_2$ are not. One proton in the latter has a *cis* spin-coupling interaction with a given F nucleus whereas the other proton has a *trans* interaction with the same nucleus. In contrast, in CH_2F_2 each proton has the same coupling to both fluorine nuclei since the bonding pathway between them is the same.

Strictly speaking, in a molecule such as CH_3CH_2Cl the three CH_3 protons are magnetically inequivalent: each one may have a different coupling to the CH_2 protons on account of the different dihedral angles between the protons. However, the three CH_3 protons are in practice made magnetically equivalent by the rapid rotation of the CH_3 group, which averages out any differences. The spectra of molecules with chemically equivalent but magnetically inequivalent sets of spins can become very complicated. For example, the proton and ^{19}F spectra of $H_2C{=}CF_2$ each consist of 12 lines. Such spectra are not considered further.

An important feature of chemically equivalent magnetic nuclei is that, although they do couple together, the coupling has no effect on the appearance of the spectrum. How this comes about can be illustrated by considering the case of an A_2 spin system. The first step is to establish the energy levels.

How is that done? 12B.1 Deriving the energy levels of an A_2 system

Consider an A_2 system of two spin-$\frac{1}{2}$ nuclei, and consider first the energy levels in the absence of spin–spin coupling. When considering spin–spin coupling, be prepared to use the complete expression for the energy (the one proportional to $\boldsymbol{I}_1{\cdot}\boldsymbol{I}_2$), because the Larmor frequencies are the same and the approximate form ($I_{1z}I_{2z}$) cannot be used as it is applicable only when the Larmor frequencies are very different.

Step 1 *Identify the states and their energies in the absence of spin–spin coupling*

There are four energy levels; they can be classified according to their *total* spin angular momentum I_{tot} (the analogue of S for several electrons) and its projection on to the z-axis, given by the quantum number M_I. There are three states with $I_{tot} = 1$, and one further state with $I_{tot} = 0$:

$$\begin{array}{lll} \text{Spins parallel, } I_{tot}=1: & M_I=+1 & \alpha\alpha \\ & M_I=0 & (1/2^{1/2})\{\alpha\beta+\beta\alpha\} \\ & M_I=-1 & \beta\beta \\ \text{Spins paired, } I_{tot}=0: & M_I=0 & (1/2^{1/2})\{\alpha\beta-\beta\alpha\} \end{array}$$

The effect of a magnetic field on these four states is shown on the left-hand side of Fig. 12B.19: the two states with $M_I = 0$ are unaffected by the field as they are composed of equal proportions of α and β spins, and both spins have the same Larmor frequency.

Step 2 *Allow for spin–spin interaction*

The scalar product in the expression $E = (hJ/\hbar^2)\boldsymbol{I}_1{\cdot}\boldsymbol{I}_2$ can be expressed in terms of the total nuclear spin $\boldsymbol{I}_{tot} = \boldsymbol{I}_1 + \boldsymbol{I}_2$ by noting that

$$I_{tot}^2 = (\boldsymbol{I}_1+\boldsymbol{I}_2)\cdot(\boldsymbol{I}_1+\boldsymbol{I}_2) = I_1^2 + I_2^2 + 2\boldsymbol{I}_1\cdot\boldsymbol{I}_2$$

Rearranging this expression to

$$\boldsymbol{I}_1\cdot\boldsymbol{I}_2 = \tfrac{1}{2}(I_{tot}^2 - I_1^2 - I_2^2)$$

and replacing the square magnitudes by their quantum-mechanical values gives:

$$\boldsymbol{I}_1\cdot\boldsymbol{I}_2 = \tfrac{1}{2}\{I_{tot}(I_{tot}+1) - I_1(I_1+1) - I_2(I_2+1)\}\hbar^2$$

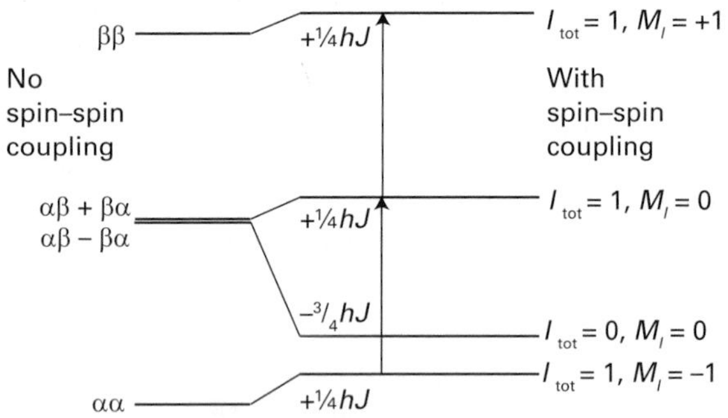

Figure 12B.19 The energy levels of an A_2 spin system in the absence of spin–spin coupling are shown on the left. When spin–spin coupling is taken into account, the energy levels on the right are obtained. Note that the effect of spin–spin coupling is to raise the three states with total nuclear spin $I_{tot} = 1$ (the triplet) by the same amount (J is positive); in contrast, the one state with $I_{tot} = 0$ (the singlet) is lowered in energy. The only allowed transitions, indicated by arrows, are those for which $\Delta I_{tot} = 0$ and $\Delta M_I = \pm 1$. These two transitions occur at the same resonance frequency as they would have in the absence of spin–spin coupling.

Then, because $I_1 = I_2 = \frac{1}{2}$, it follows that

$$E = \tfrac{1}{2}hJ\left\{I_{\text{tot}}(I_{\text{tot}}+1) - \tfrac{3}{2}\right\}$$

For parallel spins, $I_{\text{tot}} = 1$ and $E = +\frac{1}{4}hJ$; for antiparallel spins $I_{\text{tot}} = 0$ and $E = -\frac{3}{4}hJ$, as shown on the right-hand side of Fig. 12B.19.

The calculation shows that the three states with $I_{\text{tot}} = 1$ all move in energy in the same direction and by the same amount. The single state with $I_{\text{tot}} = 0$ moves three times as much in the opposite direction. In the resonance transition, the relative orientation of the nuclei cannot change, so there are no transitions between states of different I_{tot}. The selection rule $\Delta M_I = \pm 1$ also applies, and arises from the conservation of angular momentum and the unit spin of the photon. As shown in Fig. 12B.19, there are only two allowed transitions and because they have the same energy spacing they appear at the same frequency in the spectrum. Hence, the spin–spin coupling interaction does not affect the appearance of the spectrum of an A_2 molecule.

12B.4(b) Strongly coupled nuclei

The multiplets seen in NMR spectra due to the presence of spin–spin coupling are relatively simple to analyse provided the difference in chemical shifts between any two coupled spins is much greater than the value of the spin–spin coupling constant between them. This limit is often described as **weak coupling**, and the resulting spectra are described as **first-order spectra**.

When the difference in chemical shifts is comparable to the value of the spin–spin coupling constant, the multiplets take on a more complex form. Such spin systems are said to be **strongly coupled**, and the spectra are described as **second-order**. In such spectra the lines shift from where they are expected in the weak coupling case, their intensities change, and in some cases additional lines appear. Strongly coupled spectra are more difficult to analyse in the sense that the relation between the frequencies of the lines in the spectrum and the values of chemical shifts and coupling constants is not as straightforward as in the weakly coupled case.

Figure 12B.20 shows NMR spectra for two coupled spins as a function of the difference in chemical shift between the two spins. In Fig. 12B.20a (an AX species) this difference is large enough for the weak coupling limit to apply and two doublets are detected, with all lines having the same intensity. As the shift difference decreases the inner two lines gain intensity at the expense of the outer lines, and in the limit that the shift difference is zero (an A_2 species), the outer lines disappear and the inner lines converge.

If the two nuclei belong to different elements (e.g. ^{1}H and ^{13}C), or different isotopes of the same element (e.g. ^{1}H and ^{2}H), the fact that they have widely different Larmor frequencies means that the spin system will always be weakly coupled, and hence described as AX. If the two nuclei are of the same element the spin system is described as **homonuclear**, whereas if they are of different elements the system is described as **heteronuclear**.

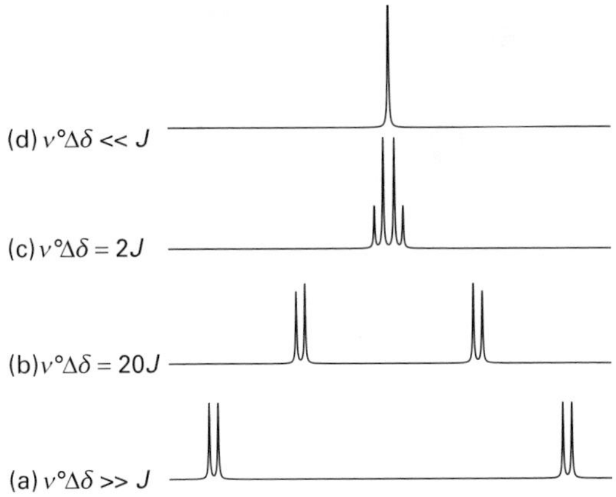

Figure 12B.20 The NMR spectra of (a) an AX system and (d) a 'nearly A_2' system are simple 'first-order' spectra (for an actual A_2 system, $\Delta\delta = 0$). At intermediate relative values of the chemical shift difference and the spin–spin coupling (b and c), more complex 'strongly coupled' spectra are obtained. Note how the inner two lines of the AX spectrum move together, grow in intensity, and form the single central line of the A_2 spectrum.

Exercises

E12B.7 Classify the ^{1}H nuclei in 1-chloro-4-bromobenzene into chemically or magnetically equivalent groups. Give your reasoning.

E12B.8 Classify the ^{19}F nuclei in PF_5 into chemically or magnetically equivalent groups. Give your reasoning.

12B.5 Exchange processes

The appearance of an NMR spectrum is changed if magnetic nuclei can jump rapidly between different environments. For example, consider the molecule *N,N*-dimethylmethanamide, $HCON(CH_3)_2$, in which the O–C–N fragment is planar, and there is restricted rotation about the C–N bond. The lowest energy conformation is shown in Fig. 12B.21. In this conformation the two methyl groups are not equivalent because one is *cis* and the other is *trans* to the carbonyl group. The two groups therefore have different environments and hence different chemical shifts.

Rotation by 180° about the C–N bond gives the same conformation, but it exchanges the CH_3 groups between the two environments. When the jumping rate of this process is low, the spectrum shows a distinct line for each CH_3 environment. When the rate is fast, the spectrum shows a single line at the mean of the two chemical shifts. At intermediate rates, the lines

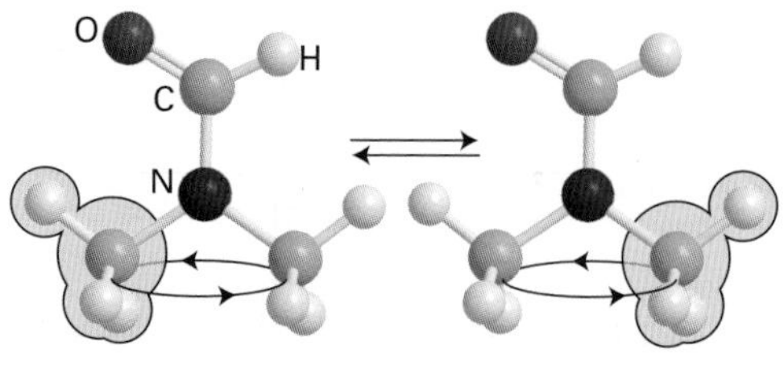

Figure 12B.21 In this molecule the two methyl groups are in different environments and so will have different chemical shifts. Rotation about the C–N bond interchanges the two groups, so that a particular methyl group is swapped between environments.

start to broaden and eventually coalesce into a single broad line. Coalescence of the two lines occurs when

$$\tau = \frac{2^{1/2}}{\pi\delta\nu} \quad \text{Condition for coalescence of two NMR lines} \quad (12B.16)$$

where τ is the lifetime of an environment and $\delta\nu$ is the difference between the Larmor frequencies of the two environments.

Brief illustration 12B.5

The NO group in *N,N*-dimethylnitrosamine, $(CH_3)_2N–NO$ (**5**), rotates about the N–N bond and, as a result, the magnetic environments of the two CH_3 groups are interchanged. The two CH_3 resonances are separated by 390 Hz in a 600 MHz spectrometer. According to eqn 12B.16,

$$\tau = \frac{2^{1/2}}{\pi \times (390\ s^{-1})} = 1.2\ ms$$

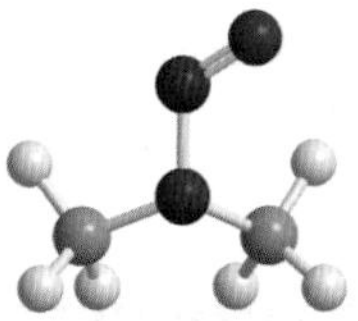

5 *N,N*-Dimethylnitrosamine

Such a lifetime corresponds to a (first-order) rate constant of $1/\tau = 870\ s^{-1}$. It follows that the signal will collapse to a single line when the rate constant for interconversion exceeds this value.

A similar explanation accounts for the loss of fine structure for protons in hydrogen atoms that can exchange with the solvent. For example, the resonance from the OH group in the spectrum of ethanol appears as a single line (Fig. 12B.22). In this molecule the hydroxyl protons are able to exchange with the protons in water which, unless special precautions are taken, is inevitably present as an impurity in the (organic) solvent. When this **chemical exchange**, an exchange of atoms, occurs, a molecule ROH with an α-spin proton (written as ROH_α) rapidly converts to ROH_β and then perhaps to ROH_α again

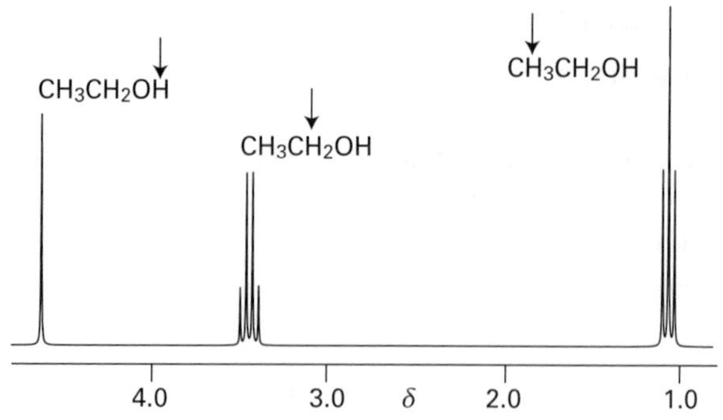

Figure 12B.22 The ^{1}H (proton) NMR spectrum of ethanol. The arrows indicate the protons giving rise to each multiplet. Due to chemical exchange between the OH proton and water molecules present in the solvent, no splittings due to coupling to the OH proton are seen.

because the protons provided by the water molecules in successive exchanges have random spin orientations.

If the rate constant for the exchange process is fast compared to the value of the coupling constant *J*, in the sense $1/\tau >> J$, the two lines merge and no splitting is seen. Because the values of coupling constants are typically just a few hertz, even rather slow exchange leads to the loss of the splitting. In the case of OH groups, only by rigorously excluding water from the solvent can the exchange rate be made slow enough that splittings due to coupling to OH protons are observed.

Exercise

E12B.9 A proton jumps between two sites with $\delta = 2.7$ and $\delta = 4.8$. What rate constant for the interconversion process is needed for the two signals to collapse to a single line in a spectrometer operating at 550 MHz?

12B.6 Solid-state NMR

In contrast to the narrow lines seen in the NMR spectra of samples in solution, the spectra from solid samples give broad lines, often to the extent that chemical shifts are not resolved. Nevertheless, there are good reasons for seeking to overcome these difficulties. They include the possibility that a compound is unstable in solution or that it is insoluble. Moreover, many species, such as polymers (both synthetic and naturally occurring), are intrinsically interesting as solids and might not be open to study by X-ray diffraction: in these cases, solid-state NMR provides a useful alternative way of probing both structure and dynamics.

There are three principal contributions to the linewidths of solids. One is the direct magnetic dipolar interaction between nuclear spins. As pointed out in the discussion of spin–spin coupling, a nuclear magnetic moment gives rise to a local magnetic field which points in different directions at different locations around the nucleus. If the only component of interest is parallel to the direction of the applied magnetic field (because only this component has a significant effect), then provided certain subtle effects arising from transformation from the static to the rotating frame are neglected, the classical expression in *The chemist's toolkit* 12B.1 can be used to write the magnitude of the local magnetic field as

$$\mathcal{B}_{\text{loc}} = -\frac{\gamma_{\text{N}}\hbar\mu_0 m_I}{4\pi R^3}(1-3\cos^2\theta) \qquad (12\text{B}.17\text{a})$$

Unlike in solution, in a solid this field is not averaged to zero by the molecular motion. Many nuclei may contribute to the total local field experienced by a nucleus of interest, and different nuclei in a sample may experience a wide range of fields. Typical dipole fields are of the order of 1 mT, which corresponds to splittings and linewidths of the order of 10 kHz for ^{1}H. When the angle θ can vary only between 0 and θ_{max}, the average value of $1 - 3\cos^2\theta$ can be shown to be $-(\cos^2\theta_{\text{max}} + \cos\theta_{\text{max}})$. This result, in conjunction with eqn 12B.17a, gives the average local field as

$$\mathcal{B}_{\text{loc,av}} = \frac{\gamma_{\text{N}}\hbar\mu_0 m_I}{4\pi R^3}(\cos^2\theta_{\text{max}} + \cos\theta_{\text{max}}) \qquad (12\text{B}.17\text{b})$$

Brief illustration 12B.6

When $\theta_{\text{max}} = 30°$ and $R = 160$ pm, the local field generated by a proton is

$$\mathcal{B}_{\text{loc,av}} = \frac{\overbrace{(3.546\ldots\times10^{-32}\ \text{T m}^3)}^{\gamma_{\text{N}}\hbar\mu_0}\times\overbrace{(\tfrac{1}{2})}^{m_I}\times\overbrace{(1.616)}^{\cos^2\theta_{\text{max}}+\cos\theta_{\text{max}}}}{4\pi\times\underbrace{(1.60\times10^{-10}\ \text{m})^3}_{R^3}}$$

$$= 5.57\times10^{-4}\ \text{T} = 0.557\ \text{mT}$$

A second source of linewidth is the anisotropy of the chemical shift. Chemical shifts arise from the ability of the applied field to generate electron currents in molecules. In general, this ability depends on the orientation of the molecule relative to the applied field. In solution, when the molecule is tumbling rapidly, only the average value of the chemical shift is relevant. However, the anisotropy is not averaged to zero for stationary molecules in a solid, and molecules in different orientations have resonances at different frequencies. The chemical shift anisotropy also varies with the angle θ between the applied field and the principal axis of the molecule as $1 - 3\cos^2\theta$.

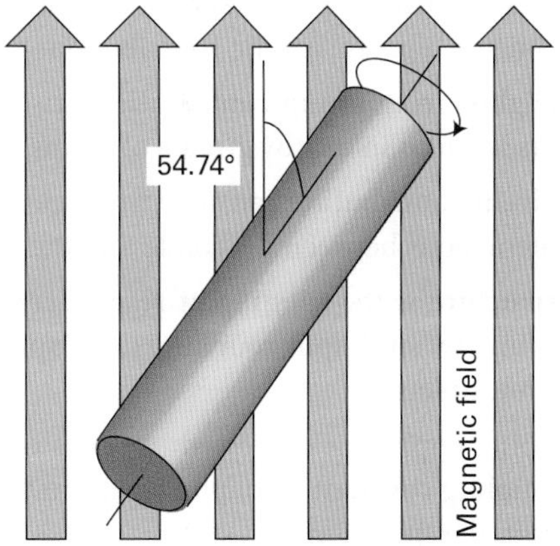

Figure 12B.23 In magic-angle spinning, the sample spins on an axis at 54.74° (i.e. arccos $1/3^{1/2}$) to the applied magnetic field. Rapid motion at this angle averages dipole–dipole interactions and chemical shift anisotropies to zero.

The third contribution is the electric quadrupole interaction. Nuclei with $I > \frac{1}{2}$ have an 'electric quadrupole moment', a measure of the extent to which the distribution of charge over the nucleus is not uniform (for instance, the positive charge may be concentrated around the equator or at the poles). An electric quadrupole interacts with an electric field gradient, such as may arise from a non-spherical distribution of charge around the nucleus. This interaction depends on the orientation of the molecule relative to the magnetic field, as in both the dipolar interaction and chemical shift anisotropy.

Fortunately, there are techniques available for reducing the linewidths of solid samples. One technique, **magic-angle spinning** (MAS), takes note of the $1 - 3\cos^2\theta$ dependence of the dipole–dipole interaction, the chemical shift anisotropy, and the electric quadrupole interaction. The 'magic angle' is the angle at which $1 - 3\cos^2\theta = 0$, and corresponds to 54.74°. In the technique, the sample is spun at high speed around an axis at the magic angle to the applied field (Fig. 12B.23). All the dipolar interactions and the anisotropies average to the value they would have at the magic angle, but at that angle they are zero. In principle, MAS therefore removes completely the line-broadening due to dipole–dipole interactions and chemical shift anisotropy. The difficulty with MAS is that the spinning frequency must not be less than the width of the spectrum, which is of the order of kilohertz. However, gas-driven sample spinners that can be rotated at up to 50 kHz are now routinely available.

Checklist of concepts

☐ 1. The **chemical shift** of a nucleus is the difference between its resonance frequency and that of a reference standard.

☐ 2. The **shielding constant** is the sum of a local contribution, a neighbouring group contribution, and a solvent contribution.

☐ 3. The **local contribution** is the sum of a diamagnetic contribution and a paramagnetic contribution.

☐ 4. The **neighbouring group contribution** arises from the currents induced in nearby groups of atoms.

☐ 5. The **solvent contribution** can arise from specific molecular interactions between the solute and the solvent.

☐ 6. **Fine structure** is the splitting of resonances into individual lines by spin–spin coupling; these splittings give rise to **multiplets**.

☐ 7. **Spin–spin coupling** is expressed in terms of the **spin–spin coupling constant** *J*; coupling leads to the splitting of lines in the spectrum.

☐ 8. The coupling constant decreases as the number of bonds separating two nuclei increases.

☐ 9. Spin–spin coupling can be explained in terms of the **polarization mechanism** and the **Fermi contact interaction**.

☐ 10. If the shift difference between two nuclei is large compared to the coupling constant between the nuclei the spin system is said to be **weakly coupled**; if the shift difference is small compared to the coupling, the spin system is **strongly coupled**.

☐ 11. **Chemically equivalent** nuclei have the same chemical shifts; the same is true of **magnetically equivalent** nuclei, but in addition the coupling constant to any other nucleus is the same for each of the equivalent nuclei.

☐ 12. Coalescence of two NMR lines occurs when nuclei are exchanged rapidly between environments by either conformational or chemical process.

☐ 13. **Magic-angle spinning** (MAS) is a technique in which the NMR linewidths in a solid sample are reduced by spinning at an angle of 54.74° to the applied magnetic field.

Checklist of equations

Property	Equation	Comment	Equation number
δ-Scale of chemical shifts	$\delta = \{(\nu - \nu^\circ)/\nu^\circ\} \times 10^6$	Definition	12B.4a
Relation between chemical shift and shielding constant	$\delta \approx (\sigma^\circ - \sigma) \times 10^6$		12B.6
Local contribution to the shielding constant	$\sigma(\text{local}) = \sigma_d + \sigma_p$		12B.8
Lamb formula	$\sigma_d = (e^2\mu_0/12\pi m_e)\langle 1/r\rangle$	Applies to atoms	12B.9
Neighbouring group contribution to the shielding constant	$\sigma(\text{neighbour}) \propto (\chi_\parallel - \chi_\perp)\{(1 - 3\cos^2\Theta)/R^3\}$		12B.10b
Karplus equation	${}^3J_{HH} = A + B\cos\phi + C\cos 2\phi$	*A*, *B*, and *C* are empirical constants	12B.14
Condition for coalescence of two NMR lines	$\tau = 2^{1/2}/\pi\delta\nu$	τ is the lifetime of the exchange process	12B.16

TOPIC 12C Pulse techniques in NMR

➤ Why do you need to know this material?

To appreciate the power and scope of modern nuclear magnetic resonance techniques you need to understand how radiofrequency pulses can be used to obtain spectra.

➤ What is the key idea?

Sequences of pulses of radiofrequency radiation manipulate nuclear spins, leading to efficient acquisition of NMR spectra and the measurement of relaxation times.

➤ What do you need to know already?

You need to be familiar with the general principles of magnetic resonance (Topics 12A and 12B), and the vector model of angular momentum (Topic 7F). The development makes use of the concept of precession at the Larmor frequency (Topic 12A).

In modern forms of NMR spectroscopy the nuclear spins are first excited by a short, intense burst of radiofrequency radiation (a 'pulse'), applied at or close to the Larmor frequency. As a result of the excitation caused by the pulse, the spins emit radiation as they return to equilibrium. This time-dependent signal is recorded and its 'Fourier transform' computed (as will be described) to give the spectrum. The technique is known as **Fourier-transform NMR** (FT-NMR). An analogy for the difference between conventional spectroscopy and pulsed NMR is the detection of the frequencies at which a bell vibrates. The 'conventional' option is to connect an audio oscillator to a loudspeaker and direct the sound towards the bell. The frequency of the sound source is then scanned until the bell starts to ring in resonance. The 'pulse' analogy is to strike the bell with a hammer and then Fourier transform the signal to identify the resonance frequencies of the bell.

One advantage of FT-NMR over conventional NMR is that it improves the sensitivity. However, the real power of the technique comes from the possibility of manipulating the nuclear spins by applying a sequence of several pulses. In this way it is possible to record spectra in which particular features are emphasized, or from which other properties of the molecule can be determined.

12C.1 The magnetization vector

To understand the pulse procedure, consider a sample composed of many identical spin-$\frac{1}{2}$ nuclei. According to the vector model of angular momentum (Topic 7F), a nuclear spin can be represented by a vector of length $\{I(I+1)\}^{1/2}$ with a component of length m_I along the z-axis. As the three components of the angular momentum are complementary variables, the x- and y-components cannot be specified if the z-component is known, so the vector lies anywhere on a cone around the z-axis. For $I=\frac{1}{2}$, the length of the vector is $3^{1/2}/2$ and when $m_I=+\frac{1}{2}$ it makes an angle of $\arccos\left\{\frac{1}{2}/(3^{1/2}/2)\right\}=54.7°$ to the z-axis (Fig. 12C.1); when $m_I=-\frac{1}{2}$ the cone makes the same angle to the $-z$-axis.

In the absence of a magnetic field, the sample consists of equal numbers of α and β nuclear spins with their vectors lying at random, stationary positions on their cones. The **magnetization**, M, of the sample, its net nuclear magnetic moment, is zero (Fig. 12C.2a). There are two changes when a magnetic field of magnitude $\mathcal{B}_0$ is applied along the z-direction:

- The energies of the two spin states change, the α spins moving to a lower energy and the β spins to a higher energy (provided $\gamma_N > 0$).

In the vector model, the two vectors are pictured as precessing at the Larmor frequency (Topic 12A, $\nu_L=\gamma_N\mathcal{B}_0/2\pi$). At 10 T, the Larmor frequency for ^1H nuclei (commonly referred to as 'protons') is 427 MHz. As the strength of the field is increased, the Larmor frequency increases and the precession becomes faster.

- The populations of the two spin states (the numbers of α and β spins) at thermal equilibrium change, with slightly more α spins than β spins (Topic 12A).

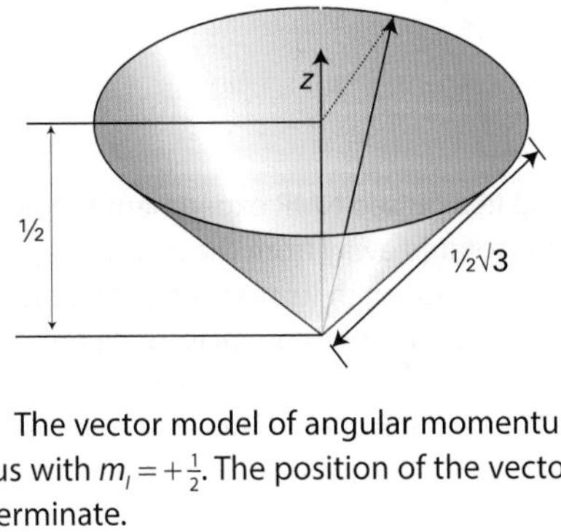

Figure 12C.1 The vector model of angular momentum for a single spin-$\frac{1}{2}$ nucleus with $m_I=+\frac{1}{2}$. The position of the vector on the cone is indeterminate.

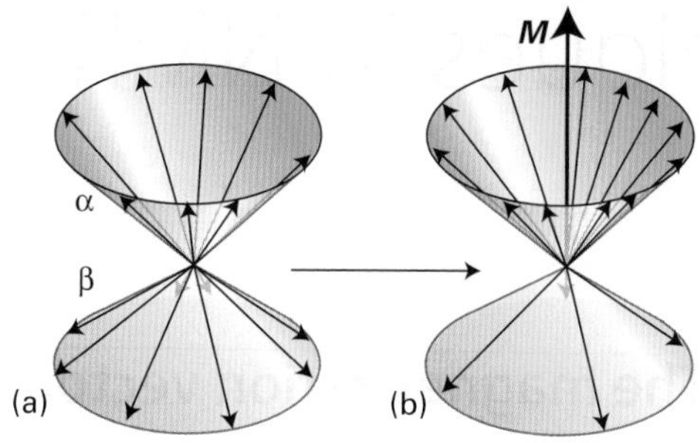

Figure 12C.2 The magnetization of a sample of spin-$\frac{1}{2}$ nuclei is the resultant of all their magnetic moments. (a) In the absence of an externally applied field, there are equal numbers of α and β spins lying at random angles around the cones: the magnetization is zero. (b) In the presence of a field there are slightly more α spins than β spins. As a result, there is a net magnetization, represented by the vector ***M***, along the *z*-axis. There is no magnetization in the transverse plane (the *xy*-plane) because the spins still lie at random angles around the cones.

This imbalance results in a net magnetization in the *z*-direction. It can be represented by a vector ***M*** lying along the *z*-axis with a length proportional to the population difference (Fig. 12C.2b). The manipulation of this net magnetization vector is the central feature of pulse techniques.

12C.1(a) The effect of the radiofrequency field

The magnetization vector can be rotated away from its equilibrium position by applying radiofrequency radiation that provides a magnetic field $\mathcal{B}_1$ lying in the *xy*-plane and rotating at the Larmor frequency (as determined by $\mathcal{B}_0$, Fig. 12C.3a). To understand this process, it is best to imagine stepping on to a **rotating frame**, a platform that rotates around the *z*-axis at the Larmor frequency: the $\mathcal{B}_1$ field is stationary in this frame (Fig. 12C.3b).

In the laboratory frame, the applied field defines the axis of quantization and the spins are either α or β with respect to that axis. In the rotating frame, the applied field has effectively disappeared and the new axis of quantization is the direction of the stationary $\mathcal{B}_1$ field. The angular momentum states are still confined to two values with components on that axis, and will be denoted α′ and β′. The vectors that represent them precess around this new axis on cones at a Larmor frequency $\nu_L' = \gamma_N \mathcal{B}_1/2\pi$. This frequency will be termed the '$\mathcal{B}_1$ Larmor frequency' to distinguish it from the '$\mathcal{B}_0$ Larmor frequency' associated with precession about $\mathcal{B}_0$.

For simplicity, suppose that there are only α spins in the sample. In the rotating frame these vectors will seem to be bunched up at the top of the α′ and β′ cones in the rotating frame (Fig. 12C.4). They precess around $\mathcal{B}_1$ and therefore migrate towards the *xy*-plane. Of course, there are β nuclei present too, which in the rotating frame are bunched together at the bottom of the α′ and β′ cones, but precess similarly. At thermal equilibrium there are fewer β spins than α spins, so the net effect is a magnetization vector initially along the *z*-axis that rotates around the $\mathcal{B}_1$ direction at the $\mathcal{B}_1$ Larmor frequency and into the *xy*-plane.

When the radiofrequency field is applied in a pulse of duration $\Delta\tau$ the magnetization rotates through an angle (in radians) of $\phi = \Delta\tau \times (\gamma_N \mathcal{B}_1/2\pi) \times 2\pi$; this angle is known as the **flip angle** of the pulse. Therefore, to achieve a flip angle ϕ, the duration of the pulse must be $\Delta\tau = \phi/\gamma_N \mathcal{B}_1$. A **90° pulse** (with 90° corresponding to $\phi = \pi/2$) of duration $\Delta\tau_{90} = \pi/2\gamma_N \mathcal{B}_1$ rotates the magnetization from the *z*-axis into the *xy*-plane (Fig. 12C.5a).

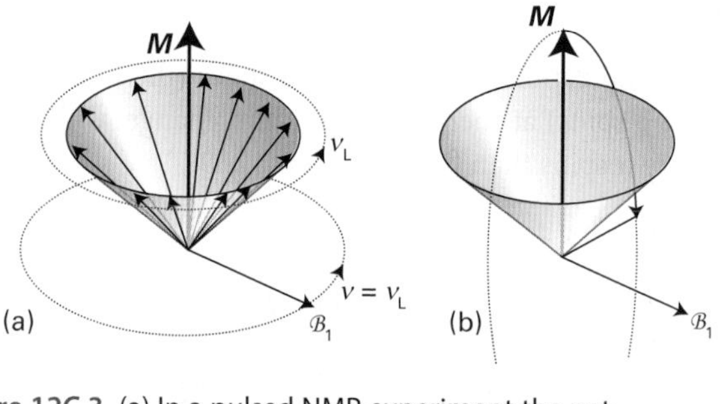

Figure 12C.3 (a) In a pulsed NMR experiment the net magnetization is rotated away from the *z*-axis by applying radiofrequency radiation with its magnetic component $\mathcal{B}_1$ rotating in the *xy*-plane at the Larmor frequency. (b) When viewed in a frame also rotating about *z* at the Larmor frequency, $\mathcal{B}_1$ appears to be stationary. The magnetization rotates around the $\mathcal{B}_1$ field, thus moving the magnetization away from the *z*-axis and generating transverse components.

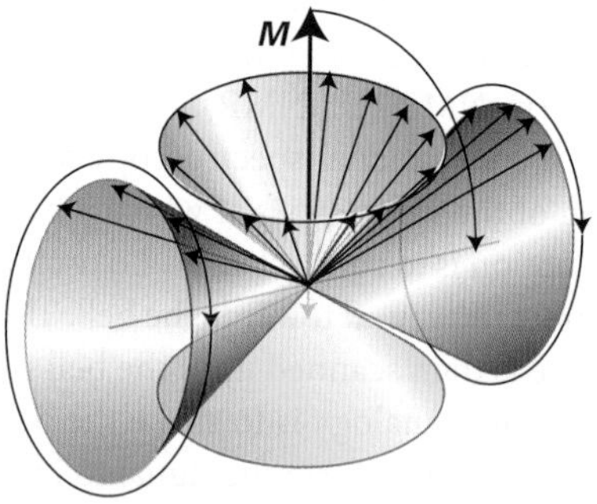

Figure 12C.4 When attention switches to the rotating frame, the vectors representing the spins are in states referring to the axis defined by $\mathcal{B}_1$, and precess on their cones. A uniform distribution of α spins in the $\mathcal{B}_0$ frame is actually a superposition of α′ and β′ states that seem to be bunched together on their cones. As the latter precess on the horizontal cones, the magnetization vector rotates into the *xy*-plane. Vectors representing β spins in the original frame behave similarly.

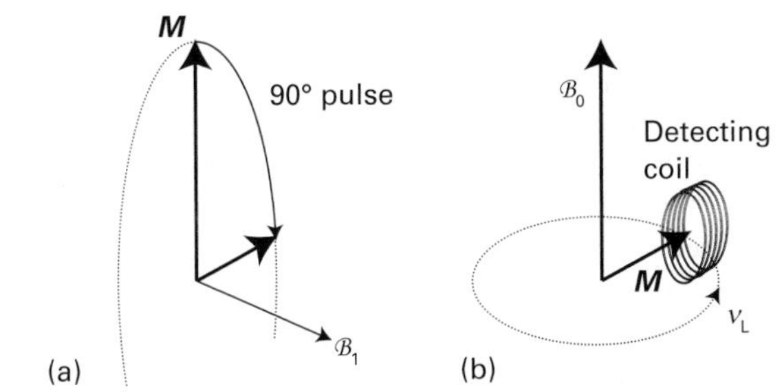

Figure 12C.5 (a) If the radiofrequency field is applied for the appropriate time, the magnetization vector is rotated into the *xy*-plane; this is termed a *90° pulse*. (b) Once the magnetization is in the transverse plane it rotates about the $\mathcal{B}_0$ field at the Larmor frequency (when observed in a static frame). The magnetization vector periodically rotates past a small coil, inducing in it an oscillating current, which is the detected signal.

Brief illustration 12C.1

The duration of a radiofrequency pulse depends on the strength of the $\mathcal{B}_1$ field. If a 90° pulse requires 10 μs, then for protons

$$\mathcal{B}_1 = \frac{\pi}{2\times\underbrace{(2.675\times10^8\ \mathrm{T^{-1}\,s^{-1}})}_{\gamma_N}\times\underbrace{(1.0\times10^{-5}\ \mathrm{s})}_{\Delta\tau}}$$
$$= 5.9\times10^{-4}\ \mathrm{T} = 0.59\ \mathrm{mT}$$

Immediately after a 90° pulse the $\mathcal{B}_1$ field is removed and the magnetization lies in the *xy*-plane. Next, imagine stepping out of the rotating frame. The magnetization vector is now rotating in the *xy*-plane at the $\mathcal{B}_0$ Larmor frequency (Fig. 12C.5b). In an NMR spectrometer a small coil is wrapped around the sample and perpendicular to the $\mathcal{B}_0$ field in such a way that the precessing magnetization vector periodically 'cuts' the coil, thereby inducing in it a small current oscillating at the $\mathcal{B}_0$ Larmor frequency. This oscillating current is detected by a radiofrequency receiver.

As time passes the magnetization returns to an equilibrium state in which it has no transverse components; as it does so the oscillating signal induced in the coil decays to zero. This decay is usually assumed to be exponential with a time constant denoted T_2. The overall form of the signal is therefore a decaying-oscillating **free-induction decay** (FID) like that shown in Fig. 12C.6 and of the form

$$S(t) = S_0\cos(2\pi\nu_L t)\mathrm{e}^{-t/T_2} \qquad \text{Free-induction decay} \qquad (12C.1)$$

So far it has been assumed that the radiofrequency radiation is exactly at the $\mathcal{B}_0$ Larmor frequency. However, virtually the same effect is obtained if the separation of the radiofrequency from the Larmor frequency is small compared to the inverse of the duration of the 90° pulse. In practice, a spectrum with several peaks, each with a slightly different Larmor frequency, can be excited by selecting a radiofrequency somewhere in the centre of the spectrum and then making sure that the 90° pulse is sufficiently short. This requirement is met by using an intense radiofrequency field, so that $\mathcal{B}_1$ is still large enough to achieve rotation through 90°.

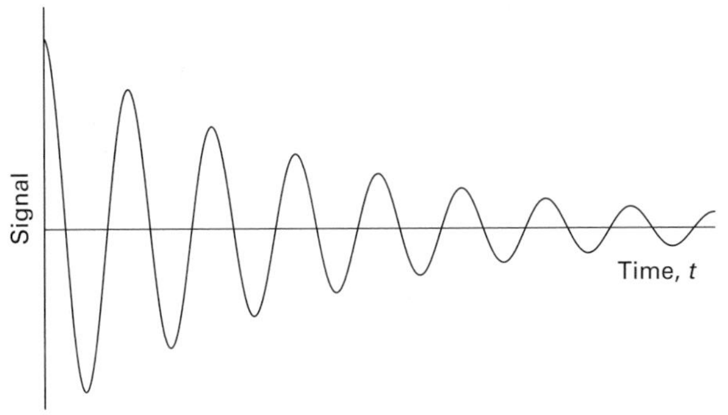

Figure 12C.6 A free-induction decay of a sample of spins with a single resonance frequency.

12C.1(b) Time- and frequency-domain signals

Each line in an NMR spectrum can be thought of as arising from its own magnetization vector. Once that vector has been rotated into the *xy*-plane it precesses at the frequency of the corresponding line. Each vector therefore contributes a decaying-oscillating term to the observed signal and the FID is the sum of many such contributions. If there is only one line it is possible to determine its frequency simply by inspecting the FID, but that is rarely possible when the signal is composite. In such cases, Fourier transformation (*The chemist's toolkit* 12C.1), as mentioned in the introduction, is used to analyse the signal.

The input to the Fourier transform is the decaying-oscillating 'time-domain' function $S(t)$, the FID. The output is the absorption spectrum, the 'frequency-domain' function, $I(\nu)$, which is obtained by computing the integral

$$I(\nu) = \int_0^\infty S(t)\cos(2\pi\nu t)\mathrm{d}t \qquad (12C.2)$$

where $I(\nu)$ is the intensity at the frequency ν. The complete frequency-domain function, which is the spectrum, is built up by evaluating this integral over a range of frequencies.

If the time-domain contains a single decaying-oscillating term, as in eqn 12C.1, the Fourier transform (see the extended

The chemist's toolkit 12C.1 The Fourier transform

A **Fourier transform** expresses any waveform as a superposition of harmonic (sine and cosine) waves. If the waveform is the real function $S(t)$, then the contribution $I(\nu)$ of the oscillating function $\cos(2\pi\nu t)$ is given by the 'cosine transform'

$$I(\nu)=\int_0^{\infty} S(t)\cos(2\pi\nu t)\mathrm{d}t \qquad (1)$$

There is an analogous transform appropriate for complex functions: see the additional information for this *Toolkit* in the enriched collection of *The chemist's toolkits*. If the signal varies slowly, then the greatest contribution comes from low-frequency waves; rapidly changing features in the signal are reproduced by high-frequency contributions. If the signal is a simple exponential decay of the form $S(t)=S_0\mathrm{e}^{-t/\tau}$, where τ is the time constant of the decay, the contribution of the wave of frequency ν is

$$I(\nu)=S_0\int_0^{\infty}\mathrm{e}^{-t/\tau}\cos(2\pi\nu t)\mathrm{d}t=\frac{S_0\tau}{1+(2\pi\nu\tau)^2} \qquad (2)$$

Sketch 1 shows a fast and slow decay and the corresponding frequency contributions: note that a slow decay has predominantly low-frequency contributions and a fast decay has contributions at higher frequencies.

If an experimental procedure results in the function $I(\nu)$ itself, then the corresponding signal can be reconstructed by forming the **inverse Fourier transform**:

$$S(t)=\frac{2}{\pi}\int_0^{\infty} I(\nu)\cos(2\pi\nu t)\mathrm{d}\nu \qquad (3)$$

Fourier transforms are applicable to spatial functions too. Their interpretation is similar but it is more appropriate to think in terms of the wavelengths of the contributing waves. Thus, if the function varies only slowly with distance, then its Fourier transform has mainly long-wavelength contributions. If the features vary quickly with distance (as in the electron density in a crystal), then short-wavelength contributions feature.

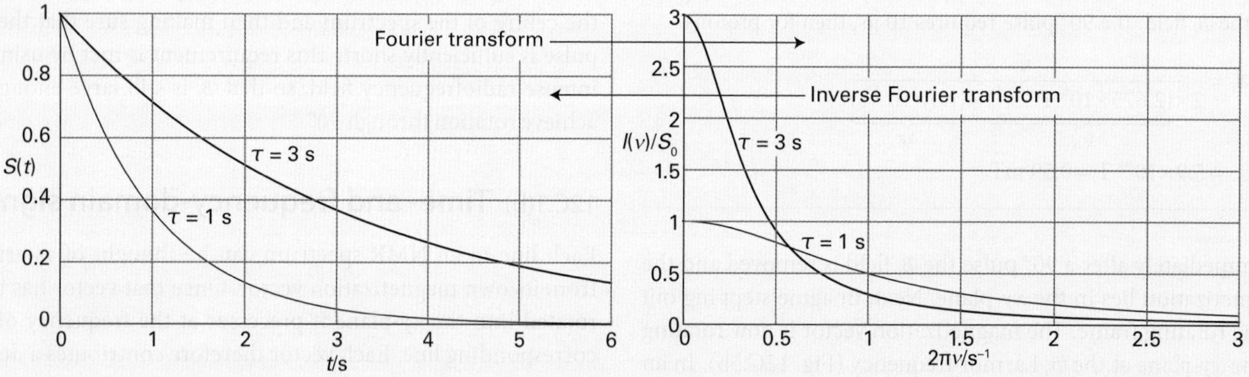

Sketch 1

The chemist's toolkit 12C.1 in the enriched collection of *The chemist's toolkits*) is

$$I(\nu)=\frac{S_0T_2}{1+(\nu_\mathrm{L}-\nu)^2(2\pi T_2)^2} \qquad (12C.3)$$

The graph of this expression has a so-called 'Lorentzian' shape, a symmetrical peak centred at $\nu=\nu_\mathrm{L}$ and with height S_0T_2; its width at half the peak height is $1/\pi T_2$ (Fig. 12C.7). If the FID consists of a sum of decaying-oscillating functions, Fourier transformation gives a spectrum consisting of a series of peaks at the various frequencies.

In practice, the FID is sampled digitally and the integral of eqn 12C.2 is evaluated numerically by a computer in the NMR spectrometer. Figure 12C.8 shows the time- and frequency-domain functions for three different FIDs of increasing complexity.

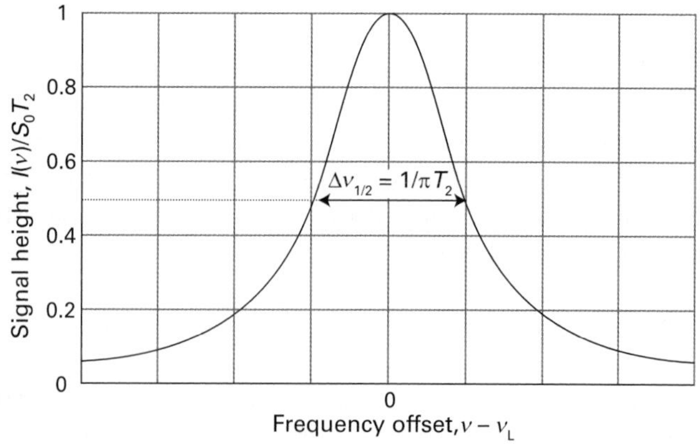

Figure 12C.7 A Lorentzian absorption line. The width at half-height depends on the time constant T_2 which characterizes the decay of the time-domain signal.

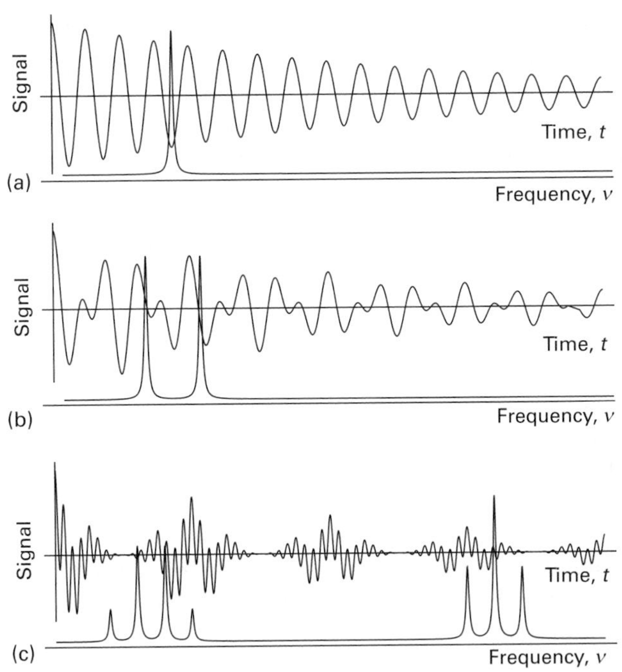

Figure 12C.8 Free-induction decays (the time domain) and the corresponding spectra (the frequency domain) obtained by Fourier transformation. (a) An uncoupled A resonance, (b) the A resonance of an AX system, (c) the A and X resonances of an A_2X_3 system. *Interact with the dynamic version of this graph in the e-book.*

Exercise

E12C.1 The duration of a 90° or 180° pulse depends on the strength of the $\mathcal{B}_1$ field. If a 180° pulse applied to ^{1}H requires 12.5 μs, what is the strength of the $\mathcal{B}_1$ field? How long would the corresponding 90° pulse require?

12C.2 Spin relaxation

Relaxation is the process by which the magnetization returns to its equilibrium value, at which point it is entirely along the z-axis, with no x- and no y-component (no transverse components). In terms of the behaviour of individual spins, the approach to equilibrium involves transitions between the two spin states in order to establish the thermal equilibrium populations of α and β. The attainment of equilibrium also requires the magnetic moments of individual nuclei becoming distributed at random angles on their two cones.

As already explained, after a 90° pulse the magnetization vector lies in the xy-plane. This orientation implies that, from the viewpoint of the laboratory quantization axis, there are now equal numbers of α and β spins because otherwise there would be a component of the magnetization in the z-direction. At thermal equilibrium, however, there is a Boltzmann distribution of spins, with more α spins than β spins (provided $\gamma_N > 0$) and a non-zero z-component of magnetization. The return of the z-component of magnetization to its equilibrium value is termed **longitudinal relaxation**. It is usually assumed that this process follows an exponential recovery curve, with a time constant called the **longitudinal relaxation time, T_1**. Because longitudinal relaxation involves the transfer of energy between the spins and the surroundings (the 'lattice'), the time constant T_1 is also called the *spin–lattice relaxation time*. If the z-component of magnetization at time t is $M_z(t)$, then the recovery to the equilibrium magnetization M_0 takes the form

$$M_z(t) - M_0 \propto e^{-t/T_1}$$ Longitudinal relaxation time [definition] (12C.4)

Immediately after a 90° pulse the fact that the magnetization vector lies in the xy-plane implies that the α and β spins are aligned in a particular way on their cones so as to give a net (and rotating) component of magnetization in the xy-plane. At thermal equilibrium, however, the spins are at random angles on their cones and there is no transverse component of magnetization. The return of the transverse magnetization to its equilibrium value of zero is termed **transverse relaxation**. It is usually assumed that this process is an exponential decay with a time constant called the **transverse relaxation time, T_2** (or *spin–spin relaxation time*). If the transverse magnetization at time t is $M_{xy}(t)$, then the decay takes the form

$$M_{xy}(t) \propto e^{-t/T_2}$$ Transverse relaxation time [definition] (12C.5)

This T_2 is the same as that describing the decay of the free-induction signal which is proportional to $M_{xy}(t)$.

12C.2(a) The mechanism of relaxation

The return of the z-component of magnetization to its equilibrium value involves transitions between α and β spin states so as to achieve the populations required by the Boltzmann distribution. These transitions are caused by local magnetic fields that fluctuate at a frequency close to the resonance frequency of the β ↔ α transition. The local fields can have a variety of origins, but commonly arise from nearby magnetic nuclei or unpaired electrons. These fields fluctuate due to the tumbling motion of molecules in a fluid sample. If molecular tumbling is too slow or too fast compared with the resonance frequency, it gives rise to a fluctuating magnetic field with a frequency that is either too low or too high to stimulate a transition between β and α, so T_1 is long. Only if the molecule tumbles at about the resonance frequency is the fluctuating magnetic field able to induce spin flips effectively, and then T_1 is short.

The rate of molecular tumbling increases with temperature and with reducing viscosity of the solvent, so a dependence of the relaxation time like that shown in Fig. 12C.9 can be expected. The quantitative treatment of relaxation times depends on setting up models of molecular motion and using, for instance, the diffusion equation (Topic 16C) adapted for rotational motion.

Transverse relaxation is the result of the individual magnetic moments of the spins losing their relative alignment as they spread out on their cones. One contribution to this

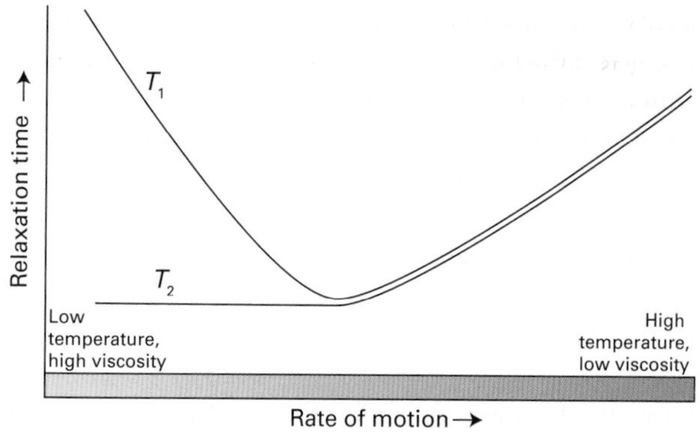

Figure 12C.9 The variation of the two relaxation times with the rate at which the molecules tumble in solution. The horizontal axis can be interpreted as representing temperature or viscosity. Note that the two relaxation times coincide when the motion is fast.

randomization is any process that involves a transition between the two spin states. That is, any process that causes longitudinal relaxation also contributes to transverse relaxation. Another contribution is the variation in the local magnetic fields experienced by the nuclei. When the fluctuations in these fields are slow, each molecule lingers in its local magnetic environment, and because the precessional rates depend on the strength of the field, the spin orientations randomize quickly around their cones. In other words, slow molecular motion corresponds to short T_2. If the molecules move rapidly from one magnetic environment to another, the effects of differences in local magnetic field average to zero: individual spins do not precess at very different rates, remain bunched for longer, and transverse relaxation does not take place as quickly. This fast motion corresponds to long T_2 (as shown in Fig. 12C.9). Calculations show that, when the motion is fast, transverse and longitudinal relaxation have similar time constants.

Brief illustration 12C.2

For a small molecule dissolved in a non-viscous solvent the value of T_2 for ^{1}H can be as long as several seconds. The width (measured at half the peak height) of the corresponding line in the spectrum is $1/\pi T_2$. For $T_2 = 3.0$ s the width is

$$\Delta\nu_{1/2} = \frac{1}{\pi T_2} = \frac{1}{\pi \times (3.0\ \text{s})} = 0.11\ \text{Hz}$$

In contrast, for a larger molecule such as a protein dissolved in water T_2 is much shorter, and a value of 30 ms is not unusual. The corresponding linewidth is 11 Hz.

So far, it has been assumed that the applied magnetic field is homogeneous (uniform) in the sense of having the same value across the sample so that the differences in Larmor frequencies arise solely from interactions within the sample. In practice, due to the limitations in the design of the magnet, the field is not perfectly uniform and is different at different locations in the sample. The inhomogeneity results in a **inhomogeneous broadening** of the resonance. It is common to express the extent of inhomogeneous broadening in terms of an **effective transverse relaxation time**, T_2^*, by using a relation like eqn 12C.5, but writing

$$T_2^* = \frac{1}{\pi \Delta\nu_{1/2}} \qquad \text{Effective transverse relaxation time [definition]} \qquad (12C.6)$$

where $\Delta\nu_{1/2}$ is the observed width at half-height of the line (which is assumed to be Lorentzian). In practice an inhomogeneously broadened line is unlikely to be Lorentzian, so the assumption that the decay is exponential and characterized by a time constant T_2^* is an approximation. The observed linewidth has a contribution from both the inhomogeneous broadening and the transverse relaxation. The latter is usually referred to as **homogeneous broadening**. Which contributions dominate depends on the molecular system being studied and the quality of the magnet.

12C.2(b) The measurement of T_1 and T_2

The longitudinal relaxation time T_1 can be measured by the **inversion recovery technique**. The first step is to apply a 180° pulse to the sample by applying the $\mathcal{B}_1$ field for twice as long as for a 90° pulse. As a result of the pulse, the magnetization vector is rotated into the $-z$-direction (Fig. 12C.10a). The effect of the pulse is to invert the population of the two levels and to result in more β spins than α spins.

Immediately after the 180° pulse no signal can be detected because the magnetization has no transverse component. The β spins immediately begin to relax back into α spins, and the magnetization vector first shrinks towards zero and then increases in the opposite direction until it reaches its thermal equilibrium value. Before that has happened, after an interval τ, a 90° pulse is applied. That pulse rotates the remaining z-component of magnetization into the xy-plane, where it generates an FID signal. The frequency-domain spectrum is then obtained by Fourier transformation in the usual way.

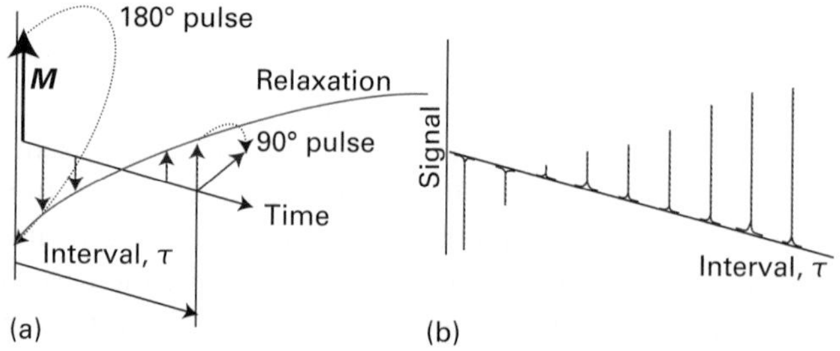

Figure 12C.10 (a) The result of applying a 180° pulse to the magnetization in the rotating frame, and the effect of a subsequent 90° pulse. (b) The amplitude of the frequency-domain spectrum varies with the interval between the two pulses because there has been time for longitudinal relaxation to occur.

The intensity of the resulting spectrum depends on the magnitude of the magnetization vector that has been rotated into the *xy*-plane. That magnitude changes exponentially with a time constant T_1 as the interval τ is increased, so the intensity of the spectrum also changes exponentially with increasing τ. The longitudinal relaxation time can therefore be measured by fitting an exponential curve to the series of spectra obtained with different values of τ.

The measurement of T_2 (as distinct from T_2^*) depends on being able to eliminate the effects of inhomogeneous broadening. The cunning required is at the root of some of the most important advances made in NMR since its introduction.

A **spin echo** is the magnetic analogue of an audible echo. The sequence of events is shown in Fig. 12C.11. The overall magnetization can be regarded as made up of a number of different magnetizations, each of which arises from a **spin packet** of nuclei with very similar precession frequencies. The spread in these frequencies arises from the inhomogeneity of $\mathcal{B}_0$ (which is responsible for inhomogeneous broadening), so different parts of the sample experience different fields. The precession frequencies also differ if there is more than one chemical shift present.

First, a 90° pulse is applied to the sample. The subsequent events are best followed in a rotating frame in which $\mathcal{B}_1$ is stationary along the *x*-axis and causes the magnetization to rotate on to the *y*-axis of the *xy*-plane. Immediately after the pulse, the spin packets begin to fan out because they have different Larmor frequencies. In Fig. 12C.11 the magnetization vectors of two representative packets are shown and are described as 'fast' and 'slow', indicating their frequency relative to the rotating frame frequency (the nominal Larmor frequency). Because the rotating frame is at the Larmor frequency, the 'fast' and 'slow' vectors rotate in opposite senses when viewed in this frame.

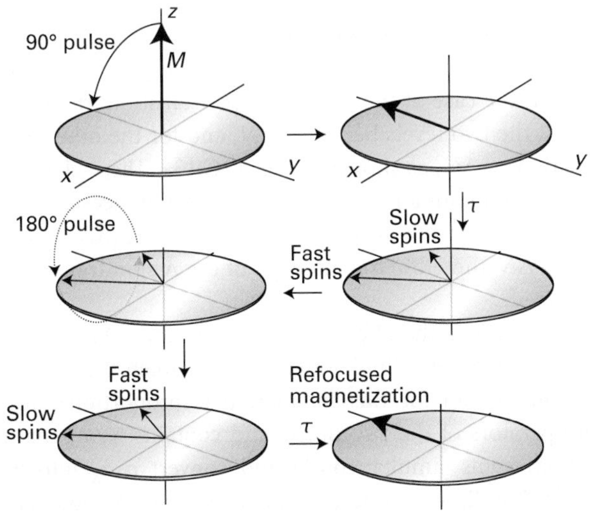

Figure 12C.11 The action of the spin echo pulse sequence 90°–τ–180°–τ, viewed in a rotating frame at the Larmor frequency. Note that the 90° pulse is applied about the *x*-axis, but the 180° pulse is applied about the *y*-axis. 'Slow' and 'Fast' refer to the speed of the spin packet relative to the rotating frame frequency.

First, suppose that there is no transverse relaxation but that the field is inhomogeneous. After an evolution period τ, a 180° pulse is applied along the *y*-axis of the rotating frame. The pulse rotates the magnetization vectors around that axis into mirror-image positions with respect to the *yz*-plane. Once there, the packets continue to move in the same direction as they did before, and so migrate back towards the *y*-axis. After an interval τ all the packets are again aligned along the axis. The resultant signal grows in magnitude, reaching a maximum, the 'spin echo', at the end of the second period τ. The fanning out caused by the field inhomogeneity is said to have been 'refocused'.

The important feature of the technique is that the size of the echo is independent of any local fields that remain constant during the two τ intervals. If a spin packet is 'fast' because it happens to be composed of spins in a region of the sample that experience a higher than average field, then it remains fast throughout both intervals, and so the angle through which it rotates is the same in the two intervals. Hence, the size of the echo is independent of inhomogeneities in the magnetic field, because these remain constant.

Now consider the consequences of transverse relaxation. This relaxation arises from fields that vary on a molecular scale, and there is no guarantee that an individual 'fast' spin will remain 'fast' in the refocusing phase: the spins within the packets therefore spread with a time constant T_2. Consequently, the effects of the relaxation are not refocused, and the size of the echo decays with the time constant T_2. The intensity of the signal after a spin echo is measured for a series of values of the delay τ, and the resulting data analysed to determine T_2.

Exercises

E12C.2 What is the effective transverse relaxation time when the width of a Lorentzian resonance line is 1.5 Hz?

E12C.3 The envelope of a free-induction decay is observed to decrease to half its initial amplitude in 1.0 s. What is the value of the transverse relaxation time, T_2?

12C.3 Spin decoupling

Carbon-13 has a natural abundance of only 1.1 per cent and is therefore described as a **dilute-spin species**. The probability that any one molecule contains more than one ^{13}C nucleus is rather low, and so the possibility of observing the effects of ^{13}C–^{13}C spin–spin coupling can be ignored. In contrast the abundance of the isotope ^{1}H is very close to 100 per cent and is therefore described as an **abundant-spin species**. All the hydrogen nuclei in a molecule can be assumed to be ^{1}H and the effects of coupling between them is observed.

Dilute-spin species are observed in NMR spectroscopy and show coupling to abundant-spin species present in the molecule. Generally speaking, ^{13}C-NMR spectra are very complex on account of the spin couplings of each ^{13}C nucleus with the numerous ^{1}H nuclei in the molecule. However, a dramatic simplification of the spectrum can be obtained by the use of **proton decoupling**. In this technique radiofrequency radiation is applied at (or close to) the ^{1}H Larmor frequency while the ^{13}C FID is being observed. This stimulation of the ^{1}H nuclei causes their spins state to change rapidly, so averaging the ^{1}H–^{13}C couplings to zero. As a result, each ^{13}C nucleus gives a single line rather than a complex multiplet. Not only is the spectrum simplified, but the sensitivity is also improved as all the intensity is concentrated into a single line.

Exercise

E12C.4 The ^{13}C NMR spectrum of ethanoic acid (acetic acid) shows a quartet centred at $\delta = 21$ with a splitting of 130 Hz. When the same spectrum is recorded using proton decoupling, the multiplet collapses to a single line. Another quartet, but with a much smaller spacing, is also seen centred at $\delta = 178$; this quartet collapses when decoupling is used. Explain these observations.

12C.4 The nuclear Overhauser effect

A common source of the local magnetic fields that are responsible for relaxation is the **dipole–dipole interaction** between two magnetic nuclei (see *The chemist's toolkit* 12B.1). In this interaction the magnetic dipole of the first spin generates a magnetic field that interacts with the magnetic dipole of the second spin. The strength of the interaction is proportional to $1/R^3$, where R is the distance between the two spins, and is also proportional to the product of the magnetogyric ratios of the spins. As a result, the interaction is characterized as being short-range and significant for nuclei with high magnetogyric ratios. In typical organic and biological molecules, which have many ^{1}H nuclei, the local fields due to the dipole–dipole interaction are likely to be the dominant source of relaxation.

The **nuclear Overhauser effect** (NOE) makes use of the relaxation caused by the dipole–dipole interaction of nuclear spins. In this effect, irradiation of one spin leads to a change in the intensity of the resonance from a second spin provided the two spins are involved in mutual dipole–dipole relaxation. The dipole–dipole interaction has only a short range, so the observation of an NOE is indicative of the closeness of the two nuclei involved and can be interpreted in terms of the structure of the molecule.

To understand the effect, consider the populations of the four levels of a homonuclear AX spin system shown in Fig. 12C.12. At thermal equilibrium, the population of the $\alpha_A\alpha_X$ level is the greatest, and that of the $\beta_A\beta_X$ level is the least; the other two levels have the same energy and an intermediate population. For

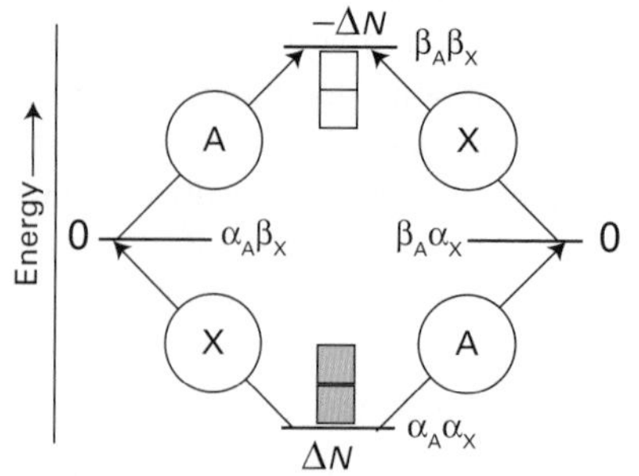

Figure 12C.12 The energy levels of an AX system and an indication of their relative populations. Each filled square above the line represents an excess population above the average, and each empty square below the line represents a lower population than the average. The labels ($\pm\Delta N$) show the deviations in population from their average value.

the purposes of this discussion it is sufficient to consider the deviations of the populations from the average value for all four levels: the $\alpha_A\alpha_X$ level has a greater population than the average, the $\beta_A\beta_X$ level has a smaller population, and the other two levels have a population equal to the average. The deviations from the average are ΔN for $\alpha_A\alpha_X$, $-\Delta N$ for $\beta_A\beta_X$, and 0 for the other two levels. These population differences are represented in Fig. 12C.12. The intensity of a transition reflects the difference in the population of the two energy levels involved, and for all four transitions this population difference (lower − upper) is ΔN, implying that all four transitions have the same intensity.

The NOE experiment involves irradiating the two X spin transitions ($\alpha_A\alpha_X \leftrightarrow \alpha_A\beta_X$ and $\beta_A\alpha_X \leftrightarrow \beta_A\beta_X$) with a radiofrequency field, but making sure that the field is sufficiently weak that the two A spin transitions are not affected. When applied for a long time, this field saturates the X spin transitions in the sense that the populations of the two energy levels become equal. In the case of the $\alpha_A\alpha_X \leftrightarrow \alpha_A\beta_X$ transition, the populations of the two levels become $\frac{1}{2}\Delta N$, and for the other X spin transition $-\frac{1}{2}\Delta N$, as shown in Fig. 12C.13a. These changes in population do not affect the population differences across the A spin transitions ($\alpha_A\alpha_X \leftrightarrow \beta_A\alpha_X$ and $\alpha_A\beta_X \leftrightarrow \beta_A\beta_X$) which remain at ΔN, therefore the intensity of the A spin transitions is unaffected.

Now consider the effect of spin relaxation. One source of relaxation is dipole–dipole interaction between the two spins. The hamiltonian for this interaction is proportional to the spin operators of the two nuclei and contains terms that can flip both spins simultaneously and convert $\alpha_A\alpha_X$ into $\beta_A\beta_X$. This double-flipping process tends to restore the populations of these two levels to their equilibrium values of ΔN and $-\Delta N$, respectively, as shown in Fig. 12C.13b. A consequence is that the population difference across each of the A spin transitions is now $\frac{3}{2}\Delta N$, which is greater than it is at equilibrium. In summary, the combination of saturating the X spin transitions and

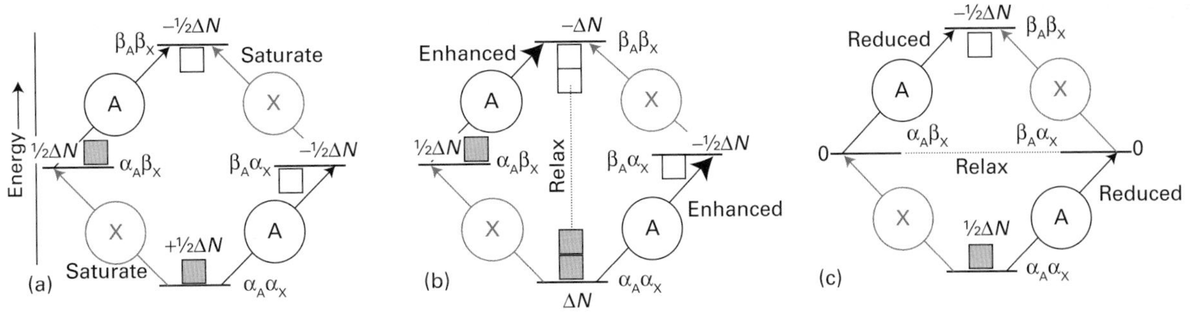

Figure 12C.13 (a) When an X transition is saturated, the populations of the two states are equalized, leading to the populations shown (using the same symbols as in Fig. 12C.12). (b) Dipole–dipole relaxation can cause transitions between the αα and ββ states, such that they return to their original populations: the result is that the population difference across the A transitions is increased. (c) Dipole–dipole relaxation can also cause the populations of the αβ and βα states to return to their equilibrium values: this decreases the population difference across the A transitions.

dipole–dipole relaxation leads to an enhancement of the intensity of the A spin transitions.

The hamiltonian for the dipole–dipole interaction also contains combinations of spin operators that flip the spins in opposite directions, so taking the system from $\alpha_A\beta_X$ to $\beta_A\alpha_X$. This process drives the populations of these states to their equilibrium values, as shown in Fig. 12C.13c. As before, the population differences across the A spin transitions are affected, but now they are reduced to $+\frac{1}{2}\Delta N$, meaning that the A spin transitions are less intense than they would have been in the absence of dipole–dipole relaxation.

It should be clear that there are two opposing effects: the relaxation-induced transitions between $\alpha_A\alpha_X$ and $\beta_A\beta_X$ which enhance the A spin transitions, and the transitions between $\alpha_A\beta_X$ and $\beta_A\alpha_X$ which reduce the intensity of these transitions. Which effect dominates depends on the relative rates of these two relaxation pathways. As in the discussion of relaxation times in Section 12C.2, the efficiency of the $\beta_A\beta_X \leftrightarrow \alpha_A\alpha_X$ relaxation is high if the dipole field oscillates close to the transition frequency, which in this case is about $2\nu_L$; likewise, the efficiency of the $\alpha_A\beta_X \leftrightarrow \beta_A\alpha_X$ relaxation is high if the dipole field is stationary (in this case there is no frequency difference between the initial and final states). Small molecules tumble rapidly and have substantial motion at $2\nu_L$. As a result, the $\beta_A\beta_X \leftrightarrow \alpha_A\alpha_X$ pathway dominates and results in an increase in the intensity of the A spin transitions. This increase is called a 'positive NOE enhancement'. On the other hand, large molecules tumble more slowly so there is less motion at $2\nu_L$. In this case, the $\alpha_A\beta_X \leftrightarrow \beta_A\alpha_X$ pathway dominates and results in a decrease in the intensity of the A spin transitions. This decrease is called a 'negative NOE enhancement'.

The NOE enhancement is usually reported in terms of the parameter η (eta), where

$$\eta = \frac{I_A - I_A^\circ}{I_A^\circ} \quad \text{NOE enhancement parameter} \quad (12C.7)$$

Here I_A° is the intensity of the signals due to nucleus A before saturation, and I_A is the intensity after the X spins have been saturated for long enough for the NOE to build up (typically several multiples of T_1). For a homonuclear system, and if the only source of relaxation is due to the dipole–dipole interaction, η lies between -1 (a negative enhancement) for slow tumbling molecules and $+\frac{1}{2}$ (a positive enhancement) for fast tumbling molecules. In practice, other sources of relaxation are invariably present so these limiting values are rarely achieved.

The utility of the NOE in NMR spectroscopy comes about because only dipole–dipole relaxation can give rise to the effect: only this type of relaxation causes both spins to flip simultaneously. Thus, if nucleus X is saturated and a change in the intensity of the transitions from nucleus A is observed, then there must be a dipole–dipole interaction between the two nuclei. As that interaction is proportional to $1/R^3$, where R is the distance between the two spins, and the relaxation it causes is proportional to the square of the interaction, the NOE is proportional to $1/R^6$. Therefore, for there to be significant dipole–dipole relaxation between two spins they must be close (for ^{1}H nuclei, not more than 0.5 nm apart), so the observation of an NOE is used as qualitative indication of the proximity of nuclei. In principle it is possible to make a quantitative estimate of the distance from the size of the NOE, but to do so requires the effects of other kinds of relaxation to be taken into account.

The value of the NOE enhancement η also depends on the values of the magnetogyric ratios of A and X, because these properties affect both the populations of the levels and the relaxation rates. For a heteronuclear spin system the maximal enhancement is

$$\eta = \frac{\gamma_X}{2\gamma_A} \quad (12C.8)$$

where γ_A and γ_X are the magnetogyric ratios of nuclei A and X, respectively.

Brief illustration 12C.3

From eqn 12C.8 and the data in Table 12A.2, the maximum NOE enhancement parameter for a ^{13}C–^{1}H pair is

$$\eta = \frac{\overbrace{2.675\times10^{8}\ \mathrm{T^{-1}\,s^{-1}}}^{\gamma_{^1\mathrm{H}}}}{2\times\underbrace{(6.73\times10^{7}\ \mathrm{T^{-1}\,s^{-1}})}_{\gamma_{^{13}\mathrm{C}}}} = 1.99$$

It is common to take advantage of this enhancement to improve the sensitivity of ^{13}C NMR spectra. Prior to recording the spectrum, the ^{1}H nuclei are irradiated so that they become saturated, leading to a build-up of the NOE enhancement on the ^{13}C nuclei.

Exercise

E12C.5 Predict the maximum NOE enhancement (as the value of η) that could be obtained for ^{31}P as a result of dipole–dipole relaxation with ^{1}H.

Checklist of concepts

☐ 1. The **free-induction decay** (FID) is the time-domain signal resulting from the precession of transverse magnetization.

☐ 2. Fourier transformation of the FID (the time domain) gives the NMR spectrum (the frequency domain).

☐ 3. **Longitudinal** (or *spin–lattice*) **relaxation** is the process by which the z-component of the magnetization returns to its equilibrium value.

☐ 4. **Transverse** (or *spin–spin*) **relaxation** is the process by which the x- and y-components of the magnetization return to their equilibrium values of zero.

☐ 5. The **longitudinal relaxation time** T_1 can be measured by the **inversion recovery technique**.

☐ 6. The **transverse relaxation time** T_2 can be measured by observing spin echoes.

☐ 7. In **proton decoupling** of ^{13}C-NMR spectra, the protons are continuously irradiated; the effect is to collapse the splittings due to the ^{13}C–^{1}H couplings.

☐ 8. The **nuclear Overhauser effect** is the modification of the intensity of one resonance by the saturation of another: it occurs only if the two spins are involved in mutual dipole–dipole relaxation.

Checklist of equations

Property	Equation	Comment	Equation number
Free-induction decay	$S(t) = S_0 \cos(2\pi\nu_L t)e^{-t/T_2}$	T_2 is the transverse relaxation time	12C.1
Width at half-height of an NMR line	$\Delta\nu_{1/2} = 1/\pi T_2$	Assumed Lorentzian	
Longitudinal relaxation	$M_z(t) - M_0 \propto e^{-t/T_1}$	T_1 is the longitudinal relaxation time	12C.4
Transverse relaxation	$M_{xy}(t) \propto e^{-t/T_2}$		12C.5
NOE enhancement parameter	$\eta = (I_A - I_A^\circ)/I_A^\circ$	Definition	12C.7

TOPIC 12D Electron paramagnetic resonance

➤ Why do you need to know this material?

Many materials and biological systems contain species with unpaired electrons and some chemical reactions generate intermediates with unpaired electrons. Electron paramagnetic resonance is a key spectroscopic tool for studying them.

➤ What is the key idea?

The details of an EPR spectrum give information on the distribution of the density of the unpaired electrons.

➤ What do you need to know already?

You need to be familiar with the concepts of electron spin (Topic 8B) and the general principles of magnetic resonance (Topic 12A). The discussion refers to spin–orbit coupling in atoms (Topic 8C) and the Fermi contact interaction in molecules (Topic 12B).

Electron paramagnetic resonance (EPR), which is also known as electron spin resonance (ESR), is used to study species that contain unpaired electrons. Both solids and liquids can be studied, but the study of gas-phase samples is complicated by the free rotation of the molecules.

12D.1 The *g*-value

According to the discussion in Topic 12A, the resonance frequency for a transition between the $m_s = -\frac{1}{2}$ and the $m_s = +\frac{1}{2}$ levels of a free electron is

$$h\nu = g_e \mu_B \mathcal{B}_0 \qquad \text{Resonance condition [free electron]} \qquad (12D.1)$$

where $g_e \approx 2.0023$. If the electron is in a radical the field it experiences differs from the applied field due to the presence of local magnetic fields arising from electronic currents induced in the molecular framework. This difference is taken into account by replacing g_e by g and expressing the resonance condition as

$$h\nu = g\mu_B \mathcal{B}_0 \qquad \text{EPR resonance condition} \qquad (12D.2)$$

where g is the **g-value** of the radical.

Electron paramagnetic resonance spectra are usually recorded by keeping the frequency of the microwave radiation fixed and then varying the magnetic field so as to bring the electron into resonance with the microwave frequency. The positions of the peaks, and the horizontal scale on spectra, are therefore specified in terms of the magnetic field.

Brief illustration 12D.1

The centre of the EPR spectrum of the methyl radical occurs at 329.40 mT in a spectrometer operating at 9.2330 GHz (radiation belonging to the X band of the microwave region). Its *g*-value is therefore

$$g = \frac{\overbrace{(6.62608\times10^{-34}\ \text{J s})}^{h}\times\overbrace{(9.2330\times10^{9}\ \text{s}^{-1})}^{\nu}}{\underbrace{(9.2740\times10^{-24}\ \text{J T}^{-1})}_{\mu_B}\times\underbrace{(0.32940\ \text{T})}_{\mathcal{B}_0}} = 2.0027$$

The *g*-value is related to the ease with which the applied field can generate currents through the molecular framework and the strength of the magnetic field these currents generate. Therefore, the *g*-value gives some information about electronic structure and plays a similar role in EPR to that played by shielding constants in NMR. A *g*-value smaller than g_e implies that in the molecule the electron experiences a magnetic field smaller than the applied field, whereas a value greater than g_e implies that the magnetic field is greater. Both outcomes are possible, depending on the details of the electronic excited states.

Two factors are responsible for the difference of the *g*-value from g_e. Electrons migrate through the molecular framework by making use of excited electronic states (Fig. 12D.1). This circulation gives rise to a local magnetic field that can add to or subtract from the applied field. The extent to which these

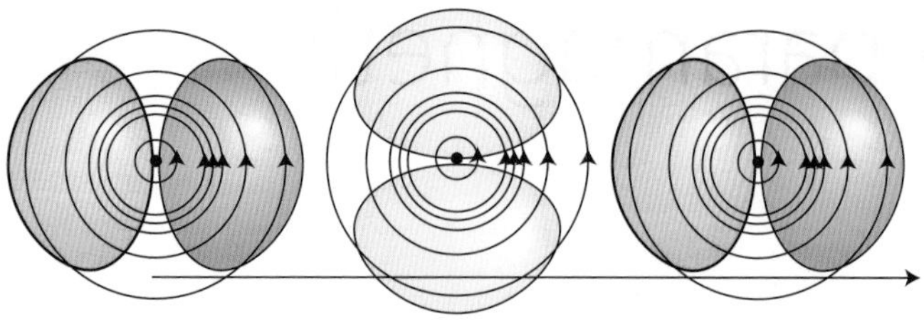

Figure 12D.1 An applied magnetic field can induce circulation of electrons that makes use of excited state orbitals (shown schematically here).

currents are induced is inversely proportional to the separation of energy levels, ΔE, in the radical or complex. Secondly, the strength of the field experienced by the electron spin as a result of these orbital currents is proportional to the spin–orbit coupling constant, ξ (Topic 8C). It follows that the difference of the g-value from g_e is proportional to $\xi/\Delta E$. This proportionality is widely observed. Many organic radicals, for which ΔE is large and ξ (for carbon) is small, have g-values close to 2.0027, not far removed from g_e itself. Inorganic radicals, which commonly are built from heavier atoms and therefore have larger spin–orbit coupling constants, have g-values typically in the range 1.9–2.1. The g-values of paramagnetic d-metal complexes often differ considerably from g_e, varying from 0 to 6, because in them ΔE is small on account of the small splitting of d-orbitals brought about by interactions with ligands (Topic 11F).

The g-value is anisotropic: that is, its magnitude depends on the orientation of the radical with respect to the applied field. The anisotropy arises from the fact that the extent to which an applied field induces currents in the molecule, and therefore the magnitude of the local field, depends on the relative orientation of the molecules and the field. In solution, when the molecule is tumbling rapidly, only the average value of the g-value is observed. Therefore, the anisotropy of the g-value is observed only for radicals trapped in solids and crystalline d-metal complexes.

Exercise

E12D.1 The centre of the EPR spectrum of atomic hydrogen lies at 329.12 mT in a spectrometer operating at 9.2231 GHz. What is the g-value of the electron in the atom?

12D.2 Hyperfine structure

The most important feature of an EPR spectrum is its **hyperfine structure**, the splitting of individual resonance lines into components. In general in spectroscopy, the term 'hyperfine structure' means the structure of a spectrum that can be traced to interactions of the electrons with nuclei other than as a result of the point electric charge of the nucleus. The source of the hyperfine structure in EPR is the magnetic interaction between the electron spin and the magnetic dipole moments of the nuclei present in the radical that give rise to local magnetic fields.

12D.2(a) The effects of nuclear spin

Consider the effect on the EPR spectrum of a single ^{1}H nucleus located somewhere in a radical. The proton spin is a source of magnetic field and, depending on the orientation of the nuclear spin, the field it generates either adds to or subtracts from the applied field. The total local field is therefore

$$\mathcal{B}_{\text{loc}} = \mathcal{B}_0 + am_I \quad m_I = \pm\tfrac{1}{2} \tag{12D.3}$$

where a is the **hyperfine coupling constant** (or *hyperfine splitting constant*); from eqn 12D.3 it follows that a has the same units as the magnetic field, for example tesla. Half the radicals in a sample have $m_I = +\tfrac{1}{2}$, so half resonate when the applied field satisfies the condition

$$h\nu = g\mu_B(\mathcal{B}_0 + \tfrac{1}{2}a), \text{ or } \mathcal{B}_0 = \frac{h\nu}{g\mu_B} - \tfrac{1}{2}a \tag{12D.4a}$$

The other half (which have $m_I = -\tfrac{1}{2}$) resonate when

$$h\nu = g\mu_B\left(\mathcal{B}_0 - \tfrac{1}{2}a\right), \text{ or } \mathcal{B}_0 = \frac{h\nu}{g\mu_B} + \tfrac{1}{2}a \tag{12D.4b}$$

Therefore, instead of a single line, the spectrum shows two lines of half the original intensity separated by a and centred on the field determined by g (Fig. 12D.2).

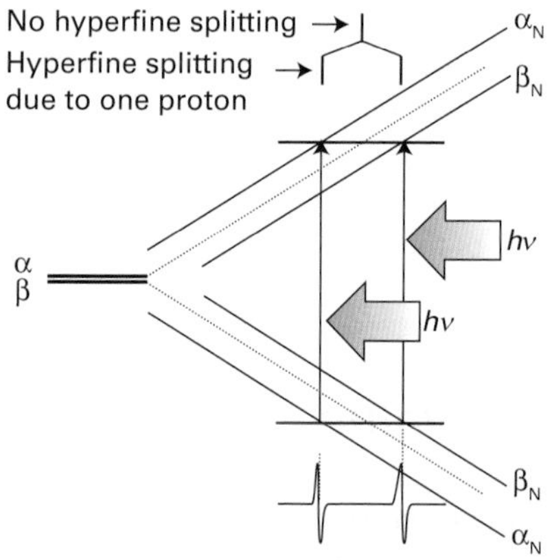

Figure 12D.2 The hyperfine interaction between an electron and a spin-$\tfrac{1}{2}$ nucleus results in four energy levels in place of the original two; α_N and β_N indicate the spin states of the nucleus. As a result, the spectrum consists of two lines (of equal intensity) instead of one. The intensity distribution can be summarized by a simple stick diagram. The diagonal lines show the energies of the states as the applied field is increased, and resonance occurs when the separation of states matches the fixed energy of the microwave photon.

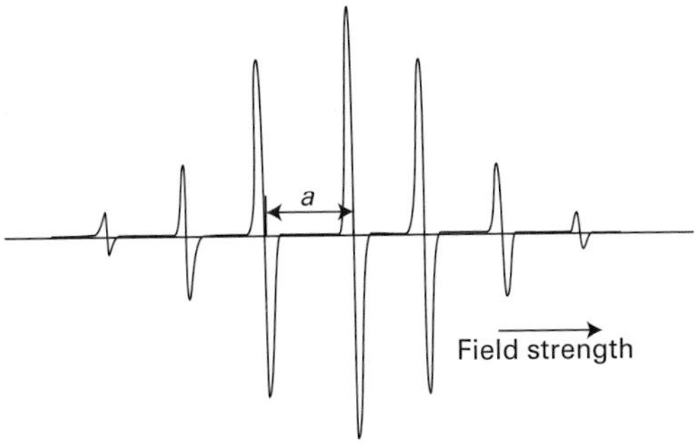

Figure 12D.3 The EPR spectrum of the benzene radical anion, $C_6H_6^-$, in solution; *a* is the hyperfine coupling constant. The centre of the spectrum is determined by the *g*-value of the radical.

If the radical contains an ^{14}N atom ($I = 1$), its EPR spectrum consists of three lines of equal intensity, because the ^{14}N nucleus has three possible spin orientations, and each spin orientation is possessed by one-third of all the radicals in the sample. In general, a spin-I nucleus splits the spectrum into $2I + 1$ hyperfine lines of equal intensity.

When there are several magnetic nuclei present in the radical, each one contributes to the hyperfine structure. In the case of equivalent protons (for example, the two CH_2 protons in the radical CH_3CH_2) some of the hyperfine lines are coincident. If the radical contains N equivalent protons, then there are $N + 1$ hyperfine lines with an intensity distribution given by Pascal's triangle (**1**). The spectrum of the benzene radical anion in Fig. 12D.3, which has seven lines with intensity ratio 1:6:15:20:15:6:1, is consistent with a radical containing six equivalent protons. More generally, if the radical contains N equivalent nuclei with spin quantum number I, then there are $2NI + 1$ hyperfine lines.

1
1 1
1 2 1
1 3 3 1
1 4 6 4 1

1

Example 12D.1 Predicting the hyperfine structure of an EPR spectrum

A radical contains one ^{14}N nucleus ($I = 1$) with hyperfine constant 1.61 mT and two equivalent protons $\left(I = \frac{1}{2}\right)$ with hyperfine constant 0.35 mT. Predict the form of the EPR spectrum.

Collect your thoughts You will need to consider the hyperfine structure that arises from each type of nucleus or group of equivalent nuclei in succession. First, split a line with one nucleus, then split each of the lines again by a second nucleus (or group of nuclei), and so on. It is best to start with the nucleus with the largest hyperfine splitting; however, any choice could be made, and the order in which nuclei are considered does not affect the conclusion.

The solution The ^{14}N nucleus gives three hyperfine lines of equal intensity separated by 1.61 mT. Each line is split into a doublet of spacing 0.35 mT by the first proton, and each line of these doublets is split into a doublet with the same 0.35 mT splitting by the second proton (Fig. 12D.4). Two of the lines in the centre coincide, so splitting by the two protons gives a 1:2:1 triplet of internal splitting 0.35 mT. Overall the spectrum consists of three identical 1:2:1 triplets.

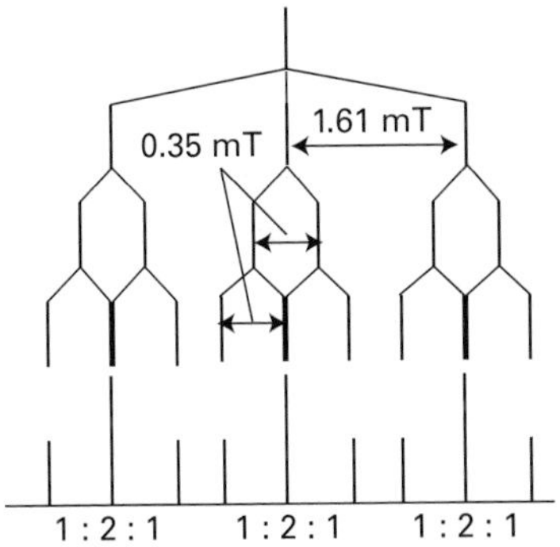

Figure 12D.4 The analysis of the hyperfine structure of a radical containing one ^{14}N nucleus ($I = 1$) and two equivalent protons.

Self-test 12D.1 Predict the form of the EPR spectrum of a radical containing three equivalent ^{14}N nuclei.

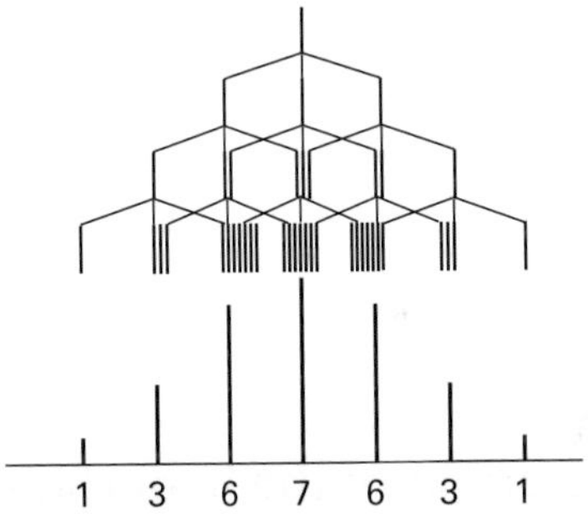

Figure 12D.5 The analysis of the hyperfine structure of a radical containing three equivalent ^{14}N nuclei.

Answer: Fig. 12D.5

12D.3 The McConnell equation

The hyperfine structure of an EPR spectrum is a kind of fingerprint that helps to identify the radicals present in a sample. Moreover, because the magnitude of the splitting depends on the distribution of the unpaired electron in the vicinity of the magnetic nuclei, the spectrum can be used to map the molecular orbital occupied by the unpaired electron.

The hyperfine splitting observed in $C_6H_6^-$ is 0.375 mT. If it is assumed that the unpaired electron is in an orbital with equal probability at each C atom, this hyperfine splitting can be attributed to the interaction between a proton and one-sixth of the unpaired electron spin density. If all the electron density were located on the neighbouring C atom, a hyperfine coupling of 6×0.375 mT = 2.25 mT would be expected. If in another aromatic radical the hyperfine coupling constant is found to be a, then the **spin density**, ρ, the probability that an unpaired electron is on the neighbouring C atom, can be calculated from the **McConnell equation**:

$$a = Q\rho \qquad \text{McConnell equation} \qquad (12D.5)$$

with $Q = 2.25$ mT. In this equation, ρ is the spin density on a C atom and a is the hyperfine splitting observed for the H atom to which it is attached.

Brief illustration 12D.2

The hyperfine structure of the EPR spectrum of the naphthalene radical anion $C_{10}H_8^-$ (2) can be interpreted as arising from interactions with two groups of four equivalent protons. Those at the 1, 4, 5, and 8 positions in the ring have a = 0.490 mT and those in the 2, 3, 6, and 7 positions have a = 0.183 mT. The spin densities obtained by using the McConnell equation are, respectively,

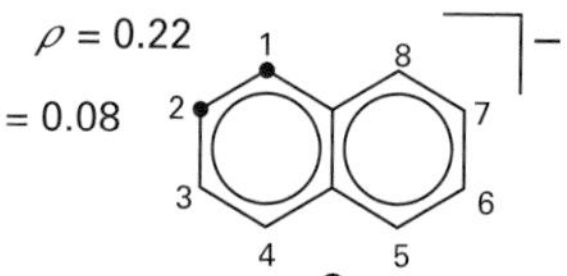

2

$$\rho = \frac{\overbrace{0.490\text{ mT}}^{a}}{\underbrace{2.25\text{ mT}}_{Q}} = 0.218 \quad \text{and} \quad \rho = \frac{0.183\text{ mT}}{2.25\text{ mT}} = 0.0813$$

12D.3(a) The origin of the hyperfine interaction

An electron in a p orbital centred on a nucleus does not approach the nucleus very closely, so the electron experiences a magnetic field that appears to arise from a point magnetic dipole. The resulting interaction is called the **dipole–dipole interaction**. The contribution of a magnetic nucleus to the local field experienced by the unpaired electron is given by an expression like that in eqn 12B.15 (a dependence proportional to $(1 - 3\cos^2\theta)/r^3$). A characteristic of this type of interaction is that it is anisotropic and averages to zero when the radical is free to tumble. Therefore, hyperfine structure due to the dipole–dipole interaction is observed only for radicals trapped in solids.

There is a second contribution to the hyperfine splitting. An s electron is spherically distributed around a nucleus and so has zero average dipole–dipole interaction with the nucleus even in a solid sample. However, because an s electron has a non-zero probability of being at the nucleus itself, it is incorrect to treat the interaction as one between two point dipoles. As explained in Topic 12B, an s electron has a 'Fermi contact interaction' with the nucleus, a magnetic interaction that occurs when the point dipole approximation fails. The contact interaction is isotropic (that is, independent of the orientation of the radical), and consequently is shown even by rapidly tumbling molecules in fluids (provided the spin density has some s character).

The dipole–dipole interactions of p electrons and the Fermi contact interaction of s electrons can be quite large. For example, a 2p electron in a nitrogen atom experiences an average field of about 4.8 mT from the ^{14}N nucleus. A 1s electron in a hydrogen atom experiences a field of about 50 mT as a result of its Fermi contact interaction with the central proton. More values are listed in Table 12D.1. The magnitudes of the contact interactions in radicals can be interpreted in terms of the s orbital character of the molecular orbital occupied by the unpaired electron, and the dipole–dipole interaction can be interpreted in terms of the p character. The analysis of hyperfine structure therefore gives information about the composition of the orbital, and especially the hybridization of the atomic orbitals.

Table 12D.1 Hyperfine coupling constants for atoms, a/mT*

Nuclide	Isotropic coupling	Anisotropic coupling
^{1}H	50.8 (1s)	
^{2}H	7.8 (1s)	
^{14}N	55.2 (2s)	4.8 (2p)
^{19}F	1720 (2s)	108.4 (2p)

* More values are given in the *Resource section*.

Brief illustration 12D.3

From Table 12D.1, the hyperfine interaction between a 2s electron and the nucleus of a nitrogen atom is 55.2 mT. The EPR spectrum of NO_2 shows an isotropic hyperfine interaction of 5.7 mT. The s character of the molecular orbital occupied by the unpaired electron is the ratio 5.7/55.2 = 0.10. For a continuation of this story, see Problem P12D.7.

Neither interaction appears to account for the hyperfine structure of the $C_6H_6^-$ anion and other aromatic radical anions. The sample is fluid, and as the radicals are tumbling the hyperfine structure cannot be due to the dipole–dipole interaction. Moreover, the protons lie in the nodal plane of the π orbital occupied by the unpaired electron, so the structure cannot be due to a Fermi contact interaction. The explanation lies in a **polarization mechanism** similar to the one responsible for spin–spin coupling in NMR. The magnetic interaction between a proton and the electrons favours one of the electrons being found with a greater probability nearby (Fig. 12D.6). The electron with

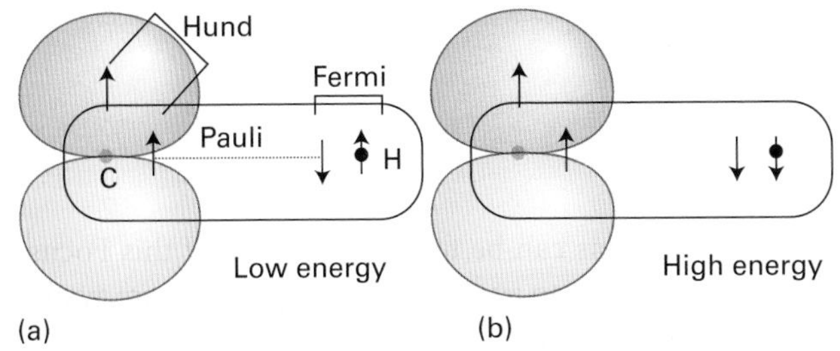

Figure 12D.6 The polarization mechanism for the hyperfine interaction in π-electron radicals. The arrangement in (a) is lower in energy than that in (b), so there is an effective coupling between the unpaired electron and the proton.

opposite spin is therefore more likely to be close to the C atom at the other end of the bond. The unpaired electron on the C atom has a lower energy if it is parallel to that electron (Hund's rule favours parallel electrons on atoms), so the unpaired electron can detect the spin of the proton indirectly. Calculation using this model leads to a hyperfine interaction in agreement with the observed value of 2.25 mT.

Exercises

E12D.2 A radical containing two equivalent ^1H nuclei shows a three-line spectrum with an intensity distribution 1:2:1. The lines occur at 330.2 mT, 332.5 mT, and 334.8 mT. What is the hyperfine coupling constant for each proton? What is the g-value of the radical given that the spectrometer is operating at 9.319 GHz?

E12D.3 A radical containing two inequivalent protons with hyperfine coupling constants 2.0 mT and 2.6 mT gives a spectrum centred on 332.5 mT. At what fields do the hyperfine lines occur and what are their relative intensities?

E12D.4 Predict the intensity distribution in the hyperfine lines of the EPR spectra of the radicals (i) $\cdot C^1H_3$, (ii) $\cdot C^2H_3$.

Checklist of concepts

- ☐ 1. The EPR resonance condition is expressed in terms of the **g-value** of the radical.
- ☐ 2. The value of g depends on the ability of the applied field to induce local electron currents in the radical and the magnetic field experienced by the electron as a result of these currents.
- ☐ 3. The **hyperfine structure** of an EPR spectrum is the splitting of individual resonance lines into components by the magnetic interaction between the electron and nuclei with spin.
- ☐ 4. If a radical contains N equivalent nuclei with spin quantum number I, then there are $2NI + 1$ hyperfine lines.
- ☐ 5. Hyperfine structure arises from **dipole–dipole interactions, Fermi contact interactions**, and the **polarization mechanism**.
- ☐ 6. The **spin density** on an atom is the probability that an unpaired electron is on that atom.

Checklist of equations

Property	Equation	Comment	Equation number
EPR resonance condition	$h\nu = g\mu_B\mathcal{B}_0$	No hyperfine interaction	12D.2
	$h\nu = g\mu_B(\mathcal{B}_0 \pm \frac{1}{2}a)$	Hyperfine interaction between an electron and a proton	12D.4
McConnell equation	$a = Q\rho$	Q = 2.25 mT	12D.5

FOCUS 12 Magnetic resonance

To test your understanding of this material, work through the *Exercises, Additional exercises, Discussion questions,* and *Problems* found throughout this Focus.

Selected solutions can be found at the end of this Focus in the e-book. Solutions to even-numbered questions are available online only to lecturers.

TOPIC 12A General principles

Discussion questions

D12A.1 Why do chemists and biochemists require spectrometers that operate at the highest available fields and frequencies to determine the structures of macromolecules by NMR spectroscopy?

D12A.2 Describe the effects of magnetic fields on the energies of nuclei and the energies of electrons. Explain the differences.

D12A.3 What is the Larmor frequency? What is its significance in magnetic resonance?

Additional exercises

E12A.6 Given that the nuclear *g*-factor, g_I, is a dimensionless number, what are the units of the nuclear magnetogyric ratio γ_N when it is expressed in tesla and hertz?

E12A.7 Given that the nuclear *g*-factor, g_I, is a dimensionless number, what are the units of the nuclear magnetogyric ratio γ_N when it is expressed in SI base units?

E12A.8 For a ^{14}N nucleus, what is the magnitude of the spin angular momentum and what are its allowed components along the *z*-axis? Express your answer in multiples of $\hbar$. What angles does the angular momentum make with the *z*-axis?

E12A.9 What is the NMR frequency of a ^{19}F nucleus in a magnetic field of 17.1 T? Express your answer in megahertz.

E12A.10 The nuclear spin quantum number of ^{33}S is $\frac{3}{2}$ and its *g*-factor is 0.4289. Calculate (in joules) the energies of the nuclear spin states in a magnetic field of 6.800 T.

E12A.11 The nuclear spin quantum number of ^{14}N is 1 and its *g*-factor is 0.404. Calculate (in joules) the energies of the nuclear spin states in a magnetic field of 10.50 T.

E12A.12 Calculate the frequency separation (in megahertz) of the nuclear spin levels of a ^{13}C nucleus in a magnetic field of 15.4 T given that its magnetogyric ratio is $6.73 \times 10^{-7}\,\mathrm{T^{-1}\,s^{-1}}$.

E12A.13 Calculate the frequency separation (in megahertz) of the nuclear spin levels of a ^{14}N nucleus in a magnetic field of 14.4 T given that its magnetogyric ratio is $1.93 \times 10^{-7}\,\mathrm{T^{-1}\,s^{-1}}$.

E12A.14 In which of the following systems is the energy level separation larger for a given magnetic field? (i) A ^{15}N nucleus, (ii) a ^{31}P nucleus.

E12A.15 Calculate the relative population differences $(N_\alpha - N_\beta)/N$ for ^{13}C nuclei in fields of (i) 0.50 T, (ii) 2.5 T, and (iii) 15.5 T at 25 °C.

E12A.16 By what factor must the applied magnetic field be increased for the relative population difference $(N_\alpha - N_\beta)/N$ to be increased by a factor of 5 for (i) ^{1}H nuclei, (ii) ^{13}C nuclei?

E12A.17 By what factor must the temperature be changed for the relative population difference $(N_\alpha - N_\beta)/N$ to be increased by a factor of 5 for ^{1}H nuclei relative to its value at room temperature? Is changing the temperature of the sample a practical way of increasing the sensitivity?

E12A.18 What is the EPR resonance frequency in a magnetic field for which the NMR frequency for ^{1}H nuclei (protons) is 500 MHz? Express your answer in gigahertz.

Problems

P12A.1 A scientist investigates the possibility of neutron spin resonance, and has available a commercial NMR spectrometer operating at 300 MHz for ^{1}H nuclei. What is the NMR frequency of the neutron in this spectrometer? What is the relative population difference at room temperature? Which is the lower energy spin state of the neutron?

P12A.2‡ The relative sensitivity of NMR spectroscopy, *R*, for equal numbers of different nuclei at constant temperature for a given magnetic field is $R \propto \{I(I+1)\}\gamma_N^3$. (a) From the data in Table 12A.2, calculate these sensitivities for ^{2}H, ^{13}C, ^{14}N, ^{15}N, and ^{11}B relative to that of ^{1}H. (b) For a given number of nuclei of a particular element, the fraction present as a particular isotope is affected by the natural abundance of that isotope. Recalculate these results taking this dependence into account.

P12A.3 The intensity of the NMR signal is given by eqn 12A.8c. The intensity can be increased further by 'isotopic labelling', which involves increasing the proportion of the atoms present that are of the desired NMR-active isotope. The degree of labelling is expressed by giving the fractional enrichment. For example, 'a 10 per cent enrichment in ^{15}N' would imply that 10 per cent of all the N atoms are ^{15}N. (a) What level of enrichment is needed for the ^{15}N signal to have the same intensity as that from ^{13}C with its natural abundance? (b) What is the intensity achievable, relative to natural abundance ^{13}C, by 100 per cent enrichment of ^{17}O?

‡ These problems were supplied by Charles Trapp and Carmen Giunta.

P12A.4 With special techniques, known collectively as 'magnetic resonance imaging' (MRI), it is possible to obtain NMR spectra of entire organisms. A key to MRI is the application of a magnetic field that varies linearly across the specimen. If the field varies in the z-direction according to $\mathcal{B}_0 + \mathcal{G}_z z$, where $\mathcal{G}_z$ is the field gradient along the z-direction, the ^{1}H nuclei have NMR frequencies given by

$$\nu_L(z) = \frac{\gamma_N}{2\pi}(\mathcal{B}_0 + \mathcal{G}_z z)$$

Similar equations may be written for gradients along the x- and y-directions. The NMR signal at frequency $\nu = \nu(z)$ is proportional to the numbers of protons at the position z. Suppose a uniform disk-shaped organ is placed in such a linear field gradient, and that in this case the NMR signal is proportional to the number of protons in a slice of width δz at each horizontal distance z from the centre of the disk. Sketch the shape of the absorption intensity for the MRI image of the disk.

TOPIC 12B Features of NMR spectra

Discussion questions

D12B.1 The earliest NMR spectrometers measured the spectrum by keeping the frequency fixed and then scanning the magnetic field to bring peaks successively into resonance. Peaks that came into resonance at higher magnetic fields were described as 'up field' and those at lower magnetic fields as 'down field'. Discuss what the terms 'up field' and 'down field' imply about chemical shifts and shielding.

D12B.2 Discuss in detail the origins of the local, neighbouring group, and solvent contributions to the shielding constant.

D12B.3 Explain why the resonance from two equivalent ^{1}H nuclei does not exhibit any splitting due to the spin–spin coupling that exists between the nuclei, but that the resonance is split by the coupling to a third (inequivalent) spin.

D12B.4 Explain the difference between magnetically equivalent and chemically equivalent nuclei, and give two examples of each.

D12B.5 Discuss how the Fermi contact interaction and the polarization mechanism contribute to spin–spin coupling in NMR.

Additional exercises

E12B.10 The ^{13}C resonance from TMS is found to occur at 125.130 000 MHz. What is the chemical shift (on the δ scale) of a peak at 125.148 750 MHz?

E12B.11 In a spectrometer operating at 500.130 000 MHz for ^{1}H, a resonance is found to occur 750 Hz higher in frequency than TMS. What is the chemical shift (on the δ scale) of this peak?

E12B.12 In a spectrometer operating at 125.130 000 MHz for ^{13}C, a resonance is found to occur 1875 Hz lower in frequency than TMS. What is the chemical shift (on the δ scale) of this peak?

E12B.13 What is the frequency separation, in hertz, between two peaks in the ^{13}C spectrum with chemical shifts $\delta = 50.0$ and $\delta = 25.5$ in a spectrometer operating at 100.130 000 MHz for ^{13}C?

E12B.14 In a spectrometer operating at 400.130 000 MHz for ^{1}H, on the δ scale what separation of peaks in the ^{1}H spectrum corresponds to a frequency difference of 550 Hz?

E12B.15 In a spectrometer operating at 200.130 000 MHz for ^{13}C, on the δ scale what separation of peaks in the ^{13}C spectrum corresponds to a frequency difference of 25 000 Hz?

E12B.16 The chemical shift of the CH_3 protons in ethoxyethane (diethyl ether) is $\delta = 1.16$ and that of the CH_2 protons is 3.36. What is the difference in local magnetic field between the two regions of the molecule when the applied field is (i) 1.9 T, (ii) 16.5 T?

E12B.17 Make a sketch, roughly to scale, of the ^{1}H NMR spectrum expected for an AX_2 spin system with $\delta_A = 1.50$, $\delta_X = 4.50$, and $J_{AX} = 5$ Hz recorded on a spectrometer operating at 500 MHz. The horizontal scale should be in hertz, taking the resonance from TMS as the origin.

E12B.18 Sketch the form of the ^{19}F NMR spectrum and the ^{11}B NMR spectrum of $^{11}BF_4^-$.

E12B.19 Sketch the form of the ^{31}P NMR spectra of a sample of $^{31}PF_6^-$.

E12B.20 Sketch the form of the ^{1}H NMR spectra of $^{14}NH_4^+$ and of $^{15}NH_4^+$.

E12B.21 Predict the multiplet you would expect for coupling to two equivalent spin-1 nuclei.

E12B.22 Use an approach similar to that shown in Figs. 12B.13 and 12B.14 to predict the multiplet you would expect for coupling to two spin-$\frac{1}{2}$ nuclei when the coupling to the two nuclei is not the same.

E12B.23 Predict the multiplet you would expect for coupling of a proton to two inequivalent spin-1 nuclei.

E12B.24 Use an approach similar to that shown in Figs. 12B.13 and 12B.14 to predict the multiplet you would expect for coupling to two equivalent spin-$\frac{5}{2}$ nuclei.

E12B.25 Predict the multiplet you would expect for coupling to three equivalent spin-$\frac{5}{2}$ nuclei.

E12B.26 Classify the ^{1}H nuclei in 1,2,3-trichlorobenzene into chemically or magnetically equivalent groups. Give your reasoning.

E12B.27 Classify the ^{19}F nuclei in SF_5^- (which is square-pyramidal) into chemically or magnetically equivalent groups. Give your reasoning.

E12B.28 A proton jumps between two sites with $\delta = 4.2$ and $\delta = 5.5$. What rate constant for the interconversion process is needed for the two signals to collapse to a single line in a spectrometer operating at 350 MHz?

Problems

P12B.1 Explain why the ^{129}Xe NMR spectrum of XeF^+ is a doublet with $J = 7600$ Hz but the ^{19}F NMR spectrum appears to be a triplet with $J = 3800$ Hz. *Hints:* ^{19}F has spin-$\frac{1}{2}$ and 100 per cent natural abundance; ^{129}Xe has spin-$\frac{1}{2}$ and 26 per cent natural abundance.

P12B.2 The ^{19}F NMR spectrum of IF_5 consists of two lines of equal intensity and a quintet (five lines with intensity ratio 1:4:6:4:1). Suggest a structure for IF_5 that is consistent with this spectrum, explaining how you arrive at your result. *Hint:* You do not need to consider possible interaction with the I nucleus.

P12B.3 The Lewis structure of SF_4 has four bonded pairs of electrons and one lone pair. Propose two structures for SF_4 based a trigonal bipyramidal coordination at S, and a further structure based on a square pyramid. For each structure, describe the expected form of the ^{19}F NMR spectrum, giving your reasons. *Hint:* You do not need to consider possible interaction with the S nucleus.

P12B.4 Refer to Fig. 12B.15 and use mathematical software or a spreadsheet to draw a family of curves showing the variation of $^3J_{HH}$ with ϕ using $A = +7.0$ Hz, $B = -1.0$ Hz, and allowing C to vary slightly from a typical value of +5.0 Hz. Explore the effect of changing the value of the parameter C on the shape of the curve. In a similar fashion, explore the effect of the values of A and B on the shape of the curve.

P12B.5‡ Various versions of the Karplus equation (eqn 12B.14) have been used to correlate data on three-bond proton coupling constants $^3J_{HH}$ in systems of the type $XYCHCHR_3R_4$. The original version (M. Karplus, *J. Am. Chem.* Soc. **85**, 2870 (1963)) is $^3J_{HH} = A\cos^2\phi_{HH} + B$. Experimentally it is found that when $R_3 = R_4 = H$, $^3J_{HH} = 7.3$ Hz; when $R_3 = CH_3$ and $R_4 = H$, $^3J_{HH} = 8.0$ Hz; when $R_3 = R_4 = CH_3$, $^3J_{HH} = 11.2$ Hz. Assuming that only staggered conformations are important, determine which version of the Karplus equation fits the data better. *Hint:* You will need to consider which conformations to include, and average the couplings predicted by the Karplus equation over them; assume that X and Y are 'bulky' groups.

P12B.6‡ It might be unexpected that the Karplus equation, which was first derived for $^3J_{HH}$ coupling constants, should also apply to three-bond coupling between the nuclei of metallic elements such as tin. T.N. Mitchell and B. Kowall (*Magn. Reson. Chem.* **33**, 325 (1995)) have studied the relation between $^3J_{HH}$ and $^3J_{SnSn}$ in compounds of the type $Me_3SnCH_2CHRSnMe_3$ and find that $(^3J_{SnSn}/\text{Hz}) = 78.86 \times (^3J_{HH}/\text{Hz}) + 27.84$. (a) Does this result support a Karplus type equation for tin nuclei? Explain your reasoning. (b) Obtain the Karplus equation for $^3J_{SnSn}$ and plot it as a function of the dihedral angle. (c) Draw the preferred conformation.

P12B.7 Show that the coupling constant as expressed by the Karplus equation (eqn 12B.14) passes through a minimum when $\cos\phi = B/4C$.

P12B.8 In a liquid, the dipolar magnetic field averages to zero: show this result by evaluating the average of the field given in eqn 12B.15. *Hint:* The relevant volume element in polar coordinates is $\sin\theta\, d\theta d\phi$.

P12B.9 Account for the following observations: (a) The ^1H NMR spectrum of cyclohexane shows a single peak at room temperature, but when the temperate is lowered significantly the peak starts to broaden and then separates into two. (b) At room temperature, the ^{19}F NMR spectrum of PF_5 shows two lines, and even at the lowest experimentally accessible temperatures the spectrum is substantially unchanged. (c) In the ^1H NMR spectrum of a casually prepared sample of ethanol a triplet and a quartet are seen. These multiplets show additional splittings if the sample is prepared with the careful exclusion of water.

TOPIC 12C Pulse techniques in NMR

Discussion questions

D12C.1 Discuss in detail the effects of a 90° pulse and of a 180° pulse on a system of spin-$\frac{1}{2}$ nuclei in a static magnetic field.

D12C.2 Suggest a reason why the relaxation times of ^{13}C nuclei are typically much longer than those of ^1H nuclei.

D12C.3 Suggest a reason why the spin–lattice relaxation time of a small molecule (like benzene) in a mobile, deuterated hydrocarbon solvent increases as the temperature increases, whereas that of a large molecule (like a polymer) decreases.

D12C.4 Discuss the origin of the nuclear Overhauser effect and how it can be used to identify nearby protons in a molecule.

D12C.5 Distinguish between homogeneous and inhomogeneous broadening.

Additional exercises

E12C.6 The duration of a 90° or 180° pulse depends on the strength of the $\mathcal{B}_1$ field. If a 90° pulse applied to ^1H requires 5 μs, what is the strength of the $\mathcal{B}_1$ field? How long would the corresponding 180° pulse require?

E12C.7 What is the effective transverse relaxation time when the width of a Lorentzian resonance line is 12 Hz?

E12C.8 If the transverse relaxation time, T_2, is 50 ms, after what time will the envelope of the free-induction decay decrease to half its initial amplitude?

E12C.9 The ^{13}C NMR spectrum of fluoroethanoic acid shows a multiplet centred at $\delta = 79$. When the same spectrum is recorded using proton decoupling, the multiplet collapses to a doublet with a splitting of 160 Hz. Another multiplet, but with much smaller splittings, is also seen centred at $\delta = 179$; this multiplet collapses to a doublet when decoupling is used. Explain these observations.

E12C.10 Predict the maximum NOE enhancement (as the value of η) that could be obtained for ^{19}F as a result of dipole–dipole relaxation with ^1H.

Problems

P12C.1 An NMR spectroscopist performs a series of experiments in which a pulse of a certain duration is applied, the free-induction decay recorded, and then Fourier transformed to give the spectrum. A pulse of duration 2.5 μs gave a satisfactory spectrum, but when the pulse duration was increased to 5.0 μs more intense peaks were seen. A further increase to 7.5 μs resulted in weaker signals, and increasing the duration to 10.0 μs gave no detectable spectrum. (a) By considering the effect of varying the flip angle of the pulse, rationalize these observations. Calculate (b) the duration of a 90° pulse and (c) the $\mathcal{B}_1$ Larmor frequency $\nu_L' = \gamma_N \mathcal{B}_1/2\pi$.

P12C.2 In a practical NMR spectrometer the free-induction decay is digitized at regular intervals before being stored in computer memory, ready for subsequent processing. Technically, it is difficult to digitize a signal at the frequencies typical of NMR, so in practice a fixed reference frequency, close to the Larmor frequency, is subtracted from the NMR frequency. The resulting difference frequency, called the *offset frequency*, is of the order of several kilohertz, rather than the hundreds of megahertz typical of NMR resonance frequencies. This lower frequency can be digitized by currently available technology. For ^1H, if this reference frequency is set at the NMR frequency of

TMS, then a peak with chemical shift δ will give rise to a contribution to the free-induction decay at $\delta \times (\nu_L/10^6)$. Use mathematical software to construct the FID curve for a set of three ^{1}H nuclei with resonances of equal intensity at $\delta = 3.2$, 4.1, and 5.0 in a spectrometer operating at 800 MHz. Assume that the reference frequency is set at the NMR frequency of TMS, that $T_2 = 0.5$ s, and plot the FID out to a maximum time of 1.5 s. Explore the effect of varying the relative amplitude of the three resonances.

P12C.3 Refer to Problem 12C.2 for the definition of offset frequency. The FID, $F(t)$, of a signal containing many frequencies, each corresponding to a different chemical shift, is given by

$$F(t) = \sum_j S_{0j} \cos(2\pi\nu_j t)\mathrm{e}^{-t/T_{2j}}$$

where, for each resonance j, S_{0j} is the maximum intensity of the signal, ν_j is the offset frequency, and T_{2j} is the spin–spin relaxation time. (a) Use mathematical software to plot the FID (out to a maximum time of 3 s) for the case

$$S_{01} = 1.0 \quad \nu_1 = 50\ \mathrm{Hz} \quad T_{21} = 0.50\ \mathrm{s}$$
$$S_{02} = 3.0 \quad \nu_1 = 10\ \mathrm{Hz} \quad T_{22} = 1.0\ \mathrm{s}$$

(b) Explore how the form of the FID changes as ν_1 and T_{21} are changed. (c) Use mathematical software to calculate and plot the Fourier transforms of the FID curves you generated in parts (a) and (b). How do spectral linewidths vary with the value of T_2? *Hint:* Most software packages offer a 'fast Fourier transform' routine with which these calculations can be made: refer to the user manual for details. You should select the cosine Fourier transform.

P12C.4 (a) In many instances it is possible to approximate the NMR lineshape by using a Lorentzian function of the form

$$I_{\text{Lorentzian}}(\omega) = \frac{S_0 T_2}{1 + T_2^2(\omega - \omega_0)^2}$$

where $I(\omega)$ is the intensity as a function of the angular frequency $\omega = 2\pi\nu$, ω_0 is the resonance frequency, S_0 is a constant, and T_2 is the spin–spin relaxation time. Confirm that for this lineshape the width at half-height is $1/\pi T_2$. (b) Under certain circumstances, NMR lines are Gaussian functions of the frequency, given by

$$I_{\text{Gaussian}}(\omega) = S_0 T_2 \mathrm{e}^{-T_2^2(\omega-\omega_0)^2}$$

Confirm that for the Gaussian lineshape the width at half-height is equal to $2(\ln 2)^{1/2}/T_2$. (c) Compare and contrast the shapes of Lorentzian and Gaussian lines by plotting two lines with the same values of S_0, T_2, and ω_0.

P12C.5 The shape of a spectral line, $I(\omega)$, is related to the free-induction decay signal $G(t)$ by

$$I(\omega) = a\ \mathrm{Re}\int_0^{\infty} G(t)\mathrm{e}^{\mathrm{i}\omega t}\,\mathrm{d}t$$

where a is a constant and 'Re' means take the real part of what follows. Calculate the lineshape corresponding to an oscillating, decaying function $G(t) = \cos \omega t\ \mathrm{e}^{-t/\tau}$. *Hint:* Write $\cos \omega t$ as $\frac{1}{2}(\mathrm{e}^{-\mathrm{i}\omega t} + \mathrm{e}^{\mathrm{i}\omega t})$.

P12C.6 In the language of Problem P12C.5, show that if $G(t) = (a \cos \omega_1 t + b \cos \omega_2 t)\mathrm{e}^{-t/\tau}$, then the spectrum consists of two lines with intensities proportional to a and b and located at $\omega = \omega_1$ and ω_2, respectively.

P12C.7 The exponential relaxation of the z-component of the magnetization $M_z(t)$ back to its equilibrium value M_0 is described by the differential equation

$$\frac{\mathrm{d}M_z(t)}{\mathrm{d}t} = -\frac{M_z(t) - M_0}{T_1}$$

(a) In the inversion recovery experiment the initial condition (at time zero) is that $M_z(0) = -M_0$, which corresponds to inversion at the magnetization by the 180° pulse. Integrate the differential equation (it is separable), impose this initial condition, and hence show that $M_z(\tau) = M_0(1 - 2\mathrm{e}^{-\tau/T_1})$, where τ is the delay between the 180° and 90° pulses. (b) Use mathematical software or a spreadsheet to plot $M_z(\tau)/M_0$ as a function of τ, taking $T_1 = 1.0$ s; explore the effect of increasing and decreasing T_1. (c) Show that a plot of $\ln\{(M_0 - M_z(\tau))/2M_0\}$ against τ is expected to be a straight line with slope $-1/T_1$. (d) In an experiment the following data were obtained; use them to determine a value for T_1.

τ/s	0.000	0.100	0.200	0.300	0.400	0.600	0.800	1.000	1.200
$M_z(\tau)/M_0$	−1.000	−0.637	−0.341	−0.098	0.101	0.398	0.596	0.729	0.819

P12C.8 Derive an expression for the time τ in an inversion recovery experiment at which the magnetization passes through zero. In an experiment it is found that this time for zero magnetization is 0.50 s; evaluate T_1. *Hint:* This Problem requires a result from Problem P12C.7.

P12C.9 The exponential relaxation of the transverse component of the magnetization $M_{xy}(t)$ back to its equilibrium value of zero is described by the differential equation

$$\frac{\mathrm{d}M_{xy}(t)}{\mathrm{d}t} = -\frac{M_{xy}(t)}{T_2}$$

(a) Integrate this differential equation (it is separable) between $t = 0$ and $t = \tau$ with the initial condition that the transverse magnetization is $M_{xy}(0)$ at $t = 0$ to obtain $M_{xy}(\tau) = M_{xy}(0)\mathrm{e}^{-\tau/T_2}$. (b) Hence, show that a plot of $\ln\{M_{xy}(\tau)/M_{xy}(0)\}$ against τ is expected to be a straight line of slope $-1/T_2$. (c) The following data were obtained in a spin-echo experiment; use the data to evaluate T_2.

τ/ms	10.0	20.0	30.0	50.0	70.0	90.0	110	130
$M_{xy}(\tau)/M_{xy}(0)$	0.819	0.670	0.549	0.368	0.247	0.165	0.111	0.074

P12C.10 In the spin echo experiment analysed in Fig. 12C.11, the 180° pulse is applied about the y-axis, resulting in the magnetization vectors being reflected in the yz-plane. The experiment works just as well when the 180° pulse is applied about the x-axis, in which case the magnetization vectors are reflected in the xz-plane. Analyse the outcome of the spin echo experiment for the case where the 180° pulse is applied about the x-axis.

P12C.11 The z-component of the magnetic field at a distance R from a magnetic moment parallel to the z-axis is given by eqn 12B.17a. In a solid, a proton at a distance R from another can experience such a field and the measurement of the splitting it causes in the spectrum can be used to calculate R. In gypsum, for instance, the splitting in the H_2O resonance can be interpreted in terms of a magnetic field of 0.715 mT generated by one proton and experienced by the other. What is the separation of the hydrogen nuclei in the H_2O molecule?

P12C.12 In a liquid crystal a molecule might not rotate freely in all directions and the dipolar interaction might not average to zero. Suppose a molecule is trapped so that, although the vector separating two protons may rotate freely around the z-axis, the colatitude may vary only between 0 and θ'. Use mathematical software to average the dipolar field over this restricted range of orientation and confirm that the average vanishes when $\theta' = \pi$ (corresponding to free rotation over a sphere). What is the average value of the local dipolar field for the H_2O molecule in Problem P12C.11 if it is dissolved in a liquid crystal that enables it to rotate up to $\theta' = 30°$?

TOPIC 12D Electron paramagnetic resonance

Discussion questions

D12D.1 Describe how the Fermi contact interaction and the polarization mechanism contribute to hyperfine interactions in EPR.

D12D.2 Explain how the EPR spectrum of an organic radical can be used to identify and map the molecular orbital occupied by the unpaired electron.

Additional exercises

E12D.5 The centre of the EPR spectrum of atomic deuterium lies at 330.02 mT in a spectrometer operating at 9.2482 GHz. What is the g-value of the electron in the atom?

E12D.6 A radical containing three equivalent protons shows a four-line spectrum with an intensity distribution 1:3:3:1. The lines occur at 331.4 mT, 333.6 mT, 335.8 mT, and 338.0 mT. What is the hyperfine coupling constant for each proton? What is the g-value of the radical given that the spectrometer is operating at 9.332 GHz?

E12D.7 A radical containing three inequivalent protons with hyperfine coupling constants 2.11 mT, 2.87 mT, and 2.89 mT gives a spectrum centred on 332.8 mT. At what fields do the hyperfine lines occur and what are their relative intensities?

E12D.8 Predict the intensity distribution in the hyperfine lines of the EPR spectra of the radicals (i) $\cdot{}^{1}\mathrm{C}{}^{1}\mathrm{H}_2{}^{1}\mathrm{C}{}^{1}\mathrm{H}_3$, (ii) $\cdot\mathrm{C}{}^{2}\mathrm{H}_2\mathrm{C}{}^{2}\mathrm{H}_3$.

E12D.9 The benzene radical anion has $g = 2.0025$. At what field should you search for resonance in a spectrometer operating at (i) 9.313 GHz, (ii) 33.80 GHz?

E12D.10 The naphthalene radical anion has $g = 2.0024$. At what field should you search for resonance in a spectrometer operating at (i) 9.501 GHz, (ii) 34.77 GHz?

E12D.11 The EPR spectrum of a radical with a single magnetic nucleus is split into four lines of equal intensity. What is the nuclear spin of the nucleus?

E12D.12 The EPR spectrum of a radical with two equivalent nuclei of a particular kind is split into five lines of intensity ratio 1:2:3:2:1. What is the spin of the nuclei?

Problems

P12D.1 It is possible to produce very high magnetic fields over small volumes by special techniques. What would be the resonance frequency of an electron spin in an organic radical in a field of 1.0 kT? How does this frequency compare to typical molecular rotational, vibrational, and electronic energy-level separations?

P12D.2 The angular NO_2 molecule has a single unpaired electron and can be trapped in a solid matrix or prepared inside a nitrite crystal by radiation damage of NO_2^- ions. When the applied field is parallel to the OO direction the centre of the spectrum lies at 333.64 mT in a spectrometer operating at 9.302 GHz. When the field lies along the bisector of the ONO angle, the resonance lies at 331.94 mT. What are the g-values in the two orientations?

P12D.3 (a) The hyperfine coupling constant is proportional to the gyromagnetic ratio of the nucleus in question, γ_N. Rationalize this observation. (b) The hyperfine coupling constant in $\cdot{}^{1}\mathrm{C}{}^{1}\mathrm{H}_3$ is 2.3 mT. Use the information in Table 12D.1 to predict the splitting between the hyperfine lines of the spectrum of $\cdot\mathrm{C}{}^{2}\mathrm{H}_3$. What are the overall widths of the multiplet in each case?

P12D.4 The 1,4-dinitrobenzene radical anion can be prepared by reduction of 1,4-dinitrobenzene. The radical anion has two equivalent N nuclei ($I = 1$) and four equivalent protons. Predict the form of the EPR spectrum using $a(\mathrm{N}) = 0.148$ mT and $a(\mathrm{H}) = 0.112$ mT.

P12D.5 The hyperfine coupling constants for the anthracene radical anion are 0.274 mT (protons 1, 4, 5, 8), 0.151 mT (protons 2, 3, 6, 7), and 0.534 mT (protons 9, 10). Use the McConnell equation to estimate the spin density at carbons 1, 2, and 9 (use $Q = 2.25$ mT).

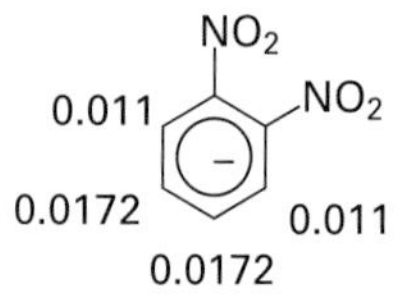

1

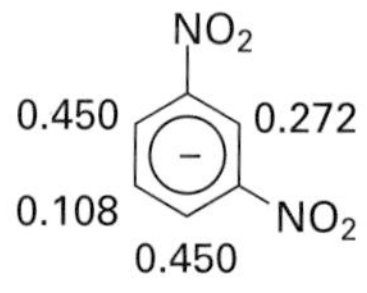

2

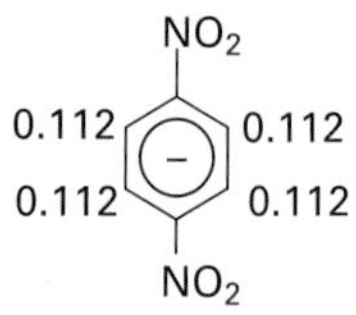

3

P12D.6 The hyperfine coupling constants observed in the radical anions (**1**), (**2**), and (**3**) are shown (in millitesla, mT). Use the value for the benzene radical anion to map the probability of finding the unpaired electron in the π orbital on each C atom.

P12D.7 When an electron occupies a 2s orbital on an N atom it has a hyperfine interaction of 55.2 mT with the nucleus. The spectrum of NO_2 shows an isotropic hyperfine interaction of 5.7 mT. For what proportion of its time is the unpaired electron of NO_2 occupying a 2s orbital? The hyperfine coupling constant for an electron in a 2p orbital of an N atom is 3.4 mT. In NO_2 the anisotropic part of the hyperfine coupling is 1.3 mT. What proportion of its time does the unpaired electron spend in the 2p orbital of the N atom in NO_2? What is the total probability that the electron will be found on (a) the N atoms, (b) the O atoms? What is the hybridization ratio of the N atom? Does the hybridization support the view that NO_2 is angular?

FOCUS 12 Magnetic resonance

Integrated activities

I12.1 Consider the following series of molecules: benzene, methylbenzene, trifluoromethylbenzene, benzonitrile, and nitrobenzene in which the substituents *para* to the C atom of interest are H, CH_3, CF_3, CN, and NO_2, respectively. (a) Use the computational method of your or your instructor's choice to calculate the net charge at the C atom *para* to these substituents in this series of organic molecules. (b) It is found empirically that the ^{13}C chemical shift of the *para* C atom increases in the order: methylbenzene, benzene, trifluoromethylbenzene, benzonitrile, nitrobenzene. Is there a correlation between the behaviour of the ^{13}C chemical shift and the computed net charge on the ^{13}C atom? (c) The ^{13}C chemical shifts of the *para* C atoms in each of the molecules that you examined computationally are as follows:

Substituent	CH_3	H	CF_3	CN	NO_2
δ	128.4	128.5	128.9	129.1	129.4

Is there a linear correlation between net charge and ^{13}C chemical shift of the *para* C atom in this series of molecules? (d) If you did find a correlation in part (c), explain the physical origins of the correlation.

I12.2 The computational techniques described in Topic 9E have shown that the amino acid tyrosine participates in a number of biological electron transfer reactions, including the processes of water oxidation to O_2 in plant photosynthesis and of O_2 reduction to water in oxidative phosphorylation. During the course of these electron transfer reactions, a tyrosine radical forms with spin density delocalized over the side chain of the amino acid. (a) The phenoxy radical shown in (5) is a suitable model of the tyrosine radical. Using molecular modelling software and the computational method of your or your instructor's choice, calculate the spin densities at the O atom and at all of the C atoms in (5). (b) Predict the form of the EPR spectrum of (5).

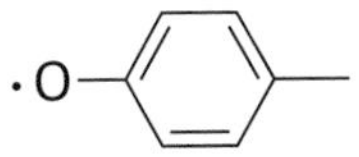

5 Phenoxy radical

I12.3 Two groups of protons have $\delta = 4.0$ and $\delta = 5.2$ and are interconverted by a conformational change of a fluxional molecule. In a 60 MHz spectrometer the spectrum collapsed into a single line at 280 K but at 300 MHz the collapse did not occur until the temperature had been raised to 300 K. Calculate the exchange rate constant at the two temperatures and hence find the activation energy of the interconversion (Topic 17D).

I12.4 NMR spectroscopy may be used to determine the equilibrium constant for dissociation of a complex between a small molecule, such as an enzyme inhibitor I, and a protein, such as an enzyme E:

$$EI \rightleftharpoons E + I \quad K_I = [E][I]/[EI]$$

In the limit of slow chemical exchange, the NMR spectrum of a proton in I would consist of two resonances: one at ν_I for free I and another at ν_{EI} for bound I. When chemical exchange is fast, the NMR spectrum of the same proton in I consists of a single peak with a resonance frequency ν given by $\nu = f_I\nu_I + f_{EI}\nu_{EI}$, where $f_I = [I]/([I] + [EI])$ and $f_{EI} = [EI]/([I] + [EI])$ are, respectively, the fractions of free I and bound I. For the purposes of analysing the data, it is also useful to define the frequency differences $\delta\nu = \nu - \nu_I$ and $\Delta\nu = \nu_{EI} - \nu_I$. Show that when the initial concentration of I, $[I]_0$, is much greater than the initial concentration of E, $[E]_0$, a plot of $[I]_0$ against $(\delta\nu)^{-1}$ is a straight line with slope $[E]_0\Delta\nu$ and y-intercept $-K_I$.

FOCUS 13

Statistical thermodynamics

Statistical thermodynamics provides the link between the microscopic properties of matter and its bulk properties. It provides a means of calculating thermodynamic properties from structural and spectroscopic data and gives insight into the molecular origins of chemical properties.

13A The Boltzmann distribution

The 'Boltzmann distribution' is used to predict the populations of states in systems at thermal equilibrium. It is an exceptionally important result because its predictions are used throughout chemistry to understand many molecular phenomena. The distribution also provides insight into the nature of 'temperature'.

13B Molecular partition functions

The Boltzmann distribution introduces a quantity known as the 'partition function' which plays a key role in statistical thermodynamics. The Topic shows how to interpret the partition function and how to calculate it in a number of simple cases.

13C Molecular energies

A partition function is the thermodynamic version of a wavefunction, and contains all the thermodynamic information about a system. In this Topic partition functions are used to calculate the mean values of the energy of the basic modes of motion of a collection of independent molecules.

13D The canonical ensemble

Molecules do interact with one another, and statistical thermodynamics would be incomplete without being able to take these interactions into account. This Topic shows how that is done in principle by introducing the 'canonical ensemble', and hints at how this concept can be used.

13E The internal energy and the entropy

This Topic shows how molecular partition functions are used to calculate and give insight into the two basic thermodynamic functions, the internal energy and the entropy. The latter is based on the definition introduced by Boltzmann of the 'statistical entropy'.

13F Derived functions

With expressions relating internal energy and entropy to partition functions, it is possible to develop expressions for the derived thermodynamic functions, such as the Helmholtz and Gibbs energies. The discussion is developed further to show how equilibrium constants can be calculated from structural and spectroscopic data.

What is an application of this material?

There are numerous applications of statistical arguments in biochemistry. One of the most directly related to partition functions is explored in *Impact* 20, accessed via the e-book: the helix–coil equilibrium in a polypeptide and the role of cooperative behaviour.

➤ **Go to the e-book for videos that feature the derivation and interpretation of equations, and applications of this material.**

TOPIC 13A The Boltzmann distribution

➤ **Why do you need to know this material?**

The Boltzmann distribution is the key to understanding a great deal of chemistry. All thermodynamic properties can be interpreted in its terms, as can the temperature dependence of equilibrium constants and the rates of chemical reactions. There is, perhaps, no more important unifying concept in chemistry.

➤ **What is the key idea?**

The most probable distribution of molecules over the available energy levels subject to certain restraints depends on a single parameter, the temperature.

➤ **What do you need to know already?**

You need to be aware that molecules can exist only in certain discrete energy levels (Topic 7A) and that in some cases more than one state has the same energy.

The problem addressed in this Topic is the calculation of the populations of the energy levels of a system consisting of a large number of non-interacting molecules. These energy levels can be associated with any mode of motion, such as vibration or rotation. Because the molecules are assumed to be non-interacting the total energy of the system is the sum of their individual energies. In a real system, a contribution to the total energy may arise from interactions between molecules, but that possibility is discounted at this stage.

The development is based on the **principle of equal a priori probabilities**, the assumption that all possibilities for the distribution of energy are equally probable. 'A priori' means loosely in this context 'as far as one knows'. In other words, there is no reason to assume that a vibrational state with a certain energy is any more or less likely to be populated than a rotational state with the same energy, provided that the system is at thermal equilibrium.

One very important conclusion will emerge from the following analysis: one distribution of the populations of the levels is overwhelmingly the most probable and this distribution depends on a single parameter, the 'temperature'. The work done here provides a molecular justification for the concept of temperature and some insight into this crucially important quantity.

13A.1 Configurations and weights

Any individual molecule may exist in states with energies ε_0, ε_1,.... For reasons that will become clear, the lowest available state is always taken as the zero of energy (that is, $\varepsilon_0 \equiv 0$), and all other energies are measured relative to that state. To obtain the actual energy from the energy calculated in this way all that is necessary is to add a term $N\varepsilon^{(0)}$, where N is the number of molecules and $\varepsilon^{(0)}$ is the true energy of the ground state. For example, when considering the vibrational contribution to the energy, $\varepsilon^{(0)}$ is the zero-point energy of vibration.

13A.1(a) Instantaneous configurations

At any instant there will be N_0 molecules in the state 0 with energy ε_0, N_1 in the state 1 with ε_1, and so on, with $N_0 + N_1 + \cdots = N$, the total number of molecules in the system. The specification of the set of populations N_0,N_1,... in the form $\{N_0,N_1,\ldots\}$ is a statement of the instantaneous **configuration** of the system. The instantaneous configuration fluctuates with time because the populations change, perhaps as a result of collisions.

Initially suppose that all the states have exactly the same energy. The energies of all the configurations are then identical, so there is no restriction on how many of the N molecules are in each state. Now picture a large number of different instantaneous configurations. One, for example, might be $\{N,0,0,\ldots\}$, corresponding to every molecule being in state 0. Another might be $\{N-2,2,0,0,\ldots\}$, in which two molecules are in state 1. The latter configuration is intrinsically more likely to be found than the former because it can be achieved in more ways. The configuration with all the molecules in the ground state can be achieved in only one way, but there are more ways of achieving the configuration $\{N-2,2,0,0,\ldots\}$ as any two of the N molecules can occupy state 1.

The number of ways the configuration $\{N-2,2,0,0,\ldots\}$ can be achieved is calculated in the following way. One molecule from the total N is chosen to migrate to state 1: there are N ways of making this choice. For the second molecule that is chosen to migrate to state 1 there are $N-1$ possible choices. Overall, the total number of ways of choosing two molecules is $N(N-1)$. However, the final configuration is the same regardless of the order in which the two molecules are chosen. Therefore, only half the choices lead to distinguishable configurations, and so the total number of ways of achieving the configuration $\{N-2,2,0,0,\}$ is $\frac{1}{2}N(N-1)$.

If, as a result of collisions, the system were to fluctuate between the configurations $\{N,0,0,\ldots\}$ and $\{N-2,2,0,\ldots\}$, it would almost always be found in the second, more likely configuration, especially if N were large. In other words, a system free to switch between the two configurations would show properties characteristic almost exclusively of the second configuration.

The next step is to develop an expression for the number of ways that a general configuration $\{N_0,N_1,\ldots\}$ can be achieved. This number is called the **weight** of the configuration and denoted $\mathcal{W}$.

How is that done? 13A.1 Evaluating the weight of a configuration

Consider the number of ways of distributing N balls into bins labelled 0, 1, 2 … such that there are N_0 balls in bin 0, N_1 in bin 1, and so on. The first ball can be selected in N different ways, the next ball in $N-1$ different ways from the balls remaining, and so on. Therefore, there are $N(N-1)\ldots 1 = N!$ ways of selecting the balls.

There are $N_0!$ different ways in which the balls could have been chosen to fill bin 0 (Fig. 13A.1). Similarly, there are $N_1!$ ways in which the balls in bin 1 could have been chosen, and so on. Therefore, the total number of *distinguishable* ways of distributing the balls so that there are N_0 in bin 0, N_1 in bin 1, etc. regardless of the order in which the balls were chosen is

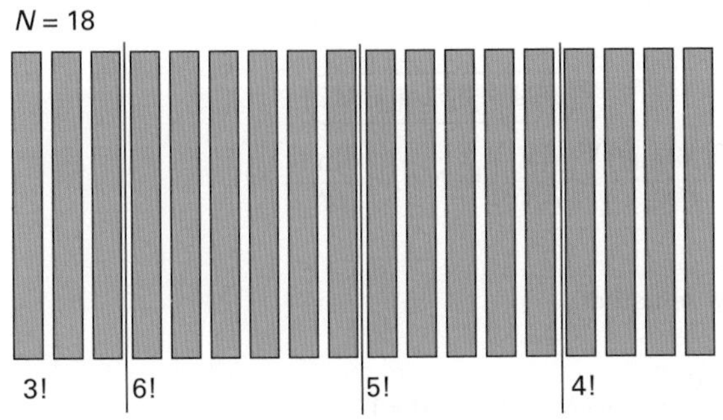

Figure 13A.1 Eighteen molecules (the vertical bars) shown here are distributed into four bins (distinguished by the three vertical lines) such that there are 3 molecules in the first, 6 in the second, and so on. There are 18! ways of selecting the balls to achieve this distribution. However, there are 3! equivalent ways of putting three molecules in the first receptacle, and likewise 6! equivalent ways of putting six molecules into the second receptacle, and so on. Hence the number of *distinguishable* ways of achieving this distribution is 18!/3!6!5!4!, or about 515 million.

$$\mathcal{W}=\frac{N!}{N_0!N_1!N_2!\cdots} \qquad (13A.1)$$

Weight of a configuration

Brief illustration 13A.1

For the configuration $\{1,0,3,5,10,1\}$, which has $N = 20$, the weight is calculated as (recall that $0! \equiv 1$)

$$\mathcal{W}=\frac{20!}{1!0!3!5!10!1!}=9.31\times10^{8}$$

It will turn out to be more convenient to deal with the natural logarithm of the weight, $\ln \mathcal{W}$, rather than with the weight itself:

$$\ln\mathcal{W}=\ln\frac{N!}{N_0!N_1!N_2!\cdots}\overset{\ln(x/y)=\ln x-\ln y}{=}\ln N!-\ln(N_0!N_1!N_2!\cdots)$$

$$\overset{\ln xy=\ln x+\ln y}{=}\ln N!-\ln N_0!-\ln N_1!-\ln N_2!-\cdots=\ln N!-\sum_i \ln N_i!$$

One reason for introducing $\ln \mathcal{W}$ is that it is easier to make approximations. In particular, the factorials can be simplified by using *Stirling's approximation*[1]

$$\ln x! \approx x\ln x - x \qquad \text{Stirling's approximation } [x >> 1] \qquad (13A.2)$$

Then the approximate expression for the weight is

$$\begin{aligned}\ln\mathcal{W}&=(N\ln N-N)-\sum_i(N_i\ln N_i-N_i)\\ &=N\ln N-N-\sum_i N_i\ln N_i+N \quad [\text{because } \textstyle\sum_i N_i=N]\\ &=N\ln N-\sum_i N_i\ln N_i \qquad (13A.3)\end{aligned}$$

13A.1(b) The most probable distribution

The configuration $\{N-2,2,0,\ldots\}$ has much greater weight than $\{N,0,0,\ldots\}$, and it should be easy to believe that there may be other configurations that have a much greater weight than both. In fact for large N there is a configuration for which the weight is overwhelmingly larger than that of any other configuration. The system will almost always be found in this dominating configuration because its weight is so much larger than other possible configurations. The properties of the system will therefore be characteristic of the dominating configuration.

This dominating configuration can be found by looking for the values of N_i that lead to a maximum value of $\mathcal{W}$. Because $\mathcal{W}$ is a function of all the N_i, this search is done by varying the N_i and looking for the values that correspond to $\mathrm{d}\mathcal{W}=0$ just as in the search for the maximum of any function. In practice it

[1] A more precise form of this approximation is $\ln x! \approx \ln(2\pi)^{1/2} + (x+\frac{1}{2})\ln x - x$.

is more convenient, and equivalent, to search for a maximum value of $\ln \mathcal{W}$.

Imagine that the configuration changes so that the population of each state i changes from N_i to $N_i + \mathrm{d}N_i$. Because $\ln \mathcal{W}$ depends on all the N_i when these change by $\mathrm{d}N_i$ the function $\ln \mathcal{W}$ changes to $\ln \mathcal{W} + \mathrm{d}\ln \mathcal{W}$, where

$$\mathrm{d}\ln \mathcal{W} = \sum_i \left(\frac{\partial \ln \mathcal{W}}{\partial N_i}\right) \mathrm{d}N_i \overset{\text{when } \mathcal{W} \text{ is a maximum}}{=} 0 \qquad (13A.4)$$

The derivative $\partial \ln \mathcal{W}/\partial N_i$ expresses how $\ln \mathcal{W}$ changes when N_i changes: if N_i changes by $\mathrm{d}N_i$, then $\ln \mathcal{W}$ changes by $(\partial \ln \mathcal{W}/\partial N_i) \times \mathrm{d}N_i$. The total change in $\ln \mathcal{W}$ is then the sum of all these changes.

Up to now it has been assumed that all the states have the same energy, which means that all the possible configurations have the same total energy. This restriction must now be removed so that the states have different energies ε_0, ε_1,.... as initially supposed. If the system is isolated it has a specified, constant, total energy and only configurations that have this total energy must be retained. It follows that the configuration with the greatest weight must also satisfy the condition

$$\sum_i N_i \varepsilon_i = E \qquad \text{Energy constraint [constant total energy]} \qquad (13A.5a)$$

where E is the total energy of the system. When the N_i change by $\mathrm{d}N_i$ the total energy must not change, so

$$\sum_i \varepsilon_i \mathrm{d}N_i = 0 \qquad \text{Energy constraint} \qquad (13A.5b)$$

The second constraint is that, because the total number of molecules present is also fixed (at N), not all the populations can be varied independently. Thus, increasing the population of one state by 1 demands that the population of another state must be reduced by 1. Therefore, the search for the maximum value of $\mathcal{W}$ is also subject to the condition

$$\sum_i N_i = N \qquad \text{Number constraint [constant total number of molecules]} \qquad (13A.6a)$$

It follows that when the N_i change by $\mathrm{d}N_i$ this sum must not change, so

$$\sum_i \mathrm{d}N_i = 0 \qquad \text{Number constraint} \qquad (13A.6b)$$

The task is to solve eqn 13A.4, that is to maximize the weight, and at the same time conform to the constraints expressed in eqns 13A.5b and 13A.6b. The way to take constraints into account was devised by the mathematician Joseph-Louis Lagrange, and is called the 'method of undetermined multipliers':

- Multiply each constraint by a constant and then add it to the main variation equation (in this case eqn 13A.4).
- Now treat the variables as though they are all independent.
- Evaluate the constants at a later stage of the calculation.

This method is the key to finding the dominant distribution.

How is that done? 13A.2 Identifying the dominant distribution

In preparation for this calculation it is useful to change eqn 13A.3 from

$$\ln \mathcal{W} = N \ln N - \sum_i N_i \ln N_i$$

to

$$\ln \mathcal{W} = N \ln N - \sum_j N_j \ln N_j$$

by using j instead of i as the label of the states. Differentiation of this expression with respect to N_i gives

$$\frac{\partial \ln \mathcal{W}}{\partial N_i} = \overbrace{\frac{\partial (N \ln N)}{\partial N_i}}^{\text{First term}} - \overbrace{\sum_j \frac{\partial (N_j \ln N_j)}{\partial N_i}}^{\text{Second term}}$$

Step 1 *Evaluate the first term in the expression*

The first term on the right is obtained (by using the product rule) as follows:

$$\frac{\partial (N \ln N)}{\partial N_i} \overset{\mathrm{d}fg/\mathrm{d}x = f\mathrm{d}g/\mathrm{d}x + g\mathrm{d}f/\mathrm{d}x}{=} \left(\frac{\partial N}{\partial N_i}\right) \ln N + N\left(\frac{\partial \ln N}{\partial N_i}\right)$$

Now note that

$$\frac{\partial N}{\partial N_i} = \frac{\partial}{\partial N_i}(N_1 + N_2 + \cdots) = 1$$

because N_i will match one and only one term in the sum whatever the value of i. Next, note that

$$\frac{\partial \ln N}{\partial N_i} \overset{\mathrm{d}\ln y/\mathrm{d}x = (1/y)\mathrm{d}y/\mathrm{d}x}{=} \frac{1}{N} \overbrace{\frac{\partial N}{\partial N_i}}^{1} = \frac{1}{N}$$

Therefore

$$\frac{\partial (N \ln N)}{\partial N_i} = \ln N + 1$$

Step 2 *Evaluate the second term*

For the derivative of the second term, first note that

$$\sum_j \frac{\partial(N_j \ln N_j)}{\partial N_i} \overset{\text{d}fg/\text{d}x = f\text{d}g/\text{d}x + g\text{d}f/\text{d}x}{=} \sum_j \left\{ \left(\frac{\partial N_j}{\partial N_i}\right) \ln N_j + N_j \left(\frac{\partial \ln N_j}{\partial N_i}\right) \right\}$$

$$\overset{\text{d}\ln y/\text{d}x = (1/y)\text{d}y/\text{d}x}{=} \sum_j \left\{ \left(\frac{\partial N_j}{\partial N_i}\right) \ln N_j + \frac{N_j}{N_j} \left(\frac{\partial N_j}{\partial N_i}\right) \right\}$$

Now note that the N_j are independent, so the only term that survives in the differentiation $\partial N_j/\partial N_i$ is the one with $j = i$. The derivative of this surviving term is $\partial N_i/\partial N_i = 1$. It follows that

$$\sum_j \frac{\partial(N_j \ln N_j)}{\partial N_i} = \sum_j \left\{ \overbrace{\left(\frac{\partial N_j}{\partial N_i}\right)}^{=1 \text{ if } i=j,\ =0 \text{ otherwise}} \ln N_j + \overbrace{\left(\frac{\partial N_j}{\partial N_i}\right)}^{=1 \text{ if } i=j,\ =0 \text{ otherwise}} \right\}$$
$$= \ln N_i + 1$$

Step 3 *Bring the two terms together*

The result of bringing the first and second terms together is

$$\frac{\partial \ln \mathcal{W}}{\partial N_i} = \ln N + 1 - (\ln N_i + 1)$$

That is,

$$\frac{\partial \ln \mathcal{W}}{\partial N_i} = -\ln \frac{N_i}{N} \tag{13A.7}$$

Step 4 *Introduce the two constraints by using Lagrange's method*

There are two constraints: a number constraint (eqn 13A.6b) and an energy constraint (eqn 13A.5b). Therefore introduce two constants α and $-\beta$ (the negative sign will be helpful later), leading to

$$\text{d}\ln \mathcal{W} = \overbrace{\sum_i \left(\frac{\partial \ln \mathcal{W}}{\partial N_i}\right)\text{d}N_i}^{\text{Main variation equation}} + \alpha \overbrace{\sum_i \text{d}N_i}^{\text{Number constraint}} - \beta \overbrace{\sum_i \varepsilon_i \text{d}N_i}^{\text{Energy constraint}}$$
$$= 0 \text{ at a maximum}$$

Substitute $\partial \ln \mathcal{W}/\partial N_i$ using eqn 13A.7 to give

$$\text{d}\ln \mathcal{W} = \sum_i \left(-\ln \frac{N_i}{N}\right)\text{d}N_i + \alpha \sum_i \text{d}N_i - \beta \sum_i \varepsilon_i \text{d}N_i$$
$$= \sum_i \left\{-\ln \frac{N_i}{N} + \alpha - \beta\varepsilon_i\right\}\text{d}N_i$$
$$= 0 \text{ at a maximum}$$

Step 5 *Treat the variables as independent*

The $\text{d}N_i$ are now treated as independent. Hence, the only way of satisfying $\text{d}\ln \mathcal{W} = 0$ is to require that for each i,

$$-\ln \frac{N_i}{N} + \alpha - \beta\varepsilon_i = 0$$

and therefore that

$$\frac{N_i}{N} = \text{e}^{\alpha - \beta\varepsilon_i} \tag{13A.8}$$

which is very close to being the Boltzmann distribution.

Step 6 *Evaluate the constant α*

The final task is to identify the values of the constants α and β. The first is found by noting that the sum of the populations N_i of the states must be equal to the total number of molecules N:

$$N = \sum_i N_i = \sum_i N\text{e}^{\alpha - \beta\varepsilon_i} = N\text{e}^{\alpha} \sum_i \text{e}^{-\beta\varepsilon_i}$$

Because the N cancels on each side of this equality, it follows that

$$\text{e}^{\alpha} = \frac{1}{\sum_i \text{e}^{-\beta\varepsilon_i}} \tag{13A.9}$$

This expression for e^{α} is used in eqn 13A.8 to give the **Boltzmann distribution**

$$\frac{N_i}{N} = \frac{\text{e}^{-\beta\varepsilon_i}}{\sum_i \text{e}^{-\beta\varepsilon_i}} \tag{13A.10a}$$

Boltzmann distribution

The Boltzmann distribution is commonly written

$$\frac{N_i}{N} = \frac{\text{e}^{-\beta\varepsilon_i}}{q} \tag{13A.10b}$$

Boltzmann distribution

where q is called the **partition function**:

$$q = \sum_i \text{e}^{-\beta\varepsilon_i} \tag{13A.11}$$

Partition function [definition]

At this stage the partition function is a convenient abbreviation for the sum, but in Topic 13B it is shown that the partition function is central to the statistical interpretation of thermodynamic properties.

Equation 13A.10a shows that the most probable populations of a given set of states with particular energies is governed by the value of a single parameter, here β. The formal deduction of the value of β depends on using the Boltzmann distribution to deduce the perfect gas equation of state (that is done in Topic 13F, and specifically in *Example* 13F.1). The result is

$$\beta = \frac{1}{kT} \tag{13A.12}$$

where T is the thermodynamic temperature and k is Boltzmann's constant. In other words:

The temperature is the unique parameter that governs the most probable populations of states of a system at thermal equilibrium.

Brief illustration 13A.2

Suppose that two conformations of a molecule differ in energy by 5.0 kJ mol^{-1} (corresponding to 8.3 zJ for a single molecule; 1 zJ = 10^{-21} J). There are therefore two states: the ground state, conformation A, with energy 0 and the higher energy state, conformation B, with energy $\varepsilon = 8.3$ zJ. At 20 °C (293 K) the denominator in eqn 13A.10a is

$$\sum_i \mathrm{e}^{-\beta\varepsilon_i} = 1 + \mathrm{e}^{-\varepsilon/kT} = 1 + \mathrm{e}^{-(8.3\times10^{-21}\,\mathrm{J})/(1.381\times10^{-23}\,\mathrm{J\,K^{-1}})\times(293\,\mathrm{K})} = 1.12\ldots$$

The proportion of molecules in conformation B at this temperature is therefore

$$\frac{N_\mathrm{B}}{N} = \frac{\mathrm{e}^{-(8.3\times10^{-21}\,\mathrm{J})/(1.381\times10^{-23}\,\mathrm{J\,K^{-1}})\times(293\,\mathrm{K})}}{1.12\ldots} = 0.11$$

or 11 per cent of the molecules.

Exercises

E13A.1 Calculate the weight of the configuration in which 16 objects are distributed in the arrangement 0, 1, 2, 3, 8, 0, 0, 0, 0, 2.

E13A.2 What are the relative populations of the states of a two-level system when the temperature is infinite?

13A.2 The relative populations of states

If only the relative populations of states are of interest the partition function in eqn 13A.10b need not be evaluated because it cancels when the ratio is taken:

$$\frac{N_i}{N_j} = \frac{\mathrm{e}^{-\beta\varepsilon_i}}{q}\frac{q}{\mathrm{e}^{-\beta\varepsilon_j}} = \frac{\mathrm{e}^{-\beta\varepsilon_i}}{\mathrm{e}^{-\beta\varepsilon_j}} = \mathrm{e}^{-\beta(\varepsilon_i-\varepsilon_j)} \qquad \text{Boltzmann population ratio} \qquad (13A.13a)$$

Note that for a given energy separation the ratio of populations N_1/N_0 decreases as β increases (and the temperature decreases). At $T = 0$ ($\beta = \infty$) all the population is in the ground state and the ratio is zero. Equation 13A.13a is enormously important for understanding a wide range of chemical phenomena and is the form in which the Boltzmann distribution is commonly employed (for instance, in the discussion of the intensities of spectral transitions, Topic 11A). It implies that the relative population of two states decreases exponentially with their difference in energy.

A very important point to note is that the Boltzmann distribution gives the relative populations of *states*, not energy *levels*. Several states might have the same energy, and each state has a relative population given by eqn 13A.13a. When calculating the relative populations of energy levels rather than states, it is necessary to take into account this degeneracy. Thus, if the level of energy ε_i is g_i-fold degenerate (in the sense that there are g_i states with that energy), and the level of energy ε_j is g_j-fold degenerate, then the relative total populations of the levels are given by

$$\frac{N_i}{N_j} = \frac{g_i\mathrm{e}^{-\beta\varepsilon_i}}{g_j\mathrm{e}^{-\beta\varepsilon_j}} = \frac{g_i}{g_j}\mathrm{e}^{-\beta(\varepsilon_i-\varepsilon_j)} \qquad \text{Boltzmann population ratio of levels} \qquad (13A.13b)$$

Example 13A.1 Calculating the relative populations of rotational levels

Calculate the relative populations of the $J = 1$ and $J = 0$ rotational levels of HCl at 25 °C; for HCl, $\tilde{B} = 10.591$ cm^{-1}.

Collect your thoughts Although the ground state is non-degenerate, you need to note that the level with $J = 1$ is triply degenerate ($M_J = 0, \pm1$); the energy of the state with quantum number J is $\varepsilon_J = hc\tilde{B}J(J+1)$ (Topic 11B). A useful relation is $kT/hc = 207.225$ cm^{-1} at 298.15 K.

The solution The energy separation of levels with $J = 1$ and $J = 0$ is $\varepsilon_1 - \varepsilon_0 = 2hc\tilde{B}$. The ratio of the population of these two levels is given by eqn 13A.13b

$$\frac{N_1}{N_0} = \frac{g_1}{g_0}\mathrm{e}^{-\beta(\varepsilon_1-\varepsilon_0)}$$

Here $g_0 = 1$ and $g_1 = 3$ hence

$$\frac{N_1}{N_0} = 3\mathrm{e}^{-2hc\tilde{B}\beta}$$

Insertion of $hc\tilde{B}\beta = hc\tilde{B}/kT = (10.591\,\mathrm{cm}^{-1})/(207.225\,\mathrm{cm}^{-1}) = 0.0511\ldots$ then gives

$$\frac{N_1}{N_0} = 3\mathrm{e}^{-2\times0.0511\ldots} = 2.708$$

Comment. Because the $J = 1$ level is triply degenerate, it has a higher population than the level with $J = 0$, despite being of higher energy.

Self-test 13A.1 What is the ratio of the populations of the levels with $J = 2$ and $J = 1$ of HCl at the same temperature?

Answer: 1.359

Exercises

E13A.3 What is the temperature of a two-level system of energy separation equivalent to $400\ \text{cm}^{-1}$ when the population of the upper state is one-third that of the lower state?

E13A.4 A certain molecule has a non-degenerate excited state lying at $540\ \text{cm}^{-1}$ above the non-degenerate ground state. At what temperature will 10 per cent of the molecules be in the upper state?

E13A.5 Calculate the relative populations of a linear rotor at 298 K in the levels with $J = 0$ and $J = 5$, given that $\tilde{B} = 2.71\ \text{cm}^{-1}$.

Checklist of concepts

- ☐ 1. The **principle of equal a priori probabilities** assumes that all possibilities for the distribution of energy are equally probable in the sense that the distribution is blind to the type of motion involved.
- ☐ 2. The **configuration** of a system of N molecules is the specification of the set of populations $N_0, N_1, \ldots$ of the energy states $\varepsilon_0, \varepsilon_1, \ldots$.
- ☐ 3. The **Boltzmann distribution** gives the numbers of molecules in each state of a system at any temperature.
- ☐ 4. The **relative populations** of energy levels, as opposed to states, must take into account the degeneracies of the energy levels.

Checklist of equations

Property	Equation	Comment	Equation number
Boltzmann distribution	$N_i/N = \text{e}^{-\beta\varepsilon_i}/q$	$\beta = 1/kT$	13A.10b
Partition function	$q = \sum_i \text{e}^{-\beta\varepsilon_i}$	See Topic 13B	13A.11
Boltzmann population ratio	$N_i/N_j = (g_i/g_j)\text{e}^{-\beta(\varepsilon_i-\varepsilon_j)}$	g_i, g_j are degeneracies	13A.13b

TOPIC 13B Molecular partition functions

> ➤ **Why do you need to know this material?**
>
> Through the partition function, statistical thermodynamics provides the link between thermodynamic data and molecular properties that have been calculated or derived from spectroscopy. Therefore, this material is an essential foundation for understanding physical and chemical properties of bulk matter in terms of the properties of the constituent molecules.
>
> ➤ **What is the key idea?**
>
> The partition function is calculated by drawing on computed or spectroscopically derived structural information about molecules.
>
> ➤ **What do you need to know already?**
>
> You need to know that the Boltzmann distribution expresses the most probable distribution of molecules over the available states (Topic 13A). You need to be aware of the expressions for the rotational and vibrational levels of molecules (Topics 11B–11D) and the energy levels of a particle in a box (Topic 7D).

The partition function $q = \sum_i e^{-\beta\varepsilon_i}$ is introduced in Topic 13A simply as a symbol to denote the sum over states that occurs in the denominator of the Boltzmann distribution (eqn 13A.10b, $p_i = e^{-\beta\varepsilon_i}/q$, with $p_i = N_i/N$). But it is far more important than that might suggest. For instance, it contains all the information needed to calculate the bulk properties of a system of independent (that is, non-interacting) particles. In this respect q plays a role for bulk matter very similar to that played by the wavefunction in quantum mechanics for individual molecules: q is a kind of thermal wavefunction.

13B.1 The significance of the partition function

The **molecular partition function** is

$$q = \sum_{\text{states } i} e^{-\beta\varepsilon_i} \qquad \text{Molecular partition function [definition]} \qquad (13B.1a)$$

where $\beta = 1/kT$. It is called the *molecular* partition function as the sum is over the states of a single molecule. As emphasized in Topic 13A, the sum is over *states*, not energy *levels*. If g_i states have the same energy ε_i (so the level is g_i-fold degenerate), then

$$q = \sum_{\text{levels } i} g_i e^{-\beta\varepsilon_i} \qquad \text{Molecular partition function [alternative definition]} \qquad (13B.1b)$$

where the sum is now over energy levels (sets of states with the same energy), not individual states. Also as emphasized in Topic 13A, the lowest available state is taken as the zero of energy, so $\varepsilon_0 \equiv 0$.

The significance of the partition function is best illustrated by calculating it for a particular set of states. Suppose a molecule has available a set of states with energies 0, ε, 2ε,... (Fig. 13B.1). The molecular partition function is

$$q = 1 + e^{-\beta\varepsilon} + e^{-2\beta\varepsilon} + \cdots = 1 + e^{-\beta\varepsilon} + (e^{-\beta\varepsilon})^2 + \cdots$$

which is the geometric series $1 + x + x^2 + \cdots$ with $x = e^{-\beta\varepsilon}$. Provided $|x| < 1$ the sum to infinity of this series is $1/(1-x)$. For any finite temperature it is indeed the case that $|x| < 1$, so the series evaluates to

$$q = \frac{1}{1 - e^{-\beta\varepsilon}} \qquad \text{Partition function [equally spaced states]} \qquad (13B.2a)$$

Figure 13B.2 shows a plot of this function against the dimensionless quantity $1/\beta\varepsilon = kT/\varepsilon$, which is proportional to temperature. At low temperatures, in the sense that $kT/\varepsilon << 1$, the partition function tends to 1. The value of the partition

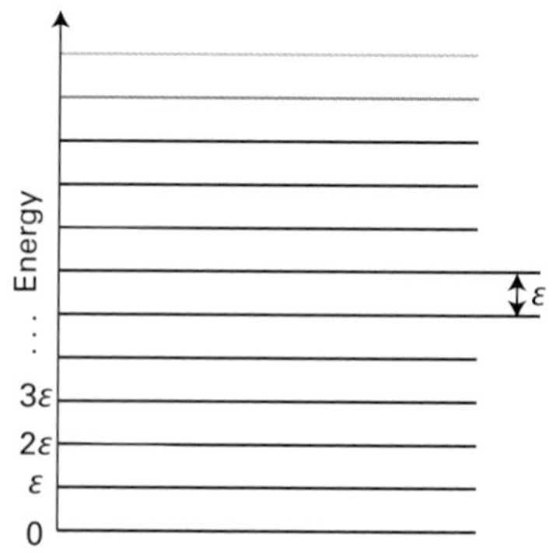

Figure 13B.1 The infinite set of states with equal spacing ε used in the calculation of the partition function in *Brief illustration* 13B.1. A harmonic oscillator has the same array of non-degenerate levels.

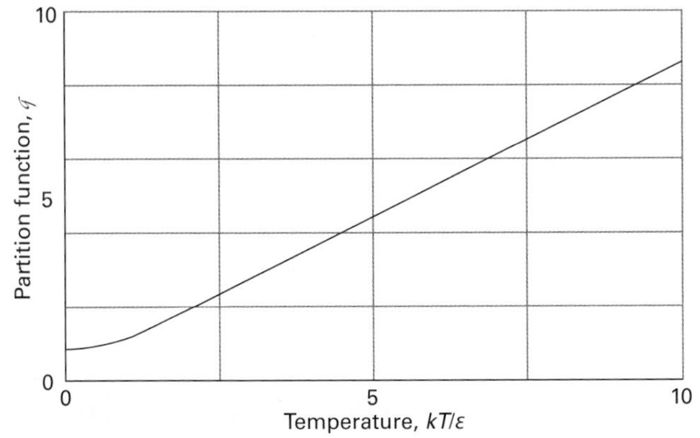

Figure 13B.2 The partition function for the system shown in Fig. 13B.1 (a harmonic oscillator) as a function of temperature.

function starts to rise significantly above 1 when $kT/\varepsilon \approx 1$, and then continues to rise indefinitely. At high temperatures, in the sense that $kT/\varepsilon >> 1$, the partition function increases linearly with the temperature.

The physical significance of a partition function can be discussed using the expression for its value for an equally spaced set of states, eqn 13B.2a. To do so, first note that the Boltzmann distribution gives the fraction, $p_i = N_i/N$, of molecules in the state with energy ε_i as $p_i = e^{-\beta\varepsilon_i}/q$. For an equally spaced set of states the partition function is given by eqn 13B.2a and the energies ε_i are given by $\varepsilon_i = i\varepsilon$. Therefore, the fractional populations are

$$p_i = \frac{e^{-\beta\varepsilon_i}}{q} = (1-e^{-\beta\varepsilon})e^{-i\beta\varepsilon} \qquad (13B.2b)$$

Figure 13B.3 shows how p_i varies with temperature. At very low temperatures ($kT/\varepsilon << 1$ or $\beta\varepsilon >> 1$), where q is close to 1, only the lowest state is significantly populated. As the temperature is raised, the population breaks out of the lowest state, and the upper states become progressively more highly populated. At the same time, the partition function rises from 1, so its value gives an indication of the range of states populated at any given temperature. The name 'partition function' reflects the sense in which q measures how the total number of molecules is distributed—partitioned—over the available states.

States that are significantly populated at a particular temperature are said to be **thermally accessible**. These are the states that contribute significantly to the sum used to compute the partition function.

The corresponding expressions for a system in which there are just two states, with energies $\varepsilon_0 = 0$ and $\varepsilon_1 = \varepsilon$ (a 'two-level system'), are

$$q = 1 + e^{-\beta\varepsilon} \qquad \text{Partition function [two-level system]} \qquad (13B.3a)$$

and

$$p_i = \frac{e^{-\beta\varepsilon_i}}{q} = \frac{e^{-\beta\varepsilon_i}}{1+e^{-\beta\varepsilon}} \qquad (13B.3b)$$

The fractional populations of the two states are therefore

$$p_0 = \frac{1}{1+e^{-\beta\varepsilon}} \qquad p_1 = \frac{e^{-\beta\varepsilon}}{1+e^{-\beta\varepsilon}} \qquad (13B.4)$$

Figure 13B.4 shows the variation of the partition function with temperature and Fig. 13B.5 shows how the fractional populations change. Notice that at very low temperatures, when $kT/\varepsilon << 1$, the fractional populations tend to $p_0 = 1$ and $p_1 = 0$, and the partition function tends to $q = 1$ (one state occupied). However, the fractional populations tend towards equality $\left(p_0 = \frac{1}{2}, p_1 = \frac{1}{2}\right)$ and $q = 2$ (two states occupied) at very high temperatures (when $kT/\varepsilon >> 1$).

A common error is to suppose that at very high temperatures, when $kT/\varepsilon >> 1$, all the molecules in the system will be found in the upper energy state. However, as seen from eqn 13B.4, in this limit the populations of states become equal. The same

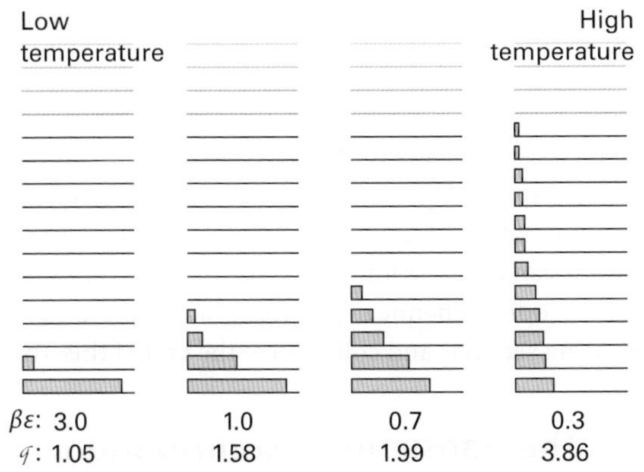

Figure 13B.3 The populations of the set states shown in Fig. 13B.1 at different temperatures, and the corresponding values of the partition function calculated from eqn 13B.2a. Note that $\beta = 1/kT$.

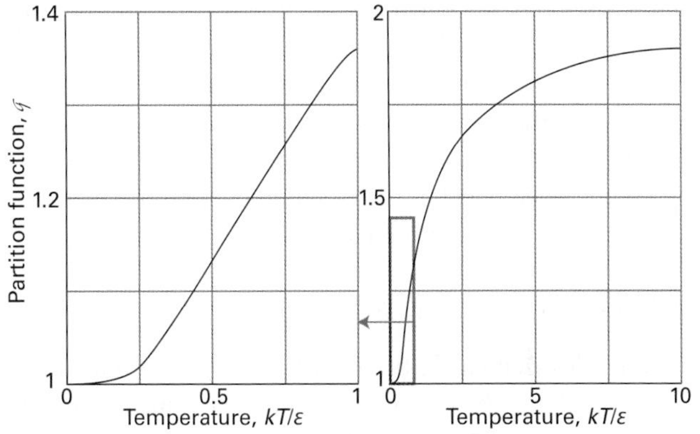

Figure 13B.4 The dependence of the partition function of a two-level system on temperature. The two graphs differ in the scale of the temperature axis to show the approach to 1 as $T \to 0$ and the slow approach to 2 as $T \to \infty$.

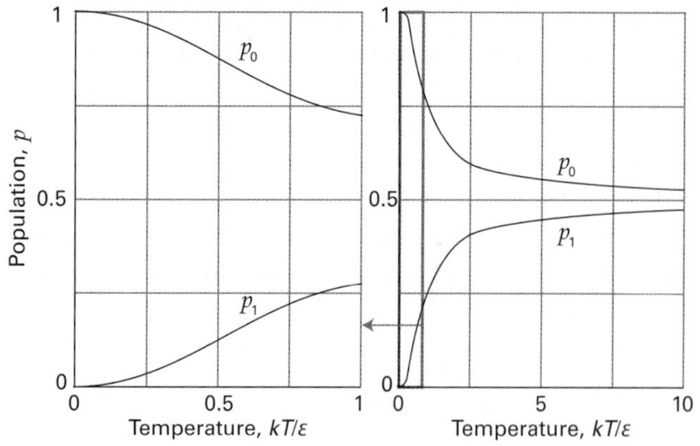

Figure 13B.5 The variation with temperature of the fractional populations of the two states of a two-level system (eqn 13B.4). Note that as the temperature approaches infinity, the populations of the two states become equal (and the fractional populations both approach 0.5).

conclusion is also true of many-level systems: if the temperature is sufficiently high, all states become equally populated.

Now consider the general case of a system with an infinite number of energy levels. The partition function is defined in eqn 13B.1b as

$$q = \sum_{\text{levels } i} g_i e^{-\beta\varepsilon_i} = g_0 + g_1 e^{-\beta\varepsilon_1} + g_2 e^{-\beta\varepsilon_2} + \cdots$$

where, as usual, the energy of the ground level is taken as zero. Now consider the limit as T approaches zero and therefore as the parameter $\beta = 1/kT$ approaches infinity. Apart from the g_0, all the terms in the expression above have the form e^{-x} and therefore in the limit of low temperature $x \to \infty$, so these terms all go to zero. The partition function therefore tends to g_0:

$$\lim_{T\to 0} q = g_0$$

That is, at $T = 0$, the partition function is equal to the degeneracy of the ground level (commonly, but not necessarily, 1).

When T is so high that for each term in the sum $\beta\varepsilon_i = \varepsilon_i/kT \approx 0$, each term in the sum now contributes g_i because $e^{-x} = 1$ when $x = 0$. It follows that the sum is equal to the sum of the degeneracies of the levels which is the total number of molecular states. In general there are an infinite number of states, so

$$\lim_{T\to\infty} q = \infty$$

In some idealized cases, the molecule may have only a finite number of states; then the upper limit of q is equal to the number of states (as for the two-level system). In summary,

> The molecular partition function gives an indication of the number of states that are thermally accessible to a molecule at the temperature of the system.

Exercises

E13B.1 Consider a system with an infinite number of doubly degenerate energy levels, evenly spaced by ε. Derive an expression for the partition function and describe its behaviour as a function of temperature. How does the behaviour compare with a system of evenly spaced non-degenerate levels?

E13B.2 Consider a two-level system in which the ground level is non-degenerate and has energy zero, and the upper level is doubly degenerate and has energy ε. At what temperature will the populations of the two levels be equal?

13B.2 Contributions to the partition function

The energy of an isolated molecule is the sum of contributions from its different modes of motion:

$$\varepsilon_i = \varepsilon_i^{\mathrm{T}} + \varepsilon_i^{\mathrm{R}} + \varepsilon_i^{\mathrm{V}} + \varepsilon_i^{\mathrm{E}} \qquad (13B.5)$$

where T denotes translation, R rotation, V vibration, and E the electronic contribution. The electronic contribution is not actually a 'mode of motion', but it is convenient to include it here. The separation of terms in eqn 13B.5 is only approximate (except for translation) because the modes are not completely independent, but in most cases the separation is acceptable.

Given that the energy is a sum of independent contributions, the partition function is a product of contributions:

$$q = \sum_i e^{-\beta\varepsilon_i} = \sum_{i(\text{all states})} e^{-\beta\varepsilon_i^{\mathrm{T}} - \beta\varepsilon_i^{\mathrm{R}} - \beta\varepsilon_i^{\mathrm{V}} - \beta\varepsilon_i^{\mathrm{E}}}$$

$$= \sum_{i(\text{translational})} \sum_{i(\text{rotational})} \sum_{i(\text{vibrational})} \sum_{i(\text{electronic})} e^{-\beta\varepsilon_i^{\mathrm{T}} - \beta\varepsilon_i^{\mathrm{R}} - \beta\varepsilon_i^{\mathrm{V}} - \beta\varepsilon_i^{\mathrm{E}}}$$

$$= \left(\sum_{i(\text{translational})} e^{-\beta\varepsilon_i^{\mathrm{T}}}\right)\left(\sum_{i(\text{rotational})} e^{-\beta\varepsilon_i^{\mathrm{R}}}\right)\left(\sum_{i(\text{vibrational})} e^{-\beta\varepsilon_i^{\mathrm{V}}}\right)$$

$$\times \left(\sum_{i(\text{electronic})} e^{-\beta\varepsilon_i^{\mathrm{E}}}\right)$$

That is,

$$q = q^{\mathrm{T}} q^{\mathrm{R}} q^{\mathrm{V}} q^{\mathrm{E}} \qquad \text{Factorization of the partition function} \qquad (13B.6)$$

This factorization means that each contribution can be investigated separately. In general, exact analytical expressions for partition functions cannot be obtained. However, approximate expressions can often be found and prove to be very important for understanding chemical phenomena; they are derived in the following sections and collected at the end of this Topic.

13B.2(a) The translational contribution

The translational partition function for a particle of mass m free to move in a one-dimensional container of length X can be evaluated by making use of the fact that the separation of energy levels is very small and that large numbers of states are accessible at normal temperatures.

How is that done? 13B.1 Deriving an expression for the translational partition function

The starting point of the derivation is eqn 7D.6 ($E_n = n^2h^2/8mL^2$) for the energy levels of a particle in a one-dimensional box. These levels are not degenerate and the quantum number n takes values 1, 2, 3 … For a molecule of mass m in a container of length X it follows that:

$$E_n = \frac{n^2h^2}{8mX^2}$$

Step 1 *Write an expression for the sum in eqn 13B.1a*

The lowest level ($n = 1$) has energy $h^2/8mX^2$, so the energies relative to that level are

$$\varepsilon_n = (n^2 - 1)\varepsilon, \quad \varepsilon = h^2/8mX^2$$

The sum to evaluate is therefore

$$q_X^{\mathrm{T}} = \sum_{n=1}^{\infty} \mathrm{e}^{-(n^2-1)\beta\varepsilon}$$

Step 2 *Make some approximations and convert the sum to an integral*

For a molecule in a container the size of a typical laboratory vessel, and at any reasonable temperature, there are an enormous number of energy levels with energies less than or equal to kT. Therefore, many terms contribute to the sum and the partition function will be very large. Negligible error is introduced by assuming that the lowest energy level has $n = 0$ because that simply adds one additional state to the very large number that contribute to the partition function. With this change the expression for the partition function becomes

$$q_X^{\mathrm{T}} \approx \sum_{n=0}^{\infty} \mathrm{e}^{-n^2\beta\varepsilon}$$

Many terms contribute to the sum, so it is well approximated by an integral

$$q_X^{\mathrm{T}} \approx \int_0^{\infty} \mathrm{e}^{-n^2\beta\varepsilon}\mathrm{d}n$$

The integral is now in a standard form.

Step 3 *Evaluate the integral*

Make the substitution $x^2 = n^2\beta\varepsilon$, implying $n = x/(\beta\varepsilon)^{1/2}$ and hence $\mathrm{d}n = \mathrm{d}x/(\beta\varepsilon)^{1/2}$. It therefore follows that

$$q_X^{\mathrm{T}} = \left(\frac{1}{\beta\varepsilon}\right)^{1/2} \overbrace{\int_0^{\infty} \mathrm{e}^{-x^2}\mathrm{d}x}^{\text{Integral G.1: } \pi^{1/2}/2} = \left(\frac{1}{\beta\varepsilon}\right)^{1/2} \frac{\pi^{1/2}}{2} \overset{\varepsilon = h^2/8mX^2}{=} \left(\frac{2\pi m}{h^2\beta}\right)^{1/2} X$$

With $\beta = 1/kT$ this relation has the form

$$q_X^{\mathrm{T}} = \frac{X}{\Lambda} \quad \Lambda = \frac{h}{(2\pi mkT)^{1/2}} \qquad (13\text{B}.7)$$

Translational partition function

The quantity Λ (uppercase lambda) has the dimensions of length and is called the **thermal wavelength** (sometimes the 'thermal de Broglie wavelength') of the molecule. The thermal wavelength decreases with increasing mass and temperature. Equation 13B.7 shows that:

- The partition function for translational motion increases with the length of the box and the mass of the particle: in each case the separation of the energy levels becomes smaller and more levels become thermally accessible.
- For a given mass and length of the box, the partition function also increases with increasing temperature (decreasing β), because more states become accessible.

The total energy of a molecule free to move in three dimensions is the sum of its translational energies in all three directions:

$$\varepsilon_{n_1n_2n_3} = \varepsilon_{n_1}^{(X)} + \varepsilon_{n_2}^{(Y)} + \varepsilon_{n_3}^{(Z)} \qquad (13\text{B}.8)$$

where n_1, n_2, and n_3 are the quantum numbers for motion in the x-, y-, and z-directions, respectively. Therefore, because $\mathrm{e}^{a+b+c} = \mathrm{e}^a\mathrm{e}^b\mathrm{e}^c$, the partition function factorizes as follows:

$$q^{\mathrm{T}} = \sum_{\text{all } n} \mathrm{e}^{-\beta\varepsilon_{n_1}^{(X)} - \beta\varepsilon_{n_2}^{(Y)} - \beta\varepsilon_{n_3}^{(Z)}} = \sum_{\text{all } n} \mathrm{e}^{-\beta\varepsilon_{n_1}^{(X)}} \mathrm{e}^{-\beta\varepsilon_{n_2}^{(Y)}} \mathrm{e}^{-\beta\varepsilon_{n_3}^{(Z)}}$$
$$= \left(\sum_{n_1} \mathrm{e}^{-\beta\varepsilon_{n_1}^{(X)}}\right)\left(\sum_{n_2} \mathrm{e}^{-\beta\varepsilon_{n_2}^{(Y)}}\right)\left(\sum_{n_3} \mathrm{e}^{-\beta\varepsilon_{n_3}^{(Z)}}\right)$$

That is,

$$q^{\mathrm{T}} = q_X^{\mathrm{T}} q_Y^{\mathrm{T}} q_Z^{\mathrm{T}} \qquad (13\text{B}.9)$$

Equation 13B.7 gives the partition function for translational motion in the x-direction. The only change for the other two directions is to replace the length X by the lengths Y or Z. Hence the partition function for motion in three dimensions is

$$q^{\mathrm{T}} = \left(\frac{2\pi m}{h^2\beta}\right)^{3/2} XYZ = \frac{(2\pi mkT)^{3/2}}{h^3} XYZ \qquad (13\text{B}.10\text{a})$$

The product of lengths XYZ is the volume, V, of the container, so

$$q^{\mathrm{T}} = \frac{V}{\Lambda^3} \qquad (13\text{B}.10\text{b})$$

Translational partition function [three-dimensional]

with Λ defined in eqn 13B.7. As in the one-dimensional case, the partition function increases with the mass of the particle (as $m^{3/2}$) and the volume of the container (as V); for a given mass and volume, the partition function increases with temperature (as $T^{3/2}$). At room temperature, $q^{\mathrm{T}} \approx 2 \times 10^{28}$ for an O_2 molecule in a vessel of volume 100 cm^3. As anticipated above, the number of thermally accessible states is very large indeed, so the approximations made in deriving the expression for the partition function are justified.

Brief illustration 13B.1

To calculate the translational partition function of an H_2 molecule confined to a $100\,\text{cm}^3$ vessel at 25 °C, use $m = 2.016m_u$. Then, from $\Lambda = h/(2\pi mkT)^{1/2}$,

$$\Lambda = \frac{6.626\times10^{-34}\,\overbrace{\text{J}}^{\text{kg m}^2\,\text{s}^{-2}}\text{s}}{\{2\pi\times(2.016\times1.6605\times10^{-27}\,\text{kg})\times(1.381\times10^{-23}\,\underbrace{\text{J}}_{\text{kg m}^2\,\text{s}^{-2}}\text{K}^{-1})\times(298\,\text{K})\}^{1/2}}$$

$$= 7.12\ldots\times10^{-11}\,\text{m}$$

Therefore,

$$q^{\text{T}} = \frac{1.00\times10^{-4}\,\text{m}^3}{(7.12\ldots\times10^{-11}\,\text{m})^3} = 2.77\times10^{26}$$

About 10^{26} quantum states are thermally accessible, even at room temperature and for this light molecule. Many states are occupied if the thermal wavelength (which in this case is 71.2 pm) is small compared with the linear dimensions of the container.

Equation 13B.10b can be interpreted in terms of the average separation, d, of the particles in the container. Because q is the total number of accessible states, the average number of translational states per molecule is q^{T}/N. For this quantity to be large, the condition $V/N\Lambda^3 >> 1$ must be met. However, V/N is the volume occupied by a single particle, and therefore the average separation of the particles is $d = (V/N)^{1/3}$. The condition for there being many states available per molecule is therefore $d^3/\Lambda^3 >> 1$, and therefore $d >> \Lambda$. That is, for eqn 13B.10b to be valid, *the average separation of the particles must be much greater than their thermal wavelength*. For 1 mol H_2 molecules at 1 bar and 298 K, the average separation is 3 nm, which is significantly larger than their thermal wavelength (71.2 pm).

The validity of eqn 13B.10b can be expressed in a different way by noting that the approximations that led to it are valid if many states are occupied, which requires V/Λ^3 to be large. That will be so if Λ is small compared with the linear dimensions of the container. For H_2 at 298 K, $\Lambda = 71$ pm, which is far smaller than any conventional container is likely to be (but comparable to pores in zeolites or cavities in clathrates). For O_2, a heavier molecule, $\Lambda = 18$ pm.

13B.2(b) The rotational contribution

The energy levels of a linear rotor are $\varepsilon_J = hc\tilde{B}J(J+1)$, with $J = 0, 1, 2, \ldots$ (Topic 11B). The state of lowest energy has zero energy, so no adjustment need be made to the energies given by this expression. Each level consists of $2J + 1$ degenerate states. Therefore, the partition function of a non-symmetrical (AB) linear rotor is computed using eqn 13B.1b as a sum over levels

$$q^{\text{R}} = \sum_J \overbrace{(2J+1)}^{g_J}\text{e}^{-\overbrace{\beta hc\tilde{B}J(J+1)}^{\varepsilon_J}} \qquad (13\text{B}.11)$$

The direct method of calculating q^{R} is to substitute the experimental values of the rotational energy levels into this expression and to sum the series numerically. (The case of symmetrical A_2 molecules is dealt with later.)

Example 13B.1 Evaluating the rotational partition function explicitly

Evaluate the rotational partition function of $^1H^{35}Cl$ at 25 °C, given that $\tilde{B} = 10.591\,\text{cm}^{-1}$.

Collect your thoughts You need to evaluate eqn 13B.11 term by term, and it is convenient to use $kT/hc = 207.225\,\text{cm}^{-1}$ at 298.15 K. The sum is readily evaluated by using mathematical software.

The solution To show how successive terms contribute, draw up the following table by using $hc\tilde{B}/kT = 0.05111$ (Fig. 13B.6):

J	0	1	2	3	4	…	10
$(2J+1)\text{e}^{-0.05111J(J+1)}$	1	2.708	3.680	3.791	3.238	…	0.076

The sum required by eqn 13B.11 (the sum of the numbers in the second row of the table) is 19.9, hence $q^{\text{R}} = 19.9$ at this temperature. Taking J up to 15 gives $q^{\text{R}} = 19.902$; adding further terms results in no further change to the value of q^{R} at this precision.

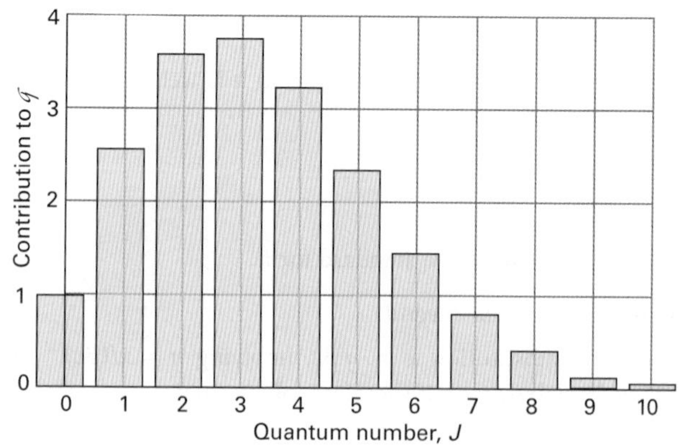

Figure 13B.6 The contributions to the rotational partition function of an HCl molecule at 25 °C. The vertical axis is the value of $(2J+1)\text{e}^{-\beta hc\tilde{B}J(J+1)}$. Successive terms (which are proportional to the populations of the levels) pass through a maximum because the population of individual states decreases exponentially, but the degeneracy of the levels increases with J.

Comment. Notice that about ten J-levels are significantly populated but the number of populated *states* is larger on account of the $(2J+1)$-fold degeneracy of each level.

Self-test 13B.1 Evaluate the rotational partition function for $^1H^{35}Cl$ at 0 °C.

Answer: 18.26

At room temperature, $kT/hc \approx 200\ \text{cm}^{-1}$. The rotational constants of many molecules are close to $1\ \text{cm}^{-1}$ (Table 11C.1) and often smaller (though the very light H_2 molecule, for which $\tilde{B} = 60.9\ \text{cm}^{-1}$, is one important exception). It follows that many rotational levels are populated at normal temperatures. When that is the case, explicit expressions for the rotational partition function can be derived.

Consider a linear rotor. If $kT >> hc\tilde{B}$ many rotational states are occupied the sum that defines the partition function can be approximated by an integral:

$$q^{R} = \int_0^{\infty} (2J+1)e^{-\beta hc\tilde{B}J(J+1)}\text{d}J$$

This integral can be evaluated without much effort by making the substitution $x = \beta hc\tilde{B}J(J+1)$, so that $\text{d}x/\text{d}J = \beta hc\tilde{B}(2J+1)$ and therefore $(2J+1)\text{d}J = \text{d}x/\beta hc\tilde{B}$. Then

$$q^{R} = \frac{1}{\beta hc\tilde{B}} \overbrace{\int_0^{\infty} e^{-x}\text{d}x}^{\text{Integral E.1: } 1} = \frac{1}{\beta hc\tilde{B}}$$

which (because $\beta = 1/kT$) is

$$q^{R} = \frac{kT}{hc\tilde{B}} \qquad \text{Rotational partition function [unsymmetrical linear molecule]} \qquad (13B.12a)$$

Brief illustration 13B.2

For $^1H^{35}Cl$ at 298.15 K, use $kT/hc = 207.225\ \text{cm}^{-1}$ and $\tilde{B} = 10.591\ \text{cm}^{-1}$. Then

$$q^{R} = \frac{kT}{hc\tilde{B}} = \frac{207.225\ \text{cm}^{-1}}{10.591\ \text{cm}^{-1}} = 19.57$$

The value is in good agreement with the exact value (19.902) and obtained with much less effort.

A similar but more elaborate approach can be used to find an expression for the rotational partition function of nonlinear molecules.

How is that done? 13B.2 Deriving an expression for the rotational partition function of a nonlinear molecule

Consider a symmetric rotor (Topic 11B) for which the energy levels are

$$E_{J,K,M_J} = hc\tilde{B}J(J+1) + hc(\tilde{A}-\tilde{B})K^2$$

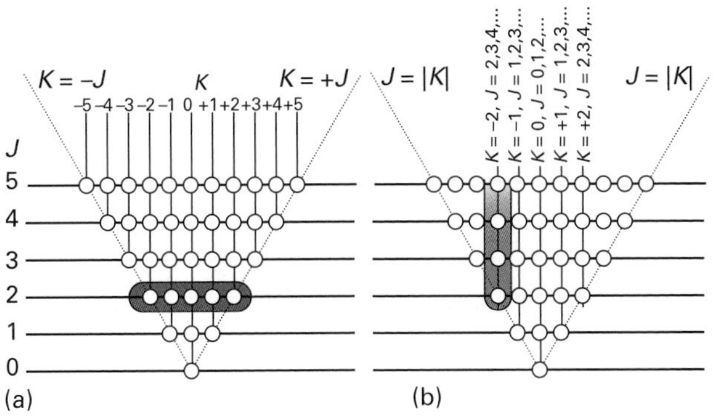

Figure 13B.7 The calculation of the rotational partition function includes a contribution, indicated by the circles, for all possible combinations of J and K. The sum is formed either (a) by allowing J to take the values 0, 1, 2,... and then for each J allowing K to range from J to $-J$, or (b) by allowing K to range from $-\infty$ to ∞, and for each value of K allowing J to take the values $|K|, |K|+1,\ldots, \infty$. The tinted areas show in (a) the sum from $K=-2$ to $+2$ for $J=2$, and in (b) the sum with $J = 2, 3 \ldots$ for $K=-2$.

with $J = 0, 1, 2,\ldots, K = J, J-1,\ldots,-J$, and $M_J = J, J-1,\ldots, -J$. Instead of considering these ranges, the same values can be covered by allowing K to range from $-\infty$ to ∞, with J confined to $|K|, |K|+1,\ldots, \infty$ for each value of K (Fig. 13B.7).

Step 1 *Write an expression for the sum over energy states*

Because the energy is independent of M_J, and there are $2J+1$ values of M_J for each value of J, each value of J is $(2J+1)$-fold degenerate. It follows that the partition function

$$q = \sum_{J=0}^{\infty}\sum_{K=-J}^{J}\sum_{M_J=-J}^{J} e^{-\beta E_{J,K,M_J}}$$

can be written equivalently as

$$q = \sum_{J=0}^{\infty}\sum_{K=-J}^{J}(2J+1)\, e^{-\beta E_{J,K,M_J}} = \sum_{K=-\infty}^{\infty}\sum_{J=|K|}^{\infty}(2J+1)\, e^{-\beta E_{J,K,M_J}}$$
$$= \sum_{K=-\infty}^{\infty} e^{-hc\beta(\tilde{A}-\tilde{B})K^2}\sum_{J=|K|}^{\infty}(2J+1)\, e^{-hc\beta\tilde{B}J(J+1)}$$

Step 2 *Convert the sums to integrals*

As for linear molecules, assume that the temperature is so high that numerous states are occupied, in which case the sums may be approximated by integrals. Then

$$q = \int_{-\infty}^{\infty} e^{-hc\beta(\tilde{A}-\tilde{B})K^2}\int_{|K|}^{\infty}(2J+1)\, e^{-hc\beta\tilde{B}J(J+1)}\text{d}J\text{d}K$$

Step 3 *Evaluate the integrals*

You should recognize the integral over J as the integral of the derivative of a function, which is the function itself, so

$$\int_{|K|}^{\infty}(2J+1)e^{-hc\beta\tilde{B}J(J+1)}\text{d}J = \frac{1}{hc\beta\tilde{B}}e^{-hc\beta\tilde{B}|K|(|K|+1)} \overset{|K|>>1 \text{ for most allowed values}}{\approx} \frac{1}{hc\beta\tilde{B}}e^{-hc\beta\tilde{B}K^2}$$

Use this result for the integral over J in Step 2:

$$q = \frac{1}{hc\beta\tilde{B}}\int_{-\infty}^{\infty} e^{-hc\beta(\tilde{A}-\tilde{B})K^2} e^{-hc\beta\tilde{B}K^2} dK = \frac{1}{hc\beta\tilde{B}}\overbrace{\int_{-\infty}^{\infty} e^{-hc\beta\tilde{A}K^2} dK}^{\text{Integral G.1}}$$

$$= \frac{1}{hc\beta\tilde{B}}\left(\frac{\pi}{hc\beta\tilde{A}}\right)^{1/2}$$

now rearrange this expression into

$$q = \frac{1}{(hc\beta)^{3/2}}\left(\frac{\pi}{\tilde{A}\tilde{B}^2}\right)^{1/2} = \left(\frac{kT}{hc}\right)^{3/2}\left(\frac{\pi}{\tilde{A}\tilde{B}^2}\right)^{1/2}$$

For an asymmetric rotor with its three moments of inertia, one of the $\tilde{B}$ is replaced by $\tilde{C}$, to give

$$q^{R} = \left(\frac{kT}{hc}\right)^{3/2}\left(\frac{\pi}{\tilde{A}\tilde{B}\tilde{C}}\right)^{1/2} \quad (13B.12b)$$

Rotational partition function [nonlinear molecule]

A useful way of expressing the temperature above which eqns 13B.12a and 13B.12b are valid is to introduce the **characteristic rotational temperature**, $\theta^{R} = hc\tilde{B}/k$. Then 'high temperature' means $T >> \theta^{R}$ and under these conditions the rotational partition function of a linear molecule is

$$q^{R} = \overbrace{\frac{k}{hc\tilde{B}}}^{1/\theta^{R}} T = \frac{T}{\theta^{R}}$$

For a nonlinear molecule a characteristic rotational temperature is defined in the same way for each of the rotational constants: $\theta^{R} = hc\tilde{X}/k$, where $\tilde{X}$ is $\tilde{A}$, $\tilde{B}$, or $\tilde{C}$. Some typical values of θ^{R} are given in Table 13B.1. The value for $^{1}H_2$ (87.6 K) is abnormally high, so the approximation must be used carefully for this molecule. However, before using eqn 13B.12a for symmetrical molecules, such as H_2, read on (to eqn 13B.13a).

The general conclusion at this stage is that

> Molecules with large moments of inertia (and hence small rotational constants and low characteristic rotational temperatures) have large rotational partition functions.

A large value of q^{R} reflects the closeness in energy (compared with kT) of the rotational levels in large, heavy molecules, and the large number of rotational states that are accessible at normal temperatures.

It is important not to include too many rotational states in the sum that defines the partition function. For a homonuclear diatomic molecule or a symmetrical linear molecule (such as CO_2 or HC≡CH), a rotation through 180° results in an indistinguishable state of the molecule. Hence, the number of thermally accessible states is only half the number that can be occupied by a heteronuclear diatomic molecule, where rotation through 180° does result in a distinguishable state. Therefore, for a symmetrical linear molecule,

$$q^{R} = \frac{kT}{2hc\tilde{B}} = \frac{T}{2\theta^{R}} \quad (13B.13a)$$

Rotational partition function [symmetrical linear rotor]

The equations for symmetrical and non-symmetrical molecules can be combined into a single expression by introducing the **symmetry number**, σ, which is the number of indistinguishable orientations of the molecule. Then

$$q^{R} = \frac{T}{\sigma\theta^{R}} \quad (13B.13b)$$

Rotational partition function [linear rotor]

For a heteronuclear diatomic molecule $\sigma = 1$; for a homonuclear diatomic molecule or a symmetrical linear molecule, $\sigma = 2$. The formal justification of this rule depends on assessing the role of the Pauli principle.

Table 13B.1 Rotational temperatures of diatomic molecules*

	θ^{R}/K
$^{1}H_2$	87.6
$^{1}H^{35}Cl$	15.2
$^{14}N_2$	2.88
$^{35}Cl_2$	0.351

* More values are given in the *Resource section*, Table 11C.1.

How is that done? 13B.3 Identifying the origin of the symmetry number

The Pauli principle forbids the occupation of certain states. It is shown in Topic 11B, for example, that $^{1}H_2$ may occupy rotational states with even J only if its nuclear spins are paired (*para*-hydrogen), and odd J states only if its nuclear spins are parallel (*ortho*-hydrogen). In *ortho*-H_2 there are three nuclear spin states for each value of J (because there are three 'parallel' spin states of the two nuclei); in *para*-H_2 there is just one nuclear spin state for each value of J.

To set up the rotational partition function and take into account the Pauli principle, note that 'ordinary' molecular hydrogen is a mixture of one part *para*-H_2 (with only its even-J rotational states occupied) and three parts *ortho*-H_2 (with only its odd-J rotational states occupied). Therefore, the average partition function for each molecule is

$$q^{R} = \tfrac{1}{4}\sum_{\text{even } J}(2J+1)e^{-\beta hc\tilde{B}J(J+1)} + \tfrac{3}{4}\sum_{\text{odd } J}(2J+1)e^{-\beta hc\tilde{B}J(J+1)}$$

The odd-J states are three times more heavily weighted than the even-J states (Fig. 13B.8). The illustration shows that approximately the same answer would be obtained for the partition function (the sum of all the populations) if each J term contributed half its normal value to the sum. That is, the last equation can be approximated as

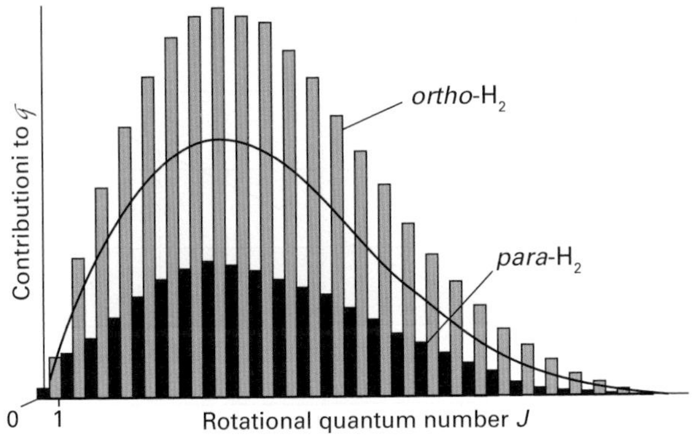

Figure 13B.8 The values of the individual terms $(2J+1)e^{-\beta hc\tilde{B}J(J+1)}$ contributing to the mean partition function of a 3:1 mixture of *ortho*- and *para*-H_2, but with a much smaller rotational constant than in reality (so as to illustrate a procedure that is in fact valid only for heavier homonuclear diatomic molecules). The partition function is the sum of all these terms. At high temperatures, the sum is approximately equal to the sum of the terms over all values of J, each with a weight of $\frac{1}{2}$. This sum is indicated by the curve.

$$q^{R} = \tfrac{1}{2}\sum_{J}(2J+1)e^{-\beta hc\tilde{B}J(J+1)}$$

and this approximation is very good when many terms contribute (at high temperatures, $T >> \theta^{R}$). At such high temperatures the sum can be approximated by the integral that led to eqn 13B.12a. Therefore, on account of the factor of $\frac{1}{2}$ in this expression, the rotational partition function for 'ordinary' molecular hydrogen at high temperatures is one-half this value, as in eqn 13B.13a.

The same type of argument may be used for linear symmetrical molecules in which identical bosons are interchanged by rotation (such as $C^{16}O_2$). As pointed out in Topic 11B, if the nuclear spin of the bosons is 0, then only even-J states are admissible. Because only half the rotational states are occupied, the rotational partition function is only half the value of the sum obtained by allowing all values of J to contribute (Fig. 13B.9).

The same care must be exercised for other types of symmetrical molecules, and for a nonlinear molecule eqn 13B.12b is replaced by

$$q^{R} = \frac{1}{\sigma}\left(\frac{kT}{hc}\right)^{3/2}\left(\frac{\pi}{\tilde{A}\tilde{B}\tilde{C}}\right)^{1/2} \quad \text{Rotational partition function [nonlinear rotor]} \quad (13B.14)$$

Some typical values of the symmetry numbers are given in Table 13B.2. To see how group theory is used to identify the value of the symmetry number, see *Integrated activity* I13.1; *Brief illustration* 13B.3 outlines the approach.

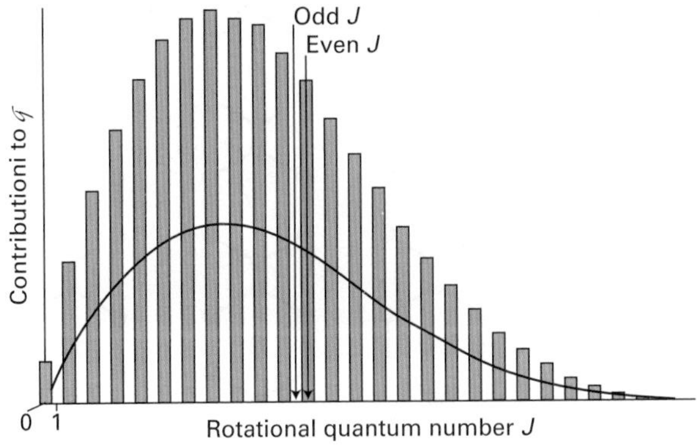

Figure 13B.9 The values of the individual terms contributing to the rotational partition function of CO_2. Only states with even J values are allowed. The full line shows the smoothed, averaged contributions of the levels.

Table 13B.2 Symmetry numbers of molecules*

	σ
$^{1}H_2$	2
$^{1}H^{2}H$	1
NH_3	3
C_6H_6	12

* More values are given in the *Resource section*, Table 11C.1.

Brief illustration 13B.3

The value $\sigma(H_2O) = 2$ reflects the fact that a 180° rotation about the bisector of the H–O–H angle interchanges two indistinguishable atoms. In NH_3, there are three indistinguishable orientations around the axis shown in (**1**). For CH_4, any of three 120° rotations about any of its four C–H bonds leaves the molecule in an indistinguishable state (**2**), so the symmetry number is $3 \times 4 = 12$.

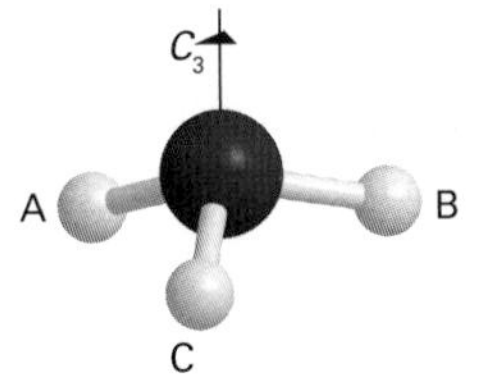

1

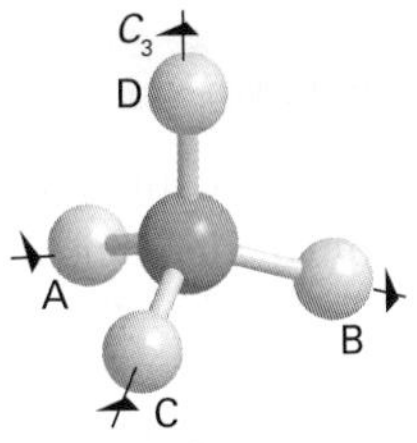

2

For benzene, any of six orientations around the axis perpendicular to the plane of the molecule leaves it apparently unchanged (Fig. 13B.10), as does a rotation of 180° around any of six axes in the plane of the molecule (three of which pass through C atoms diametrically opposite across the ring and the remaining three pass through the mid-points of C–C bonds on opposite sides of the ring).

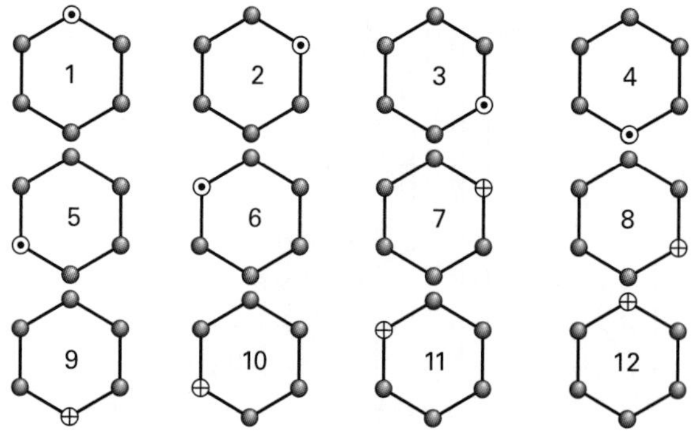

Figure 13B.10 The 12 equivalent orientations of a benzene molecule that can be reached by pure rotations and give rise to a symmetry number of 12. The plus signs indicate the underside of the hexagon after that face has been rotated into view.

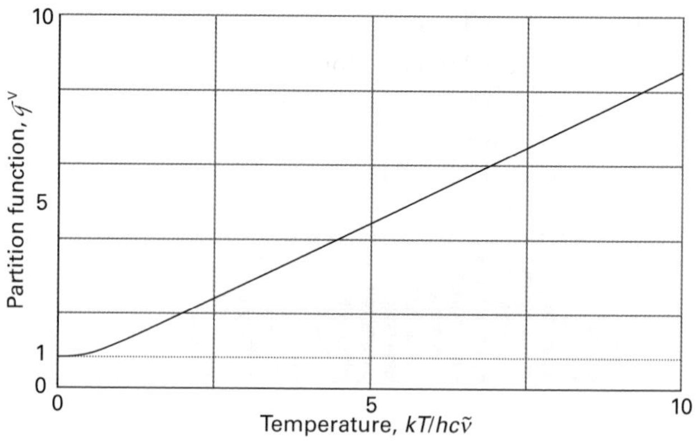

Figure 13B.11 The vibrational partition function of a molecule in the harmonic approximation. Note that the partition function is proportional to the temperature when the temperature is high ($T >> \theta^V$).

13B.2(c) The vibrational contribution

The vibrational partition function of a molecule is calculated by substituting the measured vibrational energy levels into the definition of q^V, and summing them numerically. However, provided it is permissible to assume that the vibrations are harmonic, there is a much simpler way. In that case, the vibrational energy levels form a uniform ladder of separation $hc\tilde{\nu}$ (Topics 7E and 11C), which is exactly the problem discussed earlier and leading to eqn 13B.2a. Therefore that result can be used by setting $\varepsilon = hc\tilde{\nu}$, giving

$$q^V = \frac{1}{1-e^{-\beta hc\tilde{\nu}}} \quad \text{Vibrational partition function [harmonic approximation]} \quad (13B.15)$$

This function is plotted in Fig. 13B.11 (which is essentially the same as Fig. 13B.2). Similarly, the population of each state is given by eqn 13B.2b.

Brief illustration 13B.4

To calculate the partition function of I_2 molecules at 298.15 K note that their vibrational wavenumber is 214.6 cm^{-1}. Then, because at 298.15 K, $kT/hc = 207.225$ cm^{-1},

$$\beta hc\tilde{\nu} = \frac{hc\tilde{\nu}}{kT} = \frac{214.6 \text{ cm}^{-1}}{207.225 \text{ cm}^{-1}} = 1.035\ldots$$

Then it follows from eqn 13B.15 that

$$q^V = \frac{1}{1-e^{-1.035\ldots}} = 1.55$$

From this value it can be inferred that only the ground and first excited states are significantly populated.

In a polyatomic molecule each normal mode (Topic 11D) has its own partition function (provided the anharmonicities are so small that the modes are independent). The overall vibrational partition function is the product of the individual partition functions, and so $q^V = q^V(1)q^V(2)\ldots$, where $q^V(K)$ is the partition function for normal mode K and is calculated by direct summation of the observed spectroscopic levels.

Example 13B.2 Calculating a vibrational partition function

The wavenumbers of the three normal modes of H_2O are 3656.7 cm^{-1}, 1594.8 cm^{-1}, and 3755.8 cm^{-1}. Evaluate the vibrational partition function at 1500 K.

Collect your thoughts You need to use eqn 13B.15 for each mode, and then form the product of the three contributions. At 1500 K, $kT/hc = 1042.6$ cm^{-1}.

The solution Draw up the following table displaying the contributions of each mode:

Mode	1	2	3
$\tilde{\nu}$/cm^{-1}	3656.7	1594.8	3755.8
$hc\tilde{\nu}/kT$	3.507	1.530	3.602
q^V	1.031	1.277	1.028

The overall vibrational partition function is therefore

$$q^V = 1.031 \times 1.277 \times 1.028 = 1.353$$

The three normal modes of H_2O are at such high wavenumbers that even at 1500 K most of the molecules are in their vibrational ground state.

Comment. There may be so many normal modes in a large molecule that their overall contribution may be significant even though each mode is not appreciably excited.

For example, a nonlinear molecule containing 10 atoms has $3N - 6 = 24$ normal modes (Topic 11D). If a value of about 1.1 is assumed for the vibrational partition function of one normal mode, the overall vibrational partition function is about $q^{\mathrm{V}} \approx (1.1)^{24} = 9.8$.

Self-test 13B.2 Repeat the calculation for CO_2 at the same temperature. The vibrational wavenumbers are $1388\ \mathrm{cm}^{-1}$, $667.4\ \mathrm{cm}^{-1}$, and $2349\ \mathrm{cm}^{-1}$, the second being the doubly degenerate bending mode which contributes twice to the overall vibrational partition function.

Answer: 6.79

In many molecules the vibrational wavenumbers are so great that $\beta hc\tilde{\nu} > 1$. For example, the lowest vibrational wavenumber of CH_4 is $1306\ \mathrm{cm}^{-1}$, so $\beta hc\tilde{\nu} = 6.3$ at room temperature. Most C–H stretches normally lie in the range 2850–2960 cm^{-1}, so for them $\beta hc\tilde{\nu} \approx 14$. In these cases, $\mathrm{e}^{-\beta hc\tilde{\nu}}$ in the denominator of q^{V} is very close to zero (e.g. $\mathrm{e}^{-6.3} = 0.002$), and the vibrational partition function for a single mode is very close to 1 ($q^{\mathrm{V}} = 1.002$ when $\beta hc\tilde{\nu} = 6.3$), implying that only the lowest level is significantly occupied.

Now consider the case of modes with such low vibrational frequencies that $\beta hc\tilde{\nu} << 1$. When this condition is satisfied, the expression for the partition function in eqn 13B.5 may be approximated by expanding the exponential ($\mathrm{e}^{x} = 1 + x + \cdots$) and retaining just the linear term:

$$q^{\mathrm{V}} = \frac{1}{1-\mathrm{e}^{-\beta hc\tilde{\nu}}} = \frac{1}{1-(1-\beta hc\tilde{\nu}+\cdots)}$$

That is, for low-frequency modes at high temperatures such that $\beta hc\tilde{\nu} << 1$

$$q^{\mathrm{V}} \approx \frac{kT}{hc\tilde{\nu}} \quad \text{Vibrational partition function [high-temperature approximation]} \quad (13B.16)$$

The temperatures for which eqn 13B.16 is valid can be expressed in terms of the **characteristic vibrational temperature**, $\theta^{\mathrm{V}} = hc\tilde{\nu}/k$ (Table 13B.3). The value for H_2 (6332 K) is abnormally high because the atoms are so light and the vibrational frequency is correspondingly high. In terms of the vibrational temperature, 'high temperature' means $T >> \theta^{\mathrm{V}}$, and when this condition is satisfied eqn 13B.16 is valid and can be written $q^{\mathrm{V}} = T/\theta^{\mathrm{V}}$ (the analogue of the rotational expression).

Table 13B.3 Vibrational temperatures of diatomic molecules*

	θ^{V}/K
$^{1}H_2$	6332
$^{1}H^{35}Cl$	4304
$^{14}N_2$	3393
$^{35}Cl_2$	805

* More values are given in the *Resource section*, Table 11C.1

13B.2(d) The electronic contribution

Electronic energy separations from the lowest energy level are usually very large, so for most cases $q^{\mathrm{E}} = g_0$ because only the lowest level, with degeneracy g_0, is occupied. Noble gas atoms have non-degenerate electronic ground states and so have $q^{\mathrm{E}} = 1$. In contrast, in alkali metal atoms the lowest level is doubly degenerate (corresponding to the two orientations of their electron spin), so $q^{\mathrm{E}} = 2$.

Brief illustration 13B.5

Some atoms and molecules have degenerate ground states and low-lying electronically excited degenerate states. An example is NO, which has a configuration of the form $\ldots\pi^{1}$ (Topic 11F). The energy of the two degenerate states in which the orbital and spin momenta are parallel (giving the $^{2}\Pi_{3/2}$ term, Fig. 13B.12) is slightly greater than that of the two degenerate states in which they are antiparallel (giving the $^{2}\Pi_{1/2}$ term). The separation, which arises from spin–orbit coupling, is only $121\ \mathrm{cm}^{-1}$.

If the energies of the two levels are denoted $E_{1/2} = 0$ and $E_{3/2} = \varepsilon$, then the partition function is

$$q^{\mathrm{E}} = \sum_{\text{levels } i} g_i \mathrm{e}^{-\beta\varepsilon_i} = 2 + 2\mathrm{e}^{-\beta\varepsilon}$$

This function is plotted in Fig. 13B.13. At $T = 0$, $q^{\mathrm{E}} = 2$, because only the doubly degenerate ground state is accessible. At high temperatures, q^{E} approaches 4 because all four states are accessible. At 25 °C, $q^{\mathrm{E}} = 3.1$.

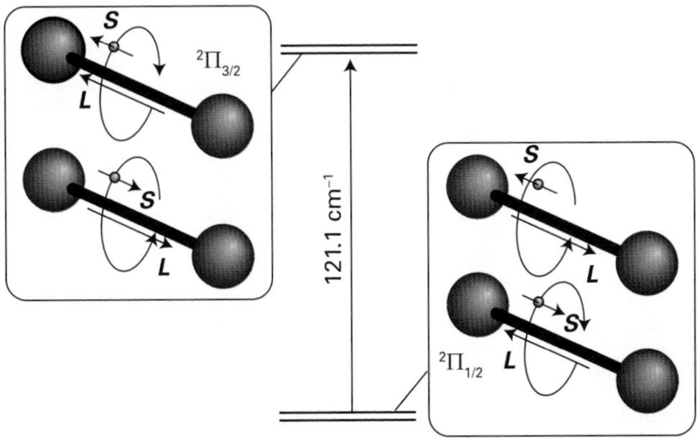

Figure 13B.12 The doubly degenerate ground electronic level of NO (with the spin and orbital angular momentum around the axis in opposite directions) and the doubly degenerate first excited level (with the spin and orbital momenta parallel). The upper level is thermally accessible at room temperature.

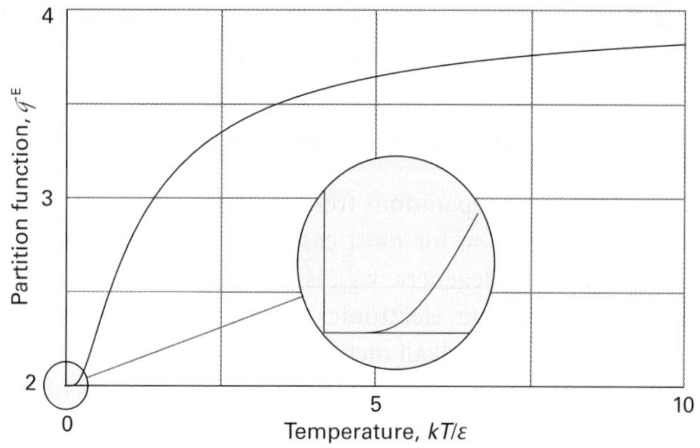

Figure 13B.13 The variation with temperature of the electronic partition function of an NO molecule. Note that the curve resembles that for a two-level system (Fig. 13B.4), but rises from 2 (the degeneracy of the lower level) and approaches 4 (the total number of states) at high temperatures.

Exercises

E13B.3 Calculate (i) the thermal wavelength, (ii) the translational partition function at 300 K and 3000 K of a molecule of molar mass 150 g mol^{-1} in a container of volume 1.00 cm^3.

E13B.4 The bond length of O_2 is 120.75 pm. Use the high-temperature approximation to calculate the rotational partition function of the molecule at 300 K.

E13B.5 Estimate the rotational partition function of ethene at 25 °C given that $\tilde{A} = 4.828$ cm^{-1}, $\tilde{B} = 1.0012$ cm^{-1}, and $\tilde{C} = 0.8282$ cm^{-1}. Take the symmetry number into account.

E13B.6 Calculate the vibrational partition function of CS_2 at 500 K given the wavenumbers of the vibrational modes as 658 cm^{-1} (symmetric stretch), 397 cm^{-1} (bend; doubly degenerate), 1535 cm^{-1} (asymmetric stretch).

E13B.7 A certain atom has a fourfold degenerate ground level, a non-degenerate electronically excited level at 2500 cm^{-1}, and a twofold degenerate level at 3500 cm^{-1}. Calculate the electronic partition function at 1900 K. What is the relative population of each level at 1900 K?

Checklist of concepts

☐ 1. The **molecular partition function** is an indication of the number of thermally accessible states at the temperature of interest.

☐ 2. If the **energy of a molecule** is given by the sum of contributions from different modes, then the molecular partition function is a product of the partition functions for each of the modes.

☐ 3. The **symmetry number** takes into account the number of indistinguishable orientations of a symmetrical molecule.

☐ 4. The **vibrational partition function** of a molecule is found by evaluating the contribution from each normal mode treated as a harmonic oscillator.

☐ 5. Because electronic energy separations from the ground state are usually very large, in most cases the **electronic partition function** is equal to the degeneracy of the lowest level.

Checklist of equations

Property	Equation	Comment	Equation number
Molecular partition function	$q = \sum_{\text{states } i} e^{-\beta\varepsilon_i}$	Definition, independent molecules	13B.1a
	$q = \sum_{\text{levels } i} g_i e^{-\beta\varepsilon_i}$	Definition, independent molecules	13B.1b
Uniform ladder	$q = 1/(1 - e^{-\beta\varepsilon})$		13B.2a
Two-level system	$q = 1 + e^{-\beta\varepsilon}$		13B.3a
Thermal wavelength	$\Lambda = h/(2\pi mkT)^{1/2}$		13B.7
Translation	$q^{T} = V/\Lambda^3$		13B.10b
Rotation	$q^{R} = kT/\sigma hc\tilde{B}$	$T >> \theta^{R}$, linear rotor	13B.13a
	$q^{R} = (1/\sigma)(kT/hc)^{3/2}(\pi/\tilde{A}\tilde{B}\tilde{C})^{1/2}$	$T >> \theta^{R}$, nonlinear rotor $\theta^{R} = hc\tilde{X}/k$, $\tilde{X} = \tilde{A}, \tilde{B}$, or $\tilde{C}$	13B.14
Vibration	$q^{V} = 1/(1 - e^{-\beta hc\tilde{\nu}})$	Harmonic approximation, $\theta^{V} = hc\tilde{\nu}/k$	13B.15

TOPIC 13C Molecular energies

➤ **Why do you need to know this material?**

For statistical thermodynamics to be useful, you need to know how to use the partition function to calculate the values of thermodynamic quantities.

➤ **What is the key idea?**

The average energy of a molecule in a collection of independent molecules can be calculated from the molecular partition function alone.

➤ **What do you need to know already?**

You need to know how to calculate a molecular partition function from computed or spectroscopic data (Topic 13B) and understand its significance as a measure of the number of accessible states. The Topic also draws on expressions for the rotational and vibrational energies of molecules (Topics 11B–11D).

A partition function in statistical thermodynamics is like a wavefunction in quantum mechanics. A wavefunction contains all the *dynamical* information about a system; a partition function contains all the *thermodynamic* information about a system. This Topic shows how one of the simplest thermodynamic properties, the mean energy of the molecules, is calculated from a knowledge of the partition function. The relations developed here are for a system composed of independent, non-interacting molecules.

13C.1 The basic equations

Consider a collection of N molecules that do not interact with one another. Any member of the collection can exist in a state i of energy ε_i measured from the lowest energy state of the molecule. The total energy of the collection, E, is

$$E = \sum_i N_i \varepsilon_i$$

where N_i is the population of state i. The mean energy of a molecule, $\langle\varepsilon\rangle$, relative to its energy in its ground state, is the total energy of the collection divided by the total number of molecules:

$$\langle\varepsilon\rangle = \frac{E}{N} = \frac{1}{N}\sum_i N_i \varepsilon_i \qquad (13C.1)$$

In Topic 13A it is shown that at temperature T the populations of the states are given by the Boltzmann distribution, eqn 13A.10b, $N_i/N = (1/q)e^{-\beta\varepsilon_i}$, so

$$\langle\varepsilon\rangle = \frac{1}{q}\sum_i \varepsilon_i e^{-\beta\varepsilon_i} \qquad (13C.2)$$

with $\beta = 1/kT$. This expression can be manipulated into a form involving only q. First note that

$$\varepsilon_i e^{-\beta\varepsilon_i} = -\frac{d}{d\beta}e^{-\beta\varepsilon_i}$$

It follows that

$$\langle\varepsilon\rangle = -\frac{1}{q}\sum_i \frac{d}{d\beta}e^{-\beta\varepsilon_i} = -\frac{1}{q}\frac{d}{d\beta}\overbrace{\sum_i e^{-\beta\varepsilon_i}}^{q} = -\frac{1}{q}\frac{dq}{d\beta} \qquad (13C.3)$$

Several points need to be made in relation to eqn 13C.3. Because $\varepsilon_0 \equiv 0$, (all energies are measured from the lowest available level), $\langle\varepsilon\rangle$ is the value of the energy relative to the actual ground-state energy. If the lowest energy of the molecule is in fact ε_{gs} rather than 0, then the true mean energy is $\varepsilon_{gs} + \langle\varepsilon\rangle$. For instance, for a harmonic oscillator, ε_{gs} is the zero-point energy, $\frac{1}{2}hc\tilde{\nu}$.

Secondly, because the partition function might depend on variables other than the temperature (e.g. the volume), the derivative with respect to β in eqn 13C.3 is actually a *partial* derivative with these other variables held constant. The complete expression relating the molecular partition function to the mean energy of a molecule is therefore

$$\langle\varepsilon\rangle = \varepsilon_{gs} - \frac{1}{q}\left(\frac{\partial q}{\partial\beta}\right)_V \qquad \text{Mean molecular energy} \qquad (13C.4a)$$

An equivalent form is obtained by noting that $dx/x = d\ln x$:

$$\langle\varepsilon\rangle = \varepsilon_{gs} - \left(\frac{\partial \ln q}{\partial\beta}\right)_V \qquad \text{Mean molecular energy} \qquad (13C.4b)$$

These two equations confirm that it is only necessary to know the partition function as a function of temperature in order to calculate the mean energy.

Brief illustration 13C.1

If a molecule has only two available states, one at 0 and the other at an energy ε, its partition function is

$$q = 1 + e^{-\beta\varepsilon}$$

Therefore, the mean energy of a collection of these molecules at a temperature T is, from eqn 13C.4a,

$$\langle\varepsilon\rangle = -\frac{1}{1+e^{-\beta\varepsilon}}\frac{d(1+e^{-\beta\varepsilon})}{d\beta} = \frac{\varepsilon e^{-\beta\varepsilon}}{1+e^{-\beta\varepsilon}} = \frac{\varepsilon}{e^{\beta\varepsilon}+1}$$

In the final step the numerator and denominator of the fraction are multiplied by $e^{+\beta\varepsilon}$. The expression for $\langle\varepsilon\rangle$ is plotted in Fig. 13C.1. Notice how the mean energy is zero when $kT << \varepsilon$, when only the lower state (at the zero of energy) is occupied, and starts to rise significantly when $kT \approx \varepsilon$. The mean energy tends to $\frac{1}{2}\varepsilon$ when $kT >> \varepsilon$; in this limit the two states become equally populated.

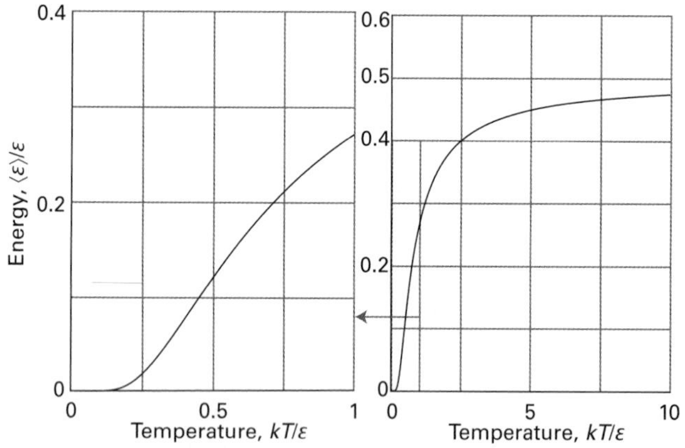

Figure 13C.1 The variation with temperature of the mean energy of a two-level system. The graph on the left shows the slow rise away from zero energy at low temperatures; the slope of the graph at $T = 0$ is 0. The graph on the right shows that as kT/ε becomes large $\langle\varepsilon\rangle/\varepsilon$ approaches 0.5. In the limit both states are equally populated.

Exercises

E13C.1 Compute the mean energy at 298 K of a two-level system of energy separation equivalent to 500 cm^{-1}; the levels are non-degenerate.

E13C.2 Derive an expression for the mean energy of a two-level system in which the ground level has energy zero and degeneracy g_0, and the upper level has energy ε and degeneracy g_1. *Hint:* start by writing an expression for the partition function and then follow the approach used in *Brief illustration* 13C.1.

13C.2 Contributions of the fundamental modes of motion

In Topic 13B it is established that each fundamental contribution to the energy, namely translation (T), rotation (R), vibration (V), electronic (E), and electron spin (S) can, to a good approximation, be treated separately.

13C.2(a) The translational contribution

For a one-dimensional container of length X, the partition function is $q_X^T = X/\Lambda$ with $\Lambda = h\beta^{1/2}/(2\pi m)^{1/2}$ (Topic 13B, with 'constant volume V' replaced by 'constant length X'). The partition function can be written as $A\beta^{-1/2}$ where A is a constant independent of temperature. This form is convenient for the calculation of the average energy by using eqn 13C.4b:

$$\langle\varepsilon_X^T\rangle = -\left(\frac{\partial \ln q}{\partial\beta}\right)_X = -\left(\frac{\partial \ln(A\beta^{-1/2})}{\partial\beta}\right)_X$$
$$= -\left(\frac{\partial(\ln A + \ln\beta^{-1/2})}{\partial\beta}\right)_X$$
$$= -\left(\frac{\partial\left(-\frac{1}{2}\ln\beta\right)}{\partial\beta}\right)_X = \frac{1}{2\beta}$$

That is,

$$\langle\varepsilon_X^T\rangle = \tfrac{1}{2}kT \qquad \text{Mean translational energy [one dimension]} \qquad (13C.5a)$$

Motion along x, y, and z are independent of each other, so each one contributes $\frac{1}{2}kT$ to the mean energy. For a molecule free to translate in three dimensions the mean energy is therefore

$$\langle\varepsilon^T\rangle = \tfrac{3}{2}kT \qquad \text{Mean translational energy [three dimensions]} \qquad (13C.5b)$$

This result is in agreement with the equipartition theorem (see *Energy: A first look*). The classical expression for the energy of translation in one dimension is $E_{k,x} = \frac{1}{2}mv_x^2$ which has one quadratic contribution. It follows from the equipartition theorem that the mean translational energy in one dimension is $\frac{1}{2}kT$, and in three dimensions the mean energy is simply three times this. The derivation of the expression for the translational partition function assumes that many translational states are occupied, so the equipartition theorem is expected to apply.

13C.2(b) The rotational contribution

The mean rotational energy of a linear molecule is obtained from the rotational partition function (eqn 13B.11):

$$q^R = \sum_J (2J+1)e^{-\beta hc\tilde{B}J(J+1)}$$

When the temperature is not high (in the sense that it is not true that $T >> \theta^{R} = hc\tilde{B}/k$) the series must be summed term by term, which for a heteronuclear diatomic molecule or other non-symmetrical linear molecule gives

$$q^{R} = 1 + 3e^{-2\beta hc\tilde{B}} + 5e^{-6\beta hc\tilde{B}} + \cdots$$

Hence, because

$$\frac{dq^{R}}{d\beta} = -hc\tilde{B}(6e^{-2\beta hc\tilde{B}} + 30e^{-6\beta hc\tilde{B}} + \cdots)$$

(q^{R} is independent of V, so the partial derivative has been replaced by a complete derivative) it follows that

$$\langle \varepsilon^{R} \rangle = -\frac{1}{q^{R}}\frac{dq^{R}}{d\beta} = \frac{hc\tilde{B}(6e^{-2\beta hc\tilde{B}} + 30e^{-6\beta hc\tilde{B}} + \cdots)}{1 + 3e^{-2\beta hc\tilde{B}} + 5e^{-6\beta hc\tilde{B}} + \cdots}$$

Mean rotational energy [unsymmetrical linear molecule] (13C.6a)

This ungainly function is plotted in Fig. 13C.2. If the temperature is much greater than the characteristic rotational temperature θ^{R} ($\theta^{R} = hc\tilde{B}/k$), q^{R} is given by eqn 13B.13b, $q^{R} = T/\sigma\theta^{R}$. The rotational partition function can be expressed in terms of β and $\tilde{B}$ as follows

$$q^{R} = \frac{T}{\sigma\theta^{R}} = \frac{kT}{\sigma hc\tilde{B}} = \frac{1}{\sigma\beta hc\tilde{B}}$$

where $\sigma = 1$ for a heteronuclear diatomic molecule. It then follows that

$$\langle \varepsilon^{R} \rangle = -\frac{1}{q^{R}}\frac{dq^{R}}{d\beta} = -\sigma\beta hc\tilde{B}\frac{d}{d\beta}\left(\frac{1}{\sigma\beta hc\tilde{B}}\right) = -\beta\overbrace{\frac{d}{d\beta}\frac{1}{\beta}}^{-1/\beta^{2}}$$

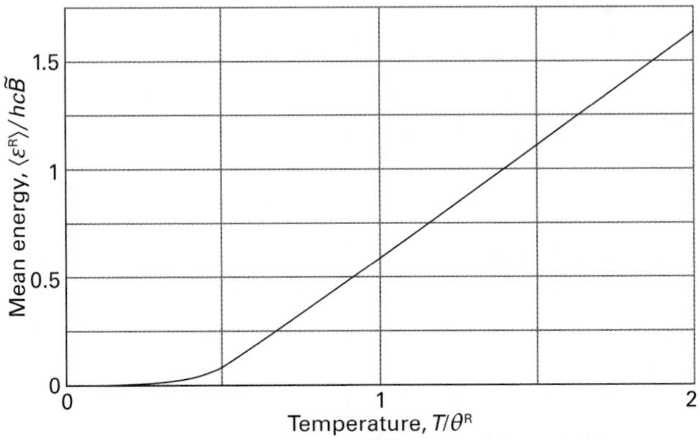

Figure 13C.2 The variation with temperature of the mean rotational energy of an unsymmetrical linear rotor. At high temperatures ($T >> \theta^{R}$), the energy is proportional to the temperature, in accord with the equipartition theorem. The limiting slope is 1, which corresponds to a mean energy of kT.

and therefore that

$$\langle \varepsilon^{R} \rangle = \frac{1}{\beta} = kT$$

Mean rotational energy [linear molecule, $T >> \theta^{R}$] (13C.6b)

The high-temperature result, which is valid when many rotational states are occupied, is also in agreement with the equipartition theorem (there are two quadratic contributions).

Brief illustration 13C.2

To estimate the mean energy of a nonlinear molecule in the high-temperature limit recognize that its rotational kinetic energy (the only contribution to its rotational energy) is $E_{k} = \frac{1}{2}I_{a}\omega_{a}^{2} + \frac{1}{2}I_{b}\omega_{b}^{2} + \frac{1}{2}I_{c}\omega_{c}^{2}$. As there are three quadratic contributions, its mean rotational energy is $\frac{3}{2}kT$. The molar contribution is $\frac{3}{2}RT$. At 25 °C, this contribution is 3.7 kJ mol^{-1}, the same as the translational contribution, giving a total of 7.4 kJ mol^{-1}. A monatomic gas has no rotational contribution.

13C.2(c) The vibrational contribution

The vibrational partition function in the harmonic approximation is given in eqn 13B.15, $q^{V} = 1/(1 - e^{-\beta hc\tilde{\nu}})$. Because q^{V} is independent of the volume, it follows that

$d(1/f)/dx = -(1/f^{2})df/dx$

$$\frac{dq^{V}}{d\beta} = \frac{d}{d\beta}\left(\frac{1}{1 - e^{-\beta hc\tilde{\nu}}}\right) = -\frac{hc\tilde{\nu}e^{-\beta hc\tilde{\nu}}}{(1 - e^{-\beta hc\tilde{\nu}})^{2}} \qquad (13C.7)$$

and hence

$$\langle \varepsilon^{V} \rangle = -\frac{1}{q^{V}}\frac{dq^{V}}{d\beta} = (1 - e^{-\beta hc\tilde{\nu}})\frac{hc\tilde{\nu}e^{-\beta hc\tilde{\nu}}}{(1 - e^{-\beta hc\tilde{\nu}})^{2}} = \frac{hc\tilde{\nu}e^{-\beta hc\tilde{\nu}}}{1 - e^{-\beta hc\tilde{\nu}}}$$

The final result, after multiplying the numerator and denominator by $e^{\beta hc\tilde{\nu}}$ is

$$\langle \varepsilon^{V} \rangle = \frac{hc\tilde{\nu}}{e^{\beta hc\tilde{\nu}} - 1}$$

Mean vibrational energy [harmonic approximation] (13C.8)

The zero-point energy, $\frac{1}{2}hc\tilde{\nu}$, can be added to the right-hand side if the mean energy is to be measured from 0 rather than the lowest attainable level (the zero-point level). The variation of the mean energy with temperature is illustrated in Fig. 13C.3.

If the temperature is much greater than the characteristic vibrational temperature θ^{V} ($\theta^{V} = hc\tilde{\nu}/k$) it follows that $T >> hc\tilde{\nu}/k$ which can be expressed as $\beta hc\tilde{\nu} << 1$. The exponential function in eqn 13C.8 can be expanded ($e^{x} = 1 + x + \cdots$) and, because

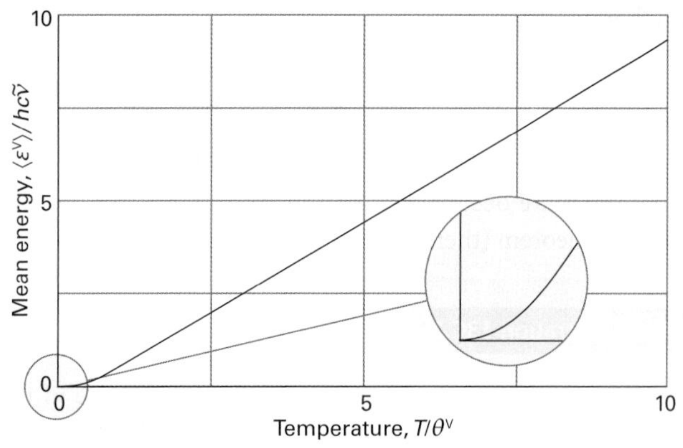

Figure 13C.3 The variation with temperature of the mean vibrational energy of a molecule in the harmonic approximation. At high temperatures ($T >> \theta^V$), the energy is proportional to the temperature, in accord with the equipartition theorem. The limiting slope is 1, which corresponds to a mean energy of kT.

$x << 1$, all but the two leading terms discarded. This approximation leads to

$$\langle\varepsilon^V\rangle = \frac{hc\tilde{\nu}}{(1+\beta hc\tilde{\nu}+\cdots)-1} \approx \frac{1}{\beta} = kT$$

Mean vibrational energy [$T >> \theta^V$] (13C.9)

This result is in agreement with the value predicted by the classical equipartition theorem (two quadratic contributions). Bear in mind, however, that the condition $T >> \theta^V$ is rarely satisfied.

Brief illustration 13C.3

To calculate the mean vibrational energy of I_2 molecules at 298.15 K note that their vibrational wavenumber is 214.6 cm^{-1} and that k/hc = 0.6950 cm^{-1} K^{-1}. The characteristic vibrational temperature is

$$\theta^V = \frac{hc\tilde{\nu}}{k} = \frac{214.6\ \text{cm}^{-1}}{0.6950\ \text{cm}^{-1}\text{K}^{-1}} = 309\ \text{K}$$

Therefore at 298 K the temperature is not large compared with θ^V so eqn 13C.9 (the equipartition value) does not apply and eqn 13C.8 must be used instead.

$$\beta hc\tilde{\nu} = \frac{hc\tilde{\nu}}{kT} = \frac{\tilde{\nu}}{(k/hc)T}$$

$$= \frac{214.6\ \text{cm}^{-1}}{(0.6950\ \text{cm}^{-1}\text{K}^{-1})\times(298.15\ \text{K})} = 1.035\ldots$$

$$\frac{\langle\varepsilon^V\rangle}{hc} = \frac{\tilde{\nu}}{e^{\beta hc\tilde{\nu}}-1} = \frac{214.6\ \text{cm}^{-1}}{e^{1.035\ldots}-1} = 118.1\ \text{cm}^{-1}$$

The addition of the zero-point energy (corresponding to $\frac{1}{2}\times 214.6$ cm^{-1}) increases this value to 225.4 cm^{-1}.

The mean molar vibrational energy (excluding the zero-point energy) is 1.41 kJ mol^{-1}, which is somewhat less than the equipartition value of 2.48 kJ mol^{-1}. For N_2 molecules, with a very much higher vibrational wavenumber of 2358 cm^{-1}, the molar vibrational energy is only 0.32 J mol^{-1}.

When there are several normal modes that can be treated as harmonic, the overall vibrational partition function is the product of each individual partition function, and the total mean vibrational energy is the sum of the mean energy of each mode.

13C.2(d) The electronic contribution

In most cases of interest, the electronic states of atoms and molecules are so widely separated that only the electronic ground state is occupied. All energies are measured from the ground state of each mode, so it follows that

$$\langle\varepsilon^E\rangle = 0$$

Mean electronic energy (13C.10)

In certain cases, there are thermally accessible states at the temperature of interest. In that case, the partition function and hence the mean electronic energy are best calculated by direct summation over the available states. Care must be taken to take any degeneracies into account, as illustrated in the following example.

Example 13C.1 Calculating the electronic contribution to the energy

The ground electronic level of a certain atom is doubly degenerate and there is an excited level that is fourfold degenerate at $\varepsilon/hc = \tilde{\nu} = 600$ cm^{-1} above the ground level. What is its mean electronic energy at 25 °C, expressed as a wavenumber?

Collect your thoughts You need to write the expression for the partition function at a general temperature T (in terms of β) and then derive the mean energy by using eqn 13C.3. Doing so involves differentiating the partition function with respect to β. Finally, substitute the data. Use $\varepsilon = hc\tilde{\nu}$, $\langle\varepsilon^E\rangle = hc\langle\tilde{\nu}^E\rangle$, and kT/hc = 207.225 cm^{-1} at 25 °C.

The solution The partition function is written as a sum over levels by using eqn 13B.1b, noting that $g_0 = 2$ and $g_1 = 4$. Therefore the partition function is $q^E = 2 + 4e^{-\beta\varepsilon}$ and the mean energy is

$$\langle\varepsilon^{\mathrm{E}}\rangle = -\frac{1}{q^{\mathrm{E}}}\frac{\mathrm{d}q^{\mathrm{E}}}{\mathrm{d}\beta} = -\frac{1}{2+4\mathrm{e}^{-\beta\varepsilon}}\overbrace{\frac{\mathrm{d}}{\mathrm{d}\beta}(2+4\mathrm{e}^{-\beta\varepsilon})}^{-4\varepsilon\mathrm{e}^{-\beta\varepsilon}}$$
$$= \frac{4\varepsilon\mathrm{e}^{-\beta\varepsilon}}{2+4\mathrm{e}^{-\beta\varepsilon}} = \frac{\varepsilon}{\frac{1}{2}\mathrm{e}^{\beta\varepsilon}+1}$$

The energy separation ε is expressed as a wavenumber $\tilde{\nu}$ using $\varepsilon = hc\tilde{\nu}$, and likewise the mean energy is expressed as a wavenumber by writing $\langle\tilde{\nu}^{\mathrm{E}}\rangle = \langle\varepsilon^{\mathrm{E}}\rangle/hc$

$$\langle\tilde{\nu}^{\mathrm{E}}\rangle = \frac{\langle\varepsilon^{\mathrm{E}}\rangle}{hc} = \frac{\varepsilon/hc}{\frac{1}{2}\mathrm{e}^{\beta\varepsilon}+1} = \frac{\tilde{\nu}}{\frac{1}{2}\mathrm{e}^{hc\tilde{\nu}/kT}+1}$$

From the data,

$$\langle\tilde{\nu}^{\mathrm{E}}\rangle = \frac{600\ \mathrm{cm}^{-1}}{\frac{1}{2}\mathrm{e}^{(600\ \mathrm{cm}^{-1})/(207.225\ \mathrm{cm}^{-1})}+1} = 59.7\ \mathrm{cm}^{-1}$$

Self-test 13C.1 Repeat the problem for an atom that has a threefold degenerate ground level and a sevenfold degenerate excited level 400 cm^{-1} above.

Answer: 101 cm^{-1}

13C.2(e) The spin contribution

An electron spin in a magnetic field $\mathcal{B}$ has two possible energy states that depend on its orientation as expressed by the magnetic quantum number m_s, and which are given by

$$E_{m_s} = g_{\mathrm{e}}\mu_{\mathrm{B}}\mathcal{B}m_s \qquad \text{Electron spin energies} \qquad (13C.11)$$

where μ_{B} is the Bohr magneton and $g_{\mathrm{e}} = 2.0023$. These energies are discussed in more detail in Topic 12A. The lower state has $m_s = -\frac{1}{2}$, so the two energy levels available to the electron lie (according to the convention that the lowest state has energy 0) at $\varepsilon_{-1/2} = 0$ and at $\varepsilon_{+1/2} = g_{\mathrm{e}}\mu_{\mathrm{B}}\mathcal{B}$. The spin partition function is therefore

$$q^{\mathrm{S}} = 1 + \mathrm{e}^{-\beta g_{\mathrm{e}}\mu_{\mathrm{B}}\mathcal{B}} \qquad \text{Spin partition function} \qquad (13C.12)$$

The mean energy of the spin is therefore

$$\langle\varepsilon^{\mathrm{S}}\rangle = -\frac{1}{q^{\mathrm{S}}}\frac{\mathrm{d}q^{\mathrm{S}}}{\mathrm{d}\beta} = -\frac{1}{1+\mathrm{e}^{-\beta g_{\mathrm{e}}\mu_{\mathrm{B}}\mathcal{B}}}\overbrace{\frac{\mathrm{d}}{\mathrm{d}\beta}(1+\mathrm{e}^{-\beta g_{\mathrm{e}}\mu_{\mathrm{B}}\mathcal{B}})}^{-g_{\mathrm{e}}\mu_{\mathrm{B}}\mathcal{B}\mathrm{e}^{-\beta g_{\mathrm{e}}\mu_{\mathrm{B}}\mathcal{B}}}$$
$$= \frac{g_{\mathrm{e}}\mu_{\mathrm{B}}\mathcal{B}\mathrm{e}^{-\beta g_{\mathrm{e}}\mu_{\mathrm{B}}\mathcal{B}}}{1+\mathrm{e}^{-\beta g_{\mathrm{e}}\mu_{\mathrm{B}}\mathcal{B}}}$$

The final expression, after multiplying the numerator and denominator by $\mathrm{e}^{\beta g_{\mathrm{e}}\mu_{\mathrm{B}}\mathcal{B}}$, is

$$\langle\varepsilon^{\mathrm{S}}\rangle = \frac{g_{\mathrm{e}}\mu_{\mathrm{B}}\mathcal{B}}{\mathrm{e}^{\beta g_{\mathrm{e}}\mu_{\mathrm{B}}\mathcal{B}}+1} \qquad \text{Mean spin energy} \qquad (13C.13)$$

This function is essentially the same as that plotted in Fig. 13C.1.

Brief illustration 13C.4

Suppose a collection of radicals is exposed to a magnetic field of 2.5 T (T denotes tesla) at 25 °C. With $\mu_{\mathrm{B}} = 9.274 \times 10^{-24}\ \mathrm{J\,T}^{-1}$,

$$g_{\mathrm{e}}\mu_{\mathrm{B}}\mathcal{B} = 2.0023\times(9.274\times10^{-24}\ \mathrm{J\,T^{-1}})\times(2.5\ \mathrm{T}) = 4.64\ldots\times10^{-23}\ \mathrm{J}$$

$$\beta g_{\mathrm{e}}\mu_{\mathrm{B}}\mathcal{B} = \frac{2.0023\times(9.274\times10^{-24}\ \mathrm{J\,T^{-1}})\times(2.5\ \mathrm{T})}{(1.381\times10^{-23}\ \mathrm{J\,K^{-1}})\times(298\ \mathrm{K})} = 0.0112\ldots$$

The mean energy is therefore

$$\langle\varepsilon^{\mathrm{S}}\rangle = \frac{4.64\ldots\times10^{-23}\ \mathrm{J}}{\mathrm{e}^{0.0112\ldots}+1} = 2.31\times10^{-23}\ \mathrm{J}$$

This energy is equivalent to 13.9 J mol^{-1} (note joules, not kilojoules).

Exercises

E13C.3 Evaluate the mean vibrational energy of Br_2 at 400 K (use $\tilde{\nu} = 323.2\ \mathrm{cm}^{-1}$). Compare your result with the equipartition value and comment on the difference.

E13C.4 A certain atom has a fourfold degenerate ground level, a non-degenerate electronically excited level at 2500 cm^{-1}, and a twofold degenerate level at 3500 cm^{-1}. Calculate the mean contribution to the electronic energy at 1900 K.

Checklist of concepts

☐ 1. The **mean molecular energy** can be calculated from the molecular partition function.

☐ 2. Individual contributions to the mean molecular energy from each mode of motion are calculated from the relevant partition functions,

☐ 3. In the high temperature limit, the results obtained for the mean molecular energy are in accord with the equipartition principle.

Checklist of equations*

Property	Equation	Comment	Equation number
Mean energy	$\langle\varepsilon\rangle = \varepsilon_{gs} - (1/q)(\partial q/\partial\beta)_V$		13C.4a
	$\langle\varepsilon\rangle = \varepsilon_{gs} - (\partial \ln q/\partial\beta)_V$	Alternative version	13C.4b
Translation	$\langle\varepsilon^T\rangle = d(kT/2)$	In d dimensions, $d = 1, 2, 3$	13C.5
Rotation	$\langle\varepsilon^R\rangle = kT$	Linear molecule, $T >> \theta^R$	13C.6b
Vibration	$\langle\varepsilon^V\rangle = hc\tilde{\nu}/(e^{\beta hc\tilde{\nu}} - 1)$	Harmonic approximation	13C.8
	$\langle\varepsilon^V\rangle = kT$	$T >> \theta^V$	13C.9
Spin	$\langle\varepsilon^S\rangle = g_e\mu_B\mathcal{B}/(e^{\beta g_e\mu_B\mathcal{B}} + 1)$	Electron in a magnetic field	13C.13

* $\beta = 1/kT$

TOPIC 13D The canonical ensemble

➤ **Why do you need to know this material?**

Whereas Topics 13B and 13C deal with independent molecules, in practice molecules do interact. Therefore, this material is essential for applying statistical thermodynamics to real gases, liquids, and solids and to any system in which intermolecular interactions cannot be neglected.

➤ **What is the key idea?**

A system composed of interacting molecules is described in terms of a canonical partition function, from which the thermodynamic properties of the system may be deduced.

➤ **What do you need to know already?**

The calculations here, which are not carried through in detail, are essentially the same as in Topic 13A. Calculations of mean energies are also essentially the same as in Topic 13C and are not repeated in detail.

The crucial concept needed in the treatment of systems of interacting particles, as in real gases and liquids, is the 'ensemble'. Like so many scientific terms, the term has basically its normal meaning of 'collection', but in statistical thermodynamics it has been sharpened and refined into a precise significance.

13D.1 The concept of ensemble

To set up an ensemble, take a closed system of specified volume, composition, and temperature, and think of it as replicated $\tilde{N}$ times (Fig. 13D.1). All the identical closed systems are regarded as being in thermal contact with one another, so they can exchange energy. The total energy of the ensemble is $\tilde{E}$. Because the members of the ensemble are all in thermal equilibrium with each other, they have the same temperature, T. The volume of each member of the ensemble is the same, so the energy levels available to the molecules are the same in each system, and each member contains the same number of molecules. This imaginary collection of replications of the actual system with a common temperature is called the **canonical ensemble**.[1]

[1] The word 'canon' means 'according to a rule'.

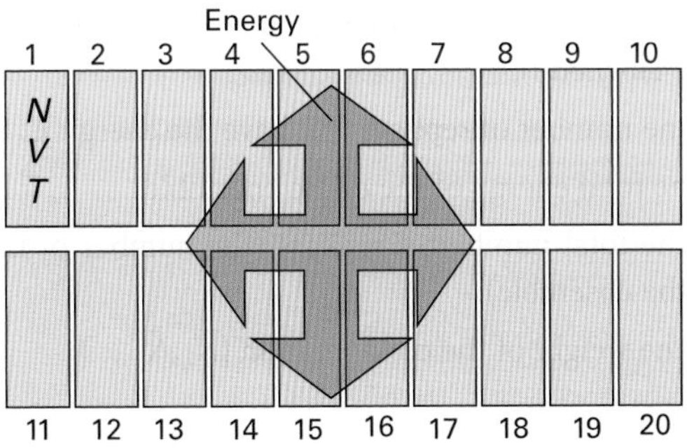

Figure 13D.1 A representation of the canonical ensemble, in this case for $\tilde{N} = 20$. The individual replications of the actual system all have the same composition and volume. They are in thermal contact with one another and so can exchange energy as heat, and because they are in thermal equilibrium with one another they all have the same temperature. The total energy of all 20 replications is a constant because the ensemble is isolated overall.

There are two other important types of ensembles. In the **microcanonical ensemble** the condition of constant temperature is replaced by the requirement that all the systems should have exactly the same energy: each system is individually isolated. In the **grand canonical ensemble** the volume and temperature of each system is the same, but they are open, which means that matter can be imagined as able to pass between them; the composition of each one may fluctuate, but now the property known as the chemical potential (μ, Topic 5A) is the same in each system. In summary:

Ensemble	Common properties
Microcanonical	V, E, N
Canonical	V, T, N
Grand canonical	V, T, μ

The microcanonical ensemble is the basis of the discussion in Topic 13A; the grand canonical ensemble will not be considered explicitly.

The important point about an ensemble is that it is a collection of *imaginary* replications of the system, so the number of members can be as large as desired; when appropriate, $\tilde{N}$ can be taken as infinite. The number of members of the ensemble in a state with energy E_i is denoted $\tilde{N}_i$, and it is possible to speak of the configuration of the ensemble (by analogy with the configuration of the system used in Topic 13A) and its weight, $\tilde{\mathcal{W}}$, the number of ways of achieving the configuration $\{\tilde{N}_0, \tilde{N}_1, \ldots\}$.

Note that $\tilde{N}$ is unrelated to N, the number of molecules in the actual system; $\tilde{N}$ is the number of *imaginary* replications of that system. In summary:

N is the number of molecules in the system, the same in each member of the ensemble.

T is the common temperature of the members.

E_i is the total energy of one member of the ensemble (the one labelled i).

$\tilde{N}_i$ is the number of replicas that have the energy E_i.

$\tilde{E}$ is the total energy of the entire ensemble.

$\tilde{N}$ is the total number of replicas (the number of members of the ensemble).

$\tilde{\mathcal{W}}$ is the weight of the configuration $\{\tilde{N}_0,\tilde{N}_1,\ldots\}$.

13D.1(a) Dominating configurations

Just as in Topic 13A, some of the configurations of the canonical ensemble are very much more probable than others. For instance, it is very unlikely that the whole of the total energy of the ensemble will accumulate in one system to give the configuration $\{\tilde{N},0,0,\ldots\}$. By analogy with the discussion in Topic 13A, there is a dominating configuration, and thermodynamic properties can be calculated by taking the average over the ensemble using that single, most probable, configuration. In the **thermodynamic limit** of $\tilde{N} \to \infty$, this dominating configuration is overwhelmingly the most probable, and dominates its properties.

The quantitative discussion follows the argument in Topic 13A (leading to the Boltzmann distribution) with the modification that N and N_i are replaced by $\tilde{N}$ and $\tilde{N}_i$. The weight $\tilde{\mathcal{W}}$ of a configuration $\{\tilde{N}_0,\tilde{N}_1,\ldots\}$ is

$$\tilde{\mathcal{W}} = \frac{\tilde{N}!}{\tilde{N}_0!\tilde{N}_1!\ldots} \qquad \text{Weight} \qquad (13\text{D}.1)$$

The configuration of greatest weight, subject to the constraints that the total energy of the ensemble is constant at $\tilde{E}$ and that the total number of members is fixed at $\tilde{N}$, is given by the **canonical distribution**:

$$\frac{\tilde{N}_i}{\tilde{N}} = \frac{e^{-\beta E_i}}{Q} \qquad Q = \sum_i e^{-\beta E_i} \qquad \text{Canonical distribution} \qquad (13\text{D}.2)$$

here the sum is over all members of the ensemble. The quantity Q, which is a function of the temperature, is called the **canonical partition function**, and $\beta = 1/kT$. Like the molecular partition function, the canonical partition function contains all the thermodynamic information about a system but allows for the possibility of interactions between the constituent molecules.

13D.1(b) Fluctuations from the most probable distribution

The canonical distribution in eqn 13D.2 is only apparently an exponentially decreasing function of the energy of the system. Just as the Boltzmann distribution gives the occupation of a single state of a molecule, the canonical distribution function gives the probability of occurrence of members in a single state i of energy E_i. There may in fact be numerous states with almost identical energies. For example, in a gas the identities of the molecules moving slowly or quickly can change without necessarily affecting the total energy. The **energy density of states**, the number of states in an energy range divided by the width of the range (Fig. 13D.2), is a very sharply increasing function of energy. It follows that the probability of a member of an ensemble having a specified energy (as distinct from being in a specified state) is given by eqn 13D.2, a sharply decreasing function, multiplied by a sharply increasing function (Fig. 13D.3). Therefore, the overall distribution is a sharply peaked function. That is, most members of the ensemble have an energy very close to the mean value.

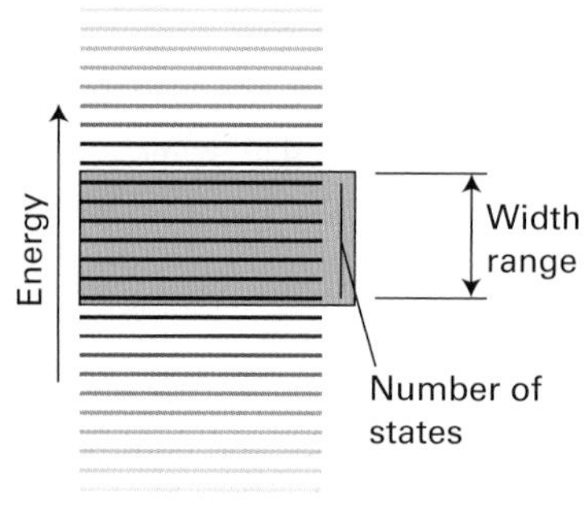

Figure 13D.2 The energy density of states is the number of states in an energy range divided by the width of the range.

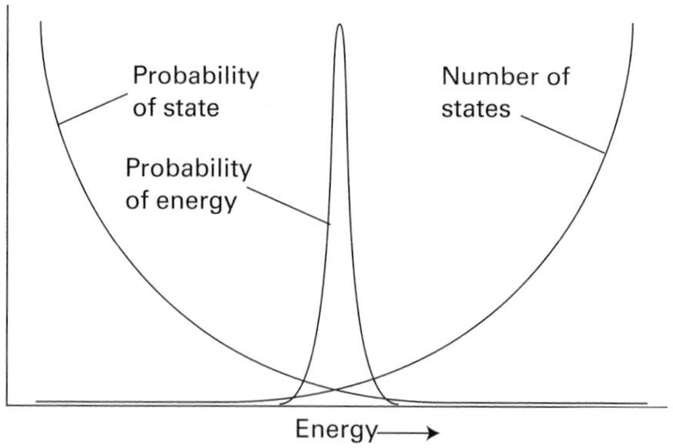

Figure 13D.3 To construct the form of the distribution of members of the canonical ensemble in terms of their energies, the probability that any one is in a state of given energy (eqn 13D.2) is multiplied by the number of states corresponding to that energy (a steeply rising function). The product is a sharply peaked function at the mean energy (here considerably magnified), which shows that almost all the members of the ensemble have that energy.

Brief illustration 13D.1

A function that increases rapidly is x^n, with n large. A function that decreases rapidly is e^{-nx}, once again, with n large. The product of these two functions, normalized so that the maxima for different values of n all coincide,

$$f(x) = e^n x^n e^{-nx}$$

is plotted for three values of n in Fig. 13D.4. Note that the width of the product does indeed decrease as n increases.

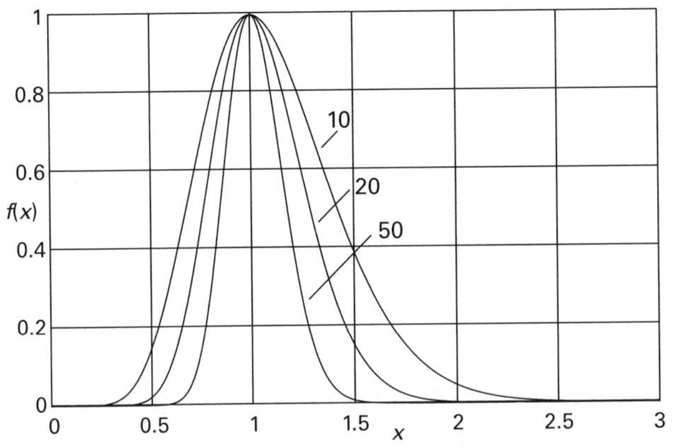

Figure 13D.4 The product of the two functions discussed in *Brief illustration* 13D.1, for three different values of n.

13D.2 The mean energy of a system

Just as the molecular partition function q can be used to calculate the mean value of a molecular property, so the canonical partition function Q can be used to calculate the mean energy of an entire system composed of molecules which might or might not be interacting with one another. Thus, Q is more general than q because it does not assume that the molecules are independent. Therefore Q can be used to discuss the properties of condensed phases and real gases where molecular interactions are important.

The total energy of the ensemble is $\tilde{E}$, and there are $\tilde{N}$ members; therefore the mean energy of a member is $\langle E\rangle = \tilde{E}/\tilde{N}$. The fraction, $\tilde{p}_i$, of members of the ensemble in a state i with energy E_i is given by the analogue of eqn 13A.10b ($p_i = e^{-\beta\varepsilon_i}/q$ with $p_i = N_i/N$) as

$$\tilde{p}_i = \frac{e^{-\beta E_i}}{Q} \qquad (13D.3)$$

Therefore it follows that

$$\langle E\rangle = \sum_i \tilde{p}_i E_i = \frac{1}{Q}\sum_i E_i e^{-\beta E_i} \qquad (13D.4)$$

By the same argument that led to eqn 13C.4a ($\langle\varepsilon\rangle = -(1/q)(\partial q/\partial\beta)_V$, when $\varepsilon_{gs} \equiv 0$),

$$\langle E\rangle = -\frac{1}{Q}\left(\frac{\partial Q}{\partial\beta}\right)_V = -\left(\frac{\partial \ln Q}{\partial\beta}\right)_V \qquad \text{Mean energy of a system} \qquad (13D.5)$$

As in the case of the mean molecular energy, the ground-state energy of the entire system must be added to this expression if it is not zero.

Exercise

E13D.1 Starting from $\langle E\rangle = (1/Q)\sum_i E_i e^{-\beta E_i}$ and the definition of Q in eqn 13D.2, derive the expression for $\langle E\rangle$ given in eqn 13D.5. (*Hint:* use the same approach as in the derivation of eqn 13C.4a.)

13D.3 Independent molecules revisited

When the molecules are in fact independent of each other, Q can be shown to be related to the molecular partition function q.

How is that done? 13D.1 Establishing the relation between Q and q

There are two cases you need to consider. The first is a system in which the particles are distinguishable (such as when they are at fixed locations in a solid). The second is a system in which the particles are indistinguishable (such as when they are in a gas and able to exchange places).

Step 1 *Consider a system of independent, distinguishable molecules*

The total energy of a collection of N independent molecules is the sum of the energies of the individual molecules. Therefore, the total energy of a state i of the system is written as

$$E_i = \varepsilon_i(1) + \varepsilon_i(2) + \cdots + \varepsilon_i(N)$$

In this expression, $\varepsilon_i(1)$ is the energy of molecule 1 when the system is in the state i, $\varepsilon_i(2)$ the energy of molecule 2 when the system is in the same state i, and so on. The canonical partition function is then

$$Q = \sum_i e^{-\beta\varepsilon_i(1)-\beta\varepsilon_i(2)-\cdots-\beta\varepsilon_i(N)}$$

Provided the molecules are distinguishable (in a sense described below), the sum over the states of the system can be reproduced by letting each molecule enter all its own individual states. Therefore, instead of summing over the states i of the system, sum over all the individual states j of molecule 1, all the states j

of molecule 2, and so on. This rewriting of the original expression leads to

$$Q = \overbrace{\left(\sum_j \mathrm{e}^{-\beta\varepsilon_j}\right)}^{q}\overbrace{\left(\sum_j \mathrm{e}^{-\beta\varepsilon_j}\right)}^{q}\cdots\overbrace{\left(\sum_j \mathrm{e}^{-\beta\varepsilon_j}\right)}^{q}$$

and therefore to $Q = q^N$.

Step 2 *Consider a system of independent, indistinguishable molecules*

If all the molecules are identical and free to move through space, it is not possible to distinguish them and the relation $Q = q^N$ is not valid. Suppose that molecule 1 is in some state a, molecule 2 is in b, and molecule 3 is in c, then one member of the ensemble has an energy $E = \varepsilon_a + \varepsilon_b + \varepsilon_c$. This member, however, is indistinguishable from one formed by putting molecule 1 in state b, molecule 2 in state c, and molecule 3 in state a, or some other permutation. There are six such permutations in all, and $N!$ in general. In the case of indistinguishable molecules, it follows that too many states have been counted in going from the sum over system states to the sum over molecular states, so writing $Q = q^N$ overestimates the value of Q. The detailed argument is quite involved, but at all except very low temperatures it turns out that the correction factor is $1/N!$, so $Q = q^N/N!$.

Step 3 *Summarize the results*

It follows that:

For distinguishable, independent molecules:
$$Q = q^N \qquad (13D.6a)$$

For indistinguishable, independent molecules:
$$Q = q^N/N! \qquad (13D.6b)$$

For molecules to be indistinguishable, they must be of the same kind: an Ar atom is never indistinguishable from a Ne atom. Their identity, however, is not the only criterion. Each identical molecule in a crystal lattice, for instance, can be 'named' with a set of coordinates. Identical molecules in a lattice can therefore be treated as distinguishable because their sites are distinguishable, and eqn 13D.6a can be used. On the other hand, identical molecules in a gas are free to move to different locations, and there is no way of keeping track of the identity of a given molecule; therefore eqn 13D.6b must be used.

Brief illustration 13D.2

For a gas of N indistinguishable, independent molecules, $Q = q^N/N!$. The energy of the system is therefore

$$\langle E\rangle = -\left(\frac{\partial \ln(q^N/N!)}{\partial\beta}\right)_V = -\left(\frac{\partial(\ln q^N - \ln N!)}{\partial\beta}\right)_V$$
$$= -\left(\frac{\partial(N\ln q)}{\partial\beta}\right)_V$$
$$= -N\left(\frac{\partial \ln q}{\partial\beta}\right)_V = N\langle\varepsilon\rangle$$

where the final step uses eqn 13C.4b, $\langle\varepsilon\rangle = -(\partial\ln q/\partial\beta)_V$. The result shows that the mean energy of the gas is N times the mean energy of a single molecule, which is the expected result for non-interacting molecules.

Exercise

E13D.2 For a system consisting of N distinguishable, independent molecules, $Q = q^N$. Use the same approach as in *Brief illustration* 13D.2 to find an expression for $\langle E\rangle$ for such a system. Comment on your answer.

13D.4 The variation of the energy with volume

When there are interactions between molecules, the energy of a collection depends on the average distance between them, and therefore on the volume that a fixed number occupy. This dependence on volume is particularly important for the discussion of real gases (Topic 1C).

To discuss the dependence of the energy on the volume at constant temperature it is necessary to evaluate $(\partial\langle E\rangle/\partial V)_T$. (In Topics 2D and 3E, this quantity is identified as the 'internal pressure' of a gas and denoted π_T.) To proceed, substitute eqn 13D.5 and obtain

$$\left(\frac{\partial\langle E\rangle}{\partial V}\right)_T = -\left(\frac{\partial}{\partial V}\left(\frac{\partial \ln Q}{\partial\beta}\right)_V\right)_T \qquad (13D.7)$$

If the molecules are non-interacting (independent) the gas will be perfect and $Q = q^N/N!$, with q the partition function for the translational and any internal (such as rotational) modes. If the gas is monatomic, only the translational mode is present and $q = V/\Lambda^3$ with $\Lambda = h/(2\pi mkT)^{1/2}$. The canonical partition function would then be $V^N/\Lambda^{3N}N!$. The presence of interactions is taken into account by replacing $V^N/N!$ by a factor called the **configuration integral**, Z, which depends on the intermolecular potentials (don't confuse this Z with the compression factor Z in Topic 1C), and writing

$$Q = \frac{Z}{\Lambda^{3N}} \qquad (13D.8)$$

It then follows that

$$\left(\frac{\partial\langle E\rangle}{\partial V}\right)_T = -\left(\frac{\partial}{\partial V}\left(\frac{\partial \ln(Z/\Lambda^{3N})}{\partial\beta}\right)_V\right)_T$$

$$= -\left(\frac{\partial}{\partial V}\left(\frac{\partial \ln Z}{\partial\beta}\right)_V\right)_T - \left(\frac{\partial}{\partial V}\left(\frac{\partial \ln(1/\Lambda^{3N})}{\partial\beta}\right)_V\right)_T$$

$\partial^2 f/\partial x\partial y = \partial^2 f/\partial y\partial x$ for the second term

$$= -\left(\frac{\partial}{\partial V}\left(\frac{\partial \ln Z}{\partial\beta}\right)_V\right)_T - \left(\frac{\partial}{\partial \beta}\overbrace{\left(\frac{\partial \ln(1/\Lambda^{3N})}{\partial V}\right)_T}^{0}\right)_V$$

$\partial \ln y/\partial x = (1/y)\partial y/\partial x$

$$= -\left(\frac{\partial}{\partial V}\left(\frac{\partial \ln Z}{\partial\beta}\right)_V\right)_T = -\left(\frac{\partial}{\partial V}\frac{1}{Z}\left(\frac{\partial Z}{\partial\beta}\right)_V\right)_T \quad (13D.9)$$

In the third line, Λ is independent of volume, so its derivative with respect to volume is zero.

For a real gas of atoms (for which the intermolecular interactions are isotropic), Z is related to the total potential energy E_p of interaction of all the particles, which depends on all their relative locations, by

$$Z = \frac{1}{N!}\int e^{-\beta E_p} d\tau_1 d\tau_2 \ldots d\tau_N \qquad \text{Configuration integral} \quad (13D.10)$$

where $d\tau_i$ is the volume element for atom i and the integration is over all the variables. The physical origin of this term is that the probability of occurrence of each arrangement of molecules possible in the sample is given by a Boltzmann distribution in which the exponent is given by the potential energy corresponding to that arrangement.

Brief illustration 13D.3

Equation 13D.10 is very difficult to manipulate in practice, even for quite simple intermolecular potentials, except for a perfect gas for which $E_p = 0$. In that case, the exponential function becomes 1 and

$$Z = \frac{1}{N!}\int d\tau_1 d\tau_2 \ldots d\tau_N = \frac{1}{N!}\left(\int d\tau\right)^N = \frac{V^N}{N!}$$

just as it should be for a perfect gas.

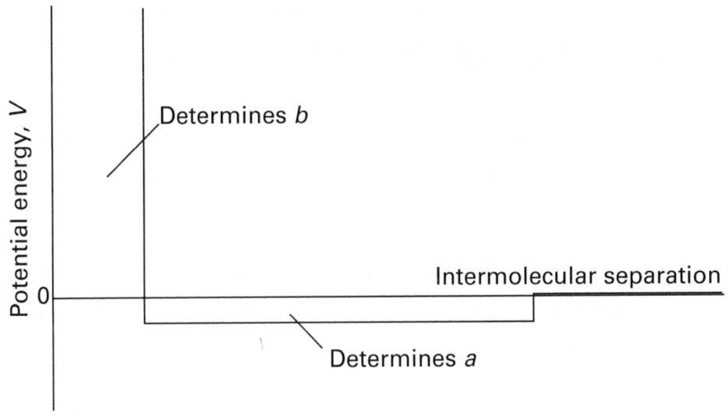

Figure 13D.5 The intermolecular potential energy of molecules in a real gas can be modelled by a central hard sphere that determines the van der Waals parameter b surrounded by a shallow attractive well that determines the parameter a. As mentioned in the text, calculations of the canonical partition function based on this are consistent with the van der Waals equation of state (Topic 1C).

If the potential energy has the form of a central hard sphere surrounded by a shallow attractive well (Fig. 13D.5), then detailed calculation, which is too involved to reproduce here (see *A deeper look* 13D.1, available to read in the e-book accompanying this text), leads to

$$\left(\frac{\partial\langle E\rangle}{\partial V}\right)_T = \frac{an^2}{V^2} \qquad \text{Attractive potential} \quad (13D.11)$$

where n is the amount of molecules present in the volume V and a is a constant that is proportional to the area under the attractive part of the potential. In *Example* 3E.2 exactly the same expression (in the form $\pi_T = an^2/V^2$) was derived from the van der Waals equation of state. The conclusion at this point is that if there are attractive interactions between molecules in a gas, then its energy increases as it expands isothermally (because $(\partial\langle E\rangle/\partial V)_T > 0$, and the slope of $\langle E\rangle$ with respect to V is positive). The energy rises because, at greater average separations, the molecules spend less time in regions where they interact favourably.

Exercise

E13D.3 Equation 13D.9 gives an expression for $(\partial\langle E\rangle/\partial V)_T$ in terms of the configuration integral Z. For a perfect gas consisting of N molecules, $Z = V^N/N!$. Use this expression for Z in eqn 13D.9 to find $(\partial\langle E\rangle/\partial V)_T$. Comment on your result.

Checklist of concepts

☐ 1. The **canonical ensemble** is a collection of imaginary replications of the actual system with a common temperature and number of particles.

☐ 2. The **canonical distribution** gives the most probable number of members of the ensemble with a specified total energy.

☐ 3. The mean energy of the members of the ensemble can be calculated from the **canonical partition function**.

Checklist of equations*

Property	Equation	Comment	Equation number
Canonical partition function	$Q=\sum_i e^{-\beta E_i}$	Definition	13D.2
Canonical distribution	$\tilde{N}_i/\tilde{N}=e^{-\beta E_i}/Q$		13D.2
Mean energy	$\langle E\rangle=-(1/Q)(\partial Q/\partial\beta)_V=-(\partial \ln Q/\partial\beta)_V$		13D.5
Canonical partition function	$Q=Z/\Lambda^{3N}$		13D.8
Configuration integral	$Z=(1/N!)\int e^{-\beta E_p}d\tau_1 d\tau_2 \ldots d\tau_N$	Isotropic interaction	13D.10
Variation of mean energy with volume	$(\partial\langle E\rangle/\partial V)_T=an^2/V^2$	Potential energy as specified in Fig. 13D.5	13D.11

* $\beta=1/kT$

TOPIC 13E The internal energy and the entropy

➤ **Why do you need to know this material?**

The importance of this discussion is the insight that a molecular interpretation provides into thermodynamic properties.

➤ **What is the key idea?**

The partition function contains all the thermodynamic information about a system and thus provides a bridge between spectroscopy and thermodynamics.

➤ **What do you need to know already?**

You need to know how to calculate a molecular partition function from structural data (Topic 13B); you should also be familiar with the concepts of internal energy (Topic 2A) and entropy (Topic 3A). This Topic makes use of the calculations of mean molecular energies in Topic 13C.

Any thermodynamic function can be obtained once the partition function is known. The two fundamental properties of thermodynamics are the internal energy, U, and the entropy, S. Once these two properties have been calculated, it is possible to calculate the derived functions, such as the Gibbs energy, G, (Topic 13F) and all the chemically interesting properties that stem from them.

13E.1 The internal energy

The first example of the importance of the molecular partition function, q, is the derivation of an expression for the internal energy.

13E.1(a) The calculation of internal energy

It is established in Topic 13C that the mean energy of a molecule in a system composed of independent (non-interacting) molecules is related to the molecular partition function by

$$\langle\varepsilon\rangle = -\frac{1}{q}\left(\frac{\partial q}{\partial \beta}\right)_V \qquad (13E.1)$$

with $\beta = 1/kT$. The total energy of a system composed of N molecules is therefore $N\langle\varepsilon\rangle$. This total energy is the energy above the ground state (the calculation of q is based on the convention $\varepsilon_0 \equiv 0$) and so the internal energy is $U(T) = U(0) + N\langle\varepsilon\rangle$, where $U(0)$ is the internal energy when only the ground state is occupied, which is the case at $T = 0$. It follows that the internal energy is related to the molecular partition function by

$$U(T) = U(0) + N\langle\varepsilon\rangle = U(0) - \frac{N}{q}\left(\frac{\partial q}{\partial \beta}\right)_V$$

Internal energy [independent molecules] (13E.2a)

In many cases, the expression for $\langle\varepsilon\rangle$ established for each mode of motion in Topic 13C can be used and it is not necessary to go back to q itself except for some formal manipulations. For instance, it is established in Topic 13C (eqn 13C.8) that the mean energy of a harmonic oscillator is $\langle\varepsilon^{V}\rangle = hc\tilde{\nu}/(e^{\beta hc\tilde{\nu}} - 1)$. It follows that the molar internal energy of a system composed of oscillators is

$$U_m^{V}(T) = U_m^{V}(0) + N_A\langle\varepsilon^{V}\rangle = U_m^{V}(0) + \frac{N_A hc\tilde{\nu}}{e^{\beta hc\tilde{\nu}} - 1}$$

Brief illustration 13E.1

The vibrational wavenumber of an I_2 molecule is 214.6 cm^{-1}. At 298.15 K, $kT/hc = 207.225$ cm^{-1}. With these values

$$\beta hc\tilde{\nu} = \frac{hc\tilde{\nu}}{kT} = \frac{214.6\ \text{cm}^{-1}}{207.225\ \text{cm}^{-1}} = 1.035\ldots$$

$$hc\tilde{\nu} = (6.626\times10^{-34}\ \text{J s})\times(2.998\times10^{10}\ \text{cm s}^{-1})\times(214.6\ \text{cm}^{-1})$$
$$= 4.26\ldots10^{-21}\ \text{J}$$

It follows that the vibrational contribution to the molar internal energy of I_2 is

$$U_m^{V}(T) = U_m^{V}(0) + \frac{(6.022\times10^{23}\ \text{mol}^{-1})\times(4.26\ldots10^{-21}\ \text{J})}{e^{1.035\ldots} - 1}$$
$$= U_m^{V}(0) + 1.41\ \text{kJ mol}^{-1}$$

An alternative form of eqn 13E.2a is

$$U(T)=U(0)-N\left(\frac{\partial \ln q}{\partial \beta}\right)_V \quad \text{Internal energy [independent molecules]} \quad (13E.2b)$$

A very similar expression is used for a system of interacting molecules (Topic 13D). In that case the mean energy $\langle E\rangle$, given by eqn 13D.5 in terms of the canonical partition function, Q, is identified as the contribution of other than the ground state to the internal energy to give

$$U(T)=U(0)-\left(\frac{\partial \ln Q}{\partial \beta}\right)_V \quad \text{Internal energy [interacting molecules]} \quad (13E.2c)$$

13E.1(b) Heat capacity

The constant-volume heat capacity (Topic 2A) is defined as $C_V=(\partial U/\partial T)_V$. It is therefore possible to find expressions for the heat capacity in terms of the partition function by differentiating an expression for U in terms of the q. The procedure is illustrated here for a system of harmonic oscillators.

The mean vibrational energy of a harmonic oscillator is given by eqn 13C.8, quoted above as $\langle\varepsilon^V\rangle=hc\tilde{\nu}/(e^{\beta hc\tilde{\nu}}-1)$. The calculation of the heat capacity involves taking the derivative with respect to T, so it is convenient to rewrite this expression for the mean energy in terms of T and the characteristic vibrational temperature $\theta^V=hc\tilde{\nu}/k$. It follows that $hc\tilde{\nu}=k\theta^V$ and $\beta hc\tilde{\nu}=k\theta^V/kT=\theta^V/T$. With these substitutions the mean energy is

$$\langle\varepsilon^V\rangle=\frac{hc\tilde{\nu}}{e^{\beta hc\tilde{\nu}}-1}=\frac{k\theta^V}{e^{\theta^V/T}-1}$$

The molar internal energy is therefore

$$U_m^V(T)=U_m^V(0)+\frac{N_A k\theta^V}{e^{\theta^V/T}-1}=U_m^V(0)+R\theta^V\boxed{\frac{1}{e^{\theta^V/T}-1}}$$

where $R=N_Ak$. To evaluate the derivative of $U_m^V(T)$ with respect to T note that only the boxed term depends on the temperature and that by using the rules for differentiating a function of a function (see Part 1 of the *Resources section*)

$$\frac{d}{dT}\boxed{\frac{1}{e^{\theta^V/T}-1}}=(-1)\times\frac{1}{(e^{\theta^V/T}-1)^2}\times e^{\theta^V/T}\times\frac{-\theta^V}{T^2}$$
$$=\frac{\theta^V}{T^2}\frac{e^{\theta^V/T}}{(e^{\theta^V/T}-1)^2}$$

Then the molar heat capacity is

$$C_{V,m}^V=\left(\frac{\partial U_m^V(T)}{\partial T}\right)_V=R\theta^V\frac{d}{dT}\frac{1}{e^{\theta^V/T}-1}$$
$$=R\theta^V\times\frac{\theta^V}{T^2}\frac{e^{\theta^V/T}}{(e^{\theta^V/T}-1)^2}=R\left(\frac{\theta^V}{T}\right)^2\frac{e^{\theta^V/T}}{(e^{\theta^V/T}-1)^2}$$

By noting that $e^{\theta^V/T}=(e^{\theta^V/2T})^2$, this expression can be rearranged into

$$C_{V,m}^V=Rf(T),\quad f(T)=\left(\frac{\theta^V}{T}\right)^2\left(\frac{e^{\theta^V/2T}}{e^{\theta^V/T}-1}\right)^2 \quad \text{Vibrational contribution to } C_{V,m} \quad (13E.3)$$

The graph in Fig. 13E.1 shows how the vibrational heat capacity depends on temperature. Note that even when the temperature is only slightly above θ^V the heat capacity is close to its equipartition value of R. Equation 13E.3 is essentially the same as the Einstein formula for the heat capacity of a solid (eqn 7A.9a) with θ^V the Einstein temperature, θ_E. The only difference is that in a solid the vibrations take place in three dimensions.

It is sometimes more convenient to convert the derivative with respect to T into a derivative with respect to $\beta=1/kT$ by using

$$\frac{d}{dT}=\frac{d\beta}{dT}\frac{d}{d\beta}=-\frac{1}{kT^2}\frac{d}{d\beta}=-k\beta^2\frac{d}{d\beta} \quad (13E.4)$$

It follows that

$$C_V=-k\beta^2\left(\frac{\partial U}{\partial \beta}\right)_V=-Nk\beta^2\left(\frac{\partial\langle\varepsilon\rangle}{\partial\beta}\right)_V\overset{\langle\varepsilon\rangle=-(\partial \ln q/\partial\beta)_V}{=}Nk\beta^2\left(\frac{\partial^2\ln q}{\partial\beta^2}\right)_V \quad \text{Heat capacity} \quad (13E.5)$$

There is a much simpler route to finding C_V when the equipartition principle can be applied, which is when many energy levels associated with a particular mode of motion are populated. Such a mode is said to be 'active'. Translational modes are

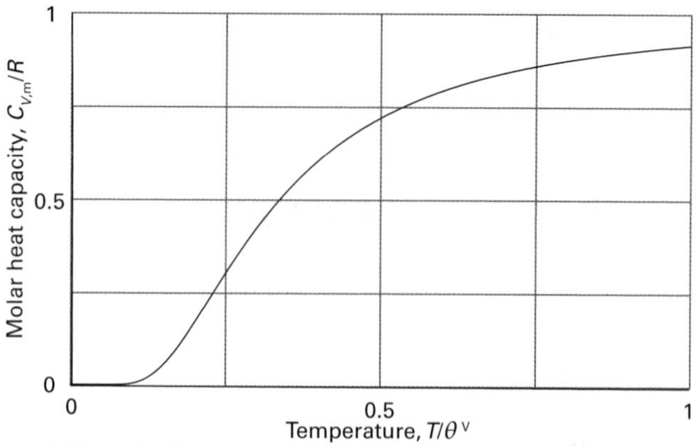

Figure 13E.1 The temperature dependence of the vibrational heat capacity of a molecule in the harmonic approximation calculated by using eqn 13E.3. Note that the heat capacity is within 10 per cent of its classical value for temperatures greater than θ^V.

always active at any reasonable temperature. A rotational mode is active when $T >> \theta^R$, where θ^R is the characteristic temperature for rotation ($\theta^R = hc\tilde{B}/k$). Similarly, a vibrational mode is active when $T >> \theta^V$ with $\theta^V = hc\tilde{\nu}/k$.

Each quadratic term contributes $\frac{1}{2}RT$ to the molar internal energy and, because $C_{V,m} = (\partial U_m/\partial T)_V$, contributes $\frac{1}{2}R$ to the molar heat capacity. In gases, all three translational modes are always active and so contribute $\frac{3}{2}R$ to the molar heat capacity. Each active rotational mode contributes $\frac{1}{2}R$, so if there are ν^{R*} such modes the contribution is $\frac{1}{2}\nu^{R*}R$. For most molecules at normal temperatures $\nu^{R*} = 2$ for linear molecules, and 3 for nonlinear molecules. Each active vibrational mode contributes R (two quadratic terms), so if there are ν^{V*} such modes their contribution is $\nu^{V*}R$. However, in most cases none of the vibrational modes is active and $\nu^{V*} \approx 0$. It follows that the total molar heat capacity of a gas is approximately

$$C_{V,m} = \tfrac{1}{2}(3 + \nu^{R*} + 2\nu^{V*})R \qquad \text{Total heat capacity [equipartition]} \qquad (13E.6)$$

Brief illustration 13E.2

The characteristic temperatures of the vibrations of H_2O are close to 5300 K, 2300 K, and 5400 K; the vibrations are therefore not active at 373 K. The three rotational modes of H_2O have characteristic temperatures 40 K, 21 K, and 13 K, so they are all active. The translational contribution is $\frac{3}{2}R = 12.5\,\mathrm{J\,K^{-1}\,mol^{-1}}$. The rotations contribute a further $12.5\,\mathrm{J\,K^{-1}\,mol^{-1}}$. Therefore, a value close to $25\,\mathrm{J\,K^{-1}\,mol^{-1}}$ is predicted. The experimental value is $26.1\,\mathrm{J\,K^{-1}\,mol^{-1}}$. The discrepancy is probably due to deviations from perfect gas behaviour.

Exercises

E13E.1 Use the equipartition theorem to estimate the constant-volume molar heat capacity of (i) I_2, (ii) CH_4, (iii) C_6H_6 in the gas phase at 25 °C.

E13E.2 Estimate the values of $\gamma = C_{p,m}/C_{V,m}$ for gaseous ammonia and methane. Do this calculation with and without the vibrational contribution to the energy. Which is closer to the experimental value of $\gamma = 1.31$ for both gases at 25 °C? *Hint:* Note that $C_{p,m} - C_{V,m} = R$ for a perfect gas.

E13E.3 The ground level of Cl is ${}^2P_{3/2}$ and a ${}^2P_{1/2}$ level lies $881\,\mathrm{cm^{-1}}$ above it. Calculate the electronic contribution to the molar heat capacity at constant volume of Cl atoms at 500 K.

13E.2 The entropy

One of the most celebrated equations in statistical thermodynamics, the 'Boltzmann formula' for the entropy, is obtained by establishing the relation between the entropy and the weight of the most probable configuration.

How is that done? 13E.1 Deriving the Boltzmann formula for the entropy

The starting point for this derivation is the expression for the internal energy, $U(T) = U(0) + N\langle\varepsilon\rangle$, which, with eqn 13C.1, $\langle\varepsilon\rangle = (1/N)\sum_i N_i\varepsilon_i$, can be written

$$U(T) = U(0) + \sum_i N_i\varepsilon_i$$

The strategy is to use classical thermodynamics to establish a relation between dS and dU, and then to use this relation to express dU in terms of the weight of the most probable configuration.

Step 1 *Write an expression for* d$U(T)$

A change in $U(T)$ may arise from either a modification of the energy levels of a system (when ε_i changes to $\varepsilon_i + \mathrm{d}\varepsilon_i$) or from a modification of the populations (when N_i changes to $N_i + \mathrm{d}N_i$). The most general change is therefore

$$\mathrm{d}U(T) = \mathrm{d}U(0) + \sum_i N_i\mathrm{d}\varepsilon_i + \boxed{\sum_i \varepsilon_i\mathrm{d}N_i}$$

Step 2 *Write an expression for* dS

Because neither $U(0)$ nor the energy levels change when a system is heated at constant volume (Fig. 13E.2), in the absence of all changes other than heating, only the third (boxed) term on the right survives. The fundamental equation $\mathrm{d}U = T\mathrm{d}S - p\mathrm{d}V$ (eqn 3E.1) relates the change in internal energy to the changes in entropy and volume. Under the constant volume conditions imposed here it follows that $\mathrm{d}U = T\mathrm{d}S$ and hence $\mathrm{d}S = \mathrm{d}U/T$. Therefore,

$$\mathrm{d}S = \frac{\mathrm{d}U}{T} = \frac{1}{T}\sum_i \varepsilon_i\mathrm{d}N_i = k\beta\sum_i \varepsilon_i\mathrm{d}N_i$$

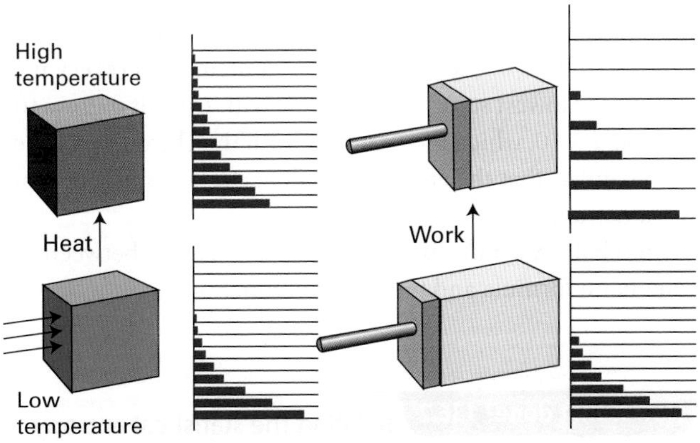

Figure 13E.2 (a) When a system is heated, the energy levels are unchanged but their populations are changed. (b) When work is done on a system, the energy levels themselves are changed. The levels in this case are the one-dimensional particle-in-a-box energy levels of Topic 7D: they depend on the size of the container and move apart as its length is decreased.

Step 3 *Write an expression for* dS *in terms of* $\mathcal{W}$

Changes in the most probable configuration (the only one to consider) are given by eqn 13A.7, $\partial(\ln\mathcal{W})/\partial N_i + \alpha - \beta\varepsilon_i = 0$. It follows that $\beta\varepsilon_i = \partial(\ln\mathcal{W})/\partial N_i + \alpha$. This expression for $\beta\varepsilon_i$ is substituted into the above expression for dS to give

$$\begin{aligned}
\mathrm{d}S &= k\beta\sum_i \varepsilon_i \mathrm{d}N_i = k\sum_i \beta\varepsilon_i \mathrm{d}N_i \\
&= k\sum_i\left(\frac{\partial\ln\mathcal{W}}{\partial N_i}+\alpha\right)\mathrm{d}N_i \\
&= k\sum_i\left(\frac{\partial\ln\mathcal{W}}{\partial N_i}\right)\mathrm{d}N_i + \boxed{k\sum_i \alpha\,\mathrm{d}N_i}
\end{aligned}$$

Because the system contains a fixed number of molecules the sum of the changes in the populations of the states $\sum_i \mathrm{d}N_i$ is zero, so the term in the box is zero. The expression $(\partial\ln\mathcal{W}/\partial N_i)\mathrm{d}N_i$ simplifies to $\mathrm{d}\ln\mathcal{W}$, leaving

$$\mathrm{d}S = k(\mathrm{d}\ln\mathcal{W})$$

This relation strongly suggests that

$$S = k\ln\mathcal{W} \qquad (13\text{E}.7)$$

Boltzmann formula for the entropy

This very important expression is the **Boltzmann formula** for the entropy. In it, $\mathcal{W}$ is the weight of the most probable configuration of the system (as discussed in Topic 13A).

13E.2(a) Entropy and the partition function

The **statistical entropy**, the entropy calculated from the Boltzmann formula, behaves in exactly the same way as the thermodynamic entropy. Thus, as the temperature is lowered, the value of $\mathcal{W}$, and hence of S, decreases because fewer configurations are consistent with the total energy. In the limit $T\to 0$, $\mathcal{W}=1$, so $\ln\mathcal{W}=0$, because only one configuration (every molecule in the lowest state) is compatible with $U(0)$. It follows that $S\to 0$ as $T\to 0$, which is compatible with the Third Law of thermodynamics, that the entropies of all perfect crystals approach the same value as $T\to 0$ (Topic 3C).

The challenge now is to establish a relation between the Boltzmann formula and the partition function.

How is that done? 13E.2 Relating the statistical entropy to the partition function

It is necessary to separate the calculation into two parts, one for distinguishable independent molecules and the other for indistinguishable independent molecules. Interactions can be taken into account, in principle at least, by a simple generalization.

Step 1 *Derive the relation for distinguishable independent molecules*

For a system composed of N distinguishable molecules, $\ln\mathcal{W}$ is given by eqn 13A.3, $\ln\mathcal{W} = N\ln N - \sum_i N_i \ln N_i$; the substitution $N=\sum_i N_i$ then gives

$$\begin{aligned}
\ln\mathcal{W} &= \overbrace{N}^{\sum_i N_i}\ln N - \sum_i N_i\ln N_i \\
&= \sum_i N_i \ln N - \sum_i N_i \ln N_i \\
&= \sum_i N_i(\ln N - \ln N_i) = -\sum_i N_i \ln\frac{N_i}{N}
\end{aligned}$$

Use this expression for $\ln\mathcal{W}$ in eqn 13E.7 to obtain the entropy as

$$S = k\ln\mathcal{W} = -k\sum_i N_i\ln\frac{N_i}{N}$$

The value of N_i/N for the most probable distribution is given by the Boltzmann distribution, $N_i/N = \mathrm{e}^{-\beta\varepsilon_i}/q$, and so

$$\ln\frac{N_i}{N} = \ln\mathrm{e}^{-\beta\varepsilon_i} - \ln q = -\beta\varepsilon_i - \ln q$$

Therefore,

$$\begin{aligned}
S &= -k\sum_i N_i\ln\frac{N_i}{N} = -k\sum_i N_i(-\beta\varepsilon_i - \ln q) \\
&= k\beta\sum_i N_i\varepsilon_i + k\sum_i N_i\ln q = k\beta\overbrace{\sum_i N_i\varepsilon_i}^{N\langle\varepsilon\rangle} + k\ln q\overbrace{\sum_i N_i}^{N} \\
&= k\beta N\langle\varepsilon\rangle + Nk\ln q
\end{aligned}$$

Finally, because $N\langle\varepsilon\rangle = U(T)-U(0)$ and $\beta = 1/kT$, it follows that

$$S = \frac{U(T)-U(0)}{T} + Nk\ln q \qquad (13\text{E}.8\text{a})$$

The entropy [independent, distinguishable molecules]

Step 2 *Derive the relation for indistinguishable molecules*

To treat a system composed of N indistinguishable molecules, the weight $\mathcal{W}$ is reduced by a factor of $N!$ because the $N!$ permutations of the molecules among the states result in the same state of the system. Therefore, $\ln\mathcal{W}$, and therefore the entropy, is reduced by $k\ln N!$ from the 'distinguishable' value. Because N is so large, Stirling's approximation, $\ln N! = N\ln N - N$, can be used to convert eqn 13E.8a into

$$\begin{aligned}
S(T) &= \frac{U(T)-U(0)}{T} + Nk\ln q - k\overbrace{(N\ln N - N)}^{\ln N!} \\
&= \frac{U(T)-U(0)}{T} + Nk\ln\frac{q}{N} + kN
\end{aligned}$$

The kN can be combined with the logarithm by writing it as $kN\ln\mathrm{e}$, to give

$$S=\frac{U(T)-U(0)}{T}+Nk\ln\frac{q\mathrm{e}}{N} \qquad (13\text{E}.8\text{b})$$

The entropy [independent, indistinguishable molecules]

Step 3 *Generalize to interacting molecules*

For completeness, the corresponding expression for interacting molecules, based on the canonical partition function in place of the molecular partition function, is

$$S=\frac{U(T)-U(0)}{T}+k\ln Q \qquad (13\text{E}.8\text{c})$$

The entropy [interacting molecules]

Equation 13E.8a expresses the entropy of a collection of independent molecules in terms of the internal energy and the molecular partition function. However, because the energy of a molecule is a sum of contributions, such as translational (T), rotational (R), vibrational (V), and electronic (E), the partition function factorizes into a product of contributions. As a result, the entropy is also the sum of the individual contributions. In a gas, the molecules are free to change places, so they are indistinguishable; therefore, for the translational contribution to the entropy, use eqn 13E.8b. For the other modes (specifically R, V, and E), which do not involve the exchange of molecules and therefore do not require the weight to be reduced by $N!$, use eqn 13E.8a.

For a system with two states, with energies 0 and ε, it is shown in Topics 13B and 13C that the partition function and mean energy are $q=1+\mathrm{e}^{-\beta\varepsilon}$ and $\langle\varepsilon\rangle=\varepsilon/(\mathrm{e}^{\beta\varepsilon}+1)$. It follows that $U(T)-U(0)=N\varepsilon/(\mathrm{e}^{\beta\varepsilon}+1)$. The contribution to the molar entropy is evaluated by using eqn 13E.8a with $N=N_\mathrm{A}$

$$\begin{aligned}S_\mathrm{m}&=\frac{U_\mathrm{m}(T)-U_\mathrm{m}(0)}{T}+N_\mathrm{A}k\ln q\\&=\frac{N_\mathrm{A}\varepsilon}{T(\mathrm{e}^{\beta\varepsilon}+1)}+N_\mathrm{A}k\ln(1+\mathrm{e}^{-\beta\varepsilon})\\&=R\left\{\frac{\beta\varepsilon}{\mathrm{e}^{\beta\varepsilon}+1}+\ln(1+\mathrm{e}^{-\beta\varepsilon})\right\}\end{aligned}$$

This awkward function is plotted in Fig. 13E.3. It should be noted that at high temperatures (when $kT>>\varepsilon$ or $\beta\varepsilon<<1$), the molar entropy approaches $R\ln 2$.

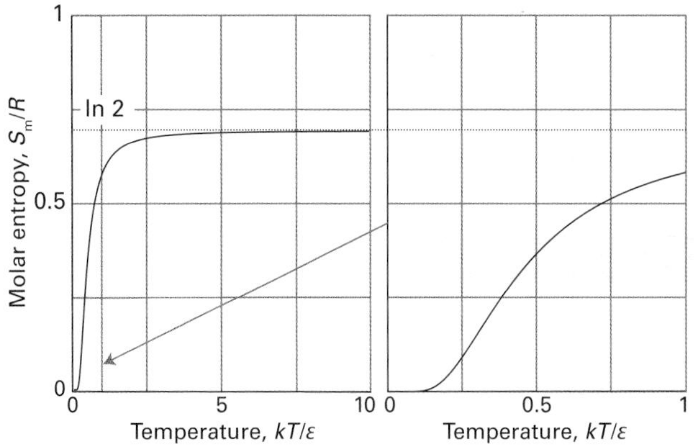

Figure 13E.3 The temperature variation of the molar entropy of a collection of two-level systems expressed as a multiple of $R=N_\mathrm{A}k$. When $kT>>\varepsilon$ the two states become equally populated and S_m approaches $R\ln 2$.

How is that done? 13E.3 Deriving the expression for the entropy of a monatomic perfect gas

You need to start with eqn 13E.8b for a collection of independent, indistinguishable atoms and write $N=nN_\mathrm{A}$, where N_A is Avogadro's constant and n is their amount (in moles). The only mode of motion for a gas of atoms is translation and application of the equipartition principle gives the contribution to the molar energy as $\frac{3}{2}RT$. It follows that $U(T)-U(0)=\frac{3}{2}nRT$. The partition function is $q=V/\Lambda^3$ (eqn 13B.10b), where Λ is the thermal wavelength (Topic 13B). Therefore,

$$\begin{aligned}S&=\frac{U(T)-U(0)}{T}+Nk\ln\frac{q\mathrm{e}}{N}=\frac{\overbrace{U(T)-U(0)}^{\frac{3}{2}nRT}}{T}+nN_\mathrm{A}k\ln\frac{q\mathrm{e}}{nN_\mathrm{A}}\\&=\tfrac{3}{2}nR+\overbrace{nN_\mathrm{A}k}^{nR}\ln\frac{V\mathrm{e}}{nN_\mathrm{A}\Lambda^3}\\&=nR\left\{\overbrace{\tfrac{3}{2}}^{\ln \mathrm{e}^{3/2}}+\ln\frac{V_\mathrm{m}\mathrm{e}}{N_\mathrm{A}\Lambda^3}\right\}=nR\ln\frac{V_\mathrm{m}\mathrm{e}^{5/2}}{N_\mathrm{A}\Lambda^3}\end{aligned}$$

where $V_\mathrm{m}=V/n$ is the molar volume of the gas and $\frac{3}{2}$ has been replaced by $\ln \mathrm{e}^{3/2}$. Division of both sides by n gives an expression for the molar entropy, the **Sackur–Tetrode equation:**

$$S_\mathrm{m}=R\ln\left(\frac{V_\mathrm{m}\mathrm{e}^{5/2}}{N_\mathrm{A}\Lambda^3}\right) \qquad (13\text{E}.9\text{a})$$

Sackur–Tetrode equation [monatomic perfect gas]

where Λ is the thermal wavelength ($\Lambda=h/(2\pi mkT)^{1/2}$). To calculate the standard molar entropy, note that $V_\mathrm{m}=RT/p$, and set $p=p^\ominus$:

$$S_\mathrm{m}^\ominus=R\ln\left(\frac{RT\mathrm{e}^{5/2}}{p^\ominus N_\mathrm{A}\Lambda^3}\right)\overset{R/N_\mathrm{A}=k}{=}R\ln\left(\frac{kT\mathrm{e}^{5/2}}{p^\ominus\Lambda^3}\right) \qquad (13\text{E}.9\text{b})$$

13E.2(b) The translational contribution

The expressions derived for the entropy are in line with what is expected for entropy as a measure of the spread of the populations of molecules over the available states. This interpretation can be illustrated by deriving an expression for the molar entropy of a monatomic perfect gas.

These expressions assume that many translational levels are occupied so that the translational partition function is given by eqn 13B.10b and the equipartition principle applies for the translational contribution to the energy. Therefore, these expressions for the entropy do not apply when T is equal to or very close to zero.

Brief illustration 13E.3

The mass of an Ar atom is $39.95m_u$. At 25 °C, the thermal wavelength of Ar is 16.0 pm and $kT = 4.12 \times 10^{-21}$ J. Therefore, the molar entropy of argon at this temperature is

$$S_m^{\ominus} = R \ln \left\{ \frac{(4.12\times10^{-21}\ \mathrm{J})\times e^{5/2}}{(10^5\ \mathrm{N\,m^{-2}})\times(1.60\times10^{-11}\ \mathrm{m})^3} \right\} = 18.6\,R = 155\ \mathrm{J\,K^{-1}\,mol^{-1}}$$

On the basis that there are fewer accessible translational states for a lighter atom than for a heavy atom under the same conditions (see below), it can be anticipated that the standard molar entropy of Ne is likely to be smaller than for Ar; its actual value is $17.6R$ at 298 K.

The physical interpretation of these equations is as follows:

- Because the molecular mass appears in the numerator (because it appears in the denominator of Λ), the molar entropy of a perfect gas of heavy molecules is greater than that of a perfect gas of light molecules under the same conditions. This feature can be understood in terms of the energy levels of a particle in a box being closer together for heavy particles than for light particles, so more states are thermally accessible.
- Because the molar volume appears in the numerator, the molar entropy increases with the molar volume of the gas, the temperature being unchanged. The reason is similar: large containers have more closely spaced energy levels than small containers, so once again more states are thermally accessible.
- Because the temperature appears in the numerator (because, like m, it appears in the denominator of Λ), the molar entropy increases with increasing temperature, the molar volume being constant. The reason for this behaviour is that more energy levels become accessible as the temperature is raised.

The expression for the entropy can be written in the form

$$S = nR \ln \frac{V_m e^{5/2}}{N_A \Lambda^3} = nR \ln \frac{V e^{5/2}}{nN_A \Lambda^3} = nR \ln aV, \quad a = \frac{e^{5/2}}{nN_A \Lambda^3}$$

where a is a constant that depends on the temperature and the atomic mass of the gas. This relation for the entropy implies that when a monatomic perfect gas expands isothermally from V_i to V_f, its entropy changes by

$$\Delta S = nR \ln aV_f - nR \ln aV_i = nR \ln \frac{V_f}{V_i} \qquad \text{Change of entropy on expansion [perfect gas, isothermal]} \qquad (13E.10)$$

This expression is the same as that obtained starting from the thermodynamic definition of entropy (Topic 3B).

13E.2(c) The rotational contribution

The rotational contribution to the molar entropy, S_m^R, can be calculated by using eqn 13E.8a once the molecular partition function is known. For a linear molecule, the high-temperature limit of q^R is $kT/\sigma hc\tilde{B}$ (eqn 13B.13b, $q^R = T/\sigma\theta^R$ with $\theta^R = hc\tilde{B}/k$) and the equipartition theorem gives the rotational contribution to the molar internal energy as RT. Therefore, from eqn 13E.8a:

$$S_m^R = \frac{\overbrace{U_m(T) - U_m(0)}^{RT}}{T} + R \ln \overbrace{q^R}^{kT/\sigma hc\tilde{B}}$$

and the contribution at high temperatures is

$$S_m^R = R\left\{1 + \ln \frac{kT}{\sigma hc\tilde{B}}\right\} \qquad \text{Rotational contribution [linear molecule, high temperature } (T >> \theta^R)] \qquad (13E.11a)$$

In terms of the rotational temperature,

$$S_m^R = R\left\{1 + \ln \frac{T}{\sigma\theta^R}\right\} \qquad \text{Rotational contribution [Linear molecule, high temperature } (T >> \theta^R)] \qquad (13E.11b)$$

This function is plotted in Fig. 13E.4. It is seen that:

- The rotational contribution to the entropy increases with temperature because more rotational states become accessible.
- The rotational contribution increases as $\tilde{B}$ decreases (θ^R decreases), because then the rotational energy levels become closer together.

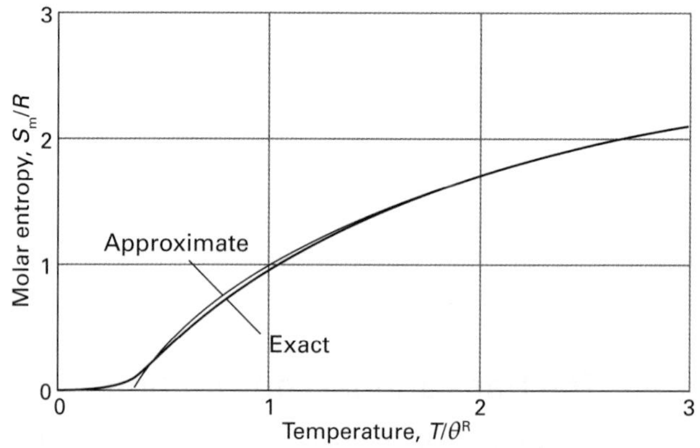

Figure 13E.4 The variation of the rotational contribution to the molar entropy of a linear molecule ($\sigma = 1$) using the high-temperature approximation (thin line) and the exact expression (thick line; for this the partition function is evaluated up to $J = 20$).

It follows that large, heavy molecules have a large rotational contribution to their entropy. As shown in *Brief illustration* 13E.4, the rotational contribution to the molar entropy of $^{35}Cl_2$ is 58.6 $J K^{-1} mol^{-1}$ whereas that for H_2 is only 12.7 $J K^{-1} mol^{-1}$. That is, it is appropriate to regard Cl_2 as a more rotationally disordered gas than H_2, in the sense that at a given temperature Cl_2 occupies a greater number of rotational states than H_2 does.

Brief illustration 13E.4

The rotational contribution for $^{35}Cl_2$ at 25 °C, for instance, is calculated by noting that $\sigma = 2$ for this homonuclear diatomic molecule and taking $\tilde{B} = 0.2441\,cm^{-1}$. The rotational temperature of the molecule is

$$\theta^R = \frac{(6.626\times10^{-34}\,J\,s)\times(2.998\times10^{10}\,cm\,s^{-1})\times(0.2441\,cm^{-1})}{1.381\times10^{-23}\,J\,K^{-1}}$$
$$= 0.351\,K$$

Therefore,

$$S_m^R = R\left\{1+\ln\frac{298\,K}{2\times(0.351\,K)}\right\} = 7.05\,R = 58.6\,J\,K^{-1}\,mol^{-1}$$

Equation 13E.11 is valid at high temperatures ($T >> \theta^R$). To track the rotational contribution down to low temperatures it is necessary to use the full form of the rotational partition function (Topic 13B). The resulting graph has the form shown in Fig. 13E.4, and it is seen that the approximate curve matches the exact curve very well for T/θ^R greater than about 1.

13E.2(d) The vibrational contribution

The vibrational contribution to the molar entropy, S_m^V, is obtained by combining the expression for the molecular partition function (eqn 13B.15, $q^V = 1/(1-e^{-\beta hc\tilde{\nu}}) = 1/(1-e^{-\beta\varepsilon})$ with $\varepsilon = hc\tilde{\nu}$) with the expression for the mean energy (eqn 13C.8, $\langle\varepsilon^V\rangle = \varepsilon/(e^{\beta\varepsilon}-1)$), to obtain

$$S_m^V = \frac{\overbrace{U_m(T)-U_m(0)}^{N_A\langle\varepsilon^V\rangle}}{\underbrace{T}_{1/k\beta}} + R\ln q^V = \frac{\overbrace{N_A k}^{R}\beta\varepsilon}{e^{\beta\varepsilon}-1} + R\ln\frac{1}{1-e^{-\beta\varepsilon}}$$
$$= R\left\{\frac{\beta\varepsilon}{e^{\beta\varepsilon}-1} - \ln(1-e^{-\beta\varepsilon})\right\}$$

That is,

$$S_m^V = R\left\{\frac{\beta hc\tilde{\nu}}{e^{\beta hc\tilde{\nu}}-1} - \ln(1-e^{-\beta hc\tilde{\nu}})\right\}$$

Vibrational contribution to the molar entropy (13E.12a)

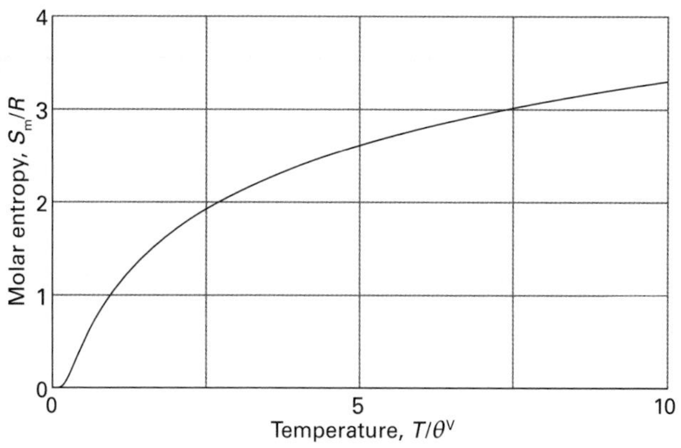

Figure 13E.5 The temperature variation of the molar entropy of a collection of harmonic oscillators expressed as a multiple of $R = N_A k$. The molar entropy approaches zero as $T \to 0$, and increases without limit as the temperature increases.

Once again it is convenient to express this formula in terms of a characteristic temperature, in this case the vibrational temperature, $\theta^V = hc\tilde{\nu}/k$:

$$S_m^V = R\left\{\frac{\theta^V/T}{e^{\theta^V/T}-1} - \ln(1-e^{-\theta^V/T})\right\}$$

Vibrational contribution to the molar entropy (13E.12b)

This function is plotted in Fig. 13E.5. As usual, it is helpful to interpret it, with the graph in mind:

- Both terms multiplying R become zero as $T \to 0$, so the entropy is zero at $T = 0$.
- The molar entropy rises as the temperature is increased as more vibrational states become accessible.
- The molar entropy is higher at a given temperature for molecules with smaller vibrational frequencies. The vibrational energy levels become closer together the smaller the vibrational frequency and so more are thermally accessible.

Brief illustration 13E.5

The vibrational wavenumber of I_2 is 214.6 cm^{-1} and so its vibrational temperature is 309 K. At 25 °C

$$S_m^V = R\left\{\frac{(309\,K)/(298\,K)}{e^{(309\,K)/(298\,K)}-1} - \ln(1-e^{-(309\,K)/(298\,K)})\right\} = 1.01\,R$$
$$= 8.38\,J\,K^{-1}\,mol^{-1}$$

13E.2(e) Residual entropies

Entropies may be calculated from spectroscopic data; they may also be measured experimentally (Topic 3C). In many cases there is good agreement, but in some the experimental entropy is less than the calculated value. One possibility is that the experimental determination failed to take a phase transition into account and a contribution of the form $\Delta_{trs}H/T_{trs}$ was incorrectly omitted from the sum. Another possibility is that some disorder is present in the solid even at $T = 0$. The entropy at $T = 0$ is then greater than zero and is called the **residual entropy**.

The origin and magnitude of the residual entropy can be explained by considering a crystal composed of AB molecules, where A and B are similar atoms (such as CO, with its very small electric dipole moment). There may be so little energy difference between … AB AB AB AB …, … AB BA BA AB …, and other arrangements that the molecules adopt the orientations AB and BA at random in the solid. The entropy arising from residual disorder can be calculated readily by using the Boltzmann formula, $S = k\ln\mathcal{W}$. To do so, suppose that two orientations are equally probable, and that the sample consists of N molecules. Because the same energy can be achieved in 2^N different ways (because each molecule can take either of two orientations), the total number of ways of achieving the same energy is $\mathcal{W} = 2^N$. It follows that

$$S = k\ln 2^N = Nk\ln 2 = nR\ln 2 \qquad (13E.13a)$$

and that a residual molar entropy of $R\ln 2 = 5.8\,\mathrm{J\,K^{-1}\,mol^{-1}}$ is expected for solids composed of molecules that can adopt either of two orientations at $T = 0$. If s orientations are possible, the residual molar entropy is

$$S_m(0) = R\ln s \qquad \text{Residual entropy} \qquad (13E.13b)$$

For CO, the measured residual entropy is $5\,\mathrm{J\,K^{-1}\,mol^{-1}}$, which is close to $R\ln 2$, the value expected for a random structure of the form …CO CO OC CO OC OC….

Similar arguments apply to more complicated cases. Consider a sample of ice with N H_2O molecules. Each O atom is surrounded tetrahedrally by four H atoms, two of which are attached by short σ bonds, the other two being attached by long hydrogen bonds (Fig. 13E.6). It follows that each of the $2N$ H atoms can be in one of two positions (either close to or far from an O atom as shown in Fig. 13E.7), resulting in 2^{2N} possible arrangements. However, not all these arrangements are acceptable. Indeed, of the $2^4 = 16$ ways of arranging four H atoms around one O atom, only 6 have two short and two long OH distances and hence are acceptable (Fig. 13E.7). Therefore, the number of permitted arrangements is $\mathcal{W} = 2^{2N}\left(\frac{6}{16}\right)^N = \left(\frac{3}{2}\right)^N$.

It then follows that the residual entropy is $S(0) \approx k\ln\left(\frac{3}{2}\right)^N = kN\ln\frac{3}{2}$, and its molar value is $S_m(0) \approx R\ln\frac{3}{2} = 3.4\,\mathrm{J K^{-1}\,mol^{-1}}$, which is in very good agreement with the experimental value of $3.4\,\mathrm{J\,K^{-1}\,mol^{-1}}$. The model, however, is not exact because it ignores the possibility that next-nearest neighbours and those beyond can influence the local arrangement of bonds.

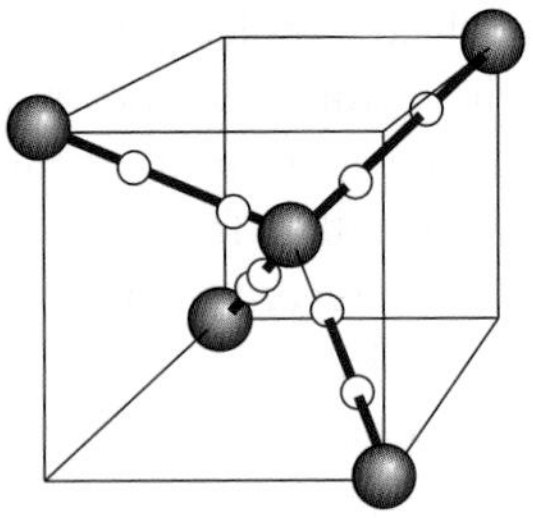

Figure 13E.6 The possible locations of H atoms around a central O atom in an ice crystal are shown by the white spheres. Only one of the locations on each bond may be occupied by an atom, and two H atoms must be close to the O atom and two H atoms must be distant from it.

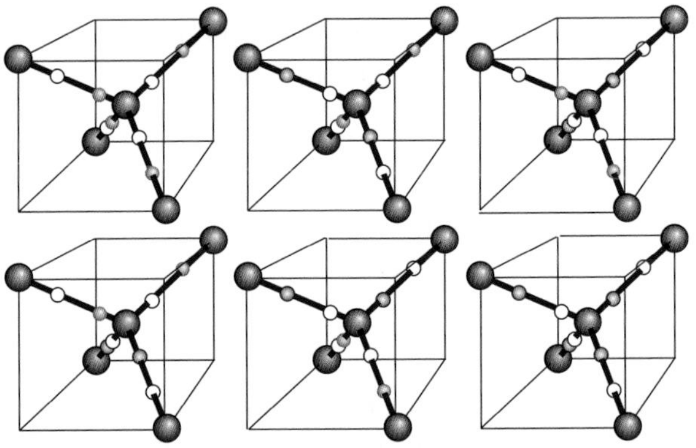

Figure 13E.7 The six possible arrangements of H atoms in the locations identified in Fig.13E.6. Occupied locations are denoted by grey spheres and unoccupied locations by white spheres.

Exercises

E13E.4 Calculate the standard molar entropy at 298 K of (i) gaseous helium, (ii) gaseous xenon.

E13E.5 At what temperature is the standard molar entropy of helium equal to standard molar entropy of xenon at 298 K?

E13E.6 Calculate the rotational partition function of H_2O at 298 K from its rotational constants $27.878\,\mathrm{cm^{-1}}$, $14.509\,\mathrm{cm^{-1}}$, and $9.287\,\mathrm{cm^{-1}}$ and use your result to calculate the rotational contribution to the molar entropy of gaseous H_2O at 25 °C.

E13E.7 The ground state of the Co^{2+} ion in $CoSO_4{\cdot}7H_2O$ may be regarded as ${}^4T_{9/2}$. The entropy of the solid at temperatures below 1 K is derived almost entirely from the electron spin. Estimate the molar entropy of the solid at these temperatures.

Checklist of concepts

☐ 1. The **internal energy** is proportional to the derivative of the logarithm of the partition function with respect to temperature.

☐ 2. The total **heat capacity** of a molecular substance is the sum of the contribution of each mode.

☐ 3. The **statistical entropy** is defined by the Boltzmann formula and expressed in terms of the molecular partition function.

☐ 4. The **residual entropy** is a non-zero entropy at $T = 0$ arising from molecular disorder.

Checklist of equations*

Property	Equation	Comment	Equation number
Internal energy	$U(T) = U(0) - (N/q)(\partial q/\partial \beta)_V = U(0) - N(\partial \ln q/\partial \beta)_V$	Independent molecules	13E.2a
Heat capacity	$C_V = Nk\beta^2(\partial^2 \ln q/\partial \beta^2)_V$	Independent molecules	13E.5
	$C_{V,\mathrm{m}} = \frac{1}{2}(3 + \nu^{\mathrm{R}*} + 2\nu^{\mathrm{V}*})R$		13E.6
Boltzmann formula for the entropy	$S = k \ln \mathcal{W}$		13E.7
Entropy	$S = \{U(T) - U(0)\}/T + Nk \ln q$	Distinguishable molecules	13E.8a
	$S = \{U(T) - U(0)\}/T + Nk \ln (q\mathrm{e}/N)$	Indistinguishable molecules	13E.8b
Sackur–Tetrode equation	$S_\mathrm{m} = R \ln(V_\mathrm{m}\mathrm{e}^{5/2}/N_\mathrm{A}\Lambda^3)$	Molar entropy of a monatomic perfect gas	13E.9a
Residual molar entropy	$S_\mathrm{m}(0) = R \ln s$	s is the number of equivalent orientations	13E.13b

* $\beta = 1/kT$

TOPIC 13F Derived functions

➤ **Why do you need to know this material?**

The power of chemical thermodynamics stems from its deployment of a variety of derived functions, particularly the enthalpy and Gibbs energy. It is therefore important to relate these functions to structural features through partition functions.

➤ **What is the key idea?**

The partition function provides a link between spectroscopic and structural data and the derived functions of thermodynamics, particularly the equilibrium constant.

➤ **What do you need to know already?**

This Topic develops the discussion of internal energy and entropy (Topic 13E). You need to know the relations between those properties and the enthalpy (Topic 2B) and the Helmholtz and Gibbs energies (Topic 3D). The final section makes use of the relation between the standard reaction Gibbs energy and the equilibrium constant (Topic 6A). Although the equations are introduced in terms of the canonical partition function (Topic 13D), all the applications are in terms of the molecular partition function (Topic 13B).

Classical thermodynamics makes extensive use of various derived functions. Thus, in thermochemistry the focus is on the enthalpy and, provided the pressure and temperature are constant, in discussions of spontaneity the focus is on the Gibbs energy. All these functions are derived from the internal energy and the entropy, which in terms of the canonical partition function, $\mathcal{Q}$, are given by

$$U(T) = U(0) - \left(\frac{\partial \ln \mathcal{Q}}{\partial \beta}\right)_V \quad \text{with } \beta = \frac{1}{kT} \qquad \text{Internal energy} \qquad (13\text{F.1a})$$

$$\begin{aligned} S(T) &= \frac{U(T) - U(0)}{T} + k \ln \mathcal{Q} \\ &= -k\beta\left(\frac{\partial \ln \mathcal{Q}}{\partial \beta}\right)_V + k \ln \mathcal{Q} \end{aligned} \qquad \text{Entropy} \qquad (13\text{F.1b})$$

There is no need to worry about this appearance of the canonical partition function (Topic 13D): the only applications in this Topic make use of the molecular partition function, q (Topic 13B). Equations 13F.1a and b can be regarded simply as a succinct way of expressing the relations for collections of independent molecules by writing $\mathcal{Q} = q^N$ for distinguishable molecules and $\mathcal{Q} = q^N/N!$ for indistinguishable molecules (as in a gas). The advantage of using $\mathcal{Q}$ is that the equations to be derived are simple and, if necessary, can be used to calculate thermodynamic properties when there are interactions between the molecules.

13F.1 The derivations

The Helmholtz energy, A, is defined as $A = U - TS$. At $T = 0$, $S = 0$ and the internal energy is $U(0)$, therefore $A(0) = U(0)$. Substitution of the expressions for $U(T)$ and $S(T)$ in eqns 13F.1a and b gives

$$\begin{aligned} A(T) &= U(T) - TS(T) \\ &= \overbrace{A(0)}^{U(0)} - \left(\frac{\partial \ln \mathcal{Q}}{\partial \beta}\right)_V - T\left\{-k\overbrace{\beta}^{1/kT}\left(\frac{\partial \ln \mathcal{Q}}{\partial \beta}\right)_V + k\ln \mathcal{Q}\right\} \\ &= A(0) - kT\ln \mathcal{Q} \end{aligned}$$

That is,

$$A(T) = A(0) - kT \ln \mathcal{Q} \qquad \text{Helmholtz energy} \qquad (13\text{F.2})$$

An infinitesimal change in conditions changes the Helmholtz energy by $\mathrm{d}A = -p\mathrm{d}V - S\mathrm{d}T$ (Table 3E.1). It follows that on imposing constant temperature ($\mathrm{d}T = 0$), the pressure and the Helmholtz energy are related by $p = -(\partial A/\partial V)_T$. It then follows from eqn 13F.2 that

$$p = kT\left(\frac{\partial \ln \mathcal{Q}}{\partial V}\right)_T \qquad \text{Pressure} \qquad (13\text{F.3})$$

This relation is entirely general, and may be used for any type of substance, including perfect gases, real gases, and liquids. Because $\mathcal{Q}$ is in general a function of the volume, temperature, and amount of substance, eqn 13F.3 is an equation of state of the kind discussed in Topics 1A and 3E.

Example 13F.1 Deriving an equation of state

Derive an expression for the pressure of a gas of independent particles.

Collect your thoughts You should suspect that the pressure is that given by the perfect gas law, $p = nRT/V$. To proceed systematically, substitute the explicit formula for Q for a gas of independent, indistinguishable molecules. Only the translational partition function depends on the volume, so there is no need to include the partition functions for the internal modes of the molecules.

The solution For a gas of independent molecules, $Q = q^N/N!$ with $q = V/\Lambda^3$:

$$p = kT\left(\frac{\partial \ln Q}{\partial V}\right)_T = kT\left(\frac{\partial \ln(q^N/N!)}{\partial V}\right)_T$$

$$= kT\left(\frac{\partial(\ln q^N - \ln N!)}{\partial V}\right)_T = kT\left(\frac{\partial(N\ln q - \ln N!)}{\partial V}\right)_T$$

$$= NkT\left(\frac{\partial \ln q}{\partial V}\right)_T - \overbrace{kT\left(\frac{\partial \ln N!}{\partial V}\right)_T}^{0}$$

$$\overset{\mathrm{d}\ln f/\mathrm{d}x = (1/f)\,\mathrm{d}f/\mathrm{d}x}{=} \frac{NkT}{q}\left(\frac{\partial q}{\partial V}\right)_T$$

$$= \frac{NkT}{V/\Lambda^3}\left(\frac{\partial(V/\Lambda^3)}{\partial V}\right)_T = \frac{NkT}{V} = \frac{nN_A kT}{V} \overset{N_A k = R}{=} \frac{nRT}{V}$$

The calculation shows that the equation of state of a gas of independent particles is indeed the perfect gas law, $pV = nRT$.

Self-test 13F.1 Derive the equation of state of a gas for which $Q = q^N f/N!$, with $q = V/\Lambda^3$, where f depends on the volume.

Answer: $p = nRT/V + kT(\partial \ln f/\partial V)_T$

If β is left undefined in eqn 13F.1, then the equation of state calculated in the same way as in *Example* 13F.1 is $p = N/\beta V$. For this result to be the perfect gas law, it follows that $\beta = 1/kT$. This is the formal proof of that relation, as anticipated and used in Topics 13A–13E. A similar approach can be used to derive an equation of state resembling the van der Waals equation: see *A deeper look* 13D.1, available to read in the e-book accompanying this text.

At this stage the expressions for U and p and the definition $H = U + pV$, with $H(0) = U(0)$, can be used to obtain an expression for the enthalpy, H, of any substance:

$$H(T) = H(0) - \left(\frac{\partial \ln Q}{\partial \beta}\right)_V + kTV\left(\frac{\partial \ln Q}{\partial V}\right)_T \qquad \text{Enthalpy} \qquad (13F.4)$$

The fact that eqn 13F.4 is rather cumbersome is a sign that the enthalpy is not a fundamental property: as shown in Topic 2B, it is more of an accounting convenience. For a gas of independent, structureless particles $U(T) - U(0) = \frac{3}{2}nRT$ and $pV = nRT$. Therefore, for such a gas, it follows directly from $H = U + pV$ that

$$H(T) - H(0) = \tfrac{5}{2}nRT \qquad (13F.5)$$

One of the most important thermodynamic functions for chemistry is the Gibbs energy, $G = H - TS$. This definition implies that $G = U + pV - TS$ and therefore that $G = A + pV$. Note that $G(0) = A(0)$, both being equal to $U(0)$. The Gibbs energy can now be written in terms of the partition function by combining the expressions for A and p:

$$G(T) = G(0) - kT\ln Q + kTV\left(\frac{\partial \ln Q}{\partial V}\right)_T \qquad \text{Gibbs energy} \qquad (13F.6)$$

This expression takes a simple form for a gas of independent molecules because pV in the expression $G = A + pV$ can be replaced by nRT:

$$G(T) = G(0) - kT\ln Q + nRT \qquad (13F.7)$$

For a gas of indistinguishable particles $Q = q^N/N!$, and therefore $\ln Q = N\ln q - \ln N!$. If Stirling's approximation, $\ln N! = N\ln N - N$, is used the expression for G simplifies as follows

$$\begin{aligned} G(T) &= G(0) - kT\ln Q + nRT \\ &= G(0) - kT(N\ln q - \overbrace{\ln N!}^{N\ln N - N}) + nRT \\ &= G(0) - kT(N\ln q - N\ln N + N) + nRT \\ &= G(0) - \overbrace{NkT}^{=nN_A kT = nRT}(\ln q - \ln N + 1) + nRT \\ &= G(0) - nRT\ln q + nRT\ln N - nRT + nRT \\ &= G(0) - nRT\ln\frac{q}{N} \end{aligned} \qquad (13F.8)$$

Note that a statistical interpretation of the Gibbs energy is now possible: because q is the number of thermally accessible states and N is the number of molecules, the difference $G(T) - G(0)$ is proportional to the logarithm of the average number of thermally accessible states available to a molecule. As this average number increases, the Gibbs energy falls further and further below $G(0)$. The thermodynamic tendency to lower Gibbs energy can now be seen to be a tendency to maximize the number of thermally accessible states.

It turns out to be convenient to define the **molar partition function**, $q_m = q/n$ (with units mol^{-1} and $n = N/N_A$). The Gibbs energy is then written in terms of q_m as

$$G(T) = G(0) - nRT\ln\frac{q_m}{N_A} \quad \text{Gibbs energy [indistinguishable, independent molecules]} \quad (13F.9a)$$

To use this expression, $G(0) = U(0)$ is identified with the energy of the system when all the molecules are in their ground state, E_0. To calculate the standard Gibbs energy, the partition function has its standard value, $q_m^{\ominus}$. This standard value is evaluated by setting the molar volume in the translational contribution equal to the standard molar volume, so $q_m^{\ominus} = (V_m^{\ominus}/\Lambda^3)q^R q^V$ with $V_m^{\ominus} = RT/p^{\ominus}$. The standard *molar* Gibbs energy is then obtained with these substitutions and after dividing through by n:

$$G_m^{\ominus}(T) = G_m^{\ominus}(0) - RT\ln\frac{q_m^{\ominus}}{N_A} \quad \text{Standard molar Gibbs energy [indistinguishable, independent molecules]} \quad (13F.9b)$$

where $G_m^{\ominus}(0) = E_{0,m}$, the molar ground-state energy of the system.

Example 13F.2 Calculating a standard Gibbs energy of formation from partition functions

Calculate the standard Gibbs energy of formation of $H_2O(g)$ at 25 °C.

Collect your thoughts Write the chemical equation for the formation reaction, and then the expression for the standard Gibbs energy of formation in terms of the standard molar Gibbs energy of each molecule; then express those Gibbs energies in terms of the molecular molar partition functions. Ignore molecular vibration as it is unlikely to be excited at 25 °C. The rotational constant for H_2 is 60.864 cm^{-1}, that for O_2 is 1.4457 cm^{-1}, and those for H_2O are 27.877, 14.512, and 9.285 cm^{-1}. Before using the approximate form of the rotational partition functions, you need to judge whether the temperature is high enough: if it is not, use the full expression. For the values of $E_{0,m}$, use bond enthalpies from Table 9C.3; for precise calculations, bond energies (at $T = 0$) should be used instead. You will need the following expressions from Topic 13B:

$$\Lambda(J) = \frac{h}{(2\pi m_J kT)^{1/2}} \qquad \overbrace{q^R = \frac{kT}{2hc\tilde{B}}}^{\text{Linear molecule, } \sigma=2} \qquad \overbrace{q^R = \frac{1}{2}\left(\frac{kT}{hc}\right)^{3/2}\left(\frac{\pi}{\tilde{A}\tilde{B}\tilde{C}}\right)^{1/2}}^{\text{Nonlinear molecule, } \sigma=2}$$

The solution The chemical reaction is $H_2(g) + \frac{1}{2}O_2(g) \rightarrow H_2O(g)$. Therefore,

$$\Delta_f G^{\ominus} = G_m^{\ominus}(H_2O,g) - G_m^{\ominus}(H_2,g) - \tfrac{1}{2}G_m^{\ominus}(O_2,g)$$

Now write the standard molar Gibbs energies in terms of the standard molar partition functions of each species J:

$$G_m^{\ominus}(J) = E_{0,m}(J) - RT\ln\frac{q_m^{\ominus}(J)}{N_A} \qquad q_m^{\ominus}(J) = q_m^{T\ominus}(J)q^R(J) = \frac{V_m^{\ominus}}{\Lambda(J)^3}q^R(J)$$

Therefore

$$\Delta_f G^{\ominus} = \left\{E_{0,m}(H_2O) - RT\ln\frac{q_m^{\ominus}(H_2O)}{N_A}\right\} - \left\{E_{0,m}(H_2) - RT\ln\frac{q_m^{\ominus}(H_2)}{N_A}\right\}$$
$$-\tfrac{1}{2}\left\{E_{0,m}(O_2) - RT\ln\frac{q_m^{\ominus}(O_2)}{N_A}\right\}$$
$$= \Delta E_{0,m} - RT\ln\frac{\overbrace{\{V_m^{\ominus}/N_A\Lambda(H_2O)^3\}q^R(H_2O)}^{q_m^{\ominus}(H_2O)/N_A}}{\underbrace{\{V_m^{\ominus}/N_A\Lambda(H_2)^3\}q^R(H_2)}_{q_m^{\ominus}(H_2)/N_A}\left[\underbrace{\{V_m^{\ominus}/N_A\Lambda(O_2)^3\}q^R(O_2)}_{q_m^{\ominus}(O_2)/N_A}\right]^{1/2}}$$
$$= \Delta E_{0,m} - RT\ln\frac{N_A^{1/2}\{\Lambda(H_2)\Lambda(O_2)^{1/2}/\Lambda(H_2O)\}^3}{V_m^{\ominus 1/2}\{q^R(H_2)q^R(O_2)^{1/2}/q^R(H_2O)\}}$$

Now substitute the data. The molar energy difference (with bond dissociation energies in kilojoules per mole indicated) is

$$\Delta E_{0,m} = \overbrace{E_{0,m}(H_2O)}^{-492-428=-920} - \overbrace{E_{0,m}(H_2)}^{-436} - \tfrac{1}{2}\overbrace{E_{0,m}(O_2)}^{-497} = -236\ \text{kJ mol}^{-1}$$

Next, establish whether the temperature is high enough for the approximate expressions for the rotational partition functions (Topic 13B) to be reliable:

	H_2O	H_2O	H_2O	H_2	O_2
$\tilde{X}$/cm^{-1}, $X = A, B,$ or C	27.877	14.512	9.285	60.864	1.4457
θ^R/K	40.1	20.9	13.4	87.6	2.1

For H_2 the characteristic rotational temperature is not much smaller than 298 K so the rotational partition function must be computed by summing terms. For all the other molecules θ^R is much smaller than 298 K and the approximate form of the partition function quoted above can be used. The data give the following values for Λ and q^R:

$$\Lambda(H_2) = 71.21\ \text{pm} \quad \Lambda(O_2) = 17.87\ \text{pm} \quad \Lambda(H_2O) = 23.82\ \text{pm}$$
$$q^R(H_2) = 1.88 \qquad q^R(O_2) = 71.67 \qquad q^R(H_2O) = 43.14$$

It then follows that

$$\Delta_f G^{\ominus} = -236\ \text{kJ mol}^{-1} - \overbrace{RT}^{2.48\ \text{kJ mol}^{-1}}\ln 0.0270 = -227\ \text{kJ mol}^{-1}$$

The value quoted in Table 2C.1 of the *Resource section* is $-228.57\,\text{kJ mol}^{-1}$.

Self-test 13F.2 Estimate the standard Gibbs energy of formation of $NH_3(g)$ at 25 °C. The rotational constants of NH_3 are $\tilde{B} = 10.001\,\text{cm}^{-1}$ and $\tilde{A} = 6.449\,\text{cm}^{-1}$, and for N_2 $\tilde{B} = 1.9987\,\text{cm}^{-1}$.

Answer: $-16\,\text{kJ mol}^{-1}$

Exercises

E13F.1 A CO_2 molecule is linear, with a rotational constant of $0.3902\,\text{cm}^{-1}$. Calculate the rotational contribution to the molar Gibbs energy at 298 K.

E13F.2 A CO_2 molecule is linear, and the wavenumbers of its vibrational modes are $1388.2\,\text{cm}^{-1}$, $2349.2\,\text{cm}^{-1}$, and $667.4\,\text{cm}^{-1}$, the last being doubly degenerate and the others non-degenerate. Calculate the vibrational contribution to the molar Gibbs energy at 298 K.

E13F.3 The ground level of Cl is ${}^2P_{3/2}$ and a ${}^2P_{1/2}$ level lies $881\,\text{cm}^{-1}$ above it. Calculate the electronic contribution to the molar Gibbs energy of Cl atoms at 500 K.

13F.2 Equilibrium constants

The following discussion considers gas-phase reactions in which the equilibrium constant is defined in terms of the partial pressures of the reactants and products.

13F.2(a) The relation between *K* and the partition function

The standard molar Gibbs energy of a gas of independent molecules is given by eqn 13F.9a, $G^{\ominus}_m(T) = G^{\ominus}_m(0) - RT\ln(q^{\ominus}_m/N_A)$, in terms of the standard molar partition function, $q^{\ominus}_m$. The equilibrium constant *K* of a reaction is related to the standard reaction Gibbs energy. The task is to combine these two relations and so obtain an expression for the equilibrium constant in terms of the molecular partition functions of the reactants and products.

How is that done? 13F.1 Relating the equilibrium constant to partition functions

You need to use the expressions for the standard molar Gibbs energies, $G^{\ominus}_m$, of each species to find an expression for the standard reaction Gibbs energy. Then find the equilibrium constant *K* by using eqn 6A.15 ($\Delta_r G^{\ominus} = -RT \ln K$).

Step 1 *Write an expression for* $\Delta_r G^{\ominus}$

From eqn 13F.9b, the standard molar reaction Gibbs energy for the reaction $aA + bB \rightarrow cC + dD$ is

$$\Delta_r G^{\ominus} = cG^{\ominus}_m(C) + dG^{\ominus}_m(D) - \{aG^{\ominus}_m(A) + bG^{\ominus}_m(B)\}$$

$$= \boxed{cG^{\ominus}_m(C,0) + dG^{\ominus}_m(D,0) - \{aG^{\ominus}_m(A,0) + bG^{\ominus}_m(B,0)\}}$$

$$-RT\left\{c\ln\frac{q^{\ominus}_{C,m}}{N_A} + d\ln\frac{q^{\ominus}_{D,m}}{N_A} - a\ln\frac{q^{\ominus}_{A,m}}{N_A} - b\ln\frac{q^{\ominus}_{B,m}}{N_A}\right\}$$

Because $G_m(J,0) = E_{0,m}(J)$, the molar ground-state energy of the species J, the first (boxed) term on the right is

$$cE_{0,m}(C,0) + dE_{0,m}(D,0) - \{aE_{0,m}(A,0) + bE_{0,m}(B,0)\} = \Delta_r E_0$$

Then, by using $a \ln x = \ln x^a$ and $\ln x + \ln y = \ln xy$,

$$\Delta_r G^{\ominus} = \Delta_r E_0 - RT\ln\frac{(q^{\ominus}_{C,m}/N_A)^c (q^{\ominus}_{D,m}/N_A)^d}{(q^{\ominus}_{A,m}/N_A)^a (q^{\ominus}_{B,m}/N_A)^b}$$

Step 2 *Write an expression for K*

Recall that $\Delta_r G^{\ominus} = -RT \ln K$ and use the above expression for $\Delta_r G^{\ominus}$ to give

$$-RT\ln K = \Delta_r E_0 - RT\ln\frac{(q^{\ominus}_{C,m}/N_A)^c (q^{\ominus}_{D,m}/N_A)^d}{(q^{\ominus}_{A,m}/N_A)^a (q^{\ominus}_{B,m}/N_A)^b}$$

This equation can now be rearranged into an expression for $\ln K$ and the $\Delta_r E_0/RT$ term taken into the logarithm:

$$\ln K = -\frac{\Delta_r E_0}{RT} + \ln\frac{(q^{\ominus}_{C,m}/N_A)^c (q^{\ominus}_{D,m}/N_A)^d}{(q^{\ominus}_{A,m}/N_A)^a (q^{\ominus}_{B,m}/N_A)^b}$$

$$= \ln e^{-\Delta_r E_0/RT} + \ln\frac{(q^{\ominus}_{C,m}/N_A)^c (q^{\ominus}_{D,m}/N_A)^d}{(q^{\ominus}_{A,m}/N_A)^a (q^{\ominus}_{B,m}/N_A)^b}$$

$$= \ln\left\{\frac{(q^{\ominus}_{C,m}/N_A)^c (q^{\ominus}_{D,m}/N_A)^d}{(q^{\ominus}_{A,m}/N_A)^a (q^{\ominus}_{B,m}/N_A)^b} e^{-\Delta_r E_0/RT}\right\}$$

The quantity within the braces on the right is therefore identifiable as the equilibrium constant *K*

$$K = \frac{(q^{\ominus}_{C,m}/N_A)^c (q^{\ominus}_{D,m}/N_A)^d}{(q^{\ominus}_{A,m}/N_A)^a (q^{\ominus}_{B,m}/N_A)^b} e^{-\Delta_r E_0/RT}$$

The equilibrium constant (13F.10a)

The quantity $\Delta_r E_0$ is the difference in molar energies of the ground states of the products and reactants and is calculated from the bond dissociation energies of the species (Fig. 13F.1).

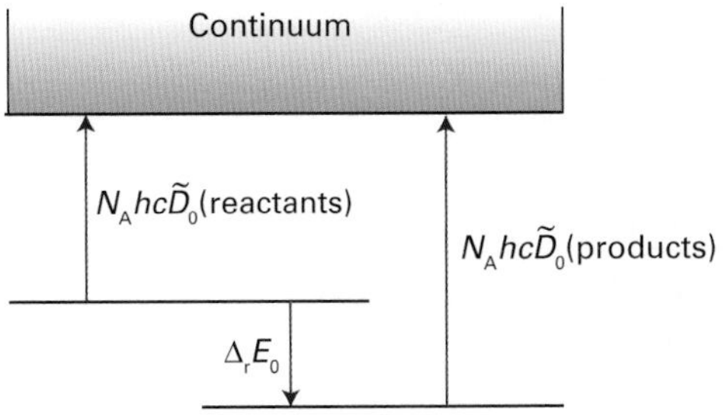

Figure 13F.1 The definition of $\Delta_r E_0$ for the calculation of equilibrium constants. The reactants are imagined as dissociating into atoms and then forming the products from the atoms.

In terms of the (signed) stoichiometric numbers introduced in Topic 2C, eqn 13F.10a can be written

$$K=\left\{\prod_J\left(\frac{q^{\ominus}_{J,m}}{N_A}\right)^{\nu_J}\right\}e^{-\Delta_r E_0/RT} \qquad (13F.10b)$$

The equilibrium constant

13F.2(b) A dissociation equilibrium

Equation 13F.10a can be used to write an expression for the equilibrium constant for the dissociation of a diatomic molecule X_2:

$$X_2(g) \rightleftharpoons 2X(g) \quad K=\frac{p_X^2}{p_{X_2}p^{\ominus}}$$

According to eqn 13F.10a (with $a=1$, $b=0$, $c=2$, and $d=0$):

$$K=\frac{(q^{\ominus}_{X,m}/N_A)^2}{q^{\ominus}_{X_2,m}/N_A}e^{-\Delta_r E_0/RT}=\frac{(q^{\ominus}_{X,m})^2}{q^{\ominus}_{X_2,m}N_A}e^{-\Delta_r E_0/RT} \qquad (13F.11a)$$

with

$$\Delta_r E_0=2E_{0,m}(X,0)-E_{0,m}(X_2,0)=N_A hc\tilde{D}_0(X-X) \qquad (13F.11b)$$

where $N_A hc\tilde{D}_0(X-X)$ is the (molar) dissociation energy of the X–X bond. The standard molar partition functions of the atoms X are

$$q^{\ominus}_{X,m}=\overbrace{g_X}^{q^E}\times\overbrace{\frac{V^{\ominus}_m}{\Lambda_X^3}}^{q^T}=\frac{g_X RT}{p^{\ominus}\Lambda_X^3} \qquad \left(V^{\ominus}_m=RT/p^{\ominus}\right)$$

where g_X is the degeneracy of the electronic ground state of X. The diatomic molecule X_2 also has rotational and vibrational degrees of freedom, so its standard molar partition function is

$$q^{\ominus}_{X_2,m}=g_{X_2}\frac{V^{\ominus}_m}{\Lambda_{X_2}^3}q^R_{X_2}q^V_{X_2}=\frac{g_{X_2}RTq^R_{X_2}q^V_{X_2}}{p^{\ominus}\Lambda_{X_2}^3}$$

where g_{X_2} is the degeneracy of the electronic ground state of X_2. It follows that

$$K=\frac{(g_X RT/p^{\ominus}\Lambda_X^3)^2}{g_{X_2}N_A RT q^R_{X_2}q^V_{X_2}/p^{\ominus}\Lambda_{X_2}^3}e^{-N_A hc\tilde{D}_0(X-X)/RT}$$

$$\overset{R=N_A k}{=}\frac{g_X^2 kT\Lambda_{X_2}^3}{g_{X_2}p^{\ominus}q^R_{X_2}q^V_{X_2}\Lambda_X^6}e^{-hc\tilde{D}_0(X-X)/kT} \qquad (13F.12)$$

All the quantities in this expression can be calculated from spectroscopic data.

Example 13F.3 Evaluating an equilibrium constant

Evaluate the equilibrium constant for the dissociation $Na_2(g) \rightarrow 2\,Na(g)$ at 1000 K. The data for Na_2 are: $\tilde{B}=0.1547\ cm^{-1}$, $\tilde{\nu}=159.2\ cm^{-1}$, and $N_A hc\tilde{D}_0=70.4\ kJ\,mol^{-1}$.

Collect your thoughts Recall that Na has a doublet ground state (term symbol $^2S_{1/2}$). You need to use eqn 13F.12 and the expressions for the partition functions assembled in Topic 13B. For such a heavy diatomic molecule it is safe to use the approximate expression for its rotational partition function (but check that assumption for consistency). Remember that for a homonuclear diatomic molecule, $\sigma=2$.

The solution The partition functions and other quantities required are as follows:

$\Lambda(Na_2)=8.14\ pm \qquad \Lambda(Na)=11.5\ pm$

$q^R(Na_2)=2246 \qquad q^V(Na_2)=4.885$

$g(Na)=2 \qquad g(Na_2)=1$

$hc\tilde{D}_0/kT=8.47$

There are many rotational states occupied, so the use of the approximate formula for $q^R(Na_2)$ is valid. Then, from eqn 13F.12,

$$K=\frac{2^2\times(1.381\times10^{-23}\ \overbrace{J}^{kg\,m^{-1}s^{-2}}K^{-1})\times(1000\ K)\times(8.14\times10^{-12}\ m)^3}{1\times(10^5\ \underbrace{Pa}_{kg\,m^{-1}s^{-2}})\times(2246)\times(4.885)\times(11.5\times10^{-12}\ m)^6}\times e^{-8.47}$$

$$=2.45$$

Self-test 13F.3 Evaluate K at 1500 K. Is the answer consistent with the dissociation being endothermic?

Answer: 52; yes

13F.2(c) Contributions to the equilibrium constant

To appreciate the physical basis of equilibrium constants, consider a simple $R \rightleftharpoons P$ gas-phase equilibrium (R for reactants, P for products).

Figure 13F.2 shows two sets of energy levels; one set of states belongs to R, and the other belongs to P. The populations of the states are given by the Boltzmann distribution, and are independent of whether any given state happens to belong to R or to P. A single Boltzmann distribution spreads, without distinction, over the two sets of states. If the spacings of R and P are similar (as in Fig. 13F.2), and the ground state of P lies above that of R, the diagram indicates that R will dominate in the equilibrium mixture. However, if P has a high density of states (a large number of states in a given energy range, as in Fig. 13F.3), then,

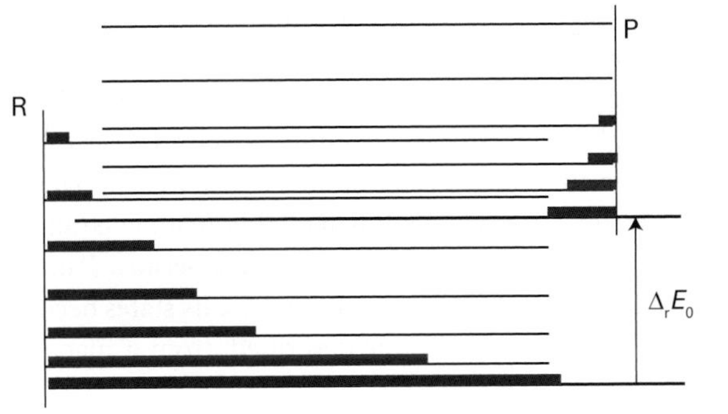

Figure 13F.2 The array of R(eactant) and P(roduct) energy levels. At equilibrium the populations of the states are given by the Boltzmann distribution, and are independent of whether any state belongs to R or to P. Here more states of R are populated than are states of P because the ground state of P lies above that of R. As $\Delta_r E_0$ increases, R becomes more and more dominant.

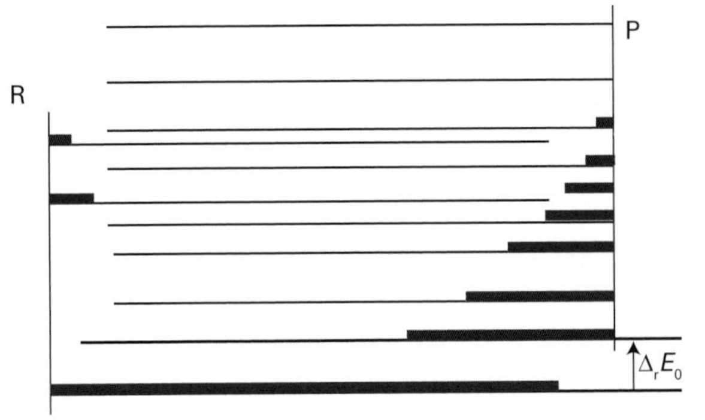

Figure 13F.3 It is important to take into account the densities of states of the molecules. Even though P might lie above R in energy (that is, $\Delta_r E_0$ is positive), P might have so many states that its total population dominates in the mixture.

even though its ground-state energy lies above that of R, the species P might still dominate at equilibrium.

It is quite easy to show that the ratio of numbers of R and P molecules at equilibrium is given by a Boltzmann-like expression.

How is that done? 13F.2 Relating the equilibrium constant to state populations

You need to start the derivation by noting that the population in a state i of the composite (R,P) system is $N_i = Ne^{-\beta\varepsilon_i}/q$, where N is the total number of molecules.

Step 1 *Write expressions for the numbers of R and P molecules*

The total number of R molecules is the sum of the populations of the (R,P) system taken over the states belonging to R; these states are labelled r with energies ε_r. The total number of P molecules is the sum over the states belonging to P; these states are labelled p with energies ε'_p (the prime is explained in a moment):

$$N_R = \sum_r N_r = \frac{N}{q}\sum_r e^{-\beta\varepsilon_r} \qquad N_P = \sum_p N_p = \frac{N}{q}\sum_p e^{-\beta\varepsilon'_p}$$

The sum over the states of R is its partition function, q_R, so $N_R = Nq_R/q$. The sum over the states of P is also a partition function, but the energies are measured from the ground state of the combined system, which is the ground state of R. If the energies of the ground states of R and P are separated by $\Delta\varepsilon_0$ (as in Fig. 13F.3), then the energies of the states of P relative to the ground state of R, ε'_p, can be written $\varepsilon'_p = \varepsilon_p + \Delta\varepsilon_0$, where ε_p are the energies of the states of P relative to the ground state of P. Writing ε'_p in this way gives N_P as

$$N_P = \frac{N}{q}\sum_p e^{-\beta(\varepsilon_p+\Delta\varepsilon_0)} = \frac{N}{q}\left(\sum_p e^{-\beta\varepsilon_p}\right)e^{-\beta\Delta\varepsilon_0} = \frac{Nq_P}{q}e^{-\beta\Delta\varepsilon_0}$$
$$= \frac{Nq_P}{q}e^{-\Delta_r E_0/RT}$$

The switch from $\Delta\varepsilon_0/kT$ to $\Delta_r E_0/RT$ in the last step is the conversion of molecular energies to molar energies. In the final expression q_P is the partition function of P with the energy of its ground state taken as zero, in the usual way.

Step 2 *Write an expression for the equilibrium constant*

The ratio of populations is

$$\frac{N_P}{N_R} = \frac{q_P}{q_R}e^{-\Delta_r E_0/RT}$$

The equilibrium constant of the $R \rightleftharpoons P$ reaction is proportional to the ratio of the numbers of the two types of molecule. Therefore,

$$K = \frac{q_P}{q_R}e^{-\Delta_r E_0/RT} \qquad (13F.13)$$

The equilibrium constant

For an $R \rightleftharpoons P$ equilibrium, the $V_m^\ominus$ factors in the partition functions cancel, so the appearance of q in place of $q^\ominus$ has no effect. In the case of a more general reaction, the conversion from q to $q^\ominus$ comes about at the stage of converting the pressures that occur in K to numbers of molecules.

The implications of eqn 13F.13 can be seen most clearly by exaggerating the molecular features that contribute to it. Suppose that R has only a single accessible level, which implies that $q_R = 1$. Also suppose that P has a large number of evenly, closely spaced levels, with spacing ε (Fig. 13F.4). The partition function for such an array is calculated in *Brief illustration* 13B.1, and is $q = 1/(1-e^{-\beta\varepsilon})$. Provided the levels are closely spaced compared to kT ($\varepsilon << kT$), the partition function is well-approximated as $q \approx kT/\varepsilon$ (Topic 13B). In this model system, the equilibrium constant is therefore given by

$$K = \frac{kT}{\varepsilon}e^{-\Delta_r E_0/RT} \qquad (13F.14)$$

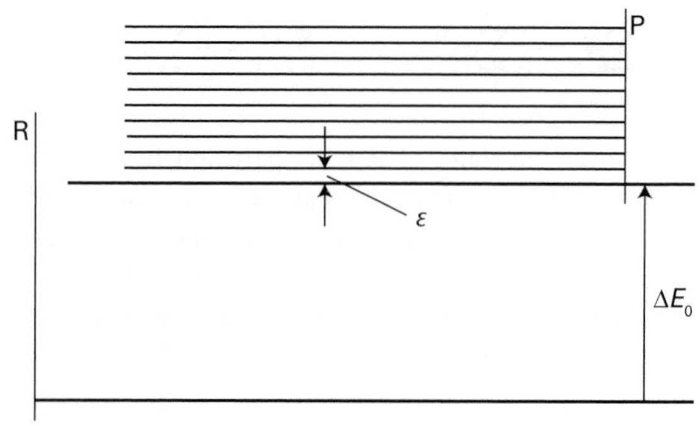

Figure 13F.4 The model used in the text for exploring the effects of energy separations and densities of states on equilibria. The products P can dominate provided $\Delta_r E_0$ is not too large and P has an appreciable density of states.

When $\Delta_r E_0$ is large compared to RT, the exponential term dominates and $K << 1$, which implies that very little P is present at equilibrium. When $\Delta_r E_0$ is smaller but still positive, K can exceed 1 if the factor kT/ε is large enough that it can overcome the exponential term (which is less than 1). The size of K then reflects the predominance of P at equilibrium on account of its high density of states.

At low temperatures $K << 1$ and the system consists entirely of R. At high temperatures, meaning $RT >> \Delta_r E_0$, the exponential function approaches 1. If ε is such that at this high temperature the factor kT/ε is large, P will be dominant at equilibrium. In this endothermic reaction (endothermic because P lies above R) a rise in temperature favours P, because its states become accessible. This is the behaviour described, from a macroscopic perspective, in Topic 6B.

The model also shows why the Gibbs energy, G, and not just the enthalpy, determines the position of equilibrium. It shows that the density of states (and hence the entropy) of each species as well as their relative energies controls the distribution of populations and hence the value of the equilibrium constant.

Exercise

E13F.4 Calculate the equilibrium constant for the reaction $A_2(g) \rightleftharpoons 2A(g)$ at 500 K using the following data: mass of A = 120 m_u; rotational temperature of A_2 = 3.00 K; $\tilde{D}_0$ = 300 cm^{-1}. Assume that the vibrational partition function = 1, and that the electronic ground states are all non-degenerate.

Checklist of concepts

☐ 1. The **thermodynamic functions** A, p, H, and G can be calculated from the canonical partition function.

☐ 2. The **equilibrium constant** can be written in terms of the partition functions of the reactants and products.

☐ 3. The equilibrium constant for dissociation of a diatomic molecule in the gas phase may be calculated from spectroscopic data.

☐ 4. The **physical basis of chemical equilibrium** can be understood in terms of a competition between energy separations and densities of states.

Checklist of equations

Property	Equation	Comment	Equation number
Helmholtz energy	$A(T) = A(0) - kT\ln Q$		13F.2
Pressure	$p = kT(\partial \ln Q/\partial V)_T$		13F.3
Enthalpy	$H(T) = H(0) - (\partial \ln Q/\partial \beta)_V + kTV(\partial \ln Q/\partial V)_T$		13F.4
Gibbs energy	$G(T) = G(0) - kT\ln Q + kTV(\partial \ln Q/\partial V)_T$		13F.6
	$G(T) = G(0) - nRT\ln(q_m/N_A)$	Perfect gas	13F.9a
Equilibrium constant	$K = \left\{\prod_J (q^{\ominus}_{J,m}/N_A)^{\nu_J}\right\} e^{-\Delta_r E_0/RT}$	Perfect gas	13F.10b

FOCUS 13 Statistical thermodynamics

To test your understanding of this material, work through the *Exercises, Additional exercises, Discussion questions*, and *Problems* found throughout this Focus.

Selected solutions can be found at the end of this Focus in the e-book. Solutions to even-numbered questions are available online only to lecturers.

Assume that all gases are perfect and that data refer to 298 K unless otherwise stated.

TOPIC 13A The Boltzmann distribution

Discussion questions

D13A.1 Discuss the relations between 'population', 'configuration', and 'weight'. What is the significance of the most probable configuration?

D13A.2 What is the significance and importance of the principle of equal a priori probabilities?

D13A.3 What is temperature?

D13A.4 Summarize the role of the Boltzmann distribution in chemistry.

Additional exercises

E13A.6 Calculate the weight of the configuration in which 21 objects are distributed in the arrangement 6, 0, 5, 0, 4, 0, 3, 0, 2, 0, 0, 1.

E13A.7 Evaluate 8! by using (i) the exact definition of a factorial, (ii) Stirling's approximation (eqn 13A.2), and (iii) the more accurate version of Stirling's approximation, $x! \approx (2\pi)^{1/2}x^{x+1/2}e^{-x}$.

E13A.8 Evaluate 10! by using (i) the definition of a factorial; (ii) Stirling's approximation (eqn 13A.2), and (iii) the more accurate version of Stirling's approximation given in Exercise E13A.7.

E13A.9 What are the relative populations of the states of a two-level system as the temperature approaches zero?

E13A.10 What is the temperature of a two-level system of energy separation equivalent to 300 cm^{-1} when the population of the upper state is one-half that of the lower state?

E13A.11 Calculate the relative populations of a spherical rotor at 298 K in the levels with $J = 0$ and $J = 5$, given that $\tilde{B} = 2.71$ cm^{-1}.

E13A.12 A certain molecule has a doubly degenerate excited state lying at 360 cm^{-1} above the non-degenerate ground state. At what temperature will 15 per cent of the molecules be in the upper state?

Problems

P13A.1 A sample consisting of five molecules has a total energy 5ε. Each molecule is able to occupy states of energy $j\varepsilon$, with $j = 0, 1, 2, \ldots$. (a) Calculate the weight of the configuration in which the molecules are distributed evenly over the available states with the stated total energy. (b) Draw up a table with columns headed by the energy of the states and write beneath them all configurations that are consistent with the total energy. Calculate the weights of each configuration and identify the most probable configurations.

P13A.2 A sample of nine molecules is numerically tractable but on the verge of being thermodynamically significant. A sample consisting of nine molecules has a total energy 9ε. Each molecule is able to occupy states of energy $j\varepsilon$, with $j = 0, 1, 2, \ldots$. Draw up a table of configurations. Before evaluating the weights of the configurations, guess (by looking for the most 'exponential' distribution of populations) which of the configurations will turn out to be the most probable. Go on to calculate the weights and identify the most probable configuration.

P13A.3 Use mathematical software to evaluate $\mathcal{W}$ for $N = 20$ for at least ten distributions over a uniform ladder of energy levels with separation ε, ensuring that the total energy is constant at 10ε. Identify the configuration of greatest weight, expressing the temperature as a multiple of ε, and compare it to the distribution predicted by the Boltzmann expression. Explore what happens as the value of the total energy is changed.

P13A.4 Suppose that two conformations A and B of a molecule differ in energy by 5.0 kJ mol^{-1}, and a third conformation C lies 0.50 kJ mol^{-1} above B. What proportion of molecules will be in conformation B at 273 K, with each conformation treated as a single energy level?

P13A.5 A certain atom has a doubly degenerate ground state and an upper level of four degenerate states at 450 cm^{-1} above the ground level. In an atomic beam study of the atoms it was observed that 30 per cent of the atoms were in the upper level, and the translational temperature of the beam was 300 K. Are the electronic states of the atoms in thermal equilibrium with the translational states? In other words, does the distribution of electronic states correspond to the same temperature as the distribution of translational states?

P13A.6 Explore the consequences of using the full version of Stirling's approximation, $x! \approx (2\pi)^{1/2}x^{x+1/2}e^{-x}$, in the development of the expression for the configuration of greatest weight. Does the more accurate approximation have a significant effect on the form of the Boltzmann distribution?

P13A.7‡ The variation of the atmospheric pressure p with altitude h is predicted by the *barometric formula* to be $p = p_0 e^{-h/H}$ where p_0 is the pressure at sea level and $H = RT/Mg$ with M the average molar mass of air and T the average temperature. Obtain the barometric formula from the Boltzmann distribution. Recall that the potential energy of a particle at height h above the surface of the Earth is mgh. Convert the barometric formula from pressure to number density, $\mathcal{N}$. Compare the relative number densities, $\mathcal{N}(h)/\mathcal{N}(0)$, for O_2 and H_2O at $h = 8.0$ km, a typical cruising altitude for commercial aircraft.

TOPIC 13B Molecular partition functions

Discussion questions

D13B.1 Describe the physical significance of the partition function.

D13B.2 What is the difference between a 'state' and an 'energy level'? Why is it important to make this distinction?

D13B.3 Why and when is it necessary to include a symmetry number in the calculation of a partition function?

Additional exercises

E13B.8 Calculate (i) the thermal wavelength, (ii) the translational partition function of a Ne atom in a cubic box of side 1.00 cm at 300 K and 3000 K.

E13B.9 Calculate the ratio of the translational partition functions of H_2 and He at the same temperature and volume.

E13B.10 Calculate the ratio of the translational partition functions of Ar and Ne at the same temperature and volume.

E13B.11 The bond length of N_2 is 109.75 pm. Use the high-temperature approximation to calculate the rotational partition function of the molecule at 300 K.

E13B.12 The NOF molecule is an asymmetric rotor with rotational constants 3.1752 cm^{-1}, 0.3951 cm^{-1}, and 0.3505 cm^{-1}. Calculate the rotational partition function of the molecule at (i) 25 °C, (ii) 100 °C.

E13B.13 The H_2O molecule is an asymmetric rotor with rotational constants 27.877 cm^{-1}, 14.512 cm^{-1}, and 9.285 cm^{-1}. Calculate the rotational partition function of the molecule at (i) 25 °C, (ii) 100 °C.

E13B.14 The rotational constant of CO is 1.931 cm^{-1}. Evaluate the rotational partition function explicitly (without approximation) and plot its value as a function of temperature. At what temperature is the value within 5 per cent of the value calculated by using eqn 13B.12a, which gives the high-temperature limit? *Hint:* Use mathematical software or a spreadsheet.

E13B.15 The rotational constant of HI is 6.511 cm^{-1}. Evaluate the rotational partition function explicitly (without approximation) and plot its value as a function of temperature. At what temperature is the value within 5 per cent of the value calculated by using eqn 13B.12a, which gives the high-temperature limit? *Hint:* Use mathematical software or a spreadsheet.

E13B.16 The rotational constant of CH_4 is 5.241 cm^{-1}. Evaluate the rotational partition function explicitly (without approximation but ignoring the role of nuclear spin) and plot its value as a function of temperature. At what temperature is the value within 5 per cent of the value calculated by using eqn 13B.12b, which gives the high-temperature limit? *Hint:* Use mathematical software or a spreadsheet.

E13B.17 The rotational constant of CCl_4 is 0.0572 cm^{-1}. Evaluate the rotational partition function explicitly (without approximation but ignoring the role of nuclear spin) and plot its value as a function of temperature. At what temperature is the value within 5 per cent of the value calculated by using eqn 13B.12b, which gives the high-temperature limit? *Hint:* Use mathematical software or a spreadsheet.

E13B.18 Give the symmetry number for each of the following molecules: (i) CO, (ii) O_2, (iii) H_2S, (iv) SiH_4, and (v) $CHCl_3$.

E13B.19 Give the symmetry number for each of the following molecules: (i) CO_2, (ii) O_3, (iii) SO_3, (iv) SF_6, and (v) Al_2Cl_6.

E13B.20 Evaluate the rotational partition function of pyridine, C_5H_5N, at 25 °C given that $\tilde{A} = 0.2014$ cm^{-1}, $\tilde{B} = 0.1936$ cm^{-1}, $\tilde{C} = 0.0987$ cm^{-1}. Take the symmetry number into account.

E13B.21 The vibrational wavenumber of Br_2 is 323.2 cm^{-1}. Evaluate the vibrational partition function explicitly (without approximation) and plot its value as a function of temperature. At what temperature is the value within 5 per cent of the value calculated from eqn 13B.16, which is valid at high temperatures?

E13B.22 The vibrational wavenumber of I_2 is 214.5 cm^{-1}. Evaluate the vibrational partition function explicitly (without approximation) and plot its value as a function of temperature. At what temperature is the value within 5 per cent of the value calculated from eqn 13B.16, which is valid at high temperatures?

E13B.23 Calculate the vibrational partition function of HCN at 900 K given the wavenumbers 3311 cm^{-1} (symmetric stretch), 712 cm^{-1} (bend; doubly degenerate), 2097 cm^{-1} (asymmetric stretch).

E13B.24 Calculate the vibrational partition function of CCl_4 at 500 K given the wavenumbers 459 cm^{-1} (symmetric stretch, non-degenerate), 217 cm^{-1} (deformation, doubly degenerate), 776 cm^{-1} (deformation, triply degenerate), 314 cm^{-1} (deformation, triply degenerate).

E13B.25 Calculate the vibrational partition function of CI_4 at 500 K given the wavenumbers 178 cm^{-1} (symmetric stretch, non-degenerate), 90 cm^{-1} (deformation, doubly degenerate), 555 cm^{-1} (deformation, triply degenerate), 125 cm^{-1} (deformation, triply degenerate).

E13B.26 A certain atom has a triply degenerate ground level, a non-degenerate electronically excited level at 850 cm^{-1}, and a fivefold degenerate level at 1100 cm^{-1}. Calculate the partition function of these electronic states at 2000 K. What is the relative population of each level at 2000 K?

‡ These problems were supplied by Charles Trapp and Carmen Giunta.

Problems

P13B.1 Consider a three-level system with levels 0, ε, and 2ε. Plot the partition function against kT/ε.

P13B.2 Plot the temperature dependence of the vibrational contribution to the molecular partition function for several values of the vibrational wavenumber. Estimate from your plots the temperature at which the partition function falls to within 10 per cent of the value expected at the high-temperature limit.

P13B.3 This problem is best done using mathematical software. Equation 13B.15 is the partition function for a harmonic oscillator. Consider a Morse oscillator (Topic 11C) in which the energy levels are given by

$$E_v = (v+\tfrac{1}{2})hc\tilde{\nu} - (v+\tfrac{1}{2})^2 hcx_e\tilde{\nu}$$

Evaluate the partition function for this oscillator, remembering to measure energies from the lowest level and to note that there is only a finite number of bound-state levels. Plot the partition function against $kT/hc\tilde{\nu}$ for values of x_e ranging from 0 to 0.1, and—on the same graph—compare your results with that for a harmonic oscillator.

P13B.4 Explore the conditions under which the 'integral' approximation for the translational partition function is not valid by considering the translational partition function of an H atom in a one-dimensional box of side comparable to that of a typical nanoparticle, 100 nm. Estimate the temperature at which, according to the integral approximation, $q = 10$ and evaluate the exact partition function at that temperature.

P13B.5 (a) Calculate the electronic partition function of a tellurium atom at (i) 298 K, (ii) 5000 K by direct summation using the following data:

Term	Degeneracy	Wavenumber/cm^{-1}
Ground	5	0
1	1	4707
2	3	4751
3	5	10 559

(b) What proportion of the Te atoms are in the ground term and in the term labelled 2 at the two temperatures?

P13B.6 The four lowest electronic levels of a Ti atom are $^3F_2$, $^3F_3$, $^3F_4$, and $^5F_1$, at 0, 170, 387, and 6557 cm^{-1}, respectively. There are many other electronic states at higher energies. The boiling point of titanium is 3287 °C. What are the relative populations of these levels at the boiling point? *Hint:* The degeneracies of the levels are $2J + 1$.

P13B.7‡ J. Sugar and A. Musgrove (*J. Phys. Chem. Ref. Data* **22**, 1213 (1993)) have published tables of energy levels for germanium atoms and cations from Ge^+ to Ge^{31+}. The lowest-lying energy levels in neutral Ge are as follows:

	$^3P_0$	$^3P_1$	$^3P_2$	$^1D_2$	$^1S_0$
(E/hc)/cm^{-1}	0	557.1	1410.0	7125.3	16 367.3

Calculate the electronic partition function at 298 K and 1000 K by direct summation. *Hint:* The degeneracy of a level J is $2J + 1$.

P13B.8 The pure rotational microwave spectrum (Topic 11B) of HCl has absorption lines at the following wavenumbers (in cm^{-1}): 21.19, 42.37, 63.56, 84.75, 105.93, 127.12, 148.31, 169.49, 190.68, 211.87, 233.06, 254.24, 275.43, 296.62, 317.80, 338.99, 360.18, 381.36, 402.55, 423.74, 444.92, 466.11, 487.30, 508.48. Calculate the rotational partition function at 25 °C by direct summation.

P13B.9 The rotational constants of CH_3Cl are $\tilde{A} = 5.097$ cm^{-1} and $\tilde{B} = 0.443$ cm^{-1}. Evaluate the rotational partition function explicitly (without approximation but ignoring the role of nuclear spin) and plot its value as a function of temperature. At what temperature is the value within 5 per cent of the value calculated by using eqn 13B.12a, which applies to the high-temperature limit? *Hint:* Use mathematical software or a spreadsheet.

P13B.10 Calculate, by explicit summation, the vibrational partition function of I_2 molecules at (a) 100 K, (b) 298 K given that its vibrational energy levels lie at the following wavenumbers above the zero-point energy level: 0, 215.30, 425.39, 636.27, 845.93 cm^{-1}. What proportion of I_2 molecules are in the ground and first two excited levels at the two temperatures?

TOPIC 13C Molecular energies

Discussion questions

D13C.1 Identify the conditions under which energies predicted from the equipartition theorem coincide with energies computed by using partition functions.

D13C.2 Describe how the mean energy of a system composed of two levels varies with temperature.

Additional exercises

E13C.5 Compute the mean energy at 400 K of a two-level system of energy separation equivalent to 600 cm^{-1}.

E13C.6 Use mathematical software or a spreadsheet to evaluate, by explicit summation, the mean rotational energy of CO and plot its value as a function of temperature. At what temperature is the equipartition value within 5 per cent of the accurate value? $\tilde{B}(CO) = 1.931$ cm^{-1}.

E13C.7 Use mathematical software or a spreadsheet to evaluate, by explicit summation, the mean rotational energy of HI and plot its value as a function of temperature. At what temperature is the equipartition value within 5 per cent of the accurate value? $\tilde{B}(HI) = 6.511$ cm^{-1}.

E13C.8 Use mathematical software or a spreadsheet to evaluate, by explicit summation, the mean rotational energy of CH_4 and plot its value as a function of temperature. At what temperature is the equipartition value within 5 per cent of the accurate value? $\tilde{B}(CH_4) = 5.241$ cm^{-1}.

E13C.9 Use mathematical software or a spreadsheet to evaluate, by explicit summation, the mean rotational energy of CCl_4 and plot its value as a function of temperature. At what temperature is the equipartition value within 5 per cent of the accurate value? $\tilde{B}(CCl_4) = 0.0572$ cm^{-1}.

E13C.10 Use mathematical software or a spreadsheet to evaluate the mean vibrational energy of Br_2 and plot its value as a function of temperature. At what temperature is the equipartition value within 5 per cent of the accurate value? Use $\tilde{\nu} = 323.2$ cm^{-1}.

E13C.11 Use mathematical software or a spreadsheet to evaluate the mean vibrational energy of I_2 and plot its value as a function of temperature. At what temperature is the equipartition value within 5 per cent of the accurate value? Use $\tilde{\nu} = 214.5\ \text{cm}^{-1}$.

E13C.12 Use mathematical software or a spreadsheet to evaluate the mean vibrational energy of CS_2 and plot its value as a function of temperature. At what temperature is the equipartition value within 5 per cent of the accurate value? Use the wavenumbers 658 cm^{-1} (symmetric stretch), 397 cm^{-1} (bend; doubly degenerate), 1535 cm^{-1} (asymmetric stretch).

E13C.13 Use mathematical software or a spreadsheet to evaluate the mean vibrational energy of HCN and plot its value as a function of temperature. At what temperature is the equipartition value within 5 per cent of the accurate value? Use the wavenumbers 3311 cm^{-1} (symmetric stretch), 712 cm^{-1} (bend; doubly degenerate), 2097 cm^{-1} (asymmetric stretch).

E13C.14 Evaluate, by explicit summation, the mean vibrational energy of CCl_4 and plot its value as a function of temperature. At what temperature is the equipartition value within 5 per cent of the accurate value? Use the wavenumbers 459 cm^{-1} (symmetric stretch, non-degenerate), 217 cm^{-1} (deformation, doubly degenerate), 776 cm^{-1} (deformation, triply degenerate), 314 cm^{-1} (deformation, triply degenerate).

E13C.15 Evaluate, by explicit summation, the mean vibrational energy of CI_4 and plot its value as a function of temperature. At what temperature is the equipartition value within 5 per cent of the accurate value? Use the wavenumbers 178 cm^{-1} (symmetric stretch, non-degenerate), 90 cm^{-1} (deformation, doubly degenerate), 555 cm^{-1} (deformation, triply degenerate), 125 cm^{-1} (deformation, triply degenerate).

E13C.16 Calculate the mean contribution to the electronic energy at 2000 K for a sample composed of the atoms specified in Exercise E13B.26.

Problems

P13C.1 Evaluate, by explicit summation, the mean rotational energy of CH_3Cl and plot its value as a function of temperature. At what temperature is the equipartition value within 5 per cent of the accurate value? Use $\tilde{A} = 5.097\ \text{cm}^{-1}$ and $\tilde{B} = 0.443\ \text{cm}^{-1}$.

P13C.2 Deduce an expression for the mean energy when each molecule can exist in states with energies 0, ε, and 2ε.

P13C.3 How much energy does it take to raise the temperature of 1.0 mol H_2O(g) from 100 °C to 200 °C at constant volume? Consider only translational and rotational contributions to the heat capacity.

P13C.4 What must the temperature be before the energy estimated from the equipartition theorem is within 2 per cent of the energy given by $\langle\varepsilon^V\rangle = hc\tilde{\nu}/(e^{\beta hc\tilde{\nu}} - 1)$?

P13C.5 Suppose a collection of species with total spin $S = 1$ is exposed to a magnetic field of 2.5 T. Calculate the mean energy of this system at 298 K. Use $g = 2.0$.

P13C.6 An electron trapped in an infinitely deep spherical well of radius R, such as may be encountered in the investigation of nanoparticles, has energies given by the expression $E_{nl} = \hbar^2 X_{nl}^2/2m_eR^2$ with X_{nl} the value obtained by searching for the zeroes of the spherical Bessel functions. The first six values (with a degeneracy of the corresponding energy level equal to $2l + 1$) are as follows:

n	1	1	1	2	1	2
l	0	1	2	0	3	1
X_{nl}	3.142	4.493	5.763	6.283	6.988	7.725

Evaluate the partition function and mean energy of an electron as a function of temperature. *Hints:* Remember to measure energies from the lowest level. Note that $\hbar^2/2m_eR^2k$ has dimensions of temperature, so can be used as the characteristic temperature θ of the system. It follows that the partition function can be expressed in terms of $E_{nl} = X_{nl}^2 k\theta$ and the dimensionless parameter T/θ. Let T/θ range from 0 to 25.

P13C.7 The NO molecule has a doubly degenerate excited electronic level 121.1 cm^{-1} above the doubly degenerate electronic ground term. Calculate and plot the electronic partition function of NO from $T = 0$ to 1000 K. Evaluate (a) the populations of the levels and (b) the mean electronic energy at 300 K.

P13C.8 Consider a system of N molecules with energy levels $\varepsilon_j = j\varepsilon$ and $j = 0, 1, 2, \ldots$. (a) Show that if the mean energy per molecule is $a\varepsilon$, then the temperature is given by

$$\beta = \frac{1}{\varepsilon}\ln\left(1+\frac{1}{a}\right)$$

Evaluate the temperature for a system in which the mean energy is ε, taking ε equivalent to 50 cm^{-1}. (b) Calculate the molecular partition function q for the system when its mean energy is $a\varepsilon$.

P13C.9 Deduce an expression for the root-mean-square energy, $\langle\varepsilon^2\rangle^{1/2}$, in terms of the partition function and hence an expression for the root-mean-square deviation from the mean, $\Delta\varepsilon = (\langle\varepsilon^2\rangle - \langle\varepsilon\rangle^2)^{1/2}$. Evaluate the resulting expression for a harmonic oscillator. *Hint:* Use $\langle\varepsilon^2\rangle = (1/q)\sum_j e^{-\beta\varepsilon_j}\varepsilon_j^2$.

TOPIC 13D The canonical ensemble

Discussion questions

D13D.1 Why is the concept of a canonical ensemble required?

D13D.2 Explain what is meant by an ensemble and why it is useful in statistical thermodynamics.

D13D.3 Under what circumstances may identical particles be regarded as distinguishable?

D13D.4 What is meant by the 'thermodynamic limit'?

Additional exercises

E13D.4 Identify the systems for which it is essential to include a factor of $1/N!$ on going from Q to q: (i) a sample of helium gas, (ii) a sample of carbon monoxide gas, (iii) a solid sample of carbon monoxide, (iv) water vapour.

E13D.5 Identify the systems for which it is essential to include a factor of $1/N!$ on going from Q to q: (i) a sample of carbon dioxide gas, (ii) a sample of graphite, (iii) a sample of diamond, (iv) ice.

Problems

P13D.1 For a perfect gas, the canonical partition function, Q, is related to the molecular partition function q by $Q = q^N/N!$. In Topic 13F it is established that $p = kT(\partial \ln Q/\partial V)_T$. Use the expression for q to derive the perfect gas law $pV = nRT$.

P13D.2 Use statistical thermodynamic arguments to show that for a perfect gas, $(\partial E/\partial V)_T = 0$

TOPIC 13E The internal energy and the entropy

Discussion questions

D13E.1 Describe the molecular features that affect the magnitudes of the constant-volume molar heat capacity of a molecular substance.

D13E.2 Discuss and illustrate the proposition that $1/T$ is a more natural measurement of temperature than T itself.

D13E.3 Discuss the relationship between the thermodynamic and statistical definitions of entropy.

D13E.4 Justify the differences between the partition-function expression for the entropy for distinguishable particles and the expression for indistinguishable particles.

D13E.5 Account for the temperature and volume dependence of the entropy of a perfect gas in terms of the Boltzmann distribution.

D13E.6 Explain the origin of residual entropy.

Additional exercises

E13E.8 Use the equipartition theorem to estimate the constant-volume molar heat capacity of (i) O_3, (ii) C_2H_6, (iii) CO_2 in the gas phase at 25 °C.

E13E.9 Estimate the value of $\gamma = C_{p,m}/C_{V,m}$ for carbon dioxide. Do this calculation with and without the vibrational contribution to the energy. Which is closer to the experimental value at 25 °C? *Hint:* Note that $C_{p,m} - C_{V,m} = R$ for a perfect gas.

E13E.10 The first electronically excited state of O_2 is ${}^1\Delta_g$ (doubly degenerate) and lies 7918.1 cm^{-1} above the ground state, which is ${}^3\Sigma_g^-$ (triply degenerate). Calculate the electronic contribution to the heat capacity of O_2 at 400 K.

E13E.11 Plot the molar heat capacity of a collection of harmonic oscillators as a function of T/θ^V, and predict the vibrational heat capacity of ethyne at (i) 298 K, (ii) 500 K. The normal modes (and their degeneracies in parentheses) occur at wavenumbers 612(2), 729(2), 1974, 3287, and 3374 cm^{-1}.

E13E.12 Plot the molar entropy of a collection of harmonic oscillators as a function of T/θ^V, and predict the standard molar entropy of ethyne at (i) 298 K, (ii) 500 K. For data, see Exercise E13E.11.

E13E.13 Calculate the translational contribution to the standard molar entropy at 298 K of (i) H_2O(g), (ii) CO_2(g).

E13E.14 At what temperature is the translational contribution to the standard molar entropy of CO_2(g) equal to that of H_2O(g) at 298 K?

E13E.15 Calculate the rotational partition function of SO_2 at 298 K from its rotational constants 2.027 36 cm^{-1}, 0.344 17 cm^{-1}, and 0.293 535 cm^{-1} and use your result to calculate the rotational contribution to the molar entropy of sulfur dioxide at 25 °C.

E13E.16 Estimate the contribution of the spin to the molar entropy of a solid sample of a d-metal complex with $S = \frac{5}{2}$.

E13E.17 Predict the vibrational contribution to the standard molar entropy of methanoic acid (formic acid, HCOOH) vapour at (i) 298 K, (ii) 500 K. The normal modes occur at wavenumbers 3570, 2943, 1770, 1387, 1229, 1105, 1033, 638, 625 cm^{-1}.

E13E.18 Predict the vibrational contribution to the standard molar entropy of ethyne at (i) 298 K, (ii) 500 K. The normal modes (and their degeneracies in parentheses) occur at wavenumbers 612(2), 729(2), 1974, 3287, and 3374 cm^{-1}.

Problems

P13E.1 An NO molecule has a doubly degenerate electronic ground state and a doubly degenerate excited state at 121.1 cm^{-1}. Calculate and plot the electronic contribution to the molar heat capacity of the molecule up to 500 K.

P13E.2 Explore whether a magnetic field can influence the heat capacity of a paramagnetic molecule by calculating the electronic contribution to the heat capacity of an NO_2 molecule in a magnetic field. Estimate the total constant-volume heat capacity using equipartition, and calculate the percentage change in heat capacity brought about by a 5.0 T magnetic field at (a) 50 K, (b) 298 K. *Hints:* Recall that NO_2 has one unpaired electron. Assume that the sample is in the gas phase at 50 K and 298 K.

P13E.3 The energy levels of a CH_3 group attached to a larger fragment are given by the expression for a particle on a ring, provided the group is rotating freely. What is the high-temperature contribution to the heat capacity and entropy of such a freely rotating group at 25 °C? *Hint:* The moment of inertia of CH_3 about its threefold rotation axis (the axis that passes through the C atom and the centre of the equilateral triangle formed by the H atoms) is 5.341×10^{-47} kg m^2.

P13E.4 Calculate the temperature dependence of the heat capacity of p-H_2 (in which only rotational states with even values of J are populated) at low temperatures on the basis that its rotational levels $J = 0$ and $J = 2$ constitute a system that resembles a two-level system except for the degeneracy of the upper level. Use $\tilde{B} = 60.864$ cm^{-1} and sketch the heat capacity curve. The experimental heat capacity of p-H_2 does in fact show a peak at low temperatures.

P13E.5‡ In a spectroscopic study of buckminsterfullerene C_{60}, F. Negri et al. (*J. Phys. Chem.* **100**, 10849 (1996)) reviewed the wavenumbers of all the vibrational modes of the molecule:

Mode	Number	Degeneracy	Wavenumber/cm^{-1}
A_u	1	1	976
T_{1u}	4	3	525, 578, 1180, 1430
T_{2u}	5	3	354, 715, 1037, 1190, 1540
G_u	6	4	345, 757, 776, 963, 1315, 1410
H_u	7	5	403, 525, 667, 738, 1215, 1342, 1566

How many modes have a vibrational temperature θ^V below 1000 K? Estimate the molar constant-volume heat capacity of C_{60} at 1000 K, counting as active all modes with θ^V below this temperature.

P13E.6 Plot the function dS/dT for a two-level system, the temperature coefficient of its entropy, against kT/ε. Is there a temperature at which this coefficient passes through a maximum? If you find a maximum, explain its physical origins.

P13E.7 Derive an expression for the molar entropy of an equally spaced three-level system; taking the spacing as ε.

P13E.8 Although expressions like $\langle\varepsilon\rangle = -d\ln q/d\beta$ are useful for formal manipulations in statistical thermodynamics, and for expressing thermodynamic functions in neat formulas, they are sometimes more trouble than they are worth in practical applications. When presented with a table of energy levels, it is often much more convenient to evaluate the following sums directly (the dots simply identify the different functions):

$$q = \sum_j e^{-\beta\varepsilon_j} \quad \dot{q} = \sum_j \beta\varepsilon_j e^{-\beta\varepsilon_j} \quad \ddot{q} = \sum_j (\beta\varepsilon_j)^2 e^{-\beta\varepsilon_j}$$

(a) Derive expressions for the internal energy, heat capacity, and entropy in terms of these three functions. (b) Apply the technique to the calculation of the electronic contribution to the constant-volume molar heat capacity of magnesium vapour at 5000 K using the following data:

Term	^1S	$^3P_0$	$^3P_1$	$^3P_2$	$^1P_1$	$^3S_1$
Degeneracy	1	1	3	5	3	3
$\tilde{\nu}/cm^{-1}$	0	21850	21870	21911	35051	41197

P13E.9 The pure rotational microwave spectrum (Topic 11B) of HCl has absorption lines at the following wavenumbers (in cm^{-1}): 21.19, 42.37, 63.56, 84.75, 105.93, 127.12, 148.31, 169.49, 190.68, 211.87, 233.06, 254.24, 275.43, 296.62, 317.80, 338.99, 360.18, 381.36, 402.55, 423.74, 444.92, 466.11, 487.30, 508.48. (a) Calculate the rotational partition function at 25 °C by direct summation. (b) Calculate the rotational contribution to the molar entropy over a range of temperature and plot the contribution as a function of temperature.

P13E.10 Calculate the standard molar entropy of N_2(g) at 298 K from its rotational constant $\tilde{B} = 1.9987$ cm^{-1} and its vibrational wavenumber $\tilde{\nu} = 2358$ cm^{-1}. The thermochemical value is 192.1 J K^{-1} mol^{-1}. What does this suggest about the solid at $T = 0$?

P13E.11‡ J.G. Dojahn et al. (*J. Phys. Chem.* **100**, 9649 (1996)) characterized the potential energy curves of the ground and electronic states of homonuclear diatomic halogen anions. The ground state of F_2^- is $^2\Sigma_u^+$ with a fundamental vibrational wavenumber of 450.0 cm^{-1} and equilibrium internuclear distance of 190.0 pm. The first two excited states are at 1.609 and 1.702 eV above the ground state. Compute the standard molar entropy of F_2^- at 298 K.

P13E.12‡ Treat carbon monoxide as a perfect gas and apply equilibrium statistical thermodynamics to the study of its properties, as specified below, in the temperature range 100–1000 K at 1 bar. (a) Examine the probability distribution of molecules over available rotational and vibrational states. (b) Explore numerically the differences, if any, between the rotational molecular partition function as calculated with the discrete energy distribution with that predicted by the high-temperature limit. (c) Plot against temperature the individual contributions to $U_m(T) - U_m(100\text{ K})$, $C_{V,m}(T)$, and $S_m(T) - S_m(100\text{ K})$ made by the translational, rotational, and vibrational degrees of freedom. *Hints:* Let $\tilde{\nu} = 2169.8$ cm^{-1} and $\tilde{B} = 1.931$ cm^{-1}; neglect anharmonicity and centrifugal distortion.

P13E.13 The energy levels of a Morse oscillator are given in Problem P13B.3. Set up the expression for the molar entropy of a collection of Morse oscillators and plot it as a function of $kT/hc\tilde{\nu}$ for a series of anharmonicity constants ranging from 0 to 0.01. Take into account only the finite number of bound states. On the same graph plot the entropy of a harmonic oscillator and investigate how the two diverge.

P13E.14 Explore how the entropy of a collection of two-level systems behaves when the temperature is formally allowed to become negative. You should also construct a graph in which the temperature is replaced by the variable $\beta = 1/kT$. Account for the appearance of the graphs physically.

P13E.15 Derive the Sackur–Tetrode equation for a monatomic gas confined to a two-dimensional surface, and hence derive an expression for the standard molar entropy of condensation to form a mobile surface film.

P13E.16 The heat capacity ratio of a gas determines the speed of sound in it through the formula $c_s = (\gamma RT/M)^{1/2}$, where $\gamma = C_{p,m}/C_{V,m}$ and M is the molar mass of the gas. Deduce an expression for the speed of sound in a perfect gas of (a) diatomic, (b) linear triatomic, (c) nonlinear triatomic molecules at high temperatures (with translation and rotation active). Estimate the speed of sound in air at 25 °C. *Hint:* Note that $C_{p,m} - C_{V,m} = R$ for a perfect gas.

P13E.17 An average human DNA molecule has 5×10^8 binucleotides (rungs on the DNA ladder) of four different kinds. If each rung were a random choice of one of these four possibilities, what would be the residual entropy associated with this typical DNA molecule?

P13E.18 It is possible to write an approximate expression for the partition function of a protein molecule by including contributions from only two states: the native and denatured forms of the polymer. Proceeding with this crude model gives insight into the contribution of denaturation to the heat capacity of a protein. According to this model, the total energy of a system of N protein molecules is $E = N\varepsilon e^{-\varepsilon/kT}/(1 + e^{-\varepsilon/kT})$, where ε is the energy separation between the denatured and native forms. (a) Show that the constant-volume molar heat capacity is

$$C_{V,m} = f(T)R \quad f(T) = \frac{(\varepsilon/kT)^2 e^{-\varepsilon/kT}}{(1+e^{-\varepsilon/kT})^2}$$

(b) Plot the variation of $C_{V,m}$ with temperature. (c) If the function $C_{V,m}(T)$ has maximum or minimum, calculate the temperature at which it occurs.

TOPIC 13F Derived functions

Discussion questions

D13F.1 Suggest a physical interpretation of the relation between pressure and the partition function.

D13F.2 Suggest a physical interpretation of the relation between equilibrium constant and the partition functions of the reactants and products in a reaction.

D13F.3 How does a statistical analysis of the equilibrium constant account for the latter's temperature dependence?

Additional exercises

E13F.5 An O_3 molecule is angular, and its vibrational wavenumbers are 1110 cm^{-1}, 705 cm^{-1}, and 1042 cm^{-1}. The rotational constants of the molecule are 3.553 cm^{-1}, 0.4452 cm^{-1}, and 0.3948 cm^{-1}. Calculate the rotational and vibrational contributions to the molar Gibbs energy at 298 K.

E13F.6 Use the information in Exercise E13E.10 to calculate the electronic contribution to the molar Gibbs energy of O_2 at 400 K.

E13F.7 Calculate the equilibrium constant of the reaction $I_2(g) \rightleftharpoons 2I(g)$ at 1000 K from the following data for I_2, $\tilde{\nu} = 214.36$ cm^{-1}, $\tilde{B} = 0.0373$ cm^{-1}, $hc\tilde{D}_e = 1.5422$ eV. The ground state of the I atoms is $^2P_{3/2}$, implying fourfold degeneracy.

E13F.8 Calculate the equilibrium constant at 298 K for the gas-phase isotopic exchange reaction $2\,^{79}Br^{81}Br \rightleftharpoons {}^{79}Br^{79}Br + {}^{81}Br^{81}Br$. The Br_2 molecule has a non-degenerate ground state, with no other electronic states nearby. Base the calculation on the wavenumber of the vibration of $^{79}Br^{81}Br$, which is 323.33 cm^{-1}.

Problems

P13F.1 Use mathematical software and work in the high-temperature limit to calculate and plot the equilibrium constant for the reaction $CD_4(g) + HCl(g) \rightleftharpoons CHD_3(g) + DCl(g)$ as a function of temperature, in the range 300 K to 1000 K. Use the following data (numbers in parentheses are degeneracies):

Molecule	$\tilde{\nu}$/cm^{-1}	$\tilde{B}$/cm^{-1}	$\tilde{A}$/cm^{-1}
CHD_3	2993 (1), 2142 (1), 1003 (3), 1291 (2), 1036 (2)	3.28	2.63
CD_4	2109 (1), 1092 (2), 2259 (3), 996 (3)	2.63	
HCl	2991 (1)	10.59	
DCl	2145 (1)	5.445	

P13F.2 The exchange of deuterium between acid and water is an important type of equilibrium, and you can examine it using spectroscopic data on the molecules. Use mathematical software and work in the high-temperature limit, calculate the equilibrium constant for the exchange reaction $H_2O(g) + DCl(g) \rightleftharpoons HDO(g) + HCl(g)$ at (a) 298 K and (b) 800 K. Use the following data:

Molecule	$\tilde{\nu}$/cm^{-1}	$\tilde{A}$/cm^{-1}	$\tilde{B}$/cm^{-1}	$\tilde{C}$/cm^{-1}
H_2O	3656.7, 1594.8, 3755.8	27.88	14.51	9.29
HDO	2726.7, 1402.2, 3707.5	23.38	9.102	6.417
HCl	2991		10.59	
DCl	2145		5.449	

P13F.3 Here you are invited to decide whether a magnetic field can influence the value of an equilibrium constant. Consider the equilibrium $I_2(g) \rightleftharpoons 2\,I(g)$ at 1000 K, and calculate the ratio of equilibrium constants $K(\mathcal{B})/K$, where $K(\mathcal{B})$ is the equilibrium constant when a magnetic field $\mathcal{B}$ is present and removes the degeneracy of the four states of the $^2P_{3/2}$ level. Data on the species are given in Exercise E13F.7. The electronic g-value of the atoms is $\frac{4}{3}$. Calculate the field required to change the equilibrium constant by 1 per cent.

P13F.4‡ R. Viswanathan et al. (*J. Phys. Chem.* **100**, 10784 (1996)) studied thermodynamic properties of several boron–silicon gas-phase species experimentally and theoretically. These species can occur in the high-temperature chemical vapour deposition (CVD) of silicon-based semiconductors. Among the computations they reported was computation of the Gibbs energy of BSi(g) at several temperatures based on a $^4\Sigma^-$ ground state with equilibrium internuclear distance of 190.5 pm, a fundamental vibrational wavenumber of 772 cm^{-1}, and a $^2\Pi$ first excited level 8000 cm^{-1} above the ground level. Calculate the value of $G_m^{\ominus}(2000\text{ K}) - G_m^{\ominus}(0)$.

P13F.5‡ The molecule Cl_2O_2, which is believed to participate in the seasonal depletion of ozone over Antarctica, has been studied by several means. M. Birk et al. (*J. Chem. Phys.* **91**, 6588 (1989)) report its rotational constants (B) as 13 109.4, 2409.8, and 2139.7 MHz. They also report that its rotational spectrum indicates a molecule with a symmetry number of 2. J. Jacobs et al. (*J. Amer. Chem. Soc.* **116**, 1106 (1994)) report its vibrational wavenumbers as 753, 542, 310, 127, 646, and 419 cm^{-1} (all non-degenerate). Calculate the value of $G_m^{\ominus}(200\text{ K}) - G_m^{\ominus}(0)$ of Cl_2O_2.

P13F.6‡ J. Hutter et al. (*J. Amer. Chem. Soc.* **116**, 750 (1994)) examined the geometric and vibrational structure of several carbon molecules of formula C_n. Given that the ground state of C_3, a molecule found in interstellar space and in flames, is an angular singlet-state species with moments of inertia $39.340m_u$ Å^2, $39.032m_u$ Å^2, and $0.3082m_u$ Å^2 (where 1 Å $= 10^{-10}$ m) and with vibrational wavenumbers 63.4, 1224.5, and 2040 cm^{-1}, calculate the value of $G_m^{\ominus}(300.0\ K) - G_m^{\ominus}(0)$ for C_3.

FOCUS 13 Statistical thermodynamics

Integrated activities

I13.1 A formal way of arriving at the value of the symmetry number is to note that σ is the order (the number of elements) of the *rotational subgroup* of the molecule, the point group of the molecule with all but the identity and the rotations removed. The rotational subgroup of H_2O is $\{E,C_2\}$, so $\sigma = 2$. The rotational subgroup of NH_3 is $\{E,2C_3\}$, so $\sigma = 3$. This recipe makes it easy to find the symmetry numbers for more complicated molecules. The rotational subgroup of CH_4 is obtained from the T character table as $\{E,8C_3,3C_2\}$, so $\sigma = 12$. For benzene, the rotational subgroup of D_{6h} is $\{E,2C_6,2C_3,C_2,3C_2',3C_2''\}$ so $\sigma = 12$. (a) Estimate the rotational partition function of ethene at 25 °C given that $\tilde{A} = 4.828$ cm^{-1}, $\tilde{B} = 1.0012$ cm^{-1}, and $\tilde{C} = 0.8282$ cm^{-1}. (b) Evaluate the rotational partition function of pyridine, C_5H_5N, at room temperature ($\tilde{A} = 0.2014$ cm^{-1}, $\tilde{B} = 0.1936$ cm^{-1}, $\tilde{C} = 0.0987$ cm^{-1}).

I13.2 A feature of the rotational molar heat capacity of H_2 is that it rises to above the classical value of R before settling back to approach that value as the temperature is increased from 0. To understand this behaviour, the heat capacity can be treated as arising from the sum of transitions between the available levels. (a) Show that the heat capacity of a linear rotor is related to the following sum:

$$\xi(\beta) = \frac{1}{q^2}\sum_{J,J'}\{\varepsilon(J) - \varepsilon(J')\}^2 g(J')g(J)e^{-\beta\{\varepsilon(J)+\varepsilon(J')\}}$$

by

$$C = \tfrac{1}{2}Nk\beta^2\xi(\beta)$$

where the $\varepsilon(J)$ are the rotational energy levels and $g(J)$ their degeneracies. Then go on to show graphically that the total contribution to the heat capacity of a linear rotor can be regarded as a sum of contributions due to transitions $0 \to 1$, $0 \to 2$, $1 \to 2$, $1 \to 3$, etc. In this way, construct the figure below for the rotational heat capacities of a linear molecule. (b) Set up a calculation like that in part (a) to analyse the vibrational contribution to the heat capacity in terms of excitations between levels and illustrate your results graphically in terms of a diagram like that in the figure below.

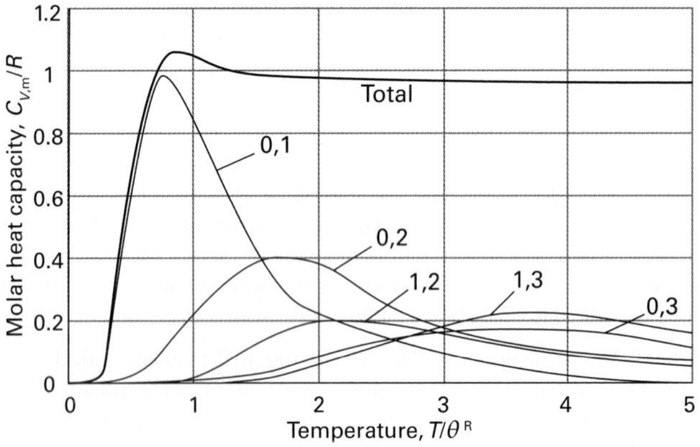

I13.3 Set up a calculation like that in Integrated activity I13.2 to analyse the vibrational contribution to the heat capacity in terms of excitations between levels and illustrate your results graphically in terms of a diagram like that in Fig. 13.1.

SOLUTIONS

Selected solutions for Exercises, *Discussion Questions*, *Additional Exercises*, and *Problems* throughout the Focuses can be found in the e-book.

AVAILABLE IN THE E-BOOK

'Impact on...' sections

'*Impact on*' sections show how physical chemistry is applied in a variety of modern contexts. They showcase physical chemistry as an evolving subject.

Go to this location in the accompanying e-book to view a list of Impacts.

'A deeper look' sections

These sections take some of the material in the text further and are there if you want to extend your knowledge and see the details of some of the more advanced derivations.

Go to this location in the accompanying e-book to view a list of Deeper Looks.

Group theory tables

A link to comprehensive group theory tables can be found at the end of the accompanying e-book.

The chemist's toolkits

The chemist's toolkits are reminders of the key mathematical, physical, and chemical concepts that you need to understand in order to follow the text.

For a consolidated and enhanced collection of the toolkits found throughout the text, go to this location in the accompanying e-book.

RESOURCE SECTION

Contents

PART 1 Mathematical resources

This brief review covers some core concepts that are used throughout the text. More specialized techniques are shown in *The chemist's toolkits*, especially the extended versions found via the e-book:

- **7B.1** Complex numbers
- **7F.1** Cylindrical coordinates
- **7F.2** Spherical polar coordinates
- **8C.1** Combining vectors
- **9D.1** Determinants
- **9E.1** Matrices
- **11A.1** Exponential and Gaussian functions
- **12B.1** Dipolar magnetic fields
- **12C.1** The Fourier transform

1.1 Integration

Integration is concerned with the areas under curves. The **integral** of a function $f(x)$, which is denoted $\int f(x)\mathrm{d}x$ (the symbol $\int$ is an elongated S denoting a sum), between the two values $x = a$ and $x = b$ is defined by imagining the x axis as divided into strips of width δx and evaluating the following sum:

$$\int_a^b f(x)\mathrm{d}x = \lim_{\delta x \to 0} \sum_i f(x_i)\delta x \quad \text{Integration [definition]}$$

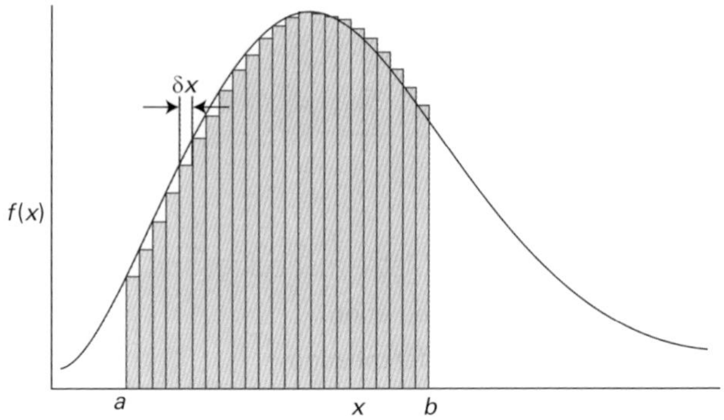

Sketch 1

As can be appreciated from Sketch 1, the integral is the area under the curve between the limits a and b. The function to be integrated is called the **integrand**. It is an astonishing mathematical fact that the integral of a function is the inverse of the differential of that function. In other words, if differentiation of f is followed by integration of the resulting function, the result is the original function f (to within a constant).

The integral in the preceding equation with the limits specified is called a **definite integral**. If it is written without the limits specified, it is called an **indefinite integral**. If the result of carrying out an indefinite integration is $g(x) + C$, where C is a constant, the following procedure is used to evaluate the corresponding definite integral:

$$\begin{aligned} I &= \int_a^b f(x)\mathrm{d}x = \{g(x)+C\}\Big|_a^b \\ &= \{g(b)+C\} - \{g(a)+C\} = g(b) - g(a) \end{aligned} \quad \text{Definite integral}$$

Note that the constant of integration disappears. The definite and indefinite integrals encountered in this text are listed in Table 1.1. They may also be calculated by using mathematical software, but sometimes it is necessary to work out an integral by hand. Two techniques are described in *The chemist's toolkit* 17B.1.

A function may depend on more than one variable, in which case it may be necessary to integrate over both the variables:

$$I = \int_a^b \int_c^d f(x,y)\mathrm{d}x\,\mathrm{d}y$$

We (but not everyone) adopt the convention that a and b are the limits of the variable x and c and d are the limits for y. This procedure is simple if the function is a product of functions of each variable and of the form $f(x,y) = X(x)Y(y)$. In this case, the double integral is just a product of each integral:

$$I = \int_a^b \int_c^d X(x)Y(y)\mathrm{d}x\ \mathrm{d}y = \int_a^b X(x)\mathrm{d}x \int_c^d Y(y)\mathrm{d}y$$

1.2 Differentiation

Differentiation is concerned with the slopes of functions, such as the rate of change of a variable with time. The formal definition of the **derivative**, $\mathrm{d}f/\mathrm{d}x$, of a function $f(x)$ is

$$\frac{\mathrm{d}f}{\mathrm{d}x} = \lim_{\delta x \to 0} \frac{f(x+\delta x) - f(x)}{\delta x} \quad \text{First derivative [definition]}$$

As shown in Sketch 2, the derivative can be interpreted as the slope of the tangent to the graph of $y = f(x)$ at a given value of x. A positive first derivative indicates that the function slopes

Table 1.1 Common integrals

	Indefinite integral*	Constraint	Definite integral
Algebraic functions			
A.1	$\int x^n \mathrm{d}x = \frac{1}{n+1}x^{n+1} + c$	$n \neq -1$	$\int_a^b x^n \mathrm{d}x = \frac{1}{n+1}(b^{n+1} - a^{n+1})$
A.2	$\int \frac{1}{x} \mathrm{d}x = \ln x + c$		$\int_a^b \frac{1}{x} \mathrm{d}x = \ln \frac{b}{a}$
A.3	$\int \frac{1}{(A-x)(B-x)} \mathrm{d}x = \frac{1}{B-A} \ln \frac{B-x}{A-x} + c$	$A \neq B$	$\int_a^b \frac{1}{(A-x)(B-x)} \mathrm{d}x = \frac{1}{B-A} \ln \frac{(B-b)(A-a)}{(A-b)(B-a)}$
Exponential functions			
E.1	$\int \mathrm{e}^{-kx} \mathrm{d}x = -\frac{1}{k}\mathrm{e}^{-kx} + c$		$\int_a^b \mathrm{e}^{-kx} \mathrm{d}x = \frac{1}{k}(\mathrm{e}^{-ka} - \mathrm{e}^{-kb})$
E.2	$\int x\mathrm{e}^{-kx} \mathrm{d}x = -\frac{1}{k^2}\mathrm{e}^{-kx} - \frac{x}{k}\mathrm{e}^{-kx} + c$		$\int_a^b x\mathrm{e}^{-kx} \mathrm{d}x = -\frac{1}{k^2}(\mathrm{e}^{-kb} - \mathrm{e}^{-ka}) - \frac{1}{k}(b\mathrm{e}^{-kb} - a\mathrm{e}^{-ka})$
E.3		$n \geq 0$ $k > 0$	$\int_0^\infty x^n \mathrm{e}^{-kx} \mathrm{d}x = \frac{n!}{k^{n+1}}$ $\quad n! = n(n-1)\ldots 1; 0! = 1$
E.4			$\int_0^\infty \frac{x^4 \mathrm{e}^x}{(\mathrm{e}^x - 1)^2} \mathrm{d}x = \frac{4\pi^4}{15}$
Gaussian functions			
G.1			$\int_0^\infty \mathrm{e}^{-kx^2} \mathrm{d}x = \frac{1}{2}\left(\frac{\pi}{k}\right)^{1/2}$
G.2	$\int x\mathrm{e}^{-kx^2} \mathrm{d}x = -\frac{1}{2k}\mathrm{e}^{-kx^2} + c$		$\int_0^\infty x\mathrm{e}^{-kx^2} \mathrm{d}x = \frac{1}{2k}$
G.3			$\int_0^\infty x^2 \mathrm{e}^{-kx^2} \mathrm{d}x = \frac{1}{4}\left(\frac{\pi}{k^3}\right)^{1/2}$
G.4	$\int x^3 \mathrm{e}^{-kx^2} \mathrm{d}x = -\frac{1}{2k^2}\mathrm{e}^{-kx^2} - \frac{x^2}{2k}\mathrm{e}^{-kx^2} + c$	$k > 0$	$\int_0^\infty x^3 \mathrm{e}^{-kx^2} \mathrm{d}x = \frac{1}{2k^2}$
G.5		$k > 0$	$\int_0^\infty x^4 \mathrm{e}^{-kx^2} \mathrm{d}x = \frac{3}{8k^2}\left(\frac{\pi}{k}\right)^{1/2}$
G.6			$\mathrm{erf}z = \frac{2}{\pi^{1/2}}\int_0^z \mathrm{e}^{-x^2} \mathrm{d}x$ $\quad \mathrm{erfc}\, z = 1 - \mathrm{erf}z$
G.7		$m \geq 0$ $k > 0$	$\int_0^\infty x^{2m+1} \mathrm{e}^{-kx^2} \mathrm{d}x = \frac{m!}{2k^{m+1}}$
G.8		$m \geq 1$ $k > 0$	$\int_0^\infty x^{2m} \mathrm{e}^{-kx^2} \mathrm{d}x = \frac{(2m-1)!!}{2^{m+1}k^m}\left(\frac{\pi}{k}\right)^{1/2}$ $n!! = n \times (n-2) \times (n-4)\ldots 1$ or 2
Trigonometric functions			
T.1	$\int \sin kx \, \mathrm{d}x = -\frac{1}{k}\cos kx + c$		$\int_0^a \sin kx \, \mathrm{d}x = \frac{1}{k}(1 - \cos kx)$
T.2	$\int \sin^2 kx \, \mathrm{d}x = \frac{1}{2}x - \frac{1}{4k}\sin 2kx + c$		$\int_0^a \sin^2 kx \, \mathrm{d}x = \frac{1}{2}a - \frac{1}{4k}\sin 2ka$
T.3	$\int \sin^3 kx \, \mathrm{d}x = -\frac{1}{3k}(\sin^2 kx + 2)\cos kx + c$		$\int_0^a \sin^3 kx \, \mathrm{d}x = \frac{1}{3k}\{2 - (\sin^2 ka + 2)\cos ka\}$
T.4	$\int \sin^4 kx \, \mathrm{d}x = \frac{3x}{8} - \frac{3}{8k}\sin kx \cos kx - \frac{1}{4k}\sin^3 kx \cos kx + c$		$\int_0^a \sin^4 kx \, \mathrm{d}x = \frac{3a}{8} - \frac{3}{8k}\sin ka \cos ka - \frac{1}{4k}\sin^3 ka \cos ka$
T.5	$\int \sin Ax \sin Bx \, \mathrm{d}x = \frac{\sin(A-B)x}{2(A-B)} - \frac{\sin(A+B)x}{2(A+B)} + c$	$A^2 \neq B^2$	$\int_0^a \sin Ax \sin Bx \, \mathrm{d}x = \frac{\sin(A-B)a}{2(A-B)} - \frac{\sin(A+B)a}{2(A+B)}$
T.6	$\int \cos Ax \cos Bx \, \mathrm{d}x = \frac{\sin(A-B)x}{2(A-B)} + \frac{\sin(A+B)x}{2(A+B)} + c$	$A^2 \neq B^2$	$\int_0^a \cos Ax \cos Bx \, \mathrm{d}x = \frac{\sin(A-B)a}{2(A-B)} + \frac{\sin(A+B)a}{2(A+B)}$
T.7	$\int \sin kx \cos kx \, \mathrm{d}x = \frac{1}{2k}\sin^2 kx + c$		$\int_0^a \sin kx \cos kx \, \mathrm{d}x = \frac{1}{2k}\sin^2 ka$
T.8	$\int \sin Ax \cos Bx \, \mathrm{d}x = \frac{\cos(B-A)x}{2(B-A)} - \frac{\cos(A+B)x}{2(A+B)} + c$	$A^2 \neq B^2$	$\int_0^a \sin Ax \cos Bx \, \mathrm{d}x = \frac{\cos(B-A)a - 1}{2(B-A)} - \frac{\cos(A+B)a - 1}{2(A+B)}$

(Continued)

Table 1.1 (Continued)

	Indefinite integral*	Constraint	Definite integral
T.9	$\int x\sin Ax\sin Bx\,dx = -\frac{d}{dB}\underbrace{\int \sin Ax\cos Bx dx}_{T.7}$	$A^2 \neq B^2$	
T.10	$\int \cos^2 kx\sin kx\,dx = -\frac{1}{3k}\cos^3 kx + c$		$\int_0^a \cos^2 kx\sin kx\,dx = \frac{1}{3k}\{1-\cos^3 ka\}$
T.11	$\int x\sin^2 kx\,dx = \frac{x^2}{4} - \frac{1}{4k}x\sin 2kx - \frac{1}{8k^2}\cos 2kx + c$		$\int_0^a x\sin^2 kx\,dx = \frac{a^2}{4} - \frac{1}{4k}a\sin 2ka - \frac{1}{8k^2}(\cos 2ka - 1)$
T.12	$\int x^2\sin^2 kx\,dx = \frac{x^3}{6} - \left(\frac{x^2}{4k} - \frac{1}{8k^3}\right)\sin 2kx - \frac{x}{4k^2}\cos 2kx + c$		$\int_0^a x^2\sin^2 kx\,dx = \frac{a^3}{6} - \left(\frac{a^2}{4k} - \frac{1}{8k^3}\right)\sin 2ka - \frac{a}{4k^2}\cos 2ka$
T.13	$\int x\cos kx\,dx = \frac{1}{k^2}\cos kx + \frac{x}{k}\sin kx + c$		$\int_0^a x\cos kx\,dx = \frac{1}{k^2}(\cos ka - 1) + \frac{a}{k}\sin ka$
T.14	$\int \cos^2 kx\,dx = \frac{x}{2} + \frac{1}{4k}\sin 2kx + c$		$\int_0^a \cos^2 kx\,dx = \frac{a}{2} + \frac{1}{4k}\sin 2ka$

* In each case, c is a constant. Note that not all indefinite integrals have a simple closed form.

upwards (as x increases), and a negative first derivative indicates the opposite. It is sometimes convenient to denote the first derivative as $f'(x)$. The **second derivative**, d^2f/dx^2, of a function is the derivative of the first derivative (here denoted f'):

$$\frac{d^2f}{dx^2} = \lim_{\delta x \to 0} \frac{f'(x+\delta x) - f'(x)}{\delta x}$$ Second derivative [definition]

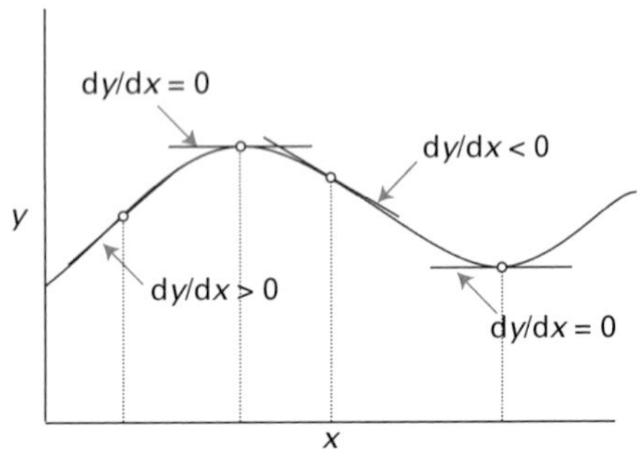

Sketch 2

It is sometimes convenient to denote the second derivative f''. As shown in Sketch 3, the second derivative of a function can be interpreted as an indication of the sharpness of the curvature of the function. A positive second derivative indicates that the function is ∪ shaped, and a negative second derivative indicates that it is ∩ shaped. The second derivative is zero at a **point of inflection**, where the curvature changes sign.

The derivatives of some common functions are as follows:

$$\frac{d}{dx}x^n = nx^{n-1} \quad \frac{d}{dx}e^{ax} = ae^{ax}$$

$$\frac{d}{dx}\sin ax = a\cos ax \qquad \frac{d}{dx}\cos ax = -a\sin ax$$

$$\frac{d}{dx}\ln ax = \frac{1}{x}$$

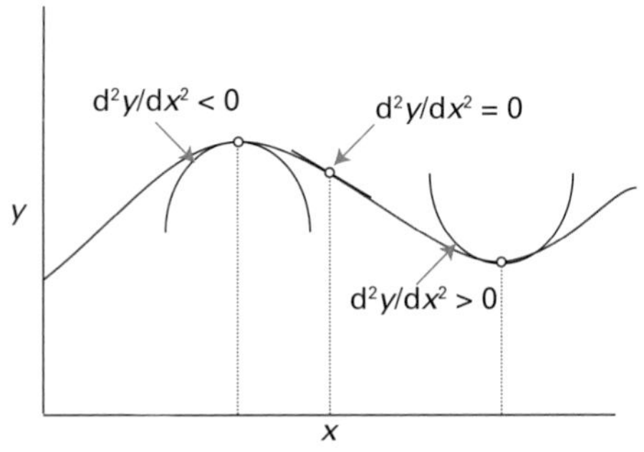

Sketch 3

It follows from the definition of the derivative that a variety of combinations of functions can be differentiated by using the following rules:

$$\frac{d}{dx}(u+v) = \frac{du}{dx} + \frac{dv}{dx}$$

$$\frac{d}{dx}uv = u\frac{dv}{dx} + v\frac{du}{dx}$$

$$\frac{d}{dx}\frac{u}{v} = \frac{1}{v}\frac{du}{dx} - \frac{u}{v^2}\frac{dv}{dx}$$

It is sometimes convenient to differentiate with respect to a function of x, rather than x itself. For instance, suppose that

$$f(x) = a + \frac{b}{x} + \frac{c}{x^2}$$

where a, b, and c are constants and you need to evaluate $df/d(1/x)$, rather than df/dx. To begin, let $y = 1/x$. Then $f(y) = a + by + cy^2$ and the derivative with respect to y is

$$\frac{df}{dy} = b + 2cy$$

Because $y = 1/x$, it follows that

$$\frac{\mathrm{d}f}{\mathrm{d}(1/x)} = b + \frac{2c}{x}$$

For further information, see *The chemist's toolkits* 2A.2 and 3E.1.

1.3 Series expansions

A function $f(x)$ can be expressed in terms of its value in the vicinity of $x = a$ by using the **Taylor series**

$$f(x) = f(a) + \left(\frac{\mathrm{d}f}{\mathrm{d}x}\right)_a (x-a) + \frac{1}{2!}\left(\frac{\mathrm{d}^2 f}{\mathrm{d}x^2}\right)_a (x-a)^2 + \cdots$$

$$= \sum_{n=0}^{\infty} \frac{1}{n!}\left(\frac{\mathrm{d}^n f}{\mathrm{d}x^n}\right)_a (x-a)^n \qquad \text{Taylor series}$$

where the notation $(\ldots)_a$ means that the derivative is evaluated at $x = a$ and $n!$ denotes a **factorial** defined as

$$n! = n(n-1)(n-2)\ldots 1, \quad 0! \equiv 1 \qquad \text{Factorial}$$

The **Maclaurin series** for a function is a special case of the Taylor series in which $a = 0$. The following Maclaurin series are used at various stages in the text:

$$(1+x)^{-1} = 1 - x + x^2 - \cdots = \sum_{n=0}^{\infty} (-1)^n x^n$$

$$\mathrm{e}^x = 1 + x + \tfrac{1}{2}x^2 + \cdots = \sum_{n=0}^{\infty} \frac{x^n}{n!}$$

$$\ln(1+x) = x - \tfrac{1}{2}x^2 + \tfrac{1}{3}x^3 - \cdots = \sum_{n=1}^{\infty} (-1)^{n+1} \frac{x^n}{n}$$

Series expansions are used to simplify calculations, because when $|x| << 1$ it is possible, to a good approximation, to terminate the series after one or two terms. Thus, provided $|x| << 1$,

$$(1+x)^{-1} \approx 1 - x$$

$$\mathrm{e}^x \approx 1 + x$$

$$\ln(1+x) \approx x$$

A series is said to **converge** if the sum approaches a finite, definite value as n approaches infinity. If it does not, the series is said to **diverge**. Thus, the series expansion of $(1+x)^{-1}$ converges for $|x| < 1$ and diverges for $|x| \geq 1$. Tests for convergence are explained in mathematical texts.

PART 2 Quantities and units

The result of a measurement is a **physical quantity** that is reported as a numerical multiple of a unit:

$$\text{physical quantity} = \text{numerical value} \times \text{unit}$$

It follows that units may be treated like algebraic quantities and may be multiplied, divided, and cancelled. Thus, the expression (physical quantity)/unit is the numerical value (a dimensionless quantity) of the measurement in the specified units. For instance, the mass m of an object could be reported as $m = 2.5$ kg or m/kg = 2.5. In this instance the unit of mass is 1 kg, but it is common to refer to the unit simply as kg (and likewise for other units). See Table 2.1 below for a list of units.

Although it is good practice to use only SI units, there will be occasions where accepted practice is so deeply rooted that physical quantities are expressed using other, non-SI units. By international convention, all physical quantities are represented by oblique (sloping) letters (for instance, m for mass); units are given in roman (upright) letters (for instance m for metre).

Units may be modified by a prefix that denotes a factor of a power of 10. Among the most common SI prefixes are those listed in Table 2.2 below. Examples of the use of these prefixes are:

$$1\,\text{nm} = 10^{-9}\,\text{m} \qquad 1\,\text{ps} = 10^{-12}\,\text{s} \qquad 1\,\mu\text{mol} = 10^{-6}\,\text{mol}$$

Powers of units apply to the prefix as well as the unit they modify. For example, $1\ \text{cm}^3 = 1\ (\text{cm})^3$, and $(10^{-2}\ \text{m})^3 = 10^{-6}\ \text{m}^3$. Note that $1\ \text{cm}^3$ does not mean $1\ \text{c}(\text{m}^3)$. When carrying out numerical calculations, it is usually safest to write out the numerical value of an observable in scientific notation (as $n.nnn \times 10^n$).

There are seven SI base units, which are listed in Table 2.3 below. All other physical quantities may be expressed as combinations of these base units. A number of these derived combinations of units have special names and symbols. For example, force is reported in the derived unit newton, $1\ \text{N} = 1\ \text{kg}\,\text{m}\,\text{s}^{-2}$ (see Table 2.4 below).

Table 2.1 Some common units

Physical quantity	Name of unit	Symbol for unit	Value*
Time	minute	min	60 s
	hour	h	3600 s
	day	d	86 400 s
	year	a	31 556 952.445 s
Length	ångström	Å	10^{-10} m
Volume	litre	L, l	1 dm^3
Mass	tonne	t	10^3 kg
Pressure	bar	bar	10^5 Pa
	atmosphere	atm	101.325 kPa
Energy	electronvolt	eV	$1.602\,176\,63 \times 10^{-19}$ J
			96.485 33 kJ mol^{-1}

* All values are exact, except for the definition of 1 eV, which depends on the measured value of e, and the year, which is not a constant and depends on a variety of astronomical assumptions.

Table 2.2 Common SI prefixes

Prefix	y	z	a	f	p	n	μ	m	c	d
Name	yocto	zepto	atto	femto	pico	nano	micro	milli	centi	deci
Factor	10^{-24}	10^{-21}	10^{-18}	10^{-15}	10^{-12}	10^{-9}	10^{-6}	10^{-3}	10^{-2}	10^{-1}
Prefix	da	h	k	M	G	T	P	E	Z	Y
Name	deca	hecto	kilo	mega	giga	tera	peta	exa	zetta	yotta
Factor	10	10^2	10^3	10^6	10^9	10^{12}	10^{15}	10^{18}	10^{21}	10^{24}

Table 2.3 The SI base units

Physical quantity	Symbol for quantity	Base unit
Length	l	metre, m
Mass	m	kilogram, kg
Time	t	second, s
Electric current	I	ampere, A
Thermodynamic temperature	T	kelvin, K
Amount of substance	n	mole, mol
Luminous intensity	I_v	candela, cd

Table 2.4 A selection of derived units

Physical quantity	Derived unit	Name and symbol of derived unit
Force	$kg\,m\,s^{-2}$	newton, N
Pressure	$kg\,m^{-1}\,s^{-2}$	pascal, Pa
	$N\,m^{-2}$	
Energy	$kg\,m^{2}\,s^{-2}$	joule, J
	$N\,m$	
	$Pa\,m^{3}$	
Power	$kg\,m^{2}\,s^{-3}$	watt, W
	$J\,s^{-1}$	

PART 3 Data

The following is a directory of all tables in the text; those included in this *Resource section* are marked with an asterisk. The remainder will be found on the pages indicated. These tables reproduce and expand the data given in the short tables in the text, and follow their numbering. Standard states refer to a pressure of $p^{\ominus} = 1$ bar. Data are for 298 K unless otherwise indicated. The general references are as follows:

AIP: D.E. Gray (ed.), *American Institute of Physics handbook*. McGraw Hill, New York (1972).

E: J. Emsley, *The elements*. Oxford University Press (1991).

HCP: D.R. Lide (ed.), *Handbook of chemistry and physics*. CRC Press, Boca Raton (2000).

JL: A.M. James and M.P. Lord, *Macmillan's chemical and physical data*. Macmillan, London (1992).

KL: G.W.C. Kaye and T.H. Laby (ed.), *Tables of physical and chemical constants*. Longman, London (1973).

LR: G.N. Lewis and M. Randall, revised by K.S. Pitzer and L. Brewer, *Thermodynamics*. McGraw Hill, New York (1961).

NBS: *NBS tables of chemical thermodynamic properties*, published as *J. Phys. Chem. Reference Data*, **11**, Supplement 2 (1982).

RS: R.A. Robinson and R.H. Stokes, *Electrolyte solutions*. Butterworth, London (1959).

TDOC: J.B. Pedley, J.D. Naylor, and S.P. Kirby, *Thermochemical data of organic compounds*. Chapman & Hall, London (1986).

Table 0.1 Physical properties of selected materials

	$\rho/(\mathrm{g\,cm^{-3}})$ at 293 K†	T_f/K	T_b/K		$\rho/(\mathrm{g\,cm^{-3}})$ at 293 K†	T_f/K	T_b/K
Elements				**Inorganic compounds**			
Aluminium(s)	2.698	933.5	2740	$CaCO_3$(s, calcite)	2.71	1612	1171[d]
Argon(g)	1.381	83.8	87.3	$CuSO_4 \cdot 5H_2O$(s)	2.284	383(−H_2O)	423(−5H_2O)
Boron(s)	2.340	2573	3931	HBr(g)	2.77	184.3	206.4
Bromine(l)	3.123	265.9	331.9	HCl(g)	1.187	159.0	191.1
Carbon(s, gr)	2.260	3700[s]		HI(g)	2.85	222.4	237.8
Carbon(s, d)	3.513			H_2O(l)	0.997	273.2	373.2
Chlorine(g)	1.507	172.2	239.2	D_2O(l)	1.104	277.0	374.6
Copper(s)	8.960	1357	2840	NH_3(g)	0.817	195.4	238.8
Fluorine(g)	1.108	53.5	85.0	KBr(s)	2.750	1003	1708
Gold(s)	19.320	1338	3080	KCl(s)	1.984	1049	1773s
Helium(g)	0.125		4.22	NaCl(s)	2.165	1074	1686
Hydrogen(g)	0.071	14.0	20.3	H_2SO_4(l)	1.841	283.5	611.2
Iodine(s)	4.930	386.7	457.5				
Iron(s)	7.874	1808	3023	**Organic compounds**			
Krypton(g)	2.413	116.6	120.8	Aniline, $C_6H_5NH_2$(l)	1.026	267	457
Lead(s)	11.350	600.6	2013	Anthracene, $C_{14}H_{10}$(s)	1.243	490	615
Lithium(s)	0.534	453.7	1620	Benzene, C_6H_6(l)	0.879	278.6	353.2
Magnesium(s)	1.738	922.0	1363	Ethanal, CH_3CHO(l)	0.788	152	293
Mercury(l)	13.546	234.3	629.7	Ethanol, C_2H_5OH(l)	0.789	156	351.4
Neon(g)	1.207	24.5	27.1	Ethanoic acid, CH_3COOH(l)	1.049	289.8	391
Nitrogen(g)	0.880	63.3	77.4	Glucose, $C_6H_{12}O_6$(s)	1.544	415	
Oxygen(g)	1.140	54.8	90.2	Methanal, HCHO(g)		181	254.0
Phosphorus(s, wh)	1.820	317.3	553	Methane, CH_4(g)		90.6	111.6
Potassium(s)	0.862	336.8	1047	Methanol, CH_3OH(l)	0.791	179.2	337.6
Silver(s)	10.500	1235	2485	Naphthalene, $C_{10}H_8$(s)	1.145	353.4	491
Sodium(s)	0.971	371.0	1156	Octane, C_8H_{18}(l)	0.703	216.4	398.8
Sulfur(s, α)	2.070	386.0	717.8	Phenol, C_6H_5OH(s)	1.073	314.1	455.0
Uranium(s)	18.950	1406	4018	Propanone, $(CH_3)_2CO$(l)	0.787	178	329
Xenon(g)	2.939	161.3	166.1	Sucrose, $C_{12}H_{22}O_{11}$(s)	1.588	457d	
Zinc(s)	7.133	692.7	1180	Tetrachloromethane, CCl_4(l)	1.63	250	349.9
				Trichloromethane, $CHCl_3$(l)	1.499	209.6	334

d: decomposes; s: sublimes; Data: AIP, E, HCP, KL. † For gases, at their boiling points.

Table 0.2 Masses and natural abundances of selected nuclides

Nuclide		m/m_u	Abundance/%
H	^{1}H	1.0078	99.985
	^{2}H	2.0140	0.015
He	^{3}He	3.0160	0.000 13
	^{4}He	4.0026	100
Li	^{6}Li	6.0151	7.42
	^{7}Li	7.0160	92.58
B	^{10}B	10.0129	19.78
	^{11}B	11.0093	80.22
C	^{12}C	12*	98.89
	^{13}C	13.0034	1.11
N	^{14}N	14.0031	99.63
	^{15}N	15.0001	0.37
O	^{16}O	15.9949	99.76
	^{17}O	16.9991	0.037
	^{18}O	17.9992	0.204
F	^{19}F	18.9984	100
P	^{31}P	30.9738	100
S	^{32}S	31.9721	95.0
	^{33}S	32.9715	0.76
	^{34}S	33.9679	4.22
Cl	^{35}Cl	34.9688	75.53
	^{37}Cl	36.9651	24.4
Br	^{79}Br	78.9183	50.54
	^{81}Br	80.9163	49.46
I	^{127}I	126.9045	100

* Exact value.

Table 8B.1 Effective nuclear charge*

	H							**He**
1s	1							1.6875
	Li	**Be**	**B**	**C**	**N**	**O**	**F**	**Ne**
1s	2.6906	3.6848	4.6795	5.6727	6.6651	7.6579	8.6501	9.6421
2s	1.2792	1.9120	2.5762	3.2166	3.8474	4.4916	5.1276	5.7584
2p			2.4214	3.1358	3.8340	4.4532	5.1000	5.7584
	Na	**Mg**	**Al**	**Si**	**P**	**S**	**Cl**	**Ar**
1s	10.6259	11.6089	12.5910	13.5745	14.5578	15.5409	16.5239	17.5075
2s	6.5714	7.3920	8.3736	9.0200	9.8250	10.6288	11.4304	12.2304
2p	6.8018	7.8258	8.9634	9.9450	10.9612	11.9770	12.9932	14.0082
3s	2.5074	3.3075	4.1172	4.9032	5.6418	6.3669	7.0683	7.7568
3p			4.0656	4.2852	4.8864	5.4819	6.1161	6.7641

* The atual charge is $Z_{eff}e$.
Data: E. Clementi and D.L. Raimondi, *Atomic screening constants from SCF functions*. IBM Res. Note NJ-27 (1963). *J. Chem. Phys.* **38**, 2686 (1963).

Table 8B.3 Ionic radii, r/pm*

Li^+(4)	Be^{2+}(4)	B^{3+}(4)	N^{3-}	O^{2-}(6)	F^-(6)		
59	27	12	171	140	133		
Na^+(6)	Mg^{2+}(6)	Al^{3+}(6)	P^{3-}	S^{2-}(6)	Cl^-(6)		
102	72	53	212	184	181		
K^+(6)	Ca^{2+}(6)	Ga^{3+}(6)	As^{3-}(6)	Se^{2-}(6)	Br^-(6)		
138	100	62	222	198	196		
Rb^+(6)	Sr^{2+}(6)	In^{3+}(6)		Te^{2-}(6)	I^-(6)		
149	116	79		221	220		
Cs^+(6)	Ba^{2+}(6)	Tl^{3+}(6)					
167	136	88					
d-block elements (high-spin ions)							
Sc^{3+}(6)	Ti^{4+}(6)	Cr^{3+}(6)	Mn^{3+}(6)	Fe^{2+}(6)	Co^{3+}(6)	Cu^{2+}(6)	Zn^{2+}(6)
73	60	61	65	63	61	73	75

* Numbers in parentheses are the coordination numbers of the ions. Values for ions without a coordination number stated are estimates.
Data: R.D. Shannon and C.T. Prewitt, *Acta Cryst.* **B25**, 925 (1969).

Table 8B.4 Ionization energies, I/(kJ mol^{-1})

H							He
1312.0							2372.3
							5250.4
Li	Be	B	C	N	O	F	Ne
513.3	899.4	800.6	1086.2	1402.3	1313.9	1681	2080.6
7298.0	1757.1	2427	2352	2856.1	3388.2	3374	3952.2
Na	Mg	Al	Si	P	S	Cl	Ar
495.8	737.7	577.4	786.5	1011.7	999.6	1251.1	1520.4
4562.4	1450.7	1816.6	1577.1	1903.2	2251	2297	2665.2
		2744.6		2912			
K	Ca	Ga	Ge	As	Se	Br	Kr
418.8	589.7	578.8	762.1	947.0	940.9	1139.9	1350.7
3051.4	1145	1979	1537	1798	2044	2104	2350
		2963	2735				
Rb	Sr	In	Sn	Sb	Te	I	Xe
403.0	549.5	558.3	708.6	833.7	869.2	1008.4	1170.4
2632	1064.2	1820.6	1411.8	1794	1795	1845.9	2046
		2704	2943.0	2443			
Cs	Ba	Tl	Pb	Bi	Po	At	Rn
375.5	502.8	589.3	715.5	703.2	812	930	1037
2420	965.1	1971.0	1450.4	1610			
		2878	3081.5	2466			

Data: E.

Table 8B.5 Electron affinities, $E_{ea}/(kJ\,mol^{-1})$

H							He
72.8							−21
Li	Be	B	C	N	O	F	Ne
59.8	≤0	23	122.5	−7	141 −844	322	−29
Na	Mg	Al	Si	P	S	Cl	Ar
52.9	≤0	44	133.6	71.7	200.4 −532	348.7	−35
K	Ca	Ga	Ge	As	Se	Br	Kr
48.3	2.37	36	116	77	195.0	324.5	−39
Rb	Sr	In	Sn	Sb	Te	I	Xe
46.9	5.03	34	121	101	190.2	295.3	−41
Cs	Ba	Tl	Pb	Bi	Po	At	Rn
45.5	13.95	30	35.2	101	186	270	−41

Data: E.

Table 9C.2 Bond lengths, R/pm

(a) Bond lengths in specific molecules

Br_2	228.3
Cl_2	198.75
CO	112.81
F_2	141.78
H_2^+	106
H_2	74.138
HBr	141.44
HCl	127.45
HF	91.680
HI	160.92
N_2	109.76
O_2	120.75

(b) Mean bond lengths from covalent radii*

H	37						
C	77(1) 67(2) 60(3)	N	74(1) 65(2)	O	66(1) 57(2)	F	64
Si	118	P	110	S	104(1) 95(2)	Cl	99
Ge	122	As	121	Se	104	Br	114
		Sb	141	Te	137	I	133

* Values are for single bonds except where indicated otherwise (values in parentheses). The length of an A–B covalent bond (of given order) is the sum of the corresponding covalent radii.

Table 9C.3a Bond dissociation enthalpies, $\Delta H^\ominus(\text{A–B})/(\text{kJ mol}^{-1})$ at 298 K*

Diatomic molecules									
H–H	436	F–F	155	Cl–Cl	242	Br–Br	193	I–I	151
O=O	497	C=O	1076	N≡N	945				
H–O	428	H–F	565	H–Cl	431	H–Br	366	H–I	299
Polyatomic molecules									
H–CH_3	435	H–NH_2	460		H–OH	492	H–C_6H_5	469	
H_3C–CH_3	368	H_2C=CH_2	720		HC≡CH	962			
HO–CH_3	377	Cl–CH_3	352		Br–CH_3	293	I–CH_3	237	
O=CO	531	HO–OH	213		O_2N–NO_2	54			

* To a good approximation bond dissociation enthalpies and dissociation energies are related by $\Delta H^\ominus = N_A hc\tilde{D}_e + \frac{3}{2}RT$ with $hc\tilde{D}_e = hc\tilde{D}_0 + \frac{1}{2}\hbar\omega$. For precise values of $N_A hc\tilde{D}_0$ for diatomic molecules, see Table 11C.1.
Data: HCP, KL.

Table 9C.3b Mean bond enthalpies, $\Delta H^\ominus(\text{A–B})/(\text{kJ mol}^{-1})$*

	H	C	N	O	F	Cl	Br	I	S	P	Si
H	436										
C	412	348(i)									
		612(ii)									
		838(iii)									
		518(a)									
N	388	305(i)	163(i)								
		613(ii)	409(ii)								
		890(iii)	946(iii)								
O	463	360(i)	157	146(i)							
		743(ii)		497(ii)							
F	565	484	270	185	155						
Cl	431	338	200	203	254	242					
Br	366	276				219	193				
I	299	238				210	178	151			
S	338	259			496	250	212		264		
P	322									201	
Si	318		374	466							226

* Mean bond enthalpies are such a crude measure of bond strength that they need not be distinguished from dissociation energies.
(i) Single bond, (ii) double bond, (iii) triple bond, (a) aromatic.
Data: HCP and L. Pauling, *The nature of the chemical bond*. Cornell University Press (1960).

Table 9D.1 Pauling (*italics*) and Mulliken electronegativities

H							He
2.20							
3.06							
Li	Be	B	C	N	O	F	Ne
0.98	*1.57*	*2.04*	*2.55*	*3.04*	*3.44*	*3.98*	
1.28	1.99	1.83	2.67	3.08	3.22	4.43	4.60
Na	Mg	Al	Si	P	S	Cl	Ar
0.93	*1.31*	*1.61*	*1.90*	*2.19*	*2.58*	*3.16*	
1.21	1.63	1.37	2.03	2.39	2.65	3.54	3.36
K	Ca	Ga	Ge	As	Se	Br	Kr
0.82	*1.00*	*1.81*	*2.01*	*2.18*	*2.55*	*2.96*	*3.0*
1.03	1.30	1.34	1.95	2.26	2.51	3.24	2.98
Rb	Sr	In	Sn	Sb	Te	I	Xe
0.82	*0.95*	*1.78*	*1.96*	*2.05*	*2.10*	*2.66*	*2.6*
0.99	1.21	1.30	1.83	2.06	2.34	2.88	2.59
Cs	Ba	Tl	Pb	Bi			
0.79	*0.89*	*2.04*	*2.33*	*2.02*			

Data: Pauling values: A.L. Allred, *J. Inorg. Nucl. Chem.* **17**, 215 (1961); L.C. Allen and J.E. Huheey, ibid., **42**, 1523 (1980). Mulliken values: L.C. Allen, *J. Am. Chem. Soc.* **111**, 9003 (1989). The Mulliken values have been scaled to the range of the Pauling values.

Table 10B.1 See Part 4.

Table 10B.2 See Part 4.

Table 11C.1 Properties of diatomic molecules

	$\tilde{\nu}/cm^{-1}$	θ^V/K	$\tilde{B}/cm^{-1}$	θ^R/K	R_e/pm	$k_f/(N\,m^{-1})$	$N_A hc\tilde{D}_0/(kJ\,mol^{-1})$	σ
$^1H_2^+$	2321.8	3341	29.8	42.9	106	160	255.8	2
$^1H_2$	4400.39	6332	60.864	87.6	74.138	574.9	432.1	2
$^2H_2$	3118.46	4487	30.442	43.8	74.154	577.0	439.6	2
$^1H^{19}F$	4138.32	5955	20.956	30.2	91.680	965.7	564.4	1
$^1H^{35}Cl$	2990.95	4304	10.593	15.2	127.45	516.3	427.7	1
$^1H^{81}Br$	2648.98	3812	8.465	12.2	141.44	411.5	362.7	1
$^1H^{127}I$	2308.09	3321	6.511	9.37	160.92	313.8	294.9	1
$^{14}N_2$	2358.07	3393	1.9987	2.88	109.76	2293.8	941.7	2
$^{16}O_2$	1580.36	2274	1.4457	2.08	120.75	1176.8	493.5	2
$^{19}F_2$	891.8	1283	0.8828	1.27	141.78	445.1	154.4	2
$^{35}Cl_2$	559.71	805	0.2441	0.351	198.75	322.7	239.3	2
$^{12}C^{16}O$	2170.21	3122	1.9313	2.78	112.81	1903.17	1071.8	1
$^{79}Br^{81}Br$	323.2	465	0.0809	10.116	283.3	245.9	190.2	1

Data: AIP.

Table 11F.1 Colour, frequency, and energy of light

Colour	λ/nm	$\nu/(10^{14}$ Hz)	$\tilde{\nu}/(10^4$ cm^{-1})	E/eV	E/(kJ mol^{-1})
Infrared	>1000	<3.00	<1.00	<1.24	<120
Red	700	4.28	1.43	1.77	171
Orange	620	4.84	1.61	2.00	193
Yellow	580	5.17	1.72	2.14	206
Green	530	5.66	1.89	2.34	226
Blue	470	6.38	2.13	2.64	254
Violet	420	7.14	2.38	2.95	285
Ultraviolet	<400	>7.5	>2.5	>3.10	>300

Data: J.G. Calvert and J.N. Pitts, *Photochemistry*. Wiley, New York (1966).

Table 11F.2 Absorption characteristics of some groups and molecules

Group	$\tilde{\nu}_{max}/(10^4$ cm^{-1})	λ_{max}/nm	ε_{max}/(dm^3 mol^{-1} cm^{-1})
C=C ($\pi^* \leftarrow \pi$)	6.10	163	1.5×10^4
	5.73	174	5.5×10^3
C=O ($\pi^* \leftarrow$ n)	3.7–3.5	270–290	10–20
–N=N–	2.9	350	15
	>3.9	<260	Strong
$-NO_2$	3.6	280	10
	4.8	210	1.0×10^4
C_6H_5-	3.9	255	200
	5.0	200	6.3×10^3
	5.5	180	1.0×10^5
$[Cu(OH_2)_6]^{2+}$(aq)	1.2	810	10
$[Cu(NH_3)_4]^{2+}$(aq)	1.7	600	50
H_2O ($\pi^* \leftarrow$ n)	6.0	167	7.0×10^3

Table 12A.2 Nuclear spin properties

Nuclide	Natural Abundance/%	Spin I	Magnetic Moment, μ/μ_N	g-value	$\gamma/(10^7$ T^{-1}s^{-1})	NMR frequency at 1 T, ν/MHz
^{1}n*		$\frac{1}{2}$	−1.9130	−3.8260	−18.324	29.164
^{1}H	99.9844	$\frac{1}{2}$	2.792 85	5.5857	26.752	42.576
^{2}H	0.0156	1	0.857 44	0.857 44	4.1067	6.536
^{3}H*		$\frac{1}{2}$	2.978 96	−4.2553	−20.380	45.414
^{10}B	19.6	3	1.8006	0.6002	2.875	4.575
^{11}B	80.4	$\frac{3}{2}$	2.6886	1.7923	8.5841	13.663
^{13}C	1.108	$\frac{1}{2}$	0.7024	1.4046	6.7272	10.708
^{14}N	99.635	1	0.403 56	0.403 56	1.9328	3.078
^{17}O	0.037	$\frac{5}{2}$	−1.893 79	−0.7572	−3.627	5.774
^{19}F	100	$\frac{1}{2}$	2.628 87	5.2567	25.177	40.077
^{31}P	100	$\frac{1}{2}$	1.1316	2.2634	10.840	17.251
^{33}S	0.74	$\frac{3}{2}$	0.6438	0.4289	2.054	3.272
^{35}Cl	75.4	$\frac{3}{2}$	0.8219	0.5479	2.624	4.176
^{37}Cl	24.6	$\frac{3}{2}$	0.6841	0.4561	2.184	3.476

* Radioactive.

μ is the magnetic moment of the spin state with the largest value of m_I: $\mu = g_I\mu_N I$ and μ_N is the nuclear magneton (see inside front cover).

Data: KL and HCP.

Table 12D.1 Hyperfine coupling constants for atoms, a/mT

Nuclide	Spin	Isotropic coupling	Anisotropic coupling
^{1}H	$\frac{1}{2}$	50.8(1s)	
^{2}H	1	7.8(1s)	
^{13}C	$\frac{1}{2}$	113.0(2s)	6.6(2p)
^{14}N	1	55.2(2s)	4.8(2p)
^{19}F	$\frac{1}{2}$	1720(2s)	108.4(2p)
^{31}P	$\frac{1}{2}$	364(3s)	20.6(3p)
^{35}Cl	$\frac{3}{2}$	168(3s)	10.0(3p)
^{37}Cl	$\frac{3}{2}$	140(3s)	8.4(3p)

Data: P.W. Atkins and M.C.R. Symons, *The structure of inorganic radicals*. Elsevier, Amsterdam (1967).

PART 4 Character tables

The groups C_1, C_s, C_i

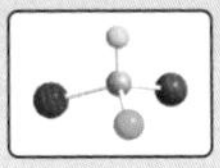

C_1 (1)	E	$h = 1$
A	1	

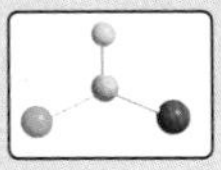

$C_s = C_h$ (m)	E	σ_h	$h = 2$	
A′	1	1	x, y, R_z	x^2, y^2, z^2, xy
A″	1	−1	z, R_x, R_y	yz, zx

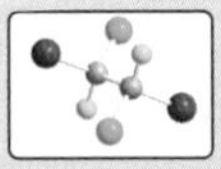

$C_i = S_2(\bar{1})$	E	i	$h = 2$	
A_g	1	1	R_x, R_y, R_z	$x^2, y^2, z^2, xy, yz, zx,$
A_u	1	−1	x, y, z	

The groups C_{nv}

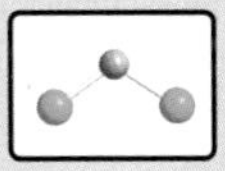

C_{2v}, $2mm$	E	C_2	σ_v	σ_v'	$h = 4$	
A_1	1	1	1	1	z, z^2, x^2, y^2	
A_2	1	1	−1	−1	xy	R_z
B_1	1	−1	1	−1	x, zx	R_y
B_2	1	−1	−1	1	y, yz	R_x

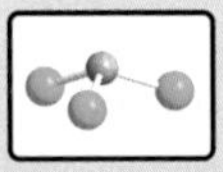

C_{3v}, $3m$	E	$2C_3$	$3\sigma_v$	$h = 6$	
A_1	1	1	1	$z, z^2, x^2 + y^2$	
A_2	1	1	−1		R_z
E	2	−1	0	$(x, y), (xy, x^2 - y^2) (yz, zx)$	(R_x, R_y)

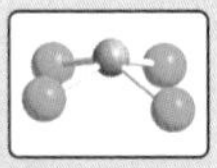

C_{4v}, $4mm$	E	C_2	$2C_4$	$2\sigma_v$	$2\sigma_d$	$h = 8$	
A_1	1	1	1	1	1	$z, z^2, x^2 + y^2$	
A_2	1	1	1	−1	−1		R_z
B_1	1	1	−1	1	−1	$x^2 - y^2$	
B_2	1	1	−1	−1	1	xy	
E	2	−2	0	0	0	$(x, y), (yz, zx)$	(R_x, R_y)

The σ_v planes coincide with the xz- and yz-planes.

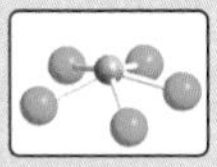

C_{5v}	E	$2C_5$	$2C_5^2$	$5\sigma_v$	$h = 10, \alpha = 72°$	
A_1	1	1	1	1	$z, z^2, x^2 + y^2$	
A_2	1	1	1	−1		R_z
E_1	2	$2 \cos \alpha$	$2 \cos 2\alpha$	0	$(x, y), (yz, zx)$	(R_x, R_y)
E_2	2	$2 \cos 2\alpha$	$2 \cos \alpha$	0	$(xy, x^2 - y^2)$	

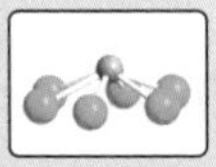

C_{6v}, $6mm$	E	C_2	$2C_3$	$2C_6$	$3\sigma_d$	$3\sigma_v$	$h = 12$	
A_1	1	1	1	1	1	1	$z, z^2, x^2 + y^2$	
A_2	1	1	1	1	−1	−1		R_z
B_1	1	−1	1	−1	−1	1		
B_2	1	−1	1	−1	1	−1		
E_1	2	−2	−1	1	0	0	$(x, y), (yz, zx)$	(R_x, R_y)
E_2	2	2	−1	−1	0	0	$(xy, x^2 - y^2)$	

$C_{\infty v}$	E	$2C_\phi^\dagger$	…	$\infty\sigma_v$	$h = \infty$	
$A_1(\Sigma^+)$	1	1	…	1	$z, z^2, x^2 + y^2$	
$A_2(\Sigma^-)$	1	1	…	−1		R_z
$E_1(\Pi)$	2	$2 \cos \phi$	…	0	$(x, y), (yz, zx)$	(R_x, R_y)
$E_2(\Delta)$	2	$2 \cos 2\phi$	…	0	$(xy, x^2 - y^2)$	
…	…	…	…	…		

† There is only one member of this class if $\phi = \pi$.

The groups D_n

D_2, 222	E	C_2^z	C_2^y	C_2^x	$h = 4$	
A_1	1	1	1	1	x^2, y^2, z^2	
B_1	1	1	−1	−1	z, xy	R_z
B_2	1	−1	1	−1	y, zx	R_y
B_3	1	−1	−1	1	x, yz	R_x

D_3, 32	E	$2C_3$	$3C_2'$	$h = 6$	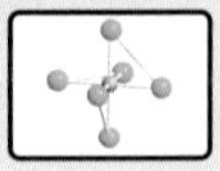
A_1	1	1	1	$z^2, x^2 + y^2$	
A_2	1	1	−1	z	R_z
E	2	−1	0	$(x, y), (yz, zx), (xy, x^2 - y^2)$	(R_x, R_y)

D_4, 422	E	C_2	$2C_4$	$2C_2'$	$2C_2''$	$h = 8$	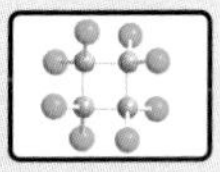
A_1	1	1	1	1	1	$z^2, x^2 + y^2$	
A_2	1	1	1	−1	−1	z	R_z
B_1	1	1	−1	1	−1	$x^2 - y^2$	
B_2	1	1	−1	−1	1	xy	
E	2	−2	0	0	0	$(x, y), (yz, zx)$	(R_x, R_y)

The groups D_{nh}

D_{2h}, mmm	E	$C_2(z)$	$C_2(y)$	$C_2(x)$	i	$\sigma(xy)$	$\sigma(yz)$	$\sigma(zx)$	$h = 8$	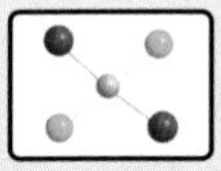
A_g	1	1	1	1	1	1	1	1		x^2, y^2, z^2
B_{1g}	1	1	−1	−1	1	1	−1	−1	R_z	xy
B_{2g}	1	−1	1	−1	1	−1	−1	1	R_y	xz
B_{3g}	1	−1	−1	1	1	−1	1	−1	R_x	yz
A_u	1	1	1	1	−1	−1	−1	−1		
B_{1u}	1	1	−1	−1	−1	−1	1	1	z	
B_{2u}	1	−1	1	−1	−1	1	1	−1	y	
B_{3u}	1	−1	−1	1	−1	1	−1	1	x	

D_{3h}, $\bar{6}2m$	E	σ_h	$2C_3$	$2S_3$	$3C_2'$	$3\sigma_v$	$h = 12$	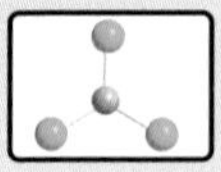
A_1'	1	1	1	1	1	1	$z^2, x^2 + y^2$	
A_2'	1	1	1	1	−1	−1		R_z
A_1''	1	−1	1	−1	1	−1		
A_2''	1	−1	1	−1	−1	1	z	
E'	2	2	−1	−1	0	0	$(x, y), (xy, x^2 - y^2)$	
E''	2	−2	−1	1	0	0	(yz, zx)	(R_x, R_y)

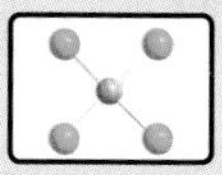

D_{4h}, 4/*mmm*	E	$2C_4$	C_2	$2C_2'$	$2C_2''$	i	$2S_4$	σ_h	$2\sigma_v$	$2\sigma_d$	$h = 16$	
A_{1g}	1	1	1	1	1	1	1	1	1	1	$x^2 + y^2, z^2$	
A_{2g}	1	1	1	−1	−1	1	1	1	−1	−1		R_z
B_{1g}	1	−1	1	1	−1	1	−1	1	1	−1	$x^2 - y^2$	
B_{2g}	1	−1	1	−1	1	1	−1	1	−1	1	xy	
E_g	2	0	−2	0	0	2	0	−2	0	0	(yz, zx)	(R_x, R_y)
A_{1u}	1	1	1	1	1	−1	−1	−1	−1	−1		
A_{2u}	1	1	1	−1	−1	−1	−1	−1	1	1	z	
B_{1u}	1	−1	1	1	−1	−1	1	−1	−1	1		
B_{2u}	1	−1	1	−1	1	−1	1	−1	1	−1		
E_u	2	0	−2	0	0	−2	0	2	0	0	(x, y)	

The C_2' axes coincide with the *x*- and *y*-axes; the σ_v planes coincide with the *xz*- and *yz*-planes.

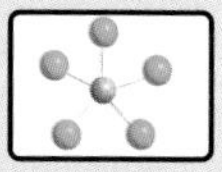

D_{5h}	E	$2C_5$	$2C_5^2$	$5C_2$	σ_h	$2S_5$	$2S_5^3$	$5\sigma_v$	$h = 20$, $\alpha = 72°$	
A_1'	1	1	1	1	1	1	1	1	$x^2 + y^2, z^2$	
A_2'	1	1	1	−1	1	1	1	−1		R_z
E_1'	2	$2\cos\alpha$	$2\cos 2\alpha$	0	2	$2\cos\alpha$	$2\cos 2\alpha$	0	(x, y)	
E_2'	2	$2\cos 2\alpha$	$2\cos\alpha$	0	2	$2\cos 2\alpha$	$2\cos\alpha$	0	$(x^2 - y^2, xy)$	
A_1''	1	1	1	1	−1	−1	−1	−1		
A_2''	1	1	1	−1	−1	−1	−1	1	z	
E_1''	2	$2\cos\alpha$	$2\cos 2\alpha$	0	−2	$-2\cos\alpha$	$-2\cos 2\alpha$	0	(yz, zx)	(R_x, R_y)
E_2''	2	$2\cos 2\alpha$	$2\cos\alpha$	0	−2	$-2\cos 2\alpha$	$-2\cos\alpha$	0		

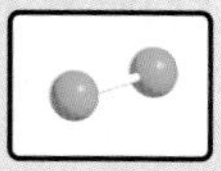

$D_{\infty h}$	E	$2C_\phi$	…	$\infty\sigma_v$	i	$2S_\infty$	…	$\infty C_2'$	$h = \infty$	
$A_{1g}(\Sigma_g^+)$	1	1	…	1	1	1	…	1	$z^2, x^2 + y^2$	
$A_{1u}(\Sigma_u^+)$	1	1	…	1	−1	−1	…	−1	z	
$A_{2g}(\Sigma_g^-)$	1	1	…	−1	1	1	…	−1		R_z
$A_{2u}(\Sigma_u^-)$	1	1	…	−1	−1	−1	…	1		
$E_{1g}(\Pi_g)$	2	$2\cos\phi$	…	0	2	$-2\cos\phi$	…	0	(yz, zx)	(R_x, R_y)
$E_{1u}(\Pi_u)$	2	$2\cos\phi$	…	0	−2	$2\cos\phi$	…	0	(x, y)	
$E_{2g}(\Delta_g)$	2	$2\cos 2\phi$	…	0	2	$2\cos 2\phi$	…	0	$(xy, x^2 - y^2)$	
$E_{2u}(\Delta_u)$	2	$2\cos 2\phi$	…	0	−2	$-2\cos 2\phi$	…	0		
…	…	…	…	…	…	…	…	…		

The cubic groups

T_d, $\bar{4}3m$	E	$8C_3$	$3C_2$	$6\sigma_d$	$6S_4$	$h = 24$	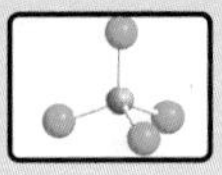
A_1	1	1	1	1	1	$x^2+y^2+z^2$	
A_2	1	1	1	−1	−1		
E	2	−1	2	0	0	$(3z^2-r^2, x^2-y^2)$	
T_1	3	0	−1	−1	1		(R_x, R_y, R_z)
T_2	3	0	−1	1	−1	(x, y, z), (xy, yz, zx)	

O_h, $m3m$	E	$8C_3$	$6C_2$	$6C_4$	$3C_2\ (=C_4^2)$	i	$6S_4$	$8S_6$	$3\sigma_h$	$6\sigma_d$	$h = 48$	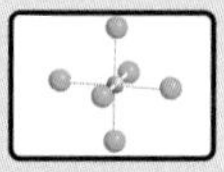
A_{1g}	1	1	1	1	1	1	1	1	1	1	$x^2+y^2+z^2$	
A_{2g}	1	1	−1	−1	1	1	−1	1	1	−1		
E_g	2	−1	0	0	2	2	0	−1	2	0	$(2z^2-x^2-y^2,$ $x^2-y^2)$	
T_{1g}	3	0	−1	1	−1	3	1	0	−1	−1		(R_x, R_y, R_z)
T_{2g}	3	0	1	−1	−1	3	−1	0	−1	1	(xy, yz, zx)	
A_{1u}	1	1	1	1	1	−1	−1	−1	−1	−1		
A_{2u}	1	1	−1	−1	1	−1	1	−1	−1	1		
E_u	2	−1	0	0	2	−2	0	1	−2	0		
T_{1u}	3	0	−1	1	−1	−3	−1	0	1	1	(x, y, z)	
T_{2u}	3	0	1	−1	−1	−3	1	0	1	−1		

The icosahedral group

I	E	$12C_5$	$12C_5^2$	$20C_3$	$15C_2$	$h = 60$	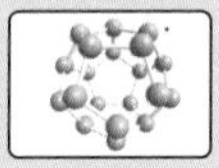
A	1	1	1	1	1	$x^2+y^2+z^2$	
T_1	3	$\frac{1}{2}(1+5^{1/2})$	$\frac{1}{2}(1-5^{1/2})$	0	−1	(x, y, z)	(R_x, R_y, R_z)
T_2	3	$\frac{1}{2}(1-5^{1/2})$	$\frac{1}{2}(1+5^{1/2})$	0	−1		
G	4	−1	−1	1	0		
H	5	0	0	−1	1	$(2z^2-x^2-y^2, x^2-y^2, xy, yz, zx)$	

* The image illustrating the group I is in fact a representation of the group I_h, which is isomorphous with a dodecahedron. The group I_h includes reflections; its order is 120.

Further information: P.W. Atkins, M.S. Child, and C.S.G. Phillips, *Tables for group theory*. Oxford University Press (1970). In this source, which is available via the e-book, other character tables such as D_{2d}, D_{3d}, D_{6h}, and I_h can be found.

INDEX

Note: Tables and figures are indicated by an italic *t* and *f* following the page number.